COMMUNICATING SCIENCE

A GLOBAL PERSPECTIVE

COMMUNICATING SCIENCE

A GLOBAL PERSPECTIVE

EDITED BY TOSS GASCOIGNE,
BERNARD SCHIELE, JOAN LEACH,
MICHELLE RIEDLINGER
WITH BRUCE V. LEWENSTEIN,
LUISA MASSARANI, PETER BROKS

Australian National University
PRESS

Published by ANU Press
The Australian National University
Acton ACT 2601, Australia
Email: anupress@anu.edu.au

Available to download for free at press.anu.edu.au

ISBN (print): 9781760463656
ISBN (online): 9781760463663

WorldCat (print): 1184001419
WorldCat (online): 1184001543

DOI: 10.22459/CS.2020

Cover design and layout by ANU Press
Cover artwork: *Acanthus* by Rosalie Gascoigne, 1995

CONTENTS

PREFACE

In 2020, one can hear the provocation that 'science is a human right' ring louder and louder.[1] At first, such an assertion seems overblown; among the human rights to which we aspire—equality, freedom from poverty, freedom of expression and freedom of assembly—'the right to science' seems quite far down the list of priorities. And yet, we find ourselves in the midst of so many problems where science might offer at least partial solutions—and understanding science also may help us avoid more of the problems science has already caused. Access to science and its benefits and opportunities is uneven across the world—as is the production of scientific knowledge. Further, the gains of scientific knowledge have been concentrated in wealthy regions of the globe while the costs of producing scientific knowledge and its application have been pushed to poorer regions. Better access to science, then, even in a modest way through science communication, has been one response to this inequality. At the Australian National Centre for Public Awareness of Science at The Australian National University (ANU), our mission for over 30 years has been to foster the 'democratic ownership of science'. So, cataloguing stories of science communication from all over the world, beginning to account for the ways that science is and could be accessible to more people as well as the ways in which scientific knowledge is held to account, is an important part of our goal. We have been proud to support the project that has produced this fine volume.

But this is a beginning.[2] And the word 'beginning' is an important one; the goal of this volume was not to produce myriad 'origin stories' of science communication (in Australia, this would be a 60,000-year-old history of Indigenous Australians sharing knowledge), but to explore the beginning of a more recent, 20th-century common project that is being realised in multiple ways around the globe. In some countries like Australia from where I write, science communication is an academic and professional activity with government support that waxes and wanes. In other places, science

1 For example, by Shabaz Khan in a seminar he gave at the Centre for the Public Awareness of Science at ANU: cpas.anu.edu.au/news-events/events/lunchtime-seminar-professor-shahbaz-khan-director-unesco-regional-science-bureau.

2 See Edward Said (1985). *Beginnings: Intention and Method*. New York: Columbia University Press.

communication is a form of activism; in still others, it is seen as part of science itself. Sometimes the goal of science communication is education, other times it is a way of sharing the creation of knowledge and yet other times science communication is about addressing items of concern—health, climate, environment, technology. And so, a beginning is a messy thing that struggles to contain all the different possibilities that a story can tell. This volume celebrates that with a proliferation of terminology, goals and aims—and the authors, while lucid, sophisticated and bringing substantial analytic skills to bear, also acknowledge a glorious mess that is science communication as it evolves from its 20th-century beginnings.

Finally, a note on methods. In some parts of the world, including Australia, empirical methods in science communication thrive and, for the most part, this is a happy advancement of the field. However, the methods of history, also empirical—of collecting evidence in an archive of sorts, of telling stories and gathering a corpus that will guide future work—are needed now more than ever. This volume, we hope, contributes to this. It brings together a corpus of stories from around the world that will be shared to provide impetus for more stories, different stories, contestation and, of course, other forms of work.

Professor Joan Leach
Director, Australian National Centre for Public Awareness of Science
The Australian National University
Canberra, Australia

ACKNOWLEDGEMENTS

We gratefully acknowledge the assistance and support of the following groups and individuals. Without their support, this book would not have been possible.

Professor Joan Leach and the Centre for Public Awareness of Science, for support and special assistance in bringing the book to fruition.

The members of the Editorial Board who guided the book through its stages: Dr Peter Broks, Professor Joan Leach, Professor Bruce V. Lewenstein, Dr Luisa Massarani, Dr Michelle Riedlinger and Professor Bernard Schiele.

The Network for the Public Communication of Science and Technology (PCST), which allowed us to test concepts, recruit authors and launch the book to the international community of science communication researchers and practitioners.

The 108 authors who wrote the 40 chapters. They were willing, responsive and cheerful in drafting, revising and re-drafting their chapters.

Bernard Schiele for his exhaustive work in compiling and analysing the data from the timelines.

Emily Tinker of ANU Press for her patience and expertise in shepherding the manuscript through the production process, and the other staff members at ANU Press who worked hard to bring the manuscript into its final form, particularly Teresa Prowse and Elouise Ball.

Ranjan Chaudhuri, for his assistance in editing the text.

Toss Gascoigne
Chief Editor

1

INTRODUCTION

A global trend, an emerging field, a multiplicity of understandings: Science communication in 39 countries

Toss Gascoigne and Bernard Schiele

This book is a comprehensive attempt to chart the history of science communication as it developed in the modern era. It tells the story from the perspective of researchers and practitioners in the field, collecting accounts of how modern science communication has developed internationally. The book contains 40 chapters: two introductory chapters, 36 chapters focusing on a single country,[1] one covering the three Scandinavian countries, and one describing the communication of health issues in a region of Africa. It involves 108 authors. The results are astounding, a unique dataset to be explored and a rich cornucopia of information.

This raises a number of questions: What knowledge can we extract from the data and how can this knowledge be shared? What are the theoretical models (either explicit or implicit) to which authors refer? What approaches have worked best, and in which context, and why? What lessons can be learned from all the experiences the chapters recount? What can we learn from the interaction between scientific knowledge and indigenous knowledge or local knowledge systems; and how can we engage people more directly with science or encourage them to participate in science? What are the best ways to counter irrational beliefs based on religion, superstition, ideology, pseudoscience or anti-science?

1 The word 'country' is used in a broad sense. It may refer to an individual nation, a group of countries or a region.

Each chapter is an in-depth introduction to a country's issues in its sociohistorical context. By comparing them, we can see connections and similarities and understand the differences and the choices made in particular situations.

Although the movement to foster the development of science culture is universal, it can only develop in the historical, cultural and social contexts of a country. This book presents the reader with a comparative analysis, not merely of quantifiable data (e.g. when the first interactive science centre was established in each country), but also the overarching dynamics at play: the narrative. What matters in each case is less the raw data than the accompanying system of significations and justifications. Most authors have taken pains to describe the sequence of events that led or prevented this or that realisation as well as the main phases, debates and issues that characterised the development of science communication. In other words, this data-rich book will reward a transversal reading that alone allows for an in-depth comparison and analysis.

The data document the different approaches taken to address practical problems: training science communicators, establishing science centres and museums, organising campaigns to lift health outcomes, science festivals and public events, media coverage of science, programs to train scientists how to communicate, and countering superstition and fake news. The chapters will reward a close examination, beyond the scope of these introductory remarks, to pivot these data to tackle cross-cutting issues in science communication, and to identify common themes, recurring challenges and potentially adaptable innovations. This is a task to be expanded in a follow-up book of analysis and comparison.

1. How the book came to be

Our book asked authors to complete a timeline for their country and nominate dates when key events took place (such as when the first university courses training science communicators were established). This is how the book began, in 2013. It was then that a simple curiosity about the nature of national experiences spurred the editor of this volume to post a message to the PCST[2] email discussion list, inviting people to compile dates of key events in science communication in their country:

2 PCST is the Network for the Public Communication of Science and Technology, an international association of science communicators: www.pcst.co/.

> The proposal is for a discussion on the emergence of science communication in different countries round the world. It would look for similarities and differences … on matters such as:
>
> - the first university courses training science communicators
> - research in science communication - the emergence of PhD courses
> - the formation of organisations and associations for science communicators
> - conferences, meetings and publications in this area
> - employment opportunities
> - the terminology (science popularisation, public understanding of science, vulgarisation, social appropriation of science etc)
>
> Have different countries followed different paths, or have we followed the same broad path and the same timelines? What were the foundation steps - how did it all begin? … [I] am interested in hearing from people who would like to be involved and can present the history of their own country.

People from 17 countries responded, each completing a timeline of up to 15 dates, and the results were presented at the PCST Conference in Brazil in 2014. Then the project went into hibernation until 2017, when the publication of an account of Ireland's science communication story (Trench, Murphy and Fahy, 2017) and a special issue of *JCOM* (Massarani, de Castro Moreira and Lewenstein, 2017) rekindled interest. Authors from the 2014 conference were asked if they wanted to expand their timelines into a chapter, and new authors were recruited to fill the gaps. An editorial board was established, a publisher signed up, and guidelines sent out to authors. The project sparked international interest and the original 17 contributing countries grew to 45 expressions of interest. What began as a set of dates in a spreadsheet had grown into full-blown chapters.

There have been earlier efforts to chart science communication. The proceedings of the PCST Conference in Montreal in 1994 featured 27 national accounts, with the 'modest aim' (in the words of the editors) 'to present a status report on the development of scientific and technological culture'. The emphasis was different—more focused on Western countries—and so was the content (not surprising given that the papers were written in the infancy of the internet, and when questions about the deficit model of science communication were just beginning). Despite this, the books taken together—the Montreal proceedings of 1994 *When Science Becomes Culture* (Schiele, 1994), and this book—represent a continuity.

It also represents a wider preoccupation to strengthen the visibility of science and technology within society—science communication being only one aspect—and the desire to better measure this visibility. In the same movement are all the surveys conducted to assess the level of science culture of given populations and the indicators developed to this end (Eurobarometer, AAAS, LES, not forgetting national surveys); and the survey of all university-based science communication programs carried out a few years ago by the EU. Sociology has demonstrated that the legitimisation of a domain is linked to its historicisation and it is not surprising to see a growing interest in the history of science communication. This book is part of a wider movement that aims to know and describe the history of the field. We hope its significance will become self-evident, because it gives a voice to the actors of the field in their own specific contexts.

2. One nagging question

During the process of commissioning the chapters, authors posed one question: what year did the 'modern era' of science communication begin? There is no simple answer to that question: it varies from country to country.

It is worth mentioning that science communication (then called science vulgarisation or popularisation) boomed in the 19th century. This period continued well into the 20th century and is seen as a golden age of science communication by many researchers. But the modes of science communication developed before World War II (WWII) were disrupted by the conflict. Scientists were not at liberty to communicate freely with the public on the grounds of national security, and after the war structural changes interrupted normal processes.

What distinguishes the post-war period from earlier times is the social driving role played by science and technology: they have become the main vehicle of an ever-accelerating movement of social, cultural and economic change, at a pace matched by the integration of the sciences and the economy. The sustainability of this movement is increasingly predicated on the training of a qualified workforce whose function is to produce new knowledge and new applications, and to acquire the skills to make use of these applications in professional and daily life. It also depends on the ability and willingness of the population to accept and adopt advances in technology. The rapid expansion of higher education during the 1960s is directly linked to this development.

The evolution of post-war society becomes incomprehensible if we do not take into account this dynamic. The narrative is borne out by successive reports of the Organisation for Economic Co-operation and Development (OECD). First, the reports stressed the fact that national competitiveness is linked to innovation: new knowledge and new applications sustain the pace of economic development. Second, they invited national governments to become key actors by creating ministries of science and technology and adopting science policies. Third, they invited states to mobilise their population through the development of science culture by the implementation of programs to promote and propagate scientific knowledge and scientific thinking. This is why the focus of this book on the post-war period is not arbitrary: it reflects this paradigm change.

The timing of the modern era of science communication varies between countries depending on their situation before WWII and the pace of change after WWII. Authors were asked to select the most appropriate date for their country, 'probably somewhere between 1945 and 1980'. The editorial board certainly recognise there were science museums, media reports and associations of science writers before that time, but the 1970s–1990s marked a step-change: new interactive science centres, new jobs, the opening of courses at universities to meet the demand for a career in science communication, the formation of journals, associations and conferences devoted to science communication and new programs to encourage the public to engage with science.

This period, as confirmed by the analysis following this introduction, can be seen as the moment when science communication asserted itself as a social necessity. It was marked by the formation of a group of social actors engaged in that field, sharing a set of practices and inventing terminologies to describe these practices and recognised as a group committed to communicating science. Over the years a second group has emerged, mostly comprised of scholars and academics whose goal is the study of science communication. These two groups are in constant interaction.

3. Diversity, optimism, doubt, challenges

3.1. Diversity

The book draws on a cross-section of countries: all continents and all cultures. The diversity is reflected across geographical location, income, religion, population, land area, democracy rating and history. Five contributing chapters are from Africa, seven from the Americas, 11 from Asia and the rest from Europe and Australasia.

Eleven countries rank outside the top 100 in per capita wealth, with Pakistan, Ghana and Uganda ranking the lowest.[3] Problems, attitudes and actions can be sharply magnified by huge disparities in wealth. There is a factor of 127 separating the richest country involved in the book from the poorest: the median figure for gross domestic product per capita for the 39 countries is US$22,928,[4] but national figures range from US$643 for Uganda to US$81,807 for Norway. This has significant implications in defining the issues countries face and where science communication would be part of the solution. It also has implications for their abilities to fund appropriate actions.

Five of the countries represented are Muslim-majority: Nigeria, Iran, Pakistan, Turkey and Malaysia. Twenty-five are Christian-majority; four have a majority of people declaring themselves 'unaffiliated'; two are Buddhist; one Hindu; and one each folk religion and Judaism[5]. Religious attitudes can shape and influence national policies on research and practice in science, with consequent implications for science communication.

Population and land area are two other variables showing great diversity in participating countries. Countries range in population from 1.3 million (Estonia) to 1.4 billion (China), a factor of over 1,000. Nine have a population of below 10 million, balanced by 10 countries with a population above 100 million. The median is 46 million.[6] National land areas vary from 709 square kilometres (Singapore) to over 16 million (Russia).[7]

Contributing countries also range widely when listed according to the Economist's Democracy Index. Twelve countries are assessed as 'full democracies'; 19 as 'flawed democracies'; 12 as 'hybrid regimes'; and three as 'authoritarian'. The Economist explains its approach as follows:

> The index rates 167 countries by 60 indicators across five broad categories: electoral process and pluralism, the functioning of government, political participation, democratic political culture and civil liberties. It is stricter than most similar indices: it concludes that just 4.5% of the world's people live in a 'full democracy'.[8]

3 GDP per capita 2018 from data.worldbank.org/indicator/ny.gdp.pcap.cd.

4 Again, see GDP per capita 2018 from data.worldbank.org/indicator/ny.gdp.pcap.cd.

5 Pew Research Centre from www.pewforum.org/2015/04/02/religious-projection-table/2020/percent/all.

6 Populations 2018 from data.worldbank.org/indicator/SP.POP.TOTL?view=chart.

7 Land areas from data.worldbank.org/indicator/AG.LND.TOTL.K2.

8 See www.economist.com/graphic-detail/2019/01/08/the-retreat-of-global-democracy-stopped-in-2018.

The book documents the way science communication has evolved and the motivations behind its development. The 38 chapters record an astounding variety of science communication modes of mediation, from drumming around the village, to traditional and social media, science cafés and festivals, and meetings with politicians. The approaches vary widely: while Germany relied on traditional media and focused its efforts in training journalists, Russia concentrated on publications and public lectures (in 1990 the Russian science journal *Argumenty i Fakty* was cited in the Guinness Book of Records as the highest weekly circulation newspaper in the world). South Korea took a whole-of-government approach that lifted the country sharply up international wealth rankings.

The science communication approaches and practices recorded for some participating countries have never been described before. But science communication is not an autonomous field. The actors always have to work and modify their work and discourse according to the context, the plays of power and the social, economic and political imperatives that constitute their environment. Thailand, for instance, was led by King Mongkut, a passionate amateur astronomer and a supporter of science communication who won great respect in his country for successfully predicting an eclipse of the sun two years before it happened in 1868 (he also provided the central character in the musical *The King and I*, a part played on Broadway and the movie by Yul Brynner).

Brazil celebrated science in the lyrics of the samba *Ciência e arte* (Science and Art) and Aotearoa New Zealand is marrying indigenous knowledge with Western science. Islamic countries have principles of Shariah (Islamic law) to consider, bearing in mind the twin tenets of halal (permissibility) and haram (forbidden) in shaping science communication activities. Nigeria tells the story of a significant science communication challenge, when a campaign to rid the country of polio was subject to an Islamic fatwa in 2003 because the Muslim community suspected it was part of a plot to sterilise children living in the northern (Muslim) parts of the country. It was only after five leaders of this community visited Egypt to witness a successful campaign that the program was allowed to continue.

3.2. Optimism

If before WWII science communication was essentially left to the discretion of scientists and enthusiasts, from the 1970s governments began to assert their role, spurred by the direct impact of science and technology on society. Since then, governments have often led moves in support of modern science communication, investing in interactive science centres, university training and public events like national science week. The British story is an example, well-documented through official reports and inquiries. So are the French and Quebec (Canada) stories. But governments with a consistent record are rare: despite public expressions of support for science communication, budgets demonstrate it is often low on the list of governmental priorities, through indifference, competition for funding, lack of commitment despite promises of support repeatedly reaffirmed, or because governments are overwhelmed by other issues.

In some countries, governments have taken a passive role and allowed external institutions and individuals to make the running. In the 1970s the German government stood back, perhaps recalling the unsavoury relationship between Nazi propaganda and science, and private foundations took the lead in funding ambitious programs to train science journalists. In the US, the absence of a strong central agency encouraged many institutions, funding bodies and societies to enter a field described variously as 'vibrant', 'jostling' and 'cacophonous' and characterised by a lack of coordination and centralisation. Reflecting this, the US authors subtitled their chapter 'It's Complicated'.

At times through history, governments have been hostile to attempts to communicate science. The fascist regimes of Italy, Spain and Portugal: the attempts of the Canadian government in 2010 to muzzle scientists on climate change, a coup d'état in Argentina that had 'devastating' effects on science, and state control and censorship in South Africa over the apartheid years—to mention some—all discouraged communication. In all cases the role of the State was a determining factor.

But generally, governments became more interested in science and science communication following WWII, when there was a growing belief in the power of science to unlock a bright new future. Governments had seen what science could do in the war years. In 1945 US President Franklin D. Roosevelt asked Vannevar Bush, Director of the US Office of Scientific Research and Development, a simple question: we have seen what science can do for us in war, but what can science do in times of peace?

Bush's response, *Science: The Endless Frontier* (1945), set out a vision for science in modern life:

> Advances in science when put to practical use mean more jobs, higher wages, shorter hours, more abundant crops, more leisure for recreation, for study … Advances in science will also bring higher standards of living, will lead to the prevention or cure of diseases, will promote conservation of our limited national resources, and will assure means of defense against aggression.

Bush's optimism was picked up by the leaders of newly independent states in Africa and Asia, and voiced by Kwame Nkrumah, prime minister of Ghana. In an address to the first conference of the Organisation of African Unity (OAU) in 1963, Nkrumah said:

> We shall accumulate machinery and establish steel works, iron foundries and factories; we shall link the various States of our continent with communications; we shall astound the world with our hydroelectric power; we shall drain marshes and swamps, clear infested areas, feed the under-nourished, and rid our people of parasites and disease. It is within the possibility of science and technology to make even the Sahara bloom into a vast field with verdant vegetation for agricultural and industrial developments. We shall harness the radio, television, giant printing presses to lift our people from the dark recesses of illiteracy ... The world is no longer moving through bush paths or on camels and donkeys.

The visions outlined by Bush and Nkrumah could only be realised if the mindset of the people was aligned to science. This was a monumental task for science communication: changing long-held beliefs and attitudes, opening the eyes of people to the possibilities of better approaches to health, agriculture and industrialisation, and preparing people for a new future. Implicit in this was the need for public acceptance, awareness and education.

Further impetus came in the 1960s. Beginning in 1963, the OECD had a central presence in asserting the fundamental role of the sciences and technologies in modern societies and insisting on the necessity for governments to develop science policies to sustain economic growth. Governments responded by establishing ministries for science and developing science policies. The second report in 1971 took a broader view of science, putting forward its social and cultural aspects. It is from that moment that science communication entered the political discourse (Piganiol et al., 1963; Brooks et al., 1971; OECD, 1981). The vision promoted by the OECD, with the exception of the United States, was quickly adopted by countries that were not members of the organisation (Henriques and Larédo, 2012).

3.3. Doubt

Eighteen countries participating in the book have a colonial history at some time in the 20th century. To that number must be added the countries in East Africa covered by the chapter on African health. This history has helped shape opportunities and attitudes to science communication.

The Nkrumah quotation above was typical of how eagerly former colonies anticipated independence and the freedom to invest in the future of their countries. Plans were formulated and optimism was strong in the newly independent countries. A lot was depending on science communication and the role it would play in changing attitudes, educating the people and modifying behaviours. But post-independence progress took time, interrupted by competing priorities, inexperience in government, inadequate resources, regional and personal conflicts and being weighed down by having to deal with immediate issues rather than the longer-term nature of investments in science. Chapters across all regions record the disappointments as well as the successes, and many authors report programs in science communication that began with enthusiasm but failed to deliver.

Faltering progress was not limited to the former colonies. Ireland has planned and lobbied for a science centre for 30 years and €13 million is still needed to complete the project. South African efforts to create a scientifically literate society are documented in the book as being hampered by inexperience and misguided politics, and the intricacies of balancing indigenous knowledge systems with modern science. In Turkey, the committee responsible for implementing the first official policy on science (and science communication) has failed to meet for many years. Mexico reports weak systems to incorporate science into the general culture of the population through collaboration and evaluation; Taiwan describes poor-quality coverage of science by mainstream media; Canada that the development of science culture has ceased to be the priority it once was; Pakistan awaits the necessary framework to communicate science to its public effectively; and Italy concludes that incentives for scientists to communicate their work are poor. Australia commissioned an inquiry into science communication in 2011 and funded national activities, but this has faded over time under a conservative government. These problems, shortcomings and issues are shared by other countries.

In addition to economic considerations, other social and cultural factors were involved. In the decades immediately after WWII, optimism was dominant in the era of Big Science, and the positive image of science reflected on scientists. Their influence was so great that although they were not in power, we could speak of a parliament of science. But although optimism was predominant (Sputnik, transistor radios, antibiotics and non-stick frying pans), doubts about the role of science and the effects of progress were emerging. Hiroshima cast a long shadow and concerns about the environment grew. Public opinion began to turn through the 1960s and 1970s. On the one hand, social movements (in Europe, the US and elsewhere), not least the peace movement, challenged authoritarianism and militarism. The continuing development of nuclear weapons was controversial, and the image of scientists was stained with the condemnation of their involvement. Environmental crises, such as the 1967 sinking of the Torrey Canyon, led to a growing awareness of the negative impacts of progress.

With the idea of progress being increasingly questioned (particularly in developed countries), the concept of risk entered collective consciousness and an ambivalence towards the development of the sciences and technologies led the UK's House of Lords to declare in 2000 that 'society's relationship with science is in a critical phase'.

3.4. Challenges

The authors have identified a number of challenges that science communication faces: fear of change, indifference, scepticism, superstition, competition for funding and resources, and cultural or religious differences. They report different experiences of a collision between 'white man's science' and indigenous knowledge: some countries have made significant progress towards resolving these issues, while others still wrestle with them. The 'system' can also work against science communication. Despite a broad expectation that scientists will discuss their work and engage with society (the 'third mission'), authors reported limited interest by institutions, funding bodies and governments in rewarding communication work. Like other parts of Europe, Italy is struggling to realise the ambitions of those advocating 'third mission' activities (broadly, 'knowledge-making and engagement with society') and extending them more generally. Scientists have busy lives and if there are no incentives to communicate, it becomes a low priority. They also can become discouraged when their communication efforts have little apparent influence on policy—scientists can be slow to realise that the kind of reliable knowledge they produce is only part of what contributes to a whole society, and that culture, economics and politics play their part in the decision-making process as well.

4. 'Science communication': Terminology and interpretation

In the course of compiling the chapters, we found that the term 'science communication' has many definitions and not all researchers or practitioners agree on its goals and boundaries. It has been variously described as an objective, goals, a process, a result and an outcome. This confusion over a definition is reflected in the terminology used internationally for the field. From the second half of the 20th century, what we have chosen to call 'science communication' for this book has flown under different headings: 'science popularisation, 'public understanding', 'vulgarisation', 'social appropriation of science and technology', 'public understanding of science' and 'scientific temper' for example. In all, the chapters mention 24 separate terms for the expression 'science communication' that we chose. We have taken note of that variety.

Most authors, consciously or unconsciously, did not define what they were writing about. Only five chapters included a definition: the US, China, Colombia, Denmark and Uganda. This suggests that an examination of the terms listed above, their evolution and an exploration of their specific meanings presents an opportunity for further research. Do the terms mean the same thing? How has usage changed over time? There appears to be no agreement on a commonly accepted definition, although a consensus is developing across the field generally that in science communication we are talking about a young transdisciplinary field, still developing and evolving, but not yet regarded as a discipline. The multiplicity of appellations implies that the field is not fixed and nor are its practices.

This observation is valid for all countries, regardless of the attempts by a number of researchers to circumscribe science communication as a field of practices and to apprehend individual practices as singular research objects. The heterogeneity of practices implies the heterogeneity of the field, and it is this heterogeneity that has led to the wide variety of terms being used for 'science communication'. In other words, the practices and discourses grouped under various different terms in different countries are the result more of the consensus of actors at a given time rather than of a genuine attempt at the objectification of these practices and discourses. One can wonder if such an objectification is feasible or even desirable.

What we can conclude from the chapters is that 'science communication' (by whatever term) is used to inform, engage, persuade, change behaviours and support better decision-making. Science communication aims to lift the social, environmental and economic standing of a nation's people. Authors report that science communication revolves around problems in diverse regional and cultural contexts: health, economic opportunity and jobs, urban resiliency, food and agriculture, clean energy, managing the development of new technologies and innovation. It may also support the participation of citizens in setting the agenda for scientific research, a democratic motivation. Italy has recently adopted initiatives to 'actively involve citizens and civil society organisations and develop a public debate'.

5. Timelines

There is a timeline at the end of every country's chapter showing key dates for that country. There is also a series of graphs and tables in the chapter following the introduction, amalgamating the national data and allowing an international comparison of (for instance) the date of the first interactive science centre or the first university courses in science communication. The collected dates give an indication as to how the field has progressed—as an academic field, as a domain of practices, as a field of practitioners and a subfield of the media.

A word of caution on the timelines. They are indicative rather than precise and show broad directions in an ecosystem of science communication. They should be regarded as markers rather than as definitive, a start to collecting meaningful episodes in the development of science communication. We realise that in some or even many cases, these records will not be the right episodes to explain the story of the development of science communication. The choice of indicators is not arbitrary although they include a degree of uncertainty. These four great generic domains are interlinked insofar as any social organisation is constituted of actors, practices and modes of communication and interaction. They should be objects of scepticism and question, not intended as a definitive account of 'the way things unfolded'. In short, they are a beginning, perhaps a controversial one, not an end.

The timelines have also posed practical difficulties, for a multitude of reasons. It was hard to identify 'firsts' because organisations changed their names, or events like science festivals stopped and started again, or regional events became national. Even the definition of what might be called a science 'festival' or a science communication degree is not always clear. People who became notable science communicators may have done their PhDs in a related area like science journalism, and what are now full courses in science communication may have begun as semester-length programs or in a different faculty. Nor are the editors or authors historians, but we do have a good grasp of the field and its evolution.

This book marks a milestone. It documents modern practice, the theories that underpin it and explains how countries have reached the position they are in today. For many participants in science communication, progress will appear to have been haphazard, characterised by interruptions, funding crises and the absence of clear and consistent policy. Recording these national stories will bring some sense of order to the process. What we have done is already useful, but it is only by a close examination and interrogation of the chapters themselves that the full value of the work of the authors will be realised.

We need to extract the lessons from the book. The questions listed at the beginning of this introduction are a starting point for the interrogation of what has been called 'a rich cornucopia of knowledge'. Understanding where we have come from and why we are here can only help the field in which we work. Our next task is to plan for an examination and interpretation of the content of the chapters.

We encourage readers to explore the book and consider the different cultural context within which each chapter was written.

References

Brooks, H., Adams, J. B., Colombo, U., Crozier, M., Kaysen, C., Kristensen, T., Okita, S., Weizäcker, C. F., Spaey, J., King, A. and Salomon, J.-J. (1971). *Science Growth and Society: A New Perspective.* Paris: OECD.

Bush, V. (1945). *Science: The Endless Frontier.* Washington DC: United States Government Printing Office.

Henriques, L. and Larédo, P. (2012). Policy-making in science policy: The 'OECD model' unveiled. *Research Policy, 42*(3), 801–816. dx.doi.org/10.1016/j.respol.2012.09.004.

Massarani, L., de Castro Moreira, I. and Lewenstein, B. (2017). A historical kaleidoscope of public communication of science and technology. *Journal of Science Communication, 16*(3), E. doi.org/10.22323/2.16030501.

Nkrumah, K. (1963, 25 May). Speech to the first conference of the Organisation of African Unity (OAU). Addis Ababa, Ethiopia.

OECD. (1981). *La politique scientifique et technologique pour les années 80.* Paris: OECD.

Piganiol, P., Herz, K., Major, R., Ramsey, N. F., Schmidt, E. I., Schultz, T. W. and Steacie, E. W. R., (1963). *Science and the Policies of Governments – The Implications of Science and Technology for National and International Affairs.* Paris: OECD.

Schiele, B. (1994). *When Science Becomes Culture. World Survey of Scientific Culture.* Éditions MultiMondes and University of Ottawa Press.

Trench, B., Murphy, P. and Fahy, D. (eds). (2017). *Little Country, Big Talk: Science Communication in Ireland.* UK: Pantaneto Press.

UK House of Lords. (2000). *Third Report: Science and Society.* Select Committee on Science and Technology.

Contributors

Toss Gascoigne is a visiting fellow at the Centre for Public Awareness of Science at The Australian National University, Canberra, Australia.

Bernard Schiele is a researcher at the Interuniversity Research Centre on Science and Technology, and Professor of Communication at the Faculty of Communication at the University of Quebec, Montreal, Canada.

2

THE TIMELINES

A broad-brush analysis

Bernard Schiele and Toss Gascoigne

The chapters compiled in this book bear witness to the development of science communication in 39 countries. The objective was twofold: to present the reader with a global understanding of the pathways different countries followed, and then to provide a means of comparison. The timelines at the end of each national chapter were constructed with this in mind, using indicators that are both comparable and recognised.

They reflect three conceptions of science communication:

- as an area of practice (indicators such as the date of establishment of an association of science writers or journalists or communicators, or the first interactive science centre)
- as a media subfield (indicators include the date when the first significant radio or television programs on science were broadcast)
- as an academic field (indicators include the dates when university courses to train science communicators were established, or when the first master's or PhD students in science communication graduated).

Other indicators were also collected, such as the date when the first national government program to support science communication was established, or the launch of the first significant initiative or report on science communication.[1]

1 At the risk of oversimplifying, science communication exists at the point of articulation of autonomous or semi-autonomous fields in interaction with one another: the academic field, the media field and domains of practice, all linked to specific fields or as sets of practices in a process of autonomisation. (Cf. Schiele, 2019; Gascoigne et al., 2010.)

In the introduction, we urged caution when considering the timelines. We said the timelines should be objects of scepticism and question, not intended as a definitive account of 'the way things unfolded'. In short, they are a beginning, perhaps a controversial one, not an end. Readers might look back at the introduction to check our reservations.

The world is complicated. Actors in science communication changed the role they played and intervened on several fronts, and any categorisation is in itself an oversimplification (as is the creation of indicators). They may, for instance, have performed in several different roles at the same time: as journalists, radio or television hosts; promoting or contributing to the institutionalisation of science communication by establishing or supporting an association; and advocating for university training (by creating programs or becoming university professors).

Therefore, the story is more complicated and cannot be reduced to 14 indicators. Trying to read the picture of science communication through these indicators makes it difficult to grasp the complexity of particular situations that governed its emergence and development. Only the chapters can do that.

However, an analysis of the timelines does show us which areas of activity have been significant from one country to another in the development of science communication, and when. It sheds light on each country's important fields of activity and their moments of significance. Therefore, we can deduce the anchor points of the actions of actors in science communication and how from there they have interacted with the other fields. And it is on this basis that a comparative approach can be undertaken.

A second limitation is the way authors interpreted the 14 questions and their knowledge of their country's history of science communication. For example, the San Francisco Exploratorium (United States) and the Ontario Science Centre (Toronto, Canada), both inaugurated in 1969 and both dedicated to visitor engagement through a hands-on approach (and thus breaking with the passive approach of traditional science museums), are commonly held to be the first interactive science centres. Or does that position belong to the Palais de la Découverte (Paris, France), which opened in 1937 and was deliberately designed without a permanent collection? Even within a country there can be debates: was the first interactive centre in France the Palais, or the Cité des Sciences et de l'Industrie, which opened in 1989 specifically with a hands-on approach?

Such questions arise on nearly all issues. Associations of science communicators are formed, dissolved and re-formed. Which date is the most significant, that of a foundation or of a re-foundation? Science communication university programs are created, cancelled and re-created. Again, which date is the appropriate one to record? In the end, we decided to acknowledge this uncertainty by accepting the dates as they were nominated in the chapters, although we are fully aware of the limitations of such an approach. This is why we regard the largely descriptive analyses that follow as offering only a general overview, an overview that could be refined with further research.

You will notice the graphs include phases of development. A 'phase' is when dates are clustered together. They are separated from the next phase by a gap of at least five years, so each cluster of dates (or 'phase') is separated by at least five years from the next one. Phases give a useful indication of concentrations of activity.

Table 2.1: Country international codes.

Aotearoa New Zealand	**NZ**	Iran	**IR**	Portugal	**PT**
Argentina	**AR**	Ireland	**IE**	Russia	**RU**
Australia	**AU**	Israel	**IL**	Singapore	**SG**
Brazil	**BR**	Italy	**IT**	South Africa	**ZA**
Canada	**CA**	Jamaica	**JM**	South Korea	**KR**
China	**CN**	Japan	**JP**	Spain	**ES**
Colombia	**CO**	Malaysia	**MY**	Sweden	**SE**
Denmark	**DK**	Mexico	**MX**	Taiwan	**TW**
Estonia	**EE**	Netherlands	**NL**	Thailand	**TH**
France	**FR**	Nigeria	**NG**	Turkey	**TR**
Germany	**DE**	Norway	**NO**	Uganda	**UG**
Ghana	**GH**	Pakistan	**PK**	United Kingdom	**GB**
India	**IN**	Philippines	**PH**	United States	**US**

1. Indicators of practice areas

1.1. First interactive science centre

All countries except Jamaica nominate a date for the creation of the *first interactive science centre*. Although the range is extensive, from 1914 in Norway to 2009 in Ireland, the trend really starts in 1975. Before that date, countries that opened science centres displayed major economic and cultural differences. How to account for the fact that although the USSR was a scientific powerhouse, symbolised by the 1957 success of Sputnik, Russia only built its first science centre in 2005?

Four stages are apparent:

- Precursors: Norway 1914, and the United States 1936
- Phase 1: Taiwan 1956, and India 1959
- Phase 2: Canada 1967, and Mexico 1970
- Phase 3: From 1975 until 2009, when all other countries developed their centres, with an average of two new science centres opening each year between 1977 and 2001.

The creation of a science centre is symbolic of the newfound importance of science and technology in a society. Although countries compete with one another to stress unique qualities of their own science centre, it is more likely that centres share strong similarities in concept, format and activities.

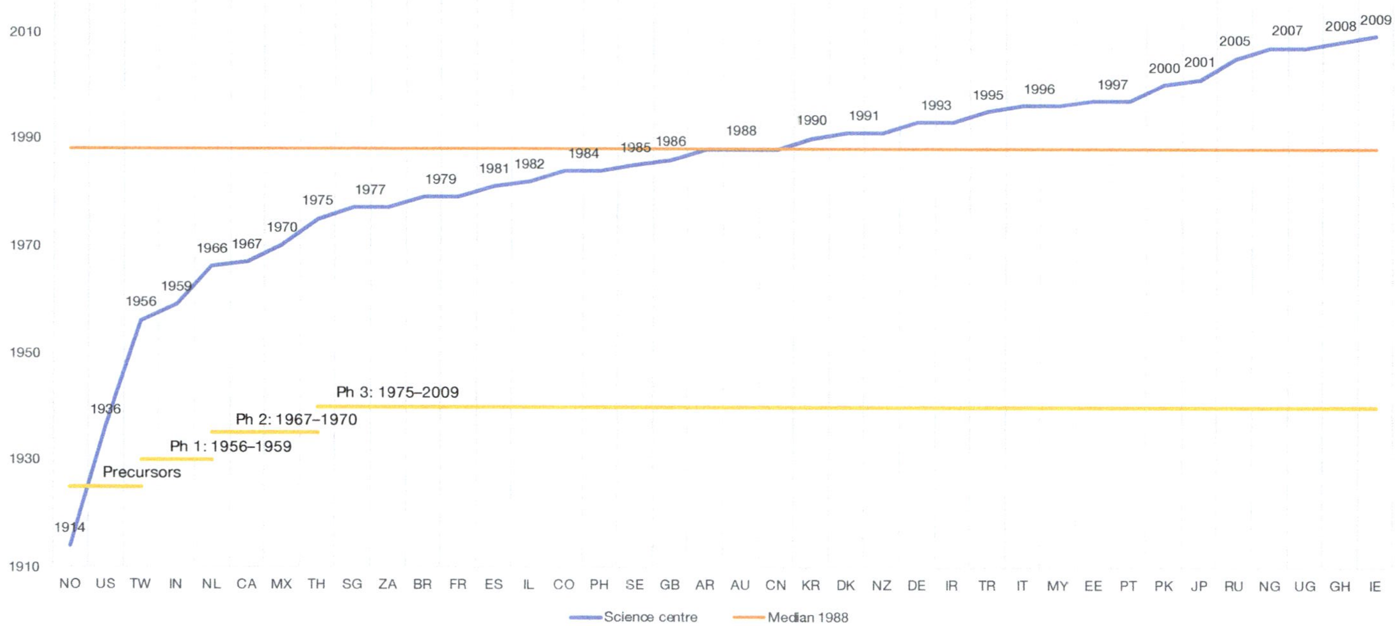

Figure 2.1: Date the first interactive science centre was established.

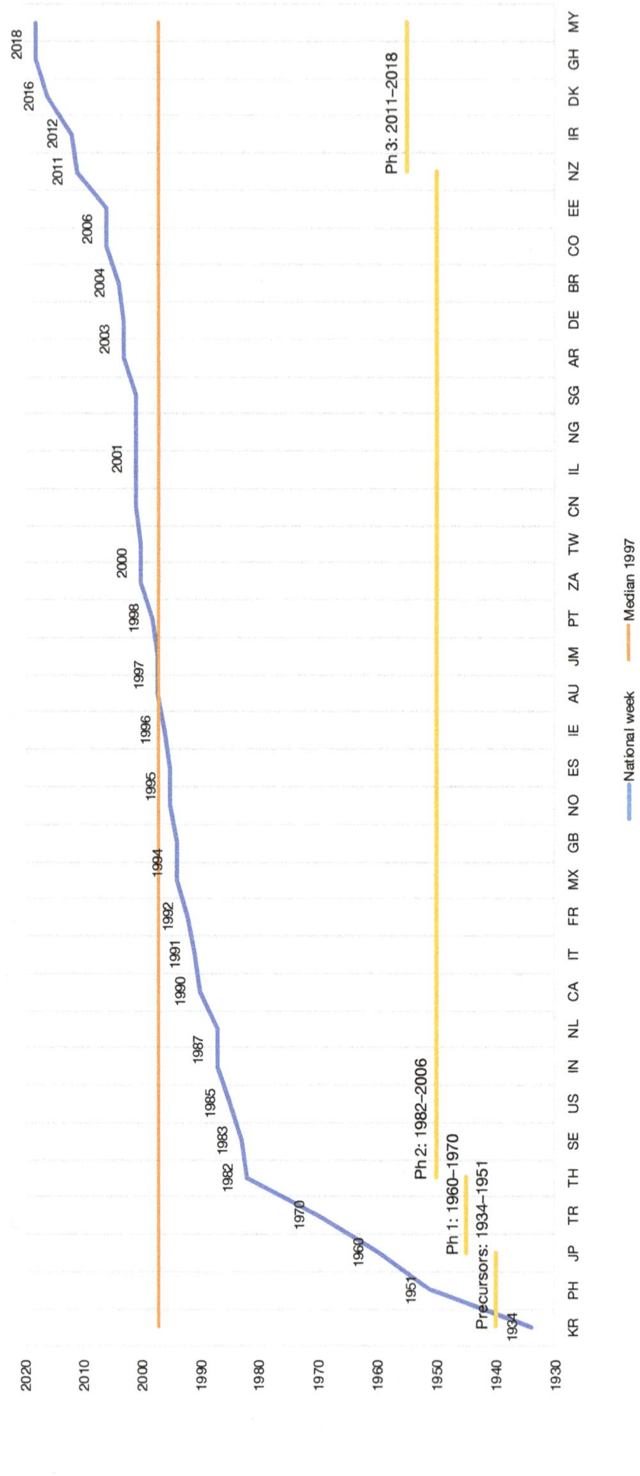

Figure 2.2: Foundation date of national science week.

1.2. National science week

Thirty-six countries report that they hold a science week, with South Korea claiming the first in 1934 and Malaysia the most recent in 2018. Pakistan, Russia and Uganda do not provide a date. The gradual creation of science weeks follows the pattern of the development of science centres.

There were four stages of activity:

- Precursors: South Korea 1934, and the Philippines 1951
- Phase 1: Japan 1960, and Turkey 1970
- Phase 2: 1982–2006: a period over which 27 countries stated a science week, and peaking between 2000–2006
- Phase 3: 2011–2018: five more countries.

The first countries to build science centres were not the first to hold a national science week. There is apparently no coordination between science communication initiatives: the development of flagship science communication initiatives like science centres does not immediately stimulate other events on a national or even regional scale. There is a time gap, of 20 years on average, between the creation of the *first science centre* and the *first national science week*, with the science centres generally being formed first. The median (mid-point) for the founding of science centres is 1988, compared to a median of 1997 for science weeks.

Most countries chose to launch a *science centre* before they inaugurated their *national week*, but eight countries began with the science week: South Korea 1934; the Philippines 1951; Japan 1960; Turkey 1970; Sweden 1984; Italy 1991; Ireland 1996; and Nigeria 2001. It is interesting to note that many of the pioneers were lower-income countries; perhaps their national science week was an expression of commitment to science communication in a context of limited resources.

1.3. First science festival

As with the science week, 36 out of 39 countries reported that they organise science festivals or science fairs, the exceptions being India, Portugal and Uganda. In some countries these coincide with the *national science week*, the two events being mutually reinforcing. With the exception of the *science fair/festival* in the United Kingdom in 1881 (a precursor and not on the graph), the movement started in the Philippines in 1951.

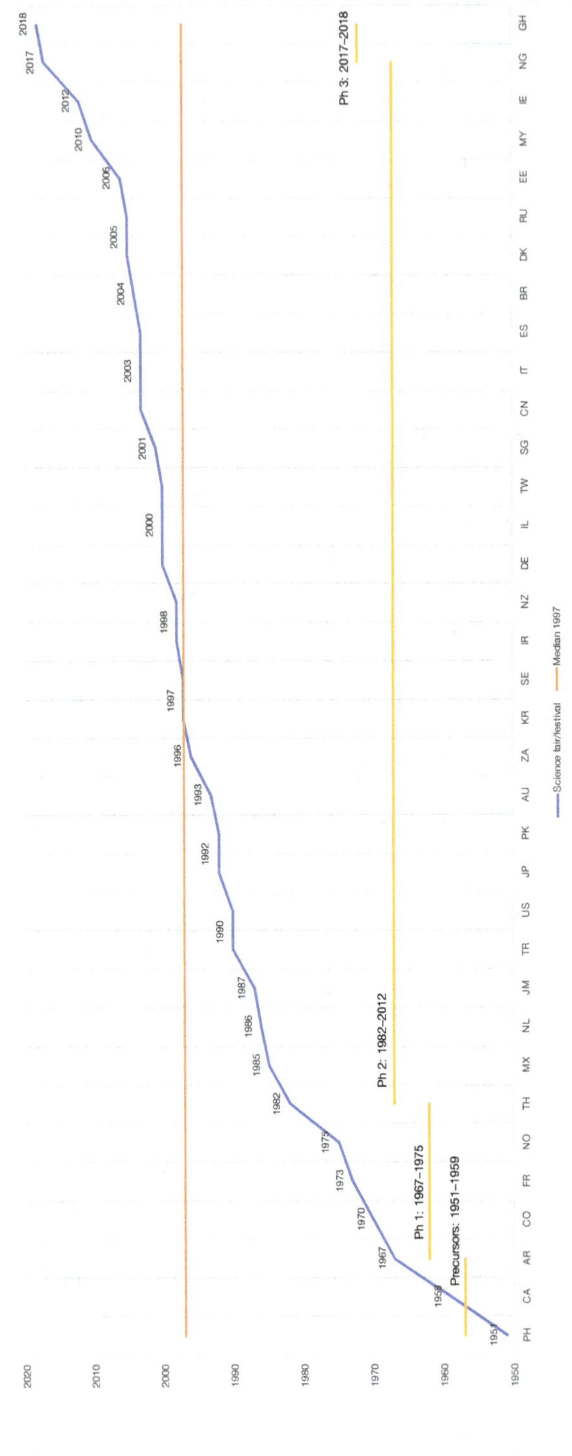

Figure 2.3: First national (or large) science fair/festival.

Science festivals progressed in four stages, following one in the UK in 1831 (not shown on graph):

- Precursors: 1951–1959: the Philippines and Canada
- Phase 1: 1967–1975: Argentina, Colombia, France and Norway
- Phase 2: 1982–2012: the longest, with 27 countries, starting in 1982 with Thailand and ending in 2012 with Ireland
- Phase 3: 2017–2018: Nigeria 2017 and Ghana 2018.

The organisation of a science festival was a marker for any country seeking to express its interests in science communication, since the event ensures science a greater visibility in the public sphere (media coverage, school trips, museum programs and so on). The creation of science festivals generally took place in a country after they had built a science centre, and the median date across the 36 countries is 1997. This is the same as the median for the creation of *national science weeks* and suggests the two events are linked or have strong ties to one another, since a science week can be the pretext for a festival. Out of the 36 countries, 15 created their science festival after the turn of the millennium, and 11 between 2000 and 2006.

1.4. Association of science writers

Thirty-three countries have an association for science writers and journalists, with the exceptions of Iran, Israel, Jamaica, Malaysia, Pakistan and Singapore. The foundation of such an association signals two things: that an emerging profession has recognised itself as such and claims its own specificity and legitimacy; and that by discussing scientific issues, a subset of writers and journalists are specialising in science.

There were four stages:

- Precursors: 1929–1934: Germany in 1929, followed by the United States in 1934
- Phase 1: 1946–1955: Seven countries founded associations (it is noteworthy that this second phase started at the end of WWII)
- Phase 2: 1962–2005: 22 countries founded associations
- Phase 3: 2014–2016: Ghana 2013, Portugal 2014, and Russia 2016.

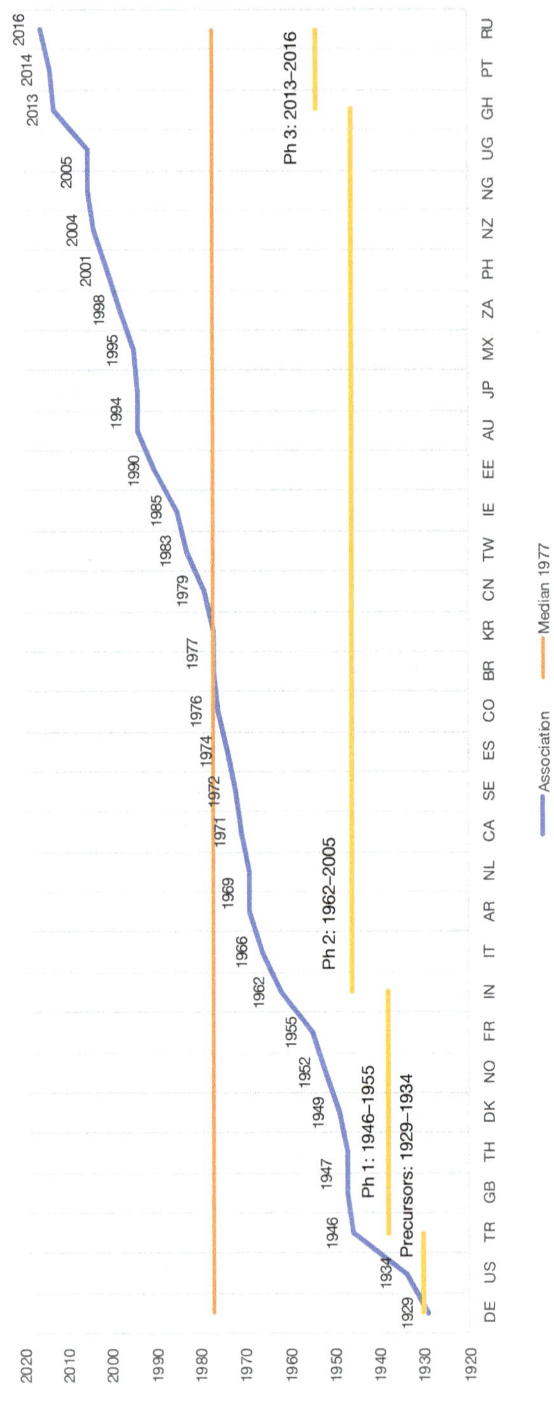

Figure 2.4: Association of science writers or journalists established.

The median year was 1977. It can be argued that the field of science communication began as a grouping of actors engaged in science writing. Popularisation or vulgarisation—expressions used to refer to science communication until the turn of the 1960s—enjoyed a golden age in the 19th century. The written medium was the dominant form, although lectures were also important (especially for natural science museums) and live demonstrations of physical or chemical phenomena were held during fairs.

The written format, governed by the rules of publishing or journalism, has left its mark on the professionalisation of science communication, if only by nominating the criteria for becoming a member. Although preceded by the first science centres, the origins of science communication are linked to the foundation of associations for science writers and journalists.

1.5. Science communication awards

Thirty-five countries have created an award for science communication by scientists, journalists and others, with only four countries not mentioning it in their national timelines (Jamaica, Japan, the Netherlands and Singapore).

Again, the chronology progressed through four stages:

- Precursor: US 1946 (Westinghouse AAAS Science Journalism Award)
- Phase 1: Sweden 1972, and Canada 1973 (Ortho-Award for print science journalism)[2]
- Phase 2: 1978–1996: involved 19 countries
- Phase 3: 2001–2018: involved 13 countries.

If we ignore the five-year gap between 1996 and 2001, new prizes are constantly being created. They signal not only a newfound appreciation for the work of science communicators and the legitimacy of the field and the role it plays in society, but also the specialisation of the field. By creating such prizes, countries want to stress the importance of science communication.

2 The *Association des communicateurs scientifiques* (Science Communicators Association, Quebec) will follow suit with the creation of the Fernand-Séguin scholarship for young science journalists. A third prize was created for the promotion of science in 2001 by the National Science and Research Council of Canada.

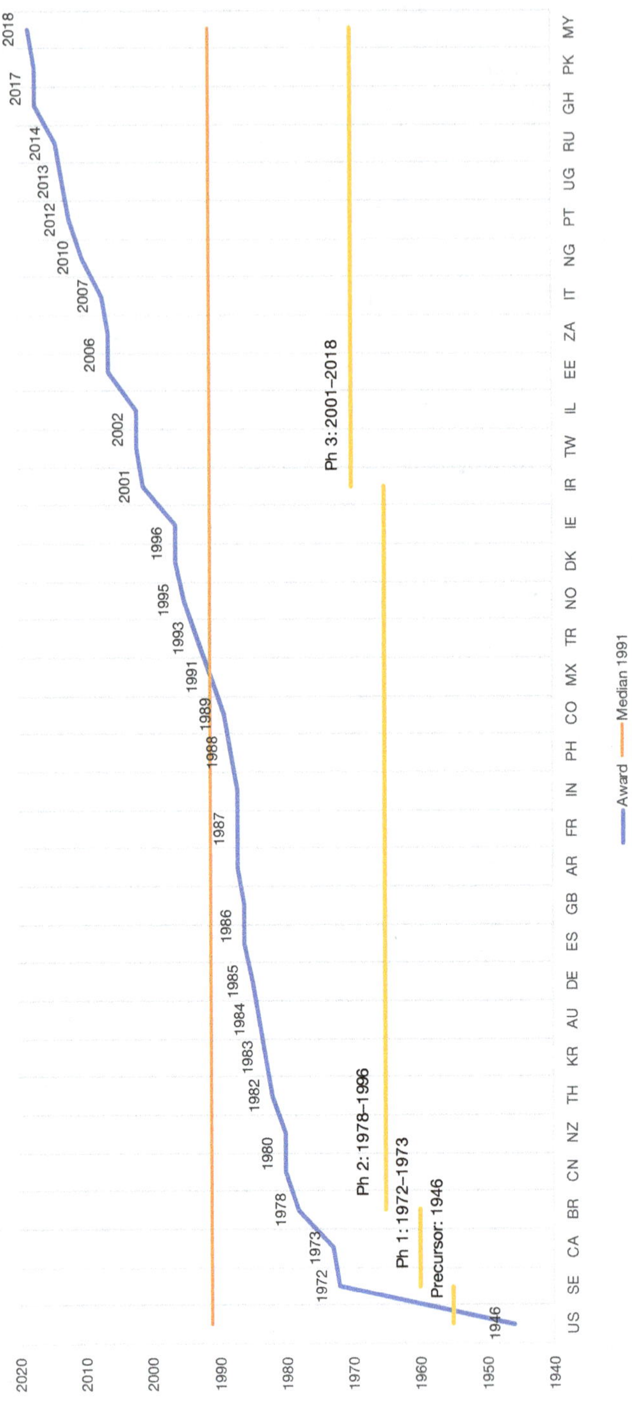

Figure 2.5: Award for scientists or journalists or others for science communication.

The median year of 1991 is slightly later than the median year for the creation of *science centres* (1988), while significantly later than the median year for creation of a *journal devoted to science communication* (1956) and of an *association for science writers and journalists* (1977). The creation of a prize signals that the constitution of a social group (science communicators) is well advanced, since the existence of such an award implies rules of communication and issues specific to this group, complementing other activities such as the journal and the association, and affirming this specific social and cultural activity.

1.6. Journals of science communication

Only 20 countries mention a journal completely or substantially devoted to science communication.[3] Although the dates for the creation of new journals can be pinpointed, they do not easily fit into phases of development. Scandinavia, whose countries, people, culture and language have close and historical ties, established its journals in 1854 (Denmark), 1877 (Norway) and 1911 (Sweden), but it cannot be said that creation of one influenced the creation of the other. The fourth journal was founded in China in 1933, 22 years later, illustrating the point that creating new journals is a slow process.

If we try to fit this activity into phases, three stages after the precursor emerge:

- Precursors: Denmark, Norway and Sweden
- Phase 1: 1938–1952: six journals were created with an average interval of three years: Australia 1938; New Zealand 1942; Argentina 1945; Thailand 1948; Brazil 1949; India 1952
- Phase 2: 1975–1979: three journals created: Mexico 1975; Colombia 1977; the United States 1979
- Phase 3: 1992–1997: the United Kingdom 1992; Spain 1995; the Netherlands 1997.

3 France mentions that examples of such journals date back to the 19th century, although it is difficult to date them accurately.

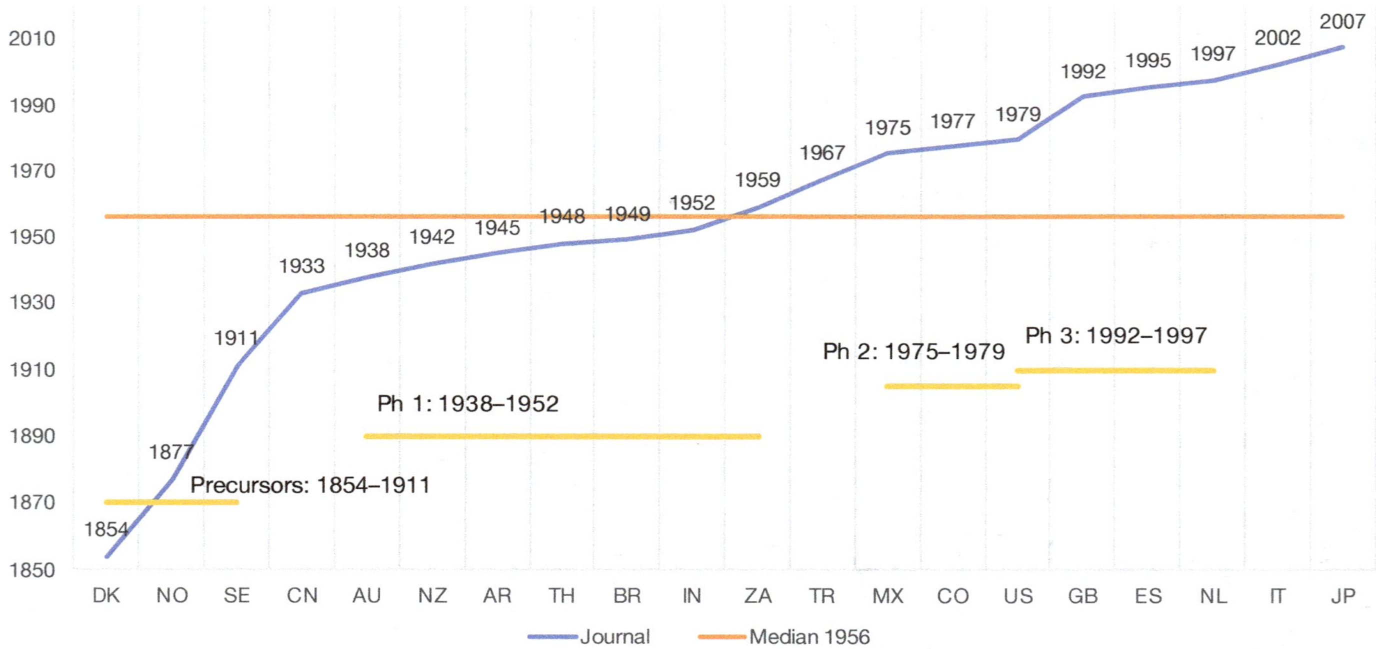

Figure 2.6: Journal completely or substantially devoted to science communication.

Other countries were scattered across the period as indicated in Figure 2.6. The median was 1956 but there does not seem to have been any general movement, as was the case for *science centres*, *national weeks*, *science fairs*, *associations* and *awards*. Half the countries surveyed (19) have no journal. While the creation of a journal may signal the beginning of the institutionalisation of science communication in any given country, the sample is too limited to draw definitive conclusions.

2. Indicators of the media subfield

2.1. Radio programs on science

The second group of indicators concerns the media field, specifically the date of the *first significant radio program on science* and the first *significant television program on science*. Radio (Figure 2.7a) came into being before television, and the majority of countries have created radio science broadcasts, or at the very least have programs that raise questions pertaining to science. Only the chapters on Japan,[4] Malaysia and Singapore do not nominate dates.

The creation of science radio broadcasts happened in seven phases extending over nearly a century, from the United States in 1920 to Ghana in 2014, with 1964 as the median year. Science radio broadcasts took off in the aftermath of WWII, and specifically from 1949 for 23 out of 39 countries.

The seven phases were:

- Phase 1: 1920–1933
- Phase 2: 1941
- Phase 3: 1949–1952
- Phase 4: 1957–1966
- Phase 5: 1976–1983
- Phase 6: 1990–2003
- Phase 7: 2010–2014.

The four most significant phases were 1, 4 and 6, each involving at least four countries.

4 The authors stated that it was very difficult to date the first science radio broadcast.

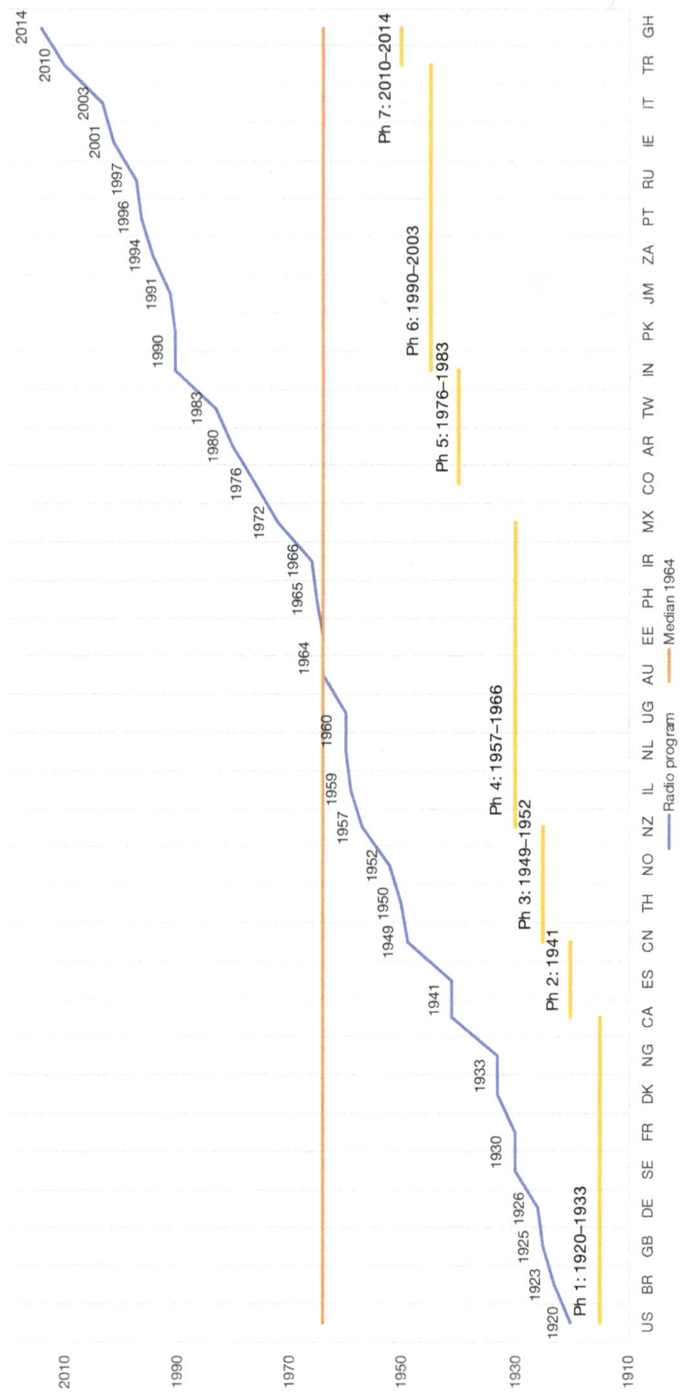

Figure 2.7a: First significant radio program on science.

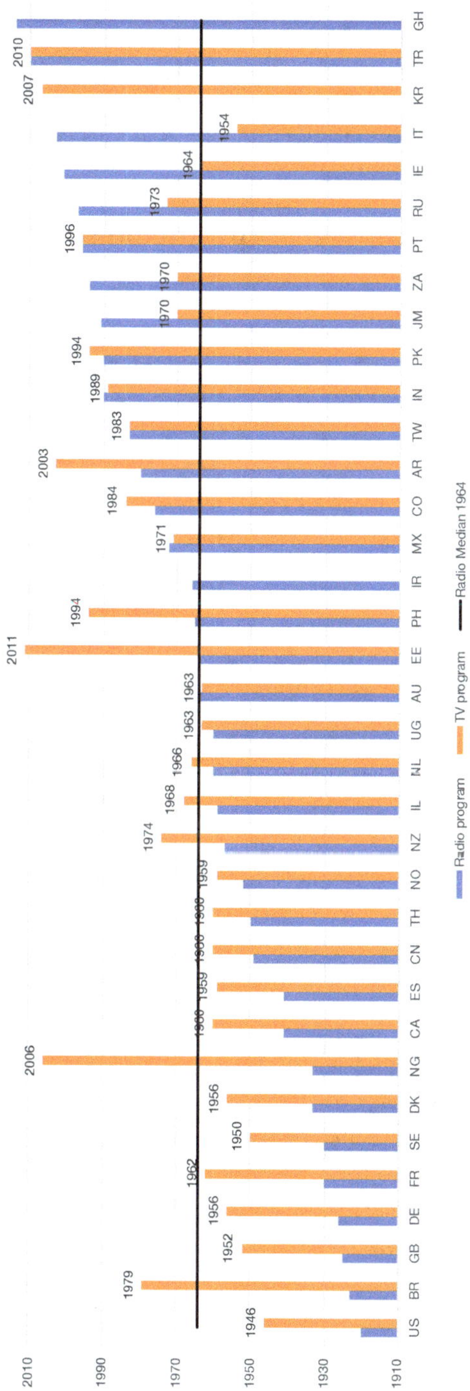

Figure 2.7b: First significant radio program on science compared to the first significant TV program on science.

The development was not continuous, nor did it seem to be evident that science topics should be raised on radio. It is significant that, although television came later, the median year for science TV broadcasts is a mere six years later than the median for science radio broadcasts, in 1970. This small difference emphasises the point that science content was not self-evident for radio producers.

2.2. Television programs on science

In contrast, the development of science TV broadcasts was more compressed. Television had huge appeal because it made it possible to show pictures of science to audiences while talking about researchers, their science and its applications. The effect was to expose viewers to worlds beyond their daily experience.

After the first burst of activity, the creation of new TV broadcasts followed at an unequal pace. There were four phases:

- Phase 1: 1946 and 1974: the first and most important phase, with 23 countries. In this period a science TV broadcast was created on average every two years
- Phase 2: 1983–1989
- Phase 3: 1994–1996
- Phase 4: 2003–2011.

Some countries did not identify dates for the first science broadcasts on either radio or TV, perhaps because they were difficult to identify or perhaps because that particular medium does not carry broadcasts on this subject. According to the authors, Singapore's (1972) and Korea's (2007) first science broadcasts were on television. Neither identified a date for radio. Neither Japan[5] nor Malaysia identified a date for the creation of TV broadcasts.

A last element: as noted, the random nature of the arrival of science on radio contrasts with the systematic nature of its arrival on television, a fact that suggests the quick realisation of television's potential (Figure 2.8a). The US, according to the authors, were the first to adopt radio (1920) and television (1946). A handful of countries even started using radio decades after adopting television, such as Jamaica (1970, 1994) and South Africa (1970, 1991).

5 The authors stressed that it was very difficult to date the first TV broadcast on science.

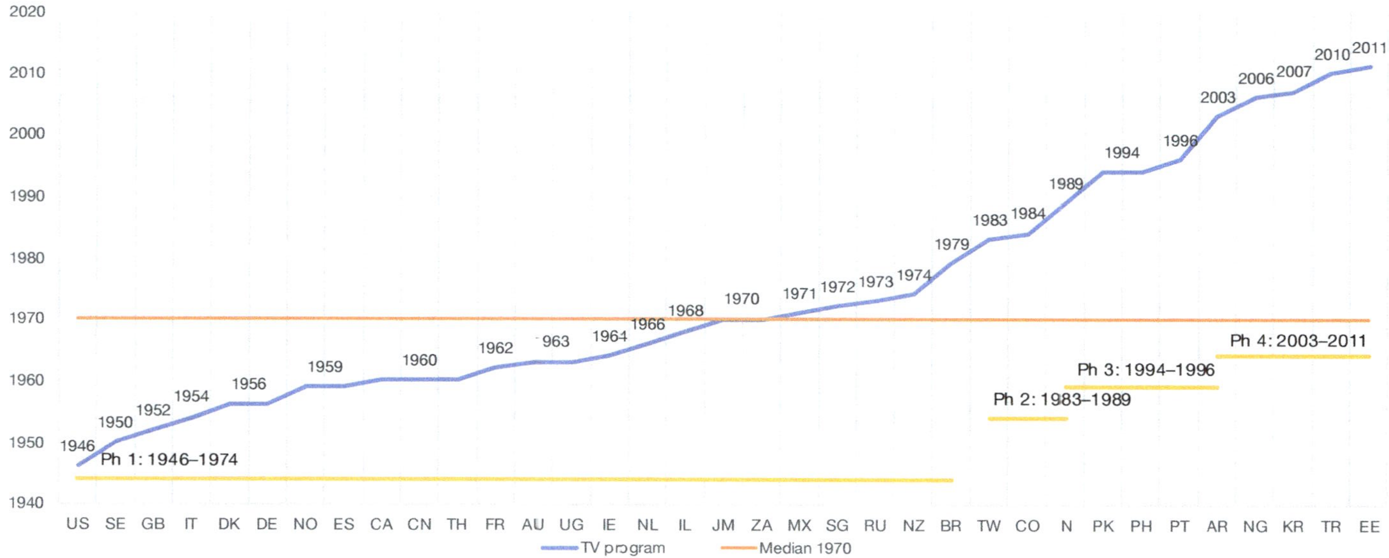

Figure 2.8a: First significant television program on science.

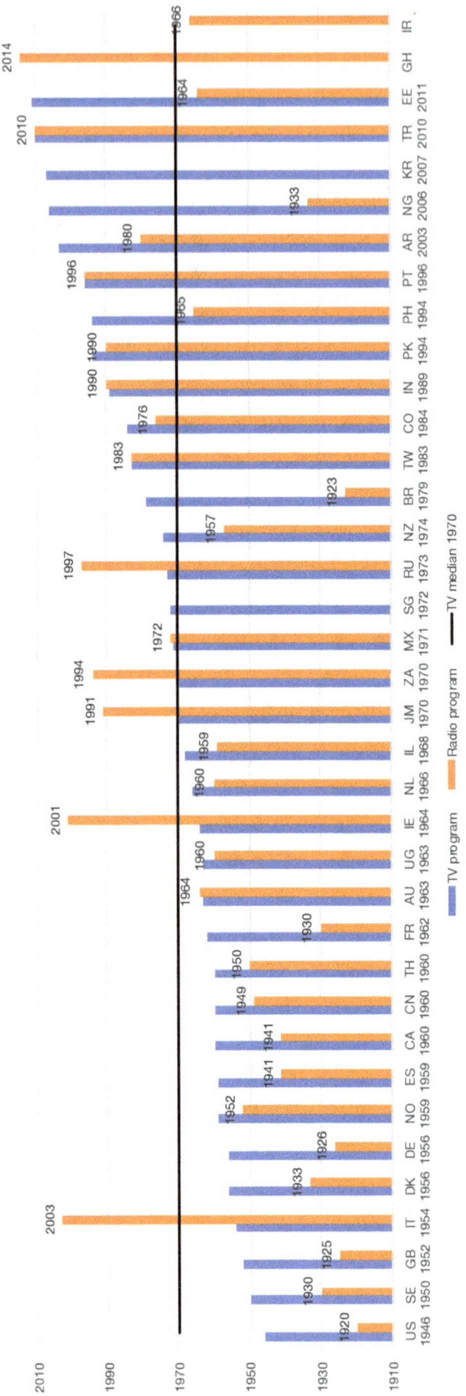

Figure 2.8b: First significant television program on science compared to the first significant radio program on science.

3. Indicators of the academic field

3.1. Courses at universities

The beginnings of systematic training in science communication, pioneered by the Philippines and the United States in 1960, are an indicator of the slow, yet continuous, professionalisation of the field over the next 60 years. Ghana (2019) is the most recent country to offer courses. Figure 2.9 shows that 35 countries offer university courses, with the exception of Malaysia, Pakistan, Singapore and Uganda.

These may be master's courses, or units in science or journalism (or other disciplinary areas) degrees, or certificate or diploma courses. Only one country identified a full undergraduate course in science communication (France).[6] In 1984, Denise Devèze-Berthet set up the first training courses (bachelor's and graduate) focused on 'communication and scientific, technical and medical information'. One year later, in Tours (France), Jean Lagoutte and colleagues created bachelor-level training at the Institute of Technology (IUT) in Tours, focused more on scientific mediation.

This development took place in one extended phase, following a lone precursor:

- Precursor: 1960
- Phase 1: 1976–2019: a new program created every 15 months on average.

The necessity of systematic training gradually became self-evident. The creation of university programs coincided with the development of academic research on questions of science communication[7] and the growing interest of states for the dissemination of scientific thought.[8]

6 We could consider this apparent absence of undergraduate courses as a possible effect of the wording of the question asked: *First university courses to train science communicators*. The dates people identified here can be seen as a combination of several things: master's degrees, or units of degrees in science or journalism or other undergraduate courses, or perhaps diplomas or certificate courses.

7 Among others: Withey, 1959; Kriegbaum, 1967; Funkhouser and Maccoby, 1970; Jurdant, 1970.

8 As a reminder: the various studies commissioned by the European Council in view of the organisation of the Strasbourg Conference of 1969 or the 1970 UNESCO report on the diffusion of sciences.

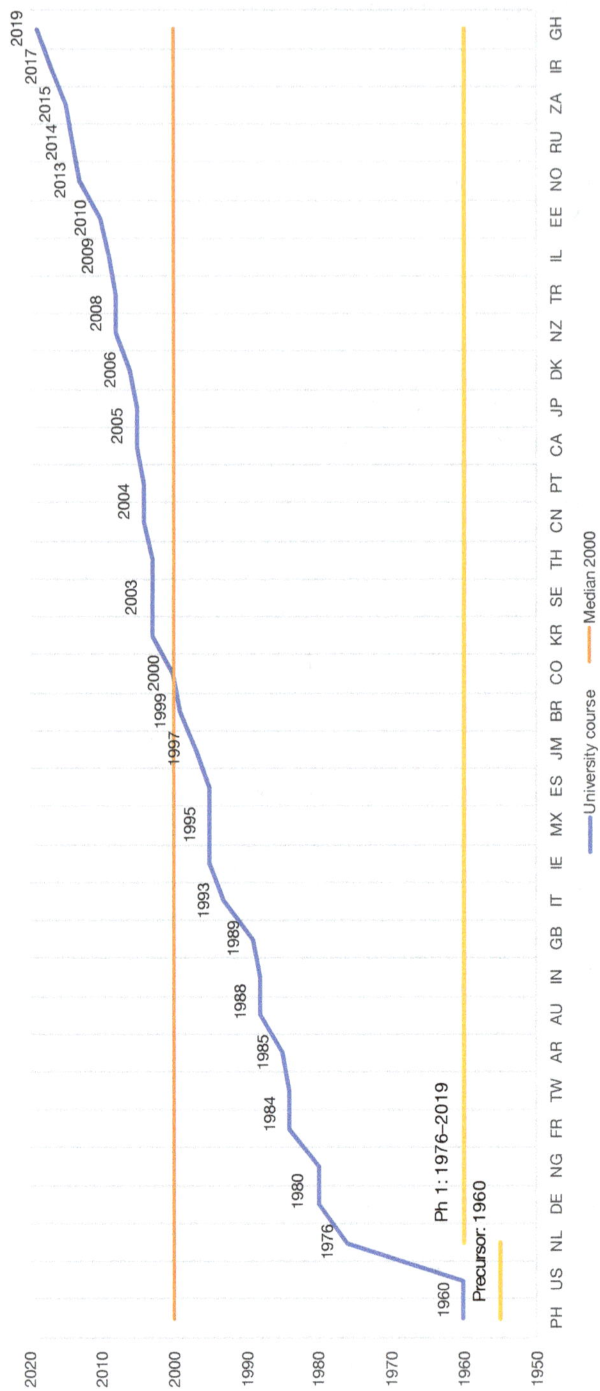

Figure 2.9: First university courses to train science communicators.

Seventeen countries began some form of training in science communication before 1999. The median year is 2000, with 18 countries commencing their training after that date.

Almost all countries report that they offer training in science communication at some level. There are five broad possibilities for this:

1. units, subjects or certificates as part of a degree in disciplines such as science or journalism
2. a diploma, as an extra qualification for a person holding a bachelor's degree
3. an undergraduate degree in science communication (we found only one report of such a degree, in France)
4. master's degree
5. PhD.

A few countries do not offer any training or research training in science communication, but people reported that they travelled overseas to undertake training (usually to earn a postgraduate qualification) before taking up positions in their home country.

3.2. First master's and PhD programs

Figures 2.10 and 2.11 should be analysed together with Figure 2.9, since they make clear not only the professionalisation of the field of science communication but its academic legitimisation. Although the development of a master's program (Figure 2.10) reflects the gradual development of criteria of competence sanctioned by training, PhD programs entail a more systematic participation in the academic field (journals, conferences, research programs and so on), and convey additional legitimacy and autonomy.

Figure 2.10 shows that the first master's in science communication was offered in the United States in 1960. There was a 24-year gap separating that from the second (France 1984). We do not have enough details to draw conclusions as to whether master's programs are an extension of undergraduate courses, which might be units in science or journalism or other degrees. Nor do we know what prerequisites were required to enter the master's program: a degree in science, social sciences or another discipline.

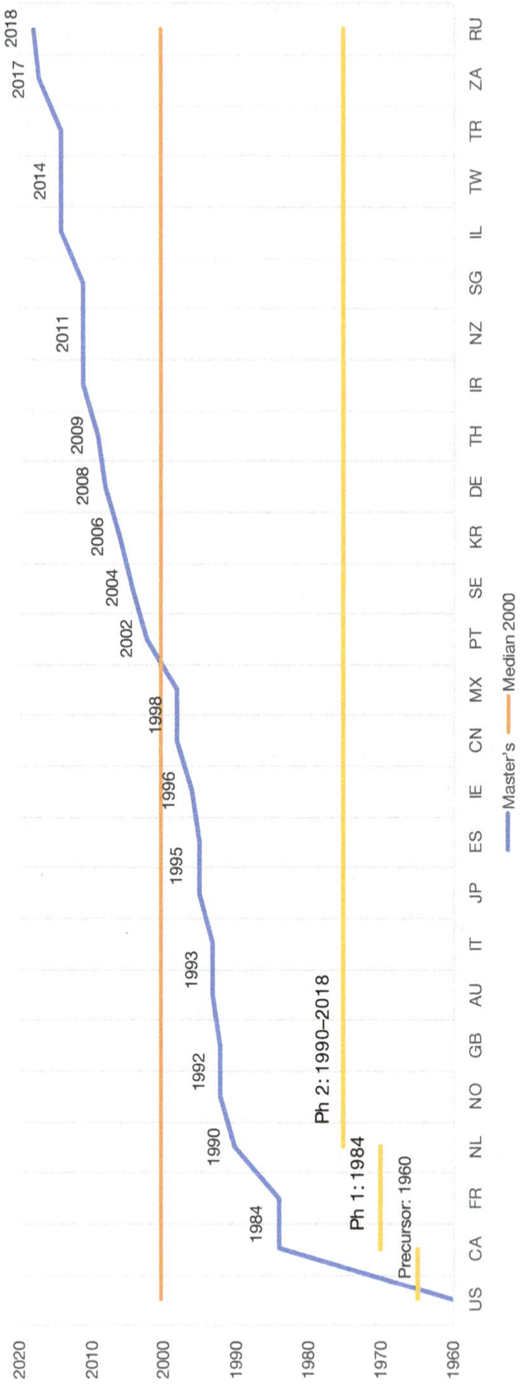

Figure 2.10: First master's degree in science communication to graduate.

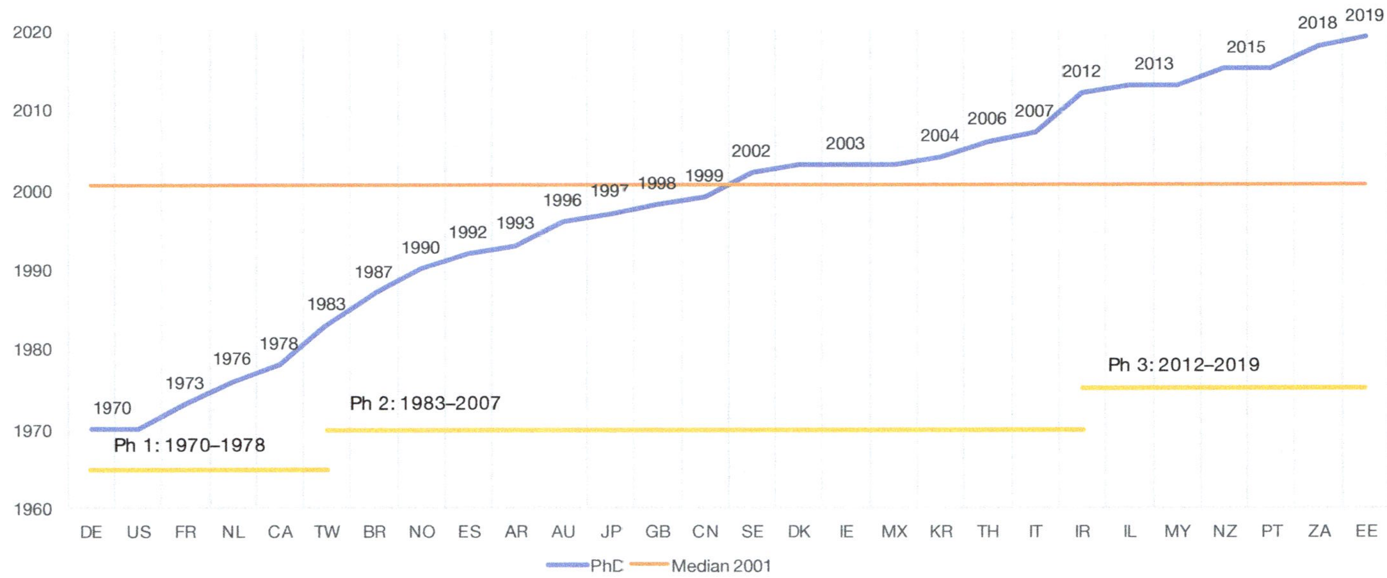

Figure 2.11: First PhD in science communication to graduate.

Twenty-six countries award master's degrees, with a median year for starting these courses of 2000. Figure 2.10 shows that it is possible to distinguish two major phases after the precursor:

- Precursor: 1960
- Phase 1: 1984
- Phase 2: 1990 on, accelerating to an average of a new country every 14 months.

Figure 2.11 shows that the development of PhDs took place in three phases:

- Phase 1: 1970–1978
- Phase 2: 1983–2007
- Phase 3: 2012–2019.

A handful of countries created PhD programs before creating a master's course. These include Canada and France (both in 1984). A comparison of the graphs tracks the patterns in the establishment of master's courses and PhDs. The differences are insignificant.

3.3. National conferences

Thirty-two countries reported having organised a national conference on science communication. This suggests a consolidation of the field since it brings together actors in this field, giving them a venue to collectively debate issues of importance. Beyond their affirmation as a group, it also gives them a visibility within society while distinguishing them from other groups of social actors.

Canada (1933) and the United Kingdom (1943) began the process, and the four subsequent phases were:

- Phase 1: Spain 1958, and Germany 1960
- Phase 2: France 1969, Colombia 1969, and US 1970
- Phase 3: 1990–1996: the movement gained momentum with six countries
- Phase 4: 2002–2019: after a six-year gap, 17 countries followed suit.

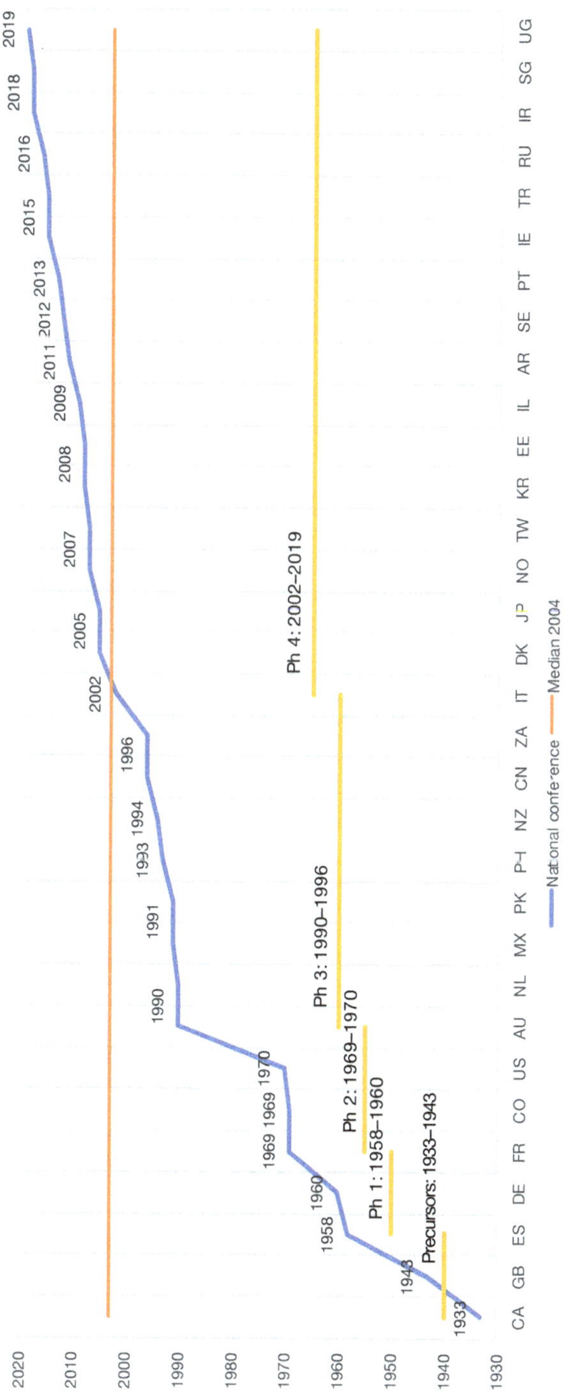

Figure 2.12: First national conference on science communication.

The first conference of the Network for the Public Communication of Science and Technology took place in Poitiers, France, in 1989. This was after the first national meetings, but perhaps the surge in national events from 1990 and 2002 owes something to having an international example to follow. The arrival of the internet and the web, to link interested parties and publicise events, made the job of organising meetings easier from the mid-1990s.

To an extent, because phases overlap with one another, we observe similarities between the institutionalisation of research and the institutionalisation of training in science communication, strengthening and affirming the community of social actors engaged in the field.

3.4. Initiative or report on science communication

At first glance, the graph on the first initiative or report on science communication differs from others. Only 28 countries make mention of an initiative or report, and Figure 2.14 shows they were discontinuously produced. Between 1933 and 1993, a 60-year period, initiatives are sporadic (one every five years on average), while the five phases that can be observed are so short that their significance may be questioned:

- Precursor: France 1937
- Phase 1: 1953–1957
- Phase 2: Colombia 1965
- Phase 3: Spain 1970
- Phase 4: 1975–1977
- Phase 5: 1985–1988
- Phase 6: Norway 1993
- Phase 7: 1999–2004
- Phase 8: 2009–2015.

The median year is 2001, and from this year the number of reports increases, with a total of 16 (one per year on average). And it is from that date on that we observe a convergence with university training and national conferences, and these dynamics may strengthen one another.

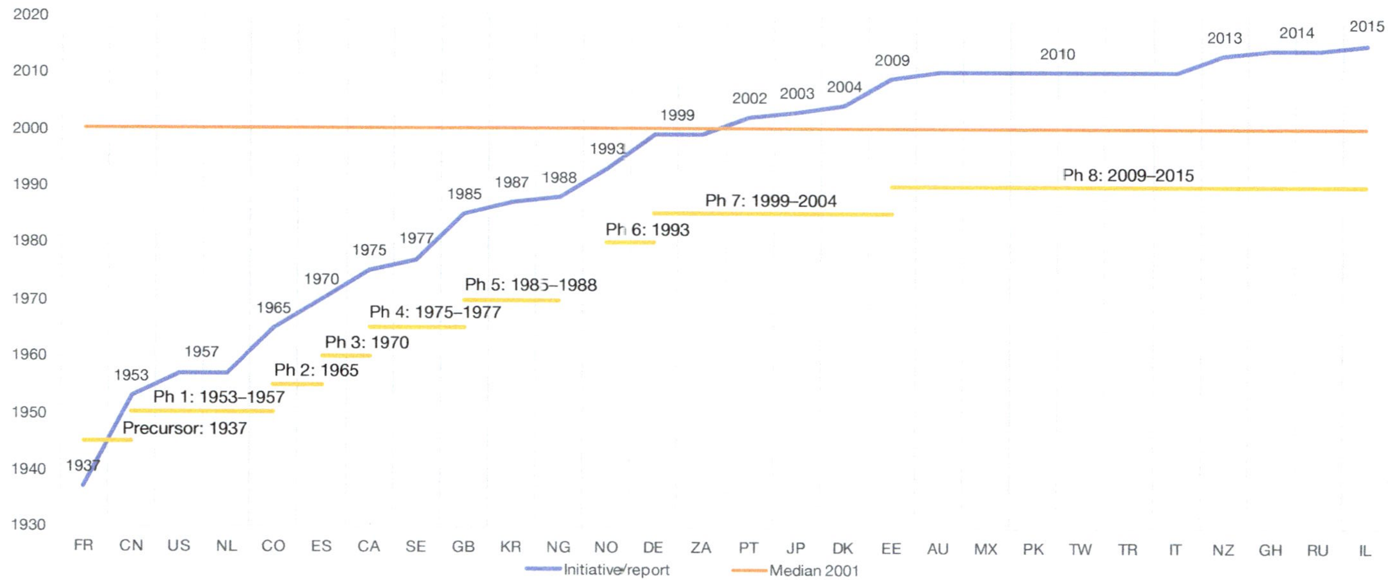

Figure 2.13: First significant initiative or report on science communication.

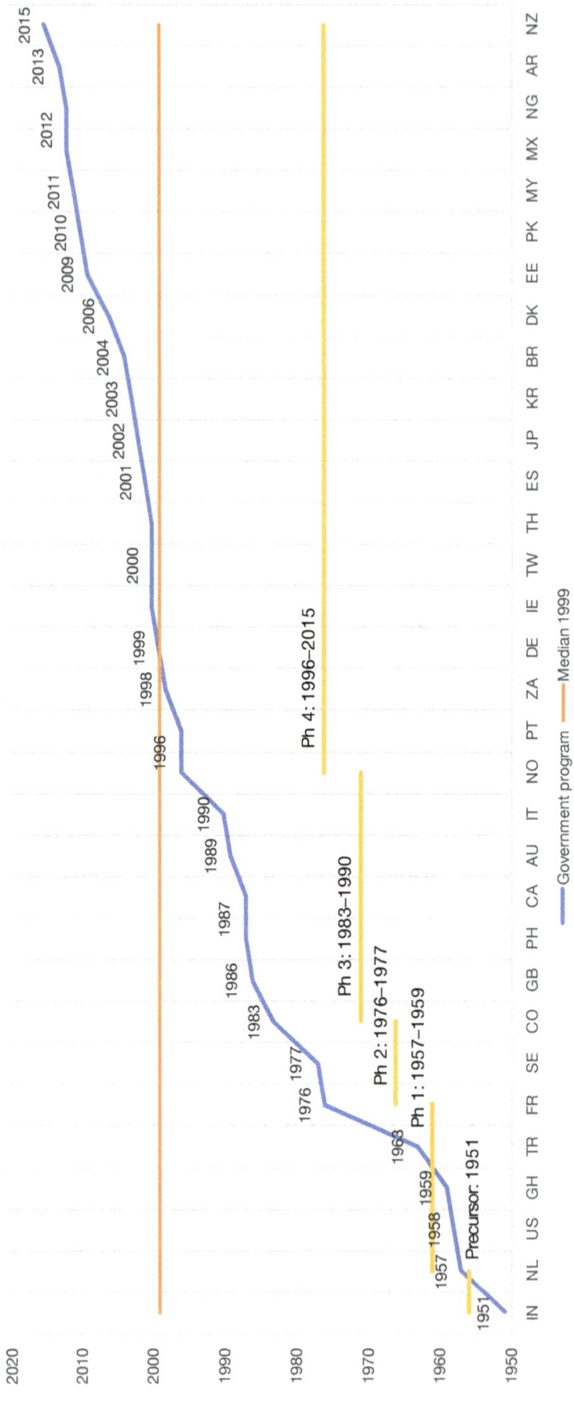

Figure 2.14: National government program to support science communication.

3.5. National programs on science communication

Thirty-two countries reported the development of national programs[9] and the median year is 1999. India 1951 was the precursor. Since 1976, 27 countries have developed national programs in three phases after the precursors:

- Precursor: 1951 India
- Phase 1: 1957–1958
- Phase 2: 1976–1977
- Phase 3: 1983–1990
- Phase 4: 1996–2015.

The 1996–2015 phase saw two national programs being created every year on average, exemplifying not only the newfound interest for science communication but also the support it gathered. Although the sample is smaller, Figure 2.12 (first national conference) and Figure 2.13 (first initiative/report on science communication) express the same newfound willingness to promote science communication through training, research or the publication of reports.

4. Which countries were the pioneers?

Below are ranked the first five countries to adopt or act in each of the 14 areas. Readers are reminded of the cautionary note posted at the beginning of this section on the accuracy and comparability of the data.

Table 2.2 presents chronologically the five first countries for each indicator.

The United States was the first to innovate or adopt in six of the 14 indicators. They were also second twice and third twice. Germany was first twice, third once and fifth twice. France, the Philippines, the United Kingdom, Canada, Norway, India, Denmark and South Korea were first on one occasion each. The Netherlands and Sweden were second twice each.

9 The authors of the chapter on China stressed that it was very difficult to date the first national program.

Table 2.2: First five countries in the 14 areas.

	Ranking				
Indicator	**1**	**2**	**3**	**4**	**5**
Science centre	NO (1914)	US (1936)	TW (1956)	IN (1959)	NL (1966)
National week	KR (1934)	PH (1951)	JP (1960)	TR (1970)	TH (1982)
Science fair/festival	GB (1831)	PH (1951)	CA (1959)	AR (1967)	CO (1970)
Association	DE (1929)	US (1934)	TR (1946)	GB (1947) TH (1947)	DK (1949)
Award	US (1946)	SE (1972)	CA (1973)	BR (1978)	CN (1980) NZ (1980)
Journal	DK (1854)	NO (1877)	SE (1911)	CN (1933)	AU (1938)
Radio program	US (1920)	BR (1923)	GB (1925)	DE (1926)	SE (1930)
Television program	US (1946)	SE (1950)	GB (1952)	IT (1954)	DK (1956)
University course	PH (1960) US (1960)	NL (1976)	DE (1980) NG (1980)	FR (1984) TW (1984)	AR (1985)
MA graduate	US (1960)	CA (1984) FR (1984)	NL (1990)	NO (1992) GB (1992)	AU (1993) IT (1993)
PhD graduate	DE (1970) US (1970)	FR (1973)	NL (1976)	CA (1978)	TW (1983)
National conference	CA (1933)	GB (1943)	ES (1958)	DE (1960)	FR (1969) CO(1969)
Initiative/report	FR (1937)	CN (1953)	US (1957)	NL (1957)	CO (1965)
National program	IN (1951)	NL (1957)	US (1958)	GH (1959)	TR (1963)

The United States appears 10 times in the table above, the Netherlands six, and Canada, France, Germany and the United Kingdom five times each.

Table 2.3 collates the results for all first-five precursor countries (the total varies from column to column because countries that innovated the same year were similarly ranked).

Table 2.3: Five first precursors.

	Ranking					
Code	**1**	**2**	**3**	**4**	**5**	**Total**
US	6	2	2			**10**
DE	2		1	2		**5**
FR	1	2		1	1	**5**
PH	1	2				**3**
GB	1	1	2	1		**5**
CA	1	1	2	1		**5**
NO	1	1		1		**3**
IN	1			1		**2**
DK	1				2	**3**
KR	1					**1**
NL		2	2	1	1	**6**
SE		2	1		1	**4**
CN		1		1	1	**3**
BR		1		1		**2**
TW			1	1	1	**3**
TR			1	1	1	**3**
ES			1			**1**
NG			1			**1**
JP			1			**1**
AR				1	1	**2**
TH				1	1	**2**
IT				1	1	**2**
GH				1		**1**
CO					3	**3**
AU					2	**2**
NZ					1	**1**
Total	**16**	**15**	**15**	**16**	**17**	**79**

4.1. Median points

Figure 2.15 takes into account all the indicators used to construct the timelines. The orange lines indicate the beginning and end of each element. For example, to the question *When did the first national (or large) science fair/festival take place*, the line stretches from 1831 to 2018. This is a 187-year period and this number is written in the column on the right side of the graph.

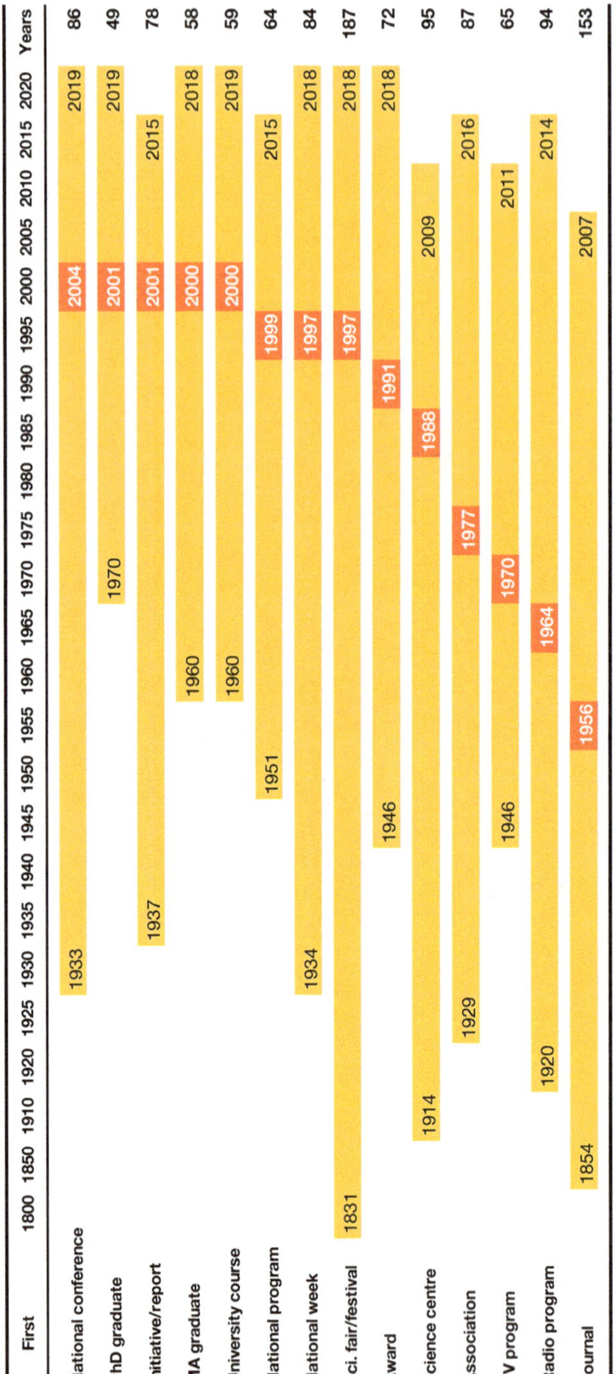

Figure 2.15: Proposed synthesis of the evolution of science communication.

The red cells indicate the median year for each question. For example, the median year for journals was 1956, which appears near the middle of the line. This means the creation of journals was more or less evenly spread, while national conferences only gained momentum in later years.

5. Conclusion

The aggregation of the indicators, ordered chronologically according to the date of the median year, gives an overall picture of the gradual development of science communication. From this data, we can assert that the creation of a journal was a foundational moment, because it anchored in society a specific type of discourse, which the later adoption of radio and television consolidated. Similarly, the award of master's or PhD degrees, located at the top of the graph, suggests that the institutionalisation process is recent and ongoing, thus lending further support to the analysis of Figure 2.13.

It is clear is that the overall pace is quickening. Countries that have not in the past invested resources into science communication increasingly see it as a pathway to prosperity, with economic, social and environmental benefits accruing to countries that apply and use science wisely. This implies an ability of its citizens to appreciate the possibilities and limitations of science. A job for the science communicator.

We conclude with a graph showing which countries have hosted the biennial PCST conference, themselves indicators of the structure and institutionalisation of science communication.

Figure 2.16: PCST conferences.

References

Funkhouser, G. R. and Maccoby, N. (1970). Communicating Science to Nonscientists, Phase 1: A Correlational Study of Textual Variables and Audience Effects, Institute for Communication Research, Stanford University.

Gascoigne, T., Cheng, D., Claessens, M., Metcalfe, J., Schiele, B. and Shi, S. (2010). Is science communication its own field? *Journal of Science Communication*, *9*(3). doi.org/10.22323/2.09030304.

Jurdant, B. (1970). *Les problèmes théoriques de la vulgarisation scientifique* [Theoretical problems raised by the popularisation of science]. France: Université Louis Pasteur de Strasbourg.

Kriegbaum, H. (1967). *Science and the Mass Media.* New York: NYU Press.

Schiele, B. (2019, 4 October). The words expressing what one does and the words expressing what one thinks he does. Communication, International School on Mind, Brain and Education, Erice, Italy.

Withey, S. B. (1959). Public opinion about science and scientists. *Public Opinion Quarterly*, *23*, 383–88.

3

AFRICA

Health communication in selected African states from colonial times

Margaret Kaseje and Verah Okeyo

1. Introduction

This chapter discusses the development and role that science communication has played in Africa, with illustrations from different countries in sub-Saharan Africa and using the field of health to demonstrate some of the challenges faced by communities and governments.

Before making a critique of science communication in Africa, it is worth noting that the discipline developed differently in Africa compared to countries in the Global North such as the US and UK. This distinction introduces an often-ignored contestation in the field: between the public understanding of science, a paradigm that anchors the justification for science communication for example in Kenya, and science communication per se. The latter may be defined as a form of contact for sharing information on science in a society using various means including institutions and communication entities such as media houses, but this is not yet settled. This is hardly surprising because it is an arena in which 'many different stakeholders battle for attention and the power of definition, because there is money in the game, there are jobs to be captured, and there are professional identities at stake' (Weingart and Guenther, 2016, p. 2), and caution is needed in its definition (Stilgoe, Lock and Wilsdon, 2014). Du Plessis (2011) defines science communication as the 'use of the impact of the media and other channels of communication to disseminate science findings' with a focus concentrated on a communication process that is reliant on multimedia through journalistic reporting on mainstream and social media, and exhibition of science in museums.

On the other hand, public understanding of science communication is, according to Bauer (2008), many activities aimed at narrowing the gap between science and the people. It also refers to research that appreciates and employs empirical methods to investigate the public's appreciation and uptake of science—or lack thereof—and how these two vary across time and context. In the West, science communication and public understanding of science exist as two different disciplines but in most countries in Africa, the lines are blurred. This can be attributed to how science was introduced in each region.

Even in developed democracies such as European countries and the United States, there is a perceived communication gap between scientists and the public, and a view that people have not the simplest grasp of basic science or the social benefits arising from science (Bensaude-Vincent, 2001). Scanty research has gone into questioning the perceived deficiency of scientific knowledge of members of the public, and most of the claims stem from an assumption of the scarce coverage of science in the mass media and other public fora (Lublinski et al., 2014; Murcott and Williams, 2013). Many reasons have been advanced to explain this gap such as the ivory towers in which science exists and the arcane nature of its practices, as well as the ostensibly dispassionate scientific discourse that scientists use (Allan, 2009). Calls for bridging this gap include efforts to make knowledge production in science open and accessible (Brenner, 1998)—the rise of many open-access journals—and also funding ways to improve science communication: training scientists to be media savvy and improving the quality of science journalism. In Africa, the gap in science communication has been wide, particularly where access to scientific journal publications is restricted by fees and membership requirements. This gap has been particularly evident in the health domain because information is restricted to populations who in the end depend on their personal active engagement with health providers.

The utilitarian reason for science education was to enable populations to improve their living conditions, and communicators of science adopted this approach when working with schools and the media. Science education aimed to improve production practices in agriculture and animal husbandry, and populations were expected to boost food production for their use and for sale and export, and to improve health and life expectancy. Science education ultimately influenced attitudes and behaviours of the population, where practices were seen to be 'civilised' if emanating from science education and 'uncivilised' if they came from people's beliefs and practices.

2. Colonisation of Africa

Colonisation of Africa reached its peak in the early 1900s when land, the most prized asset for the people living on the continent was taken by colonisers. Colonising activities slackened off after World Wars I and II, due to new priorities and worsening economies of the colonial powers in Europe, together with a growing sense of nationalism and the wave of independence in African nations. The need to revive Europe's economy, devastated by two world wars, led to a focus on what the African continent could provide. The major benefits to colonial powers were the land they seized from indigenous populations and gave to settlers to farm, and mineral resources. South Africa, in particular, became attractive due to availability of these two commodities: land and minerals. This was exemplified by the establishment of the British South African Company (Wilson, 2011).

With growing awareness of self-rule, African countries such as Kenya, Angola, Algeria and Mozambique fought battles of freedom from colonial rule. Colonial governments, most notably Britain, France and Portugal, made little effort to prepare their colonies for independence, and instead tried to absorb them. Britain perhaps was an exception in one way, investing in the education of the population in its colonies, which gave rise to African leaders such as Kwame Nkrumah (the first President of Ghana); these leaders eventually fought for independence for their countries.

The colonial governments' attitude towards science was from the standpoint of providing science education to their territories. The curriculum was limited to general knowledge of the basic sciences of biology, chemistry and physics taught at elementary levels in primary schools and later elaborated in secondary schools. Post–secondary school science education required more advanced skills and this was provided to the initial group of indigenous students. A few students advanced to higher levels of diploma and degrees, some offered in Europe through scholarships provided by the colonisers and largely to fulfil the need for trained workforces for the post-independence era in the continent.

When African countries became independent, science communication content consisted of basic sciences in these newly emerging nations where workforces needed to be equipped with necessary skills to take over the governance of their new nations. Priorities for governments in the emerging independent countries were for development and economic growth, propelled through education, technology and human resources. Thus, science communication became a key priority for the population, filling the need for knowledge and skills to increase food production and levels of literacy for improved livelihoods.

With the departure of the colonisers, remnants of their rule were apparent in the colonies' legal, educational, political, security and health systems, some of which remain today. When the newly independent countries took over the reins, politics played a big role and politicians propagated messages to rally people and consolidate their rule. Science communication was not in the forefront; yet it could have offered an alternative view of and evidence-based solutions to problems. Instead, research institutions were the custodians of communication and were more accepted within the education system than in the public domain.

3. Science communication and the public in post-colonial times

Since the growth of science communication in the 1970s in Africa, a pocket of communication researchers has argued that the gap between science and the public was created by popularisers for their own self legitimisation as brokers of science communication (Brenner, 1998; Jurdant, 1969). Far from exploring the lack of knowledge in the public about science, this chapter takes as a starting point the acceptance of 'third parties' in the development and practice of science communication in Africa in general. It presents the interests, methods and effect of the third parties in the media as an arena. These, as we will illustrate later, include the influence of public relations and funding on science journalism and the choreographed public presentation by scientists. This chapter situates science communication in terms of Jurgen Habermas' notion of the 'public sphere'. Habermas is one of the most widely read social theorists in the post-WWII era and his writings have deeply influenced humanities and social science scholarship. His argument was that all speech acts have an inherent purpose: the goal of mutual understanding, and that human beings possess the communicative competence to bring about such understanding (see Habermas et al., 1974; Calhoun, 1992; Goode, 2005).

Arguments have been put forward for the right of citizens to have accessible science information to inform their choices on issues. The importance of science communication is not reflected in the editorial space given to science issues: in South Africa, for instance, less than 2 per cent of mainstream newspapers' space was devoted to science-related topics such as technology, environmental affairs and medicine/health (Van Rooyen, 2004). South Africa has a long record of defined science communication events bringing scientists and media together and highlighting the roles of science in society (Joubert, 2001). On the other hand, du Plessis (2017) cites the influence of politics in a society where a population divided by colonisation and the apartheid system muzzled the development, research and use of science communication in higher learning institutions. Despite such constraints, science communication is a topic of research in institutions of higher learning, like other disciplines. Science communication research in Africa has focused on the practice of science communication—for example, science cafés in Kenya (Mutheu and Wanjala, 2009); providing health information through the radio in Malawi (Nyirenda et al., 2016), use of internet for health education in South Africa (Coleman, 2012); communicating science subjects through musical shows in South Africa (Fish et al., 2016); and media coverage of science information as exemplified in the coverage of genetically modified (GM) crops in Kenya (DeRosier et al., 2015).

But this work (like all forms of research in Africa) is mainly the domain of research and academic institutions, and these have been hampered by lack of resources unless they have partnered with counterparts from the Western world in getting proposals funded.

4. Research, the media and public relations

Since the 1970s, an increasing amount of scientific research is conducted under private patronage, particularly by major companies; and researchers operate increasingly in a commercial climate, with the imperatives of reputation-management and securing market share guiding the development of knowledge products. This puts pressure on science communication. Science writing is less interested in public information and education as communication staff employed by research institutes, universities and companies focus on securing public attention for particular scientists, products, research groups and scientific institutions. The model of professional public relations for science, though nothing new, turns into a generalised and domineering practice.

Traditional modes of science communication, especially with the public, have been through print (newspapers), television, radio and science cafés, but the advent of social media platforms such as blogs, Twitter and Facebook has diminished resources to media organisations in their previously commercial model of engaging with the public. This increases the dual risks of: (a) 'scientific fraud' because of higher production pressures on scientists

(Cookson, 2009; van Noorden, 2011; Schulz, 2016); and (b) lower quality in the societal conversation of science, because of the publicity imperative for research and researchers (Nelkin, 1987).

One implication is that journalists have fewer resources to check their stories. In order not to turn into a festival of hyperbole and misinformation, science reporting requires the structures of a public sphere capable of scrutinising the process of knowledge production outside science itself and supporting the peer review process. For science communication, this amounts to a paradigm change (see Bauer, 2008). Where there is diminished public participation in discussions on health issues and challenges faced by populations, there is less likelihood that any negative factors or issues of supervision will be unearthed; and much less likelihood that sustained long-term solutions to health problems will be introduced.

The scientific communities' perspectives were reflected in the developing field of science journalism, a specialism among professional journalists that has grown in strength and presence alongside the broader field of science communication. Then, and to a large extent now, a background in the natural sciences was practically an entry requirement. With the gradual professionalisation of science communication, courses have proliferated in universities or as part of professional development. Until recently, most have been accommodated in science faculties or professional societies, targeted at science students, graduates or professionals, and often delivered by 'converted' scientists.

5. Public perceptions and the gap between researchers and their subjects

The current research environment can widen the gap between researchers and their subjects. Research can be a means to further a crystallised form of communication and has played a major role in the African scene in the quest for improved food production, better health practices and knowledge for informed decision-making. However, the research has been in the domain of researchers and academic institutions with little information given to the people on whom the research is being carried out. The people provide information and data to the researchers, but in few cases are the research outcomes provided to the people in a language or format that is comprehensible to them. Nor is their feedback sought on the research process and content. Some universities are an exception in providing research science information, especially in South Africa (Stellenbosch and Rhodes universities) where rural communities have been involved in providing their perceptions of science research projects and science communication.

This gaping chasm between researchers and the public, and other structural and cultural issues, have been implicated in the failure to adopt programs after monies have been spent on the research. An example where the chasm caused problems is a community's interruption of a study in western Kenya where researchers were conducting a verbal autopsy, which involved asking the relatives of the deceased what had caused the deaths of their loved ones. The oral interviews with the villagers were accompanied by conducting an actual medical autopsy to match the answer given by the villagers (Interview, 2018). Unaware of the nature and the processes of the research, the villagers accused the researchers of harvesting organs from their dead relatives for rituals. These interruptions and general ignorance of the public (perhaps caused by the researchers not explaining what they were doing in the experiment) has motivated research funders to allocate a Public Engagement Fund to the consortia and the scientists that they are funding to conduct public and community engagement events to the public.

6. Access to public debates and the media is limited

In Kenya, research institutions were established in the 1970s as centres for knowledge generation and transfer, leading to Kenya hosting international research institutions such as the International Centre of Insect Physiology and Ecology (ICIPE), the World Agroforestry Centre—also known as the International Centre for Research in Agroforestry (ICRAF)—and the International Livestock Research Institute (ILRI). Some of the research in these institutions addressed GM crops; however, researchers limited their work to seeking people's perceptions and knowledge about such crops but not making the research findings available to the general public. While debates and dialogue in the west in the 1980s concerning healthy foods and GM crops were informed by research, in contrast public debate and decisions in Africa took place without the public having access to research findings. There was eventual passive acceptance of GM crops due to constraining factors of food shortages (see Kimenju, 2011, for related discussion).

Universities in the post-colonial era have played significant roles in science communication, where exchange of scholars and professors, conferences and publications have contributed to an exchange of knowledge that surpassed geographical boundaries. Science communication in Africa claims a universality that does not appreciate the inequalities that exist in society, which public relations creates. The notion of universality is based on the idea of the *public sphere*, which allows discussions and debate, and can influence public policy. It may be carried out in the media, social media or at meetings (see Butsch, 2007 and 2011 for discussions of the public sphere specifically relating to the media). However, full engagement of the public has been lacking due to a barrier between the public and scholars.

Social differences have barred certain individuals from accessing important conversation spaces. As an example, the Nations Leadership Forum organised by Nation Media Group, East and Central Africa's largest media house, is one of the few media-backed forums where scientists are invited as panellists in discussions touching on science-related issues, such as the Sustainable Development Goals, health, technology and food security. Invitations to panellists are not on merit but depend on a fee being paid to the corporate communication department of the media house (Interview, 16 October 2018). The topics selected are of interest to those capable of paying the fee, and includes the large funding agencies such as the Bill and Melinda Gates Foundation. Fraser (1992) argues that this public sphere's 'rational deliberation' is a bourgeois individualistic social practice where only the moneyed, privileged and educated in society meet to pursue their individualistic needs, but there is some evidence of governments attempting to include all concerned in the spirit of collectivism (*ujamaa*), as evidenced in Tanzania.

There could not have been a better time to talk about selected communication than now in Africa, when funding for science communication and journalism is derived from the very people that bankroll the research such as philanthropic organisations like the Bill and Melinda Gates Foundation (Bunce, 2016; Downie and Schudson, 2009; Wright et al., 2018). It is a fraction of the monies the Foundation invests into biomedical research and agriculture, but it is involved in Africa's media in training and funding the actual news production process. Kenya's *Daily Nation,* South Africa's *Bhekisisa* hosted in the *Mail and Guardian* are examples of such funding and dedicated centres of health journalism; and non-government organisations (NGOs) are also active in health communication.

Declining advertising, the biggest sources of revenue for media houses, has placed pressure on traditional media and provided an entry point for the foundations to finance journalism. In Africa, the Bill and Melinda Gates Foundation is quite visible in science journalism and communication. The money is not just for the actual production of science journalism but also capacity-building in African journalists through training and fellowships (Mayonzo, 2012) to ameliorate gaps in science journalism such as coverage that lacks depth or context (Ainslie, 1966); outright partisanship; lack of professionalism (Schiffrin, 2010); laziness (Owuor, 2008); and lack of ethics (Nyamnjoh, 2005). The capacity-building is offered by organisations like the International Centre for Journalists (ICFJ) as well as the Africa Science Desk, which is also funded by the Bill and Melinda Gates Foundation.

These activities are debated, because foundations have strong interests and fund activities and areas of communication of interest to them (Scott, Bunce and Wright, 2017):

> Private foundations that support media in order to change the world have views about how journalism can make the world a better place, what the world's problems are and, in some cases, what solutions should be implemented. By using journalism to promote coverage of health or governance or corruption or elections or criminal justice, these foundations are making decisions for all of us about what problems the public should know about and even pressure governments to fix. That involvement in agenda-setting and public policy affects everyone, regardless of where they live, and so it matters how donors and journalists negotiate and implement such agreements. (Schiffrin, 2017, p. 3)

In Kenya, as in other African countries, the education system focuses on science subjects, while institutes for agricultural skills produce graduates skilled in improving food production. Improving health care was through a trained health workforce, largely nurses and clinical officers and to a lesser extent physicians. Science communication therefore was available to a select group of students in institutions, with the ordinary people having less access to science communication. While certain sections of the population have been trained in science, there is still a chasm on effective communication with the ordinary citizens.

Apart from the aforementioned challenges of lacklustre coverage, science journalism has caused social change, especially in positive health-seeking behaviour (Westoff and Rodriguez, 1995).

7. Science communication against the backdrop of science events

During the late 1950s and early 1960s, African countries experienced a wave of historical events encompassing independence, economic, social and agricultural production as well as health care. These changes were reflected in the existing means of communication, largely radio and newspapers. Colleges and universities played a key role, and the three East African countries combined their resources to form joint higher learning institutions for sciences at the University of Nairobi in Kenya, economic studies at Makerere University in Uganda and law at Dar es Salaam University in Tanzania. However, there were no formal courses in science communications in these institutions. Kenya was to (later in the 1970s) develop an institute of mass communications for journalists reporting mainly for radio and newspapers.

Science communication in Africa has taken place against the backdrop of science events and advances in the Western world such as space exploration and computer development. The education system in Africa reflected those in the west where science subjects (mathematics, biology, physics and chemistry) were taught for a workforce to take on jobs in post-colonial countries. The focus on science became intense in Kenya where a number of institutes of science and technology were established in the 1970s with the aim of building a more informed population and providing skills needed to propel these post-colonial countries into the 21st century. In health, there was a need to inform populations and encourage positive health actions that would ameliorate low life-expectancy rates and causes of morbidity and mortality among vulnerable groups such as mothers and children. In order to address the need for health communication in Africa, there emerged a primary health care approach, following the Alma-Ata Declaration[1] of 1978. This declaration was adopted by African countries and led to the dissemination of health communication through trained community health volunteers who reached populations at the household level.

Secondary school performance in science in the 1960s reached a peak where high-performing schools were those that excelled in science subjects. In Kenya, Francis Carey, a British educationist was renowned for his science teaching skills and having been taught by him was viewed as a mark of excellence. While the focus on examination performance in science subjects reigned in

1 The Alma-Ata Declaration of 1978 emerged as a major milestone of the 20th century in the field of public health, and it identified primary health care as the key to the attainment of the goal of Health for All.

most secondary schools, the outcomes over the years in Kenya have been positive, where technology advancements in mobile money transfer and other practical phone applications for businesspeople and farmers has emerged and contributed globally to mobile technology.

The emergence of HIV and AIDS placed the health agenda at a different level, because this was a disease that appeared to affect not just a small population in a given geographical area but had global impact. The means of spread and the effect of HIV and AIDS on populations were devastating, pushing the need to share information in innovative and open ways that had not been previously experienced for a health issue. Counselling, testing and sharing one's HIV status required knowledge about the disease and open communication among health and non-health professionals.

Early HIV campaigns during the 1990s used the 'shock and fear' approach in attempting to change people's behaviour. Billboard and radio messages depicted thin and wasted 'victims' of the disease. However, this approach was counterproductive and only strengthened fear and stigma in communities and resulted in the isolation of AIDS patients. Rumours and misbeliefs abounded, regarding for example touching infected people, sharing eating utensils with them or being bitten by mosquitoes that were believed to transmit the HIV virus. With no impact on infection rates, high mortality rates and expensive treatment regimes, program interventions changed their approach to one of positive living, integration of affected and infected people, disclosure, counselling and stigma reduction. This approach was spearheaded by TASO (The AIDS Support Organisation) in Uganda and eventually adopted globally and by WHO as an effective approach to HIV and AIDS reduction.

In Africa, politics has played a significant role in the health of people. In elections of political leaders every five or so years, politicians become the communication media for health information and, depending on their agendas, can also be sources of science communication. The health issue of HIV and AIDS has been used by politicians to push their agendas for a healthy nation. They have demonstrated that HIV testing and counselling is an effective strategy to address the issue, through talks and speeches and by undergoing tests themselves to prove that they ascribe to the desired actions. With the emergence of democratic space, partly spurred on by the globalisation of information, there has been more tolerance resulting in freer communication.

8. Science communication and public sentiments

While the Western world debated nuclear power in the 1970s and recombinant DNA and GM crops in the 1980s, these issues emerged later in Africa, but only as research topics. The public were not engaged in debates as in the west. Economic development was a priority and any science information and technology that was perceived as being useful for this purpose was provided to government extension workers and volunteer community workers trained and motivated by NGOs. The model of science communication retained the two distinct groups—those that possessed the science, and those who were recipients of this information. With the emergence of a global world through a burst of communication channels, public opinion is now reflected in debates on the role of agriculture and the quest for an agricultural green revolution in Africa similar to that which dramatically changed crop production in Asia and Latin America. Public discourse using social media continues on topics such as climate change and global warming, drought, and the consequent demands for irrigation.

In some southern African countries, governments attempted to involve farmers in increasing food production. Towards this end, governments provided subsidies to farmers consisting of seeds and digging implements accompanied by widely disseminated information on farming practices through radio, newspapers and by government agricultural extension workers. While food production increased temporarily, dependency of farmers on government subsidies also increased during the same period. With other factors of unpredictable rainfall, price fluctuations and long-held beliefs on farming practices such as timing of planting and which family member should plant their field first before other family members, farming practices did not change over time, resulting in late planting and poor harvests.

Science communication has used traditional means such as radio, television and newspapers to communicate science information, with radio reaching the widest audience. The *baraza* or community meeting in East Africa has been a forum used by NGOs to educate and inform people on development initiatives that includes science information. This method is relevant to the intended rural audience since it promotes analysis of issues affecting communities and solutions. The *baraza*, a regular communal meeting presided over by community leaders, is an open forum for all members of the village to voice their opinions on priority issues that include health, social events and security. A *baraza* employs storytelling, proverbs and role-playing that highlight messages for community action, including science messages in a culturally acceptable manner.

A health issue in rural Kenya is the high numbers of babies delivered by unskilled people, leading to high mortality of mothers and babies. A solution is for mothers to reach hospitals in time for delivery by skilled health providers. In one *baraza* session in western Kenya, communities discussed how to assist pregnant women to reach a health facility in time for skilled delivery by health workers. However, in rural settings, transport to health facilities is largely unavailable when needed, or is unaffordable. Community members discussed the issue and agreed to train and enlist local motorcycle riders to transport mothers to hospitals for delivery. The *baraza* session used role plays to depict the late arrival of a mother to a health facility, reactions from the health providers, eventual safe delivery of the baby, and mother and baby riding back home safely.

Different emotions and reactions characterised the *baraza* session: humour and laughter during the role-play scene where the pregnant mother was arguing with the husband that the baby was due, while the husband disagreed because he had competing commitments; apprehension (will the pregnant woman arrive at the hospital in time, or will the baby be delivered by the roadside?); empathy with the situation; and the final happy resolution (role play of crying baby and women and husband smiling happily). The approach of using motorcycle riders to transport pregnant women to hospital was adopted by rural communities and led to more women delivering safely in hospitals, thus addressing one high cause of deaths of pregnant women and their babies.

For primary and secondary school children, visits to museums, animal reserves and animal orphanages in Kenya have served to provide science information to the youth.

9. Science communication through events and social media

South Africa is noted for holding science communication events. ScienceLink holds science communication events for international participants consisting of students, researchers and NGO workers as part of building a community of practice in science communication. Current science communication methods are blogs—for example, the South African SciBraai, which was initiated in 2013 as a means of social media communication and includes Facebook, Twitter and Instagram feeds and exists to promote science reporting and communication.

Most recently, with the advent of social media, science communication has taken on a new dimension in Africa. The Association of South African Women in Science and Engineering (SAWISE) has a public Facebook platform for science communication for those concerned with science activities and research. Online communication with readily available information—for example on HIV and AIDS and gender health issues—has enabled more people to be reached with science information and for people to voice their opinions and concerns while getting immediate feedback. Online communication has resulted in greater shared health information with a wider audience that has no geographical boundaries.

Some health issues that are sensitive to communities require diverse approaches—for example, issues of sexual and reproductive health that include female genital mutilation (FGM) or female circumcision require dialogue and in-depth discussions with those involved with the practice. The procedure of FGM is mostly carried out on girls in their babyhood to adolescence and before reaching the age of 15 years. In Kenya between 25 and 50 per cent of girls and women aged 15 to 49 years have undergone FGM (UNICEF, 2013). According to international standards and requirements, FGM violates the human rights of girls and women and is known to lead to health complications, some of which are extremely severe. FGM is performed for various sociocultural reasons including the need to be accepted and to conform to social norms, initiation into adulthood and marriage, notions of cleanliness after FGM, and the stand that a cultural practice is not to be argued against. Despite negative consequences, FGM is endorsed by local leaders and is carried out by respected community members who also have traditional roles such as serving as birth attendants. Through our continuous dialogue with a certain group of community members and leaders, their attitudes changed over time and by giving birth attendants alternative roles as birth companions who accompany pregnant women to hospital, they have ceased to perform FGM.

However, the spread of information has been hampered by poor internet and electricity connectivity and high costs of phone and laptop devices, particularly in rural and remote regions. While the Western world has experienced a vast expansion of communication channels and methods (such as science events), Africa has seen restricted information flow and is largely influenced by political systems where information is state-controlled and vetted for political correctness, and in cases where learning institutions and media houses have been under state control.

The issue of HIV exemplifies state-controlled information in instances where any information portrayed depicts the government in a negative light. During early 2000, when HIV treatment was not available in public facilities, health systems in sub-Saharan Africa were burdened with high treatment costs—for example, in South Africa and Uganda where the burden of the HIV infection and AIDS stretched health systems and services. In such contexts, governments controlled information made available to the public largely due to leaders' objectives of staying in favour with the populace. However when treatment became affordable, and governments demonstrated action to improve the lives of people by ensuring universal access to treatment, there was less information restriction and even over-exposure of government information about their actions to provide people with treatment and their global partnerships in ending the AIDS epidemic and contributing to Sustainable Development Goals.

While in the recent past, science communication has been restricted to providing information on science as opposed to inviting discussion and exchange of ideas, most recent communication in science has leveraged the emergence of the internet and social media. These have revolutionised how information is retrieved, shared and disseminated to intended audiences. Communication applications such as WhatsApp enable the teaching of science in a virtual manner, which is a shift from traditional classroom interaction or state-controlled media where the content was highly regulated and focused on the leadership rather than on information demanded or needed by the population.

The emergence of the internet and social media that includes Google and websites, Twitter, Instagram, Facebook and YouTube have had both positive and negative effects on science communication during a period of rapid expansion of global communication. On the positive side, more information is shared with a wider audience across different platforms. However, interaction is less personal and may be prone to misinterpretation. The emergence of 'fake news' and misinformation has diminished trust in this channel of information.

Some forms of science communication took place in the interface between researchers and users of technologies in the 1990s in Africa. This process did not reflect what communicators would term today as 'true' communication, since these were research contexts that often did not allow feedback on the process or content covered in a research study. For example, research on HIV and AIDS, family planning and malaria had outcomes that required behaviour change; however, communicating research findings in a manner that supports community behaviour change has had challenges due to the gap between researchers and end-users of the research findings.

People resisted scientific findings when they could not connect the science to their health conditions and where science ran against socioeconomic and sociocultural beliefs and practices. For example, using bed nets to prevent mosquito bites required the ability to afford insecticide-treated bed nets. In the case of HIV infection, the intervention of limiting sexual partners to one uninfected partner ran against the cultural requirement to inherit a woman whose husband had died of HIV infection in order to ensure social stability of the deceased's household. While progress has been made in improving positive health behaviour, impoverishment and cultural beliefs and practices persist, leading to preventable malaria and HIV infections in sub-Saharan Africa.

10. Sources of information for populations in Africa

In the post-independence period, radio became a versatile tool to communicate important issues of the day. It played a major role in reaching especially rural populations with scientific information. Literacy levels affect the reading population—for example, Kenya has 79 per cent literacy (and numeracy), and this influences any science communication in the print media. In the early 1970s, a program by a non-government organisation provided health information to remote villages through a radio program paraphrased as 'doctor health'. During the same period, the national radio station transmitted a weekly children's program that discussed health actions for children at home and school. Newspapers were second to radio in disseminating information with their special reports on health topics.

The country has moved from traditional means such as print, photography, radio and television broadcasts to embrace advancements in new media, social networks and videos (YouTube, Twitter, Skype, Facebook, Blogger, LinkedIn), where science information is readily available for those with access to phones and internet.

While most countries in sub-Saharan Africa have achieved improved health indicators, including life expectancy and education, challenges continue to emerge with emerging diseases and conditions that include climate change. However, Africa is a resilient continent and will continue to face future challenges in science communication, technologies and livelihoods as resolutely as in the past.

References

Ainslie, R. (1966). *The Press in Africa: communications past and present.* London: Gollancz.

Allan, S. (2009). Making Science Newsworthy: Exploring the Conventions of Science Journalism. In R. Holliman (ed.), *Investigating Science Communication in the Information Age: Implications for Public Engagement and Popular Media* (pp. 149–65). Oxford: Oxford University Press.

Bauer, M. W. (2008). Public Understanding of Science. In M. Bucchi and B. Trench (eds), *Handbook of public communication of science and technology* (pp. 111–30). London: Routledge.

Bensaude-Vincent, B. (2001). A genealogy of the increasing gap between science and the public. *Public Understanding of Science, 10*, 99–113. doi.org/10.1088/0963-6625/10/1/307.

Brenner, S. (1998). The impact of society on science. *Science, 282*(5393), 1411–12. doi.org/10.1126/science.282.5393.1411.

Bunce, M. (2016). Foundations, philanthropy and international journalism. *Ethical Space: The International Journal of Communication Ethics, 13*(2/3), 6–15.

Butsch, R. (2011). Audiences and Publics, Media and Public Spheres. In V. Nightingale (ed.), *The Handbook of Media Audiences* (pp. 147–68). Wiley-Blackwell. doi.org/10.1002/9781444340525.ch7.

Butsch, R. (ed.). (2007). *Media and Public Spheres.* Hampshire: Palgrave Macmillan.

Calhoun, C. (ed.). (1992). *Habermas and the Public Sphere.* Cambridge: Massachusetts Institute of Technology.

Coleman, A. (2012). Towards delivering e-health education using Public Internet Terminals (PIT) systems in rural communities in South Africa. *Engineering, 11*(4). doi.org/10.22323/2.11040201.

Cookson, C. (2009, 23 May). The Real Lab Rats. *Financial Times.* Retrieved from www.ft.com/content/72985c24-465d-11de-803f-00144feabdc0.

DeRosier, C., Sulemana, I., James, H. S., Jr, Valdivia, C., Folk, W. and Smith, R. D. (2015). A comparative analysis of media reporting of perceived risks and benefits of genetically modified crops and foods in Kenyan and international newspapers. *Public Understanding of Science, 24*(5), 563–81. doi.org/10.1177/0963662514568061.

Downie, L., Jr and Schudson, M. (2009). The Reconstruction of American Journalism. *Columbia Journalism Review.* Retrieved from archives.cjr.org/reconstruction/the_reconstruction_of_american.php.

du Plessis, H. (2011). Science and Society: Public Understanding of Science (PUS) in South Africa and Other Collaborations. HRSC Seminar series, January. Pretoria. Retrieved from www.hsrc.ac.za/uploads/pageContent/166/Science&studyHSRC SpeechJan2011.pdf.

du Plessis, H. (2017). Politics of science communication in South Africa. *Journal of Science Communication 16*(03), A03. doi.org/10.22323/2.16030203.

Fraser, N. (1992). Rethinking the Public Sphere: A Contribution to the Critique of Actually Existing Democracy. In C. Calhoun (ed.), *Habermas and the Public Sphere* (pp. 109–42). Cambridge: Massachusetts Institute of Technology.

Goode, L. (2005). *Jürgen Habermas. Democracy and the Public Sphere.* London: Pluto Press.

Habermas, J., Lennox, S. and Lennox, F. (1974). The Public Sphere: An Encyclopedia Article (1964). *New German Critique, 3*(3), 49–55. doi.org/10.2307/487737.

Joubert, M. (2001). Report: Priorities and Challenges for Science Communication in South Africa. *Science Communication*, *22*(2), 316–33. doi.org/10.1177/1075547001022003008.

Jurdant, B. (1969). Vulgarisation scientifique et idéologie. *Communications, 14*, 150–61. doi.org/10.3406/comm.1969.1203.

Kimenju, S. C., Groote, H. D., Bett, C. and Wanyama, J. (2011). Farmers, consumers and gatekeepers and their attitudes towards biotechnology. *African Journal of Biotechnology, 10*(23), 4767–76.

Lublinski, J., Reichert, I., Denis, A., Fleury, J. M., Labassi, O. and Spurk, C. (2014). Advances in African and Arab science journalism: Capacity building and new newsroom structures through digital peer-to-peer support. *Ecquid Novi: African Journalism Studies, 35*(2), 4–22. doi.org/10.1080/02560054.2014.919945.

Mayonzo, L. (2012). *Media, Communication and Development: Three Approaches.* SAGE Publications India.

Murcott, T. H. L. and Williams, A. (2013). The challenges for science journalism in the UK. *Progress in Physical Geography*, *37*(2), 152–60. doi.org/10.1177/030913331 2471285.

Mutheu, J. and Wanjala, R. (2009). The public, parasites and coffee: The Kenyan Science Café concept. *Trends in Parasitology*, *25*(6), 245. doi.org/10.1016/j.pt. 2009.03.002.

Nelkin, D. (1987). Risk and the Press. *Industrial Crisis Quarterly, 1*(2), 3–9. doi.org/10.1177/108602668700100202.

Nyamnjoh, F. (2005). *Africa's Media: Democracy and the Politics of Belonging.* Zed Books.

Nyamnjoh, F. (2015). Journalism in Africa: Modernity, Africanity. *African Journalism Studies, 36*(1), 37–43. doi.org/10.1080/23743670.2015.1008128.

Nyirenda, D., Chipasula Makawa, T., Chapita, G., Mdalla, C., Nkolokosa, M., O'byrne, T., Heyderman, R. and Desmond, N. (2016). Public engagement in Malawi through a health-talk radio programme '*Umoyo nkukambirana*': A mixed-methods evaluation. *Public Understanding of Science*, 27(2), 229–42. doi.org/10.1177/0963662516656110.

Owuor, O. (2008). *The Challenges of Communicating Science: A Journalist's Perspective.* Nairobi.

Schiffrin, A. (2010). Not Really Enough. *Journalism Practice, 4*(3), 405–16. doi.org/10.1080/17512781003643244.

Schiffrin, A. (2017). *Same Beds, Different Dreams? Charitable Foundations and Newsroom Independence in the Global South.* Retrieved from www.cima.ned.org/wp-content/uploads/2017/02/CIMA-Media-Philanthropy_Schiffrin.pdf.

Schulz, J. B., Cookson, M. R. and Hausmann, L. (2016). The impact of fraudulent and irreproducible data to the translational research crisis – solutions and implementation. *Journal of Neurochemistry, 139*(S2), 253–70. doi.org/10.1111/jnc.13844.

Scott, M., Bunce, M. and Wright, K. (2017). Donor Power and the News: The Influence of Foundation Funding on International Public Service Journalism. *International Journal of Press/Politics, 22*(2), 163–84. doi.org/10.1177/1940161217693394.

Stilgoe, J., Lock, S. J. and Wilsdon, J. (2014). Why should we promote public engagement with science? *Public Understanding of Science, 23*(1), 4–15.

UN Children's Fund (UNICEF). (2014). *UNICEF Annual Report 2013.* Retrieved from www.unicef.org/publications/files/UNICEF_Annual_Report_2013_web_26_June_2014.pdf.

Van Noorden, R. (2011). Science publishing: The trouble with retractions. *Nature, 478*, 26–8. doi.org/10.1038/478026a.

Van Rooyen, C. (2004). Science not a priority for SA press. *Science in Africa.* Retrieved from www.scienceinafrica.com/science-not-priority-sa-press.

Weingart, P. and Guenther, L. (2016). Science communication and the issue of trust. *Journal of Science Communication, 15*(5). doi.org/10.22323/2.15050301.

Westoff, C. F. and Rodríguez, G. (1995). The Mass Media and Family Planning in Kenya. *International Family Planning Perspectives, 21*(1), 26. doi.org/10.2307/2133602.

Wilson, F. (2011). Historical Roots of Inequality in South Africa. *Economic History of Developing Regions, 26*(1), 1–15, doi.org/10.1080/20780389.2011.583026.

Wright, K., Scott, M. and Bunce, M. (2018). Foundation-funded Journalism, Philanthrocapitalism and Tainted Donors. *Journalism Studies, 20*(5), 675–95. doi.org/10.1080/1461670X.2017.1417053.

Contributors

Dr Margaret Kaseje is a community health, health policy and public health specialist based in Kenya. She is Professor and Director of Research and Programmes, Tropical Institute of Community Health and Development, Kisumu, Kenya; and Visiting Professor for Health Policy and Planning and International Health at Université Libre des Pays des Grands Lacs, Goma DRC.

Verah Okeyo is a global health reporter from Kenya with seven years of experience at the region's largest media house. She has seven journalism awards, five from Kenya and two from the continent on health, environmental and gender reporting in East Africa.

4

AOTEAROA NEW ZEALAND

Participatory science and bicultural knowledge communication

Jean S. Fleming, Nancy Longnecker, Rhian A. Salmon, and Daniel C. H. Hikuroa

1. Introduction

Science communication continues to evolve internationally as a field of study (Gascoigne et al., 2010; Trench, 2012). In Aotearoa[1] New Zealand (NZ), there have been increased opportunities in education, jobs, funding and prizes related to science communication. NZ's bicultural status, as defined by the Treaty of Waitangi (Hudson and Russell, 2009; Mohi and Roberts, 2009; Orange, 2011), provides a unique cultural context within which scientific research and science communication occur. Fleming and Star (2017) have previously documented the history of emergence and development of Western science communication in NZ. An overview is presented in the timeline at the end of the chapter. In this chapter we further explore some specific aspects of the science communication ecosystem in NZ, as well as drivers behind a notable shift towards more participatory science and science communication.

New Zealanders have a strong history of acting firmly and independently, as demonstrated by the banning of nuclear-powered or armed ships in 1984 despite the country's strong alliance with the United States. Aotearoa New Zealand's strong *kaitiaki* (guardianship) ethic, especially amongst Māori, but also amongst Pākehā (non-Māori) New Zealanders, has empowered environmental activism. For example, the successful Save Manapouri

1 Aotearoa is a Māori name for New Zealand's North Island and is often used as a name for the entire country.

Campaign ran from 1969 to 1972; not only did it prevent the raising of the level of Lake Manapouri for construction of the Manapouri Power Project (Mark and Johnson, 1985; Mark, 2001, 2015), it also influenced the results of a federal election.

This independent streak is reflected in the way NZ has moved to accept the value of indigenous knowledge ahead of other countries. The past decades have seen a significant shift in the way in which indigenous knowledge, knowledge systems and engagement processes are respected and incorporated into nationwide funding, research practice and public engagement. As discussed in more detail later, NZ has a Vision Mātauranga policy, which recognises the potential of *mātauranga* (Māori knowledge, culture, values and world view) and its value to current research projects. The Ministry of Business, Innovation and Employment holds that Vision Mātauranga 'unlocks the science and innovation potential of Māori knowledge, resources and people' (MBIE, 2018).

Independence has made the country more cautious about scientific advances. The development of new technologies such as genetic modification and nanotechnology has led to an increase in public mistrust of science (Hipkins et al., 2002). The initial response of scientists and science institutions to this caution about new technologies was to provide more information, to fill a perceived 'deficit' in knowledge in the public. This was driven perhaps by the traditions of the Royal Society of London after the release in 1985 of the Bodmer Report on the Public Understanding of Science (Collins and Bodmer, 1986; Pieczka and Escobar, 2013). In NZ and around the world, scientists, educators and policy makers worked to increase public awareness and acceptance of evidence-based science and knowledge (Bucchi, 1998; Wilsdon and Willis, 2004; Munshi et al., 2016; Smallman, 2016).

In the early part of the 21st century, science communicators began to stress the importance of dialogue with the public (Bucchi, 2008; Cronin, 2008; France et al., 2012; Green and Rohan, 2012; Zorn et al., 2012). In NZ, public consultation played an important role in the government's establishment of 11 National Science Challenges in 2013 (described below). A potential opportunity for further research into science communication specific to the NZ context was missed when the proposed 12th National Science Challenge, which would have specifically addressed science in society, was not funded in the same way as the other challenges. Nevertheless, the value of science communication was recognised and each of the National Science Challenges has an imperative to engage with a variety of stakeholders, including the public.

By 2015, the value of participatory science projects was accepted by scientists, policy makers and funders (Peters, Eames and Hamilton, 2015; Galbraith et al., 2016; Peters et al., 2016; Sullivan and Molles, 2016; Storey and Wright-Stow, 2017; Blake et al., 2018). A relatively small but important new national funding initiative, Unlocking Curious Minds,[2] was announced in 2014 with the intent of fostering participatory science and stimulating engagement of all New Zealanders with science, particularly those who may otherwise be cut off from access to science education and outreach (New Zealand Government, 2018b). Partially as a result of these initiatives, participatory science and science communication has developed rapidly over the past five years in NZ, especially in the area of environmental monitoring and data gathering, as scientists' trust of data gathered by community participants increases and the pool of funding expands (Peters, Eames and Hamilton, 2015; Peters et al., 2016).

2. Human settlement of Aotearoa New Zealand and early practice of science

The most recent Māori settlers arrived in NZ around 750 years ago (McWethy et al., 2009; Seersholm et al., 2018), equipped with unparalleled ocean-voyaging technology and navigation and observation skills. Mātauranga includes knowledge generated using techniques consistent with the scientific method, but explained according to a Māori worldview (Hikuroa, 2017). The arrival of Māori in NZ rapidly introduced large-scale changes, with systematic burning of about half of the forest cover and hunting of the large, flightless moa (Aves: Dinornithiformes) to extinction within a few hundred years (McWethy et al., 2014).

The first European visitors came to NZ in 1642 and by the late 18th century the country was regularly visited by explorers, scientists and naturalists. They were keen to collect new species of plant or animal or to observe the transit of Venus (Fleming, 1987; Priestley, 2010). The arrival of colonisers from the Northern Hemisphere in the 18th and 19th centuries led to further changes in land use, including additional clearance of native forest cover and systematic draining of wetlands. Commercialised hunting of whales led to localised extinctions, and the collection of museum specimens in the

2 The Unlocking Curious Minds contestable fund supports the objective of A Nation of Curious Minds – He Whenua Hihiri I te Mahara to encourage and enable better engagement with science and technology across all of New Zealand, see www.mbie.govt.nz/science-and-technology/science-and-innovation/funding-information-and-opportunities/investment-funds/unlocking-curious-minds-contestable-fund/.

pursuit of science led to further extinctions, including the huia (wattlebird) (Lambert et al., 2009). The signing of the Treaty of Waitangi in 1840 (Bess, 2011; Orange, 2011) was supposed to ensure that the property rights of Māori remained after the Crown acquired sovereignty over their territories (Bess, 2011). Some of the first visitors from the Northern Hemisphere were scientists who studied the geology and natural history of NZ (Fleming, 1987; Priestley, 2008), and science communication in its current Western manifestation started with these pioneer scientists as they reported their findings both to the NZ Government and back 'home' to Great Britain. Mātauranga remained largely in the private domain of Māori, being regarded as superstition or myth by non-Māori scientists (Broughton and McBreen, 2015) until the end of the 20th century, when different perspectives on science became more valued (Hikuroa, 2017).

3. Scientific institutions and government policy

As the European population grew, institutes and museums were quickly established, and public lectures and meetings on the new geological or natural history findings were common (Priestley, 2008). The first colonial scientific and philosophical institutions, along with museums, were established in NZ in the mid to late 19th century (Priestley, 2010; Fleming, 1987). The Royal Society Te Apārangi (updated name officially announced in 2017) was founded in 1868 and published its *Transactions* from that date. Publicly acknowledged scientific research in the late 19th century was largely performed and self-funded by 'Victorian gentlemen', who earned income from professions such as the law or medicine and 'had a goal of demonstrating the power of science to further the common good' (Martin, 2017). As the universities were established (the University of Otago in 1869 and Canterbury College, part of the University of New Zealand, in 1873), salaried scientists and technical staff were more common (Martin, 2017).

The NZ Government established the Department of Scientific and Industrial Research (DSIR) in 1926, to drive research and innovation in natural sciences, agriculture and industry. As a new workforce grew, the New Zealand Association of Scientific Workers (later the NZ Association of Scientists; NZAS) was formed in 1942. The first issue of their journal *New Zealand Science Review* soon followed. This group also established NZ's first award for science communication in 1990 (Gregory, 2017). By the 1970s, many NZ scientists worked directly for the government as public servants in the DSIR (Galbreath, 1998).

In the early 1990s, DSIR was disestablished and the National Government restructured the science scene, creating in its place the Ministry of Research, Science and Technology, the Foundation for Research, Science and Technology (both of which have subsequently been further restructured) and the Crown Research Institutes (CRIs), of which there were originally 10 (Martin, 2017). There was more emphasis placed on obtaining external funding for research and on research for commercialisation. The push for modern developments in science communication gradually gained support from the scientific community, as many perceived a need to gain acceptance for new ideas or technologies (Douglas, 1988; Levitin, 2015; Csiszar, 2017), as well as a need to communicate clearly during natural disasters, such as earthquakes and volcanoes (Orchiston, 2010; Blake et al., 2018) or biosecurity emergencies (Bewsell et al., 2012; Warner, 2012; Muellner et al., 2018).

A Public Good Science Fund and a number of Centres of Research Excellence (CoREs) were established in the early 2000s (McCarthy and Rands, 2013; Martin, 2017). As commercial imperatives increased, the CRIs started producing their own public relations material, often bypassing the scientists themselves (Ashwell, 2016). Given NZ's geography and relatively small population size, there were few major daily newspapers and no public broadcasting television channels. Newsroom restructuring exacerbated a drop in the number of science journalists, diminishing independent science reporting (Ashwell, 2016). Ashwell (2016) revealed differences in opinion about the standard of science journalism between scientists and the communication advisors: scientists maintained that science reporting was poor, while communication advisors said their organisations were generally reported well. Furthermore, it has been noted that communication teams from CRIs, universities and science industries are often more interested in messaging, fundraising or creating a particular public image than reporting objectively (Salmon and Priestley, 2015). However, both scientists and communication advisors brought up issues of newsroom restructuring and increased pressures on science journalists, with cutbacks to personnel resulting in an increased use of press release material by journalists, often verbatim (Ashwell, 2016).

A publicly funded Science Media Centre was established in 2008 (Salmon and Priestley, 2015) to better inform journalists of current scientific research results. Concerns remain about the ability for publicly funded scientists to speak openly about their research results and expertise (Griffin, 2014; Hendy, 2016), as 40 per cent of those surveyed by the NZAS indicated in 2014 (Griffin, 2014).

The position of the NZ Prime Minister's Chief Science Advisor was established in 2009 to provide scientific advice and inform government policy with scientific evidence. Professor Sir Peter Gluckman held the inaugural position until mid-2018, when Professor Juliet Gerrard was appointed. The roles of the position now include raising the profile of science in NZ, especially amongst young people, making science more accessible to the public and encouraging the science community to build trusted relationships with communities (OPMCSA, 2018). The success of the Prime Minister's Science Advisor role also led to the creation of further science advisor roles being embedded within several government ministries (or departments) (MBIE, 2016).

In 2013, the government established a National Science Challenges panel, consisting of members of the public as well as science stakeholders and chaired by Sir Peter Gluckman, to prioritise research funding related to important national issues (Gluckman, 2013). The findings of this panel led to a major restructuring of the national funding of scientific research in 2013 (Salmon and Priestley, 2015). The National Science Challenges initiative aimed to promote greater commercial applications of scientific knowledge, reflecting ongoing neoliberal reforms by the National Party Government in NZ (Prussing and Newbury, 2016). They were also meant to lead to greater public engagement with science (Leitch et al., 2014). Eleven National Science Challenges were announced, focusing on environmental issues (land and water, resilience to nature, climate change, biological heritage, sustainable seas); health issues (child health, ageing populations, healthy living, nutrition); technological issues (science for technological innovation); or a combination of these (building better homes, towns and cities). These all became large, 'mission-led', interdisciplinary research programs with a substantial public or sector engagement component (Gluckman, 2013; MBIE, 2013).

The conclusions of the National Science Challenge panel's report stated that 'deficits' in the public's understanding of science needed addressing (Salmon and Priestley, 2015), suggesting sympathy for a 'deficit' approach to science communication. The proposed 'remedy' to this was the establishment of a 12th challenge, focused on 'Science and Society'. Rather than outsourcing this to research organisations, as occurred for the other 11 challenges, the government chose to manage this component internally.

As a result, in 2014, the previously mentioned strategy (A Nation of Curious Minds – He Whenua Hihiri I te Mahara) was officially launched (New Zealand Government, 2018a). It was designed to fund projects bringing science to society, thereby enabling better engagement with science and technology for all New Zealanders. Two funding initiatives were launched to support the strategy: A Nation of Curious Minds and three pilot participatory science platforms (launched in 2015) (New Zealand Government, 2018a), indicating the desire of the government to move towards more dialogic, participatory science communication. The resultant support from the public, scientists and government for participatory science appears to have informed the views of significant decision-makers in the science establishment who previously tended towards a deficit approach to science communication.

While there is a place for experts sharing their knowledge in a deficit-style approach, a broader participatory approach to science and its communication is now widely thought to be more effective (Bucchi, 2008; Salmon and Priestley, 2015; Longnecker, 2016a). A NZ example of ineffective use of a deficit, non-dialogic approach is described in a section below, 'Seeing science differently: Indigenous science and community engagement'.

4. A new kind of science communication in Aotearoa New Zealand

As discussed in more detail by Fleming and Star (2017), much of NZ's recent public engagement with scientific issues has been fuelled by health-related issues and by grassroots environmental movements. As in other modern societies with ageing populations, socioeconomic inequities and significant indigenous populations, NZ faces numerous public health challenges. The desire to protect the environment is also strong and multi-pronged, as tourism is a major contributor to the local economy and the natural environment is a key attraction for visitors (Fiedler et al., 2008). Science communication has a significant role to play in addressing these challenges.

However, the increase in communication of science has not just been driven by health-related issues and conservation of native plants and animals. Increasing frequency of physical disasters such as storms, flooding events and coastal erosion has hit society hard, from the smallest communities to our largest cities, and the growing cost of climate change to citizens and governments is becoming clear (Gifford et al., 1996; Ministry for the Environment, 2008; Roper et al., 2016; Royal Society of New Zealand, 2016). Furthermore, the rise in awareness of single-use plastic pollution (Gregory, 1978; Klein, 2018), the need to reduce the amount of waste going to landfill (Davies, 2009), the deterioration of NZ's freshwater and the decline of native freshwater fish (Joy et al., 2018), and the spread of pathogens such as myrtle rust and *Mycoplasma bovis* have all contributed to fierce debate of these environmental issues.

Changes in New Zealanders' attitudes to science were observed through the 2000s (Fleming and Star, 2017). The most recent survey, released in March 2018 and entitled *Public Engagement with Science & Technology*, shows 90 per cent of New Zealanders are interested in learning about science and 60 per cent feel well informed about science. However, New Zealanders are less likely to feel that science is important in their lives and almost a third feel science has become too specialised, with too much conflicting information making it hard to know what to believe (Nielsen Research, 2018).

In spite of diverse support, there are challenges to unfettered science communication. Hendy (2016) proposes that the lack of depth in some fields in NZ means that scientists cannot retreat from public communication but have an obligation to engage. He goes further to suggest that NZ scientists may need to adopt the potentially controversial role of advocate. He makes an eloquent case for greater communication to the NZ public about the practice of science (Longnecker, 2016b).

Below we discuss three aspects of science communication that play a significant role in influencing the national science communication ecosystem of NZ. These are environmental issues, hazards and disasters, and indigenous science and community engagement. These are by no means representative of the full scope of either the scientific issues that drive this work or the science communication activities that occur; for example, climate change communication is increasingly important in NZ (Salmon et al., 2017). Similarly, an exploration of the many health-related communication initiatives would warrant an article of its own and is beyond the scope of this chapter. We conclude this chapter with an overview of the increase in science communication training and education that has occurred over recent years.

5. Environmental issues, communication and citizen science initiatives

NZ has many species listed as in danger of extinction from habitat loss and predation by introduced predators (Dowding and Murphy, 2001; St Clair, 2011; Norbury and Jones, 2015; Ruffino et al., 2015; Thoresen et al., 2017). In the 1980s and 1990s, there was an upsurge in conservation initiatives by community groups in response to the decline in native bird numbers. Amongst other initiatives, this led to the establishment of fenced ecosanctuaries, such as Zealandia (originally the Karori Wildlife Sanctuary) near Wellington (Campbell-Hunt, 2002) and Orokonui Ecosanctuary near Dunedin (Tanentzap and Lloyd, 2017). There are now well over 20 ecosanctuaries in

NZ and its offshore islands, which act to exclude introduced mammalian predators (possum, rat, cat, dog, ferret, weasel and stoat) and browsers and grazers (possum, rabbit, pig, goat and deer) to enable native ecologies to re-establish and sustain local flora, bird, lizard and insect populations (Campbell-Hunt and Campbell-Hunt, 2013).

Figure 4.1: A volunteer checks a stoat trap in dense bush near Wellington, New Zealand.

Source: Photo courtesy of Jean Fleming.

Figure 4.2: Participants in the Marine Metre Squared activity use their guides to identify the organisms they have found within the square metre.

Source: Photo courtesy of Chris Paulin.

The success of ecosanctuaries has reawakened a vision of a NZ free of the many introduced pest species (Sullivan and Molles, 2016) and led to the Predator Free 2050 initiative, supported by government, non-government organisations and philanthropic trusts, and run by the Department of Conservation (Department of Conservation, 2018). This initiative sparked numerous predator-trapping initiatives throughout the country, perhaps epitomised by the group Predator Free NZ (Predator Free NZ, 2018). This grassroots movement in conservation has led to significant government and community engagement, and community-to-community engagement, both in cities and rurally, to help protect, restore and conserve NZ's natural heritage. These successful restoration and pest removal projects (Tanentzap and Lloyd, 2017) have brought New Zealanders face-to-face with the reality of their love of domestic cats and dogs, which contribute to the death of native birds and lizards (Morgan et al., 2009; Farnworth et al., 2010; Gordon et al., 2010; van Heezik et al., 2010; Farnworth et al., 2011; Coughlin and van Heezik, 2014; Aguilar et al., 2015; Harrod et al., 2016; Twardek et al., 2017; Walker et al., 2017).

Participatory science communication initiatives issues also experienced a significant boost through the arrival of the aforementioned Curious Minds funding, which launched a flurry of applications from all over the country. Extra funding was added to the pool at the end of the first year to keep up with demand. Since 2015, 114 projects have been funded, including those on the Participatory Science Platforms (New Zealand Government, 2018b). Around 40 per cent of these projects involved studies in conservation, environmental restoration or natural history, including studies on native birds, fish and bats. A similar proportion of projects involved *mātauranga*, ranging from a program to design Māori digital learning games to one creating marine learning environments through Māori knowledge to inform *kaitiakitanga* (guardianship) of the oceans and *Ahi Pepe Mothnet*, a study investigating moth numbers and species throughout the country (New Zealand Government, 2018b).

The goal of Predator Free 2050 (Department of Conservation, 2018) has highlighted the value of NZ's native species, as well as sparking debate on the best way to eradicate introduced predator species. Citizen science initiatives such as NatureWatch (which soon merged with the global iNaturalist (iNaturalist, 2018)), Marine Metre Squared (Fleming et al., 2017), the New Zealand Garden Bird Survey (Spurr, 2012; Liberatore et al., 2018) or *Ahi Pepe Mothnet* (Manaaki Whenua Landcare Research, 2018) have transformed science engagement across the country, by increasing participation and engagement of young and old. Volunteers are important for the success of many of the environmental projects. When asked why they volunteered, people often said they enjoyed meeting 'like-minded' or 'interesting' new friends. Volunteering was seen as a social activity, bringing better physical and mental health through doing, learning, seeing, contributing and being active outdoors (Fleming, 2017).

6. Hazards and disasters

NZ is a country prone to seismic events. In 2010 and 2011 a series of magnitude 6 and 7 earthquakes hit the city of Christchurch in the South Island (Kaiser et al., 2012). The second major shake struck at lunchtime on a working day, causing catastrophic damage to the city, and resulting in 185 deaths (Potter et al., 2015). The government of the day was focused on reporting the events accurately and appeared to be afraid to communicate information that might be wrong or alarm the community (Bryner, 2017). A full analysis of the poor communication of risk and the major players in the suppression of knowledge about the chance of aftershocks can be found in Gorman (2017). The lack of scientific information led to an information void that was filled with a claim by weather forecaster Ken Ring that he had predicted the second earthquake (The National Business Review, 2011). Geologist Dr Mark Quigley, at the time from the University of Canterbury, Christchurch, stepped up to show New Zealanders (and in particular the people of Christchurch) what had happened in his own backyard. He emphasised the very complex nature of the seismic event, gave scenarios with ranges of possible outcomes instead of absolutes, wrote about his experience in a deeply personal way (Quigley, 2012) and was honest and open on public media (Quigley, 2018a). Quigley was awarded the 2011 Prime Minister's Prize for Science Media Communication and the NZAS Science Communication Award for his work in communicating earthquake science to the public in the aftermath of these earthquakes (Quigley, 2018b).

Figure 4.3: Dr Mark Quigley, winner of the Prime Minister's Prize for Science Media Communication in 2011, in Riccarton, near Christchurch in 2015.

Source: Photo courtesy of Candice Egan.

NZ experienced a second major earthquake event when a magnitude 7.8 earthquake hit Kaikōura and the upper South Island in November 2016, causing landslides and raising parts of the land more than 4 metres (Guo et al., 2018), resulting in considerable damage to roads and property from north Canterbury to Wellington (Bradley et al., 2017; Kaiser et al., 2017). The response of the government and the scientific community was improved from the Christchurch experience, with better public engagement (Blake et al., 2018), including workshops to assess the effectiveness of the recovery response and the lessons learnt (Hatton et al., 2017). The Māori disaster management response received stronger recognition in the media after the Kaikōura earthquake, in that media stories highlighted the effectiveness of community-led initiatives as well as the importance of maintaining a unified and well-integrated approach to recovery management.

7. Seeing science differently: Indigenous science and community engagement

> Māori concerns about risks to their culture have been consistently marginalized by classing them as 'intangible' or spiritual positions and therefore unable to be evaluated alongside empirical scientific risk assessments. (Hudson et al., 2012)

In a succinct video (Goodall, 2016), Dr John Perrott points out that Māori 'belong' (*I belong therefore I am*), whereas Western scientists are trained to think as individuals (*I think therefore I am*). In Māori *whakapapa* (genealogy and cosmology), relationships with the land, flora and fauna are fundamental and all life is valued, as is collaboration and nurturing, all from a position of subjectivity not objectivity (Goodall, 2016). It has taken a long time to reach this open understanding of *mātauranga* in the NZ science world.

In the late 1980s, in response to the poor health status of Māori, nursing in NZ embarked on a process of improving nurse 'cultural safety' though a process of self-examination and change using Māori nurses. The objective was to challenge trainee nurses to see the world through the eyes of the patients. Cultural safety became a requirement for state examinations in 1992 (Papps and Ramsden, 1996). These changes can be seen as the beginning of 'Western' science engaging with *te Ao* (the Māori world). *Ngā Pae o te Māramatanga* (Horizons of Insight), NZ's Māori Centre of Research Excellence, funded by the Tertiary Education Commission (TEC) and hosted by the University of Auckland, was established in 2002 as one of the foundational Centres of Research Excellence. The centre's current research themes are *Whai Rawa* (The Māori Economy), *Te Tai Ao* (The Natural Environment), *Mauri Ora* (Human Flourishing) and *Te Reo me Ngā Tikanga* (Māori Language and Protocols).

In 2007, the Vision Mātauranga Policy was launched by the Ministry of Research, Science and Technology (restructured as the Ministry of Business, Industry and Employment in 2012). The Vision Mātauranga policy recognised that while there were many opportunities for Māori communities to make distinctive contributions to research, science and technology, and that many opportunities lay in the innovation potential of Māori knowledge, people and resources, they were not being realised (MBIE, 2018). In 2010, the Minister of Science and Innovation approved the integration of Vision Mātauranga across investment priority areas and established the Vision Mātauranga Capability Fund. In 2011, Vision Mātauranga policy was incorporated into the Statements of Core Purpose of the CRIs. CRIs are now required to enable the innovation potential of Māori knowledge, resources and people as part of their operating principles. Vision Mātauranga thus allowed communities, and the knowledge and potential therein previously disconnected from the science sector, to become fully engaged in science and technology. Vision Mātauranga is now also an integral part of all National Science Challenges, as well as a consideration in all government-led scientific research funding processes (MBIE, 2018).

The arrival of MAI Review (now *MAI Journal*[3]) in 2006 marked another turning point. This refereed academic journal is part of the Capability Building Programme of *Ngā Pae o te Māramatanga*. *MAI Journal* contains a wealth of quality material contributing to the body of knowledge about Māori and indigenous development, and help to advance the capabilities of Māori and Indigenous people engaged in research and scholarly training. A growing awareness of *mātauranga* followed in the science institutions of NZ. A values-based process of cross-cultural dialogue was proposed (Wilcox et al., 2008) and explored (Hudson et al., 2012). Engagement of the Māori community with research on the human brain helped to realise the potential of such research to Māori communities (Bohannon, 2007). Huntington disease research embarked on by Dr Melanie Cheung (Ngāti Rangitihi) for her doctoral project struck a potential barrier: the need to use human brains (from cadavers), considered by Māori to be *tapu* or sacred, and not to be touched (Bohannon, 2007). Engagement with tribal elders, who came to realise the importance of such research on the human brain to help Māori communities, was key in finding a solution to the impasse. Although the elders could not change the brain's *tapu* status, they created *tikanga* (ritual protocols), that became part of everyday lab routine. These *tikanga* enabled Dr Cheung to undertake the work in a culturally safe way (Bohannon, 2007).

This approach pioneered a way for many more researchers, providing a culturally safe working environment for subsequent cohorts of Māori brain researchers. Furthermore, Dr Cheung's work and attention to values such as *tapu* and the creation of *tikanga* has provided confidence for Māori who suffer from any brain disease, who likely otherwise would never have come forward. In a similar way, a special issue of the *New Zealand Journal of Zoology* was devoted to discussion of a seabird harvest in Aotearoa controlled solely

3 *MAI Journal* is an open-access journal that publishes multidisciplinary peer-reviewed articles that critically analyse and address indigenous and Pacific issues in the context of Aotearoa New Zealand.

by Māori (muttonbirding). These discussions explored the intersection of *mātauranga* and science as related ways of guiding sustainable harvest management (Moller, 2009). That special issue prompted the Royal Society of New Zealand to facilitate a 'cross-cultural environmental research and management' Challenges and Progress Forum, with local and international contributions published in the *Journal of the Royal Society of New Zealand* (Vol. 39, No. 4). Special Mātauranga Māori issues were recently published in the *NZ Journal of Marine and Freshwater Research* 52(4) 2018, *NZ Journal of Ecology* 43(3) 2019 and *NZ Science Review* 75(4) 2019.

Table 4.1: Some differences between *mātauranga* and science.

Mātauranga	**Science**
Knowledge as belonging	Knowledge for control and knowing
Explicit intrinsic values	Implicit instrumental values
Intuition as method	Intuition rarely acknowledged
Participatory 'experiencers' of systems	Detached observers of 'systems'
Inclusion of facts and values	Facts and values separated
Holistic worldview	Nature and culture separate
Everything is interconnected	Everything physical is interconnected

Source: Hikuroa (2017).

Hikuroa (2017) demonstrated that some *mātauranga* was and is generated using techniques consistent with the scientific method but differs by being explained according to a Māori world view. Some further differences between *mātauranga* and science are detailed in Table 3.1. An understanding of the relevance of *mātauranga* is crucial in encouraging young Māori into careers in science in NZ.

An insight into the gulf between *mātauranga* and the thinking of earlier Western government and industry can be gained from the following description of forestry contamination of an important site (Hikuroa et al., 2011).

> Until the early 1960s Rotoitipaku was a shallow lake fed by the spring Te Wai U o Tūwharetoa with active hot springs and sinter terraces on its southern shore, and a prominent feature of Te Kete Poutama … It was integral for traditional food gathering practices (*mahinga kai*) of the local Māori residents (*tangata whenua*), the home for waterborne guardians (*kaitiaki*) and was the hub for community activity. You would not recognize Te Kete Poutama if you searched for that idyllic scene today. In 1954 the New Zealand Government passed the Tasman Pulp and Paper Company Enabling Act 1954 that essentially gave the Tasman company carte blanche to do everything necessary to construct

> and operate a pulp and paper mill in Kawerau. In effect, the Act removed the tangata whenua's *manawhenua* and therefore ability to act as kaitiaki … Over 600,000 m^3 of waste containing toxic material has been dumped on the site … No longer does the lake teem with wildlife (indeed, there is no longer a lake), no longer do the hot springs provide warmth and relief for weary and aching bones, no longer do the ancestors of *Ngāti Tūwharetoa ki Kawerau* rest in peace, comforted by the constant companionship of their offspring—the mauri of Te Kete Poutama has been significantly compromised and the mana of the tangata whenua significantly impacted. (Hikuroa et al., 2011, pp. 1–2)

In subsequent years, projects were developed by Māori researchers to reflect community concern at the framing of *mātauranga* as relevant only in the traditional context and not in the modern science world (Hudson et al., 2012). Thus, Māori engaged the science community to broaden perceptions and used *mātauranga* to inspire collaborations that might lead to new avenues of scientific exploration. Some examples include the Kaitiaki Geothermal Development Model (e.g. Pryor, 2010), Te Awaroa (Hikuroa et al., 2018) and Te Ao Mārama – Centre for Fundamental Inquiry research collaboration, University of Auckland.

The development of the National Science Challenges was seen by some to marginalise participation by Māori researchers, in part through constructing 'Māori' and 'science' as separate, yet simultaneously recognising culturally distinctive forms of Māori knowledge. Others saw greater opportunity for *mātauranga* by the emphasis on Vision Mātauranga. By advocating for the validity of their *mātauranga*, Māori health researchers contested Pākehā values and priorities, reasserted the validity of *mātauranga*, and began the process of changing the production of knowledge in NZ across the board (Prussing and Newbury, 2016).

Māori advocacy against the values and priorities held by the majority of scientists and science institutions in NZ challenged trends to undermine collective rights and undercut support for culturally diverse worldviews by pointing out the ways in which market-driven priorities produce and sustain environmental, social and health inequities. One basis for the advocacy stems from the inclusion of values in *mātauranga*. By 2016, Sciblogs, the blog site of NZ's Science Media Centre, had recognised and defined the values of *mātauranga* (Goodall, 2016). Ocean Ripeka Mercier and her co-writers were discussing the potential marriage of Western science and *mātauranga* and proclaiming the 'veritableness' of Māori science (Macfarlane, 2016, 2017). Unsurprisingly, the marriage does not always involve smooth sailing—there is always potential for miscommunications between parties coming to a problem or project with different world views (Longnecker and Scott, 2018).

Figure 4.4: Researchers discussing the *mauri* of Te Kete Poutama within Waitaha Ariki Kore, ancestral house of Tohia o te Rangi marae, Kawerau. Left to right: Colleen Skerrett-White, Tomairangi Fox and Dan Hikuroa, November 2011.

Source: Photo courtesy of Ngā Pae o te Māramatanga.

Nonetheless, in recent years we have seen this marriage in action. Indigenous communities are increasingly taking the lead in river restoration, using the process as an opportunity to re-engage deeply with their rivers, while revealing sociocultural and political dimensions of restoration under-reported in ecological and social science literatures (Fox et al., 2017). Mātauranga solutions to some of NZ's toughest problems, including kauri dieback disease (Chetham and Shortland, 2013), are increasingly being sought (Williams, 2017; Darragh, 2018; Harrison, 2018; Hurihanganui, 2018; New Zealand's Biological Heritage, 2018; Pitama et al., 2018).

8. Learning to communicate science

One approach to improve understanding and practice of science communication, as well as fostering a culture with greater critical feedback about contemporary science communication, is to provide education and training to aspiring communicators and scientists. Early academic courses in communication studies, including the communication of science, were present in NZ from the late 1970s, but training mainly focused on science journalism until the end of 20th century (Fleming and Star, 2017).

Early in the 21st century, public debate about science was exemplified by the New Zealand Royal Commission on Genetic Modification (RCGM), which engaged thousands of ordinary New Zealanders with developments in biotechnology (Eichelbaum et al., 2001). Ten thousand public submissions were made to the commission and over 90 per cent of these were strongly against genetic modification, although knowledge of the technology was limited. Many submitters were concerned by the control of science and scientists by large corporations such as Monsanto rather than the biotechnology itself (Fleming, 2003). The RCGM showed the government of the day that NZ needed to move on from the deficit model of science communication and to include New Zealanders in difficult scientific decisions (Fleming, 2003). As the Chair of the RCGM, Sir Thomas Eichelbaum, said: 'Few minds may have been changed in the process but everyone emerged better informed and more willing to listen to each other' (Fleming, 2003).

The AC Nielsen survey conducted by the RCGM showed the majority of New Zealanders were interested in and engaged with aspects of science and technology, especially where personal or societal benefits were most obvious (such as health). Most New Zealanders surveyed in 2002 held strongly realistic views of science, with a significant proportion appearing to hold the view that 'seeing was believing' and not inclined to take scientific claims on trust, possibly because of a lack of understanding about how scientific evidence was generated (Hipkins et al., 2002). There is evidence that knowledge and understanding of GM has changed since 2002 and that a large majority of New Zealanders would now accept at least some forms of GM (Hope, 2014).

It took until 2008 for the first tertiary training centre, the University of Otago's Centre for Science Communication (University of Otago, 2018), to be established in NZ. This successful centre has grown over the past decade to be one of the biggest centres for postgraduate training in science communication in the world. Science communication is taught at four other universities in NZ and is a key focus in undergraduate and postgraduate programs established in 2013 by the Centre for Science in Society at Victoria University of Wellington.

An increasing number of research degrees are being awarded in science communication in NZ. While the first PhD in science communication was awarded as recently as 2015 by the University of Otago (Cade, 2015), at time of writing its Centre for Science Communication has 14 PhD students and 60 postgraduate science communication coursework and MSciComm students enrolled. Government-funded research centres are increasingly engaged with science communication research. For example, in 2018, Te Pūnaha Matatini (a Centre of Research Excellence) provided a fully funded PhD scholarship to explore the process of science communication in NZ.

NZ's first science communication conference in 2001 was run by the Royal Society of New Zealand Te Apārangi, which has played a key role in colonial and post-colonial NZ science (Fleming and Star, 2017). In 2004, agricultural journalist Peter Burke formed the national Science Communicators' Association of New Zealand (SCANZ). He had attended an international science communication conference (PCST) in 2002 in Cape Town, South Africa, and noted the value of organisations for science communicators that existed in many countries around the world. The growing number of science communicators, along with a decreasing number of specialised science journalists employed by national newspapers (Hendy, 2016), has seen SCANZ grow substantially, with the annual conference becoming a popular place for presentation of science communication research (SCANZ, 2018). In 2018, the international biennial PCST conference was hosted in Dunedin, NZ, bringing together colleagues from around to the world for this important opportunity to share knowledge and practices.

Today, the value of science communication for NZ society is recognised not only through the NZAS's Science Communication Award launched in 1990 (now the Cranwell Medal), but also prizes awarded by the Royal Society Te Apārangi (Callaghan Medal, launched 2011), the Science Communicators' Association of New Zealand (launched 2015), and the prestigious Prime Minister's Science Communication Prize (launched 2009) worth NZ$100,000.

9. Conclusions

In NZ, the beginning of the 21st century has seen an increase in the collaboration of *mātauranga* with Western science at a time of heightened awareness about environmental issues. Māori researchers have pushed to widen the perspectives of the scientific community, using dialogue initially, but more and more by engaging communities with *mātauranga*. *Mātauranga* and new dialogic approaches have complemented and added depth to established practices in science communication, such as conferences, science festivals and social media. Māori have taken science communication in Aotearoa New Zealand in new directions, with an increasing emphasis on the inherent values of the science being communicated.

There have also been significant increases in training opportunities, formal education, funding and jobs in science communication over the past decade. Linked to this has been a substantial rise in the number and diversity of participatory science programs in NZ, a good proportion concerning environmental protection and restoration (Peters et al., 2016), assisted by the establishment of a source of funding (New Zealand Government, 2018b).

As a result, NZ's communities are engaging more with science, hopefully creating more science-literate publics and a more public-literate and bicultural science community.

Acknowledgement

The authors gratefully acknowledge and thank Associate Professor Rebecca Priestley (Victoria University of New Zealand) for her advice on this chapter.

References

Aguilar, G. D., Farnworth, M. J. and Winder, L. (2015). Mapping the stray domestic cat (*Felis catus*) population in New Zealand: Species distribution modelling with a climate change scenario and implications for protected areas. *Applied Geography*, *63*, 146–54. doi.org/10.1016/j.apgeog.2015.06.019.

Ashwell, D. J. (2016). The challenges of science journalism: The perspectives of scientists, science communication advisors and journalists from New Zealand. *Public Understanding of Science*, *25*(3), 379–93. doi.org/10.1177/09636625 14556144.

Bess, R. (2011). New Zealand's Treaty of Waitangi and the doctrine of discovery: Implications for the foreshore and seabed. *Marine Policy*, *35*(1), 85–94. doi.org/ 10.1016/j.marpol.2010.08.007.

Bewsell, D., Bigsby, H. and Cullen, R. (2012). Using involvement to understand individual responses to an issue: The case of New Zealand biosecurity. *New Zealand Journal of Agricultural Research*, *55*(1), 73–88. doi.org/10.1080/ 00288233.2011.647354.

Blake, D., Johnston, D., Leonard, G., McLaren, L. and Becker, J. (2018). A Citizen Science Initiative to Understand Community Response to the Kaikoura Earthquake and Tsunami Warning in Petone and Eastbourne, Wellington, Aotearoa/New Zealand. *Bulletin of the Seismological Society of America, 108*(3B), 1807–17. doi.org/10.1785/0120170292.

Bohannon, J. (2007). Michael Walker – A home for Maori science. *Science*, *318*(5852), 907. doi.org/10.1126/science.318.5852.907.

Bradley, B. A., Razafindrakoto, H. N. T. and Polak, V. (2017). Ground-Motion Observations from the 14 November 2016 M-w 7.8 Kaikoura, New Zealand, Earthquake and Insights from Broadband Simulations. *Seismological Research Letters*, *88*(3), 740–56. doi.org/10.1785/0220160225.

Broughton, D. and McBreen, K. (2015). Mātauranga Māori, tino rangatiratanga and the future of New Zealand Science. *Journal of the Royal Society of New Zealand*, *45*(2), 83–88. doi.org/10.1080/03036758.2015.1011171.

Bryner, V. (2017). Communicating the sciences of disaster risk reduction: media stories surrounding the Canterbury earthquakes of 2010–2011 (unpublished PhD thesis). University of Otago, Dunedin, New Zealand.

Bucchi, M. (1998). The future of science has begun. The communication of science to the public: Science and the media. Proceedings of the Fifth International Conference. *Public Understanding of Science*, *7*(4), 349–50.

Bucchi, M. (2008). Of deficits, deviations and dialogues: Theories of public communication of science. In M. Bucchi and B. Trench (eds), *Handbook of Public Communication of Science and Technology*. Abingdon, UK and New York, USA: Routledge.

Cade, O. J. (2015). Communicating science through poetry from 1780 to the present (PhD thesis). University of Otago, Dunedin, New Zealand.

Campbell-Hunt, D. (2002). *Developing a Sanctuary: The Karori Experience.* Wellington: Victoria Link Ltd.

Campbell-Hunt, D. and Campbell-Hunt, C. (2013). *Ecosanctuaries: Communities building a future for New Zealand's threatened ecologies*. Dunedin: Otago University Press.

Chetham, J. and Shortland, T. (2013). *Kauri cultural health indicators – Monitoring.* Auckland: Kauri Dieback Joint Agency Response, Tangata Whenua Roopu.

Collins, P. M. D. and Bodmer, W. F. (1986). The Public Understanding of Science. *Studies in Science Education*, *13*(1), 96–104.

Coughlin, C. E. and van Heezik, Y. (2014). Weighed down by science: Do collar-mounted devices affect domestic cat behaviour and movement? *Wildlife Research*, *41*(7), 606–14. doi.org/10.1071/wr14160.

Cronin, K. (2008). The privatization of public talk: A New Zealand case study on the use of dialogue for civic engagement in biotechnology governance. *New Genetics and Society*, *27*(3), 285–99. doi.org/10.1080/14636770802326950.

Csiszar, A. (2017). How lives became lists and scientific papers became data: Cataloguing authorship during the nineteenth century. *British Journal for the History of Science*, *50*(1), 23–60. doi.org/10.1017/s0007087417000012.

Darragh, T. A. (2018). Finding New Zealand's Scientific Heritage: From Mātauranga Māori to Augustus Hamilton. *Journal of the Royal Society of New Zealand*, *29*(2), 207. doi.org/10.1080/03036758.2016.1265995.

Davies, A. R. (2009). Clean and green? A governance analysis of waste management in New Zealand. *Journal of Environmental Planning and Management*, *52*(2), 157–76. doi.org/10.1080/09640560802666503.

Department of Conservation. (2018). Predator Free 2050. Wellington: Department of Conservation. Retrieved from www.doc.govt.nz/nature/pests-and-threats/predator-free-2050/.

Douglas, G. L. (1988). History of science and technology resources at the Linnean Society of London. *British Journal for the History of Science, 21*(71), 489–93.

Dowding, J. E. and Murphy, E. C. (2001). The impact of predation by introduced mammals on endemic shorebirds in New Zealand: A conservation perspective. *Biological Conservation, 99*(1), 47–64. doi.org/10.1016/s0006-3207(00)00187-7.

Eichelbaum, T., Allan, J., Fleming, J. and Randerson, R. (2001). *Report of the Royal Commission on Genetic Modification*. Wellington: Department of Internal Affairs, New Zealand.

Farnworth, M. J., Campbell, J. and Adams, N. J. (2010). Public awareness in New Zealand of animal welfare legislation relating to cats. *New Zealand Veterinary Journal, 58*(4), 213–17. doi.org/10.1080/00480169.2010.68624.

Farnworth, M. J., Campbell, J. and Adams, N. J. (2011). What's in a Name? Perceptions of Stray and Feral Cat Welfare and Control in Aotearoa, New Zealand. *Journal of Applied Animal Welfare Science, 14*(1), 59–74. doi.org/10.1080/10888705.2011.527604.

Fiedler, A. K., Landis, D. A. and Wratten, S. D. (2008). Maximizing ecosystem services from conservation biological control: The role of habitat management. *Biological Control, 45*(2), 254–71. doi.org/10.1016/j.biocontrol.2007.12.009.

Fleming, C. A. (1987). *Science, Settlers and Scholars*. Wellington: The Royal Society of New Zealand.

Fleming, J. S. (2003). Inform, educate or influence? New Zealand's experience of the debate on genetic modification. *The Round Table: The Commonwealth Journal of International Affairs, 371*, 487–99. doi.org/10.1080/0035853032000126174.

Fleming, J. S. (2017). Why do we do it? *Forest & Bird Magazine, Te Reo o te Taiao, 364*, 28–9.

Fleming, J. S., Broni, B., Copeland, S., Hunt, D. and Newburn, R. (2017). Thirty years of science outreach at the University of Otago. *New Zealand Science Review, 74*(1), 10–13.

Fleming, J. S. and Star, J. (2017). The emergence of science communication in Aotearoa New Zealand. *Journal of Science Communication, 16*(3), A02. doi.org/10.22323/2.16030202.

Fox, C. A., Reo, N. J., Turner, D. A., Cook, J., Dituri, F., Fessell, B., Jenkins, J., Johnson, A., Rakena, T. M., Riley, C., Turner, A., Williams, J. and Wilson, M. (2017). 'The river is us; the river is in our veins': Re-defining river restoration in three indigenous communities. *Sustainability Science, 12*(4), 521–33. doi.org/10.1007/s11625-016-0421-1.

France, B., Mora, H. A. and Bay, J. L. (2012). Changing Perspectives: Exploring a pedagogy to examine other perspectives about stem cell research. *International Journal of Science Education, 34*(5), 803–24. doi.org/10.1080/09500693.2011.630427.

Galbraith, M., Bollard-Breen, B. and Towns, D. R. (2016). The Community–Conservation Conundrum: Is Citizen Science the Answer? *Land, 5*(4). doi.org/10.3390/land5040037.

Galbreath, Ross (1998). *DSIR: Making Science Work for New Zealand: Themes from the History of the Department of Scientific and Industrial Research, 1926–1992.* Wellington: Victoria University Press.

Gascoigne, T., Cheng, D., Claessens, M., Metcalfe, J., Schiele B. and Shi, S. (2010). Is science communication its own field? , *Journal of Science Communication, 9*(3), C04. doi.org/10.22323/2.09030304.

Gifford, R. M., Campbell, B. D. and Howden, S. M. (1996). Options for adapting agriculture to climate change: Australian and New Zealand examples. In W. J. Bouma, G. I. Pearman and M. R. Manning (eds), *Greenhouse: Coping with climate change* (pp. 399–416). Melbourne: CSIRO Publishing.

Gluckman, P. (2013). *Report of the National Science Challenges Panel. Ministry of Science and Innovation.* Wellington: New Zealand Government.

Goodall, L. (2016). What is Mātauranga Māori? SciBlogs. Retrieved from sciblogs.co.nz/matau-taiao/2016/02/26/what-is-matauranga-maori/.

Gordon, J. K., Matthaci, C. and van Hcczik, Y. (2010). Bellcd collars rcducc catch of domestic cats in New Zealand by half. *Wildlife Research, 37*(5), 372–78. doi.org/10.1071/wr09127.

Gorman, P. (2017). *Portacom City. Reporting on the Christchurch and Kaikōura earthquakes.* Wellington: Bridget Williams Books.

Green, W. and Rohan, M. (2012). Opposition to aerial 1080 poisoning for control of invasive mammals in New Zealand: risk perceptions and agency responses. *Journal of the Royal Society of New Zealand, 42*(3), 185–213. doi.org/10.1080/03036758.2011.556130.

Gregory, G. (2017). Service to science: History of the New Zealand Association of Scientists. *Journal of the Royal Society of New Zealand, 47*(2), 175–80. doi.org/10.1080/03036758.2017.1291435.

Gregory, M. R. (1978). Accumulation and distribution of virgin plastic granules on New Zealand beaches. *New Zealand Journal of Marine and Freshwater Research, 12*(4), 399–414.

Griffin, P. (2014). When scientists don't speak out – gag orders and funding fears revealed. SciBlogs. Retrieved from sciblogs.co.nz/griffins-gadgets/2014/11/03/when-scientists-dont-speak-out-gag-orders-and-funding-fears-revealed/.

Guo, P., Han, Z., Ran, H., Luo, J.-H., Wu, G. and Yuan, R.-M. (2018). Co-seismic surface rupture of Papatea fault and reactivation mechanism of the Clarence landslide during the 2016 Mw7.8 Kaikoura earthquake, New Zealand. *Bulletin of Engineering Geology and the Environment, 5*. doi.org/10.1007/s10064-018-1338-y.

Harrison, R. (2018). *Potential whakapapa Māori solution for kāuri dieback outbreak.* Māori Television. Retrieved from www.teaomaori.news/potential-whakapapa-maori-solution-kauri-dieback-outbreak.

Harrod, M., Keown, A. J. and Farnworth, M. J. (2016). Use and perception of collars for companion cats in New Zealand. *New Zealand Veterinary Journal, 64*(2), 121–24. doi.org/10.1080/00480169.2015.1110064.

Hatton, T., Kipp, B., Brown, C. and Seville, E. (2017). *Kaikoura earthquake social science research workshop report.* QuakeCoRE publication number 0177.

Hendy, S. (2016). *Silencing Science: Science that is not heard is not science at all.* Wellington: Bridget Williams Books.

Hikuroa, D. (2017). Mātauranga Māori – the ūkaipō of knowledge in New Zealand. *Journal of the Royal Society of New Zealand, 47*(1), 5–10. doi.org/10.1080/03036758.2016.1252407.

Hikuroa, D. (2018). *Te Awaroa: Voice of the river.* Te Pūnaha Matatini. Retrieved from www.tepunahamatatini.ac.nz/2017/09/11/wainz2/.

Hikuroa, D., Slade, A. and Gravley, D. (2011). Implementing Māori indigenous knowledge (mātauranga) in a scientific paradigm: Restoring the mauri to Te Kete Poutama. *MAI Review, 3*. Retrieved from www.review.mai.ac.nz/mrindex/MR/article/viewArticle/433.html.

Hipkins, R., Stockwell, W., Bolstad, R. and Baker, R. (2002). *Commonsense, trust and science: How patterns of beliefs and attitudes to science pose challenges for effective communication.* Ministry of Research, Science and Technology. Wellington: New Zealand Government.

Hope, K. M. (2014). Socioscientific issues: A survey of New Zealanders' attitudes towards GE (unpublished MSciComm thesis). University of Otago, Dunedin, New Zealand.

Hudson, M., Roberts, M., Smith, L., Tiakiwai, S. J. and Hemi, M. (2012). The art of dialogue with indigenous communities in the new biotechnology world. *New Genetics and Society, 31*(1), 11–24. doi.org/10.1080/14636778.2011.597979.

Hudson, M. L. and Russell, K. (2009). The Treaty of Waitangi and research ethics in Aotearoa. *Journal of Bioethical Inquiry, 6*(1), 61–8. doi.org/10.1007/s11673-008-9127-0.

Hurihanganui, T. A. (2018). *Kauri dieback fight: Scientist warns of Māori exclusion.* Radio New Zealand. Retrieved from www.rnz.co.nz/news/te-manu-korihi/365776/kauri-dieback-fight-scientist-warns-of-maori-exclusion.

iNaturalist. (2018). *iNaturalist NZ – Mātaki Taiao*. Wellington. Retrieved from inaturalist.nz/.

Joy, M. K., Foote, K. J., McNie, P. and Piria, M. (2018). Decline in New Zealand's freshwater fish fauna: Effect of land use. *Marine and Freshwater Research, 70*(1) 114–24. doi.org/10.1071/mf18028.

Kaiser, A., Balfour, N., Fry, B., Holden, C., Litchfield, N., Gerstenberger, M., D'Anastasio, E., Horspool, N., McVerry, G., Ristau, J., Bannister, S., Christophersen, A., Clark, K., Power, W., Rhoades, D., Massey, C., Hamling, I., Wallace, L., Mountjoy, J., Kaneko, Y., Benites, R., Van Houtte, C., Dellow, S., Wotherspoon, L., Elwood, K. and Gledhill, K. (2017). The 2016 Kaikoura, New Zealand, Earthquake: Preliminary Seismological Report. *Seismological Research Letters, 88*(3), 727–39. doi.org/10.1785/0220170018.

Kaiser, A., Holden, C., Beavan, J., Beetham, D., Benites, R., Celentano, A., Collett, D., Cousins, J., Cubrinovski, M., Dellow, G., Denys, P., Fielding, E., Fry, B., Gerstenberger, M., Langridge, R., Massey, C., Motagh, M., Pondard, N., McVerry, G., Ristau, J., Stirling, M., Thomas, J., Uma, S. R. and Zhao, J. (2012). The Mw 6.2 Christchurch earthquake of February 2011: Preliminary report. *New Zealand Journal of Geology and Geophysics, 55*(1), 67–90. doi.org/10.1080/00288306.2011.641182.

Klein, A. (2018). New Zealand becomes the latest country to ban plastic bags. *New Scientist, 239*(3191), 5. doi.org/10.1016/s0262-4079(18)31453-2.

Lambert, D. M., Shepherd, L. D., Huynen, L., Beans-Picôn, G., Walter, G. H. and Millar, C. D. (2009). The molecular ecology of the extinct New Zealand huia. *PLoS One, 4*(11), e8019. doi.org/10.1371/journal.pone.0008019.

Leitch, S., Motion, J., Merlot, E. and Davenport, S. (2014). The fall of research and rise of innovation: Changes in New Zealand science policy discourse. *Science and Public Policy, 41*(1), 119–23. doi.org/10.1093/scipol/sct042.

Levitin, D. (2015). 'Made Up from Many Experimental Notions': The Society of Apothecaries, Medical Humanism, and the Rhetoric of Experience in 1630s London. *Journal of the History of Medicine and Allied Sciences, 70*(4), 549–87. doi.org/10.1093/jhmas/jru027.

Liberatore, A., Bowkett, E., MacLeod, C., Spurr, E. and Longnecker, N. (2018). Social Media as a Platform a Citizen Science Community of Practice. *Citizen Science: Theory and Practice, 3*(1), 3. doi.org/10.5334/cstp.108.

Longnecker, N. (2016a). An integrated model of science communication: More than providing evidence. *Journal of Science Communication, 15*(5). doi.org/10.22323/2.15050401.

Longnecker, N. (2016b). Book Review: Silencing Science by Shaun Hendy. *Journal of the Royal Society of New Zealand, 46*(3–4), 238–9. doi.org/10.1080/03036758.2016.1230553.

Longnecker, N. and Scott, C. (2018). Challenges of cross-cultural communication in production of a collaborative exhibition: Wai ora, Mauri ora. *Journal of Science Communication, 17*(04), C05. doi.org/10.22323/2.17040305.

Macfarlane, A. (2016). Enhancing Matauranga Maori and Global Indigenous Knowledge. *New Zealand Journal of Educational Studies, 51*(1), 129–32.

Macfarlane, A. (2017). Enhancing Matauranga Maori and Global Indigenous Knowledge. *New Zealand Journal of Educational Studies, 52*(1), 205.

Manaaki Whenua Landcare Research. (2018). Ahi Pepe: Mothnet. Dunedin: Manaaki Whenua Landcare Research. Retrieved from www.landcareresearch.co.nz/information-for/citizen-science/mothnet.

Mark, A. (2001). 'Integrating conservation with hydro-electric development of Lakes Manapouri and Te Anau, New Zealand: an exercise in complexity'. *Complexity International, 8*(1), 1–19.

Mark, A. F. (2015). *Standing My Ground: A voice for nature conservation.* Dunedin: Otago University Press.

Mark, A. F. and Johnson, P. N. (1985). Ecologically derived guidelines for managing two New Zealand lakes. *Environmental Management, 9*(4), 355–64. doi.org/10.1007/bf01867306.

Martin, J. E. (2017). Reflections on the history of the Royal Society Te Apārangi. *New Zealand Journal of Social Sciences, 12*(2), 277–88. doi.org/10.1080/1177083x.2017.1391301.

MBIE. (2013). *National Science Challenges: Key messages regarding challenge principles.* MBIE-MAKO-10986337, Ministry of Business, Innovation and Employment. Wellington: New Zealand Government.

MBIE. (2016). *Our Science Board.* Ministry of Business, Innovation and Employment. Wellington: New Zealand Government. Retrieved from www.mbie.govt.nz/about/who-we-are/our-external-boards/our-science-board/.

MBIE. (2018). *Vision Mātauranga.* Ministry of Business, Innovation and Employment. Wellington: New Zealand Government. Retrieved from www.mbie.govt.nz/info-services/science-innovation/agencies-policies-budget-initiatives/vision-matauranga-policy.

McCarthy, D. and Rands, M. (2013). Learned societies: A bridge between research, policy making and funding. *Studies in Higher Education, 38*(3), 470–83. doi.org/10.1080/03075079.2013.773216.

McWethy, D. B., Whitlock, C., Wilmshurst, J. M., McGlone, M. S. and Li, X. (2009). Rapid deforestation of South Islands, New Zealands, by early Polynesian fires. *Holocene, 19*(6), 883–97. doi.org/10.1177/0959683609336563.

McWethy, D. B., Wilmshurst, J. M., Whitlock, C., Wood, J. R. and McGlone, M. S. (2014). A High-Resolution Chronology of Rapid Forest Transitions following Polynesian Arrival in New Zealand. *PLoS One*, *9*(11). doi.org/10.1371/journal.pone.0111328.

Ministry for the Environment. (2008). *Preparing for Climate Change: A guide for local government in New Zealand*. Wellington: Ministry for the Environment.

Mohi, J. H. and Roberts, W. D. (2009). Delivering a strategy for working with Māori, and developing responsiveness to an increasingly multicultural population: A perspective from the National Library of New Zealand. *IFLA Journal*, *35*(1), 48–58 and 79–82. doi.org/10.1177/0340035208102037.

Moller, H. (2009). Mātauranga Māori, science and seabirds in New Zealand. *New Zealand Journal of Zoology*, *36*(3), 203–10. doi.org/10.1080/03014220909510151.

Morgan, S. A., Hansen, C. M., Ross, J. G., Hickling, G. J., Ogilvie, S. C. and Paterson, A. M. (2009). Urban cat (Felis catus) movement and predation activity associated with a wetland reserve in New Zealand. *Wildlife Research*, *36*(7), 574–80. doi.org/10.1071/wr09023.

Muellner, P., Hodges, D., Ahlstrom, C., Newman, M., Davidson, R., Pfeiffer, D., Marshall, J. and Morley, C. (2018). Creating a framework for the prioritization of biosecurity risks to the New Zealand dairy industry. *Transboundary and Emerging Diseases*, *65*(4), 1067–77. doi.org/10.1111/tbed.12848.

Munshi, D., Kurian, P. A., Morrison, T. and Morrison, S. L. (2016). Redesigning the architecture of policy-making: Engaging with Māori on nanotechnology in New Zealand. *Public Understanding of Science*, *25*(3), 287–302. doi.org/10.1177/0963662514548629.

New Zealand Government. (2018a). Origins of Curious Minds. Wellington: Ministry of Business, Innovation and Employment. Retrieved from www.curiousminds.nz/about/history/.

New Zealand Government. (2018b). Projects funded by Curious Minds. Wellington: Ministry of Business, Innovation and Employment. Retrieved from www.curiousminds.nz/projects/.

New Zealand's Biological Heritage. (2018). Stopping kauri dieback in its tracks. National Science Challenges. Retrieved from bioheritage.nz/research/stopping-kauri-dieback/.

Nielsen Research. (2018). *Report on Public Engagement with Science & Technology*. Wellington: Ministry of Business, Innovation and Employment.

Norbury, G. and Jones, C. (2015). Pests controlling pests: does predator control lead to greater European rabbit abundance in Australasia? *Mammal Review*, *45*(2), 79–87. doi.org/10.1111/mam.12034.

Office of the Prime Minister's Chief Science Advisor (OPMCSA). (2018). Our Office – Prime Minister's Chief Science Advisor. Wellington. Retrieved from www.pmcsa.ac.nz/who-we-are/our-office/.

Orange, C. (2011). *The Treaty of Waitangi*. Wellington, NZ: Bridget Williams Books.

Orchiston, C. (2010). Tourism and seismic risk: Perceptions, preparedness and resilience in the zone of the Alpine Fault, Southern Alps, New Zealand (PhD thesis). University of Otago, Dunedin, New Zealand.

Papps, E. and Ramsden, I. (1996). Cultural safety in nursing: The New Zealand experience. *International Journal for Quality in Health Care, 8*(5), 491–97. doi.org/10.1093/intqhc/8.5.491.

Peters, M. A., Eames, C. and Hamilton, D. (2015). The use and value of citizen science data in New Zealand. *Journal of the Royal Society of New Zealand, 45*(3), 151–60. doi.org/10.1080/03036758.2015.1051549.

Peters, M. A., Hamilton, D., Eames, C., Innes, J. and Mason, N. W. H. (2016). The current state of community-based environmental monitoring in New Zealand. *New Zealand Journal of Ecology, 40*(3), 279–88. doi.org/10.20417/nzjecol.40.37.

Pieczka, M. and Escobar, O. (2013). Dialogue and science: Innovation in policy-making and the discourse of public engagement in the UK. *Science and Public Policy, 40*(1), 113–26. doi.org/10.1093/scipol/scs073.

Pitama, S. G., Palmer, S. C., Huria, T., Lacey, C. and Wilkinson, T. (2018). Implementation and impact of indigenous health curricula: a systematic review. *Medical Education, 52*(9), 898–909. doi.org/10.1111/medu.13613.

Potter, S. H., Becker, J. S., Johnston, D. M. and Rossiter, K. P. (2015). An overview of the impacts of the 2010–2011 Canterbury earthquakes. *International Journal of Disaster Risk Reduction, 14*, 6–14. doi.org/10.1016/j.ijdrr.2015.01.014.

Predator Free NZ. (2018). Predator Free NZ. Retrieved from predatorfreenz.org/.

Priestley, R. (ed.). (2008). *The Awa Book of New Zealand Science*. Wellington, NZ: Awa Press.

Priestley, R. (2010). A Survey of the History of Science in New Zealand, 1769–1992. *History Compass, 8*(6), 474–90. doi.org/10.1111/j.1478-0542.2010.00684.x.

Prussing, E. and Newbury, E. (2016). Neoliberalism and indigenous knowledge: Maori health research and the cultural politics of New Zealand's 'National Science Challenges'. *Social Science & Medicine, 150*, 57–66. doi.org/10.1016/j.socscimed.2015.12.012.

Pryor, R. (2010). *A kaitiaki approach to geothermal development: Encompassing the Māori worldview in New Zealand's growing renewable energy industry*. Retrieved from frontiersabroad.com/wp-content/uploads/2012/09/RebeccaPryor.pdf.

Quigley, M. C. (2012). Mark Quigley's long, cold year. *The Press*. Retrieved from www.stuff.co.nz/the-press/news/christchurch-earthquake-2011/12-51/the-earth/6413953/Mark-Quigleys-long-cold-year.

Quigley, M. C. (2018a). Dr Quigs. Retrieved from www.drquigs.com/.

Quigley, M. C. (2018b). About. Retrieved from www.drquigs.com/?page_id=13.

Roper, J., Ganesh, S. and Zorn, T. E. (2016). Doubt, Delay, and Discourse: Skeptics' Strategies to Politicize Climate Change. *Science Communication*, *38*(6), 776–99. doi.org/10.1177/1075547016677043.

Royal Society of New Zealand. (2016). *Climate change implications for New Zealand* (Online). Wellington, New Zealand: Royal Society of New Zealand. doi.org/10.1080/03036758.2018.1476389.

Ruffino, L., Zarzoso-Lacoste, D. and Vidal, E. (2015). Assessment of invasive rodent impacts on island avifauna: methods, limitations and the way forward. *Wildlife Research*, *42*(2), 185–95. doi.org/10.1071/wr15047.

Salmon, R. A. and Priestley, R. K. (2015). A future for public engagement with science in New Zealand. *Journal of the Royal Society of New Zealand*, *45*(2), 101–7. doi.org/10.1080/03036758.2015.1023320.

Salmon, R., Priestley, R., Fontana, M., and Milfont, T. (2017). Climate Change Communication in New Zealand. *Oxford Research Encyclopedia of Climate Science*. doi.org/10.1093/acrefore/9780190228620.013.475.

SCANZ. (2018). Science Communicators' Association of New Zealand. Auckland. Retrieved from www.scanz.co.nz/.

Seersholm, F. V., Cole, T. L., Grealy, A., Rawlence, N. J., Greig, K., Knapp, M., Stat, M., Hansen, A. J., Easton, L. J., Shepherd, L., Tennyson, A. J. D., Scofield, R. P., Walter, R. and Bunce, M. (2018). Subsistence practices, past biodiversity, and anthropogenic impacts revealed by New Zealand-wide ancient DNA survey. *Proceedings of the National Academy of Sciences of the United States of America*, *115*(30), 7771–6. doi.org/10.1073/pnas.1803573115.

Smallman, M. (2016). Public Understanding of Science in turbulent times III: Deficit to dialogue, champions to critics. *Public Understanding of Science*, *25*(2), 186–97. doi.org/10.1177/0963662514549141.

Spurr, E. B. (2012). New Zealand Garden Bird Survey – analysis of the first four years. *New Zealand Journal of Ecology*. *36*(3), 1–13.

St Clair, J. J. H. (2011). The impacts of invasive rodents on island invertebrates. *Biological Conservation*, *144*(1), 68–81. doi.org/10.1016/j.biocon.2010.10.006.

Storey, R. G. and Wright-Stow, A. (2017). Community-based monitoring of New Zealand stream macroinvertebrates: agreement between volunteer and professional assessments and performance of volunteer indices. *New Zealand Journal of Marine and Freshwater Research*, *51*(1), 60–77. doi.org/10.1080/00288330.2016.1266674.

Sullivan, J. J. and Molles, L. E. (2016). Biodiversity monitoring by community-based restoration groups in New Zealand. *Ecological Management & Restoration*, *17*(3), 210–17. doi.org/10.1111/emr.12225.

Tanentzap, A. J. and Lloyd, K. M. (2017). Fencing in nature? Predator exclusion restores habitat for native fauna and leads biodiversity to spill over into the wider landscape. *Biological Conservation*, *214*, 119–26. doi.org/10.1016/j.biocon.2017.08.001.

The National Business Review. (2011). Scientists back Campbell in Moon Man quake prediction row. *The National Business Review* (Online). Retrieved from www.nbr.co.nz/article/scientists-side-campbell-moon-man-quake-prediction-dispute-ck-87208.

Thoresen, J. J., Towns, D., Leuzinger, S., Durrett, M., Mulder, C. P. H. and Wardle, D. A. (2017). Invasive rodents have multiple indirect effects on seabird island invertebrate food web structure. *Ecological Applications*, *27*(4), 1190–98. doi.org/10.1002/eap.1513.

Trench, B. (2012). Vital and vulnerable: Science communication as a university subject. In B. Schiele, M. Claessens, and S. Shi (eds), *Science Communication in the World–practices, theories and trends* (pp. 241–57). Springer. doi.org/10.1007/978-94-007-4279-6_16.

Twardek, W. M., Peiman, K. S., Gallagher, A. J. and Cooke, S. J. (2017). Fido, Fluffy, and wildlife conservation: The environmental consequences of domesticated animals. *Environmental Reviews*, *25*(4), 381–95. doi.org/10.1139/er-2016-0111.

University of Otago. (2018). The Centre for Science Communication (Online). Dunedin: University of Otago. Retrieved from www.otago.ac.nz/science-communication/about/index.html, doi.org/10.26473/atlaanz.2015.1.1/001.

van Heezik, Y., Smyth, A., Adams, A. and Gordon, J. (2010). Do domestic cats impose an unsustainable harvest on urban bird populations? *Biological Conservation*, *143*(1), 121–30. doi.org/10.1016/j.biocon.2009.09.013.

Walker, J. K., Bruce, S. J. and Dale, A. R. (2017). A Survey of Public Opinion on Cat (Felis catus) Predation and the Future Direction of Cat Management in New Zealand. *Animals*, *7*(7), 49. doi.org/10.3390/ani7070049.

Warner, K. D. (2012). Fighting pathophobia: how to construct constructive public engagement with biocontrol for nature without augmenting public fears. *Biocontrol*, *57*(2), 307–17. doi.org/10.1007/s10526-011-9419-x.

Wilcox, P. L., Charity, J. A., Roberts, M. R., Tauwhare, S., Tipene-Matua, B., Kereama-Royal, I., Hunter, R., Kani, H. M. and Moke-Delaneyz, P. (2008). A values-based process for cross-cultural dialogue between scientists and Maori. *Journal of the Royal Society of New Zealand, 38*(3), 215–27. doi.org/10.1080/03014220809510555.

Williams, J. (2017). Some approaches to traditional Maori knowledge. *Reviews in Anthropology, 46*(4), 180–96.

Wilsdon, J. and Willis, R. (2004). *See-through science: Why public engagement needs to move upstream.* London, Demos.

Zorn, T. E., Roper, J., Weaver, C. K. and Rigby, C. (2012). Influence in science dialogue: Individual attitude changes as a result of dialogue between laypersons and scientists. *Public Understanding of Science, 21*(7), 848–64. doi.org/10.1177/0963662510386292.

Timeline

Event	Name	Date	Comment
First interactive science centre established.	Otago Museum's Discovery World in Dunedin	1991	
First national (or large regional) science festival.	International Science Festival, Dunedin	1998	
An association of science writers or journalists or communicators established.	Science Communicators Association of New Zealand (SCANZ)	2004	
First university courses to train science communicators.	Centre for Science Communication, Dunedin	2008	
First master's students in science communication graduate.	Centre for Science Communication, Dunedin	2011	
First PhD students in science communication graduate.	Octavia Cade	2015	University of Otago, Dunedin
First national conference in science communication.	Assoc. Scientific and Technical Communicators (NZ)	1994	
National government program to support science communication established.	'Unlocking Curious Minds'	2015	

Event	Name	Date	Comment
First significant initiative or report on science communication.	Report of the National Science Challenges Panel	2013	
National Science Week founded.	National Primary Science Week	2011	Run by The Science Learning Hub
A journal completely or substantially devoted to science communication established.	NZ Science Review	1942	Published by the New Zealand Association of Scientific Workers (NZAScW), later NZAS
First significant radio programs on science.	Science Report, Radio NZ	1957	
First significant TV programs on science.	Neil Harraway and Robert Brown's documentary on takahe	1974	The takahe is a flightless bird indigenous to New Zealand
First awards for scientists or journalists or others for science communication.	NZAS Science Journalism award	1980	
Date hosted a PCST conference.	PCST 2018, Dunedin	2018	2013: PCST-SCANZ Symposium, Christchurch
Other significant events.	William Doherty, first PhD student communicating mātauranga (indigenous science)	2010	Date of graduation, University of Auckland

Contributors

Professor Jean S. Fleming is Professor Emerita of Science Communication at the University of Otago, Dunedin, New Zealand. Jean has led the Popularising Science stream at the Centre for Science Communication for six years.

Professor Nancy Longnecker is Professor of Science Communication at the University of Otago. Before moving to New Zealand, she developed and delivered the science communication course at the University of Western Australia.

Dr Rhian A. Salmon is a senior lecturer and co-leader of the Centre for Science in Society at Victoria University of Wellington in New Zealand. She works as both a practitioner and researcher in public engagement.

Dr Dan C. H. Hikuroa is senior lecturer, Māori Studies, Te Wananga o Waipapa, University of Auckland. He is a world expert on integrating indigenous knowledge and science.

5

ARGENTINA

Contexts, agents and practices in science communication

Carina Cortassa and Cecilia Rosen

1. Introduction

To tell the story of an ongoing process it is necessary to take some risks, though these are well worth the intellectual challenge they present. Under that premise, in this chapter we will analyse the emergence and expansion of science communication in Argentina. We will emphasise the incipient advancements accomplished since the beginning of this century in two areas: the field of practice and the academic field.

Our work is inspired by the 'cultural cartography' approach proposed by Gieryn (1999), which entails exploring the evolving dynamics of different domains of knowledge and practice whose boundaries are drawn and redrawn in the context of both epistemic and extra-epistemic factors. Rather than presenting an exhaustive but meaningless chronology of events, we aim to identify key moments in the development of the local field and to place them in the broader context of its interactions with other spheres: scientific organisations and communities, public policies, the mass media, and the cultural industry.

Although the first attempts to popularise science in Argentina can be traced back to the very beginnings of the country,[1] our analysis in Section 1 starts around the mid-20th century. It is in this period that the national system of research and development acquired its current form and when some of its most prominent institutions and members started to spread science to society.

1 Argentina became formally independent from Spain in 1816.

The Argentinean Association for the Advancement of Science (AAPC), for example, played a leading role in promoting communication practices among its fellows with an explicit goal: to increase citizens' scientific literacy as a way to improve their attitudes towards local science.

Decades later, a professional scientist was the driving force behind another milestone for science communication. In 1984, chemist Enrique Belocopitow (1926–2007) envisaged the Program for Science and Technology Popularisation at the Campomar Institute of Research in Biochemistry.[2] In addition to promoting the creation of press areas at research centres, the program launched the first course exclusively dedicated to train science reporters. As described in Section 2, these and other achievements, accomplished between the mid-1980s and the early years of this century, set the future direction for the development of the local field.

Section 3 focuses on the last decade, which is certainly the richest and liveliest part of this story. A variety of dimensions converge during this period: from the increased interest of public policies in the diversification of popularisation agents and venues, to the growth of institutional press offices in scientific institutions, to a decline of specialised journalism in traditional media. As a result, a displacement of agents among different professional niches is currently occurring, visibly redefining the field's borders.

The last past of the chapter briefly describes the evolution of local public perceptions of science obtained through four national surveys given on this issue since 2004.

2. What is past is prologue

According to Cazaux (2010), science communication activities began to be developed in Argentina in the post-colonial period. Besides the establishment of local institutions devoted to promoting and spreading scientific knowledge—universities, museums, academies and societies[3]—nascent journals and other local publications were already including scientific and technical news among other news of cultural and political interest. However, as Nowak (2008) states, the available information is fragmented and insufficient to portray an accurate account of this embryonic stage of specialised journalism in the country.

2 Since 2001, the Leloir Institute Foundation.

3 Among others, the first Museum of Natural History (1823, currently the Argentinean Museum of Natural Sciences); the Argentinean Scientific Society (1872); the Archaeological and Anthropological Museum of Buenos Aires (1877, moved to La Plata in 1884); the National Astronomical Observatory in Córdoba Province (1881); the Zoological and the Botanical Gardens of Buenos Aires (1888 and 1898).

During the last decades of the 19th century, under positivist ideas that identified science and technology as engines of progress, influential newspapers such as *La Prensa* [The Press] (1869) and *La Nación* [The Nation] (1870) started to include regular sections devoted to these topics (Cazaux, 2010, pp. 94–95). This trend was imitated by many popular outlets that appeared early in the new century to meet public demand. Though the role was far from being recognised as such, figures who might be regarded as scientific journalists emerged on editorial staffs. This is the case of physician and journalist Jacobo Brailovsky (1906–2005), who started working for *La Nación* in 1924 and soon after began covering science and related issues. Years later, he would start writing a weekly column called 'Science in a Few Traces'. Brailovsky led the first Argentinean Association of Science Journalism (1969) and received several awards for his contributions to national science communication.

The strengthening of an articulated national system of science and technology in Argentina took place between 1950 and 1960 as part of a process of political and economic modernisation. Systematic policies for research and development were implemented for the first time during this period, following the international trend initiated at the end of World War II (Albornoz, 2007). It was also at that time that the first institutions outside universities emerged, driving the growth and consolidation of local science.

Beyond government purposes, the AAPC played a leading role in that process. The association was created in 1934 by some of the most respected scholars from several different disciplines, together with a journalist. Although anecdotal, the AAPC's origins are worthy of mention in these pages. In 1933, a senator publicly regretted the shortage of people in the country who were 'exclusively dedicated to the study of philosophy and sciences, and to disseminate their research outcomes among their students' (AAPC, 2017). Besides provoking the scientific community's fierce reaction, his statement outraged a journalist at the popular magazine *El Hogar* [Home], which was the widest-circulating magazine at that time in Buenos Aires. In response to the politician's statement, reporter Carlos Silva not only published a long series of articles devoted to highlighting the work of local researchers, he also became actively involved with them to create the AAPC, making him the only founding member who was not a scientist.[4]

4 The AAPC's first chair was Bernardo Houssay (Nobel Prize for Physiology and Medicine, 1974), who would also head the National Council of Science and Technology created in 1958.

The AAPC has published the magazine *Ciencia e Investigación* [Science and Research] since 1945, and its has resisted all adverse conditions faced since then.[5] Despite being a publication addressed to a limited public—presumably with certain skills and a previous interest in science—von Stecher (2017) highlights the strong focus on popularisation given by its creators and successive editors. An editorial note in 1950 clearly expresses the conception and values underlying the magazine's aims:

> To make science comprehensive, to inform the public about its advancements and discoveries, to orientate the people and governments in the application of scientific knowledge for the common good, and not for the destruction nor slavery of man, and to seek the diffusion of scientific thinking. (von Stecher, 2017, p. 200)

Several decades before scientific communities were challenged to 'learn to communicate with the public, be willing to do so, and indeed consider it their duty to do so' (Bodmer, 1985, p. 6), the AAPC had already placed this commitment among the main *Duties of the Science Man*.

In 1966, the fragile Argentinean democracy suffered a new onslaught when a coup d'état placed the country under a civic-military government whose effects on scientific development would be devastating in the short and long term.[6] After a brief democratic interregnum (1973–76), the next dictatorship (1976–83) widened and deepened the decay of the academic and scientific system.

3. The true beginning (1985–2000)

The return to democracy in 1983 brought with it the need to rebuild the country in all its dimensions: political, economic, social, cultural and scientific. In this last sense, although the ideological persecutions that affected institutions and individuals were surpassed, both the magnitude of the brain-drain as well as the restrictions due to economic instability did not allow for a full recovery to the scientific and technological development levels of the 1960s. However, even with many ups and downs, between the end of the 20th century and the early 2000s, it is possible to identify a series of events that served as the basis for the future growth of the field.

5 *Ciencia e Investigación* is currently published digitally on a quarterly basis. See aargentinapciencias.org/publicaciones/revista-cei.

6 On 29 July 1966, students, researchers and authorities from the University of Buenos Aires involved in the resistance against the dictatorship were fiercely repressed by police forces. This event—known as 'The Night of the Long Sticks'—started a process of the dismantling of labs, groups and research centres of international prestige. Hundreds of scholars resigned or were dismissed, and a large number of them went into exile.

The process followed two main paths. The first was that during this period, formal efforts to professionalise human resources in science communication appeared together with the creation of specific areas within research institutions. The second direction saw science issues gaining increased visibility and recognition both in the mass media and the cultural industry.

3.1. Training and professionalisation of human resources

The systematic and institutionalised training of science communicators in Argentina did not begin with journalism or communication degrees, but with a life sciences research centre. In 1984, driven by his concerns about the weakness of scientific culture in local society, chemist Enrique Belocopitow created the first Program for Science and Technology Popularisation at the Campomar Institute of Research in Biochemistry, and activities started the following year. In tune with classical issues of the Public Understanding of Science movement (Thomas and Durant, 1987), Belocopitow was convinced that better strategies of science teaching and massive popularisation through the media were fundamental tools to increase the public's interest in and engagement with science and technology—hence his interest in training professionals to best carry out these functions (Belocopitow, 1998, p. 145).

As Neffa (2014) states, the original program consisted of two parts:[7] the creation of institutional communication offices and science journalism courses.

The first popularisation centre was implemented within the Campomar Institute with the purpose of sending news about its research activities to the mass media. From 1992 onwards, the initiative was expanded through an agreement with the University of Buenos Aires—the biggest university in the country—until all its faculties had a network of specialised communication areas.[8] Publications and other activities promoted from public research bodies and addressing the lay public grew steadily from then on, and some of them—such as *EXACTAmente* magazine, edited by the Exact and Natural Sciences Faculty since 1994, became pillars of the field and remain so to the present.

Despite the relevance of the communication areas built at the behest of the program, Belocopitow's initiative would succeed mainly due to its emphasis on training human resources in the first courses on science journalism in the country. The plan consisted of two parts: over one semester, participants

7 The third element is a science news agency created in 2006.

8 Between 1989 and 1992, the science communication centres network produced coverage for *Diarios and Noticias* [Daily and News], the main private news agencies in the country at that time.

obtained theoretical and practical training; later, those who wanted to improve their skills received a full-time scholarship to work at a research centre and experience the nature of scientific practices 'from the inside'. The program had a multiplying effect among other institutions, generating the emergence of similar initiatives elsewhere.[9]

Emphasising the practical aspects of science communication and addressing itself to graduates of scientific and humanistic programs, the course 'aims to provide basic elements of discourse adaptation, textual organisation and journalistic style' (Loewy and Calabrese, 2016, p. 3). Other specific goals are that students learn to write quality news articles and acquire a fluent handling of information sources, as well as reading and synthesising scientific papers. The program played a central role in building and delimitating the field, and also promoted efforts to secure its autonomy from other spheres, such as general journalism. It came to be the seedbed of science journalists and communicators for more than two decades. Not all of the approximately 1,000 students who participated in the seminar over the years have worked professionally in science communication. But, conversely, a large number of the local field's agents have been trained at the Campomar/Leloir program and other more recently developed training initiatives (Rosen, 2018).

Nowadays, founding a science communication program in a research institution would by no means be regarded as an eccentricity. But 30 years ago, the situation in Argentina was quite different. As Neffa (2014) points out, the emergence and continuity of the enterprise have been possible mainly due to Belocopitow's legitimacy, which allowed him to overcome the internal challenges generated by his initiative. It was his renowned epistemic reputation that permitted him to introduce and sustain such a 'foreign' matter in terms of the goals and functions of a life sciences centre. In the words of a former fellow of the program:

> He [Belocopitow] incarnated the association between the two points [scientists and communicators]. No other person could have sustained this project at the heart of one of the most prestigious institutes … No one else was conceivable for that task. (in Neffa, 2014, p. 228)

9 Some of those experiences, such as the popularisation seminars in the faculties of Pharmacy and Biochemistry and of Exact and Natural Sciences, still exist. Moreover, the latter inspired a postgraduate program on science communication launched in 2015.

3.2. Science in the media and the cultural industry

Along with the ups and downs of economic instability, science journalism and communication witnessed some flourishing moments during the middle of the 1980s and the late 1990s.

By then, the three main national newspapers—*Clarín, La Nación* and *Página 12* [Page 12]—had fixed pages for science news. In the first case, after minor changes of name and format, the space 'eventually became a supplement dedicated to research news, having as a reference the traditional Science Times sections at *The New York Times*' (Nowak, 2008, p. 2). The section was replaced in 1997 for a technology supplement, moving science coverage to the general news pages. In the early to mid-1990s, traditional science coverage in *La Nación* alternated between a weekly supplement and an unfixed segment, along with a monthly health section. Both spaces converged by the end of the decade under the title 'Science and Health', until its disappearance in 2011. Since then, science news has been spread over different sections of the newspaper. It is worth mentioning here the distinguished journalist Nora Bär: she has coordinated, edited and reported on science, health and technology for the newspaper since 1980, and has been widely recognised as one of the most important science journalists in the country.

Página 12 entered the local market in 1987 and soon incorporated the supplement *Futuro* [Future] into its Saturday edition. Its editor was Leonardo Moledo (1947–2014), a physicist, mathematician and writer who also covered stories for the main section of the newspaper. With a special sensitivity to link together science, arts and humanities, Moledo became another emblematic figure in the field due to his large and diversified range of work, which included articles, popular books and novels, as well as television scripts and plays. He introduced scientific cafés to the country and led the *Galileo Galilei Planetarium* from 2000 until 2007. He is well remembered by his colleagues and disciples for his most repeated phrase: 'Popularisation is the continuance of science by other means'.[10]

Popular science magazines also experienced a successful, although short period at the same time. Three monthly publications aimed at the general public were launched in the mid-1980s to the early 1990s by commercial editorial houses—*Muy Interesante* [Very Interesting] (1985), *Descubrir* [Discover] (1989) and *Conozca Más* [Know More] (1989)—and the non-profit association *Ciencia Hoy* [Science Today] released its own self-titled periodical

10 Immediately after Moledo's death, *Página 12* cancelled the supplement. Daily science news continued to be included in different sections of the newspaper.

in 1988.[11] Between 1991 and 1993, in what de Vedia (1998) calls 'the golden triennium', the first three together reached an average monthly circulation of 450,000, something quite unusual for these products in Argentina. However, the trend did not last: five years later, the figure had dropped drastically by 75 per cent, partly due to economic considerations, and partly due to the gradual increase of science popularisation products in journals and specialised television channels (de Vedia, 1998). As a result, *Descubrir* and *Conozca Más* disappeared from the market, and the local edition of *Muy Interesante* was replaced by the Mexican version in 2007. *Ciencia Hoy*, for its part, is still published thanks to subsidisation.

A similar downward trend affects the current journalistic context as part of a parallel process of expansion and contraction of the field described below.

3.3. Interactive science centres

The emergence of the first interactive science museums and centres marked another important milestone in the local scene between the 1980s and the end of the 20th century, echoing the earlier stage of these initiatives throughout Latin America (Cambre, 2015). A number of pioneering institutions of different scales appeared at this time in Argentina under the motto 'Forbidden not to touch!'—this in overt contrast to the silent and contemplative attitude expected of visitors to traditional museums.

With the slogan 'For the curious, from ages 4 to 100', the Participatory Museum of Science opened its doors in 1988 in Buenos Aires, providing inspiration for subsequent experiences. Soon after came the New World Museum (1990), within the framework of the National University of La Plata's popularisation program;[12] the Exploratorium Interactive Sciences and Arts Centre (1995), also in Buenos Aires; and, lastly, a project outside the capital was founded in 1996—the Puerto-Ciencia at the National University of Entre Ríos.

The number of such interactive science centres certainly grew in the 2000s, and the Argentinean Association of Centres and Museums of Science (AACeMuCyT) has actively worked to strengthen them since 2007. The outcome of a recent survey of Latin American countries (Massarani et al., 2015) shows that in 2015, Argentina possessed 18 institutions of this kind—a figure that might have slightly risen since then—with most

11 Based on its Brazilian equivalent, *Ciencia Hoy*'s profile of 'high popularisation' makes it more similar to the aforementioned *Ciencia e Investigación*.

12 This museum is a founding member of the Latin American and Caribbean Network for the Popularisation of Science and Technology (RED POP).

concentrated in the capital city and surrounding areas. This puts the country in third place in the region, followed closely by Colombia, El Salvador and Uruguay, but places them far from the 58 and 272 institutions found in Mexico and Brazil, respectively.

4. Current trends in the field (2005–18)

In 2001, Argentina was hit by an economic and financial crisis of an unprecedented magnitude, followed by a slow and arduous period of recovery in all its dimensions. National policies for research and development were re-appraised and that process culminated with the creation of the Ministry of Science, Technology and Innovation (MINCyT) in 2007. In a decision of highly symbolic and practical value, the scientific area reached the highest position in the government structure for the first time in the country's institutional history.[13]

Since the beginning, MINCyT has strongly promoted science communication initiatives among public institutions and in the media, and has also had a bandwagon effect on other actors' involvement. However, in the current context of reconfiguration faced by the commercial mass media, the other side of the coin is a clear retraction of science journalism spaces. Together, the attracting of governmental attention and the progressive decline of private editorial interest has led to a visible mobility of practitioners among different work scenarios. The growing hybridisation of the spaces for practice and professional roles are among the most salient features that characterise Argentinean science communication's current state of affairs (Rosen, 2018).

4.1. Public policies for science communication and scientific culture

In developed countries, policies devoted to improving scientific literacy were consolidated during the second half of the 20th century as a means of enhancing public interest and awareness of scientific and technological development. This original purpose has been significantly broadened during recent decades, both due to the emergence of new social demands and the increasing influence of the discipline's academic rhetoric on official discourse. Besides highlighting governmental efforts to ensure civil society's support, public policies on the issue currently foster new goals. Among others, these include: to democratise access to knowledge, stimulate the development

13 In a heavily resisted political decision, in 2018, the ministry was reduced to its previous position of Secretariat.

of an innovative culture, promote scientific vocations and extend public participation in controversial issues (see, *inter alia*, the essays compiled by Felt, 2003, pp. 47–108).

Generally speaking, Latin American countries did not pay much attention to the matter until the beginning of the 21st century, when allusions to popularisation aims and activities started appearing more systematically in the framework of policy papers (Polino and Cortassa, 2015). With regard to Argentina, Neffa claims that, before this period, governments used to approach the issue of popularisation by means of some sort of 'commonsense voluntarism', making a series of vague statements like 'building a knowledge society', 'promoting the cultural assimilation of science' and 'stimulating scientific vocations', but without further reference to the means by which these objectives would be met (Neffa, 2014, p. 146).

The first document to explicitly embrace the topic was the *National Plan for Science, Technology, and Innovation* (2003), which was written within the framework of the aforementioned broader process of research and development reappraisal. In addition to introducing the first National Science Week, which has been held annually since then, the plan included a series of actions aimed at the training of human resources in science communication and improving cooperation with the educational system. The implementation of the first National Survey on the Public Perception of Science was also announced, underlining the importance of accurate statistical information as the basis for the formulation of scientific policies with major explicit and conscious support by the citizenry (SECyT, 2002).

The creation of MINCyT brought with it the strengthening of strategies aimed at the public circulation of science (particularly local science) on at least three levels as follows.

At the level of concrete actions, between 2013 and 2015, the ministry supported a broad spectrum of initiatives to promote scientific culture:[14] competitions for audio-visual media, photography and science literature; events such as National Science Week; scientific cafés; events organised along with the education sector (such as fairs, the Scientific Olympic Games and science clubs); the launching of TEC-TV (a science channel owned by MINCyT, which also produced popularisation content for other public TV stations); and subsidies for public or private projects. Additionally, in

14 A survey among Ibero-American countries shows that, at this time, Argentina was part of the most dynamic group in this sense—together with Spain, Portugal, Chile and Brazil—with 15 or more actions carried out by governmental agencies, above the average of nine actions in the total sample (Polino and Cortassa, 2015).

2001, the first large-scale fair for science, technology and the arts, *Tecnópolis,* was inaugurated, and in 2015 opened the Science Culture Center (C3), an interactive space of popularisation. Its first director was scientist and science communicator Diego Golombek, one of the country's most prominent figures in science popularisation.[15]

At the institutional level, from 2013 onwards all those actions were unified in the National Program of Science and Innovation Popularisation, contributing to visualising and prioritising the field within the ministerial structure.[16]

At the discursive level, the *National Plan of Science, Technology, and Innovation 2012–2015* continued the trend to include specific courses of action to 'expand popularisation, culture, and literacy actions in science and technology, and promote an innovation culture in society, and create and/or strengthen territorial structures (museums, agencies, directions, etc.) of scientific culture'. Its main goal is to 'bring activities and products of science and technology closer to society in order to enhance community participation and the social appropriation of knowledge as well as to promote scientific vocations in children and young people' (MINCyT, 2013, pp. 88, 103).

The government's interest in scientific culture policies had a visible effect on other social actors and sectors, contributing to the experimental expansion of the field as we will describe in the next section.

4.2. Expanding frontiers

Three dimensions synthesise the recent growth of science communication in Argentina: a) the development of institutional communication; b) a new boom in the science cultural industry; and c) the diversification of professional career options and the socialisation spaces of the agents.

At a global level, one of the most utilised indicators for exemplifying the reach of science mediatisation—the 'communicative turn'—is the growth of communication areas in research institutions (Polino and Castelfranchi, 2012). As we previously said, a handful of organisations in Argentina have had stable areas or programs since the 1980s; however, the sustained trend of developing specific spaces is an ongoing process (Cortassa, Andrés and Wursten, 2017; Ruggiero and Bello, 2015). Papers presented at the Public Communication of Science Congress (COPUCI) show that stories and

15 Golombek has authored several science books; he has also created and hosted acclaimed science television shows. He received several national and international prizes for his work, including the Kalinga/UNESCO Award in 2015.

16 Although the program no longer exists as such, some of its actions are still active.

analyses of experiences in institutional spaces have almost doubled between the first edition (2011) and the most recent one (2017). It can be said that, currently, the main universities and research organisations all have a communication office; some of them have also created their own science news agencies online. The first one, launched in 2006, was the Scientific and Technological News Agency at the Leloir Institute Foundation, created as part of the popularisation program already mentioned. More recently it was joined by the Science, Technology and Society Agency (National University of La Matanza, 2010), South–South Technology (National University of San Martín, 2013) and UNScience (National University of Córdoba, 2014).

A second feature that reflects the local field's expansion is the reinvigoration of popularisation products in diverse platforms and formats. The editorial market, for instance, has experienced a noticeable growth in recent years. One of the most relevant projects has been the collection *Ciencia que Ladra* [Barking Science, Siglo XXI Editorial], which published more than 70 titles and sold more than a million copies in 15 countries by 2010 (Bacher, 2010). More recently, the University of Buenos Aires editorial house launched the book series *Ciencia Joven* [Young Science], while other commercial houses have strongly promoted scientific titles to take advantage of the increasing demand.[17]

Popularisation programs on television also experienced an accentuated heightening in this period, especially on public and university networks, in an overt contrast to the marginalised status of science topics on newscasts that will be discussed below. Besides the creation of TEC-TV, the MINCyT channel, other state broadcasters have included science shows of great popularity, such as *Proyecto G* [G Project], *El Cerebro y Yo* [The Brain and Me] and *Conversaciones* [Conversations], all presented by Diego Golombek. Perhaps the most significant example of this flowering—and its current uncertain state—is the flagship program *Científicos Industria Argentina* [Argentinean Industry Scientists] presented by Adrián Paenza, a mathematician and another leading figure in local science communication. After 13 years on air, the show ceased broadcasting near the end of 2016.[18]

The third indicator of the local field's expansion during the last decade has been the strengthening of the training options for specialised communicators. Besides the specific subjects included in journalism and mass communications

17 For example, in 2014, four popular books on neuroscience were among the 10 most read in the non-fiction category (Berdichevsky, 2015).

18 A producer's farewell letter suggests that the show ended due to the presenter's ideological differences with current government science policies.

undergraduate degrees, four postgraduate programs were recently launched[19] at the National Universities of Córdoba, Río Negro, Buenos Aires and Buenos Aires Province. These courses are heterogeneous as regards their goals, ranging from those that seek to provide practical tools for practitioners in various areas, to others more focused on academic research (such as the longstanding master's in Science, Technology and Society at the University of Quilmes, the first of its kind in the country).

A recent study by Rosen (2018) explores the agents' 'entry to the field', as well as the different training opportunities that allow them to accumulate the background to be recognised as 'professional science journalists/communicators'. Through interviews with experienced practitioners, the author found that although most of them affirm that they obtained their expertise through working practice, they also complemented this empirical training with specialised courses throughout their careers. Many respondents agree that these courses marked a 'before' and 'after' in their respective personal pathways—that is where they became aware of the full potential of science journalism as a career opportunity.

According to Vara (2015), the expansion of formal training is a clear signal of the professionalisation of the sector, as is the emergence of two active spaces that bring together agents and interests: the Argentinean Network of Science Journalists (RADPC, created in 2007)[20] and the Public Communication of Science Congress (COPUCI), organised in 2011.

The RADPC now has more than 100 members, including communicators, journalists and even researchers who work in mass media and scientific institutions. Some of its objectives are 'to promote the debate on practices and professional ethics as well as the exchange of experiences, knowledge, and concerns with colleagues of other countries; to develop and promote activities and opportunities for professional training; and to allow the exchange and appropriation of research on the relationship between science, technology, society, and the mass media'.[21] The association frequently organises or participates in events to train and update journalists, having also an active presence in international networks such as the World Federation of Science Journalists. It also edits an annual volume containing selected pieces published by their members.

19 According to Massarani, Reynoso, Murriello and Castillo (2016), 65 per cent of current Latin American courses were created during the last decade.

20 As previously mentioned, the first association of this kind was created in 1969, and another attempt emerged during the 1990s. Nevertheless, unlike the current one (RADPC), 'these associations never had more than a small group of fellows and were dissolved for different reasons, that we must investigate' (Vara, 2015, p. 2).

21 RADPC website: radpc.org/.

The multiplication of venues and the growing number of practitioners have been accompanied by the strengthening of the vernacular academic research and reflection on science communication. COPUCI conferences emerged as a relevant realm to articulate the several groups of actors committed to the matter. These events—organised yearly between 2011 and 2015, and biennially since then—have allowed the establishment of a rich and sustained dialogue, and partnership between journalists, communicators, researchers, stakeholders and policymakers, boosting a critical and reflective vision on their respective practices, goals and responsibilities.

4.3. The decline of science journalism and new job market scenarios

The local development of science communication in Argentina has followed a similar pattern to other countries, where the diversification of actors, spaces and problems constitutes a distinctive feature of the field (Bucchi and Trench, 2008, p. 3). Nevertheless, its recent expansion did not follow a regular pattern; on the contrary, science journalism in the country appears to be in a worrying decline.

With the turn of the century, science seemed to be gaining notoriety in the media in contrast to what has been registered in previous years (SECyT, 2006). The position was summarised by Calabrese, Geller and Loewy (2013, p. 5): 'news on science and technology and especially in medicine, appear more frequently in the daily agenda of traditional media and there is an explosion of science popularisation in digital media'.

One of the communicators interviewed by Rosen (2018) perceived a comparatively more receptive attitude among editors in recent years: '20 or 25 years ago all the editors, everyone, thought science didn't attract the public ... In the last decade the amount and quality of science coverage in the media has grown'. However, the same study shows that this view is not frequently shared by her colleagues, who are much less positive about the current state of science journalism in the country. While acknowledging the recent expansion of science communication, the interviewees also express a clear concern about the setback in job opportunities for full-time science journalists. That is especially true among those seeking to cover science independently from science institutions or governmental agencies—currently two of the most prominent job providers, and both more interested in conveying a non-neutral, highly positive view of the subject.

In this context, many testimonies express worries about the progressive loss of science sections in the national media. *La Nación* cancelled its weekly section in 2011 and so did *Página 12* (in 2014) with *Futuro*, another longstanding symbol of local science journalism.[22] Although in both cases science coverage continues to be found in the general pages, the cutbacks or closure of spaces are perceived as symptoms of a larger retraction process. In the same vein, in contrast to the growth of popularisation initiatives, science news on Argentinean television is still marginal: by 2014, only 0.8 per cent of all news content covered these topics, occupying the 15th of 20 places among covered subjects (Halpern, 2015).

Although most of the practitioners interviewed by Rosen held a pessimistic view of the current state and the future of local science journalism, their perceptions cannot be fully confirmed considering the lack of systematic studies about either science coverage in the media or the evolution of the job market. However, some of her findings are suggestive: of 21 informants, only four worked as full-time science journalists at the time the data were collected (2016), 12 were sporadic collaborators or freelancers, four worked at institutional press offices, and the remaining four were employed by governmental agencies. These data somehow support the generalised opinion about the existence of better work opportunities outside the traditional journalistic niches and justifies the gradual displacement of many reporters to other realms in pursuit of better wages, job stability and greater prospects for professional advancement.

5. The public perception of science and technology

Four national surveys on public perception of science and technology were carried out in Argentina between 2003 and 2015 (SECyT, 2004, 2007; MINCyT, 2014, 2015),[23] with their frequency depending on budgetary considerations. The questionnaires follow the standard model that measures attitudes on general topics related to the public's degree of interest and information, information sources, attitudes towards science, and images of scientists and public support for research, avoiding from the beginning any questions regarding scientific literacy (SECyT, 2004). Except for the most

22 A testimony gathered by Rosen (2018) states categorically: 'Everybody says 'science sells' … but on the other hand you see that, at least in the print press, there is less and less science content. If science is so attractive, what's going on with science journalism?'

23 The data was gathered, respectively, in 2003, 2006, 2012 and 2014. A fifth survey was supposed to be held in 2019, but it has not been implemented thus far.

recent one, each survey has included questions on the perception of issues related to a current public policy interest: GMOs in 2003, nuclear power and software development in 2006, and scientific vocations in 2012.

A cross-temporal comparison of the data shows that, broadly speaking, positive attitudes towards science grew 18 per cent between 2012 and 2015, reaching 74 per cent of the respondents to the survey. Since the second edition, the state has been more progressively seen as the main source of funding for research and development[24] and more than half of respondents believed that support should increase. It is not a coincidence that these figures started rising at the same time the government began heavily promoting its sectorial policies and efforts to the public. Even so, seven years after its creation, by 2015 MINCyT was still unknown by nearly half of those surveyed, and *Tecnópolis*—the high-profile science and technology exhibition opened in 2011—generated great interest (75 per cent of the sample in 2012, 83 per cent in 2015), but few visitors (15 per cent of the sample in 2012, 17 per cent in 2015).[25]

In line with international trends, scientists rank among the best-reputed professionals in Argentina. In 2015, eight out of 10 respondents highly appreciated their work, a proportion that had increased through successive surveys. The majority considered them to be the most reliable source of information in cases of uncertainty and/or controversial public issues. Nevertheless, the percentage of people who could not identify a single national scientific institution rose from 60 per cent in the first survey to 70 per cent in the last. One might say that almost everyone likes scientists, but only a few know where they can be found.

Finally, with regards to the interest in and consumption of scientific news, the variations in the questions over the survey's editions hamper an accurate comparison of the data. However, a general trend shows that, even though most of the population positively values the social and individual importance of scientific information, in the informants' self-reported information levels the topic ranks lower than others such as sports, arts and culture, politics and religion—except, perhaps, for medical and health issues, on which more than half of respondents considered themselves to be 'well' or 'very well' informed (MINCyT, 2015). In the first three surveys, television programs and documentaries were the most cited sources of information, followed by the print media, internet sites, radio programs, popular magazines and books, in

24 Although national expenditure on research and development is scarcely 0.6 per cent of the GDP, the government funds around 70 per cent of the overall spending against less than 20 per cent that comes from the private sector (Ibero-American Network for Science and Technology Indicators: www.ricyt.org/?layout=blog).

25 That is understandable given that the surveys are carried out nationwide and *Tecnópolis* is located in Buenos Aires. Since 2016, the fair has had thematic itinerant exhibitions.

that order. The last poll's available data only reflects that in both 2012 and 2015, only three out of 10 people knew about the existence of the public channel TEC-TV but, among those who did, viewership increased by 12 percentage points (41 per cent of the sample in 2012, 53 per cent in 2015).

6. Concluding remarks

Even though science communication in Argentina has a relatively short history, during the last 15 years the field has grown slowly but steadily in importance and visibility. This is especially so between 2003 and 2015, when public policies focused on strengthening the national system of science and technology as a whole were formulated, encouraging not only knowledge production and application but its social dissemination as well. It is still unknown whether this positive evolution will continue under the current government, which, since 2015, has shown not quite favourable signs.

One of the main features that characterises the local field's expansion has been the creation and further consolidation of areas devoted to science communication and the general promotion of scientific culture at universities, research centres and government agencies. The emergence of more and better training options can be considered another positive indicator, as well as the growing number of specialised research groups spread throughout several universities in the country.

With regards to science journalism, some actors perceive the ups and downs experienced from 2000 onwards as worrying symptoms of a rapid and systematic decline, while others see them as part of the current trends and transformations taking place in the global press. In any case, these changes put an additional strain on science journalists' already fragile identities, which must deal with a high level of job insecurity and the consequent need to adapt to distinct roles, values and practices. The multiplication of *hybrid* professionals—who work simultaneously in journalism and press or public relations in institutional areas—raises new challenges for the shaping of the field's boundaries; for instance, when practitioners working for research centres or governmental agencies display values more related to journalism, such as objectivity, critical scrutiny and professional autonomy. To what extent this situation involves a potential or actual conflict of interest is at the core of a lengthy internal debate between community members.

The current state of development of science communication in Argentina is the result of a rapid evolution during which the local field's dynamic has been marked by a constant hybridisation of practices, actors and values. After decades of generally successful initiatives, through advances and retreats, the field today faces several exciting challenges posed both by internal and external factors. Hopefully this account of a recent but rich history will stimulate further and deeper reflections, debates and dialogues towards the domain's consolidation.

Acknowledgements

The authors want to thank the information and help of Alicia Calabrese, Matias Loewy, Ana Maria Vara and Susana Gallado.

References

Albornoz, M. (2007). Los problemas de la ciencia y el poder [The problems of science and power]. *Revista CTS*, *8*(3), 47–65.

Argentinean Association for the Advancement of Science (AAPC). (2017). AAPC Review. Retrieved from aargentinapciencias.org/institucional/historia/.

Bacher, S. (2010). El boom de la divulgación científica ¿Por qué la ciencia vende tantos libros? [A boom of science popularisation. Why does science sell so many books?]. *Clarín*, *23*(March). edant.revistaenie.clarin.com/notas/2010/03/23/_-02163038.htm (site discontinued).

Belocopitow, E. (1998). ¿Por qué hacer Divulgación Científica en la Argentina? [Why is it necessary to do Science Popularisation in Argentina?]. *Redes*, *5*(11), 141–63.

Berdichevsky, E. (2015). Aproximaciones a la poesía de la mente [Approaches to the Mind's Poetry]. *Clarín*, *6*(February). doi.org/10.12795/ph.2005.v19.i01.07.

Bodmer, W. F. (1985). *The Public Understanding of Science*. Royal Society, London, royalsociety.org/~/media/Royal_Society_Content/policy/publications/1985/10700.pdf.

Bucchi, M. and Trench, B. (eds) (2008). *Handbook of Public Communication of Science and Technology*. Routledge, New York. doi.org/10.4324/9780203928240.

Calabrese, A., Geller, B. and Loewy, M. (2013). Del laboratorio a los medios [From the laboratory to the media]. *Ciencia Hoy*. cienciahoy.org.ar/2013/10/del-laboratorio-los-medios.

Cambre, M. (2015). Museos interactivos de ciencia y tecnología en América Latina. [Interactive museums of science and technology in Latin America]. In L. Massarani, (org.), *RedPOP: 25 años de popularisación de la ciencia en América Latina* (pp. 41–50). Museu da Vida, Rio de Janeiro. doi.org/10.22323/3.01010201.

Cazaux, D. (2010). *Historia de la Divulgación Científica Argentina* [History of Science Popularisation in Argentina]. Teseo, Buenos Aires.

Cortassa, C., Andrés, G. and Wursten, A. (eds) (2017). *Comunicar la Ciencia: escenarios y prácticas. Memorias del COPUCI 2015* [Communicating Science: scenarios and practices. Proceedings of COPUCI 2015]. Editorial de la Universidad Nacional de Entre Ríos, Paraná.

de Vedia, M. (1998). Sufrieron una abrupta caída las revistas de divulgación científica [Popular science magazines suffer from an abrupt decline]. *La Nación*, 14 April. www.lanacion.com.ar/93512-sufrieron-una-abrupta-caida-las-revistas-de-divul gacion-cientifica.

Felt, U. (ed.) (2003). *O.P.U.S. Optimizing Public Understanding of Science and Technology. Final Report*, sts.univie.ac.at/fileadmin/user_upload/i_sts/Forschung/Projekte_abgeschlossen/final_report_opus.pdf.

Gieryn, T. (1999). *Cultural boundaries of science. Credibility on the line*. Chicago and London: University of Chicago Press.

Halpern, G. (2015). *Informe ante consulta del INTI sobre monitoreo de ciencia en medios* [Report at request by INTI about the monitoring of science in the media]. Defensoría del Público de Servicios de Comunicación Audiovisual, Buenos Aires.

Loewy, M. and Calabrese, A. (2016). ¿De qué sirve formarse en periodismo científico cuando no se trabaja como tal? Encuesta entre graduados del Programa de Ciencia y Técnica de la Fundación Instituto Leloir [What is the point of being trained in science journalism when you don't work in it? Survey of graduates from the Science and Technique Program at the Leloir Institute]. *RIHUMSO Journal*, *2*(8), 1–11.

Massarani, L. et al. (eds) (2015). *Guía de Centros y Museos de Ciencia de América Latina y el Caribe* [Guide on Science Centers and Museums in Latin America and the Caribbean]. Museu da Vida/Casa de Oswaldo Cruz/Fiocruz: RedPOP, UNESCO, Montevideo and Rio de Janeiro.

Massarani, L., Reynoso, E., Murriello, S. and Castillo, A. (2016). Posgrado en Comunicación de la Ciencia en América Latina: un mapa y algunas reflexiones [Postgraduate Programs in Science Communication in Latin America: a map and some reflections]. *Journal of Science Communication*, *15*(5), A03. jcom.sissa.it/sites/default/files/documents/JCOM_1505_2016_A03_es.pdf.

Ministerio de Ciencia, Tecnología e Innovación Productiva de la Nación (MINCyT) [Ministry of Science, Technology, and Productive Innovation]. (2013). *Plan Nacional de Ciencia, Tecnología e Innovación Productiva. Lineamientos generales 2012–2015* [National Plan of Science, Technology, and Productive Innovation. General Guidelines 2012–2015]. MINCyT, Buenos Aires. cdi.mecon.gov.ar/bases/docelec/ac1030.pdf.

Ministerio de Ciencia, Tecnología e Innovación Productiva de la Nación (MINCyT) [Ministry of Science, Technology, and Productive Innovation]. (2014). *La Percepción de los Argentinos sobre la Investigación Científica en el País. Tercera Encuesta Nacional* [The Perception of Argentineans about Scientific Research. Third National Survey]. MINCyT, Buenos Aires. www.argentina.gob.ar/sites/default/files/est_inst_tercera-encuesta-nacional-de-percepcion-de-los-argentinos-sobre-la-investigacion-cientifica.pdf.

Ministerio de Ciencia, Tecnología e Innovación Productiva de la Nación (MINCyT) [Ministry of Science, Technology, and Productive Innovation]. (2015). *Cuarta Encuesta Nacional de Percepción Pública de la Ciencia. La evolución de la percepción pública de la ciencia y la tecnología en la Argentina, 2003-2015* [Fourth National Survey on Public Perception of Science. The evolution of the public perception of science and technology in Argentina, 2003–2015], MINCyT, Buenos Aires. www.mincyt.gob.ar/estudios/cuarta-encuesta-nacional-de-percepcion-publica-de-la-ciencia-11656.

Neffa, G. (2014). La Comunicación Pública de la Ciencia en las Instituciones Cientificas nacionales [Public Science Communication at National Scientific Institutions] (PhD thesis). Buenos Aires University at Buenos Aires, Argentina. doi.org/10.24193/cbb.2018.22.05.

Nowak, C. (2008). Una historia de ciencia y periodismo en la Argentina [A history of science and journalism in Argentina]. *Circunstancia, 15*(January). dialnet.unirioja.es/servlet/articulo?codigo=2559750.

Polino, C. and Castelfranchi, Y. (2012). The 'communicate turn' in contemporary techno-science: Latin American approaches and global tendencies. In B. Schiele, M. Claessens and S. Shunke (eds), *Science communication in the world: Practices, theories and trends* (pp. 3–17). London and New York: Springer. doi.org/10.1007/978-94-007-4279-6_1.

Polino C. and Cortassa, C. (2015). *La promoción de la cultura científica. Un análisis de las políticas públicas en los países iberoamericanos* [The promotion of scientific culture, An analysis of public policies in Ibero-American countries]. Observatorio Iberoamericano de la Ciencia, la Tecnología y la Sociedad de la Organisación de Estados Iberoamericanos (OCTS-OEI) [Ibero-American Observatory of Science, Technology, and Society of the Organisation of Ibero-American States], Buenos Aires.

Rosen, C. (2018). Prácticas y valores del periodismo de ciencias en la Argentina. Un análisis exploratorio del campo y los comunicadores [Practices and values of science journalism in Argentina. An exploratory analysis of the field and the communicators] (PhD thesis). Latin American School of Social Sciences, Buenos Aires. hdl.handle.net/10469/13936.

Ruggiero, G. and Bello, M. (comps.) (2015). *Las universidades frente al problema de comunicar la ciencia* [Universities facing the problem of communicating science], Editorial de la Universidad Nacional de General. Sarmiento, Buenos Aires. doi.org/10.19137/an1409.

Secretaría de Ciencia y Tecnología de la Nación (SECyT) [Secretary of Science and Technology]. (2002). *Plan Nacional de Ciencia y Tecnología 2002–2003* [National Plan of Science and Technology 2002–2003]. SECyT, Buenos Aires. doi.org/10.3989/arbor.2003.i690.678.

Secretaría de Ciencia y Tecnología de la Nación (SECyT) [Secretary of Science and Technology]. (2004). *Los Argentinos y su visión de la Ciencia y la Tecnología: Primera Encuesta Nacional de Percepción Pública de la Ciencia* [Argentineans and their vision of science and technology: First National Survey of Public Perception of Science]. MINCyT-SECyT, Buenos Aires.

Secretaría de Ciencia y Tecnología de la Nación (SECyT) [Secretary of Science and Technology]. (2006). *Análisis de la oferta informativa sobre ciencia y tecnología en los principales diarios argentinos. Reporte Final* [Analysis of news on science and technology in the main Argentinean newspapers. Final Report]. Observatory of Science, Technology and Productive Innovation, SECyT, Buenos Aires.

Secretaría de Ciencia y Tecnología de la Nación (SECyT) [Secretary of Science and Technology]. (2007). *La Percepción de los Argentinos sobre la Investigación Científica en el País. Segunda Encuesta Nacional* [Scientific Research Perception by Argentineans. Second National Survey]. Observatory of Science, Technology and Productive Innovation, SECyT, Buenos Aires. doi.org/10.15517/psm.v13i1.19290.

Thomas, G. and Durant, J. (1987). Why should we promote the Public Understanding of Science? *Scientific Literacy Papers,* Summer, 1–14.

Vara, A. M. (2015). Periodismo científico: entre la profesionalisación y los desafíos del cambio tecnológico [Science journalism: between profesionalisation and the challenges of technological change]. In S. Espinosa (ed.), *Ciencia, arte y tecnología. Enfoques plurales para un abordaje multidisciplinar, Escritos sobre Tecnología* [Science, art, and technology. Plural approaches for a mutidisciplinary focus]. Editorial de la Universidad Nacional de Lanús, *1*, 167–84.

von Stecher, P. (2017). El lenguaje de la ciencia y de su divulgación en la revista argentina Ciencia e Investigación (1945–1955) [The language of science and its popularisation in the Argentinean magazine, Science and Research (1945–1955)]. *Logos: Revista de Lingüística, Filosofía y Literatura, 27*(2), 198–210. doi.org/10.15443/RL2715.

Timeline

Event	Name	Date	Comment
First interactive science centre established.	Participative Science Museum	1988	See www.mpc.org.ar/home.htm
First national (or large regional) science festival.	Science and Technology Youth Fair	1967	
Association of science writers or journalists or communicators established.	Argentinean Association of Science Journalism	1969	Currently the Argentinean Network of Science Journalism (2007). Website: radpc.org/

Event	Name	Date	Comment
First university courses to train science communicators.	Program for Science and Technology Popularisation – Campomar Institute of Research In Biochemistry	1985	The first courses to train science communicators did not start at a University but at a life science's research centre
First PhD students in science communication graduate.		1993	
First national conference in science communication.	Public Communication of Science and Technology Congress	2011	
National government program to support science communication established.	National Program of Science and Innovation Popularisation	2013	
National Science Week founded.		2003	
A journal completely or substantially devoted to science communication was established.	*Science and Research*	1945	
First significant radio programs on science.		1980s	Mostly on health issues
First significant TV programs on science.	TN Science: *Todo Noticias*	2003	
First awards for scientists or journalists or others for science communication.	Konex Foundation Awards for Communication and Journalism	1987	(Special Mention for Science Journalism) 2007: National Award on Science Journalism 2010: National Award for PCST

Contributors

Dr Carina Cortassa is head of the Research and Postgraduate Studies at the School of Educational Sciences, National University of Entre Ríos, Argentina.

Cecilia Rosen is a science journalist and science communication researcher based in Mexico City.

6

AUSTRALIA

The five stages of development of science communication

Toss Gascoigne and Jenni Metcalfe

1. Introduction

Modern science communication in Australia as a discipline emerged in five distinct stages. Some of these steps were influenced by events in other Western countries, especially the UK and the US, while others were influenced by Australia's own particular environment and culture.

By the time European settlers arrived in Australia in 1788, it had been occupied for 80,000 years by the Aboriginal people. They used oral traditions such as dance and stories to pass on knowledge about the natural world. This was the first stage.

White settlement heralded the second stage, characterised by scientific interest in Australia's unique natural history and the need to establish a source of food in an agricultural environment far removed from European conditions. This stage lasted through the two world wars.

US President Roosevelt was advised that the driver of economic growth after World War II (WWII) would be the application of science, and that only a community sympathetic and informed about science could benefit (Bush, 1945). This view was picked up by other countries, including Australia: this was the dawn of the third stage, right after WWII.

The fourth stage, which ushered in the modern era of science communication, arrived with a new emphasis on the public communication of science through increased media coverage of science, the creation of interactive science centres, government programs to support science communication and an increase in science communication employment.

The start of the fifth can be associated with the advent of the first dedicated, professional Australian science communicators, and the provision of university courses and training for an expanding cohort of workers. Academic research increased and science communication developed greater cooperation with industry through the Cooperative Research Centre program, and greater connections with politicians through the Science Meets Parliament initiative.

Many institutions and individuals contributed to these five stages, and this chapter describes their roles and the contributions they made to the emergence of modern science communication in Australia.

2. Historical endeavours

The earliest drivers of recorded science communication in Australia were the needs of European settlers on their arrival from 1788. They considered Australia a harsh country, with poor soil, a difficult climate, strange animals and crops that stubbornly refused to grow in seasonal conditions that were the reverse of the northern hemisphere.

The relationship of European settlers with the Aboriginal communities in various areas was difficult and they were not able to use Aboriginal knowledge to improve their situation.

> For thousands of years Aboriginal people survived in the Australian landscape relying on their intricate knowledge of the land and its plants and animals. Tracking and hunting, digging soakages and maintaining surface waters were just some of the ways that people survived. (Central Land Council, 2019)

The settlers relied on their own worldview of Western science. Lindy Orthia (2016) documents the new colony's fascination with science from its earliest days in a study of Sydney's mass media and popular culture. This is hardly surprising, given that every expedition to explore Australia included a person with scientific interests.

Cultural and scientific institutions appeared 40 years after white settlement: the Philosophical Society of 1821 and Sydney's Australia Museum in 1827 were among the first. Other state (regional) museums followed. Traditions of scientific inquiry were extended in the 19th century by mechanics institutes,[1] botanic gardens, learned societies, public libraries and universities:

> By the 1870s it was clear that the programme that had unfolded in these [learned] societies was one largely committed to the collection, description and classification of Australian natural history, phenomena and resources, combined with a discussion of practical matters involved in colonial development. This reflected the mood of the times, which had little patience with abstract theorizing. (Inkster and Todd in Home, 1989, p. 113)

Publication of Australian newspapers began in 1803. Science stories were primarily lifted from American and British publications and often covered agricultural topics.[2] But there was input from domestic journalists, including the Reverend W. B. Clarke and James S. Bray. Bray described himself as a 'naturalist and science journalist' (Burns, 2014b) and was perhaps the first to use this term in Australia. He wrote for the *Sydney Morning Herald* and other antipodean journals on subjects including venomous snakes:

> The season, popularly known as the 'snake season' for 1894–95 has passed away. There is little chance of any person being bitten by a reptile, in a state of nature, for the next five months to come. All our venomous reptiles have already retired, or are about to retire, into their winter quarters. (Bray, 1895)

Bray examined NSW Government records 'for recorded instances of deaths and bites from venomous reptiles' and noted where victims were bitten (most frequently on the finger); the most dangerous snake (black snake); and popular treatments (strychnine, scarifying, or tying a ligature and sucking the wound).

The inauguration of the Australasian and New Zealand Association for the Advancement of Science (ANZAAS) in 1888 was a major advance in science communication. Building on a Victorian-era fascination with science and modelled on the British Association for the Advancement of Science (Macleod, 1988, p. 19), ANZAAS was a meeting place for scientists and the public, to make 'science more widely understood, more generously and wisely

1 The objective of the typical Mechanics Institute was 'the diffusion of scientific, literary and other useful knowledge among its members and the community generally and particularly among the young as well as the operative classes'. See R. W. E. Wilmot, quoted in Home (1989).

2 See Burns (2014a, pp. 72–6).

supported, more directly beneficial to the nation, and more accountable to the public interest' (ibid.). ANZAAS conferences attracted large crowds, as the 'only national forum for science' in Australia (Fenner, 2005, p. 2).[3]

The results of science were taken more directly to the people by agricultural extension officers. Established in the 1860s, extension services helped farmers develop agricultural systems where traditional European approaches had failed to cope with the unfamiliar local soils and conditions (Hunt et al., 2012, p. 11). The demand for these services grew after World War I, when 40,000 demobilised soldiers took up land grants under the Soldier Settlement Scheme. Many had no previous farming experience and much of the land they were offered had only marginal agricultural potential (ibid., p. 12), so a problem–solution exchange from a visiting government extension officer was very helpful.

From 1900 to 1926 the population of Australia increased from 3.7 million to 6 million. Agriculture was a mainstay of the economy and the national government wanted to make greater use of science to develop the country. In 1926 it created the Council for Scientific and Industrial Research (CSIR), a body that originally targeted problems and productivity in the agricultural sector—sheep blow-fly, pizzle disease and seed testing. CSIR worked with the state governments to offer extension services, initially to farmers but over the years as it expanded into other research areas, it employed people for the specific purpose of communicating research results to the public and industry.

On 1 January 1901 the six independent colonies of the Australian continent united to form a single country. By 1939, the population had doubled to 7 million. Nine universities in 1900 had become 17 by 1939. Attendance at ANZAAS conferences rose from 693 in 1900 to 1,200 in 1939. Science journalism became more visible, with a sharp increase in science reports in newspapers. There was a national push to communicate research results to farmers. And then WWII arrived.

3. Leading into the modern era: post–World War II

The war changed everything: new industries, different jobs and new employment opportunities emerged, and people moved to the cities to take advantage. Employment in rural industries dropped from 24 per cent of the

3 The colonial period of science communication is discussed in the paper 'The Emergence of Modern Science Communication in Australia' (Gascoigne and Metcalfe, 2017).

workforce in 1901 to 14 per cent in 1954, and then to under 3 per cent in 2011 (Pollard, 2000; Abjorensen, 2015). Agricultural jobs based on manual drudgery morphed into city work requiring different skills and, increasingly, an educated workforce.

Science caught the imagination of the world through transistor radios, the moon landings, hovercraft and the polio vaccine. In Australia, plagued by rabbits,[4] three scientists won fame when they injected themselves with the myxoma virus in order to demonstrate the safety to humans of this rabbit disease.

The government recognised the benefits of scientific research in the shape of the potential for new industries, new employment and solutions to environmental problems. Research was shifting into unfamiliar territory. Before the war most research related to agriculture, important to Australia's economy and familiar to its citizens: diseases and pests affecting animals and plants, food, forest products and fuel (Rivett, 1972, p. 88). But in the years after 1945, less familiar topics emerged: lasers, quasars and pulsars, genetic manipulation and computing science. Science was increasingly outside the everyday experience of Australians, and the comprehension gap between scientists and the general population grew.

The growing interest in science by the federal government led to a tussle for control with the science community and a debate about restrictions applying to the communication of science. Scientists had a philosophical attachment to the free exchange of ideas, but the Australian Government felt restrictions should apply to research with military or commercial implications. Complicating this was a distrust of scientists in conservative political circles, and the shameful debate in the Australian Parliament beginning on 30 September 1948 (complete with McCarthyist overtones[5]) about who should control the scientific agenda (Rouse, 2002, p. 166).

The love affair with science went through phases in Australia as it did internationally: fascination at the machines of war; awe mixed with revulsion at the atomic bomb; and wonder and optimism at Sputnik in the 1950s. But concerns about the side effects of science also emerged: the anti-nuclear demonstrations of the 1950s; the publication of Rachel Carson's *Silent Spring* in 1962; and napalm and Agent Orange in the Vietnam War. These concerns amplified workers' suspicions about science: how would new technologies affect traditional jobs?

4 Rabbits are a pest in Australia, estimated to cost the economy AU$200 million each year.

5 For details, see Rivett (1972).

Stephen Hill charts the course of this love affair in *The Future for Sale*. Science was a source of 'international prestige and emancipation' in the 1960s but that soured in the 1970s when people saw the potential downside, sparking 'a shift in public consciousness towards disenchantment with science's intrinsic promise' (Hill, 1993, p. 63).

The post-war years saw the opening of university courses in the history and philosophy of science. These began at the University of Sydney in 1945 and the University of Melbourne in 1946; it is claimed they were among the first departments in the world to tackle such subjects. Their discussions pre-empted questions of subsequent concern to science communication (History and Philosophy of Science, 2019).

The foundations were now set, and the role of science communication and the science communicator unfolded over the next 50 years as a profession, an area of training and a field of study.

4. The modern era

There is no set date for the beginning of the modern era of science communication. It began some time in a 20-year envelope from the mid-1960s. Media activity was marked by the establishment of the Radio Science Unit at the Australian Broadcasting Corporation (ABC) in 1964. Questacon, the first interactive science centre, was forming in the mind of Dr Michael Gore (its inaugural director) by 1980 and opened in 1988. The Commonwealth Scientific and Industrial Research Organisation (CSIRO)[6] accelerated the engagement of staff to manage communication activities in the 1980s.

Universities made tentative steps into science communication courses from the late 1980s, and the first national government program to support science communication followed the release of the Bodmer Report in the UK in 1985. Bodmer was an international watershed that reverberated around the world. It recommended actions for scientists, educators, the media, industry, government and museums, aiming to increase overall awareness of science and the way it pervades modern life:

> Science and technology play a major role in most aspects of our daily lives both at home and at work. Our industry and thus our national prosperity depend on them. Almost all public policy issues

6 CSIR was renamed CSIRO in 1949.

> have scientific or technological implications. Everybody, therefore, needs some understanding of science, its accomplishments and its limitations. (Royal Society, 1985, p. 6)

The influence Bodmer had on science communication policy in Australia can be inferred. As Simon Lock points out, UK institutions have a long history in science communication and 'initiatives in this area, particularly institutional programmes in the public understanding of science, have frequently become exemplars for other countries when developing their own' (Lock, 2011, p. 18). Australia as a former colony and member of the Commonwealth borrowed freely from Britain in setting up many of its institutions, including ANZAAS and the Science Media Centre.

We argue that all these influences—greater expenditure on science, the quickening pace of change, international influences and new government awareness of the power of technology—created new demands on communication. There were demands for accountability, for awareness and for education so the population could take advantage of new opportunities. Together, they triggered a demand for the science explainer, a person who could translate the complexities of science into language comprehensible by a layperson.

In part, this was to counter concern about public understanding and attitudes to science (Eckersley and Woodruff, 1984). After reviewing six surveys of popular attitudes to science and technology in Australia, Richard Eckersley concluded that:

> Australians applaud technological process and fear it … we generally regard science and technology as a good thing, but feel threatened by their growing and seemingly uncontrolled power … this anxiety may be heightened by the fact that few of us feel we are very well informed about science and technology. (Eckersley, 1987, p. 1)

The rationale for science communication activities in Australia followed a pattern familiar to other Western countries (Gascoigne, 2001):

- We want a modern, knowledge-based economy.
- Only this sort of economy will deliver the satisfying, high-paying, sustainable jobs that will ensure our national prosperity.
- We believe it would assist us to achieve this sort of economy if we had a population that understands and appreciates science.
- We need a population that understands health and safety issues, like AIDS.
- We also want to ensure the next generation of scientists and technologists, and to stimulate students to do science at school and university, particularly in the 'hard' sciences like mathematics, physics and chemistry.

5. Science in the media: Julius Sumner Miller, Peter Pockley and the ABC Science Unit

Television was introduced to Australia in 1956. The most memorable early science program was *Why is it so?*, where 'the blazing-eyed American Professor Julius Sumner Miller performed like a magician simple experiments demonstrating laws of nature: Groucho Marx PhD, someone called him' (Inglis, 1983, p. 215). The program ran for 20 years and is remembered with affection by many older Australians, not least because Miller never held back on explaining the science. (A measure of his impact was that Miller was commissioned to make TV commercials for Cadbury chocolate, where he emphasised the health benefits of 'a glass and a half of full-cream dairy milk'.)

In 1964, the ABC Science Unit was formed at the urging of the Australian Academy of Science, which 'wanted the ABC to take science more seriously' (ibid., p. 215). Peter Pockley was appointed Talks Assistant (Special Duties) with responsibility for both radio and television broadcasts on science. He began a new science program *Insight* in 1965, and his arrival coincided with moves at ABC Radio to sharpen the style of presenters: less stuffy, more sparkle and more attuned to the growing market of people travelling in cars.

> Pockley was responsible for science in both media, but it was television he really cared about; he itched to work up programs of the kind he had seen done on the BBC. For a start he was given Science Question Time, a fortnightly program … of a panel answering questions sent in by viewers. (Inglis, 1983, p. 215)

Robyn Williams joined Pockley and the ABC Science Unit in 1972 and broadcast his first *Science Show* in 1975. It is the longest-running radio science program in Australia and Williams has been officially designated a National Living Treasure.[7] Apart from interviewing countless scientists and serving as President of Australian Science Communicators, part of the Williams legend is that early in his career he made guest appearances in *The Goodies*, *Monty Python's Flying Circus* and *Doctor Who* (ABC News, 2018).

7 In 1997, the National Trust of Australia (NSW) invited the public to nominate Australians they most valued.

Figure 6.1: Peter Pockley.

Source: Made available by Cultural Collections, Auchmuty Library, University of Newcastle.

Figure 6.2: Robyn Williams.

Source: Made available by ABC RN.

Pockley was his boss; in an obituary Williams described Pockley as a pioneer and an innovator with a marvellous on-air presence:

> He turned out to be a magnificent mentor, offering the most thorough going training any budding broadcaster could wish for. And it was clear from the start that the Unit was everywhere, and pioneering on all fronts. Our Apollo mission coverage (I was hired to prepare for Apollo 16) was across all radio networks and also on TV. And when it came to craft skills we were the first in the ABC to try talk-back[8] (using an Army field telephone, which would have been more suitable for Rommel in North Africa) and we were secretly doing our own editing. (Williams, 2013)

Pockley set up radio programs at the ABC such as *Insight* and *The World Tomorrow*, but the ABC hesitated about letting him loose on television. He was widely regarded as a prickly character, and this led to monumental battles with management. He was the first senior officer to be charged with breaking staff rules about public comments (spectacularly at a packed ANZAAS Conference, when he responded frankly to a question about ABC plans for science on television).

8 When listeners phone in to ask questions of the presenters, live to air.

The formation of the ABC Science Unit was a major development in the reporting of science. Pockley claimed it 'catalysed newspapers into appointing their first science writers' (Pockley, 2004). It was certainly a significant step and may have sparked interest in science by the print media—certainly all major newspapers had journalists dedicated to covering science and technology by 1990 (Metcalfe and Gascoigne, 1995). Burns, however, records a history of science journalists stretching back well before the Science Unit was formed (Burns, 2014b).

Pockley persuaded the Australian Academy of Science to become more involved in public communication. The Academy had been formed in 1954 and the first object in its charter is 'to promote, declare and disseminate scientific knowledge' (Fenner, 2005, p. 314). But faced with a multitude of tasks, limited resources and high expectations, the Academy concentrated on scientific publications and educational materials, with public activities such as lectures, exhibitions (e.g. *Illusion and Reality* in 1983) and forums regarded as supplementary.

The academy did accept one of Pockley's suggestions in establishing the Australian Science and Technology Information Service (ASTIS) in 1985, a database of expert scientists willing to talk to the media. Pockley chaired media briefings at the University of Technology Sydney (UTS), assembling panels of experts to address topics of media interest, briefing them on media expectations and leading a discussion before an audience of journalists.

The trial briefings were successful, and the idea attracted significant private sponsorship, but the government chose to discontinue funding and the briefings ended (Fenner, 2005, p. 275). Ten years later, they reappeared, this time run by the Australian Science Media Centre (AusSMC), to inform media coverage of science issues. It is no coincidence that the inaugural (and current) director of SMC is Susannah Eliott, a former colleague of Peter Pockley at UTS.

6. The first science centre opened in 1988

Australia's first interactive science centre was the vision of one man, Dr Michael Gore, a physicist at The Australian National University (ANU). Like others, he was inspired by Frank Oppenheimer's *Exploratorium* in San Francisco. Gore's centre began modestly in 1980: it 'opened on a shoestring budget, and every science organisation in Canberra helped build the displays. Based in an old inner-city infants' school, it grew and began to attract private sponsorship' (Gascoigne and Metcalfe, 1994, p. 412).

Gore piloted his home-grown science centre into Questacon, the National Science and Technology Centre. His cause was aided by a bicentennial gift from the Japanese Government, which paid half the cost of the new building. Questacon aimed to change science museums from 'dusty, static, even dead emporia of esoteric mysteries ... [they] started to come alive, to involve and question their millions of visitors; they started to explain, entertain and educate' (ibid., p. 410).

Professor Chris Bryant takes up the story:

> At once, it attracted large crowds that needed to be managed, so Mike [Gore] hit on the idea of training student explainers to give science shows in the side rooms, thus easing the press of people on the main floor. This pragmatic response to an organizational problem had a far-reaching effect, as these young men and women eventually formed the foundation of the Questacon Science Circus ... the most extensive outreach project of any science centre in the world, with the whole of Australia as its territory. (Bryant et al., 2015, p. 129)

Questacon and the new Centre for Public Awareness of Science (CPAS) at ANU formed a partnership to train the 'student explainers' in both the theory and practice of science communication. The training evolved from a Graduate Certificate into a Graduate Diploma and a master's degree program.

Australia also had other science centres and travelling science educational activities, and they have a patchy history. Bryant lists nine centres that opened in Australia between 1980 and 1992, from Wollongong to Hobart. The passion of 'local champions' was integral to their success, but this proved a fragile base: several subsequently closed, went through funding crises or moved to other venues. (Bryant et al., 2015).

7. Government programs to support science communication

From the 1960s, both sides of government had concerns about science, driven by low investment in research and development by industry, and collapses in the prices of agriculture and mining products. In 1981, the Liberal (conservative) Minister for Science and Technology David Thomson urged scientists to speak out and to carry the public with them. Similar calls were made by both his predecessors and his successors, with ALP minister

Barry O. Jones famously calling scientists 'wimps' in 1984 for remaining silent while he was fighting for the science budget in cabinet[9] (New Scientist, 1989, p. 20).

It was dawning on government that scientists willing to be more vocal about their research and its benefits (as part of a broad push into science communication) might help achieve national objectives: informing the public, changing behaviours and stimulating interest in science. The Labor Party announced a policy on science communication in 1982:

> [to] initiate a continuous public information campaign in an attempt to demystify scientific processes, to raise levels of community understanding about science and technology so that the Australian people and their political representatives can be directly involved in choosing between options and determining priorities. (Bhatal, 1985, p. 1)

But government support for science communication was minimal. There were no policies and no institutions to achieve government objectives, and any moves were experimental. In 1984 the Labor Government took a tentative step, establishing the Commission for the Future:

> Its emphasis will be on explaining future scenarios pointing out the range of options opened by new technology and then saying to the people, you must choose for yourselves. (Jones, quoted in Bhatal, 1985, p. 1)

Five years later, after the under-performing commission had closed, the government made a major announcement on science. Urged by his polymath science minister Barry O. Jones, Prime Minister Bob Hawke delivered the 'Clever Country' speech, a landmark address leading into the 1990 election. It set out new expectations of science: 'No longer content to be just the lucky country, Australia must become the clever country' (Hawke, 1990). Influential was an OECD report critical of Australia's National Science and Technology Policy: such an 'external international perspective can be a strong motivator for action'.[10] Hawke unveiled spending measures, a new rhetoric about science, and a program to support science communication: the Science and Technology Awareness Program (STAP).

9 Jones is an extraordinary figure, a member of all four Academies in Australia, prolific author and visionary for science. He later denied he had called scientists 'wimps' but admitted to accusing them of 'wimpish behaviour' (Personal correspondence, c. 2002).

10 Personal communication, Professor Sue Stocklmayer, 2017.

STAP was Australia's first national science communication program. Created in 1989, it had seven staff and a budget in 1991–92 of AU$0.7 million, rising to AU$1.7 million in 1992–93 (10 cents per head of the population of Australia). These were modest resources given its ambitious aims:

> The Government's Science and Technology Awareness Program aims to increase public awareness of the central role that science and technology play in national life, including economic and social development. The contribution of science and technology to industry, and the contribution of our manufacturing and services industries to national development, are not widely recognised by Australians. This lack of recognition appears to be one reason for the reluctance of Australians to adopt new technologies and innovative practices in the workplace. (Commonwealth of Australia, 1991, p. 4)

The program's five target groups were young people and their teachers; women; industry and business leaders; scientists; and journalists. STAP funded about 40 special projects annually, including briefing forums for the media, an annual register of science communicators,[11] and co-hosting the first national conference for science communicators in August 1990.

The Budget Statement for the next year (1992–93) built on this base. It renamed the Science Council 'The Prime Minister's Science and Engineering Council' and gave it a new task: 'to enhance awareness in community of the importance of science, technology and engineering for Australia's economic and social development' (Commonwealth of Australia, 1992, p. 2.91).

So in quick succession through the 1980s, Questacon was launched; STAP was established; and the 'Clever Country' speech meant more support for programs aiming to lift awareness of science and technology (ibid., p. 1.20–1.22).

The achievements of STAP were modest, reflecting its experimental nature and limited funding. Over the next 15 years, successive governments tinkered with the program, not satisfied it was making much of a difference. The Australian community was not engaged, student numbers in science were falling, and investment in research by industry was among the lowest in the OECD. STAP was modified, renamed and eventually replaced by *Inspiring Australia*, a new strategy launched in February 2010 by science minister Carr (and still running):

11 The Register of Australian Science and Technology Communicators was first published in 1990. By the time of the Third Register in 1994, there were 407 individuals listed (70 academia; 72 education; 31 electronic media; 31 print media; 36 government; 5 industry; 47 museums and science centres; 52 research institutions; 44 in service industries and 19 other).

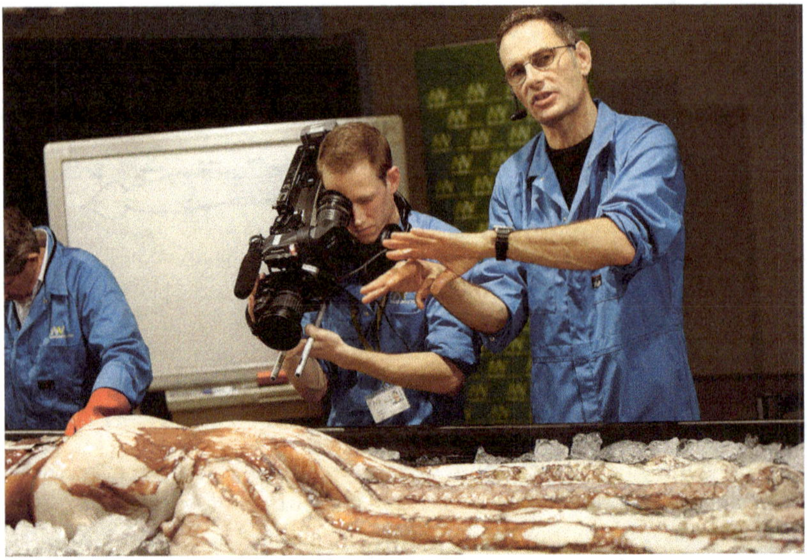

Figure 6.3: Dr Mark Norman dissecting a Giant Squid at Melbourne Museum in front of the media and streamed live on the internet, 2008.

Source: © Museums Victoria. Photographer: Jon Augier.

Again the Government had lofty aims:

> we must communicate and engage the wider community in science. Australia aspires to an innovative society with a technologically skilled workforce, a scientifically literate community and well informed decision makers. The *'Inspiring Australia'* strategy aims to build a strong, open relationship between science and society, underpinned by effective communication of science and its uses. (Commonwealth of Australia, 2010, p. xiii)

Science minister Carr was blunter: 'If we are serious about giving people a real voice in how we run this high-tech world of ours, we have to be serious about science communication' (Carr, 2010).

Inspiring Australia introduced new approaches: a long-overdue recognition that the humanities and social sciences play a part in science discussions, and an attempt to co-ordinate national and regional activities. It established working groups to examine specific issues: the working group on 'Science and the Media', for instance, produced a 60-page report with 26 recommendations. It aimed to bring a new energy to the sector, targeting priority groups (youth, industry, women).

After 10 years, the impact of *Inspiring Australia* is hard to gauge. Evaluating science communication is notoriously difficult, but its small initial budget and a lack of enthusiasm combined with declining funding from the current government suggest tokenism rather than a serious attempt to tackle an ambitious agenda.

8. The birth of Australian Science Communicators 1994

In the early 1990s, people working as science communicators were isolated. Every science research organisation, every museum and each of the 37 divisions of CSIRO had at least one communicator, but there were few mechanisms to enable them to exchange ideas. An association had been formed in about 1987 (with Pockley playing a lead role) but it died after 18 months under the weight of logistical challenges. A national conference of science communicators (with a schoolteacher focus) was convened in 1990 by Questacon and the government, and STAP's registers of science and technology communicators were preliminary steps to events of 1994.

The impetus to form an association came to a head in 1993, when the authors of this chapter (Gascoigne and Metcalfe) were asked to present a paper at the Public Communication of Science and Technology (PCST) Conference in Montreal. Their paper was a summary: which organisations were communicating science in Australia, what were they doing, and how much were they spending?

The authors collected information by asking organisations for a copy of their annual report and interviewing communication staff, and it quickly became clear that people in these roles felt isolated. There was little opportunity for them to share experiences. Tertiary courses in science communication were at an embryonic stage. Publishing articles on science communication was a fanciful idea and attending international meetings to discuss professional issues was almost unknown. The internet as a means of communication was in its infancy.

Clearly there was a need for an association. A meeting at the National Press Club in Canberra in 1994 was the first step. Twenty-three people attended: from the science media, research organisations, science-based institutions and government departments. Alison Leigh, then executive producer of the ABC's national television science show, chaired the meeting; and the only contentious discussion was whether membership should be limited to working journalists or open to anyone prepared to pay the subscription. The latter view prevailed.

The next step was a general invitation to communicators to register as Foundation Members: 375 people paid AU$25, and these funds together with donations from the Academy of Science, the Institution of Engineers, CSIRO and the Department of Science provided Australian Science Communicators with working capital. The inaugural general meeting at the ANZAAS Conference in Geelong on 22 September 1994 voted to adopt the draft constitution and elected science journalists Julian Cribb (*The Australian* newspaper) and Ian Anderson (Australasian editor of *New Scientist*) as president and vice-president.

The major achievement of Australian Science Communicators (ASC) has been to create a network where none existed before. Members have discovered colleagues and common interests, and worked together to organise events such as National Science Week.[12] ASC hosted the PCST conference in 1996 and the World Conference of Science Journalists in 2007, and runs annual national conferences and a newsletter. None of this would have been possible without an association. The ASC website, discussion list and new media provide a mechanism for advertising conferences and publications. The result has been a large increase in Australian participation in journals and meetings internationally (aided by the advent of the internet and the rise across the world in science communication activities).

9. University courses in science communication

An ASC survey in Australia in 1996 recorded 16 universities offering or planning to offer courses in science communication. These were a mixture: short courses for scientists in writing and handling the media; a bachelor's degree in science and the media; electives for science and journalism students in writing science; and postgraduate and short courses in technical writing (ASC, 1996).

> This initial focus of university science communication courses was on training scientists rather than catering to the newly-emerging profession of 'science communicators', and many of the teachers of science communication had scientific rather than communication qualifications and experience. (Metcalfe and Gascoigne, 2012, p. 23)

12 The first national event was held in 1997.

Figure 6.4: Chris Bryant, Mike Gore and Sue Stocklmayer at ANU campus, 2019.

Source: Photo by Toss Gascoigne.

The first universities offering master's courses in science communication were the University of Central Queensland (UCQ) in 1993 and ANU in 1996. Emeritus Professor Lesley Warner designed the UCQ course in 1992:

> The idea to offer the courses came from my contacts with Chris Bryant and others at Questacon in the early 1990s and I went to the … workshop on media skills that you[13] offered in Rockhampton in August 1992 (I value my attendance certificate most highly). I was looking for niche-market courses that could be offered in the external mode, and at the time we thought there was an opportunity going begging. The interest in sci-comm was developing and there were no post grad courses in Australia. I also found an interesting initiative that was being offered across Universities in Belfast and Dublin so I met with them as well. Basically though the final format was put together by an interdisciplinary group across the university … There was no model to follow.[14]

13 Authors Gascoigne and Metcalfe.

14 Personal communication, email from Emeritus Professor Lesley Warner, 20 February 2014.

An influential factor in the ANU course was the partnership with Questacon's Science Circus. Shortly after the foundation of the Circus in 1985, Questacon Director Mike Gore asked ANU Dean of Science Chris Bryant if the newly formed Centre for the Public Awareness of Science (CPAS) could provide theoretical training and a formal qualification to back up the Circus' more practical approach. The graduate certificate program began in 1988 and was converted into a graduate diploma in 1990.[15] Demand from outside the Circus grew, and Sue Stocklmayer was appointed as lecturer in science communication. Master's and PhD programs followed in due course.

These courses were pioneering. A preliminary survey records only China (1989), the UK (1990) and India (1993) running master's or research degrees in science communication any earlier (Gascoigne, 2014). The work of Mike Gore, Chris Bryant and Sue Stocklmayer blazed a trail in the Antipodes.[16]

CPAS and other courses responded to the need to train people to match new job opportunities in science communication. In 1990, CSIRO employed communicators in each of its 37 divisions: to write articles, issue media releases, organise exhibitions, assist researchers with presentations and papers, organise demonstration days for farmers and manage relationships with funding bodies and industry. Staff at museums and the new science centres had similar roles, with greater focus on educational activities. Managing organisational websites later became an important part of a communicator's work.

The normal prerequisite to enter a science communication course in Australia was a degree in science. This contrasts with European and American courses likely to be offered in journalism, arts or social sciences contexts.[17] CPAS was established in 1996, and its courses were advertised as allowing 'a new generation of highly qualified scientists to become skilled communicators who can engage people with the science, technology, or medical information that is most relevant to them'.[18]

Requirements have changed over time. University of Queensland's courses were designed 'for science graduates, or those with strong science backgrounds, who wish to communicate effectively with scientists and

15 Personal communication, email from Emeritus Professor Chris Bryant, 2017.

16 Gore, Bryant and Stocklmayer worked together to in initiate and develop the courses: Gore as founder and first Director of Questacon; Bryant as Dean of Science at ANU; and Stocklmayer as the first Director of the Centre for the Public Awareness of Science, ANU.

17 See, for instance, Directory of Science Communication Courses and Programs by Sharon Friedman and Sharon Dunwoody et al., published in years from 1978 and listing and describing science communication courses in the US.

18 As published on CPAS's website in March 2015, site now discontinued.

professionals in business, industry, government, and the media'. This wording was subsequently modified: 'for graduates from disciplines engaging with science'.[19] The University of Western Australia required a strong mathematics background.

Since 1996, when 16 universities offered courses, there has been a consolidation. The 16 are now half a dozen and CPAS has emerged as the powerhouse. But this is a volatile area and new courses are liable to rise and fall as personnel, funding and vice-chancellors change.

10. Research in science communication

As science communication courses grew and became concentrated in a handful of Australian universities, interest in research at postgraduate level increased. It started at CPAS in 1996, and the number of PhD students has grown steadily to about 30 today. Two other universities have had a significant presence in science communication: the University of Queensland and the University of Western Australia. In 2011 they had 10 PhD students and 20 master's students between them, but numbers declined after staff changes.

Publication of articles on science communication was virtually non-existent in Australia before 1990. The reasons were simple: science communication was emerging as a field of study; *Knowledge: Creation, Diffusion, Utilization* (now published as *Journal of Science Communication*) was the only specialised journal in the field (although related journals might carry articles); and there was no culture of publication. A search of 22 journals relevant to science communication discovered only two articles by Australian authors published before 1990 (Metcalfe and Gascoigne, 2012).

But the trend is upwards. Metcalfe's 2012 study identified a total of 73 articles, 23 published in the period 1990–99, and a further 48 since 2000. In the last five years, authors attached to CPAS have published four books, 18 book chapters and 49 refereed papers.[20] A study of the three international journals devoted to science communication found that Australians were fourth in producing peer-reviewed papers, behind the US, UK and Canada (Guenther and Joubert, 2017).

19 As published on University of Queensland Science Communication Field of Study website in 2015, site now discontinued.

20 A list of publications by all CPAS staff and students is published at: cpas.anu.edu.au/research/publications/archive (accessed 6 November 2018).

Metcalfe's review (2012) of Australian science communication scholarly papers notes other trends in publishing over the last 15 years:

- a more interdisciplinary approach
- a greater diversity of topics
- a move in the disciplinary background of researchers, away from natural sciences and towards the humanities/social sciences
- more research into models of science communication, and engagement.

11. Communication with industry: Cooperative Research Centres, 1990

Communication with industry (apart from agriculture) has been a problematic area in Australia. Part of this is the nature of the commercial sector. Industrial investment in research is lower than in comparable countries, partly because many companies in Australia are branch offices. To increase the industrial take-up of research, the government moved to encourage collaborative research: consortiums of research organisations, industrial companies, government departments and the end users. In 1990 it established the Cooperative Research Centre (CRC) program.

The CRC program brought industry and researchers into networked groups bound by a common interest. Groups worked in areas including renewable energy, viticulture, freshwater ecology and asthma. The program involved public good issues as well, notably in the environmental area where the 'industry partners' were government agencies or organisations aiming to improve matters such as air or water quality. A notable example was the CRC for Freshwater Ecology run by the redoubtable Peter Cullen. Cullen pioneered the use of 'knowledge brokers' to link research with end-users and drew up a 'Top Hundred' list of all the people he had to communicate with. He said that as long as these people knew what was happening at his CRC, it was achieving its aim.

Consolidation of research was an advantage, but communication was a challenge, as Riedlinger et al. notes:

> There are many advantages to functional diversity, including increased innovation, renewal, and creativity. The main disadvantage of functionally diverse organizations is that diversity makes communication and cooperation difficult. (Riedlinger et al., 2004, p. 56)

By 1994 more than 50 CRCs had been established, each with up to 100 partners. The challenge of communicating across a cooperative venture of this diversity was initially underestimated, with the communication role often carried out by the administrator of the CRC, outsourced to a communicator working with one of the partners, or devolved to the researchers. A review of the CRC program in 2003 (Howard Partners, 2003) concluded that a successful CRC requires a major commitment to regular communication among all participants (researchers, the board, industry bodies, government and the community), but observed this area was often under-resourced:

> For many CRCs external communication does not appear to be a high priority. Yet communication is an important path to adoption. … CRCs spend, on average, only 1.9 per cent of their total expenditure on communication. Given the purpose of the CRC Programme in promoting adoption, this low level of commitment to communication is of concern. (Howard Partners, 2003, p. 91)

As the program has matured, so has the role of communication. The focus on industry and making partnerships work across institutional cultures has re-orientated the communicator's normal focus on communicating with the public. The belated recognition of the importance of communication is demonstrated by a guide published by the CRC Association in 2010, emphasising the central importance of internal and external communication and listing 20 different approaches (CRC Association, 2010).

12. Communicating with politicians: Science meets parliament, 1999

Australians are proud of their scientists: in 1966 when the country moved to decimal currency, half the new bank notes featured portraits of scientists. But in political terms, science is a low priority, shuffled around from department to department and low on the ministerial totem pole. Supporting science is not seen a vote-winning policy.

Parliamentarians are a difficult target group for science communication. Only a handful of the 227 national parliamentarians have university-level qualifications in science. Their lack of knowledge is apparent when parliament discusses science-based issues like water, energy and genetic engineering. There is a vocal distrust of science by some parliamentarians, part of an inclination by conservative forces to dismiss scientists as 'chardonnay socialists and latte-sipping liberals'. One result has been a long-running and rancorous debate on energy and climate policy.

In 1999, a national advocacy body for science (the Federation of Australian Scientific and Technological Societies (FASTS)) tackled this issue head-on with a new event, 'Science meets Parliament' (SmP).[21] Loosely modelled on the US 'Congressional Visits Day', it brought 160 scientists to the national parliament to meet MPs and talk about science and its benefits. Half of the national parliamentarians agreed to meetings with a pair of scientists.

SmP was an instant success and galvanised enthusiasm across the science community. Its stated purpose was to inform MPs about science, but it also served to educate scientists on the policymaking process. It was a trail-blazer in Australia: no other organisation or sector had attempted to influence parliamentary opinion by 'invading' Parliament House *en masse*. SmP is firmly established as an annual event and is much copied by organisations from other sectors and regions. It has spawned a European event with a similar name: Science meets Parliaments.[22]

SmP had a controversial start. At the initial event FASTS employed Robbie Swan, a well-known lobbyist, to tutor participating scientists in advocacy. One senator was not impressed:

> Senator Harradine: So I was rather surprised to learn of the methods that the Federation of Australian Scientific and Technological Societies used yesterday to assist scientists in their communications with parliamentarians. Some lobbyists go about their work in a very professional way. They might say they apply similar systematic and rigorous methods as do scientists. By and large, they are a reputable group. We all know who they are, and there is no need to mention them by name. But whom did the organisation decide to pick to tutor the scientists in how to approach lobbyists? None other than the spokesman for the porn industry.
>
> Interjection: Oh, no!
>
> Senator Harradine: Yes, the spokesman of the porn industry, Robbie Swan, or should I say Caroline Sweetly, whichever alias he goes by. (Hansard, 2001, p. 26332)

21 For a description of the event, see Gascoigne (2005).

22 See europa.eu/newsroom/events/science-meets-parliaments_en.

Figure 6.5: *Canberra Times*, Monday 11 October 1999.
Source: Published by permission of the artist, Pat Campbell.

An integral part of SmP is an address to the National Press Club (NPC). FASTS had earlier pioneered the use of the NPC to make the public case for science, hoping to profit from the fact that major speeches are televised live. But in the 55-year history of the NPC, no scientist had ever delivered a televised address before FASTS organised an address by Professor Ian Lowe in 1997. FASTS used the NPC to influence the policy debate and connect with a wider spectrum of interests in Australia, including industry. The Academy of Science has followed suit, organising its own speakers from 2001.

13. Conclusion

The story of science communication in Australia began with the Aboriginal oral traditions of passing on knowledge about the natural world. Today there is a rich diversity of science communicators in Australia, including those working for research organisations like CSIRO, in state government departments, for industry and in private consultancy businesses. In the last decade, we have seen science communication put more emphasis on the need to 'engage' the public. One of these publics is the Aboriginal community, as the scientific establishment tries to incorporate indigenous knowledge gained over 80,000 years into its Western science model.

All the institutions and events documented above have been important influences on the emergence of modern science communication in Australia. This account omits (for reasons of space) other organisations: the Academy of Technological Sciences and Engineering; the Royal Institute of Australia; the communication efforts of agriculturally based research and development organisations; the Powerhouse and Australian Museums; the environmental movement from the 1960s; and national awards including the Eurekas and the Australia Prize.

Science communication has had its successes and failures in the modern era. Successes include the growth of courses and research at universities, Questacon, the birth of Australian Science Communicators and the engagement of communicators in international conversations. The 'Clever Country' speech and a national report on science communication (*Inspiring Australia*) were significant policy steps. Science communication has reached out to difficult audiences in the worlds of politics and business. It continues to grapple with hard questions of audiences and publics, and research is coming up with more sophisticated communication tools. There is a lively and well-connected national community actively engaged in practice and research.

But it is also fighting a continuing battle to be effective. What is the best way of engaging with the public, finding a balance between deficit, dialogue and participatory activities? How should communicators act to moderate a discussion between researchers, interested parties and political interests? In the domain of climate change, science communication has been both a success and a failure: polling shows a strong revival of public support (59 per cent) for Australia to take action,[23] but the obstructive stance of the current national government is a source of frustration.

Measured against the government's stated objectives ('We must communicate and engage the wider community in science. Australia aspires to an innovative society with a technologically skilled workforce, a scientifically literate community and well-informed decision makers' (Commonwealth of Australia, 2010, p. xiii)), there has been limited progress. Enrolments in STEM subjects at universities hit a 20-year low in 2017 (Wood, 2017) and decision-makers continue to resist action on climate change and managing Australia's major river system. CSIRO has cut funding for communication activities.

The challenge for Australian science communicators and those who support and employ them is to find the resources to support long-term science communication programs that can help scientists, policy makers and different publics better address the environmental, social and economic issues facing the country.

23 The Lowy poll: 59 per cent of respondents agreed with the statement: 'climate change is a serious and pressing problem. We should begin taking steps now even if this involves significant costs'. See lowyinstitutepoll.lowyinstitute.org/climate-change-and-energy/.

References

ABC News. (2018). Robyn Williams. *ABC Radio National*. Retrieved from www.abc.net.au/radionational/robyn-williams/2913842.

Abjorensen, N. and Docherty, J. C. (2015). *Historical Dictionary of Australia* (4th edition). Appendix E, p. 438. Australia: Rowman and Littlefield. doi.org/10.1108/rr-08-2015-0203.

Australian Science Communicators. (ASC). (1996). *Vulcan* newsletter. Canberra.

Bhatal, R. S. (1985). Science Centres and/or Science Museums for Australia. *Journal of Proceedings, Royal Society of New South Wales, 118*.

Bray, J. S. (1895, 8 June). The Snake Season in New South Wales. *Supplement to the Evening Post*, Wellington, New Zealand. Retrieved from paperspast.natlib.govt.nz/imageserver-newspapers/EP18950608.1.5.pdf.

Bryant, C., Gore, M. and Stocklmayer, S. (2015). The Australian Science Centre Movement 1980–2000: Part 1 – Questacon, the National Science and Technology Centre. *Historical Records of Australian Science, 26*, 122–32. doi.org/10.1071/HR15008.

Burns, M. (2014a). A Brief History of Science Communication. *Media International Australia, 150*(1), 72–6. doi.org/10.1177/1329878x1415000116.

Burns, M. (2014b). *A Companion to the Australian Media*. Melbourne: Australian Scholarly Publishing.

Bush, V. (1945). *Science: The Endless Frontier*. Washington DC: United States Government Printing Office.

Carr, K. (2010). Address to the Australian Science Communicators National Conference, at ANU Canberra, 8 February. Retrieved from parlinfo.aph.gov.au/parlInfo/search/display/display.w3p;query=Id%3A%22media%2Fpressrel%2F1TEW6%22;src1=sm1.

Central Land Council. (2019). *Indigenous Ecological Knowledge*. Retrieved from www.clc.org.au/articles/info/indigenous-ecological-knowledge.

Commonwealth of Australia. (1991). *Science and Technology Budget Statement 1991–92*. Canberra: Australian Government Publishing Service. Retrieved from www.industry.gov.au/sites/default/files/1991-92-science-technology-budget-statement.pdf.

Commonwealth of Australia. (1992). *Science and Technology Budget Statement 1992–93*. Canberra: Australian Government Publishing Service. Retrieved from www.industry.gov.au/sites/g/files/net3906/f/1992-93-science-technology-budget-statement.pdf.

Commonwealth of Australia. (2010). *Inspiring Australia: A national strategy for engagement with the sciences.* Canberra, ACT: Department of Innovation. Retrieved from www.industry.gov.au/sites/default/files/2018-10/inspiring_australia_-_a_national_strategy_for_engagement_with_the_sciences_2010.pdf.

CRC Association. (2010). Communication in Cooperative Research Centres. Retrieved from crca.asn.au/wp-content/uploads/2012/05/CRCGuideH-Communication-in-CRCs.pdf.

Eckersley, R. (1988). *Australian Attitudes to Science and Technology and the Future: A report for the Commission for the Future.* Canberra: Australian Government Publishing Service.

Eckersley, R. M. and Woodruff, B. J. (1984). Public Perceptions of CSIRO: A Staff Viewpoint. *Report to the CSIRO Executive*, 22 March.

Fenner, F. (ed.) (2005). *The Australian Academy of Science: The First Fifty Years.* Canberra, ACT: Australian Academy of Science.

Gascoigne, T. H. (2005). Science meets Parliament. In M. Claessens (ed.), *Communicating European Research.* Dordrecht: Springer.

Gascoigne, T. H. (2014). How did modern science communication emerge in different countries round the world? *PCST Conference.* 5–8 May, Salvador, Brazil.

Gascoigne, T. H. and Metcalfe, J.E. (1994). Public communication of science and technology in Australia. In B. Schiele (ed.), *When Science Becomes Culture.* Ottawa, Canada: University of Ottawa Press.

Gascoigne T. H. and Metcalfe, J. E. (2001). Report: The evaluation of national programs of science awareness. *Science Communication*, *23*(1), 66–76. doi.org/10.1177/1075547001023001007.

Gascoigne, T. H. and Metcalfe, J. E. (2017). The Emergence of Modern Science Communication in Australia. *Journal of Science Communication, Special issue: History of Science Communication, 16*(3). doi.org/10.22323/2.16030201.

Guenther, J. and Joubert, M. (2017). Science communication as a field of research: identifying trends, challenges and gaps by analyzing research papers. *Journal of Science Communication, 16*(2), A02. doi.org/10.22323/2.16020202.

Hansard. (2001). Proceedings of the Senate, 22 August. parlinfo.aph.gov.au/parlInfo/search/display/display.w3p;db=CHAMBER;id=chamber/hansards/2001-08-22/0065;query=Id:%22chamber/hansards/2001-08-22/0000%22.

Hawke, R. J. (1990). Election speech, 8 March. Retrieved from electionspeeches.moadoph.gov.au/speeches/1990-bob-hawke.

History and Philosophy of Science. (2019). *Overview.* University of Melbourne. Retrieved from arts.unimelb.edu.au/school-of-historical-and-philosophical-studies/study/history-and-philosophy-of-science.

Hill, S. (1993). The Future for Sale. *Search 24*(3), 63–7.

Home, R. W. (ed.). (1989). *Australian science in the making*. Cambridge, UK: Cambridge University Press.

Howard Partners. (2003). *Evaluation of the Cooperative Research Centres Programme*. Department of Education, Science and Training, July.

Hunt, W., Birch, C., Coutts, J. and Vanclay, F. (2012). The Many Turnings of Agricultural Extension in Australia. *The Journal of Agricultural Education and Extension, 18*(1), 9–26. doi.org/10.1080/1389224X.2012.638780.

Inglis, K. S. (1983). *This is the ABC: The Australian Broadcasting Commission 1932–1983*. Melbourne: Melbourne University Press.

Lock, S. (2011). Deficits and Dialogues: Science communication and the public understanding of science in the UK. In W. Bodmer (Author) and D. Bennett and R. Jennings (eds), *Successful science communication: Telling it like it is*, pp. 17–30, Cambridge: Cambridge University Press. doi.org/10.1017/CBO97805117 60228.004.

MacLeod, R. (ed.) (1988). *The Commonwealth of Science: ANZAAS and the Scientific Enterprise in Australia 1888–1988*. Australia: Oxford University Press.

Metcalfe, J. and Gascoigne, T. (1995). Science journalism in Australia. *Public Understanding of Science, 4*, 411–28. doi.org/10.1088/0963-6625/4/4/005.

Metcalfe, J. and Gascoigne, T. (2012). The Evolution of Science Communication Research in Australia. In B. Schiele, M. Claessens and S. Shi (eds), *Science Communication in the World: Practices, Theories and Trends* (pp. 19–32). Dordrecht, The Netherlands: Springer. doi.org/10.1007/978-94-007-4279-6_2.

New Scientist. (1989). Shear Indifference. Editorial, 5 August.

Orthia, L. A. (2016). 'Laudably Communicating to the World': Science in Sydney's Public Culture, 1788–1821. *Historical Records of Australian Science, 27*, 1–12. doi.org/10.1071/HR15018.

Pockley, P. (2004). 40th Anniversary of the ABC Science Unit. 19 September. Retrieved from www.abc.net.au/radionational/programs/ockhamsrazor/40th-anniversary-of-the-abc-science-unit/3425990#transcript.

Pollard, J. (2000). A Hundred Years of Agriculture. *Year Book Australia, 2000*. Retrieved from www.abs.gov.au/ausstats/abs@.nsf/94713ad445ff1425ca25682 000192af2/3852d05cd2263db5ca2569de0026c588!OpenDocument.

Riedlinger, M. E., Gallois, C., McKay, S. and Pittam, J. (2004). Impact of Social Group Processes and Functional Diversity on Communication in Networked Organizations. *Journal of Applied Communication Research, 32*(1), 55–79. doi.org/ 10.1080/0090988042000178130.

Rivett, R. (1972). *David Rivett: Fighter for Australian Science*. Melbourne: Rivett. doi.org/10.1071/9780643109964.

Rowse, T. (2002). *Nugget Coombs: A reforming life*. Melbourne: Cambridge University Press.

Royal Society (1985). *The Public Understanding of Science*. London, UK.

Williams, R. (2013, 28 September). Peter Pockley: Champion for science in a wakening world. *Sydney Morning Herald*.

Wood, P. (2017, 30 March). STEM enrolments hit 20-year low, but scientists have an idea to stop the slide. *ABC News Breakfast*. Retrieved from www.abc.net.au/news/2017-03-30/science-maths-enrolments-hit-20y-low-but-scientists-have-a-plan/8395798.

Timeline

Event	Name	Date	Comment
First interactive science centre established.	National Science and Technology Centre (Questacon)	1988	Questacon began in an inner-city school in 1980, and was renamed Questacon in 1988
First national (or large regional) science festival?	Regional science festival in Canberra (the national capital)	1993	In 1997, the festival was expanded into National Science Week
Association of science writers or journalists or communicators established.	Australian Science Communicators	1994	Inaugural AGM held at ANZAAS meeting in Geelong
First university courses to train science communicators.	Graduate certificate in science communication, ANU	1988	1990: Converted to a graduate diploma There were earlier short courses and diplomas
First master's students in science communication graduate.	University of Southern Queensland	1993	1996: Master's offered by CPAS, ANU
First PhD students in science communication graduate.	CPAS offered a PhD in science communication	1996	First PhD graduate in science communication from CPAS
First national conference in science communication.	In Canberra. Primarily for teachers	August 1990	1996: Melbourne, part of the PCST Conference
National government program to support science communication established.	Science and Technology Awareness Program (STAP)	1989	2010: STAP was replaced by a new program after release of the *Inspiring Australia* report

Event	Name	Date	Comment
First significant initiative or report on science communication.	*Inspiring Australia: A national strategy for engagement with the sciences*	February 2010	
National Science Week founded.	National Science Week established	1997	Runs annually in August, with 1,000 events
A journal completely or substantially devoted to science communication established.	*Australasian Journal of Science* first published	1938	'To publish news on scientific topics of general interest' (name change to *Search*, now defunct)
First significant radio programs on science.	Radio Science Unit formed at the ABC	1964	Peter Pockley was first reporter and producer
First significant TV programs on science.	*Why is it so?* Iconic TV program hosted by Julius Sumner Miller	1963	
First awards for scientists or journalists or others for science communication.	Michael Daley Awards for excellence in science journalism	1984	1997: The Daley Awards were rolled into the Eureka Awards, run by the Australia Museum
Date hosted a PCST conference.	Melbourne hosted the fourth PCST Conference	1996	
Other significant events.	Australasian Association for the Advancement of Science founded	1886	1888: renamed the Australia and New Zealand Association for the Advancement of Science
	New decimal currency, had portraits of seven scientists on bank notes	1966	
	First Science meets Parliament (SMP)	1999	Scientists meet national politicians to explain the value of their work
	Australian Science Media Centre (AusSMC) formed	2005	An independent service to help media report science accurately
	5th World Conference of Science Journalists	2007	Conference in Melbourne

Contributors

Toss Gascoigne is a visiting fellow at the Centre for Public Awareness of Science at The Australian National University. He is interested in the interface between science and policy.

Dr Jenni Metcalfe is a science communicator, journalist and educator. She is director of Econnect Communication, and lectures and publishes internationally in science and media communication and is the co-editor of several books on science communication.

7

BRAZIL

History, significant breakthroughs and present challenges in science communication

Luisa Massarani and Ildeu de Castro Moreira

1. Introduction

From the 16th to the 18th centuries, systematic scientific activities and the communication of ideas of modern science were almost non-existent in Brazil, which at that time was a Portuguese colony of exploitation. The country had a small population, most of whom were illiterate. Furthermore, in the 18th century, the press and the publication of books were prohibited in Brazil. New scientific knowledge was accessible only to a few individuals belonging to an elite of people who had been educated abroad.

The first, yet still very limited, consistent expression of science communication in Brazil occurred in the early 19th century. It was due to an overriding political motive: the Portuguese Court[1] had arrived in Brazil and it was necessary to create suitable conditions for the administration of the metropolis and the colony. Ports were opened, the ban on printing was lifted, and the first institutions linked to science and technology (S&T) were created. These included the Real Horto [Royal Garden] (1808), the Real Academia Militar [Royal Military Academy] (1810) and the Museu Real [Royal Museum] (1818). All these institutions were located in Rio de Janeiro, which was the capital of Brazil until 1960, when the capital moved to Brasília.

1 The court moved to Brazil in 1807, when Napoleon threatened to invade Portugal. It remained there until 1821.

The Museu Real, later called Museu Nacional [National Museum] and whose images collapsing in fire in September 2018 shocked the world, was created by the Emperor Dom João VI to stimulate scientific knowledge in Brazil. Initially it hosted collections of stuffed animals, biological materials and machines, among other objects.

Since then, science communication activities have taken place in Brazil, with higher or lower intensities according to different times and initiatives, including stories in the mass media, science communication magazines and popular conferences.[2] The 1920s in particular should be highlighted, because this was when science communication was used as a tool by the embryonic scientific community (led by the then recently created Brazilian Academy of Sciences) to promote the basic sciences (see Massarani and Moreira, 2016). In 1923, in the saloons of the Academy of Science, Radio Sociedade (Society Radio) was born. It was the very first radio station in Brazil, set up only a couple of years after the first radio broadcasts in the world, and aimed to broadcast information, science content and music.

2. After World War II

New spaces for science communication emerged in Brazil after World War II (WWII) as part of a global movement in which science gained prominence. An expression of the concerns of the time was the creation of the Brazilian Society for the Advancement of Science (SBPC, from the acronym in Portuguese) in 1948. Public engagement was an important part of its activities.

Some newspapers (mainly in Rio de Janeiro and São Paulo) created science sections (Esteves, 2011), such as *A Noite* and the newspapers of the Folha Group. In the latter, José Reis—an icon of science communication in Brazil—was a columnist for about six decades (Massarani et al., 2018). Many of the stories that appeared in the newspapers described activities of the Brazilian research institutions and advocated for better conditions for science practice in the country: funding, resources, infrastructure, status, recognition.

By the 1950s, public interest in the physical sciences was awakened due to debates about the use of nuclear energy for military and civilian purposes, and also because the Brazilian scientist Cesar Lattes participated in the discovery and identification of the pi-meson in the years 1947–48. Popular newspapers and magazines with wide circulations, such as *O Cruzeiro* and *Manchete*,

2 It is worth mentioning the Popular Conferences of Gloria in the 19th century (see Correia, 1876).

published science stories highlighting activities of Brazilian institutions and researchers and the advances made in the field of nuclear energy. An example of the popular interest in Cesar Lattes's work can be found in the lyrics of the samba *Ciência e arte* [Science and Art], written by Cartola and Carlos Cachaça, for the Mangueira samba school (1948), in which Lattes and the painter Pedro Américo are honoured:

> You are my Brazil everywhere.
> Whether in science or art
> Portentous and towering ...
> There are sages like Pedro Américo and Cesar Lattes.[3]

The influence of the atomic bomb and its consequences led to many references in the Brazilian literature of this period, particularly in the poetry of Carlos Drummond and Vinícius de Moraes.

An example of best practice was the supplement *Ciência para todos* [Science for All]. It was published monthly by *A Manhã*, between 1948 and 1953. Having as editor Fernando de Souza Reis—brother of the abovementioned José Reis—it had the participation of key scientists, such as the biologist Oswaldo Frota-Pessoa (Esteves et al., 2006). Its editorial line was renowned for its appreciation of Brazilian science, for encouraging research activities and for its particularly positive outlook on science and scientists.

The *Jornal do Commercio* took a similar approach on its Sunday page *Ciência* [Science], published between 1958 and 1962 (Moreira and Massarani, 2011). Coordinated by the scientist Walter Oswaldo Cruz and supported by other scientists and professors, it emphasised the role of science for the development of Brazil. Its presentation shows the mentality of the time:

> This supplement, inaugurated today, is a mild gesture to awaken the country on this morning of its industrialization era. Its soft villager's sleep of citizens from a few centuries ago, needs to end since the machines wait them for multiplying the richness to be shared, by means of the industry. Brazil will not develop without technicians, and technicians are the human product of science. The development, the prestige, the understanding and the advancement of science will open the only path towards the growth of real economic independence of our country.[4]

3 www.vagalume.com.br/cartola/ciencia-e-arte.html.

4 *Jornal do Commercio*, 16 November 1958, Third Section, p. 3.

3. Science museums and centres: The post-war years

Science museums and science centres are important to Brazil. For more than a century, the science museums in Brazil were mostly natural history, such as the Museu Nacional in Rio de Janeiro, destroyed by fire in September 2018, and the Goeldi Museum in Pará, both created in the 19th century. They fell within the typical model of museums endowed with material objects intended for preservation, static displays and no hands-on activities.

Since the 1920s, science museums with more dynamic characteristics have been proposed. For example, Edgard Roquette-Pinto, one of the most enthusiastic science-communicator scientists of the 1920s (his voice was frequently heard on the first Brazilian radio programs) proposed a museum similar to the Deutsches Museum (Venâncio Filho, 1995). New attempts were made in the 1950s and 1960s in Rio de Janeiro and São Paulo: illustrative of those attempts is a story in a newspaper about the proposal of a Science Museum and a Planetarium in the Ibirapuera Park, in 1954.[5] The Planetarium of São Paulo, the first in Brazil, however, was set up in January 1957.

In Rio de Janeiro, the scientist Carlos Chagas Filho systematically tried to persuade stakeholders (such as the dean of the Federal University of Rio de Janeiro and state authorities) to create a designed space based on the Palais de la Découverte (France) and Chicago Museum (US), with the objective of 'spreading scientific knowledge into the public and enhancing secondary education, by providing students with basic ideas mainly with regards to practical demonstration'.[6] The museum was intended to 'present issues related to physics, genetics, nuclear power, tropical diseases and a few aspects related to oil'.[7] A working group with representatives from the Brazilian Centre for Physics Research (CBPF) and the Municipality of the Federal District was created in Rio de Janeiro and an agreement signed in November 1956 for the building of a Science Museum in Guanabara.[8] It was also to have a planetarium and an aquarium. But none of these attempts succeeded (Valente, 2008). The Planetarium of Rio de Janeiro was finally opened in 1970 by the State Secretary of Science and Technology, but the very first science museum of Brazil was created in Bahia, the Museu da Ciência e Tecnologia [Museum

5 See, for example, *Diário da Tarde*, 6 February 1954.
6 *Jornal do Commercio*, 14 May 1961, Third Section, p. 1.
7 ibid.
8 *Jornal do Brasil*, 24 May 1956; *Diário da Tarde*, 3 November 1956.

of Science and Technology] in 1979.[9] Still, due to the fact that it is not very interactive, some scholars and practitioners prefer to consider Ciência Viva [Science Alive] the first hands-on science museum. It was created in 1982 in Rio de Janeiro, in a project led by a group of scientists and educators, some of them linked to the Brazilian Society for the Advancement of Science, and was the result of a partnership with the Exploratorium in the United States.

The movement that lead to the emergence of science centres throughout the country is part of a bigger movement in the 1960s, triggered by the influence of transformations in science education in the United States and is based on the importance of experimentation for science teaching.

4. Dictatorship and resistance

In 1964, a military coup occurred. This had profound social, economic, educational and scientific effects on the country and greatly decreased the momentum of several initiatives in public engagement in science. The dictatorship severely hit sectors of the scientific community, forcing many people into exile, including scientists and students.[10]

Within that scenario, the SBPC took on an important role in government resistance. Its annual meetings clearly became the basis of an opposition movement and had political impact on the public and on the mass media (Fernandes, 1990). In favour of democracy and in defence of an alternative development for Brazil, the SBPC supported the idea of science being an important tool to overcome underdevelopment and to address the social issues faced by the country at that time.

The annual meetings of the SBPC began to play an important role in science communication, attracting thousands of students, scientists, teachers and other participants—over 20,000 people per meeting. The mobilisation around SBPC in the 1970s and 1980s resulted in groups of scientists, professors, teachers and students starting movements in several parts of the country. They organised seminars and science communication events to promote the implementation of scientific-cultural places and to create new science communication tools.

9 See www.guiadasartes.com.br/bahia/salvador/museu-de-ciencia-e-tecnologia.

10 See the special issue of *Ciência e Cultura*, 'University in dictatorship', *66*(4), 20–53, available at cienciaecultura.bvs.br/scielo.php?script=sci_issuetoc&pid=0009-672520140004&lng=pt&nrm=iso.

5. Science journalism: A post-war Latin American movement

In Brazil, the very first newspapers in the 19th century published science stories, and since then media have provided (at different levels) room for science. An important development in science journalism as a field occurred in the 1960s and 1970s, inspired by the Spaniard Manuel Calvo Hernando, together with a group of Latin Americans including Jacobo Brailovsky (Argentina), José Reis (Brazil), Arístides Bastidas (Venezuela), Sergio Prenafeta (Chile) and Antonio Cacua Prada (Colombia) (Massarani et al., 2012).

This movement led to the creation of science journalism associations in the region, in countries such as Argentina (1969), Venezuela (1971), Chile (1976), Colombia (1976) and Brazil (1977). Calvo Hernando played an important role in Brazil in the training of science journalists, delivering a course in 1972 at the University of São Paulo. During the hard days of the dictatorship, science journalism kept a low profile and did not come to the attention of the authorities as individuals in an association. According to Julio Abramczyk, who chaired the Brazilian Association after its first president José Reis, the efforts of this association in its first decade were largely geared towards attracting professional journalists.[11]

As an expression of the value of the field and in honour of José Reis, the Brazilian government created the José Reis Award for Science Communication in 1978. It is an annual award through the National Council for Scientific and Technological Development (CNPq), and is awarded alternately between journalists, scientists and institutions.

6. The recent decades

The military dictatorship ended in 1985, and Brazil reverted to a democracy with a new constitution and free elections in 1988. The science community was once again free to express its views, and the years since then have been a period of growth in science communication, particularly rich in diversified experiences. Despite this, Brazil is still far from developing an extensive and qualified program that would reflect consistent public policy support for this field.

11 Interview granted by Julio Abramczyk to Luisa Massarani on 4 January 2012.

7. Media in a post-coup era

It has been a tumultuous era for the media in the past 30 years. Threats to traditional media have come from the internet and the rise of new media, and later from the global financial crisis. Brazil has been affected like other countries. A prominent action was the creation of the *Ciência Hoje* [Science Today] magazine in 1982[12] by the SBPC. *Ciência Hoje* later expanded to include printed and online publications and a magazine for children (*Ciência Hoje das Crianças*), which had widespread circulation.[13] In the following years, other science communication magazines emerged. From 1981 to 1984, Abril Publishing launched a Brazilian version of the *Science Illustrated* magazine, published by Reader's Digest. In 1987, the same publisher created the *Superinteressante* [Super interesting] magazine, following the model of the Spanish magazine *Muy Interesante.* In 1991, the magazine *Globo Ciência,* now called *Galileu,* was launched. However commercial interests together with a distorted vision regarding what science communication really is have led to an unfortunate situation today: the scientific credibility of many of the publications is poor and often favours pseudoscience.

In 2002, the magazine *Scientific American Brazil* was launched, and it includes articles by Brazilian scientists and journalists. Recently, several research-supporting State-Foundations (FAPs or Fundação de Amparo à Pesquisa) such as Fapesp (State of São Paulo), Faperj (Rio de Janeiro), Fapemig (Minas Gerais) and Fapeam (Amazonas) began publishing science magazines.

Despite its extensive reach throughout Brazil, radio broadcasts still rarely cover S&T issues. There were innovative initiatives such as the *E por falar em ciência* [And speaking of science] program, broadcast by Radio MEC between 1992 and 1997 (Werneck, 2002). The total number of science programs is about three dozen, most from short-range university stations. The Federal University of Minas Gerais has excelled in the production of science communication programs for its radio and TV stations.

The first attempt at creating a science program for television occurred in 1979, when *Nossa Ciência* [Our Science] was broadcast by the government channel in Rio de Janeiro. It lasted only 10 episodes. *Globo Ciência* [Globo Science], a television program on the Globo channel supported by the Roberto Marinho Foundation, was more successful: it was created in 1984 and lasted

12 It is not a coincidence that the interactive science centre Ciência Viva was created in the same year; it represents the efforts and concerns of scientists and educators in engaging with society.

13 In recent years, however, the publications have been facing economic problems and challenges in keeping up to date with the new context for science communication.

until 2014. In the beginning, *Globo Ciência* had a more journalistic format, but since 1984 its ratings and audience numbers have gone up and down as it has rethought its goals and formats and the best way to communicate science.

The public TV station Cultura of São Paulo has also broadcast several science communication programs. *Tome Ciência* [Take Science] ran news and interviews on S&T issues from 1987 to 1990 on the public TVE channel. It resumed in 2004 and is currently broadcast by several university channels. Some of these programs were produced in Brazil, such as the *Minuto Científico* [Scientific Minute] (1996–97) or *Ver Ciência* [Watching Science] (2002–05), while other programs aimed at a younger public (such as *The World of Beakman,* 1994–2002) were imported and translated into Portuguese. Some state or university channels, such as the TV Educativa of Espírito Santo or the UFMG TV, have also broadcast science communication programs.

Even though there are only a few science communication–focused programs on TV, science does find space in other programs. Television is present in about 97 per cent of Brazilian households,[14] so Brazilians see science on programs such as newscasts, where science stories occupied an average of 7.3 per cent of the daily time of *Jornal Nacional* from April 2009 to March 2010 (Ramalho et al., 2012) and 3.8 per cent of the daily broadcast of *Repórter Brasil,* broadcast by the public channel TV Brazil over the same period of time (Reznik et al., 2014). Other television programs such as the variety Sunday program *Fantástico* also convey issues of S&T (Medeiros et al., 2013).

The International Festival of Science in TV, *Ver Ciência* [View Science], an annual festival of TV science programs produced in Brazil and abroad, was created in 1994. From 2004, the festival became part of the National Week of Science and Technology and had spread throughout all Brazil. Several cable channels have programs related to S&T but the access to these programs is limited to a relatively small proportion of the Brazilian population.

Although some best practices can be observed, the quality of the science presented in the mass media does not reach adequate levels, even though it has improved over recent years. Science is often presented as a spectacular occurrence, where scientific discoveries are episodic and made by particularly gifted individuals. Real or imagined applications of science are given great emphasis, but the processes of its production, its contexts, its limitations and its uncertainties are usually ignored, and simplified conceptual models of the relationship between science and the public prevail. Quality varies significantly, but in general it can be said that Brazilian science journalism is

14 National survey carried out by IBGE, in 2011.

still largely based on a limited vision of scientific activity, with little mention of the contexts of its production and of its social impacts. The image of the scientist in the media is predominantly male, and we have stereotypical roles for men and women scientists: while men go out to literally explore other worlds, women take care of health and the body. This is the case, for example, of the *Jornal Nacional* and the *Fantástico* program (Castelfranchi et al., 2014).

In the 1980s, new sections devoted to science appeared in the pages of the daily newspapers. In the following years they underwent intermittent questioning processes, that resulted in reduced or increased content. National newspapers with the largest print runs, such as *O Globo*, *Jornal do Brasil*, *Folha de São Paulo* and *O Estado de São Paulo*, and local newspapers such as the *Jornal do Commercio* in Recife, *Correio Brasiliense* in Brasília and *Zero Hora* in Porto Alegre, have provided space for the science communication and may have expert journalists writing science stories (Almeida et al., 2011). At a national level, space provided in newspapers for science stories is generally limited and there are very few journalists with adequate competence in the field.

However, some subjects have generated general public interest due to their major impact on the public (such as genetically modified food, climate change or embryonic cells) and have led to an increase of space dedicated to S&T in different sections of newspapers (Massarani et al., 2003). A significant portion of the coverage of S&T issues follows the agenda of developed countries. Several newspapers provide little space for national scientific production and do not give room for science from other Latin America countries. This emphasis on non-local science does not seem to be as prominent on television, especially on television news, which often addresses local scientific production (Ramalho et al., 2012). Discussions about S&T policies and communication about the results of public actions related to science communication are often excluded from news coverage. This causes the public sphere to be less influential than it could be in the discussion about public policies in this field.

8. Internet and new media

The use of the Internet for science communication in Brazil mainly occurs when science centres and museums, scientific institutions, research groups on science communication and a few governmental agencies become involved. There have been an increased number of bloggers, and we have observed the growing engagement of scientists and science communicators similar to Europe and United States, yet at a lower level. The *ScienceBlogs Brasil* gateway is associated with the largest gateway of science blogs in the world and has about 40 blogs.[15] The use of social media networks such as Facebook has grown and dominates communication with quick information, but its full potential is still unexplored. Recently, YouTubers such as Atila Iamarino have attracted followers on the channel Nerdologia, and Pirula, with Canal do Pirula. In 2016, the network ScienceVlogs Brasil (SvBr) was created by combining 47 channels of science videos.

The quantitative use of the Internet to search for information about S&T is already close to the use of TV. Webvideo has the potential to have a significant impact on the communication of information about S&T, and new communication tools and procedures on the horizon will certainly arise.

An important feature of the new science centres and science museums created in the country, following an international trend, was their interactivity with the public. Currently there are about 260 museums registered in the 2015 Brazilian guide of museums and science centres (Almeida et al., 2015). Most are small or medium in size and have very low levels of interactivity, and very few are big enough to receive over 100,000 visitors a year. They are mostly financed by public funds with very limited private participation.

One of the country's largest science museums is the Museu de C&T [Museum of Science and Technology] at the Pontifical Catholic University of Rio Grande do Sul in Porto Alegre. Its pre-history starts in the year 1967, with a collection of animals, rocks and minerals collected by biologist Jeter Bertoletti. In 1993, the museum moved to a new building, with a display area of 17,500 square metres. It became one of the most visited museums in the country. More recently in São Paulo, the Catavento Cultural e Educacional (2010) and Museu do Amanhã [Museum of Tomorrow] (2015) were created. The latter attracted more than 3 million visitors in less than two years. In the meantime, one of the first and most important science museums in Brazil, Estação Ciência [Science Station], closed its doors. It was created in 1987 in the city of São Paulo and linked to the University of São Paulo.

Considering the size of the country and its population, the number of scientific-cultural spaces in Brazil is still low in comparison to more advanced countries. The geographical and social distribution of museums in Brazil is very uneven, with a greater concentration in the wealthier areas in the southeastern part of the country. These institutions have a limited capacity to boost science communication, considering the size of the Brazilian population. Attendance is small, although increasing: surveys show the percentage of Brazilians who claim to have visited a science museum in the last 12 months is 12 per cent, lower than the average of European countries but triple the rate of 10 years ago, according to a 2015 survey (Moreira et al., 2015).

15 See sciencevlogsbrasil.com.br.

The survey indicated that a large number of Brazilians (41 per cent) visit locations for contact with nature, through botanical gardens, zoos and environmental parks. Very few of these spaces, however, promote science communication and environmental education programs. Policies and public actions in favour of these would engage a large number of Brazilians in activities focused on environmental education and science communication.

9. Organisations, events and public policy

From the point of view of professionals and institutions related to science communication, some organisations and events stand out. In addition to the Brazilian Association of Science Journalism, created in 1977, there exist representative bodies for zoos and aquariums (1977), botanical gardens (1991) and planetariums (1996). The Brazilian Association of Science Centres and Museums was created in 1999. At a larger scale, the Red de Popularización de la Ciencia y la Tecnología en América Latina y el Caribe (RedPOP) [Network for the Communication of Science and Technology of Latin America and the Caribbean] was created in 1990, as a network that links science communication centres and programs of the region (Massarani, 2015). The activities of some prominent Brazilian science communicators have been recognised with the Kalinga Prize, an international prize granted by the United Nations Educational, Scientific and Cultural Organization (UNESCO) for science communication. Brazilian recipients of this prize are José Reis (in 1974), Oswaldo Frota-Pessoa (1982), Ennio Candotti (1988), Ernest Hamburger (2000) and Jetter Bertoletti (2005).

From 2003, institutions, scientific organisations and stakeholder groups have urged the establishment of broader public policies in science communication. This occurred when organised movements and activities focused on valuing and strengthening communication actions.

The Department of Popularization and Diffusion of Science and Technology was established as one of the tools of this public policy in 2004. It was located within the Ministry of Science, Technology and Innovation (MCTI). This led to a significant increase in the incentives for science communication, including new programs and financial support, and national coordination. Over the following 10 years, nearly three dozen calls were made through the National Council for Scientific and Technological Development (CNPq) to support science communication projects, including the creation and development of science centres and museums, the development of science Olympiads, science fairs and exhibitions.

There are now about 10 national science Olympiads, in subjects including astronomy and astronautics, physics, mathematics, chemistry, history, health and the environment. They are organised by scientific societies and/or research institutions. The Brazilian Mathematics Olympics of the Public Schools is noteworthy for its huge scope and educational impact, since it reaches nearly 20 million students, rendering it the largest event of its kind in the world.

In recent years, many celebrations regarding the international years proclaimed by the United Nations/UNESCO have been supported by the government and carried out by scientific societies and educational and research institutions. Many initiatives were developed in the World Year of Physics (2005), the International Year of Astronomy (2009) and the International Year of Chemistry (2011), reaching millions of people throughout the country. Another program supports 'mobile science', with trucks, buses and vans designed to take science out to remote areas of the country.

Several state funding agencies launched calls for science communication over the past years and promoted initiatives within their states, in line with the federal actions and often linked to them. Among these states are Minas Gerais, Rio de Janeiro, São Paulo, Bahia, Amazonas, Rio Grande do Norte, Espírito Santo, Ceará, Maranhão and Rio Grande do Sul. But in recent years such calls are less frequent or have even been discontinued. This is in a general context where the science sector has been losing funds and status within the Brazilian government: less than 24 hours after the Temer government took over the Presidency in Brazil in 2016, the science ministry was collapsed with the communication ministry and the funds for science were cut to one-third of their 2013 levels in the budget. In 2019, a further reduction of about 40 per cent in the budget of the ministry was announced. The Department of Popularization and Diffusion of S&T had its status increasingly downgraded in the structure of the ministry; at present, the activities related to science communication and science education are split across different sectors.

The National Week of Science and Technology was created by presidential decree in 2004 and nationally coordinated by the science ministry. Its objective is to get the public, especially children and teenagers, engaged in activities related to science and to encourage creativity, scientific attitude and innovation. Universities, research institutions, schools, science centres and museums, funding agencies, the media, non-government organisations and businesses have increasingly been participating in the National Week of Science and Technology. A decentralised structure stimulates the involvement and participation of the local-level public institutions of research and education.

Another significant political milestone was reached when science communication was included in three key documents for the formulation of public policies, namely the National Plans for Science, Technology and Innovation (ST&I) for years 2007–10 and for 2011–15 as well as in the so-called *Blue Book*, a summary of the discussions of the Fourth National Conference on ST&I held in May 2010 (Livro Azul, 2010). Major challenges for the country identified by the conference were to promote a radical change in science education and to improve the actions focused on communicating science to the general public. However, the program was not put into practice.

In 2009, the CNPq, one of the leading governmental funding agencies, created a Science Communication Advisory Committee to support science communication projects. CNPq also created the Lattes Platform, an online curriculum vitae (CV) platform that records the scientific production and researchers' activities throughout the country, with a specific section for activities focused on education and popularisation of S&T.

Another important indicator of how activities of science communication are increasingly being valued is that they are now present in the program of the National Institutes of Science and Technology, created by the science ministry with several state funding agencies in 2008. The program aims to articulate and mobilise researchers, boost internationally competitive basic and fundamental research, encourage scientific and technological research, promote innovation and entrepreneurship, and establish programs that contribute to science education and public engagement with science.

Over the last few years, the interest in science communication in universities and research institutions had increased significantly. The Ministry of Education created a National Outreach Programme geared towards public institutions of higher education in 2003 to support university outreach, including science communication activities. Today in most cases only isolated individuals or small groups are still active, with little institutional support and little interaction with the institutions themselves. When we analyse the higher education offered to specialists in this field, the situation reveals itself to be even more fragile with a great dispersion of the courses, even though it is improving. On the other hand, the academic field of science communication has been growing in Brazil: a recent study identified that 51 per cent of the 609 scientific papers in science communication published in Latin America are written by Brazilians (Massarani et al., 2017).

In contrast to this general background of science communication—growing, but yet still fragile—Brazilians declare considerable interest in science issues. According to a 2015 survey on public perceptions of S&T in Brazil, the population is interested in science (61 per cent), a percentage similar to those interested in sports or economics. However, there is a great lack of knowledge regarding Brazilian scientists and Brazilian institutions: the vast majority (87 per cent) of people interviewed could not name any research institution or any important Brazilian scientist (94 per cent).

Overall, the survey showed a noticeably positive and optimistic vision with regards to the role of S&T. Castelfranchi et al. (2013) indicate that optimistic attitudes about S&T do not generally depend on the educational level or the information declared or accessed by those people. Individuals with little schooling and low levels of information generally have positive attitudes; by contrast, people with higher education and greater access to information often have diverse attitudes, being optimistic in some respects, but more critical in others. The overall positive view of the role of S&T and the expressed confidence in scientists do not prevent the public from being aware of the importance of having the whole society participating in the definition of S&T policies. However, this awareness has not lead to a more active engagement of authorities in chasing a greater social participation in decision-making.

10. Current challenges of science communication in Brazil

We have shown that, in recent years, there was an expansion of science communication activities in Brazil, despite the many major limitations.

Although rare in the country, there already exist some diagnostics on the current framework of these activities, including their scope as well as their impact (Moreira, 2006), but we do not have the in-depth analyses that would enable the drafting of consistent public policies. This would require expanded data and perspectives and more in-depth research and collective reflections, and we will not attempt to carry out such a task here. An excellent example of this type of diagnosis—which could serve as an inspiration and a model for a more global and prospective assessment of the field in Brazil—was made in the United States by the Committee on Learning Science in Informal Environments, of the National Research Council of the National Academies. This committee produced a comprehensive study on the state of the art of public communication in S&T (Bell et al., 2009).

We will limit ourselves to indicating general challenges that should be faced decisively, with adequate resources and enthusiasm from researchers and professionals, from educational and research institutions, and from federal and local authorities. Science communication activities are strongly correlated with the quality of basic education, especially science education, and here the Fourth National Conference on ST&I indicated the need for a real revolution, given the enormous shortcomings in the scientific education of most young Brazilians (Livro Azul, 2010). Such activities can contribute to improving the teaching of science, with emphasis on methods and practices that enhance and promote creativity, experimentation and inter-discipline.

The main challenges for science communication can be divided into three parts. The first one is to reach the entire Brazilian population. The objective would be to reach middle sectors and poor and excluded sectors—both in urban and rural areas: a total of over 150 million people. This challenge requires time, resources and a lot of trained people. An important contributing action would be to promote science communication in a network. This would improve and integrate science centres and museums and other stakeholders. It would tackle the problem of an unequal distribution of effort, in both a regional and social sense, and encourage a closer interaction with the formal education system. Public events for science communication, such as the National Week of S&T, should be extended to all corners of the country: those involved with the public communication of science should go where the people are. Another aspect, within the goal of a greater social engagement, is to use the mass media in a much more intense and qualified way, particularly the Internet and social networks. There is the need to greatly expand science communication in the public channels and within the research institutions.

Public policies constitute the second part of the challenge for the development of science communication. Despite the advances of the last decade public policies are still vulnerable to change in political attitudes. Existing structures need to be strengthened and new stimulatory policies created, and to do this both continuity and expansion of means and resources are needed.

Science communication is already part of several government agendas and many education and research institutions. Politically, however, this has not been done to the necessary extent, and the level of investment is not enough to get a socially broad, inclusive and qualified performance (Moreira, 2006; Ferreira, 2014). A drastic decrease in the stifling bureaucracy that plagues S&T and its communication in the country is not a minor challenge. Administrative processes need to be simpler and clearer.

The third aspect of the challenge is to improve the quality of today's science communication. One goal is to increase and improve the training of science journalists, science communicators and scientists, as well as academically valuing these activities. Undergraduate students, particularly those in the fields of science and engineering, should be given more encouragement to participate in public communication actions of S&T. With regards to science centres and museums, some of the permanent challenges are to incorporate new technologies into their practices, to increase accessibility, to encourage public engagement in big issues of science and society, and to improve the training of their staff (Marandino, 2005). It is hoped that innovative initiatives are stimulated and supported, such as the new path for natural science museums, linked with environmental and social concerns.[16]

We need more studies and analyses to guide science communication activities. Outreach activities of S&T are still in their early stages, and greater understanding of the best strategies, practices and impacts of the outreach activities are required. Similar research should examine the characteristics, attitudes and expectations of the audience and on public perceptions of S&T. Academic assessment of education activities, particularly in the field of science communication, is still low, despite recent actions of universities and funding agencies; these actions should be encouraged. Cooperation with other countries, through partnerships with groups and researchers from abroad who conduct research and high-quality science communication can also provide advances and significant improvements for this field in Brazil, particularly in the training of qualified personnel.

Cultural aspects are important in any communication process, and an improved interface between art, science and society should be taken into consideration in science communication activities. Another goal is to recognise the value of the cultural and humanistic aspects of science, as well as the acknowledgement of popular and traditional knowledge. Citizen science, which favours the collective learning and construction of knowledge and its social appropriation, has significant potential, indicating the role of scientific education integrated with a public engagement with science.

One of the most effective ways to educate the young and the public about science is to put them in the role of researchers and make them use, even at a restricted level, the methods of science in an effective dialogue between science, its actors and the public.

11. Final considerations

There is still a long way to go until we can declare the existence of high-level science communication and an adequate social appropriation of scientific and technological knowledge encompassing all of Brazilian society. But the journey began long ago, with faltering steps in previous decades. One of the biggest hurdles in science communication in Brazil—actually, in science

16 A nice example is the Museum of Amazonia (MUSA), in Manaus.

in general and other aspects—is the vulnerability of the initiatives, actions and policies. Much of what was built has been destroyed. Our field requires a continuity of action and purpose in which tiredness is not an option. Overcoming the great challenges outlined here is a fundamental collective task, depending on the creation of effective public policies. But also, and most importantly, it depends on the collective action of scientists, teachers, science communicators, journalists, museum experts, students, and all the people involved with scientific work and its communication.

As João Cabral de Melo Neto beautifully poeticised: 'A rooster on its own does not weave a morning; it will always need other roosters ... so that the morning, starting from a fine web, can be woven among all the roosters'.[17]

References

Almeida C., Brito, F., Ferreira, R., Massarani, L. and Amorim, L. (2015). *Centros e Museus de Ciência do Brasil.* Rio de Janeiro: ABCMC, Casa da Ciência/UFRJ e Museu da Vida.

Almeida, C., Ramalho, M., Buys, B. and Massarani, L. (2011). La cobertura de la ciencia en América Latina: estudio de periódicos de elite en nueve países de la región. In Moreno C. (ed.), *Periodismo y divulgación científica. Tendencias en el ámbito iberoamericano* (pp. 75–97). Madrid: OEI and Biblioteca Nueva. doi.org/10.22323/3.01010201.

Bell, P., Lewenstein, B., Shouse, A. and Feder, M. (eds). (2009). *Learning Science in Informal Environments: People, Places, and Pursuits.* Washington DC: National Academies Press. doi.org/10.17226/12190.

Castelfranchi, Y., Massarani, L. and Ramalho, M. (2014). War, anxiety, optimism and triumph: a study on science in the main Brazilian TV news. *Journal of Science Communication, 13*(3), A1. doi.org/10.22323/2.13030201.

Castelfranchi, Y., Vilela, E., Moreira, I. and Massarani, L. (2013). As opiniões dos brasileiros sobre C&T: o paradoxo da relação entre informação e atitudes. *História, Ciências, Saúde-Manguinhos, 20*(supl.), 1163–83. doi.org/10.1590/S0104-59702013000400005.

Correia, C. M. F. (1876). *Conferências Populares. 1876.* Rio de Janeiro: Typ Imp Const J Villeneuve & C.

17 In Portuguese: 'um galo sozinho não tece uma manhã, ele precisará sempre de outros galos ... para que a manhã, desde uma teia tênue, se vá tecendo, entre todos os galos'.

Esteves, B. (2011). Os cientistas vão à imprensa: divulgação científica nos jornais brasileiros (1945–1964). In L. Massarani, C. Jurberg and L. De Meis (eds), *Um gesto ameno para acordar o país: a ciência no 'Jornal do Commercio' [1958–1962]* (pp. 13–23). Rio de Janeiro: Museu da Vida / Fundação Oswaldo Cruz.

Esteves, B., Massarani, L. and Moreira, I. (2006). Ciência para Todos e a divulgação científica na imprensa brasileira entre 1948 e 1953. *Rev Soc Bras Hist Ciênc, 4*(1), 62–85.

Fernandes, A. M. (1990). *A construção da ciência no Brasil e a SBPC.* Brasília: Editora UnB.

Ferreira, J. R. (2014). Popularização da Ciência e Tecnologia e as Políticas Públicas no Brasil (2003–2012) (PhD thesis). Programa de Pós-Graduação em Ciências Biológicas – Biofísica, IBCCF / UFRJ.

Livro Azul. (2010). *IV Conferência Nacional de Ciência, Tecnologia e Inovação.* Brasília: CGEE and MCT. Retrieved from www.cgee.org.br/documents/10195/734063/livroAzul_digital_18jan2011_6990.pdf/68c79d24-d589-42f5-ac66-4d6d728f9691?version=1.3.

Marandino, M. (2005). A pesquisa educacional e a produção de saberes nos museus de ciência. *História, Ciências, Saúde – Manguinhos, 12*(supl.), 161–81. doi.org/10.1590/S0104-59702005000400009.

Massarani, L. (ed.). (2015). *RedPOP: 25 años de popularización de la ciencia en América Latina.* Rio de Janeiro: RedPOP, Unesco, Museu da Vida.

Massarani, L., Amorim, L., Bauer, M. and Montes de Oca, A. (2012). Periodismo científico: reflexiones sobre la práctica en América Latina. *Chasqui, 120*, 73–7.

Massarani, L., Burlamaqui, M. and Passos, J. (2018). *José Reis, caixeiro-viajante da ciência.* Rio de Janeiro: Casa de Oswaldo Cruz.

Massarani, L. and Moreira, I. (2016). Science communication in Brazil: A historical review and considerations about the current situation. *Anais da Academia Brasileira de Ciências, 88*(3), 1577–95. doi.org/10.1590/0001-3765201620150338.

Massarani, L., Moreira, I. and Magalhães, I. (2003). Quando a genética vira notícia: Um mapeamento da genética nos jornais diários. *Ciênc Ambient*, 26, 141–48.

Massarani, L., Rocha, M., Pedersoli, C., Almeida, C., Amorim, L., Cambre, M., Nepote, A. C., Rocha, J. N., Aguirre, C., Gonçalvez, J. C., Cordioli, L. and Ferreira, F. (2017). *Aproximaciones a la investigación en divulgación de la ciencia en América Latina a partir de sus artículos académicos.* Rio de Janeiro: RedPOP e Casa de Oswaldo Cruz.

Medeiros, F., Ramalho, M., Caldas, C. and Massarani, L. (2013). Ciência e tecnologia em um programa de infotainment: uma análise de conteúdo da cobertura do Fantástico. *InterCom: Rev Bras Ciênc Comum, 36*(1), 127–47. doi.org/10.1590/S1809-58442013000100007.

Moreira, I. (2006). A inclusão social e a popularização da ciência e tecnologia no Brasil. *Inclusão Social, 1*(2), 11–16.

Moreira, I., Castelfranchi, Y., Falcão, D., Badaró, A. and Massarani, L. (2015). *A ciência e a tecnologia no olhar dos brasileiros – Percepção pública da C&T no Brasil – 2015.* Brasília: Centro de Gestão e Estudos Estratégicos.

Moreira, I. and Massarani, L. (2011). A divulgação científica no Jornal do Commercio (1958–1962). In L. Massarani, C. Jurberg and L. De Meis (eds), *Um gesto ameno para acordar o país: a ciência no 'Jornal do Commercio' [1958–1962]* (pp. 43–62). Rio de Janeiro: Museu da Vida / Fundação Oswaldo Cruz.

Ramalho, R., Polino, C. and Massarani, L. (2012). From the laboratory to prime time: science coverage in the main Brazilian TV newscast. *Journal of Science Communication 11*(2), A1. doi.org/10.22323/2.11020202.

Reznik, G., Massarani, L., Ramalho, M. and Amorim, L. (2014). Ciência na televisão pública: uma análise do telejornal Repórter Brasil. *Alexandria, 7*(1), 157–78.

Valente, M. E. (2008). Museus de Ciências e Tecnologia no Brasil: uma história da museologia entre as décadas de 1950 – 1970 (PhD thesis). Universidade Estadual de Campinas. doi.org/10.20396/revpibic2720191726.

Venâncio Filho, A. (1995). *Francisco Venâncio Filho: Um educador brasileiro 1894–1994.* Rio de Janeiro: Nova Fronteira.

Werneck, E. (2002). E por falar em ciência … no rádio' In L. Massarani, I. Moreira and F. Brito (eds), *Ciência e público: caminhos da divulgação científica no Brasil* (pp. 79–88). Rio de Janeiro: Casa da Ciência/UFRJ.

Timeline

Event	Name	Date	Comment
First interactive science centre established.	Museu da Ciência e Tecnologia [Museum of Science and Technology] in Bahia	1979	The more interactive Ciência Viva [Science Alive] was created in 1982 in Rio de Janeiro
First national (or large regional) science festival.	National Week of Science and Technology	2004	
Association of science writers or journalists or communicators established.	Brazilian Association of Science Journalism	1977	
First university courses to train science communicators.	A one-year diploma course in science journalism at the State University of Campinas	1999	There were earlier short courses

Event	Name	Date	Comment
First PhD students in science communication graduate.	Probably Wilson Bueno	1987	
National government program to support science communication established.		2004	
A journal completely or substantially devoted to science communication established.	*Ciência e cultura*	1949	
First significant radio programs on science.	*Radio Sociedade*	1923	
First significant TV programs on science.	*Nossa Ciência* [Our Science] was broadcast in Rio de Janeiro	1979	The more successful *Globo Ciência* [Globo Science], ran from 1984 until 2014
First awards for scientists or journalists or others for science communication.	José Reis Award for Science Communication, CNPq	1978	
Date hosted a PCST conference.	Salvador, Brazil	2014	

Contributors

Dr Luisa Massarani is coordinator of the Brazilian National Institute of Public Communication of Science and Technology, and researcher and science communicator at the Oswaldo Cruz Foundation.

Dr Ildeu de Castro Moreira teaches a postgraduate program in the History of Science and Physics Education at the Federal University of Rio de Janiero.

8

CANADA

One country, two cultures: Two routes to science communication

Michelle Riedlinger, Alexandre Schiele and Germana Barata

1. Introduction

This chapter provides an account of modern science communication in Canada, including historical factors influencing its development, and the development of the distinct Province of Quebec. Canada is a bilingual country. Over 24 million Canadians (74 per cent of the population) speak English at home, and nearly 8.2 million (23 per cent) speak French. Over 5.7 million people (71 per cent of the population) in Quebec speak only French at home (Statistics Canada, 2017). Britain's conquest of New France[1] in the 1760s and Canada's Westminster democratic governance structures and institutions and membership of the Commonwealth significantly shape the country's activities. The 10 Canadian provinces retain several exclusive rights that the Canadian parliament and the federal government cannot infringe upon, and provinces and territories actively work to preserve their cultural distinctiveness through these rights. Quebec's efforts for political, social, economic and cultural emancipation, starting in the interwar years, continue to influence the lives of Canadians in and outside Quebec. Canada also continues to feel the social, cultural, economic and political impacts of its largest neighbour and trade partner, the United States (Dubas and Martel, 1973; Levere, 1988).

1 New France was the area colonised by France in North America from 1534. It ended with the cession of New France to Great Britain and Spain in 1763 under the Treaty of Paris.

The chapter is divided into two main parts, the history of science communication in Canada, followed by an account of its development in Quebec. Science communication in Canada cannot be understood without considering the 'Two Cultures', with few influences or contact between Quebec and the rest of Canada. Science communication in Quebec largely developed on its own, and later influenced the development of science communication in France and other Francophone nations. Significant space will be devoted in this chapter to its development.

2. Science communication approaches in Canada

2.1. Early approaches to science communication in Canada

Like many other industrialised societies, the first half of the 19th century in Canada was marked by the activities of a commercial and professional bourgeoisie steeped in the cultural and intellectual ideas of the Enlightenment. The thrust of science communication initiatives came from learned societies and a collection of individual initiatives. Among these societies were the Literary and Historical Society of Quebec (1824), the Natural History Society of Montreal (1827) and the Mechanics' Institute (1830). While aimed at the elite, these societies contributed to science communication through talks at libraries and natural history museums, aiming to complement humanist education with science culture (Schiele et al., 1994). Since Montreal was the economic capital of Canada at the time, it is difficult to distinguish Canadian science communication initiatives from specifically Quebec initiatives during those early years. In the early part of the 19th century, the political capital of Canada shifted often, before being fixed at Ottawa on the now Ontario–Quebec border. However, Montreal remained the economic capital well into the 20th century.

Canada's learned societies were created following the reorganisation of British North America as a confederation in 1867. Governor-General, the Marquess of Lorne, founded the Royal Society of Canada in 1882. He suggested it could bring Canada's intellectual centres together if they were modelled on the British Royal Society and the French Académie des Sciences (Levere, 1988). In 1916, the Canadian government created the National Research Council (NRC) of Canada, to promote Canadian scientific progress to the public, along with making investments in science and technology research. World War I provided a push for physics and engineering development, funded in part by university and industry research scholarships for postgraduate studies.

In its aftermath, Canada experienced a rise in the number of young researchers trained in Canadian universities, and publications and communication through scientific events supported by the Royal Society of Canada (Gingras, 1991). The Royal Canadian Geographical Society published the first issue of *Canadian Geographic* in 1930 (RCGS, 2018).

During World War II, Canada collaborated in building atomic bombs, and was caught up in the international movement to simultaneously develop, promote and regulate science. As intervention by federal and provincial governments in scientific research grew, science policy became an important issue for the nation (Gingras, 1991).

2.2. National and provincial government policies and programs

In Canada, national policy is made by the Canadian parliament. However, Canadian provinces retain some responsibilities and enjoy a number of exclusive rights upon which the Canadian parliament and federal government cannot intrude: property and civil rights, administration of justice, natural resources and the environment, education, health, welfare, and municipalities. Provincial activities exist in tandem with efforts stemming from federal government programs and national entities, and provincial and territorial governments often resist federal interference. Each province has its own objectives, interests and practices when it comes to science communication.

Eighty-two per cent of government resources devoted to science communication are expended in the four most populous provinces in Canada (Ontario, Quebec, British Columbia and Alberta). Quebec was the first province to create a science policy, in the 1970s (Dufour and Gringas, 1988); and in 1980 was the first province to set goals for the diffusion of science and technology, and support democratic public debates on important issues (Dufour and De la Mothe, 1993).

A 2011 inventory completed by Schiele, Landry and Schiele of informal science communication activities in Canada catalogued over 700 programs or organisations managing initiatives associated with science outreach, science journalism or science education. According to the Council of Canadian Academies (2014), Quebec, Yukon and the Northwest Territories governments have established visions for science promotion and activities aimed at engaging youth and adults with science. Ontario, Saskatchewan and Alberta focus their financial support on skills development to build workforce capacity and encourage innovation for regional industries. Manitoba and the Atlantic Provinces support science outreach programs that complement their provincial science curricula.

Provincial programs have expanded or been adapted at the national level. For example, Canada's National Science and Technology Week, founded in 1990, was modelled on events already happening in Quebec and British Columbia. *Science Rendezvous*, a community science festival, started in the Greater Toronto Area in 2002 and is now the largest science festival in Canada, with events in over 23 cities (Science Rendezvous, 2018).

Federal and provincial government support for science and science communication has waned during the 21st century. Scientific literacy programs and activities can still be found in all Canadian provinces and territories, but with reduced federal and provincial government funding. Science communication in Canada has diversified and relies more heavily on industry, university and wider public support mechanisms. A number of independent research institutions are now well-respected contributors to science communication in Canada.

2.3. The emergence of professional science journalism

After World War II, scientists and scientific societies in Canada increased their roles as science popularisers and Canada followed the United States in professionalising science journalism. The Canadian section of the US-based National Association of Science Writers (NASW) gained its autonomy in 1970 as the Canadian Science Writers Association (CSWA) (Visser-deVries, 2015).

The 1970s through to the early 1980s saw different publishers entering the market with science-based magazines. Prominent examples include *Science Forums, OWL* and *Science Dimensions*, but none managed to establish themselves on a commercial basis and all were relatively short-lived. The Quebec-based French-language *Québec Science* (1962) and *Les Petis Débrouillards* (1981) are the only science magazines to have found commercial success. Canada's science writing efforts were recognised internationally in 2002 when science writer and broadcaster Véronique Morin was elected as the first President of the World Federation of Science Journalists.

Canada has two well-recognised and long-running television series devoted to science that first aired on Canadian television in 1960: Canada Broadcasting Corporation's (CBC) *Nature of Things* and its French-language equivalent *Aux frontières de la science* (Council of Canadian Academies, 2014). Canada's longest-running radio show *Quirks and Quarks* first aired on CBC Radio in 1975 and is currently broadcast in over 40 countries. Three well-known Canadian science communicators have hosted the show: David Suzuki (1975–79), Jay Ingram (1979–91), and Bob McDonald (1992–present).

Michael Ryan and Sharon Dunwoody conducted one of the first surveys of professional Canadian science communicators in 1975. They surveyed 152 members of the National Association of Science Writers who resided in the US or Canada. Respondents recommended 'aspiring science writers' take more classes in science (in preference to journalism) and practice science writing through part-time jobs or summer jobs, which might suggest a lack of science writing courses that balanced both skills. The survey did not report on where respondents were employed or the kind of work they did.

2.4. Science literacy efforts in Canada and creating a science culture

Like many industrialised countries during the 1980s and 1990s, Canada's investment in science communication was heavily influenced by the Organisation for Economic Co-operation and Development (OECD), and OECD work could be considered one of the main drivers of modern science communication in Canada (Pitre, 1994). In 1984, the Canadian government introduced its first national science and technology policy, the Public Awareness Programme for Science and Technology, subsequently redefined and renamed as Science and Culture Canada in 1987 (Dufour and De la Mothe, 1993). The associated Science and Culture Canada Grants scheme had dual aims of supporting economic development and national unity through scientific literacy efforts. Commentators have argued that this policy was only possible because it recognised provincial autonomy: the focus was provincial and national cooperation, and it emphasised province-building initiatives (Dufour and Gringas, 1988).

Two notable national surveys of science literacy and attitudes towards science have been conducted in Canada. In 1989, Edna Einsiedel conducted a telephone survey of over 2,000 Canadian adults assessing their scientific literacy and attitudes towards science and technology (Einsiedel, 1994). Einsiedel was one of the first researchers to investigate the relationship between knowledge and attitudes using survey data (Bauer, Allum and Miller, 2007). She found that high scientific literacy levels in the Canadian survey population were positively correlated with positive attitudes towards science, particularly trust in science and feelings of efficacy in relation to science. In 2014, a second survey on science culture with 2,004 respondents was conducted by EKOS Research Associates Inc. on behalf of the Council of Canadian Academies (CCA). It showed that, compared with other countries, Canadians' attitudes towards science and technology are generally positive and that people have few reservations about science. Ninety-three per cent of the Canadians surveyed reported an interest in science and technology.

Canadian's factual knowledge about science has improved since Einsiedel's 1989 study. Both surveys used the same seven knowledge-related questions, and findings demonstrated an improvement in the science knowledge of survey respondents for each question. Authors of the 'science culture' report for 2014 attribute this to an increase in the level of education. However, it is important to consider the impact of public attitudes on levels of scientific literacy. Findings of a subsequent survey conducted in 2016 by the Ontario Science Centre show that, while most of the 1,578 respondents were confident about their level of science knowledge, their understanding of controversial topics that divide the Canadian population such as climate change, vaccinations and genetically modified organisms was variable (Montgomery, 2016; Ontario Science Centre, 2016). For example, 89 per cent claimed they understood the science behind vaccinations, but 19 per cent also said (erroneously) that there were links between vaccinations and autism. The Quebec Provincial Government has also produced several studies and reports on the state of science culture in Quebec, and these are discussed in the section on Quebec in this chapter.

2.5. Popular science: Journalists, scientists and the media

Early national and provincial government policies recognised the economic potential of science and technology and guided the science communication research accordingly (Schiele and Landry, 2012). One of the earliest studies, *A Research Study on Science Communication*, conducted in 1973 by Orest Dubas and Lisa Martel, focused on science media reporting. The researchers found that Canada had around two dozen full-time science and technology reporters and as many as 200 reporters who covered at least some stories related to science and technology. The researchers used data compiled by Margaret Brasch from the Journalism Department at Carleton University, and information provided by CSWA (Dubas and Martel, 1973). Building on traditions of media research conducted in the US (see, for example, Friedman, 1986; Dunwoody, 1980; and Nelkin, 1995), Canadian researchers identified how science news develops, how it is reported through various channels, and the various audiences for science news. Through content analyses of Canadian newspapers and surveys with Canadian journalists, researchers have found that 'hard' science news stories in Canada were mostly framed internationally, positively and drawn from wire services (Einsiedel, 1992; Saari, Gibson and Osler, 1998).

Raymond Duchesne (1981) was one of the first Canadian researchers to critique studies of science media reporting in Quebec, arguing that the media representations of science in Quebec served dominant interests and lacked critical perspective. A few years later, Chris Dornan came to the same conclusion for Canadian Anglophone media (Dornan, 1988, 1990).

But writing for the popular media did not always achieve the policy impact that scientists desired. In more recent research, Bentley and colleagues (2011) analysed academic attitudes to popular science publishing. In a comparison of academic authors in 13 countries, they found that Canadians were some of the most productive, publishing more popular science articles in newspapers and magazines than academics in most other countries. But it did not always work for the scientist-authors: John Besley and Kathryn O'Hara drew on survey work to investigate the attitudes of 1,142 Canadian researchers who had received a federal research grant. Using survey questions administered previously in the US with American Association for the Advancement of Science (AAAS) members, they found that Canadian researchers were extremely concerned that policymakers were not using scientific evidence in their decision-making. Canadian researchers wanted to see an impact from their own work rather than just seeing it communicated (Besley and O'Hara, 2018).

2.6. Canadian journalism awards

Canada has a number of awards dedicated to excellence in science media and communication practices. In 1973, the CSWA launched the Ortho Award for print science journalism, and over the years the program has expanded to include awards for radio, television, magazine, newspaper, trade publication, books and the Herb Lampert student writing award. In 1981, the awards were renamed the Science in Society Journalism Awards and were sponsored by government and the private sector. In 1981, Association des communicateurs scientifiques (ACS) created the Fernand-Séguin scholarship to encourage reporting excellence in young science journalists. In 1982, the Royal Canadian Institute created the annual Sandford Fleming Award to recognise Canadians who make outstanding contributions to the public understanding of science. The Royal Society of Canada established the McNeil Medal in 1991, awarded to an individual for the ability to promote and communicate science to students and the public within Canada. Finally, the Natural Sciences and Engineering Research Council of Canada launched the annual Awards for Science Promotion in 2001, to honour individuals and groups who make outstanding contributions to the promotion of science in Canada.

2.7. From scientist to science broadcaster: Fernand Séguin and David Suzuki

Fernand Séguin (1922–86) in Québec and David Suzuki (1936–) in English Canada are examples of scientists turned science broadcasters who achieved national prominence, in different contexts and one generation apart. Séguin started his professional life as a biochemist at Université de Montréal in 1945, and Suzuki as a geneticist at the University of British Columbia in 1963.

Very early in his career (1947), Séguin joined the radio program *Radio-Collège* at Radio-Canada, continuing after its transition to television in 1954. Putting an end to his academic career and devoting himself full time to journalism from the mid-1950s, Séguin scripted his radio and television appearances (*L'école buissonière*, 1955; *La joie de connaître*, 1955–57; *Le roman de la science*, 1957–60; *Les frontières de la science*, 1960–61; *Sciences réalité*, 1975–78). From the 1960s, he developed his own radio and television programs (*Connaissances d'aujourd'hui*, 1965–66; Sel *de la Terre*, 1965–70; *Magazine Science*, 1970–71; *La Science et vous*, 1971–79).

By comparison, Suzuki waited nearly 10 years to make the transition to broadcasting, taking sabbatical leave to host the TV program *Science Magazine* at CBC and the radio program *Quirks and Quarks*. Suzuki was criticised by some peers for taking sabbatical leave to pursue activities unrelated to advancing scientific knowledge (Dornan, 1987). In 1979 he moved over to television to host *The Nature of Things*, motivated to move because of the reach of television and the need for an informed Canadian public that could engage in democratic decision-making (Suzuki, 2006). He worked with the British Broadcasting Corporation (BBC) and the Public Broadcasting Service (PBS) on *The Secret of Life*, which aired in 1993, and with the Discovery Channel on *The Brain*, which aired in 1994. He has authored or co-authored more than 50 books, including 20 books for children. Suzuki is also a well-known critic of government inaction on environmental protection. He remained an academic until his retirement in 2000.

Séguin and Suzuki have received numerous awards for their communication work. Both received the title of Officer of the Order of Canada, and both were elevated to Companions of the Order of Canada (Séguin in 1978, elevated in 1988; and Suzuki in 1977, elevated in 2005). They also received the international UNESCO Kalinga Prize for the Popularisation of Science for their careers as broadcast communicators; Séguin in 1977 and Suzuki in 1986.

Figure 8.1: Canadian science broadcaster and environmental activist David Suzuki.

Source: David Suzuki Foundation, davidsuzuki.org.

Figure 8.2: Canadian science broadcaster Fernand Séguin, 1961.

Source: Archives UdeM, Fonds Fernand Séguin, cote P0241/1fp,05156.

While Séguin remained a committed science educator, Suzuki became an environmental activist. In 1990, he co-founded the David Suzuki Foundation to provide 'evidence-based research, education and policy analysis' in order to 'work to conserve and protect the natural environment, and help create a sustainable Canada' (David Suzuki Foundation, 2018, para. 4). But Suzuki attracted significant criticism in Canada: he announced in 2012 that he was stepping down from the board of his own foundation because of political and media pressure that was putting the charitable status of his organisation at risk (Stoymenoff, 2012). Most recently, business leaders, donors, faculty members and Premier of Alberta Rachel Notley criticised the University of Alberta for awarding Suzuki an honorary degree because he was a critic of the Alberta tar sands development (Bennett, 2018).

2.8. Science centres and museums

In tandem with a focus on science media reporting, Canada has a long tradition of communicating science through science centres. The Council of Canadian Academies (2014) notes that most provinces support science centres, parks and museums to some degree. In 1967 the Canada Museum of Science and Technology opened in Ottawa; in 1969, the Ontario Science Centre opened in Toronto. The Ontario Science Centre was Canada's first interactive science centre, modelled on San Francisco's Exploratorium and Detroit's Museum of Science and Technology. Science centres in other provinces and territories followed. The Canadian Council of Science Centres (CCSC) (now the Canadian Association of Science Centres (CASC)), was established in 1985 as a national network of science centres in Canada. It represents more than 40 science centres, museums, aquariums and planetariums. Canadian research into the management and design of these centres, audience reception and appropriate of content has made a valuable contribution to the field internationally (see, for example, Schiele, 2008, 2014; Schiele and Koster, 2000).

2.9. Training and research at Canadian universities

Before formal courses in science communication were established, academic research on science communication was supported through other disciplinary programs. Canada's first PhD thesis on a 'science communication' topic was written by Bernard Schiele, a graduate student in education at the University of Montreal in 1978, under the title: *Incidence télévisuelle sur la diffusion des connaissances scientifiques vulgarisées: Science-Réalité, un cas particulier de vulgarisation scientifique* [Televisual incidences on the diffusion of vulgarised scientific knowledge: Sciences-Réalité, a specific case of scientific vulgarisation]. The first master's thesis was produced in 1984 by Suzanne Champagne, who graduated from the Faculty of Social Sciences at Laval University, Quebec City. Her thesis was titled: *La vulgarisation scientifique, ses agents, ses adeptes: le cas du magazine Québec science* [Scientific vulgarisation, its agents, its devotees: The case of the magazine *Quebec science*].

The first accredited science communication qualification in Canada is a multidisciplinary graduate diploma offered in partnership between Science North and Laurentian University in Sudbury, Ontario, since 2005. In 2010, Mount Saint Vincent University created the first undergraduate degree in science communication. The first master's in science communication was launched in 2017 at Laurentian University. This points to the very recent emergence of science communication as an umbrella term for a number of recognised professions. Universities in other provinces, including Quebec, Nova Scotia and British Columbia, offer science communication training as part of other accredited programs. Since 1994, the University of Toronto has offered a master's program in Biomedical Communications, while Carleton University recently acquired a Chair in Digital Science Journalism. Science journalism research and teaching programs are well established at Concordia University (since 2008) and the University of British Columbia (since 2010).

Canadian researchers actively participate in international debates on science communication. An analysis of English-language science communication research output worldwide, distributed through the Web of Science database

(1976–2015), shows that Canada is among the top five countries publishing on science communication and the public, along with the US, the UK, Germany and Australia. A similar pattern is identified in contributions to three of the most relevant international English-language journals in science communication—*Science Communication, Public Understanding of Science and Technology* and *Journal of Science Communication*—where Canadian researchers are the fifth most prolific, with 64 papers published between 1994 and 2015 (Barata, Caldas and Gascoigne, 2018). In terms of general science research output, Canada is in seventh place for the number of publications distributed through Scopus Database (1996–2017) and the Web of Science (Schnell, 2017).

2.10. Citizen involvement in science communication

Communication researchers have recorded a shift in science communication to more dialogic models of communication, starting at the beginning of the 20th century (Einsiedel, 2008). Canada has taken part in this shift, but responded later than other industrial societies to calls for policies to support greater citizen participation in decision-making processes or 'upstream engagement'. Canada's first consensus conference was held in Alberta in March 1999 on the topic of food safety (Einsiedel and Eastman, 2000). This event was thought to be too controversial to receive federal ministry support (Einsiedel, Jelsøe and Breck, 2001) but was funded through a range of other sources: Canada's national social sciences research grant body, a provincial government community grant, the University of Calgary and two non-government organisations (the National Institute of Nutrition, and the Food Biotechnology Communications Network).

During the 2000s, governments at various levels supported commissions, citizen conferences and science cafés on health risks including genetic modification and assisted human reproduction. Canadian governments, research institutions and non-profit organisations have come to recognise lay expertise and the co-production of knowledge through government and citizen science initiatives supported by non-profit organisations. For example, Canada's *NatureWatch* web portal, launched in 2000, helps recruit citizen scientists to track changes in the environment through a number of programs: *FrogWatch, PlantWatch, WormWatch, IceWatch* and *MilkweedWatch*. Canada's national web-based portal for citizen science was recently launched on Science.gc.ca. It contains links to 20 Canadian projects requiring citizen input including earthquake monitoring, flu tracking, agriculture and climate condition changes, tree-disease outbreaks, and shark and whale sightings (Government of Canada, 2018).

2.11. Government attempts to muzzle scientists and grassroots resistance

In 2010, Canada came to world prominence for the federal government's attempts to muzzle scientists on controversial issues such as climate change (Turner, 2013). Organisations including the CSWA were instrumental in raising awareness about this issue and advocating for scientists' freedom to speak (see, for example, O'Hara, 2010). In the same year, the Canadian Science Media Centre (SMC) opened as a not-for-profit entity to address the media's lack of access to Canadian scientists and to provide expert support for non-specialist reporters. The non-profit science advocacy group, Evidence for Democracy, was formed in 2012 after thousands of Canadians marched in Death to Evidence rallies calling for more evidence-based decision-making for a strong Canadian democracy (Evidence for Democracy, 2013; Turner, 2013). Evidence for Democracy organised Stand Up for Science rallies in September 2013 in 17 Canadian cities. In 2013, the Professional Institute of the Public Service of Canada (PIPSC) surveyed its scientist members employed in federal institutions and found that 90 per cent of them felt they were not allowed to speak freely to the media about their work (PIPSC, 2013). In 2014, PIPSC and the United States Union of Concerned Scientists (UCSUSA) coordinated a letter signed by 815 scientists from 32 countries to Prime Minister Stephen Harper urging him to lift media communication restrictions on federal researchers and stop drastic funding cuts to the federal science budget (An Open Letter on Science to Canadian Prime Minister Stephen Harper, 2014). According to commentators, science was a major campaign issue prior to the 2015 federal election (see, for example, Halpern, 2015) and attracted extensive media coverage and all-party debates on science policy (CBC, 2015). In 2017, PIPSC launched a new survey and concluded that while there had been some progress, the change of government had not rectified the 10 years of damage to science communication (PIPSC, 2017).

2.12. The impact of new and emerging technologies on science communication

Coupled with government ambivalence towards science communication over the last decade, Canada has experienced the impact of new and emerging technologies and changing economic conditions. These changes have reshaped the mainstream media landscape in many parts of the world, including Canada, and the effects have been exacerbated by neoliberal agendas. The changes and their impacts on Canadian journalism were captured in the Canadian survey report *The Shattered Mirror* (2017). The survey found that Canadians prefer to be informed through the media but on their own timelines and with little or no cost to themselves.

Canada's science media have responded to new media in many ways. For example, in 2005, CBC's *Quirks and Quarks* became the first major CBC radio show to be made available as a free podcast. Canada's very active blogging community has been developing from the early 2000s, and recent digital initiatives are helping redefine what independent science communication looks like. These initiatives include *Science Borealis*, launched in 2013 (Science Borealis, 2018), *Hakai Magazine* launched in 2015 (Hakai Magazine, n.d.), and *The Conversation Canada* launched in 2017 (The Conversation Canada, 2018). Twitter, Instagram and YouTube are also supporting a growing number of science communicators engaging a diverse range of publics in digital spaces. For example, in 2013 Canadian astronaut Chris Hadfield used social media (Twitter and YouTube) to communicate about life on board the International Space Station. Over 392,000 people followed him on Twitter while he was aboard the Space Station (Strauss, 2018), and in 2018 he had over 2.3 million followers.

In 2016, the Canadian Science Writers Association changed its name to the Science Writers and Communicators of Canada Association (SWCC) to reflect the new diversity of its membership as well as the declining number of full-time journalists in mass media organisations. SWCC now describes itself as a national alliance of professional science communicators in all media, to reflect the blurring boundaries between journalism, science communication and public relations activities (SWCC, 2017). In 2017, SWCC launched the People's Choice Awards for Canada's favourite science site and Canada's favourite blog to reflect the inclusion of new media.

3. Science communication in Quebec: National emancipation and the centrality of the state

The Province of Quebec is the second most populous and second richest in Canada after Ontario. To many, while Ontario is the successor of the Province of Upper Canada, Quebec is the successor of Lower Canada. Rivalries between the two provinces run deep. The Province of Quebec is the reorganisation of the Colony of New France following the British Conquest in 1763. By comparison, Ontario was founded by the Loyalists, supporters of the British Crown, system of government and way of life. They had resisted the drive for American independence and fled to British-ruled territories in 1783. The two societies, the 'two solitudes', find their origins in these events.

To some extent, the modern tensions between Quebec and the rest of Canada are the continuation of the opposition between 'Canadiens', the conquered French-speaking subjects of a Catholic absolute monarchy, and the 'Empire Loyalists', the defeated English-speaking supporters of the British Crown during the American War of Independence who were forced to resettle in great numbers in Quebec and Ontario. Both the 'Canadiens', the future Québécois and the 'Empire Loyalists' remained attached to their respective religions and institutions and rejected rule by others. The British Empire could only consolidate its hold by ensuring that 'Empire Loyalists' were ruled by the British Common Law, and 'Canadiens' by the French Civil Code. Thus, the creation of Upper and Lower Canada, which only reinforced the distinctiveness of each.

As the British population expanded across Canada, the 'Canadiens', who were at first the largest population and remained so in Quebec, became a minority in Canada as a whole. While English Canadians rallied for King and Empire at the onset of the Great War, French Canadians resisted conscription. As noted above, Canadian provinces retain several exclusive rights upon which the Canadian parliament and federal government cannot intrude; and successive Quebec governments have actively used these prerogatives to preserve and develop French Canada's and then Quebec's cultural distinctiveness.

3.1. The first trends in science communication

The Quebec government brought support to existing learned societies before taking over science communication efforts in the 1840s, becoming the impetus for science communication and the development of science education in general (Chartrand et al., 2008). Yet government activities remained secondary to the efforts of other groups and individuals until the pull of industrialisation from the late 1850s. This led to a full reorganisation of the education system and, from that time on, the sciences were considered to be a necessary condition for economic and industrial development. The second industrial revolution, from the late 19th century, was even more impactful than the first, and prompted the development of vocational education in Quebec and the creation in Montreal of professional schools such as the Surveying School (1907) and the School of Forestry (1910), both affiliated to Quebec City–based Laval University.

In those years, clerics played important and overlapping roles in scientific research and the diffusion of science. They debated publicly in the exploding number of short-lived print journals (Carle and Guédon, 1988): *Le naturaliste canadien* (1868), *La Science populaire illustrée* (1886), *La Science pour tous* (1891), etc. In parallel, the press followed emerging scientific controversies very closely. The Catholic Church retained control over education and knowledge in general in Quebec, publicly denouncing Darwin's thesis while refraining from formally banning it (Chartrand et al., 2008).

3.2. The catalyst of national affirmation

Because of the political and social context in Quebec, World War II had an impact but it was not the catalyst for science communication as in the rest of Canada and many other countries. What is now known as the public communication of science and technology (PCST) started in Quebec in the interwar period as part of a movement for political, social, economic and cultural emancipation, and led by Brother Marie-Victorin (1885–1944). The objective was to refute the dominant Anglophone discourse that French-Canadians were too 'Latin' and Catholic to be scientifically minded. Vulgarisation not only played a major role to that end, it acquired its contemporary legitimacy (Chartrand et al., 2008).

The founding of the University of Montreal in 1920 revolutionised Francophone post-secondary education in Quebec. Science remained secondary and a majority of graduates were clerics, but it opened administrative and teaching positions to laymen and also opened the first Faculty of Science. In parallel, as a way to increase the number of graduates entering the workforce, the Quebec government created a scholarship program for students wishing to pursue post-secondary education in France. By ensuring a francophone study and work environment in Quebec, the conditions were set for the development of a francophone scientific community. As secondary education underwent reform from 1923, scientists such as Brother Marie-Victorin advocated in magazines and newspapers for greater inclusion of science in the school curriculum. Supporters of French-Canadian economic nationalism, these advocates argued that economic and intellectual independence was impossible without the mastery of knowledge and the control of natural resources (Gingras, 1996). It is with this specific purpose that the Association Canadienne-Française pour l'Avancement des Sciences (ACFAS) [French-Canadian Association for the Advancement of Science] was founded in 1923 (Gingras, 1994). There is no English-speaking equivalent in Canada. Five years later, the Institut Franco-Canadien (IFSC) was founded with the express purpose of improving the exchanges between Quebec and French scientists.

The Great Depression of the 1930s could have led to the shrinking of already limited resources and facilities, potentially plaguing not only PCST initiatives but science education and scientific research in general. However, the active resistance and mobilisation of the scientific community, notably Brother Marie-

Victorin, ensured their preservation and expansion (Gingras, 1996). In 1931, the scout movement-inspired Cercles des Jeunes Naturalistes (CJN) [Circles of Young Naturalists] and the Zoological Garden of Quebec City were founded. In 1933, the first genuine ACFAS conference was held (Gingras, 1994); a scientific exhibition designed by the CJN attracted 100,000 persons in two weeks; and the construction of the Montreal Zoological Garden started. These events demonstrated the growing interest of the public in science and scientific issues. The scientific community brought its support to the Union Nationale party during the 1936 election, and its election gave a voice to the scientific community and ensured its institutionalisation. One of the results of the election was the establishment of ACFAS as the main structure of the scientific community (Duchesne, 1978).

Figure 8.3: Brother Marie-Victorin circa 1920, photographer Albert Dumas.

Source: Collection Centre d'archives de Québec, Bibliothèque et Archives nationales du Québec, cote :P1000,S4,D83,PM39.

Yet universities and research institutes did not receive the funding and the support to match the development of science and the growing numbers of students. Fundamental research was stifled. In parallel, the Union Nationale government favoured staunch conservative policies in all domains and upheld the continued domination of the Catholic Church on educational matters. Although the 1949 Royal Commission on National Developments in the Arts, Letters and Sciences recommended financing universities, the Union Nationale government refused all subsidies, denouncing federal interference in provincial prerogatives. The 1953 Royal Commission of Inquiry on Constitutional Problems, called by the Quebec government, made similar recommendations (Gouvernement du Québec, 1956), but they failed to meet the demands of the scientific community (Duchesne, 1978).

This period came to be known as the Great Darkness. Growing dissatisfaction with the situation in Quebec was expressed as early as 1948 by the artist Paul Émile Borduas (1905–60) in the *Manifeste du Refus global* [The Global

Rejection Manifesto], the publication of which was the catalyst for the growing radicalisation of the movement for secularisation and national emancipation. The success of PCST, as for other conditions of this emancipation, was seen as resting on the successful taking and exercising of political power by nationalist and progressive actors.

3.3. The Quiet Revolution and the pull of independence

The 1960 election of the Liberal Party became the catalyst for the modernisation of all aspects of Quebec life. Now the State, regardless of the political party in power, was the motor behind the development of science and technology. Although individual initiatives remained a significant feature, the role of the State in the development and evolution of Quebec since the 1960s (and not least of all in PCST) cannot be understated (Duchesne, 1978). The 1961 Royal Commission on Education in the Province of Quebec, commonly known as the Parent Commission, advocated and guided a deep reform of the education system. Although scientific research was still not seen as a fundamental issue and the business community wished to tie it to economic development, the commission prioritised fundamental research (Gouvernement du Québec, 1965). Although the CJN started to wane, the government promoted scientific leisure activities, such as science fairs, through the *Conseils du loisir scientifique* and the *Association des jeunes scientifiques*. In 1962, ACFAS created *Le Jeune Scientifique*, which remains today Canada's oldest and only science magazine for adults. This French-language magazine was renamed *Québec Science* in 1970 and is published today. In 1968, ACFAS created the *Conseil de la jeunesse scientifique* (Gingras, 1994).

It was not until the 1970s that PCST became a priority for the State. The lagging of research in Quebec became a critical issue to universities, the scientific community and the business community. At this time, a large proportion of Québécois were reaching adulthood. This generation was not only literate and critical but they were also active participants in the debates of the time. Independence was the main issue. The Quebec nation had come of age and, thus, the more nationalist the party in power, the more its PCST initiatives were aimed at independence. In 1971, the new liberal government published *Les principes de la politique scientifique du Québec* [The Principles of the Quebec Science Policy] (Gouvernement du Québec, 1971), which emphasised fundamental research and science training. The same year, an interdepartmental committee on science policies was set up, followed in 1972 by the Science Policy Council. In 1974, ACFAS set up the Fédération du loisir scientifique (Gingras, 1994). During those years, a new generation of science communicators and journalists came of age, which notably manifested itself with the renaming of the *Le Jeune Scientifique* to *Québec Science* (1970).

With the 1977 election, the Parti Québécois government prepared for independence, imposing French as the only official language in the Province. The Quebec branch of the CSWA broke away to reorganise itself as the ACS. A new press agency solely dedicated to PCST was founded the very next year, Agence Science-Presse, launching its own science magazine: *Hebdo-Science*. The Parti Québécois launched the most thorough reform of education and PCST to date. In 1978, the government published a white paper on cultural development (Gouvernement du Québec, 1978); in 1979, a green paper on science policy (Gouvernement du Québec, 1979); and, in 1980, another white paper entitled *Un projet collectif: énoncé d'orientations et plan d'action pour la mise en oeuvre d'une politique québécoise de la recherche scientifique* [A Collective Project: An Orientation Statement and Action Plan for the Implementation of a Quebec Scientific Research Policy] (Gouvernement du Québec, 1980). In this last paper, the government announced a significant allocation of resources for university and industrial research and the sponsorship of vulgarisation and scientific leisure, including a Maison des sciences [House of Science] to be built in Montreal. In the same vein, the Semaine des Sciences [Science Week], the Petits débrouillards movement and the magazine *La puce à l'oreille* (1982) were launched. In 1981, the Conseil du développement du loisir science [Science Leisure Development Council] replaced the ACFAS-promoted Conseil de la jeunesse scientifique. However, to the great disappointment of the Parti Québécois government and the nationalist intellectuals, the Quebec electorate rejected independence by a large majority in the 1980 referendum.

3.4. The 1980s economic crisis and the neoliberal shift

In the 1980s, Quebec felt the full thrust of the economic crises that had been brewing since the mid-1970s. From 1982 on, governments around the world were compelled to improve their countries' competitiveness through reform. The Parti Québécois was no exception. In this new context, and at the behest of the OECD (1980, 1981), the reform of PCST and its promotion was seen as crucial to what the Parti Québécois government called the 'technological turn' (Gouvernement du Québec, 1982). Now State policy would promote PCST for economic growth, regardless of the party in power, and often to the detriment of other ends (Pitre, 1994).

Contrary to what may be expected, there was little to no direct influence of ideas coming from the American tradition of 'scientific literacy', British 'public understanding of science' or French 'culture scientifique, technique et industrielle', which were all emerging in parallel. In fact, it was the development of PCST in Quebec that influenced the development of PCST in France during those years (Santerre, 2008).

Without doubt, the 1980s were the high point of State intervention in PCST: the State published a number of official papers promoting PCST and enacted numerous policies aimed at supporting, developing and providing resources for PCST. In 1983, the Ministry of Higher Education launched a program designed to support the development of PCST and, in 1985, the ministry itself was reorganised as the Ministry of Higher Education and Science. Although the government put a stop to the creation of a Maison des sciences et des techniques, it published in 1986 a report entitled *La diffusion de la culture scientifique et technique au Québec* [Science and Technical Culture Diffusion in Quebec] (Gouvernement du Québec, 1986), and in the following year it inaugurated an International Science Fair, an exhibition entitled *Images du futur* [Images of the future] dedicated to high-tech applications in the arts, and *Expotec*, a yearly exhibition on a science topic.

The year 1988 was particularly momentous. First, the government published the *Énoncé d'orientations et plan de développement de la culture scientifique et technique au Québec* [Policy Declaration and Development Plan of Science and Technology Culture in Quebec] (Ministère de l'Enseignement supérieur et de la Science, 1988), which identified as the main dissemination channels educational institutions, media, leisure, exhibitions, science camps, museum institutions and interpretation centres. Second, the city of Montreal launched its first Cultural Municipal Policies in science, technology and heritage. Third, the government started to support PCST magazines in order to consolidate this new sector and to help new magazines establish themselves. Fourth, the government set up programs to sponsor national (i.e. Quebec-based) PCST organisations and major national PCST events: Agence Science-Presse, La Société de la Semaine des sciences and, from 1992 on, the Festival du international du film scientifique [International Science Film Festival] and the Expo-sciences panquébécoise [Panquebec Science fair]. Fifth, the government created support programs for temporary and traveling science and technology exhibitions. Finally, the Musée de la Civilisation [Civilisation Museum] was inaugurated in Quebec City.

The first half of the 1990s was especially dynamic as well. In 1992, an internal government document entitled *Les défis de la culture scientifique et technique au Québec* [Challenges of Science and technical culture in Quebec] (Ministère de l'Enseignement supérieur et de la Science, 1992) acknowledged the existence of PCST infrastructure and made recommendations to make science learning more attractive, stimulate the involvement of scientists, promote a positive image of science, increase the resources of local organisations and promote the collaboration of existing public networks. Among notable events were the transformation of the *Semaine des sciences* into the *Quinzaine des sciences* [Science Fortnight] and the renaming of the *Société de la Semaine des science* to

Société de la promotion de la science et de la technology, and the enlargement of its mandate. The Petits débrouillards movement rapidly expanded and spread internationally, not only to Francophone countries, but also to the Czech Republic and Slovakia as well (Les Petits Débrouillards, n.d.). The launch of *Les Petits Débrouillards* book series and the inauguration of the Montreal Biodôme, an indoor zoo with the recreation of natural habitats, all happened around that time.

The fiscal year of 1994–95 saw the opening of the Montreal Biosphere (an environmental museum), the Botanical Garden Complex (a botanical and entomological museum), the Armand Frappier Museum (a human health museum) and the Cosmodome (a space museum). The same year, the government published a report on the importance of PCST entitled *Miser sur le savoir* [Bet on Knowledge] (Gouvernement du Québec, 1994). The third (and largest to date) international conference of the PCST Network took place in Montreal alongside an international science exhibition, with exhibits coming from world-class institutions and with the support of both the federal and provincial governments. This conference became the basis for the first global survey of science culture: *When science becomes culture: World survey of scientific culture*. Finally, in 2000, the Montreal Science Centre or, more accurately, the Interactive Science Centre (iSci), was inaugurated.

And yet, at the turn of the 1990s, neoliberalism and its state counterpart, new public management, finally took precedence over all other priorities of the State. As a result, PCST was left to individual, associative and community actors while the State largely disengaged itself. In 1994, the Science Development Directory was transferred from the Ministry of Education to the Ministry of Commerce, Science and Technology. The next year, the narrow defeat of the second referendum broke the momentum for independence and, with it, for state- and nation-building. The government still ordered a number of studies on the state of PCST in Quebec at the turn of the 2000s (Gouvernement du Québec, 2002a, 2002b), but the State ceased to be the main driving force behind PCST. The last study ordered by the Quebec government on PCST was published in 2004 (Gouvernement du Québec, 2004).

4. The future for modern science communication in Canada

Recent surveys of Canadian science communicators identified though Twitter and Instagram show that, compared to traditional science communication professionals, social media communicators are younger, paid less (or not at all) for their science communication activities, and have been communicating science for fewer years than other kinds of science communicators (Riedlinger, Barata and Schiele, 2019). They are more likely to have a science background (rather than communication, journalism or education background) and are less likely to be members of professional associations. These communicators tend to be based in Ontario, Quebec and British Columbia, and communicate with each other through their own informal networks. Canadian social media science communicators are primarily located in the provinces identified by Schiele and Landry (2012) as the most prolific regions for science communication in Canada, where Canada's most prestigious and traditional universities are located, and where the bulk of Canada's population is concentrated. While some science journalists and communicators in Canada mourn the perceived loss of control over science communication as a loss of quality and accuracy, others welcome digital technology for the public engagement potential it offers. For example, Canadian science Instagram communicator Samantha Yammine was recently criticised in a *Science* magazine op-ed piece for trivialising scientific endeavours on social media (Wright, 2018). However, supporters of Yammine argued that she was successfully responding to the Instagram medium in her communication (see, for example, Lougheed, 2018; Marks, 2018). *Science* has subsequently published an article by Yammine and other social media communicators on the benefits of social media for science communication (Yammine, Liu, Jarreau and Coe, 2018). Social media platforms are allowing space for sociopolitically motivated communicators in Canada to work productively. The impact of these social media science communication efforts is difficult to assess; yet open science for consensus building and support for science in society efforts are needed in Canada now more than ever.

Canada has seen increased investments in science as described by the Naylor Report and the Global Young Academy, but science communication and outreach efforts are still needed to support science culture nationally (Boon, 2017a). Funding for activities happens at the federal level through agency funding; however, Canadian scientists, science communicators and science policymakers have criticised some recent initiatives for being primarily aimed at youth rather than adults, supporting mainly traditional and established organisations rather than innovative science communication initiatives, and having limited connection with the current and broader community of science communicators in Canada. While some science communicators are actively advocating for greater institutional support for a wider range of science communication initiatives (see Boon, 2017b), governments and scientific communities have been slow to respond.

Austerity continues to dominate public policy in Quebec, and science culture has ceased to be a priority. The Society for the Promotion of Science and Technology dissolved in 2010 and State-sponsored PCST in Quebec has come to an end. PCST actors and networks in Quebec persevere although they face difficulties in achieving an online presence in a global, yet overwhelmingly Anglophone, social media environment. However, the European Union program *Horizon 2020* may very well encourage a new period of renewed government interest in science communication.

References

An Open Letter on Science to Canadian Prime Minister Stephen Harper. (2014). Retrieved from s3.amazonaws.com/ucs-documents/science-and-democracy/canada-letter-word-by-country.pdf.

Association des communicateurs scientifiques de Quebec (ACS). (2018). Our association. Retrieved from www.acs.qc.ca/notre-association.html.

Barata, G., Caldas, G. and Gascoigne, T. (2018) Science communication research: National and international contributions. *An Acad Bras Cienc.*, *90*(2.1), 2523–42. doi.org/10.1590/0001-3765201720160822.

Bauer, M. W., Allum, N. and Miller, S. (2007). What can we learn from 25 years of PUS survey research? Liberating and expanding the agenda. *Public Understanding of Science*, *16*(1), 79–95. doi.org/10.1177/0963662506071287.

Bennett, D. (2018, 25 April). Notley disagrees with University of Alberta honorary degree for David Suzuki. *Globe and Mail.* Retrieved from www.theglobeandmail.com/canada/article-notley-disagrees-with-university-of-alberta-honorary-degree-for-david/.

Bentley, P. and Kyvik, S. (2011). Academic staff and public communication: a survey of popular science publishing across 13 countries. *Public Understanding of Science*, *20*(1), 48–63. doi.org/10.1177/0963662510384461.

Besley, J. and O'Hara, K. (2018, 14 April). Scientists hit the streets to command policy-makers attention. *NatureIndex*. Retrieved from www.natureindex.com/news-blog/scientists-hit-the-streets-to-command-policy-makers-attention.

Boon, S. (2017a, 13 September). Science funding in Canada should include science communication. *Watershed Notes* [Blog post]. Retrieved from snowhydro1.wordpress.com/2017/09/13/science-funding-in-canada-should-include-science-communication/.

Boon, S. (2017b, Sept 11). Canadian science needs more than funding: it needs public champions. *Science Borealis*. Retrieved from blog.scienceborealis.ca/canadian-science-needs-more-than-funding-it-needs-public-champions/.

Borduas, P.-E. (1948). *Refus global.* Montréal: Mithra-Mythe.

Carl, P. and Guédon, J.-C. (1988). Vulgarisation et développement des sciences et des techniques: le cas du Québec (1850–1950). In D. Jacobi and B. Schiele (eds), *Vulgariser la science. Le procès de l'ignorance* (p. 201). Seysell: Champ Vallon. doi.org/10.7202/1002104ar.

Carpentier, J.-M., and Ouellet, D. (1994). *Fernand Seguin: le savant imaginaire*, Montréal: Libre Expression.

CBC. (2015, 10 October). All-party election science debate. *Quirks and Quarks*. Retrieved from www.cbc.ca/radio/quirks/quirks-quarks-for-oct-10-2015-1.3264916/all-party-election-science-debate-1.3264954.

Chartrand, L., Duchesne, R. and Gingras, Y. (2008). *Histoire des sciences au Québec de la Nouvelle-France à nos jours*, Montréal: Boréal. doi.org/10.7202/045475ar.

Council of Canadian Academies. (2014). *Science culture: Where Canada stands*. Retrieved from www.scienceadvice.ca/en/assessments/completed/science-culture.aspx.

David Suzuki Foundation (2018). About us. Retrieved from davidsuzuki.org/about/.

Dornan, C. (1987). Science as ideology: The problem of science and the media reconsidered (PhD thesis). McGill University. Retrieved from www.mcgill.ca/ahcs/graduate/grads.

Dornan, C. (1988). The 'problem' of science and the media: A few seminal texts in their context, 1956–1965. *Journal of Communication Inquiry*, *12*(2), 53–70. doi.org/10.1177/019685998801200205.

Dornan, C. (1990). Some problems in conceptualizing the issue of 'science and the media'. *Critical Studies in Media Communication*, *7*(1), 48–71. doi.org/10.1080/15295039009360163.

Dubas, O. and Martel, L. (1973). *Media impact: A research study on science communication* (Vol. 2). Ministry of State, Science and Technology.

Duchesne, R. (1978). *La science et le pouvoir au Québec (1920–1965)*, Québec: Éditeur officiel.

Duchesne, R. (1981). L'état et la culture scientifique des québécois. *Les Cahiers du Socialismes*, *8*(Fall), 109–38.

Dufour, P. and De la Mothe, J. (1993). *Science and technology in Canada*. Detroit: Gale/Cengage Learning.

Dufour, P. and Gingras, Y. (1988). Development of Canadian science and technology. *Science and Public Policy*, *15*(1), 13–18.

Dunwoody, S. (1980). The science writing inner club: A communication link between science and the lay public. *Science, Technology, & Human Values*, *5*(30), 14–22. Retrieved from www.jstor.org/stable/689304.

Einsiedel, E. F. (1992). Framing science and technology in the Canadian press. *Public Understanding of Science, 1*, 89–102. doi.org/10.1088/0963-6625/1/1/011.

Einsiedel, E. F. (1994). Mental maps of science: Knowledge and attitudes among Canadian adults. *International Journal of Public Opinion Research, 6*(1), 35–44. doi.org/10.1093/ijpor/6.1.35.

Einsiedel, E. F. (2008). Public participation and dialogue. In M. Bucchi and B. Trench (eds), *Handbook of public communication of science and technology* (pp. 173–84). London: Routledge.

Einsiedel, E. F., Jelsøe, E. and Breck, T. (2001). Publics at the technology table: The consensus conference in Denmark, Canada, and Australia. *Public Understanding of Science, 10*(1), 83–98.

Evidence for Democracy. (2013). About E4D. Retrieved from evidencefordemocracy.ca/en/about.

Friedman, S. M. (1986). The journalist's world. In S. M. Friedman, S. Dunwoody and C. L. Rogers (eds), *Scientists and journalists: Reporting science as news* (pp. 17–41). New York: The Free Press.

Gingras, Y. (1991). Physics and the rise of scientific research in Canada. Montreal: McGill-Queen's University Press.

Gingras, Y. (1994). *Pour l'avancement des sciences: histoire de l'ACFAS, 1923–1993*, Montréal: Boréal. doi.org/10.7202/305406ar.

Gingras, Y. (1996). Marie-Victorin intellectuel. In *Frère Marie-Victorin, Science, culture et nation*, Montréal: Boréal.

Gouvernement du Québec. (1956). *Commission royale d'enquête sur les problèmes constitutionnels*. Québec: Éditeur officiel.

Gouvernement du Québec. (1965). *Rapport Parent*. Québec: Éditeur officiel.

Gouvernement du Québec. (1971). *Les principes de la politique scientifique du Québec*. Québec: Éditeur official. doi.org/10.7202/040482ar.

Gouvernement du Québec. (1978). *La politique québécoise du développement culturel* (vols 1 and 2). Québec: Éditeur officiel. doi.org/10.7202/1054250ar.

Gouvernement du Québec. (1979). *Pour une politique québécoise de la recherche scientifique*. Québec: Éditeur officiel. doi.org/10.7202/1054250ar.

Gouvernement du Québec. (1980). *Un projet collectif: énoncé d'orientations et plan d'action pour la mise en oeuvre d'une politique québécoise de la recherchée scientifique*. Québec: Éditeur official. doi.org/10.7202/1054250ar.

Gouvernement du Québec. (1982). *Le virage technologique. Bâtir le Québec – Phase 2. Programme d'action économique 1982–1986*. Québec: Éditeur officiel. doi.org/10.7202/600896ar.

Gouvernement du Québec, Conseil de la Science et de la Technologie. (1986). *La diffusion de la culture scientifique et technique au Québec*. Québec: Éditeur officiel. doi.org/10.3138/cmlr.44.4.739a.

Gouvernement du Québec, Conseil de la Science et de la Technologie. (1994). *Miser sur le savoir*. Québec: Éditeur officiel. doi.org/10.3138/cmlr.44.4.739a.

Gouvernement du Québec, Conseil de la Science et de la Technologie. (2002a). *La culture scientifique et technique au Québec: un bilan*. Québec: Éditeur officiel. doi.org/10.3138/cmlr.44.4.739a.

Gouvernement du Québec, Conseil de la Science et de la Technologie. (2002b). *Enquête sur la culture scientifique et technique des Québécoises et des Québécois*. Québec: Éditeur officiel. doi.org/10.7202/1055295ar.

Gouvernement du Québec, Conseil de la Science et de la Technologie. (2004). *La culture scientifique et technique: une interface entre sciences, la technologie et la société*. Québec: Éditeur officiel. doi.org/10.3138/cmlr.44.4.739a.

Government of Canada. (2018). Citizen Science Portal. Retrieved from science.gc.ca/eic/site/063.nsf/eng/h_97169.html.

Hakai Magazine. (n.d.). About us. Retrieved from www.hakaimagazine.com/about-us/.

Halpern, M. (2015, 27 October). How science helped to swing the Canadian election. *The Guardian*. Retrieved from www.theguardian.com/science/political-science/2015/oct/21/how-science-helped-to-swing-the-canadian-election.

Les Petits Débrouillards (n.d.). Histoire. Retrieved from www.lespetitsdebrouillards-idf.org/qui-sommes-nous.

Levere, T. H. (1988). The history of science of Canada. *British Journal of Historical Science*, *71*(4), 419–25.

Lougheed, T. (2018, 27 March). ICYMI. Social media sells science short. Sad! *Science Writers and Communicators of Canada*. Retrieved from sciencewriters.ca/6002325.

Macleod, D. (1996). A small band of devotees: Recent history of Canadian science and technology. *Journal of Canadian Studies*, *31*(2), 183–189.

Marks, G. S. (2018, 20 March). A scientist responds to that 'Science' Instagram essay. *Massive Science*. Retrieved from massivesci.com/articles/instagram-science-sam-women-representation/.

Ministère de l'Enseignement supérieur et de la Science. (1988). Énoncé d'orientations et plan de développement de la culture scientifique et technique au Québec, May. doi.org/10.4000/books.pum.18232.

Ministère de l'Enseignement supérieur et de la Science, Direction du développement scientifique. (1992). Les défis de la culture scientfique et technique au Québec, réflexions sur les pratiques et les strategies ministèrielles en mantière de culture scientifique et technique, May. doi.org/10.4000/books.pum.18232.

Montgomery, M. (2016). Canadians and science literacy: We're No 1…um…maybe not. Retrieved from www.rcinet.ca/en/2016/09/23/canadians-and-science-literacy-were-no-1um-maybe-not/.

Natural Sciences and Engineering Council of Canada. (2015). *NSERC 2020: A strategic plan*. Retrieved from www.nserc-crsng.gc.ca/NSERC-CRSNG/NSERC 2020-CRSNG2020/NSERC2020-2020CRSNG_eng.pdf.

Nelkin, D. (1995). *Selling science: How the press covers science and technology*. New York: Freeman.

O'Hara, K. (2010, 29 September). Canada must free scientists to talk to journalists. *Nature, 467*. doi.org/0.1038/467501a.

OECD. (1980). *Changement technique et politique économique: la science et la technologie dans le nouveau contexte économique et social*. Paris: OECD. doi.org/10.3406/estat.1991.5574.

OECD. (1981). *La politique scientifique et technologique pour les années 1980*. Paris: OECD.

Ontario Science Centre. (2016). *Ontario Science Centre survey reveals gap in public understanding of critical scientific issues* [News release]. Retrieved from www.ontariosciencecentre.ca/Media/Details/432/.

PIPSC. (2013). *The big chill: Silencing public interest science, a survey*. Retrieved from: www.pipsc.ca/portal/page/portal/website/issues/science/bigchill.

PIPSC. (2017). *Defrosting public science: A survey report on federal government efforts to reverse the effects of The Big Chill*. Retrieved from www.pipsc.ca/sites/default/files/comms/Defrosting-report-e_v4%202_1.pdf.

Pitre, R. (1994). La culture scientifique et technique et les politiques scientifiques au Québec: le rôle de l'OCDE. In B. Schiele (ed.), *When Science Becomes Culture*, Proceedings II, Sainte-Foy: Multimondes.

RCGS. (2018). About us. Retrieved from www.rcgs.org/about/history/rcgs_cg1930.asp.

Riedlinger, M., Barata, G. and Schiele, A. (2019). The landscape of science communication in contemporary Canada: A focus on anglophone actors and networks. *Cultures of Science, 2*(1), 51–63. Retrieved from cos.cnais.org.cn/html/collection-full-text.html?id=2083a2b0e95d49fb9651d432bac21665&num=5.

Ryan, P. M. and Dunwoody, S. L. (1975). Specialized training and the science writer. *Journal of Technical Writing and Communication, 5*(2), 82–90. doi.org/10.2190/98ug-a2hw-w96w-dk33.

Saari, M.-A., Gibson, C. and Osler, A. (1998) Endangered species: science writers in the Canadian daily press. *Public Understanding of Science, 7*, 61–81. doi.org/10.1177/096366259800700105.

Santerre, L. (2008). From democratization of knowledge to bridge building between science, technology and society. In D. Cheng, M. Claessens, T. Gascoigne, J. Metcalfe, B. Schiele and S. Shi (eds), *Communicating science in social contexts: New models, new practices*, New York: Springer. doi.org/10.1007/978-1-4020-8598-7_17.

Schiele, B. (2008). Science museums and science centres. In M. Bucchi and B. Trench (eds), *Handbook of public communication of science and technology* (pp. 41–54). London: Routledge. doi.org/10.4324/9780203483794.ch4.

Schiele, B. (2014). Science museums and centres: evolution and contemporary trends. In M. Bucchi and B. Trench (eds), *Routledge handbook of public communication of science and technology* (pp. 56–73). London: Routledge. doi.org/10.4324/9780203483794.ch4.

Schiele, B., Amyot, M. and Benoit, C. (1994). Quebec: Overview of scientific and technological culture and assessment of government action. In B. Schiele (ed.), *When science becomes culture* (pp. 13–72). Ottawa: University of Ottawa Press.

Schiele, B. and Koster, E. H. (2000). *Science centers for this century*. Quebec: Editions Multimondes.

Schiele, B. and Landry, A. (2012). The development of science communication studies in Canada. In B. Schiele, M. Claessens and S. Shi (eds), *Science communication in the world* (pp. 33–63). Springer, Dordrecht. doi.org/10.1007/978-94-007-4279-6_3.

Schiele, B., Landry, A. and Schiele, A. (2011). *Science communication in Canada: An inventory for the major PCST initiatives carried out in Canada*. Research report to the Korean Foundation for the Advancement of Science and Creativity. Montreal: CIRST-UQAM, University of Quebec at Montreal.

Schnell, J. (2017, 19 June). Canada at 150: The Canadian research landscape. *Clarivate Analytics*. Retrieved from clarivate.com/blog/science-research-connect/canada-150-canadian-research-landscape/.

Science Borealis. (2018). About us. Retrieved from scienceborealis.ca/about-us/.

Science Rendezvous. (2018). About Science Rendezvous. Retrieved from www.science rendezvous.ca/about/.

Statistics Canada (2017). English, French and official language minorities in Canada. Retrieved from www12.statcan.gc.ca/census-recensement/2016/as-sa/98-200-x/2016011/98-200-x2016011-eng.cfm.

Stoymenoff, A. (2012, 17 April). David Suzuki forced to resign from board of foundation by 'vicious' media attacks. *Vancouver Sun*. Retrieved from www.vancouverobserver.com/sustainability/2012/04/17/david-suzuki-forced-resign-board-foundation-vicious-media-attacks.

Strauss, S. (2018). Bye-bye moonwalks: Hello Tweets. *Science Writers and Communicators of Canada*. Retrieved from sciencewriters.ca/.

Suzuki, D. T. (2006). *David Suzuki: The autobiography*. Vancouver: Greystone Books.

SWCC. (2017). Mandate. Retrieved from sciencewriters.ca/Mandate.

The Conversation Canada. (2018). Who we are. Retrieved from theconversation.com/ca/who-we-are.

Turner, C. (2013). *War on science: Muzzled scientists and wilful blindness in Stephen Harper's Canada*. Vancouver/Berkeley: Greystone Books.

Visser-deVries, A. (2015). CSWA/ACRS celebrates 45th anniversary. History. Retrieved from sciencewriters.ca/History.

Weigold, M. F. (2001). Communicating science: A review of the literature. *Science Communication*, *23*(2), 164–93. doi.org/10.1177/1075547001023002005.

Wright, M. (2018). Instagram won't solve inequality. *Science*, *359*(6381), 1294. doi.org/10.1126/science.359.6381.1294.

Yammine, S. Z., Liu, C., Jarreau, P. B. and Coe, I. R. (2018). Social media for social change in science. *Science*, *360*(6385), 162–63. doi.org/10.1126/science.aat7303.

Canada and Quebec timeline

Event	Name	Date	Comments
First interactive science centre established.	The (traditional) Canada Museum of Science and Technology in Ottawa	1967	1969: The Ontario Science Centre opens in Toronto. This was more interactive
First national (or large regional) science festivals.	Science fairs in Winnipeg, Edmonton, Toronto, etc.	1959	2008: Science Rendezvous is now the largest science festival in Canada
Association of science writers or journalists or communicators.	Canadian Science Writers Association (CSWA) formed. Now renamed (SWCC)	1970	1977: The Association des communicateurs scientifiques de Quebec (ACS) formed
First university courses to train science communicators.	A graduate diploma by Science North and Laurentian University	2005	2008: First undergraduate degree at Mount Saint Vincent University, Halifax
First master's students in science communication to graduate.	First master's thesis by Suzanne Champagne, Faculty of Social Sciences, Laval University	1984	1997: The English-language master's by Chantelle Bita Kudsi at McGill University, Montreal
First PhD students in science communication graduate.	First French-language PhD was Bernard Schiele, Faculty of Education, University of Montreal	1978	1987: Canada's first English-language PhD: Chris Dornan, Communications Program, McGill University, Montreal

Event	Name	Date	Comments
First national conferences in science communication.	Conference Association Canadienne Française pour l'Avancement des Sciences	1933	1971: The first annual meeting of the Canadian Science Writers' Association
National government programs to support science communication.	National science and technology strategy (InnovAction) released	1987	2007: A federal strategy to foster a culture that values entrepreneurship and gets Canadians excited about science
First significant initiatives or reports on science communication.	Symons Royal Commission on Canadian Studies acknowledges the role of science in society in Canadian culture	1975	1980: Quebec government white paper, *Un projet collectif*, plus other reports 2014: Academies report, *Science Culture*
National Science Week founded.	Semaine des Sciences [Science Week] launched in Quebec	1980	1990: First National Science and Technology Week
First significant radio programs on science.	*Radio-Collège* at Radio-Canada	1941	1975: *Quirks and Quarks* airs on CBC Radio
First significant TV programs on science.	CBC's *Nature of Things* airs on television	1960	1960: Fernand Séguin hosts *Aux frontières de la science* 1979: David Suzuki hosts *The Nature of Things*
First awards for scientists or journalists or others for science communication.	The Canadian Science Writers Association (CSWA) launched the Ortho Award for print science journalism	1973	1981: The Fernand-Séguin scholarship to encourage young science journalists 2001: NSERC Awards for Science Promotion
Date hosted a PCST conference.	Third PCST conference held in Montréal, 'When Science Becomes Culture'	1994	2004: The 4th World Conference of Science Journalists in Montreal 2017: 'Science and You' Conference in Montréal
Other significant events.	Report by Orest Dubas and Lisa Martel, *Media impact: A research study on science communication*	1973–75	1989: Edna Einsiedel conducts surveys on public perceptions of science
	The first consensus conference held on the topic of food safety	1999	
	The Canadian Science Media Centre opens	2010	

Contributors

Dr Michelle Riedlinger is an associate professor at the University of the Fraser Valley, British Columbia, Canada, and secretary of the PCST Network and her career spans the practical and theoretical sides of science communication.

Dr Alexandre Schiele holds a PhD in communication science (Sorbonne) and another in political science (University of Quebec). He is working on a project 'Mapping the New Science Communication Landscape in Canada'.

Dr Germana Barata is a science communication researcher at the Laboratory of Advanced Studies in Journalism (Labjor) at the State University of Campinas, Brazil, and a member of the Scientific Committee of the PCST Network.

9

CHINA

Science popularisation on the road forever

Yin Lin and Li Honglin

This chapter outlines modern science popularisation across a span of nearly 70 years. The authors take the timeline as the guide, unfolding the story of modern science popularisation development in three stages. Considering the predominant role that the Chinese government has played, this chapter puts more emphasis on government policies and guiding principles for science popularisation and the effects they have brought about.

1. Sailing: The starting point and context of modern science popularisation in China (1949–78)

Since the late 19th century, science was integrated into Chinese culture, but this changed radically in 1949 with the founding of the new China. Before 1949, science was invited to China by a small number of groups and oriented towards the elite,[1] which opened the door for science and science popularisation in China (Ren, Yin and Li, 2012). After 1949, science popularisation in China started to be institutionalised, with the government as the main driving force and the public as the target. This closely related to

1 There are two main ways and stages for science entering China. One is Western missionaries spreading Western science and technology to Chinese scholar-bureaucrats along with Christianity before the 20th century. The second is the new culture movement in the early 20th century when Chinese avant-garde intellectuals advocated democracy and science and promoted establishing many science and technology associations within the intellectuals' communities with the purpose of developing scientific research and popularisation to save the country.

the specific political, economic and cultural background at that time. This government-led model has contributed to the rapid development and success of science communication in China (Office for Implementing the Action Plan, 2018), though with many problems.

1.1. The institutionalisation of science popularisation in China

When the People's Republic of China (PRC) was founded in 1949 there were a thousand things to be done, and science popularisation was located in a specific context. On the one hand, science and technology attracted worldwide attention during World War II, and the Chinese government and elites were fully aware of its importance for national development and international competition. On the other hand, the new China faced multiple development dilemmas: in politics, it was important to consolidate power and stabilise social order; in the economy, it was urgent to achieve industrial development; in culture, it was necessary to improve people's education levels and eliminate feudal superstitions. All of this greatly strengthened the position of science and technology, regarded as important means to achieve economic development, promote social stability and enlighten the people. The government, scientists and the public were full of expectations for science and technology, especially the public that maintained a high level of interest in scientific knowledge. 'There has never been such urgent demand for science in China as today … Scientists, technicians, skilled modern workers and literate farmers are needed' (Guo, 1950). These demands provided internal impetus for the development of science popularisation.

In such a context, science popularisation in China developed rapidly. It changed from government-led to government-driven, with the driving force being a combination of government, and science and technology associations.

1.1.1. Government-leading mode: The Science Popularisation Bureau of central government

In September 1949, scientists suggested that we should 'work hard to develop natural science so as to serve the construction of industry, agriculture and national defense. Encourage scientific developments and inventions, popularise scientific knowledge' (Chinese People's Political Consultative Conference Committee, 1949). This was written into the *Common Program of Chinese People's Political Consultative Conference* (taken as the temporary constitution). Taking this as the key tenet, the central government set up the Science Popularisation Bureau (SPB) in November 1949 to explore science popularisation in China. That began the government-leading mode of science popularisation, in which resources were managed by administrative means. After two years' exploration and practice, when workers, farmers and cadres were the main target, and lectures, exhibitions, book editing and technical training were the main forms, it was realised that 'it is not enough to rely on the government alone. There must be mass organisations, and the scientific community and all sectors of society mobilised to form a mass science popularisation campaign' (Zhang, 2001). 'It's not necessary to have a special science popularisation institution in government, and the new science popularisation association could undertake its work' (China Association for Science and Technology, 2002). Thus, the SPB was disbanded in October 1951, and the government-leading mode of science popularisation with an administrative bureau in charge came to its end.

1.1.2. Science popularisation mode 'driven by government, operated by associations': The establishment of CAST

In August 1950, the Chinese Natural Scientific Representatives' Conference convened in Beijing. At the conference, the National Federation of Natural Science Associations and the National Association for Science and Technology Popularisation came into existence, with the former taking scientific research development as its core task, and the latter taking scientific knowledge popularisation. It was pointed out that scientists should serve the people and national economy, and science associations should coordinate their work towards national economic and cultural construction. In September 1958, the National Federation of Natural Science Associations and National Association for Science and Technology Popularisation merged into one organisation named China Association for Science and Technology (CAST), to adapt to the scientific and technological innovation movement in the Great Leap Forward,[2] and satisfy public demands for science and technology. CAST began to take on the responsibility of popularising science in China and the political function of participating in the political consultation system of the Chinese Communist Party (CPC) as a people's organisation. The responsibility was closely related to the political function.

At that time, the model for science popularisation in China was 'driven by government, operated by associations', and the institutionalisation of modern science popularisation was set (Yin, 2008). China's science and technology associations and science popularisation entered a new era. Lots of local science associations were established in provinces, cities, autonomous districts,

2 'Great Leap Forward' refers to the nationwide movement in China from 1958–60. Unrealistic economic production plans were made and put into effect despite of the low productivity level at that time.

counties, industrial and mining corporations, enterprises, communes, schools and units. They were initiated by scientists under the guidance of the government to promote scientific research and popularisation. A large stable network for science popularisation was formed and carried out by government and science associations at national or local levels, pushed forward by the combination of government administration and associations (Shen, 2002).

1.2. Formation of the top-down mass science communication model

Through continuous exploration, China had formed a top-down, large-scale and mass-oriented science communication mode. During this period, China's science popularisation was 'combined with production, combined with reality and make every effort to be extensive', and 'serve[d] production through vigorously mass scientific and technological professional activities' (Deng, 2008). 'The science popularisation first must be a massive movement ¼ must accommodate the urgent demands then and there ¼ must be concentrated rather than dispersed so as to get better effects' (Zhang, 2001).

These mass science popularisation activities could be divided into two levels. The first was an anti-illiteracy campaign, popularising basic knowledge of natural sciences to the public and encourage them to know science and learn science. The second level focused on the state and the actual conditions of industrial and agricultural production, popularising practical knowledge and imparting skills training to raise productivity. The latter, in particular, was the focus of science popularisation in China during this period. For example, 1959 to 1961 were continuous lean years in agriculture in China due to unrealistic economic production plans, bad seasons and poor farming techniques, and the country suffered severe food shortages. In this context, CAST took 'serving agriculture as its long-term principal task' (Deng, 2008), and organised staff to help farmers with experimental activities. It performed agricultural science popularisation and training, and played an important role in supporting national industrial and agricultural production. Science popularisation was developing from basic knowledge popularisation to the combination of basic knowledge popularisation and technology promotion, and with practicality as its core concern.

Forms of science popularisation were mainly speeches, lectures, broadcasts, exhibitions and periodicals and magazines (i.e. popular forms that could directly connect to people). The statistic showed that there had been 72 million speeches, 170,000 exhibits and 130,000 films and slides during the period 1950–58, with an attendance of 1 billion (Ren and Zhai, 2012).

Science in the media was dominated by popular science books, followed by periodicals and newspapers. Statistics of the National Bibliography[3] from 1949 to 1965 listed 24,036 popular science books. Basic knowledge of natural science accounted for about 9 per cent, medicine and health care 11 per cent, transportation 40 per cent, agriculture and forestry 37 per cent, and children's books about 3 per cent (Liu, 2010). Science popularisation for agriculture and industry were also important categories for popular books. The *100,000 Whys* series created a miracle in Chinese popular science book publishing, introducing scientific knowledge in the form of questions and answers on common natural phenomena in daily life. From 1961 to 1964, 5.8 million copies were published. By 1978, over 10 million copies had been printed and they became best-selling popular science books. There were two main reasons: a lot of scientists got involved in revising the series to increase the readability; and the education level for all was generally low and *100,000 Whys* could be easily read by both adults and teenagers. Most important, of course, was the Chinese public's hunger for scientific knowledge.

At the same time, science periodicals, magazines and newspapers for the public developed fast. By 1965, the number of popular science periodicals and magazines had grown from less than 10 in the early days of the new China to 55 (Liu, 2010): *Knowledge is Power, Amateur Astronomer, Aviation Enthusiasts* and *Aerospace Knowledge* are examples. In 1958, 19 local associations for science and technology launched science and technology newspapers. Closely following the national political and economic development situation, these newspapers communicated basic knowledge to the public and became quite popular.

Radio and television were welcomed as new forms of communication and loved by the public. In September 1949, Bei Ping XinHua radio station (renamed China National Radio in December 1949) established the first popular science radio program in China, *Popular Natural Science Lecture*. The broad direction was to 'popularise science education, broadcast all sorts of natural science common sense systematically', and the program committed 'service to production and construction; improve the scientific and technological knowledge of the audience; propagate dialectical materialism and helping the audience to establish Marxist world outlook; popularise physiological health knowledge' (Liu et al., 1991). In 1953, on the basis of the earlier *Popular Natural Science Lecture*, China National Radio launched the program *Science Knowledge*, which established the model of science broadcasting in China. Since

3 The National Bibliography is the only yearbook catalogue in China. It has been compiled year by year since 1949 and purports to be a comprehensive enumeration of all books published in that year. It is a necessary reference book for publishing houses, libraries, information materials, scientific research and teaching departments.

then, radio stations successively launched science radio programs, forming a huge popular science radio network covering a vast area. According to China National Radio, popular science programs had a great impact on listeners. By 1966, *Science Knowledge* was one of the top 10 programs.

In 1960, China established the first science TV programs *Scientific Knowledge* and *Medical Consultant*, aiming to 'spread basic knowledge and serve life and production' (Yang, 1998). The programs usually asked scientists to impart scientific knowledge to the audience, and the topics generally involved people's daily production and life, such as 'electric shock first aid', 'the use of household appliances', 'breeding the long-haired rabbit' and so on. Although the form was relatively monotonous, it laid a foundation for the development of science and education TV programs. The social influence of early TV science programs was limited, mainly because China's TV infrastructure was inadequate and people could not afford expensive TV sets. However, TV became a 'teacher' to impart knowledge, providing the psychological conditions for the audience to respect and accept the programs unconditionally.

1.3. Discussion and reflection on the governmental position

Science popularisation in China during this period followed a typical governmental position. As in other countries, the public was regarded as needing to be educated and informed (Liu, 2004). To some extent, the government could justify using this approach: the deficit model of science communication was applied to China and succeeded (Fan, 2004). In 1949, the illiteracy rate of the Chinese public reached 80 per cent, and feudal superstitions and pseudoscientific beliefs were very common. The public was very homogeneous: lack of scientific knowledge, scientific attitude and scientific spirit, but holding a general attitude of love and support for science and technology and expecting to improve their living conditions through science and technology. The government recognised the urgent needs of the public and the benefit of using the 'deficit model' approach (Lewenstein, 2003) to popularise science, eliminate scientific illiteracy and help the public understand the world. The government concentrated resources and power as well as its organisational and social mobilisation ability to promote the establishment of science popularisation. This laid a foundation for its long-term development.

This government position also posed problems. First, the political environment played a decisive role in the development of science popularisation. Science popularisation had not concerned itself with the environment or laying a foundation for sustainable development, and political turbulence would have a substantial impact on it. During the 'cultural revolution' from 1966 to 1976, science popularisation in China was almost at a standstill. Second, from the standpoint of the government, science popularisation had an pragmatic aim of political construction and economic development. It paid more attention to natural science knowledge, practical technology, and skills training than scientific attitude and scientific spirit. The target groups for science popularisation were workers, farmers and civil servants, while teenagers, as important talents for future development, did not receive much attention. Furthermore, as important actors in science communication, scientists failed to play their roles fully. For political reasons, intellectuals and scientists were often at a loss about what to do. During the 'cultural revolution', a large number lost their jobs and their enthusiasm for science popularisation. For a long time afterwards, intellectuals and scientists were cautious, and their motives and positions in science popularisation were suppressed or covered up, and this hindered its development. The individual attributes and expressions of the public were often neglected, and science popularisation found it difficult to meet the personal and diversified needs of the public.

2. Developing: Upspring of multiple communication forms (1978–2006)

At the National Science Conference of March 1978, the paramount leader Deng Xiaoping addressed the principle that 'science and technology are productive forces'. This aroused the enthusiasm of scientists, technologists and educators and stimulated their passion. This was a turning point and ushered in a new stage for science popularisation.

For nearly 30 years, science popularisation as a term has been consolidated in the social culture and accepted by lay people. They are willing to spend time in visiting science museums, opening laboratories in universities and joining in activities during the National Science Popularisation Day (known as the Science Festival). They accept science as part of their cultural life and important in changing their lives. There are now more options for science engagement to choose from: as well as television, magazines, books and public lectures, on-site activities and interactive communication began to emerge. By 2000, as science popularisation become more visible as a social and cultural behaviour, science communication research emerged as a subfield of science and technology studies in China and gradually became a relatively independent domain of research (Xu, Huang and Wu, 2015).

2.1. Science popularisation infrastructure: From the first S&T museum to a system of S&T museums with Chinese characteristics

Represented by the opening of Bengbu Science and Technology Museum (1984) and the last-stage project of the China Science and Technology Museum (1988) in the 1980s, the first science and technology museums in China emerged. Defined as 'comprehensive multifunctional places to hold science and technology activities', they served as places for exhibitions, science education and communication, S&T training and consultation, S&T activities for youth, etc. (Cheng, 2014).[4] But the education function of S&T museums was weak and, by the end of 2000, only 11 out of more than 320 S&T museums took exhibition education as their predominant function (Zhu et al., 2011).

The science popularisation investments and venues could not meet the demands in different regions for a considerable time due to general economic conditions. Nor could residents in outlying regions and remote rural places afford to go to cities to visit museums because of the geographical and traffic conditions. In order to satisfy demands of districts that could not afford to construct S&T museums, or those that could afford neither exhibits nor the maintenance of museums even if they had the buildings, CAST launched the Science Wagon project in 2000, sending science exhibits to villages as travelling displays. By the end of 2012, more than 600 science wagons have provided travelling exhibitions at 90,000 sites.

CAST started another project named Mobile Science Museums in 2010, sending science popularisation exhibits, show boards, video materials and scientific experiment tool kits to small cities and counties on demand. These exhibits can be displayed anywhere appropriate for public visits, not necessarily in museums. Grassroot infrastructures for science popularisation had been constructed to expand the coverage of the public science popularisation service. By the end of 2010, there were 212,500 science popularisation galleries, 68,000 science and technology popularisation sites in urban communities and 370,000 in rural villages (Ministry of Science and Technology, 2010). In addition, since the creation of the China Digital

4 The functions and tasks of S&T museums were defined as 'comprehensive multifunctional places for science and technology activities' in both the work report and the *Proposed Regulations on Management of S&T Museums* issued at the Symposium for Management of S&T Museums in May 1989 (Cheng, 2014).

Science and Technology Museum (CDSTM) in 2006, more than 40 S&T museums have constructed a sub-website of CDSTM. Some have science popularisation columns on their websites, forming a platform for the public.

The different layers of science museum resources use the design concept of a 'modern S&T museum system with Chinese characteristics' (Cheng, 2014), to meet different public demands in regions with unbalanced economic, educational and cultural levels. The China Science and Technology Museum is the central designer and allocates exhibition resources in the national museum system. It provides S&T education and demonstration services for people visiting the museum in Beijing, and duplicates its resources and sends them to remote places. Similarly, the large and medium-sized museums in provincial capitals provide services for local citizens and the small museums that cannot afford to update their exhibitions regularly. The mobile science museums and science wagons radiate out across the country. In remote places, galleries and rural libraries are like S&T information distribution centres. For netizens there are digital S&T museums. Most S&T museums' materials are shared and cyclically utilised: larger museums provide small grassroots centres with technical services such as exhibition content and personnel training.

Businesses also began to pay attention to science popularisation. The China Aerospace Industry Corporation established the China Youth Space Science Popularisation Fund; the Sony Corporation set up the first experiential S&T museum in Asia, Seek Dream in Sony at the Oriental Plaza in Beijing; the Shenyang Aircraft Industry (Group) Co. Ltd. created the Aviation Science Exhibition Area; the Qingdao Haier group established an open S&T museum. The categories and layers of the science popularisation infrastructure system have become multidimensional.

2.2. Public-engaging science popularisation activities

Besides science communication in museums, activities like S&T Week, Science Popularisation Day, S&T festivals and so on were reaching more people. For example, activities implemented by larger cities in China in 1994 ran for two months with more than 40 cities participating. The International Week of Science and Peace, and summer camps like World Population Day, World Environment Day and International Ocean Year, all attract extensive participation. CAST launched 85,000 science popularisation activities from 1995 to 2000, with an attendance of more than 70 million (China Association for Science and Technology, 2002). The following activities have the most extensive influences:

- Science and Technology Week: mass S&T activity approved by Chinese government in 2001 and carried out nation-wide in the third week of May. The organising committee comes from 10 or more departments and organisations like the Ministry of Science and Technology, the Propaganda Department of CPC Central Committee, and CAST.
- National Science Day: science popularisation activities held initially on 29 June 2003 by CAST for the implementation of the *Law of PRC on Science and Technology Popularisation* to celebrate its anniversary (NPC, 2002). The date was changed to the third weekend in September in 2005, so that all people (especially students) could participate.

2.3. Further development of the institutionalisation of science popularisation

In the period 1978–2006, science communication and popularisation experienced great development. At the macro level, it was the urgent demand for science and technology development that drove progress, and the rapid development of science popularisation was closely supported by the policies of the Chinese government.

With the development of the socialist market economy in China in the 1990s, the importance of science popularisation increased as all areas had urgent demands for science and technology. At the Fourth National Representative Conference of CAST in 1991, Jiang Zemin, then the General Secretary of CPC Central Committee, pointed out that the economic growth of China should rely on science and technology development and the promotion of workers' scientific literacy. This was a major requirement for reinforcing science popularisation work in China. It was proposed in the work report by CAST to CPC Central Committee in 1993 that:

> Science and technology popularisation is a public activity involving the whole society. There should be laws to regulate its importance, status, tasks, as well as the rights and duties of government, social entity and personnel on it. We suggest that the *Law of the People's Republic of China on Popularisation of Science and Technology* should to be formulated and issued. (Cui, 2010)

Since it takes a long time to produce a law, the *Instructions on Strengthening Engagement in Science and Technology Popularisation* (hereinafter the *Instructions*) was issued by the State Council on 5 December 1994 to give instructions for the general principles, central tasks and main measures of science popularisation (Central Committee of the Communist Party of China and the State Council, 1994). The idea was to put science popularisation on a strategic level for the country. It was the first programmatic document on science popularisation issued jointly by CPC Central Committee and State Council after the founding of the new China and the document guiding science popularisation work before the issue of the *Law of PRC on Science and Technology Popularisation* (NPC, 2002).

Facing the challenges of science popularisation losing its priority among some local governments and superstitions gathering strength since the 1980s, the *Instructions* call for the governments at each level to place science popularisation on their working agendas. Science popularisation engagement was to be included in the State's ninth 5-year plan and in local, economic, science and technology programs. The construction of science museums, science centres and public spaces for science popularisation activities were to be supported. It demanded that sensationalist media reporting on superstitions and pseudoscience were to be opposed. In this document, science popularisation is regarded as the approach to advancing the 'material civilisation' as well as 'mental civilisation' of the nation and an indispensable way to fostering the new generation (Shi and Zhang, 2012).

The legislation of science popularisation law did not stop at that. Investigations and research identified the current challenges of science popularisation in China:

1. there is not enough recognition of the strategic status and importance of science popularisation
2. the administrative mechanism of science popularisation does not work efficiently due to less smooth coordination between CAST (with its branches) and the relevant government departments
3. the funding of science popularisation was low and not enough to satisfy the needs of realistic work
4. there were not enough science popularisation venues and infrastructures with exhibition and education measures were lagging behind
5. science popularisation organisations were not sound, and the literacy of working staff was not high (Cui, 2010).

The legislation of the *Law of the People's Republic of China on Popularisation of Science and Technology* (hereafter the *Law*) aimed to solve these problems. The Law has six parts and 34 items, stipulating 'science popularisation is one of the long-term tasks of the nation'; and that government departments, social organisations, enterprises and institutions 'should carry out science popularisation'. In the *Law*, words like 'the country supports … science popularisation' and 'the country protects … legal rights' appear many times, bringing the importance and legality of science popularisation to national attention. It is stated that while CAST is the main social force undertaking science popularisation, the administrative S&T departments of the State Council should be responsible for compiling work plans at national level and pushing the work forward by issuing guiding policies and supervision. On the whole though, CAST is responsible for the implementation and organisation of science popularisation. The *Law* gives macro guidance on the duties of different institutions. In the third chapter, 'Social Responsibilities', guidance is given to educational institutions, schools, S&T museums, mass media, enterprises and other related entities. It is written that: 'Kinds of rural economic organisations, agricultural technology spreading institutions and rural technology associations should perform scientific and technological knowledge popularisation when spreading advanced and applicative technologies to peasants'; and 'Enterprises should perform science popularisation in their technological innovation and technician trainings, establish and open science popularisation venues and infrastructures to public if condition allows'.

Funding is crucial. If there is a shortage of funds, the *Law* would lack force and difficulties in science popularisation could not be settled. It is regretful that the *Law* does not give a clear plan on levels of funding. Unprecedentedly, though, the *Law* includes science popularisation funding in the national financial budget and requires the financial investment level to be gradually increased. This provides a legal basis for the budget and is considered a great step forward. According to the CAST *Statistical Yearbook*, the funds in 2000 and 2001 were ¥1.365 billion (about US$200 million) and ¥1.683 billion respectively, while in 2002 and 2003 it was ¥2.537 billion and ¥2.650 billion respectively (China Association for Science and Technology, 2000, 2001, 2002, 2004), reflecting strong support.

Science popularisation during this period reached a new national level because of government attention to the development of S&T. Institutionalisation became more consolidated because of legislation and policymaking. Consequently, the consciousness and capabilities of social organisations for science popularisation have been stimulated and fostered. Science popularisation infrastructure increased, the science media was well funded, large-scale science popularisation activities have been carried out, more social forces and materials have been invested, and more members of the public are benefiting.

The model of government as the 'pushing hands' has its pros and cons. The investment in science popularisation is huge but the engagement of public and science community in a real sense is questioned by some researchers. They usually criticise the 'deficit model' approach standing behind popularising science and appeal for more 'constructive' involvement from different stakeholders of S&T (Jia and Liu, 2014). Although the policies have promoted the social recognition of the importance of science popularisation, there is usually a lack of follow-up regulations that make sure the policies work in practice (Chen, 2015). But at this stage, science popularisation activities have been merged with the social life of Chinese citizens via many channels and things are shifting from serving national economic development to a more independent domain, both in the sense of a national strategy and in the form of cultural life.

3. Current and future: Social mobilisation and participation (after 2006)

Social media platforms are gradually replacing traditional media like books, articles and TV programs. The young generation are accustomed to receiving S&T information via cell phone apps like TikTok, Quick and WeChat news. Science popularisation is often combined with entertainment.

Since science and technology not only impact the economy and society, but are also embedded in the daily life of people; their popularisation is leading people to another understanding of the relationship between S&T and society in which people are the core concern. They are striving for robust interactions between the development of S&T and society, to ensure the greatest benefit to society in the sense of sustainable development. Thus the scientific literacy of citizens and their attitudes towards the development of science-in-society relations are becoming very important to science popularisation.

3.1. Outline of the National Scheme for Scientific Literacy: Network and cooperation

In the past 30 years, science popularisation in China has mainly developed in a government-driven way, and CAST has been the main body that organised activities. Since the promulgation of the *Law*, more and more government departments and social institutions have devoted themselves to science popularisation, and a lack of coordination and guidance to the work is emerging. To resolve the problem, in 2006 the State Council issued the *Outline of National Scheme for Scientific Literacy (2006–2010–2020)* (hereinafter referred to as the *Outline*), which put forward new concepts, guidelines, plans and measures for the direction of science popularisation for the next 15 years (State Council, 2006). In order to mobilise social participation, a 'broad alliance and cooperation' framework has been adopted. More than

30 national ministries, research institutes and non-government organisations such as the Ministry of Science and Technology, Ministry of Culture, Ministry of Finance, Ministry of Agriculture, Ministry of Education, Chinese Academy of Sciences and CAST have become member units of the *Outline*. The *Outline* is a long-term plan to be promoted in different regions, groups and stages. The standards of national scientific and technological literacy at different stages of development, as well as objectives, key tasks and measures have been studied, and action plans and programs drawn up.

The implementation of the *Outline* is in the form of action plans and capacity-building projects. The key social groups include teenagers, farmers, the urban workforce, leading cadres and civil servants. It is expected that the improvement of scientific literacy of the key groups will promote that of the whole nation. Capacity-building projects are targeted at the weak points of science popularisation, and include projects to improve science education and training in residential communities, informatisation, infrastructure, popular science industry and human resource training.

In the 10 years since the issue of the *Outline*, science popularisation in China has indeed witnessed a rapid growth in human, financial and material input, as well as great achievements. For example, according to the data from the national surveys on civic scientific literacy conducted by CAST, the proportion of Chinese citizens with basic scientific literacy shows a rapidly growing trend: from 3.27 per cent in 2010 to 6.20 per cent in 2015, and projected to grow to 8.47 per cent in 2018 (Office for Implementing the Action Plan, 2018). A simple change of number cannot explain the characteristics of science popularisation in China, and we still need to look into it from a more multidimensional perspective.

3.2. Science communicator training

Since these matters became more important, professional training for communicators and researchers has gained urgency. It can be traced back to the 1980s when universities began to set up science communication majors, including training programs for undergraduates, master's students and a few doctorates. Most universities focus more on theoretical research aimed at cultivating students with knowledge of natural sciences, humanities and social sciences, equipping them with skills of science writing and using the media. Graduates are mostly science journalists, editors and freelancers involved in science books and magazines, as well as researchers. However, in recent years, facing a shortage of science communicators, universities have invested more in science communication majors. Many universities have increased majors and the number of students enrolled and improved the training plans. At present, most universities have classified science communication majors as secondary disciplines under journalism, communication and philosophy.

In China's '985 Project' universities,[5] there are two types with professional education related to science communication. One runs their program independently, such as Peking University, University of Science and Technology of China, Beijing Institute of Technology, Hunan University, University of Chinese Academy of Sciences, Fudan University and China Agricultural University. The other seeks to cooperate with CAST and other organisations and includes universities such as Beijing Normal University and Tsinghua University. The training programs of science communicators in these universities are mostly for undergraduate and master's degrees, and only a few universities have doctoral training programs majoring in philosophy of science and technology. In local colleges and universities, there are a few like Zhengzhou University and Zhongyuan University of Technology that also have master's programs (Mo, 2006).

In 2012, CAST and the Ministry of Education initiated their training program for a master's degree in science communication. Ten key universities were selected and, for the past six years, they have focused their training programs in science education, popular science product design and professionals for the science media, mostly to meet the needs of S&T museums. These programs intend to use the resources of pilot universities and museums in designing training programs and materials, sharing teachers and innovating the training model.

Up to July 2017, 571 postgraduates have been enrolled in six pilot universities. In 2015 and 2016, 306 students graduated from these universities with an employment rate of 94.38 per cent. They mostly went to enterprises, S&T museums, primary and secondary education institutions, government departments, universities and some went on to doctoral degrees. Among them, 14.1 per cent of the graduates went to S&T museums, while 13.7 per cent went to work in S&T enterprises (Department of Science Popularisation, 2017).

3.3. Science popularisation model

During the first 10 years of the 21st century, many scholars of science communication discussed and debated the concepts and models of science popularisation and compared them to developed countries. They believed that the concept and model in China was something between traditional

5 This refers to universities that are included in the national plan to be given resources to become high-level universities.

science popularisation and the public understanding of science—that is to say, between the top-down model (which represented the national standpoint) and the deficit model (which represented the standpoint of the science community). In some cases, the science communication model shows the traits of the dialogue model (which represented the standpoint of the public) (Liu, 2009). It is not easy to give a definitive answer to this question. In fact, the ideas and approaches of science communication of these three models are not hierarchically ordered or ranked and are no better or worse in an absolute sense. They simply focus on different approaches in line with different communication groups, conditions and objects.

In China today, these three models coexist, and the concept of public engagement is integrated into policymaking and the concept design of science popularisation projects and activities. For example, in February 2017, the Ministry of Education issued new standards for the science curriculum that made science education compulsory from the primary school. The Ministry of Environmental Protection approved the 'Measures on Public Participation in Environmental Protection' on 2 July 2015, aimed at providing legal rights for citizens to get access to environmental information, to participate in environmental protection and to open channels for social participation (Ministry of Environmental Protection, 2015). Not only are the government departments and science associations the main bodies of science popularisation, more stakeholders are getting involved. The science community and the media play a more active and influential role. Citizens' awareness of participating in science-related social affairs is growing, as shown in the protests of Xiamen PX project,[6] debates on genetically modified food and so on.

Compared with the developed countries, Chinese people utilised less critical thinking in considering what science and technology might bring to them. However, under the influence of the world campaign of science communication, the government encourages people to get involved and requests people to develop their scientific literacy in more sophisticated ways (Ren, Yin and Li, 2012).

At present, public scientific literacy refers to the following abilities:

- To possess knowledge of scientific content, method, thoughts and ethos.
- To apply this knowledge to resolve practical problems and participate in public affairs concerning science and technology.

6 In 2007, citizens in Xiamen initiated a series of demonstrations and protests against the new factory construction of the P-Xylene Project that could be harmful to their health. Several hearings on this matter have been held with the participation of policymakers, government officials, scientists and the citizens in Xiamen.

The multidimensional perception of science and technology and their functions in society expands the public's understanding of science and technology. Chinese people have come to realise the interactive relationship between science and the world, and the way it changes with a more comprehensive perspective.

Although public scientific literacy is being given more importance in the overall development of society by the government, the Chinese science community seems to fall into a dilemma when it comes to science popularisation. They usually recognise science popularisation as their social responsibility and an important part of their job in addition to their research, but not very many scientists like to communicate what they are doing to the public, especially when invited to do so in formal channels like TV interviews or newspaper reports. They prefer to act as 'informal risk communicators' in situations such as conversations between friends, neighbours or fellow travellers, in order to earn more trust from the public (Zhang, 2015). In May 2016, the National Conference on Science and Technology Innovation, the General Assembly of Academicians of Chinese Academy of Sciences and Chinese Academy of Engineering, and the Ninth National Congress of CAST were held together. On this occasion, Chinese leaders put forward the statement that 'science popularisation should be attached the same importance as that of science and technology innovation' (Xi, 2016). As the second largest economy in the world at present, if China wants to realise the transformation and upgrading of its economic development, it will not only depend on the driving force of science and technology innovation but also on the understanding, support and participation from the public in the process of innovation. To make the innovation process, economic growth and the public's quality of life work in a coordinated and sustainable way, science popularisation can obviously be of great help. But how to go about this, how science popularisation and science innovation can really become the two wings of the same bird, how the science community might be motivated to become powerful and active social actors for science popularisation, these are the crucial issues to be further studied.

References

Central Committee of the Communist Party of China and the State Council. (1994). Instructions on Strengthening Engagement in Science and Technology Popularisation. *People's Republic of China Yearbook, 1995* (pp. 335–37).

Chen, T. (2015). Looking into science popularisation system from the perspective of governmental policies and social cooperation. *Science Popularisation Study, 1*, 49–55.

Cheng, D. (ed.). (2014). *Research on modern S&T museum system with Chinese characteristics*. Beijing: China Science and Technology Press.

China Association for Science and Technology. (2000). *Statistical Yearbook*. Beijing: China Statistics Press.

China Association for Science and Technology. (2001). *Statistical Yearbook*. Beijing: China Statistics Press.

China Association for Science and Technology. (2002). *Statistical Yearbook*. Beijing: China Statistics Press.

China Association for Science and Technology. (2004). *Statistical Yearbook*. Beijing: China Statistics Press.

Chinese People's Political Consultative Conference Committee. (1949). The Common Program of the Chinese People's Political Consultative Conference. The first plenary session of the Chinese People's Political Consultative Conference. Official statement published in *People's Daily* (30 September), p. 2.

Cui, J. (2010). Looking into the background and course of promulgation of Law of PRC on Popularisation of Science and Technology. *CAST Forum, 281*(12), 2–3.

Deng, N. (2008). *Development and Responsibility: the 50 years of CAST* (pp. 57–73). Beijing: China Science and Technology Press.

Department of Science Popularisation. (2017). Work Report on Master Degree Programme Pilot (unpublished).

Fan, H. (2004, 9 January). Interpretation of 'traditional science popularisation'. *Science Times*, p. 3.

Guo, M. (1950). Science Popularisation and Improvement. *Newsletter of Science Popularisation* (restricted publications of Science Popularisation Bureau).

Jia, H. and Liu, L. (2014). Unbalanced progress: The hard road from science popularisation to public engagement with science in China. *Public Understanding of Science, 23*(1), 32–37.

Lewenstein, B. V. (2003). Editorial. *Public Understanding of Science, 12*, 357–58. doi.org/10.1177/0963662503124002.

Liu, G., Cai, Y.,Wang, H., Song, G. and Yang, Y. (1989). *Selected works of popular science broadcasting*. Beijing: Academic Publishing House.

Liu, H. (2004). On the Perspectives in Science Popularisation and the Tenets of Science Communication [in Chinese]. *Studies in Dialectics of Nature, 8*, 76–80.

Liu, H. (2009). Three models and three stages of science communication in China. *Science Popularisation Study, 2*, 10–18.

Liu, X. (2010). A study on the history of science popularisation of Peoples' Republic of China (PhD dissertation). University of Science and Technology of China.

Ministry of Environmental Protection. (2015). *Measures on Public Participation in Environmental Protection*. Retrieved from www.mee.gov.cn/gkml/hbb/bl/201507/t20150720_306928.htm.

Ministry of Science and Technology. (2010). *Chinese Popular Science Statistics* (pp. 49–52). Beijing: Scientific and Technical Documents Publishing House.

Mo, Y. (2006). Science communication major setting in Chinese universities. *Science Popularisation Study*, *2*, 31–35.

National People's Congress of the People's Republic of China (NPC). (2002). *Law of the People's Republic of China on Popularization of Science and Technology*. Beijing: Popular Science Press.

Office for Implementing the Action Plan Outline for Improving People's Science Literacy. (2018). *Report on the Construction of Public Science Literacy in China* (17 September). Beijing.

Ren, F., Yin, L. and Li, H. (2012). Science Popularisation Studies in China. In B. Schiele, S. Shi and M. Claessens (eds), *Science communication in the world: Practices, theories and trends* (pp. 65–79). Springer.

Ren, F. and Zhai, J. (2012). *Introduction to Science and Technology Communication and Popularisation*. Beijing: China Science and Technology Press.

Shen, Z. (2002). An investigation into the history of China's science popularisation [in Chinese]. *Science Popularisation*, supplementary issue, 5–60.

Shi, S. and Zhang, H. (2012). Policy perspective on science popularisation in China. In B. Schiele, S. Shi and M. Claessens (eds), *Science communication in the world: Practices, theories and trends* (pp. 81–94). Springer. doi.org/10.1007/978-94-007-4279-6_5.

State Council. (2006). *Outline of the National Scheme for Scientific Literacy (2006–2010–2020)*. Beijing: Popular Science Press.

Xi, J. (2016). *Strive to build the world scientific and technological power*. Speech at the National Conference on Scientific and Technological Innovation, the Conference of Academicians of the Chinese Academy of Sciences and the Ninth National Congress of the China Association for Science and Technology. Beijing: People's Publishing House.

Xu, L., Huang, B. and Wu, G. (2015). Mapping science communication scholarship in China: Content analysis on breadth, depth and agenda of published research. *Public Understanding of Science*, *24*(8). doi.org/10.1177/0963662515600966.

Yang, W. (1998). *The history of CCTV*. Beijing: Beijing Press.

Yin, C. (2008). The Differences Between Popular Science Bureau and China Association for Science and Technology. *Science Popularization Study*, *4*, 65–71.

Zhang, D. (2001, 2 March). Science popularisation in China: A brief review of a century. *Times of Science and Technology*, p. 12.

Zhang, J. (2015). The 'credibility paradox' in China's science communication: views from scientific practitioners. *Public Understanding of Science, 24*(8), 913–27. doi.org/10.1177/0963662515598249.

Zhu, Y., Cai, W. and Zhao, Y. (2011). Report on the Development of S&T Museums. In F. Ren and Z. Li (ed.), *Report on Development of Science Popularisation Infrastructures in China* (p. 53). Beijing: Social Sciences Academic Press.

Timeline

Event	Name	Date	Comment
First interactive science centre established.	China science and technology museum	1988	
First national (or large regional) science festival.	The first national science festival was held on 29 June 2003	June 2003	Since 2005, the date has been changed to the third weekend of September every year
An association of science writers or journalists or communicators established.	China Science Writers Association	August 1979	May 1988: Chinese Society for Science and Technology Journalism was established
First university courses to train science communicators.	University of Chinese Academy of Sciences established the first master's degree in education of science communication	2004	The university then set up a series of courses
First master's students in science communication graduate.	First master's students graduated from the University of Chinese Academy of Sciences	2007/1998	From the late 1990s master's students in philosophy, journalism, management etc. took science communication as their research direction. The first thesis on science communication was in 1998
First PhD students in science communication graduate.	There is no formal PhD degree in education of science communication in China	1999	Some PhD students gained degrees from 1999 in philosophy, journalism, management etc., but taking science communication as their research direction
First national conference in science communication.	Conference organised by Ministry of Science and Technology	February 1996	Hosted by the Publicity Department, Ministry of Science and Technology

Event	Name	Date	Comment
National government program to support science communication established.	Too hard to identify		These were many national programs but hard to find accurate records
First significant initiative or report on science communication.	*Instructions on Strengthening the Leadership to the Associations for the Popularisation of Science and Technology*	April 1953	This is the first document issued by the government in regard to science popularisation
National Science Week founded.		May 2001	Occurring in the third week of May
A journal completely or substantially devoted to science communication established.	*Science Pictorial*	August 1933	
First significant radio programs on science.	*Popular Natural Science Lecture*	September 1949	
First significant TV programs on science	*Science Knowledge* and *Medical Consultant*	January 1960	
First awards for scientists or journalists or others for science communication.	National Outstanding Popular Science Works Award	1980	
Date hosted a PCST conference	PCST Working Symposium, Beijing	June 2005	
Other significant events	First national press on popular science publishing	July 1956	Popular Science Press
	First law on science communication	June 2002	*Law of the People's Republic of China on Science Popularisation*

Contributors

Dr Yin Lin graduated from the Graduate School of the Chinese Academy of Social Sciences. She is a senior researcher and Deputy Director of the Division of Science Popularisation Policy Research at CRISP.

Dr Li Honglin is associate professor at the China Research Institute of Science Popularisation (CRISP), and Director of the Research and Communication Department at the China Science Writers Association (CSWA).

10

COLOMBIA

Stories in the history of science communication

Sandra Daza-Caicedo, Luisa Barbosa-Gómez, Tania Arboleda-Castrillón and Marcela Lozano-Borda

This chapter describes some of the main events of science communication during the last 60 years in Colombia. According to some authors, this history can be traced back to the mid-19th century. However, much research and archival work is needed to confirm this earlier point of origin. We have chosen the 1960s as a starting point for two reasons. First, it was the beginning of the institutionalisation of scientific policy in the country and consequently of government support to science communication activities. Second, this period has been widely studied by several authors and its activities have been documented (Fog, 2004; Lozano, 2005; Daza-Caicedo and Arboleda, 2007; Hermelin, 2011; Daza-Caicedo and Lozano-Borda, 2013).

We have reviewed official documents (scientific policy, public office for science communication files, official websites, etc.), as well as consulting books and papers on this topic. We have considered previous research including interviews and archive reviews. The history told here relies on the most visible and central actions and those that have had support from the government: actions of the Administrative Department of Science, Technology and Innovation (Colciencias),[1] which was created in 1968 and commissioned in 1969 to design and execute the country's scientific policy as well as to promote and finance the development of science, technology and innovation in Colombia.

1 In 2009, this institute changed its name to Administrative Department of Science, Technology and Innovation. However, since its creation in 1968 it has always been recognised as Colciencias, which is the term we will use in this chapter.

Our story has many gaps. For instance, we do not consider science communication activities carried out by the universities, the main producers of scientific knowledge in the country. This overlooks some radio and television stations, magazines and activities that researchers carry out in co-production with communities. Nor have we accounted for activities of community organisations, non-government organisations and other socially based organisations. We offer just a piece of 'the whole' history, some stories in the history of science communication.

The play on words we make between *History* and *stories* looks to emphasise not just the partial character of History, but the fact that science communication is part of a socioeconomic context, with social actors and their conflicts, political processes and different comprehensions about what science and technology mean. In other words, while great events occur—such as the creation of institutions, science festivals and museums, TV programs, awards, academic training—there are tensions, excluded actors and negotiations between stakeholders about the purposes and practices of science communication. In Colombia there has been a thoughtful debate on this topic for the last 20 years. As a result, some sectors refer to 'social appropriation of science and technology' (ASCyT for its acronym in Spanish) instead of science communication.

The meaning of ASCyT has changed over the years. Initially it intended to differentiate itself from deficit models, but it has been changing as researchers from different fields (critical studies of communication, social studies of science and technology, communication for development) conduct research on the practices, actors and objectives of science communication activities. Although there is still no agreement on its definition, the academic and political debate on ASCyT has allowed a critical reflection on its activities and objectives.

This chapter is composed of three sections. The first one is devoted to a *history of events*. We use a set of periods suggested by previous works to give an account of the changes in public policies as well as in the terminology used to address science communication. We will go through five different periods that can be differentiated by the way science communication has been done and what it has been called. As we cover more than a half century, the narrative can sometimes get 'overcrowded': too many names, institutions, activities and projects. We encourage the reader to let themselves dig quickly into this history of events. The second section deals with the *stories* of science communication, centred in the debate around the local concept of social appropriation of science and technology. Finally, we offer some conclusions that try to summarise the trends of the many diverse actions done in the field of science communication in Colombia in the last decades.

1. The history of events

1.1. From newspapers to informal education

In the second half of the 20th century, after World War II, science and technology (S&T) arose as a promise of development. Latin America witnessed the emergence of S&T programs intended to 'modernise' countries. Development agencies such as the United Nations, the Organisation of American States (OEA) and the United Nations Educational, Scientific and Cultural Organization (UNESCO) offered financial and conceptual support in the areas of education, science and technology. These agencies and some intellectuals of the region started promoting the objective of 'closing the gap' between the underdeveloped Latin American countries and the developed world by the means of science and technology. This gave birth during the 1960s and 1970s to the National Councils and Ministries of Science and Technology in various countries, including Colombia (Franco-Avellaneda and von Linsingen, 2011). The new panorama required the implementation of technology and knowledge transfer processes accompanied by communication strategies in the fields of health and agriculture, among others, with the promise of 'modernisation'. Such promises set the conditions and interests for the communication of S&T through the media.

In this context, during the 1960s an enthusiasm and an institutionalisation of science and technology emerged, including science communication activities. A series of courses and workshops were held in Chile, Ecuador and Argentina between 1962 and 1965. In 1967, Ibero-American leaders agreed on a regional program in which education and scientific and technological dissemination stood out among the priority objectives (Declaración de Punta del Este, 1967; Calvo Hernando, 1999).

In Colombia, the first science communication activities in the 1960s were in the form of science education. The program Scientific Activities for Youth was funded by the MIT–Harvard Club of Colombia, the Bank of the Republic and the Ford Foundation (Posada et al., 1995); and focused on supporting science teaching and promoting scientific vocations (Fog, 1995).

In 1968, Colciencias was founded as the government organisation in charge of leading the policy and the development of science, technology and innovation activities in the country. Two years after that, the Colombian Association for the Advancement of Science (ACAC) emerged, a non-profit private organisation whose mission was to create consciousness of the importance of science, technology and innovation (STI) to society. These two events constituted a breakthrough in the history of science communication in Colombia, as they created the central institutions that have determined the policies and the dynamics of the field in the country.

Colciencias stated for the first time in an official document (Colciencias, 1971) the necessity to 'create consciousness on the important social role of science and technology in the mind of every Colombian'. However, for a long time the actions to achieve this goal remained focused on science education for Colombian youth and the program had a very limited budget.

In 1969 another milestone occurred in this history: a round table in the capital city Bogotá about the value of education, science and culture for national progress. The event was convened by the Ministry of Education and the OEA (Massarani et al., 2012). As a result, the Inter-American Centre for the Production of Educational and Scientific Material for the Press (CIMPEC) was created. This institution had the important task of producing scientific and technological material for the media in Colombia and all other member countries (Fog, 2004). That same year, the first National Congress of Science Journalism was held in the city of Medellín, which motivated the first ideas for the subsequent creation of the Ibero-American Association of Science Journalism (Massarani et al., 2012).

During this period and until 1976, a weekly scientific supplement appeared in *El Tiempo*, the largest-circulation national newspaper. In addition, the national radio station had a half-hour program on Saturdays dedicated to scientific dissemination (Muñoz Quevedo, 1986). The first science TV show *Naturalia* was broadcast in 1974, presenting documentaries of explorers and adventurers such as Jacques Cousteau and David Attenborough (Cortés-Fonnegra, 2014). Within this context of enthusiasm for the development of science and technology, in 1976 the Colombian Association of Science Journalism (ACPC)[2] was created.

In this early phase of science communication in Colombia, multilateral organisations played a crucial role. For instance, the Inter-American Bank (IDB) lent money to Colciencias in 1982 to promote science and technology and, by doing so, settled the line of action around these topics. The activities relating to science communication were shaped according to what was occurring in the Anglo-Saxon countries: magazines and TV shows promoting scientific activity were created. The discourse was also affected when Colombia adopted the term 'science popularisation' instead of 'dissemination' (Daza-Caicedo and Lozano-Borda, 2013).

2 See acpc.com.co/acpc-asociacion-colombiana-periodismo-cientifico/.

With the IDB loan, a new Science and Technology Policy was launched by Colciencias (1983). The policy stated three areas of action for the strengthening of science popularisation in the country: science journalism, the scientific activities for young people and the use of mass communication media. By including mass media and science journalism, the activities of that period reached several publics from different ages and regions instead of just the young.

The first written publications completely devoted to science communication were established with the newsletter *Letter from Colciencias* and the magazine *Colombia, Science and Technology* in 1982. The first science TV show was launched with the name *Diffusion and Scientific-technological Formation* in 1984. It was co-produced by Colciencias and the Colombian Institute for the Promotion of Higher Education (ICFES) and showed the advances Colombia had made in S&T, as well as introducing the leading professionals in the field to the public.

By this time, science popularisation was perceived as an effort arising from institutions, the media and individuals related to science teaching and information, as well as researchers who actively participated in outreach of their research. With the involvement of the mass media—first cinema and radio and then TV—a new era of professionalisation arose.

In 1986, the Executive Secretariat of the Andrés Bello Agreement[3] and the Konrad Adenauer Foundation developed a program to promote the specialisation of journalists in science and technology; as a result, two books were published in Bogotá in 1986 and 1988. They contributed to the scarce bibliography in the Spanish language available on science topics (Calvo Hernando, 2002).

In 1987, following the line of informal science education for youth, the national program Cuclí Cuclí started. Supported by the National Ministry of Education and the National University, the program promoted science by linking it with primary and secondary educational curricula. This program was active until 1997 and laid the foundations for other initiatives such as Nautilius (1995–96), the Pléyade project (1997–98), Cuclí-Pléyade (1998–2001) and Ondas. All these programs aimed at promoting scientific vocations and interesting children and young persons in science and technology (Lozano-Borda, 2013).

3 The Andrés Bello Agreement is an inter-governmental organisation for educational, scientific, technological and cultural integration in Ibero-America. It was created in order to contribute to the equitable, sustainable and democratic development of the member states through a treaty signed in Bogotá on 31 January 1970. Members are: Bolivia, Chile, Colombia, Cuba, Ecuador, Spain, México, Panamá, Paraguay, Perú, the Dominican Republic and Venezuela.

That same year, the first S&T fair Expociencia was organised by the ACAC with the support of Colciencias. This fair is still organised annually, although there have been gap years.

In 1990, the Network of Popularisation of Science and Technology for Latin America and the Caribbean (RedPOP) was established. Colciencias played a major role in the creation of the network, and its objective was to contribute to the enhancement, exchange and active cooperation between the teams, programs and institutions devoted to the popularisation of S&T in the region (Massarani, 2015; Pabón, 2017).

In 1991, the second loan from the IDB allowed the restructuring of Colciencias and its transformation into a division fully dedicated to the organisation of activities targeting a non-scientific public. The newly created department elaborated a proposal for a National Plan for the Popularisation of Science and Technology, a first attempt to develop a policy on this topic—although it was not officially accepted.

2. The boom of interactive museums, science journalism courses and more

Another milestone in the history of science communication in Colombia was the creation of the Mission of Science, Education and Development in 1993. The mission was a group of academics commissioned by the president, and they worked together to elaborate a set of recommendations on the future course of science, education and development in the country. The mission wrote the report *Colombia, at the edge of opportunity* (Posada et al., 1995) and used for the first time the term 'social appropriation of science and technology' in an official document with political intention. The report did not evolve into a policy but constituted an important step towards an S&T system with cultural, ethical and democratic basis.

Even though a new term was proposed in the document, the aims and activities described did not differentiate from the traditional ideas of mass popularisation. This could be in part due to the fact that 'appropriation' was proposed as a means of raising awareness of the importance of promoting science in culture, without having fully developed a research-based conceptual framework around its meaning. As a result, the term often got merged with others used in public policy, such as 'popularisation', 'dissemination', 'science journalism' and 'science education'.

During this decade, Colciencias supported initiatives aimed at boosting science communication through the media, and public universities played an important role. For instance, in 1993 the Universidad de Valle created the first news agency for science and technology in Colombia. The University Agency of Science Journalism (AUPEC) gave rise in 1994 to the first science journalism award, which received entries from 86 journalistic works from newspapers, TV and radio. This university also produced the TV show *Eureka* (1996), and the National University of Colombia followed with *Mente Nueva* [New Mind] in 2000. Colciencias promoted other television shows in these years, many of which faced big problems of funding and acceptance by the private mass media, such as the series *Universes* (1997) and *For Science* (2003).

The Museo de la Ciencia y el Juego [Museum of Science and Play] founded in 1984 was one of the pioneers in the country, using interactive, playful and itinerant activities to explore scientific concepts. The network of Small Interactive Museums of Science (Liliput) in Colombia and Ecuador was consolidated during these years. Another initiative promoted by the museum was Maletas viajeras [Scientific Suitcases] that inspired the concept of itinerant museums and allowed many Colombian towns to access science communication activities.

In 1998, and promoted by ACAC, the interactive museum Maloka was founded with funding from a third loan of the IDB. Maloka was the first cultural and educational megaproject focused on science and technology, using much of the economic resources devoted by Colciencias to science communication. Since its creation, Maloka has a cultural program that offers many entertaining activities, hosts activities such as science cafés and works with teachers to explore the pedagogical possibilities of the centre and to collect opinions and suggestions.

With the new century, a new program of Scientific Activities for Youth started. Ondas [Waves] was launched in 2001 thanks to the collaboration between Colciencias and the Foundation for Education and Social Development (FES). It became the longest-lasting science communication initiative for children in Colombia. The program carries out research projects based on the interest and curiosity of the children, designed and developed by the children together with their teachers.

In 2002, the Colombian Association of Science Journalism set its focus on three projects: training of science journalists around the country, research on the work of science and technology communicators, and setting up a S&T news agency (NotiCyT). Even though the agency was not a pioneer in the field, it did have an impact on local media as well as Latin American newspapers (Vélez Lopera, 2013). Despite this, it only lasted five years.

Other institutions also initiated efforts to offer training in science communication. In 2002 and then 2004, the ACPC organised a series of seminars in different regions of Colombia (Daza-Caicedo et al., 2006). Some years later, the ACAC together with the Javeriana University offered training courses on the social appropriation of sciences.

During this period one 'innovative' activity called *regional encounters* took place. It was developed by Colciencias with the support of the ACAC. The encounters were events that, for the first time, created spaces for direct dialogue between researchers, industry and journalists all over the country. Their objective was to socialise the results of the activities promoted by Colciencias among several actors of the different regions.

In 2003, theatre came into play. The National Theatre Company's Theatre of Science attempted to promote science and technology through artistic language. Two plays resulted from this, and were presented in big events in Bogotá, like the Book Fair and Expociencia.

Institutions started recognising the value of dialogue around the course of science communication. In 2004, Maloka organised the first academic forum Conciencia abierta.[4] It gathered together communicators, academics and researchers from different Latin American countries to reflect upon the process of appropriation of S&T in the region. The organisation of the forum was supported by Colciencias and the Andrés Bello Agreement.

3. Appropriation becomes a policy

In Colombia, the National Council of Socio-Economic Policy (CONPES) is the highest planning authority. It advises the national government in all aspects of socioeconomic development by coordinating and monitoring the study and approval of public documents that describe different policies to be applied in the country. In 2005, CONPES approved the *National Policy for the Social Appropriation of Science, Technology and Innovation* (ASCTeI) (Colciencias, 2005).

4 The literal translation is 'Open consciousness'. In Spanish, the name is a play on words that also means 'open with science'.

This sought to better define the terminology involved, using an extensive theoretical and epistemological approach, as well as describing a strategy for implementation. It constituted an inflection point by considering the population as active participants with decision-making capacity. Therefore, it challenged the deficit model that had been implemented until that time.

One year after the CONPES, the Colombian Science and Technology Observatory (OCyT) published a report on the assessment of activities for the public communication of S&T. With this report, Colciencias manifested the need to migrate towards a more democratic model 'in which the publics are not conceived as mere receptors of scientific information' and are instead recognised as active participants in the process of knowledge production (Daza-Caicedo et al., 2006).

For the first time, authorities, institutions and actors in the national system of S&T were urged to 'democratise' scientific knowledge. 'New' strategies came into play, and they took into consideration citizen participation, public opinion and the interests and needs of society. Some activities focused on participation rather than mass media, promoting forums, open debates and the creation of networks.

Following that line of thought, the National Science and Technology Week was founded in 2006. This initiative created a space for the different sectors of society to engage in the process of scientific knowledge production. With National Science Week another important goal was achieved: to decentralise the activities that mostly occurred in Bogotá. Science Week was organised biennially until 2014. Colciencias supported it financially and technically, and this initiative has presence in 25 (out of 33) territorial entities of the country (32 departments plus the capital district). Science Week helped shift activities away from conferences and talks, and towards more participatory events with the integration of different actors, including universities, research centres, companies, students, community agents of different kinds and indigenous and Afro-descendant communities.

In 2008, 10 years after the opening of Maloka, the second science centre Parque Explora opened in Medellin. From the start, this centre developed a participatory approach much in line with the idea of appropriation. In 2010, Maloka published the book *Relocating the Social Appropriation of Science and Technology* (Pérez-Bustos and Tafur Sequera, 2010) aimed at recognising the effort of actors in different regions of the country acting in favour of the social appropriation of knowledge.

Figure 10.1: Parque Explora, Medellin.

Source: Guía de Viajes Oficial de Medellín.

The public policy of this period shows a misalignment between the goals and the strategies of ASCTeI in Colombia. Although the theory is clear about the importance of the relationship between science, technology and society, a great part of the actions still focused on a vision of science, technology and innovation that is external and independent from their cultural and social contexts.

4. The appropriation policy gets a strategy

The National Strategy of ASCTeI was launched in 2010 as a complement to the National Policy of 2005 (Colciencias, 2010). It led to one of the most important changes at the organisational level: the design and realisation of national open calls instead of financing projects evaluated one by one. This allowed a greater control in the execution of the policy as well as opening up and introducing a greater democratisation of the resources, as the regions and types of organisations diversified.

The strategy differentiates from previous public policy by recognising several types of actors, not just scientists, involved in the processes of the generation and use of knowledge. It distances itself from the assumption of a gap between 'producers' of knowledge and 'receptors–users', on the basis that science is a social construction. As the civil society's agency is recognised, the development of initiatives acquires a democratic and participatory perspective.

The strategy consists of four lines of action:

- Promote citizen participation in the construction of public policy in STI.
- Communicate S&T from the perspective of society, to favour the development of reflective and contextualised communication projects for the understanding, dialogue and opinion formation on the relations of STI and society.
- Exchange and transfer knowledge, to stimulate initiatives with an effective integration into specific local and social contexts that contribute to development.
- Develop training and measurement mechanisms to generate knowledge about the different ways in which scientific and technological knowledge is appropriated in Colombian society.

These lines of action are expressed to a greater or lesser degree in the programs and initiatives financed by Colciencias between 2010 and 2018.

In 2011 the Science, Technology and Innovation Fund was created to increase scientific and technological capacities, as well as the innovation and competitiveness of Colombian regions. It was financed by royalties from natural-resource extraction (Article 29, Law 1530, 2012 ; Official Journal of the Republic of Colombia, 2012). Within five years (2013–18), around 50 ASCTeI projects were financed with these resources. More than half of these belonged to the Ondas program, and the rest was attributed to ASCTeI, science centres and scientific and technological vocations. Some ASCTeI projects were aimed at strengthening the citizens' culture in STI through, for example, research projects to promote the capacities, skills and attitudes towards science and technology research in children and young people linked to primary and secondary education in the Caribbean island of San Andrés. Another aimed at reaching the Buen Vivir [Good Living] and the territorialisation of peace in the ecoregion of Perijá.

'Territorialisation of peace' means to develop peace-building processes in diverse Colombian territories. Even though the administrative division in Colombia is the department (there are 33, including the capital district of Bogotá), the concept of 'territory' is related to the identity of various groups whose life and permanence in these lands are constantly threatened. Groups affected by this include peasants, indigenous people and the Afro-Colombian people, a result of the colonial legacy and inequity in land distribution that is concentrated in very few rich families. The underlying tension since the 'discovery' of the continent and the transfer by force of the African people to these lands and subsequent enslavement of them and the indigenous people were the main reason for 60 years of warfare.

There are difficulties with the ASCTeI projects, and many actors report it is impossible to spend the money because of the logistics and paperwork, ending with an absurd budget in which 'two of every five pesos in the fund went unspent' (Bajak, 2018).

In 2011 the first public forum, Replantémonos, was held in the Congress, aimed at promoting an effective forest policy that responded to a social, economic and ecological crisis. In 2012, a National Water Meeting held regional workshops to collect and systematise experiences on water as a factor of economic and social development, as a basic human need and as a risk in the face of droughts, floods and pollution. And finally, in 2014, the National Forum on Social Appropriation of Science and Technology elaborated recommendations to the strategy (Aguirre et al., 2011). Unfortunately, a change in the director of Colciencias that year did not give continuity to the forums, as the new director did not agree to accept these recommendations.

In 2012, the OCyT organised the Survey on the Public Perception of Science and Technology. Its results revealed the general public's lack of knowledge of local science and scientists, and the influence of the media on the stereotypical misrepresentation of S&T among Colombians. For instance, it was evidenced that people thought science was confined to natural and basic science; that it was all about discoveries and technology and that it was done by men who were above average intelligence (Daza-Caicedo et al., 2014). This was the third official survey of its kind carried out in the country. The first survey was published in 1995 by Colciencias and the ACAC (Misión de Ciencia, Educación y Desarrollo, 1995) and the second one was published in 2005 (Aguirre, 2005), carried out by Colciencias and the National Consulting Centre.

Between 2011 and 2013, more than 20 communication projects were funded to produce content for mass media broadcasts (including radio and virtual channels) and to develop strategies to contextualise science, technology and innovation in a critical perspective. Some of these projects were developed in local communities with the active participation of members of the communities, research groups and organisations that promote ASCTeI processes.

With the purpose of encouraging collaborative work among science, technology and innovation experts and communities or social organisations, in 2012 the program Ideas for Change was launched.[5] In the first phase, poor communities postulated needs that could be addressed through scientific-technological knowledge. Then, universities and other organisations put forward possible solutions. After technical evaluation and prioritisation, solutions are then implemented by negotiation, exchange and knowledge transfer between experts and communities. So far, the program has made four biennial calls (2012–18). 'Water and poverty' was the theme of the first call, and it was to address water problems in the communities of three territorial entities. Ten innovative scientific-technological solutions for access to drinkable water were proposed. In 2014, the call Pacific Pure Energy addressed energy solutions for clean and renewable sources and, as a result, 14 collaborative solutions to supply energy in community spaces such as schools and health centres were implemented. In 2016, the third call sought to promote the preservation and conservation of the environment through the design and application of clean technologies contributing to the sustainable use of natural resources; 14 solutions were proposed. The fourth called 'Science and ICT for Peace' is under development. It aims to support processes of social appropriation

5 See minciencias.gov.co/cultura-en-ctei/ideas-para-el-cambio.

of science, technology and ICT in collaborative construction with surviving communities from the armed conflict in Colombia, contributing to human and sustainable development in a post-conflict context.

A Ciencia Cierta[6] is another initiative developed during this period. It takes the form of a contest that invites communities to share experiences in which they have applied scientific and technological knowledge. Those with the greatest impact and relevance in the social field then participate in learning processes and knowledge exchange in sessions involving scientists, technologists and communities. Up to now, three versions have been developed (2013, 2015, 2017), supporting 36 experiences regarding the agricultural sector, food security and micro-enterprises for the use of biodiversity. A more recent call (2018) has been developed around community projects for the conservation of ecosystems.

In 2014, Colciencias implemented Virtualia,[7] an online training program with support of the United Nations Development Program (UNDP) Virtual School, to provide tools for the ASCTeI to assist strategic actors and institutions. The program aimed at strengthening capacities and processes and boosting social practices with a focus on human development. The training included the development of written projects based on proposals for solving local problems. Courses were run for community leaders of local organisations interested in the development of participatory processes involving the ASCTeI—namely professionals working in public libraries and museums (science museums, botanic gardens, planetariums, zoos, aquariums).

From 2014, a change in the chief executive officer of Colciencias also changed the approach of the ASCTeI as proposed in the strategy: the key priority became to strengthen the development of museums and science centres. With this turn, Colciencias provided guidelines and supported the creation, recognition and strengthening of science centres and promoted the National Network of Science Centres[8] (Colciencias, 2015). Likewise, a general model of good sustainability practices in science centres was developed to respond to the need for these organisations 'to remain valid in times of uncertainty and thrive under changing economic, legislative, cultural and environmental conditions' (Colciencias, 2016).

6 See www.acienciacierta.gov.co.

7 See www.colciencias.gov.co/cultura-en-ctei/apropiacion-social/virtualia.

8 See www.colciencias.gov.co/cultura-en-ctei/apropiacion-social/centros-ciencia/por-que-los-centros-ciencia.

Parallel to the National Strategy of ASCTeI, Colciencias currently supports a strategy called Todo es Ciencia [All is science]. It mainly follows the deficit model of science communication through the production and issue of web series, news, opinion articles and documentaries 'in order to show a diverse panorama to inspire people to appropriate science as an engine of a better world'.[9]

Furthermore, local processes are developing in regions of Colombia like Antioquia where the book *Social Appropriation of Knowledge, the Role of Communication* (Aguirre et al., 2013) gathered some support with critical views on the topic and proposing a more regional approach. In the Caribbean coast, the University Jorge Tadeo Lozano has organised living laboratories and researched on the appropriation carried out in universities (Hernández et al., 2016). In the Coffee Region, the Quindío University developed the Centre for Scientific Culture and risk management programs.

5. From science communication to social appropriation of science of technology

In the field of science communication, it is very important to tell a story listing milestones, main characters and the consolidation of institutions and policies. This history allows for comparisons and lets readers analyse the creation and strengthening of this field. However, to the 'history of events' one should add a 'social history'. Therefore, one must consider the tensions, the power games, the different conceptions about communication, the silent actors and so on.

We do not intend to write down the complete story in these few pages. However, we would like to emphasise that S&T communication in Colombia has been a dynamic field, always involved in conceptual disputes—sometimes enriching, sometimes not. A first element in this complex dynamic is set out in the previous section: there is a close link between science communication, public policy and transnational agencies and their agendas. Very few activities have been done without the financial support of Colciencias. This is probably due to the scarce resources devoted to S&T in the country, but also because the first visible activities in the 20th century occurred thanks to IBD loans and with the support of other transnational institutions like UNESCO, OEA and the Andrés Bells Agreement. This limited the type of actions that could be conceived and executed (Daza-Caicedo et al., 2006).

9 See www.todoesciencia.gov.co/todo_es_ciencia/presentes-en-docmontevideo-uruguay-2017.

The IDB loan allowed the growth and development of some activities, like the museums and interactive science centres, but was not flexible enough to encourage activities of 'alternative communication' or local initiatives that did not follow international models. Moreover, the lack of continuity in the availability of resources translated into the quick death and disappearance of many activities.

So the relationship between communication for development—carried out through public health campaigns, agricultural transfer, alphabetisation campaigns[10]—and science communication is yet to be pointed out. We left out of the analysis to what extent the development agenda favoured or hindered local processes of scientific production and communication and how the ideals of development biased ways of understanding the science–society relationship.[11]

Between the 1980s and 1990s, many academics started pointing out the need of society to appropriate science. By the beginning of the 21st century—and after some ground-breaking actors and communication activities—an academic debate started around the dominant communication model. The discussion was inspired by the debates from Roqueplo (1983), Raichvarg and Jaques (1991), Durant (1993), Lewenstein (2003), Irwin and Michael (2003), Michael (2002) and Felt (2003) about the 'deficit' and 'democratic' models in the public communication of science. The core of the discussion orbited around four questions: Why communicate science? From whom? To whom? And how?

Those debates stimulated the consolidation and use of the concept 'social appropriation of science and technology' (ASCyT) and a new idea to get away from the deficit model of communication. The new approach fostered dialogue between techno-scientific knowledge and local perspectives, promoting a scientific practice concentrating on the solution of local problems and with the inclusion of civil society. Such debate was encouraged by social and human sciences researchers, most of them focused on the social studies of science and technology. Due to their backgrounds, they insisted on considering science as a social construct (Woolgar, 1988; Latour, 1992, 2007; Hess, 1997; Bloor, 2003).

10 Campaigns to teach adults to read and write.

11 Some of these issues have been addressed by Escobar (1996), Gómez-Morales (2005), Pérez-Bustos (2009) and Franco-Avellaneda and von Linsingen (2011).

The term 'social appropriation of science and technology' was quickly embraced by politics and actors in the field of science communication, but its meaning has not been stable. Some use it as synonym of words as 'popularisation', 'dissemination', 'outreach' and 'communication'. Others try to fill it with sense, as in the definition by Franco-Avellaneda and Pérez-Bustos (2010):

> An intentional social process in which diverse actors in a reflexive manner articulate to exchange, combine, negotiate and dialogue knowledge, motivated by their needs and interests to use, apply and enrich such knowledge in their contexts and concrete realities. We understand that this intentional social process happens through mediations of recognition, information, teaching-learning, circulation, transfer, transformation and production of knowledge, among others, of which science and technology are its main object.

In the ASCyT policy the concept is defined as (Colciencias, 2010):

> A process of understanding and intervention of the relationships between techno science and society, built upon active participation of the various social groups that create knowledge.

These two definitions seek to account for the complexity and diversity of mediation devices that comprise ASCyT, emphasising the need to think of it as a critical, context-dependent activity that involves the participation of different actors. This conception considers communication as a process that must be understood 'in the context of historical, social, geographical, political, cultural conditions' (Huergo, 2001). In that sense, there is no ASCyT strategy that always applies uniformly to all cultures.

Why is there no consensus on the meaning of appropriation? There are multiple interests that range from stakeholders looking for scarce resources and maintaining the management of activities, to promoting certain imaginaries about science and technology or specific ways of giving continuity to policies. On the other hand, because science communication and appropriation remain as secondary areas in the macro policies of science and technology, they have not obtained important resources or clear instruments for their implementation.

Finally, there are voices that still have very little participation and power: communities could use techno-scientific knowledge to negotiate with their own traditional knowledge (indigenous, Afro-Colombian, rural, etc.) to solve contextual problems. The questions that arose in the 1990s are still open: what kind of science and for whom?

6. Conclusions

The history of science communication in Colombia shows a slow development of the field in at least five sorts of activities: science journalism and communication through mass media; scientific activities for youth; museums and science centres; citizen participation; and research on social appropriation of knowledge.

We have documented a significant number of mass media productions made in Colombia, mainly focused on children. In most cases, they have been short-term strategies with weak alliances that have not facilitated co-production with public channels. Science journalism has lost the momentum of the late 1960s and it has been difficult to keep active the Colombian Association of Science Journalism. As an important stakeholder, the institution should have been centrally placed to position and mobilise the issue in the country. The association has been boosted recently by a new generation of young science journalists and communicators with interesting projects.[12] On the other hand, we evidence the low prestige of science journalism in the Colombian mass media: in TV news, science and technology has a very low key presence compared to Latin American countries like Brazil. Even when these topics are covered, there is little contextualisation and use of scientific sources, with a bias for presenting science in a positive way and rarely mentioning scientific controversies that limit public debate (Arboleda Castrillón et al., 2015; Ramalho, Arboleda and Hermelin, 2017). This situation is intertwined with the limited professional training on offer for mediators and scientific journalists.

Many science communication experiences in the country target children and young people in schools—this was particularly so until 2005. They aim at motivating future generations for science. Slowly, the regionalisation of this topic allows the appearance of new actors, and options to communicate science and opportunities to build more dialogical initiatives. This means we require not only scientific journalists, but other types of 'mediators'.

Another important group of initiatives are the S&T museums and interactive centres. In Colombia they arose through the initiative of scientific communities interested in developing effective translation exercises that show science as relevant and 'fun' to diverse audiences, with the aim of achieving

12 Some of these new science communication projects are Shots de Ciencia (soundcloud.com/shots-de-ciencia), cientificamente.net (www.facebook.com/pg/Cientificamente.net/about/?ref=page_internal), or Ciencia Café Pa' Sumercé (cienciacafesumerce.wordpress.com, www.youtube.com/watch?v=w1lkk5CqAuQ).

a greater appreciation of their practices and implications. These strategies to seduce the public gained great relevance in the allocation of Colciencias' budget and other national public entities. However, the museums' offerings tend to quickly fall short in front of a population eager for novelty and expecting to find new wonders at each visit. Their renewal and maintenance are expensive and difficult to sustain over time, especially when only a few interactive centres in the country have significant support from private companies. Because of this, and because national goals are prioritised, public investment in S&T interactive museums and centres has been very variable, going from periods of great support to periods of almost none.

Despite the above, interactive museums are significant in Colombia and they have become a reference point in public communication of S&T. Likewise, they have been laboratories of new types of experiences, which seek in an innovative manner to impact the field of informal education and the social integration with the environment in which they are located.

As an alternative to the traditional models of communication of science, citizen participation initiatives have emerged. These initiatives have been strongly promoted in the last few years and emerge from the need to have dialogical models to contribute to the social appropriation of knowledge. Hence, forums and public debates about controversial issues of local interest are organised and new calls encourage co-production of knowledge and knowledge-sharing. Although it has not been possible to fully consolidate this, it shows an interest for constructing strategies that break the mould of deficit models of communication. It is noteworthy that the constant change of direction in Colciencias during the last period has had an impact on its strategic approach, going vertiginously from participatory to deficit models and ending abruptly with ongoing proposals.

In the last 20 years, there is growing interest for evaluating strategies related to the identification, differentiation and understanding of audiences. This idea is motivating research and evaluation in Colombians' perceptions of science. The discussion has also been framed theoretically, encouraging an academic community beginning to engage with this interest, feeding a reflective and critical view on ASCyT processes and creating evaluation systems. The ASCyT is gradually consolidating not only as a field of practice but also as a relevant field of studies.

Universities have been an important player in Latin American scientific and technological development and are considered the centre of knowledge production in the region (Albornoz, Barrere and Sokil, 2017). Thus, it is necessary to identify the role that Colombian universities have in the communication and appropriation of S&T. They should open the debate on how the knowledge science produces is useful to society, to its environment and to local communities. In brief, their role in ASCyT should go beyond training future mediators. Universities can encourage processes of encounter, communication, negotiation and exchange for new forms and places of knowledge production that enable the resolution of social problems. But this requires researchers to work on local research agendas, carry out processes of dialogue with communities and turn their interests not only to the production of papers in indexed journals but to other forms and formats of communication. It is urgent to recognise the work of researchers who devote time and effort to these activities and try to connect scientific production with local necessities and agendas.

For years we thought that the mission of science communication was to conquer citizen's hearts, but decades of activities have taught us that we also need to conquer the heart of researchers and policymakers. First, because we need more resources, both economic and human, for science communication to develop new and innovative activities. Second, because it is still necessary to understand that citizens can be involved in science and technology activities and policy design. Colombia has made a great effort in elaborating a discourse that recognises the citizens' right to participate, but the materialisation of it is still a work in progress.

References

Aguirre, C., Arboleda Castrillón, T., Cassiani, S., Daza-Caicedo, S., Guivant, J., Hermelin Bravo, D., Hilgartner, S., Jasanoff, S., Lozano-Borda, M., Lozano Hincapié, M., Maldonado Castañeda, O. J., Metcalfe, J., Olive, L., Raigoso, C., Sutz, J. and Tafur Sequera, M. (2011). *Ciencia, Tecnología y Democracia: Reflexiones en torno a la apropiación social del conocimiento*. M. Lozano-Borda and T. Pérez-Bustos (eds). Medellín: Universidad Eafit / Colciencias.

Aguirre, C., Dominguez, E., Echeverry, J., Castaño, M., Raichvarg, D., Trelles, I., Jaillier, E. Velez, N., Jaramillo, F., Acevedo, C., Marin, B., Marquez, J., Domínguez, E., Echeverry, J. and Castaño, M. (2013). *Apropiación social del conocimiento, el papel de la comunicación*. E. Domínguez, J. Echeverry and M. Castaño (eds). Medellín: Universidad de Antioquia.

Aguirre, J. (ed.) (2005). *La percepción que tienen los Colombianos sobre la Ciencia y la Tecnología*. Bogotá, Colombia: Colciencias.

Albornoz, M., Barrere, R. and Sokil, J. (2017). 2.1. Las Universidades lideran la I+D en América Latina. In RICYT, *El estado de la Ciencia*. Retrieved from www.ricyt.org/wp-content/uploads/2017/10/files_Estado-de-la-Ciencia-2017_E_2017_2_1_Las_Universidades_lideran_id.pdf.

Arboleda Castrillón, T., Hermelin, D., Massarani, L. and Reznik, G. (2015). Ciencia y Tecnología en los telediarios colombianos: sobre lo que se cubre y lo que no se cubre. *Revista Ensaio, 17*(1), 208–29. doi.org/10.1590/1983-21172017517011 0.

Bajak, A. (2018). Science in Colombia on the cusp of change. *Nature, 562,* 7728. doi.org/10.1038/d41586-018-07113-6.

Bloor, D. (2003). *Conocimiento e imaginario social.* Barcelona: Gedisa Editorial.

Calvo Hernando, M. (1999). *El nuevo periodismo de la ciencia.* Quito, Ecuador: Ediciones CIESPAL.

Calvo Hernando, M. (2002). La divulgación de la ciencia: ambiciones y objetivos. El periodismo científico -PC-, no contribuye a satisfacer las necesidades básicas de América Latina en este campo. *Colombia Ciencia y Tecnología, 20*(1), 10–17.

Colciencias. (1971). Política científica y tecnológica: consideraciones preliminares. Documento de trabajo. Santa Fé de Bogotá: Colciencias. Retrieved from repositorio.colciencias.gov.co/bitstream/handle/11146/1269/1987-V11-N3-4-Articulos-Art%203-4.4.pdf?sequence=1&isAllowed=y.

Colciencias. (1983). Plan de Concertación Nacional en Ciencia y Tecnología. Santa Fé de Bogotá: Colciencias. Retrieved from repositorio.colciencias.gov.co/bitstream/handle/11146/1398/1983-V1-N4-Articulos-Art%204.3.pdf?sequence=1&isAllowed=y.

Colciencias. (1998). *Colciencias 30 años: Memorias de un compromiso.* Santa Fé de Bogotá: Colciencias.

Colciencias. (2005). *Política de apropiación social de la ciencia, la tecnología y la innovación.* Bogotá. Retrieved from www.colciencias.gov.co/sncyt/ley29.html.

Colciencias. (2010). *Estrategia nacional de apropiación social de la ciencia, la tecnología y la innovación.* Bogotá: Colciencias.

Colciencias. (2015). *Lineamientos para el reconocimiento de centros de ciencia en Colombia.* Bogotá, Colombia. Retrieved from colciencias.gov.co/sites/default/files/ckeditor_files/lineamientos_centrosdeciencia.pdf.

Colciencias. (2016). *Modelo general de buenas prácticas de sostenibilidad en centros de ciencia.* Bogotá, Colombia. Retrieved from colciencias.gov.co/sites/default/files/ckeditor_files/modelo-buenas-practicas.pdf.

Cortés-Fonnegra, L. M. (2014). Naturalia: una huella en la historia de la divulgación científica en Colombia. *Desde la Biblioteca, 48,* 25–33.

Daza-Caicedo, S., Arboleda, T., Rivera, A., Bucheli, V. and Alzate, J. (2006). *Evaluación de las actividades de comunicación pública de la ciencia y la tecnología en el Sistema Nacional de Ciencia y Tecnología Colombiano. 1990–2004.* Report. Bogotá DC: Observatorio Colombiano de Ciencia y Tecnología.

Daza-Caicedo, S. and Arboleda, T. (2007). Comunicación pública de la ciencia y la tecnología en Colombia: ¿políticas para la democratisación del conocimiento? *Signo y pensamiento*, *26*(50), 100–25.

Daza-Caicedo, S. and Lozano-Borda, M. (2013). Actividades hacia 'otros públicos'. Entre la difusión, la apropiación y la gobernanza de la ciencia y la tecnología. In M. Salazar (ed.), *Colciencias cuarenta años. Entre la legitimidad, la normatividad y la práctica* (pp. 280–353). Bogotá DC: Observatorio Colombiano de Ciencia y Tecnología.

Daza-Caicedo, S., Lozano-Borda, M., Bueno Castellanos, E., Gómez-Morales, Y. J., Salazar Acosta, M., Jaime, A., Aguirre, J., Rueda Ortíz, R., Franco-Avellaneda, M., Rincón, O., Pérez-Bustos, T., Farias, D., Suárez, R. and Osorio, C. (2014). *Percepciones de las ciencias y las tecnologías en Colombia.* S. Daza-Caicedo and M. Lozano-Borda (eds). Bogotá DC: Observatorio Colombiano de Ciencia y Tecnología.

Declaración de Punta del Este (1967). Declaración de los Presidentes de América, 12 al 14 de abril de 1967. Punta del Este, Uruguay.

Durant, J. (1993). What is scientific literacy? In J. Durant and J. Gregory (eds), *Science and Culture in Europe* (pp. 129–37). London: Science Museum.

Escobar, A. (1996). *La invención del Tercer Mundo Construcción y deconstrucción del desarrollo.* Santa Fé de Bogotá: Editorial Norma. Retrieved from cronicon.net/paginas/Documentos/No.10.pdf.

Felt, U. (2003). *Optimising Public Understanding of Science and Technology.* Vienna.

Fog, L. (1995). *La A.C.A.C., 25 años creando futuro.* ACAC. Retrieved from catalogo.cenda.edu.co/cgi-bin/koha/opac-detail.pl?biblionumber=1898.

Fog, L. (2004). El Periodismo científico en Colombia un lento despegue. *Quark*, *34*, 59–65.

Franco-Avellaneda, M. and von Linsingen, I. (2011). Popularisaciones de la ciencia y la tecnología en América Latina: mirando la política científica en clave educativa. *Revista mexicana de investigación educativa.* Consejo Mexicano de Investigación Educativa A.C., *16*(51), 1253–72.

Franco-Avellaneda, M. and Pérez-Bustos, T. (2010). Tensiones y convergencias en torno a la apropiación social de la ciencia y la tecnología en Colombia. In CyT, C. de I. para la A. S. de la (ed.), *Deslocalizando la apropiación social de la ciencia y la tecnología en Colombia. Aportes desde prácticas diversas.* Bogotá: Maloka-Colciencias.

Gómez-Morales, Y. J. (2005). Política científica colombiana y bibliometría: usos. *Nómadas*, *22*, 241–54.

Hermelin, D. (2011). Un contexto para la comunicación pública en Colombia: de las herencias eurocéntricas a los medios para la acción. *Co-herencia. Revista de Humanidades, 14*(8), 231–60.

Hernández, Y., Puello, K., Perafán, A. and Martínez-Dueñas, W. (2016). Aproximación a las relaciones entre Ciencia, Tecnología, y Sociedad en la Dinámica Investigativa de la Universidad del Magdalena. *Diálogos en Mercosur – Rede Acadêmica Série Diálogos en Red – Nuestra América, 2*, 7–23.

Hess, D. (1997). *Science studies. An advanced introduction.* New York and London: New York University Press.

Huergo, J. (2001). La popularización de la ciencia y la tecnología: interpretaciones desde la comunicación. *Seminario Latinoamericano Red-POP-Cono Sur, 14 al 17 de mayo de 2001*. La Plata.

Irwin, A. and Michael, M. (2003). *Science, Social Theory and Public Knowledge.* Maidenhead, Philadelphia: Open University Press.

Latour, B. (1992). *Ciencia en acción.* Barcelona: Editorial Labor.

Latour, B. (2007). *Nunca fuimos modernos. Ensayo de antropología simétrica.* Siglo XXI editores.

Lewenstein, B. (2003). *Models of Public Communication of Science & Technology.* Ithaca. Retrieved from edisciplinas.usp.br/pluginfile.php/43775/mod_resource/content/1/Texto/Lewenstein2003.

Lozano, M. (2005). *Programas y experiencias en popularisación de la ciencia y la tecnología: panorámica desde los países del Convenio Andrés Bello.* Bogotá DC: Convenio Andrés Bello.

Lozano-Borda, M. (2013). Una mirada hacia la medición de la apropiación social de la ciencia y la tecnología a la luz del programa Ondas. In J. Lucio (ed.), *Observando el Sistema Colombiano de Ciencia, Tecnología e Innovación: sus actores y sus productos* (pp. 311–43). Bogotá DC: Observatorio Colombiano de Ciencia y Tecnología.

Massarani, L., Amorm, L., Bauer, M. and Montes de Oca, A. (2012). Periodismo científico: reflexiones sobre la práctica en América Latina. Chasqui. *Revista Latinoamericana de Comunicación, 120*, 73–77.

Massarani, L. (2015). *RedPOP: 25 años de popularisación de la ciencia en América Latina.* Rio de Janeiro: Museu da Vida/ Casa de Oswaldo Cruz / Fiocruz / Unesco.

Michael, M. (2002). Comprehension, Apprehension, Prehension: Heterogeneity and the Public Understanding of Science. *Science, Technology & Human Values, 27*(3), 357–78. doi.org/10.1177/016224390202700302.

Misión de Ciencia, Educación y Desarrollo (1995). *Colombia: Al filo de la oportunidad.* Colección documentos de la Misión, Tomo 1. Bogotá: Presidencia de la República-Colciencias.

Muñoz Quevedo, J. (1986). El estado actual del periodismo científico en Colombia. *Periodismo científico en los países del Convenio Andres Bello*, pp. 51–72. Bogotá, Colombia: Secretaría del Convenio Andrés Bello.

Official Journal of the Republic of Colombia. (2012). Article 29, Law 1530. Bogotá, DC, Colombia. Retrieved from svrpubindc.imprenta.gov.co/diario/index.xhtml.

Pabón, R. (2017). Apropiación social del conocimiento: una aproximación teórica y perspectivas para Colombia. *Educación y Humanismo*, *20*(34), 116–39. doi.org/10.17081/eduhum.20.34.2629.

Pérez-Bustos, T. (2009). Tan lejostan cerca. Articulaciones entre la popularisación de la ciencia y la tecnología y los sistemas educativos en Colombia. *Interciencia*, *34*(11). doi.org/10.1097/AJP.0b013e3182290d90.

Pérez-Bustos, T. and Tafur Sequera, M. (2010). *Deslocalizando la Apropiación Social de la Ciencia y la Tecnología*. Aportes desde Prácticas Diversas, Bogotá, Maloka-Colciencias.

Posada, E., Hoyos, N., Pantoja, A., Carvajal, C. and Marín, M. (1995). Apropiación social de la ciencia y la tecnología. In Z. Periodismo (ed.), *Ciencia y Educación para el Desarrollo, Informe de Comisionados III* (pp. 9–224). Bogotá DC: Imprenta Nacional.

Raichvarg, D. and Jacques, J. (1991). *Savants et ignorants : Une histoire de la vulgarisation des sciences*. Paris: Seuil.

Ramalho, M., Arboleda, T. and Hermelin, D. (2017). A cobertura de ciência em telejornais do Brasil e da Colômbia. *História, Ciências, Saúde – Manguinhos*, *24*(1), 223–42. doi.org/10.1590/s0104-59702016005000019.

Roqueplo, P. (1983). *El reparto del saber*. Gedisa. Retrieved from www.casadellibro.com/libro-el-reparto-del-saber/9788474321807/33327.

Vélez Lopera, N. (2013). Vida y obra de Lisbeth Fog : entender la ciencia para divulgarla. *Desde la Biblioteca*, *46*, 15–21.

Woolgar, S. (1988). *Ciencia: abriendo la caja negra*. Barcelona: Anthropos Ltd.

Timeline

Event	Name	Date	Comment
First interactive science centre established.	Museo de la Ciencia y el Juego – Universidad Nacional de Colombia	1984	
First national (or large regional) science festival.	Science fairs	1970s	Organised by the National Education Ministry of Colombia
An association of science writers or journalists or communicators established.	Asociación Colombiana de Periodismo Científico (ACPC) [Colombian Association of Science Journalism]	1976	
First university courses to train science communicators.	Science journalism course at the University Corporation Minuto de Dios	2000s	Taught by members of the Colombian Association of Science Journalism.
First master's students in science communication graduate.	There is no specific master's degree in science communication	No	2003: Master's at the National University of Colombia, but in appropriation and communication of science
First national conference in science communication.	Roundtable 'Creation of a public awareness of education, science and culture', Bogota	1969	1969: National Congress of Science Journalism took place in Medellín 1969: Creation of the Ibero-American Association of Science Journalism
National government program to support science communication established.	This was enunciated in the policy for Colciencias called *Plan for National Consensus on Science and Technology*	1983	It was based on five working areas, one of which was the popularisation of science
First significant initiative or report on science communication.	*Scientific Activities for Youth*	1965	
National Science Week founded.	National Science Week	2006	10 regions joined this initiative
A journal completely or substantially devoted to science communication established.	*Science, Technology and Development Journal*	1977	
First significant radio programs on science.	National radio program on Saturdays dedicated to scientific dissemination	1976	

Event	Name	Date	Comment
First significant TV programs on science.	*Dissemination and scientific-technological training*	1984	A co-production of Colciencias and ICFES on research projects, technological development
First awards for scientists or journalists or others for science communication.	National Award for Scientific Merit	1989	Organised by the Colombian Association for the Advancement of Science and Technology
Other significant events.	The term 'social appropriation of science and technology' first used in science policy	1995	
	National policy for social appropriation of science, technology and innovation	2005	
	National Strategy of Social Appropriation of Science and Technology	2010	

Contributors

Dr Sandra Daza-Caicedo leads research in the social appropriation of science and technology. She was co-editor of the Colombian science and technology indicators yearbook and was in charge of the national surveys on public understanding of science and technology.

Luisa Barbosa-Gómez is a researcher at the Studies Centre on Science, Communication and Society at Universitat Pompeu Fabra (CCS-UPF) in Barcelona and the academic coordinator of the master's degree in scientific, medical and environmental communication at UPF – Barcelona School of Management.

Dr Tania Arboleda-Castrillón is an independent consultant and researcher on the social appropriation of science and technology. She recently worked as a consultant for the National Administrative Department of Science and Technology in Colombia.

Marcela Lozano-Borda is a professor of the Department of Communication of the Pontificia Universidad Javeriana in Bogotá. She coordinated the Colciencias (National Department of Science and Technology) social appropriation of science area, where she led the design and formulation of the national strategy.

11

DENMARK, NORWAY AND SWEDEN

Share, make useful and critically discuss: Science communication

Per Hetland, Dick Kasperowski and Kristian H. Nielsen

1. Science communication in Scandinavia from the 15th century to the mid-20th century

Scandinavia is the geographical region roughly comprising present-day Denmark, Norway and Sweden. The three Scandinavian countries have a shared cultural history, and the three national languages belong to the same dialect continuum and are mutually intelligible. Although the name Scandinavia can be traced back to the Roman natural philosopher Pliny the Elder, it only became popular during the 18th century when ideas about common identity and historical roots became prominent.

Danish, Norwegian and Swedish are all North Germanic languages. All three languages include direct translations of the German term *wissenschaft*, which traditionally is used to denote all spheres of knowledge from theology through the humanities to the social, natural, medical and practical sciences. Historically, the term *populärwissenschaft* and the related Scandinavian translations arose in the 19th century to denote communication of all sciences to non-academic audiences (Daum, 2002). Today, Danes, Norwegians and Swedes also talk about *forskningsinformation*, *forskningsformidling* and *forskningskommunikation* (literally, 'research information', 'research dissemination' and 'research communication'). Again, these terms imply all disciplines.

King Christian I, who for a brief period in the 15th century restored the Kalmar Union with Denmark, Norway and Sweden under a single monarch, established the University of Copenhagen in 1479. During the Enlightenment period, scholars founded royal societies in Stockholm (1739), Copenhagen (1740) and Trondheim (1760) to advance and communicate scientific and practical knowledge. The late 18th and early 19th centuries also saw the establishment of royal societies dedicated to agriculture, forestry and mining.

Priests played an important role in this early phase of science communication, as they typically were the only persons in their local communities who had received higher education. A well-known example from the Scandinavian countries are the so-called 'potato priests' promoting potato cultivation among peasants (Brenna, 2011). The potato priests understood that potatoes, introduced into Scandinavia during the 18th century, were vital to the health and wellbeing of their congregations. They had to fight resistance to the new crop and superstitious beliefs about the potato causing disease and low yields in other crops. In the end, the public potato campaign was quite successful since potatoes constitute an important part of contemporary national diets; yet we have to remember that it also helped that people increasingly began to produce alcohol (aquavit and schnapps) from potatoes.

Science communication in Scandinavia really took off during the 19th century with the formation of societies and magazines dedicated to science popularisation, the opening of natural history museums and botanical gardens for the general public and, not least, the so-called rural awakening. The latter term indicates not only the spread of enlightened Christian thought, but also widespread interest in rural development and social stability. In the late 19th century and into the 20th, the newly formed national governments in the Scandinavian countries, partly inspired by the US agricultural extension model for diffusing the results of agricultural research to farmers, established agricultural test stations and other extension services. The rise of the Folk high school movement, providing adult education to the rural population, also meant that many more people were able to seek more or less science-based reasoning and information.

Figure 11.1: Scientific explorations are a source of national pride. Dr Nansen and the first Fram expedition (1893–96).

Source: The Fram Museum (used with permission).

During the 19th and 20th centuries, science communication played an important role in the formation of national identity. As mentioned, the very idea of a common cultural heritage in the three Scandinavian countries builds partly on literary and historical research. Scientific exploration, too, has had high visibility by way of the media and popular books. Explorers such as the Danish Knud Rasmussen, the Norwegian Fridtjof Nansen and the Finnish-Swedish Adolf Erik Nordenskiöld all became national celebrities because, for many people, they combined scientific discovery with national pride, understanding and unity. The explorer-as-national-hero tradition continues up until today, although the distinction between scientific and popular exploration seems to have become sharper.

Industrialisation during the 19th century proceeded at an uneven pace, with Sweden leading the way followed by Denmark and Norway. Primary industries such as farming, forestry, fishing and mining, however, remained the mainstay of the Scandinavian economies up until the 1930s. It was in the 1930s and 1940s that all three Scandinavian countries established the prevailing, but largely implicit, social contract between science and society, relying on widespread ideas of social corporatism, consensus and citizenship. The state agreed with various corporate groups such as agricultural, labour, trade and scientific associations that science and technology were key factors in driving economic growth and welfare. This political consensus lent broad support for collaborations between public and private research, but also for higher education, lifelong learning and public communication of science and technology. At the core of society was the informed citizen, responsible for their own self-improvement and enlightenment.

In what follows we will treat the emergence of contemporary science communication in each of the three Scandinavian countries from roughly the end of World War II onwards. Although Denmark, Norway and Sweden fared very differently during the war, the post-war period for all countries to a high degree revolved around the development of the welfare state. The welfare state model combined internationalism and liberal capitalism with a large public sector organised into many subsectors such as education, health, military, research, etc. Science policy therefore emerged as 'a proper concern of government', as Vannevar Bush put it in his 1945 report to the US president, *Science: The Endless Frontier* (Bush, 1945). In Scandinavia, the national governments established many public sector research organisations to produce and communicate knowledge relevant to sectorial governance. More or less at the same time, the Nordic countries, comprising the three Scandinavian countries as well as Finland and Iceland, established the Nordic Council in 1952, driven by the desire to make the Nordic region one that people would want to live and work in.

2. The emergence of modern science communication in Denmark

Compared to other European countries, Denmark remained relatively untouched by the damages from World War II. For most of the war, Denmark continued to cooperate politically and economically with the German occupation forces and, due to the cooperative atmosphere and the availability of dairy products, the German occupation soldiers soon nicknamed Denmark *die Sahnefront* (literally, the Cream Front) (Poulsen, 1991). Following the war, Denmark immediately sought to re-establish its foreign relations and made the defence of the realm a top priority. Denmark joined the United Nations in June 1945 and the North Atlantic Treaty Organization (NATO) in April 1949. In addition, Denmark's share of the European Recovery Program (the Marshall Plan) was relatively higher than its two neighbouring Scandinavian countries, Norway and Sweden. By the late 1950s, the industrial sector overtook the agricultural sector in relative size in terms of national employment, and from the 1960s onwards public sector investments led to the development of a strong welfare state.

In the immediate post-war period, scientists and journalists were optimistic about science communication. Communicating science, they believed, was important not only for general education but also for the public appreciation and understanding of science. Already during the war, Børge Michelsen, one of the first Danish science journalists, had begun campaigning for improved working conditions for younger scientists. Warning against what he called the 'proletarianisation of science', Michelsen together with a few scientists petitioned the government for the establishment of a committee for the advancement of science in Denmark. Due to the collapse of the protectorate government in August 1943 when the German occupation forces took full control of Denmark, the government did not establish the committee until after the war (Nielsen, 2008a).

Børge Michelsen established the Association of Danish Science Journalists in 1949. The association was short-lived, having little impact on the field of science journalism in Denmark. In 1951, one of the other founding members, Niels Blædel, played an important role in the campaign for the advancement of science in Denmark that Michelsen had helped set in motion. By this time, Michelsen had left Denmark to accept the position of head of the Division for Science and its Popularisation at UNESCO (Nielsen, 2018). The campaign culminated when thousands of scientists and students marched the streets of Copenhagen in favour of increased government support for scientific research and higher education. Blædel took part in the organisation of the demonstration, covering it extensively in his newspaper, *Politiken*. Because of the mounting pressure to advance science, the government established the first public research foundation in 1952. The foundation soon became known as the State's Carlsberg Foundation in reference to the fact that before 1952 most support for basic research had come from private foundations such as the Carlsberg Foundation (Knudsen, 2006).

When these events took place in Copenhagen, a large science–media–government partnership was well underway, namely the Galathea Deep Sea Expedition 1950–52. Like the campaign to advance science, preparations for the expedition had begun during the war where explorers were planning to launch a Danish expedition foundation after the war. Their purpose was dual. They wanted not only to support three Danish scientific expeditions—one to Greenland, one to Central Asia and one circumnavigation of the globe, i.e. the Galathea Expedition—but also to communicate to audiences at home and abroad that the collaboration with Nazi Germany during the war had been a complete mistake since Denmark really stood for peace, international collaboration and scientific research. The Galathea Expedition had its own media section, consisting of the head of the section journalist-explorer Hakon

Mielche, one other journalist and three photographers. The media section reported on the expedition through telegrams, magazine articles, books, films and radio features (Nielsen, 2008b).

Throughout the 1950s and 1960s, the Danish Broadcasting Corporation (DR) increased its coverage of science and technology. There were regular TV and radio shows about science, most aimed at adult education. Still, the most popular science broadcasts seem to have been radio lectures, a continuation of a tradition established in the interwar years. In 1957, DR launched its Sunday University, a weekly radio show on Sunday from 11 am –12 pm, where scientists and lecturers from all disciplines gave lectures on historical, literary, philosophical, scientific, technological and other issues. The show became hugely popular, continuing into the 1970s. DR also produced a book series to accompany the lectures. One of the reasons for the show's popularity was probably its so-called 'radiophonic' adaptation, where the DR journalists would carefully select proper subject matters and edit the presenter's lecture notes so that listeners with no academic education would also see the lectures as interesting and be able to understand them (Nielsen and Nielsen, 2006).

Unprecedented economic growth, but also cultural upheaval and sociopolitical change, characterised 1960s and early 1970s Denmark. The anti-nuclear and environmental movements along with Marxist-inspired social critique led to public debate about societal inequalities as well as the health and environmental risks produced by capitalism, science and technology. When the Association of Danish Science Journalists again formed in 1976, its members included 'red' science journalists who saw critical science journalism, quite often with a Marxist bent, as a form of counter-expertise. They strongly believed that scientists and science journalists would have a crucial role to play in transforming the industrial, capitalist society to a socialist one (Waneck, 2005).

The renewable energy movement became strong in the course of the 1970s, enabling new formats of science communication. Wind energy in particular was an important renewable energy source for a country like Denmark with its long traditions in wind turbine development and scarce energy resources in terms of sun, rivers, forests and fossil fuels, but with a lot of wind. To promote wind energy as a viable alternative to coal, oil and nuclear power, engineers, researchers, renewable energy activists, green politicians, wind turbine producers and owners, science journalists and many others created informal, yet strongly interconnected, grassroots communication networks supporting a high degree of learning-by-interaction and public participation in sociotechnical issues relating to wind energy. Regular informal meetings, magazines, booklets, television programs, political activism and professional engineering communication were all means to promote knowledge sharing and gain public visibility (Nielsen and Heymann, 2012).

Figure 11.2: Leading up to COP15 in Copenhagen in 2009, the Danish Board of Technology initiated the first World Wide Views (WWViews) consultation on global warming, which was the first-ever global citizen consultation with roughly 4,000 citizens in 38 countries. The picture shows citizens in São Paulo, Brazil, deliberating some of the core issues at stake.

Source: The Danish Board of Technology (used with permission).

The Danish Parliament established the Danish Board of Technology (DBT) in 1986 to communicate knowledge about new technology and its societal consequences. The parliament and the board understood technology assessment, the board's remit, as involving not only expert assessments, but also public deliberation and public participation. The board developed and employed many deliberative and participatory methods such as scenario workshops, citizen juries and consensus conferences. The first consensus conference in Denmark predates the establishment of the DBT, taking place in 1983 on the theme of breast cancer. In contrast to the model developed in the United States by the National Institutes of Health, where the expert panel would prepare the final consensus document, Danish consensus conferences typically result in a consensus statement authored by the citizen panel. The DBT have been instrumental in making the Danish-style consensus conferences popular around the world. Scholars have argued that the work of the DBT demonstrates the strength of integrative political processes with respect to scientific and technological issues by institutionalising a close link between political and participatory aspects of democracy, thus enforcing the Danish culture of consensus (Andersen and Jæger, 1999; Horst and Irwin, 2010).

Despite—or maybe because of—Denmark's strong tradition of public participation in science, the Danish Government introduced the third mission for Danish universities rather later than other European countries (Zomer and Benneworth, 2011). The 2003 Act on Universities introduced the idea that the Danish universities (all universities are public) are obligated to 'collaborate with the surrounding society and contribute to the development of international collaboration'. In May 2003, the Minister of Science, Helge Sander, established a think tank on the public understanding and appreciation of science that included mostly people working with journalism and communication. The think tank propounded the term 'research communication' as a way to provide citizens with enough knowledge and competencies to be able to enter into 'a democratic dialogue about research, its results and processes, its benefits and opportunities, its consequences, dilemmas and risks' (Tænketanken for forståelse for forskning, 2004). The think tank recommended allocating 2 per cent of all research grants to research communication and funds for research in research communication, but did not specify if this meant 2 per cent of the total pool of funds, or 2 per cent of each individual grant. Although many Danish scientists were positive towards the 2 per cent recommendation, the ministry never enacted it (Nielsen et al., 2007). The ministry did, however, announce two funding schemes for research in research communication and established the national Festival of Research, another one of the think tank's recommendations.

The third mission of the Danish universities introduced in 2003 really has two elements (Aagaard and Mejlgaard, 2012). First, the universities should contribute to growth, welfare and development in society. Second, as important institutions of knowledge and culture, they should exchange knowledge and competencies with actors in society, disseminate knowledge about scientific results and methods to the public and encourage their employees to take part in public debate. Studies indicate that most scientists and engineers have limited engagement in any kind of third mission activities, but those that do tend to engage in public sector service, industrial collaboration or executive involvement (Mejlgaard and Ryan, 2017). Researchers from the humanities, however, are overwhelmingly engaged in disseminating knowledge to the public through books, newspapers, magazines, blogs and social media (Johansson et al., 2018).

3. The emergence of modern science communication in Norway

At the beginning of World War II, the Norwegian academic system remained relatively untouched. However, in 1942, a large number of schoolteachers were sent to labour camps in northern Norway, and the University of Oslo was closed in November 1943. At the beginning of 1944, 'most of the academic leaders, both students and staff, were imprisoned, hiding, or in exile' (Fure, 2018, p. 113). Many of those held in concentration camps in Germany, and still living there at the end of the war, were saved by a rescue operation organised by the Swedish Red Cross and Danish Government only weeks before the collapse of the Third Reich. During the war one notable form of science communication was a large number of articles, pamphlets and books telling people how to secure their own subsistence. Following the war, Norway, similarly to Denmark, joined the United Nations in June 1945 and NATO in April 1949. As with Denmark, Norway received important assistance through the US's Marshall Plan, and a number of scientists and engineers travelled to the US to learn. Quite likely this aid led to the Norwegian academic system being strongly inspired by the Anglo-American one, and it has had long-lasting effects.

However, to explain the position of Norway it is necessary to take one step back. Due to the country serving as the junior partner in two unions, first with Denmark (1380–1814) and then with Sweden (1814–1905), academic institutions arrived late in Norway. What is now known as the University of Oslo was established in 1811. The University of Oslo played an important role in meeting the intellectual needs of the emerging independent nation of Norway. In fact, it has been said that the University of Oslo has gone through three distinct periods of development, with each period having a marked influence on Norwegian society (Myhre, 2018):

1. A national civil servant university (until approximately 1880) that educated the professionals needed to serve the state.
2. The growth of a research university with a definite national flavour from about 1880. Norwegian scientists mostly 'excelled—internationally speaking—in disciplines connected to practical and national interests in fisheries, shipping and polar exploration: meteorology, geology, oceanography, marine biology and studies of the northern lights' (Myhre, 2018, p. 23).
3. An internationalised mass university from approximately 1960 onward.

These three periods of development also reflect changing understandings of science communication. Science communication delivered as an element of formal education was the dominant task during the first period of the university's life, while popular science communication became part of the process of national identity formation and modernisation from the 1880s until the 1960s. Finally, during the third period, science communication in Norway evolved in parallel with that seen in the other Nordic countries.

During the 19th and 20th centuries, polar exploration was important for developing a national identity. This tradition was revived after World War II. The best-known Norwegian explorers of recent times are arguably Helge Ingstad (1899–2001), who is internationally recognised for mapping Viking settlements in Canada (Ingstad, 2009, 2010); and Thor Heyerdahl (1914–2002), who is internationally renowned for his experimental archaeology and numerous expeditions (Kvam, 2005, 2008, 2013). Both explorers produced a great number of popular books, presentations and documentary films, and Heyerdahl in particular had a large international audience, with his books being translated into more than 70 languages. The documentary film chronicling the Kon-Tiki expedition won a number of international awards. Both Ingstad and Heyerdahl also used the new mass media of the time, namely radio and TV, to spread news of their achievements, while the Kon-Tiki Museum was established in 1949. Polar studies are still important, and Norway has several research facilities in the Arctic and Antarctica. Climate change has spurred science communication with both stakeholders and different publics.

Although several authors have attempted to analyse science and technology communication in Norway from a broader perspective (Andersen and Hornmoen, 2011; Bentley and Kyvik, 2011; Kyvik, 2005; Løvhaug, 2011), there have been only a very few analyses of the Norwegian science and technology communication policy framing those communication activities (Hetland, 2014). Approximately 50 per cent of all faculty members published at least one popular science article during a three-year period (1989–91 and 1998–2000), although just 6 per cent of faculty members published half of all popular science articles (Kyvik, 2005).

The third mission was first mentioned in the law governing the University of Bergen in 1948, then in the revised law governing the University of Oslo in 1955. In 1995, the law governing all public higher education institutions adopted the third mission, as did the revised law governing both public and private higher education institutions in 2005. Finally, in 2013, the expanded and strengthened Act Relating to Universities and University Colleges declared that higher education institutions have three key missions:

education, scientific research, and science and technology communication. Consequently, science and technology communication is sometimes referred to as 'the third mission'. This states that the universities should: 1) contribute to the public communication of science and technology; 2) contribute to innovation; and 3) ensure the participation of higher education staff in public debate. One important condition in relation to engaging in the third mission is academic freedom (Underdal et al., 2006), which from 2007 was explicitly included in the Act Relating to Universities and University Colleges.

In this context, it is also important to note that the Norges forskningsråd (NFR, the Research Council of Norway) is the major player funding research and promoting science communication on a larger scale, as private money plays only a minor role. The overall objective of the NFR's national science communication strategy is: 'Through a general dissemination of research, the aim is to encompass that part of the general public who are not traditional users of research simultaneous to including research in the public debate' (Norges forskningsråd, 1997, p. 7). Three main groups of people are defined as being especially important in this regard:

1. Children and young people, who will form the basis for the recruitment of future researchers.
2. Teachers, who serve as disseminators of research results to their pupils.
3. Journalists, who disseminate research and, due to their position, can influence the science policy debate.

Several different activities promoted science communication in Norway. The Norwegian Contest for Young Scientists was launched as a private initiative in 1968, although it quickly grew in terms of both its ambition and its scope. The Nysgjerrigper Science Knowledge Project for children attending primary school was established in 1990, and the Norwegian Science Week was inaugurated in 1995, the same year in which the Science Channel was established as a joint project conducted between the largest universities and university colleges, resulting in weekly transmissions by the Norwegian Broadcasting Corporation. This activity took place alongside the launch of a number of other projects directed toward children and young people, including TV series such as *Newton*. *Forskning.no* was established in 2002 as an online newspaper devoted to Norwegian and international science, included several possibilities for feedback and debate. *ScienceNordic* was launched in 2011 with science news from the Nordic countries in English.

The Science Centres Program was established in 2003 as an important project, not least in terms of stimulating the interest of young people in the STEM fields (science, technology, engineering and mathematics). By 2018, Norway had nine regional science centres, as well as five additional specialised science centres. It was considered important to strengthen museum-based activities aimed at schools. Both the museums and the science centres adopted the use of social media platforms in order to increase the possibilities for inquiry-based learning. Finally, the Researcher Grand Prix was established in 2010 and proved an exciting opportunity for doctoral candidates to learn valuable skills in communicating their research, and at the same time participate in a national competition.

The concept of environmental citizenship has grown in importance since 1900, although it had its first significant breakthrough during the 1970s, both as a popular movement and as a growing concern within Norwegian universities (Anker, 2011; Berntsen, 2011). Hydropower accounts for almost all of Norway's electricity production. Yet the fight against the extensive building of hydroelectric power plants broadly lasted throughout the 1970s and culminated in debates about the Alta Hydroelectric Power Scheme in the early 1980s. This conflict caused a number of environmental organisations to engage in activities intended to strengthen their collaboration with the indigenous people of the north, the Sami. Science communication was used extensively to argue for a more sustainable environmental policy. Such ecological arguments relied on the natural sciences, as well as on the social sciences and humanities. A key intellectual and inspirational figure within the environmental movement was Professor Arne Næss from the University of Oslo. He coined the term 'deep ecology', and he had a long-lasting influence across various disciplines both within and outside academia. A more traditional approach to environmental issues was presented within the established political regime by former Prime Minister Gro Harlem Brundtland, chair of the Brundtland Commission, which led to the publication of the Brundtland Report on sustainable development in 1987. Within just two decades, environmental concerns had become a matter of general concern within the international policy arena.

The Norwegian authorities recognised the need for new forums to encourage active public debate concerning technology and in 1999, inspired by the Danes, they established the Norwegian Board of Technology. This was tasked with determining the possibilities and consequences associated with new technologies, for both society and individual citizens. Its findings were to be made known to the parliament, as well as to other authorities and to the public. The board was allowed to determine both the specific areas for discussion and its working methods. However, importance was to be attached to the methods by which laypeople could become engaged in its activities. Today, the Norwegian Board of Technology employs several working methods, such as hosting laypeople's conferences, workshops, citizens' panels and hearings.

In other words, the Norwegian authorities wish to engage the general public in a more comprehensive debate concerning technology. A little before this, the NFR launched a policy for user-oriented science communication in 1996. The policy underlines that user-oriented science communication requires two-way communication between the researcher and the user. The aim is to empower the user to act.

Two distinct trends are currently apparent within the field of Norwegian science communication, namely a stronger emphasis on open science (OS), and the growing professionalisation of science communication, including more public relations within academia. Within the long-term Norwegian plan for research and higher education (Kunnskapsdepartementet, 2018), OS is assigned a high priority. In most documents concerning OS, transparency and knowledge-sharing are emphasised. Scientific publications based on research and development projects funded wholly or partially by the Research Council are to be made openly accessible to all interested parties to the greatest extent possible. At the same time, participation in the actual performance of science is to be opened up.

It is often taken for granted that OS and citizen science (CS) are two sides of the same coin; however, this issue is not as straightforward as might initially be expected (Kennedy, 2016). Several initiatives work toward achieving greater openness, and in Norway, most CS projects, for example, in relation to biodiversity mapping, encourage openness and inclusiveness. For instance, the Ministry of Climate and Environment, through the Norwegian Biodiversity Information Centre, launched a new service known as Artsobservasjoner [Species Observations] in 2008. It is a digital reporting system that is open to everybody. Between 2008 and 2018, about 20 million observations have been recorded, mostly by amateur naturalists. These observations are crucial in many respects, with one application being the species map service used by planners and the like. Consequently, CS represents one emerging method of participatory science communication (Hetland, 2011). In 2013, Farbrot published a guidebook concerning practical science communication that remains much in use (Farbrot, 2013). The book illustrates how science communication is increasingly converging with PR. However, more empirical studies of both phenomena in the Norwegian context are required as academic institutions increasingly establish their own communication units, on all levels, to promote science communication.

4. The emergence of modern science communication in Sweden

Sweden remained neutral during World War II, and neutrality required allocating substantial state funds to national defence. In general, due to a national focus on self-sufficiency requiring that Sweden itself produce and provide as much as possible of what the country needed, research received generous funding. Particularly, money went to forestry, mining and food production, which helped boost the Swedish economy in the post-war period. During the war, initiatives such as the establishment of the so-called Kursverksamheten KV (literally, course activity), which later became Folkuniversitetet (the People's University, one of many adult education activities known as *studieförbund*, or study unions) served to strengthen general education and serve as intellectual resources to resist Nazism and fascism (SOU, 1946:68, p. 28).

In the 1960s, initiatives in research policy started to move beyond a rhetoric on *bildung* and democracy. The longstanding ideal of *bildung* in Sweden was inspired by the German tradition of self-cultivation, where philosophy, personal experiences, formal and informal education are envisioned as resulting in personal and cultural maturation. This is often described as a harmonisation of the individual's mind and heart and a unification of selfhood and identity with the national culture or society. However, this ideal, as an individual project, was now questioned in calls for concrete practices and incentives for the public communication of science. Industry had since long been the prioritised recipient for research results and the natural partner for cooperation (in Swedish, *samverkan*), with the universities. However, now a more intense discussion took place over why the communication of science was to be directed at the public.

The general argument was that Swedish society had now reached a state of 'scientification' realising the need for every citizen to understand and communicate scientific knowledge. Without this information, citizens would not be able to take part in societal decision-making, and, ultimately, this would lead to a democratic deficit (Universitetskanslersämbetet, 1972). Science communication providing citizens with access to research information, therefore, was more than entertainment, education and formation; it had become a civil right and democratic virtue.

The wider context for this discussion was science policy initiatives during the 1950s–1960s in creating the Sectorial Principle in Swedish research. This stated that state authorities with responsibility for sectors such as environment, defence, education, etc. also had the responsibility to produce the knowledge necessary to fulfil their politically defined mission. Usually these tasks were commissioned from researchers at the universities, but who were now criticised for producing knowledge below scientific standards. The criticism of the Sectorial Principle was two-fold: that the science it generated was substandard, and that, importantly, it created a dislocation of power in Swedish society, from elected members of parliament to experts in the administration of state authorities.

During the latter part of the 1970s and into the 1980s and 1990s, the belief that the Swedish state could control and steer societal development with an increasingly instrumentally motivated research policy lost ground to arguments based on research quality and excellence. The allocation of resources for scientific research should no longer be left to civil servants in government, but rather should be performed by the universities where peer review and other mechanisms for scientific quality could be guaranteed. Research was to be brought back to the universities, so to speak, yet at the same time it was also important to ensure the public were involved in discussions about science.

In the latter part of the 1970s, long-established relations between industry and university research largely defined the official Swedish policy on science communication and the formal relationships between universities and society. The organisation of science communication outside government and industry, mainly defined by concepts such as *bildung* and democracy, was in a different situation. Although sometimes resources were shared with the industrial networks, other actors, particularly the organisations involved in adult education, were responsible for communicating science to the public. In spite of this, the Higher Education Law of 1977 stated that university researchers now had a 'third mission' to fulfil, and it was their responsibility to inform and collaborate with society, and from 2009 to make their knowledge useful.

The Law was largely the result of discussions following several public inquiries and a product of university administrators and politicians (Universitetskanslersämbetet, 1972; Centrala organisationskommittén för högskolereformen, 1975). Traditionally, the Law has been taken as evidence for the opening up of the Swedish university. It encouraged researchers to communicate their work and results to the public, on the basis that this work was funded by taxpayers' money. Thus, research was brought back to the universities from the sectorial sphere, assigning the task of communicating high-quality scientific results to the researcher responsible for creating those results.

The Law was constituted on 10 February 1977. The sixth paragraph stated:

> That work at the universities shall also include to disseminate knowledge on research and development. Knowledge shall also be disseminated on what experiences and knowledge have been gained and how this knowledge shall be applied (Regeringens proposition om utbildning och forskning inom högskolan m m., 1977).

The Law prescribes a third task for the Swedish universities, in addition to the well-established tasks of education and research. It is not evident if the third mission is associated with any new economic resources allocated to the universities. In the referral process preceding the Law, no institutions made any enquires on the possibilities of increased resources from the state for the universities in their work with the third mission. It appears as if the universities did not perceive of the third mission as something more, new or requiring extra funding. In the formulation of the law there is no information about the intended recipients of science communication. It is not self-evident that it is the general public as such or individual or groups of citizens that are targeted for the efforts in the dissemination of research.

Forskningsrådsnämnden (FRN, Swedish Council for Planning and Co-ordination of Research) was established in February 1977 as a state authority to oversee the third mission. Sweden is unique in having a state authority from 1977 to 2000, to initiate, investigate and oversee public communication of science. Swedish researchers have to a large extent been considered as insufficient or inadequate in their assigned task to communicate science to the public. This is largely because little value or importance has been assigned to the third task by the Swedish research community, in spite of many attempts since the late 1970s (both in terms of legislation and education) to encourage public communication of science.

However, it would be a mistake to believe that Swedish researchers do not engage in such activities. Later studies have shown that researchers communicate their research extensively, legislation or not, and create strategies to accommodate public communication of science in already established and recognised value systems of merit.

The standard tale is that Swedish scientists do not engage in public communication of science. However, this is a result built upon evaluations and surveys, not looking into the practices of the research community. In an upcoming book on the history of the third mission in Sweden, applications for professorships at Swedish universities have been systematically analysed. The data show that researchers are doing much of what can be subsumed under the third mission, but they categorise these efforts as 'education' or 'research' using traditional categories of merit. Publishing in a popular science magazine can be called 'research overview', giving public lectures called 'education', etc. (Kasperowski, Bragesjö and Kullenberg, in progress).

FRN's assumption was that science communication and the third mission would not happen by itself: researchers must be encouraged, trained or even coerced into communicating with the public. The 'societal side' of the university has to be evoked by laws and, most often in the history of FRN, supplemented with different stimulus packages such as workshops in writing popular science, courses for PhD students, journalist exchange programs, etc., all of which encourage and support researchers in their efforts to fulfil the third mission obligations. Several FRN reports have lamented the fact that researchers are not rewarded for third mission activities as they gain relatively little merit for these activities compared to research and teaching, and minimal economic resources have been made available for public communication of science (Ramström, 1986, p. 74).

Even proposals to establish courses in science communication for researchers have been met with scepticism from university boards, because they thought that other types of training and qualifications, particularly for PhD students, were more important. This led to several recommendations by FRN, among them a proposal to the government of issuing a complementary higher education regulation in 1985. The Swedish Government accepted the FRN recommendations, stating that individual researchers should be rewarded and promoted for their 'research information' activities together with scientifically meritorious outputs, pedagogical work and other skills (Högskoleförordning, 1985).

The government assigned the universities the task to coordinate their communication and information activities. However, no funds were allocated for this, and the local universities had to prioritise existing funding. Together with the Swedish Office for Higher Education, FRN was assigned the task of overseeing that the new regulation was actually applied in employment procedures at the universities. However, an analyst contracted by the FRN concluded that Swedish researchers were well prepared to deal with the third mission and therefore did not need training or coercion by law. To be a researcher is also to be able to communicate your work to students and the public (Ramström, 1986, p. 46). The problem cannot be fought focusing on the individual researcher or PhD student; it is of a structural kind. When advice is issued to FRN in the report it is stated that 'research information should be directed to a general public not to a narrowly defined group of professionals for their use in their everyday work' (Ramström, 1986, p. 25).

FRN was critical of the report since communicating research information to professionals is also part of activities it promotes, and here the longstanding conflict of science communication in Sweden becomes visible. Public communication of science is resting upon an image of diffusion from researchers at the higher seats of learning, coupled with a very strong rhetoric on the needs of the modern knowledge-based representative democracy, including citizens with a high degree of trust in science as an institution. However, this rhetoric has seldom been realised in any larger coordinated projects or programs. Communication of science to professionals in industry, policy and business has, since the 1920s, been an integral part of the cooperation between universities and 'the surrounding society', enjoying infrastructural and economic resources. This tradition does not produce a strong rhetoric of democracy, but is often clad in innovation and global competition, and has at times affected the trust for science and research negatively in Sweden.

During the late 1990s, FRN as a state authority was questioned by prominent natural scientists regarding the distribution of large grants for the procurement of advanced technologies for research in the natural and medical sciences. In the late 1990s, FRN was shut down following a dispute over the handling of large research grants, and its responsibilities (including science communication) were handed over to the Swedish Research Council. The shutdown of FRN was thus not motivated by any arguments concerning the public communication of science. Now, the responsibility for developing new initiatives in science communication was to be carried out by the research council and the universities themselves. Whereas the limited distribution of funds to public communication of science by FRN had been directed to projects initiated by individual researchers and on the training of researchers in science communication, the Swedish Government now wants the research councils to take responsibility for increasing researchers' options for reaching out to the public (Forskning för kunskap och framsteg, 1993).

What is the image of the Swedish researcher as a public communicator of science in the light of ideals and practices employed by FRN? The general answer is that Swedish researchers cannot be steered, trained or encouraged to fulfil the third mission—they simply fail to adhere to the different initiatives taken by FRN. In particular, this is the fact concerning communication of science to the general public carried out with the ideals of *bildung* and democracy.

None of the strategies for solving the problem of the low value placed on public communication of science by FRN seem to result in any significant change among Swedish researchers, with some exceptions. The first successful example dates from the late 1970s and is characterised by the creation of a new public for science communication (Dewey, 1991; Marres, 2005). In 1980, after several years of public debate Sweden held a public referendum on nuclear power. FRN early on produced a popular science booklet series called *Källa*. The referendum on nuclear power in Sweden gave FRN the opportunity to produce several issues on nuclear power, many of them debates between researchers from different disciplines and moderated by a 'neutral' researcher. The articles on nuclear power in *Källa* were considered an important contribution to the public debate. Over 50 volumes of *Källa* were published during 1980–2003, presenting new and socially relevant research, debated and moderated by the researchers themselves. Usually research identified by the FRN as important 'from a societal perspective' was debated. Environment and energy, health and medicine, the relation between technology, nature and culture, digitalisation and, during the later 1990s, biotechnology were recurrent themes. This initiative has almost paradoxical qualities: given the right circumstances, Swedish researchers are not reluctant to fulfil the third mission and they also seem to have the communicative skills necessary.

A further example concerns calls and applications to the position of professor at Swedish universities. Historical data show that despite the low merit value of third mission activities, they are actually being practised. Document analysis of applications to the position of professor (n=126) on all faculties from 1976–1982 and 1996–2002 at the University of Gothenburg shows that applicants do communicate their science to the public; however, they manage the low merit value by constructing their efforts as 'education' or 'research'. The low value merit value of third mission activities means that they do not attract rewards, do not help secure positions at universities, and even that you can be accused of producing 'popular science' instead of original research. Public communication of science is thus constructed or subsumed under already recognised categories of merit. Calls for applications, however, very seldom put public communication as a requirement for the position as professor, third mission or not (Kasperowski, Bragesjö and Kullenberg, in progress).

The third example is a demand put on researchers by funders from the late 1990s onwards in Sweden in relation to research applications. Sweden is a country where it is necessary to obtain external funding for research, since positions as lecturers are usually primarily funded only to fulfill educational tasks. The incentives to apply for external funding are therefore high and competitive, since research and publications are the hard currency for career movement. From the late 1990s onwards, all Swedish major funders of research have requested researchers to draft popularised versions of their research proposals. This is true for the Swedish Research Council, established in 2001 and supporting research of the highest quality within all scientific fields, and previously for the specialised, disciplinary research councils as well as FRN.[1]

Swedish researchers do publicly communicate science almost independently of the Higher Education Law. The construction of the Swedish researcher as not fulfilling a role as a communicator of science to the public is, at least in view of our limited data, a narrative decoupled from the practices of Swedish researchers. Today, discussions on public communication of science and research in Sweden are to a large extent defined by open science and the possibilities of opening science further 'upstream' in the research process itself, and the communication efforts needed for such a change. Universities are increasingly taking advantage of social media and digital resources as a large cadre of communicators are now developing a new ecology of science communication in Sweden.

Figure 11.3: Forum for Forskningskommunikation [Forum for Research Communication] is an annual event held in relation to Vetenskapfestivalen Göteborg [International Science Festival Gothenborg]. The forum brings together science communicators, scientists and others to discuss pertinent topics in relation to science communication. The topic of the 2019 event was 'Impact: Good, bad or dangerous?'.

Source: International Science Festival Gothenborg (used with permission).

1 Interview of Anna-Maria Fleetwood, Senior Adviser, External Relations, Swedish Research Council, by Dick Kasperowski, 4 October 2018.

5. Concluding remarks

Denmark, Norway and Sweden all have a rich history of science communication. Popularisation efforts by the scientific community have co-existed and co-developed with efforts to make science communication useful for the purposes of democracy, education, farming, environmental protection, industrial development, public health, social welfare and more. One of the challenges faced in all three countries is how to match the demands of the academic community, particularly attuned to specialist, technical communication, with the demands of society (including most academics), hoping to share, make useful and critically discuss the fruits of research. The enactment of the third mission for all public universities serves as a modern example of such a match, but also shows the difficulties involved. There are many similarities across the three countries covered in this chapter, such as an ongoing emphasis on the role of science communication in enforcing citizenship, public deliberation and social responsibility, but also many differences. The making of modern science communication in Scandinavia also testifies to the fact that Denmark, Norway and Sweden are—and always have been—firmly embedded in international developments.

As open economies, the three Scandinavian countries have a long tradition of international collaboration. Denmark joined the present EU in 1973, Sweden in 1995 and Norway joined the European Economic Area in 1994. As such, they are all three collaborating partners of Horizon 2020, and national science policy is strongly influenced by EU policy. An important policy objective is to ensure that the results of publicly funded research are made open and available to the different publics. Furthermore, responsible research and innovation is high on the agenda. One crucial question for each of the three countries is how they manage to build bridges between international ambitions and national publics facilitating science communication that build identity and citizenship.

References

Aagaard, K. and Mejlgaard, N. (eds) (2012). *Dansk forskningspolitik efter årtusindskiftet*. Aarhus, Aarhus University Press.

Andersen, E. K. and Hornmoen, H. (2011). Mediating Science in Norway: Practices and Transformations in Major Newspapers. *Media Transformations*, *5*, 88–104.

Andersen, I.-E. and Jæger, B. (1999). Scenario workshops and consensus conferences: towards more democratic decision-making. *Science and Public Policy*, *26*, 331–40. doi.org/10.3152/147154399781782301.

Anker, P. (2011). Den store økologiske vekkelsen som har hjemsøkt vårt land. In P. Anker, M. Gulbrandsen, E. Larsen, J. W. Løvhaug and B. S. Tranøy (eds), *Universitetet i Oslo 1811–2011: Samtidshistoriske perspektiver* (pp. 103–71). Oslo, Unipub.

Bentley, P. and Kyvik, S. (2011). Academic Staff and Public Communication: A Survey of Popular Science Publishing Across 13 Countries. *Public Understanding of Science, 20*(1), 48–63. doi.org/10.1177/0963662510384461.

Berntsen, B. (2011). *Grønne linjer: Natur- og miljøvernets historie i Norge.* Oslo: Unipub.

Brenna, B. (2011). Clergymen Abiding in the Fields: The Making of the Naturalist Observer in Eighteenth-Century Norwegian Natural History. *Science in Context, 24*, 143–66. doi.org/10.1017/S0269889711000044.

Bush, V. (1945). *Science: The Endless Frontier.* Washington DC: United States Government Printing Office.

Centrala organisationskommittén för högskolereformen (H 75). (1975). *Högskolelag och andra författningar för högskolan.* Stockholm, LiberLäromedel/Utbildningsförl. för Utbildningsdep.

Daum, A. W. (2002). *Wissenschaftspopularisierung im 19. Jahrhundert: Bürgerliche Kultur, Naturwissenschaftliche Bildung und die Deutsche Öffentlichkeit, 1848–1914.* München, Oldenbourg. doi.org/10.1515/9783486832501.

Dewey, J. (1991). *The public and its problems.* Athens, OH: Swallow Press.

Farbrot, A. (2013). *Forskningskommunikasjon: Praktisk håndbok for forskere og kommunikasjonsrådgivere.* Oslo: Cappelen Damm Akademisk.

Forskning för kunskap och framsteg. (1993). *Regeringen Proposition 1992/93:170.*

Fure, J. S. (2018). A university under siege. In K. Helsvig and J. E. Myhre (eds), *Making a Modern University: The University of Oslo 1811–2018* (pp. 11–39). Oslo: Scandinavian Academic Press.

Hetland, P. (2011). Science 2.0: Bridging Science and the Public. *Nordic Journal of Digital Literacy, 6*, 326–39.

Hetland, P. (2014). Models in Science Communication Policy: Formatting Public Engagement and Expertise. *Nordic Journal of Science and Technology Studies, 2*(2), 5–17. doi.org/10.5324/njsts.v2i2.2144.

Högskoleförordning. (1985). Svensk författningssamling, SFS nr 1977: 26, kapital 19, § 30.

Horst, M. and Irwin, A. (2010). Nations at Ease with Radical Knowledge: On Consensus, Consensusing and False Consensusness. *Social Studies of Science, 40*, 105–26. doi.org/10.1177/0306312709341500.

Ingstad, B. (2009). *Eventyret: en biografi om Helge Ingstad.* Oslo: Gyldendal.

Ingstad, B. (2010). *Oppdagelsen: en biografi om Anne Stine og Helge Ingstad*. Oslo: Gyldendal.

Johansson, L. G., Pedersen, D. B. and Stjernfelt, F. (2018). *Humanistiske universitetsforskeres vidensformidling og videnssamarbejde*. Copenhagen: Humanomics, Aalborg University.

Kasperowski, D., Bragesjö, F. and Kullenberg, C (in progress). *Den tredje uppgiften: Lagarna, folket och forskarna* [The third mission: The laws, the people and the researchers].

Kennedy, E. B. (2016). When Citizen Science Meets Science Policy. In D. Cacalier and E. B. Kennedy (eds), *The Rightful Place of Science: Citizen Science* (pp. 21–50). Tempe, AZ: Consortium for Science, Policy & Outcomes.

Knudsen, H. (2006). Politik, penge og forskningsvilkår. In H. Nielsen and K. H. Nielsen (eds), *Viden uden grænser, 1920–70. Dansk Naturvidenskabs Historie, bind 4* (pp. 21–50). Aarhus: Aarhus University Press.

Kunnskapsdepartementet. (2018). *Langtidsplan for forskning og høyere utdanning 2019–2028*. Oslo: Kunnskapsdepartementet.

Kvam, R. (2005). *Thor Heyerdahl: Mannen og havet*. Oslo: Gyldendal.

Kvam, R. (2008). *Thor Heyerdahl: Mannen og verden*. Oslo: Gyldendal.

Kvam, R. (2013). *Thor Heyerdahl: Mannen og mytene*. Oslo: Gyldendal.

Kyvik, S. (2005). Popular Science Publishing and Contributions to Public Discourse Among University Faculty. *Science Communication*, *26*(3), 288–311. doi.org/10.1177/1075547004273022.

Løvhaug, J. W. (2011). Overlæreren i rampelyset: Universitetet i mediesamfunnet 1970–2011. In P. Anker, M. Gulbrandsen, E. Larsen, J. W. Løvhaug and B. S. Tranøy (eds), *Universitetet i Oslo 1811–2011: Samtidshistoriske perspektiver* (pp. 367–459). Oslo: Unipub.

Marres, N. S. (2005). No issue, no public: democratic deficits after the displacement of politics (PhD thesis). Amsterdam, Ipskamp Printpartners. Retrieved from pure.uva.nl/ws/files/3890776/38026_thesis_nm_final.pdf.

Mejlgaard, N. and Ryan, T. (2017). Patterns of third mission engagement among scientists and engineers. *Research Evaluation*, *26*, 326–36. doi.org/10.1093/reseval/rvx032.

Myhre, J. E. (2018). The University of Oslo through 200 years: From national civil servant school to mass university. In K. Helsvig and J. E. Myhre (eds), *Making a Modern University: The University of Oslo 1811–2018* (pp. 11–39). Oslo: Scandinavian Academic Press.

Nielsen, K. H. (2008a). Enacting the social relations of science: historical (anti-) boundary-work of Danish science journalist Borge Michelsen. *Public Understanding of Science, 17*, 171–88. doi.org/10.1177/0963662506067661.

Nielsen, K. H. (2008b). In quest of publicity: the science-media partnership of the Galathea Deep Sea Expedition from 1950 to 1952. *Public Understanding of Science, 18,* 464–80. doi.org/10.1177/0963662507083529.

Nielsen, K. H. (2018). 1947–1952: UNESCO's Division for Science & Its Popularisation. *Public Understanding of Science*, E-Pub Ahead of Print, 12 November. doi.org/10.1177/0963662518787572.

Nielsen, K. H. and Heymann, M. (2012). Winds of change: communication and wind power technology development in Denmark and Germany from 1973 to ca. 1985. *Engineering Studies, 4,* 11–31. doi.org/10.1080/19378629.2011.649921.

Nielsen, K. H., Kjaer, C. R. and Dahlgaard, J. (2007). Scientists and science communication: a Danish survey. *Journal of Science Communication, 6*(1). doi.org/10.22323/2.06010201.

Nielsen, K. H. and Nielsen, H. (2006). Formidling af naturvidenskab i presse, radio og tv. In H. Nielsen and K. H. Nielsen (eds), *Viden uden grænser, 1920–70. Dansk Naturvidenskabs Historie, bind 4* (pp. 405–22). Aarhus: Aarhus University Press.

Norges forskningsråd. (1997). *Nasjonal strategi for allmennrettet forskningsformidling.* Oslo: Norges forskningsråd.

Poulsen, H. (1991). Die Deutsche Besatzungspolitik. In R. Bohn et al. (eds), *Neutralität und totalitäre Aggression: Nordeuropa und die Grossmächte im Zweiten Weltkrieg* (pp. 369–79). Stuttgart: Franz Steiner Verlag.

Ramström, D. (1986). *Att dela med sig av kunskapens glädje: en utvärdering av FRNs stöd till högskolans forskningsinformation. Rapport 86:8.* Stockholm: Forskningsrådsnämnden.

Regeringens proposition om utbildning och forskning inom högskolan m.m. (1977). 1976/1977:59.

Statens offentliga utredningar (SOU). (1946:68). *Betänkande och förslag angående det fria och frivilliga folkbildningsarbetet.* D.1, Allmänt folkbildningsarbete: 1944 års folkbildningsutredning. Stockholm.

Tænketanken for forståelse for forskning. (2004). *Forsk og fortæl!* Copenhagen: Ministry of Science.

Underdal, A., Ullenius, C., Storsul, T., Olsen, J. P., Kjenndalen, K., Hagesæter, G. Giertsen, J., Bernt, J. F. and Rigmor, A. (2006). *Akademisk frihet: Individuelle rettigheter og institusjonelle styringsbehov.* Oslo: Kunnskapsdepartementet/Norges offentlige utredninger.

Universitetskanslersämbetet. (1972). *Förslag till försöksverksamhet med forskningsinformation*. Stockholm: SAMSUS.

Waneck, M. (2005). Rød naturvidenskab (Master's thesis). Aalborg, Aalborg University. Retrieved from projekter.aau.dk/projekter/files/6145654/Rød%20Naturvidenskab.pdf.

Zomer, A. and Benneworth, P. (2011). The Rise of the University's Third Mission. In J. Enders, H. F. de Boer and D. F. Westerheijden (eds), *Reform of Higher Education in Europe* (pp. 81–101). Rotterdam: Sense Publishers. doi.org/10.1007/978-94-6091-555-0_6.

Timeline

Event	Denmark	Norway	Sweden
First interactive science centre established	1991 Experimentarium	1914 Norwegian Museum of Science and Technology	1985 Teknorama
First national (or large regional) science festival	2005	1975	1997
An association of science writers or journalists or communicators established	1949/1976	1952	1972
First university courses to train science communicators	2006	2013	2003
First master's students in science communication graduate		1992	2004
First PhD students in science communication graduate	2003	1990	2002
First national conference in science communication	2005	2007	2012
National government program to support science communication established	2006	1996	1977–2000
First significant initiative or report on science communication	2004	1993	1977
National Science Week founded	2016	1995	1983

Event	Denmark	Norway	Sweden
A journal completely or substantially devoted to science communication established	1854	1877	1911–(1916) 1906–
First significant radio programs on science	1933	1952	1930/1937
First significant TV programs on science	1956	1959	Late 1950s/1971
First awards for scientists or journalists or others for science communication	1996/2004	1995	1972
Date hosted a PCST conference	2008		2008
Other significant events: Law on third mission for the universities	2003	1995/2005/2013	1977/1997/2009

Contributors

Dr Per Hetland is professor at the University of Oslo, Department of Education. He holds a Dr Philos. in science communication from the University of Oslo and a PhD in innovation studies from Roskilde University.

Dick Kasperowski is associate professor of Theory of Science at the University of Gothenburg. Informed by current perspectives in science and technology studies, his main interests include citizen science, governance of science, participatory practices in science and the humanities and open collaborative projects in scientific work.

Dr Kristian Nielsen is associate professor and head of the Centre for Science Studies, Aarhus University. He is currently working on the emergence of public understanding of science.

12

ESTONIA

Science communication in a post-Soviet country

Arko Olesk

Estonia, with just 1.3 million people, is one of the smallest countries in the world to use its own language as the primary language in all areas of social life, including media and all levels of education. Today, the country also has a modern science communication landscape with science centres, science festivals and other regular events, established science journalism, and a national program to foster science communication. This modern setting is mostly a product of rapid developments during the last 15 years, when Estonia's accession to the European Union functioned as a major catalyst.

Estonia belongs to a group of Central and Eastern European countries that underwent at least three significant transitions during the 20th century: first, gaining independence following World War I, then being incorporated into the Soviet Union or its sphere of influence during World War II, and finally, returning to democracy within a capitalist market structure in the late 1980s and early 1990s. Each transition brought with it a disruption that has made a steady development impossible: each time, old societal structures were dismantled or radically reshaped.

Various statistical indicators place Estonia as one of the most successful of the post-Soviet group of countries making the transition to the Western world. This also concerns science and science communication where the comparison with similar post-Soviet countries demonstrates that the development towards modern science communication is by no means a given. Therefore, the Estonian example helps describe and explain both the characteristics of science communication under the Soviet regime and what forces and factors lead to the establishment of a modern science communication system.

1. Historical background

The first Estonian-language periodical publication, the magazine *Lühhike öppetus* [Brief instruction] aimed to provide Estonian peasants with practical medical advice, both for themselves and for their cattle. The magazine was published in 1766–67 by the Baltic-German Estophile Peter Ernst Wilde and was part of the Enlightenment-inspired efforts of the German nobility who were convinced that 'if peasants' virtues were developed and proper education provided, their social circumstances would improve' (Lauk et al., 1993).

A similar focus on education and cultural development was promoted in the mid-19th century by the emerging Estonian elite leading the national awakening movement. For example, Friedrich Reinhold Kreutzwald, the doctor who penned the Estonian national epic *Kalevipoeg*, also published the widely read popular science magazine *Ma-ilm ja mõnda* [The world and other things] (1848), the first illustrated Estonian magazine (Peegel, 1994).

The contribution of Tartu University (founded in 1632 by Sweden and re-opened in 1802 under Russian Czarist rule) became more important towards the end of the 19th century when students' organisations established themselves as venues to bring science to the public. This complemented the opening to the public of the university's natural history museum and Botanical Gardens in 1802 and 1803 respectively.

During the first period of independence (1918–40), fundamental sciences were considered impractical for a small nation like Estonia and the emphasis was on 'national' sciences (i.e. those dealing with Estonian history, culture, nature, etc.) or applied sciences such as agriculture (Kalling and Tammiksaar, 2008). Scientific societies became leading communicators by publishing books and magazines, including *Eesti Loodus* [Estonian Nature], which is still published today, and *Loodusevaatleja* [Nature's Observer]. The initiator of the latter, botanist Gustav Vilbaste, considered it crucial that the publications avoided academic language and were written in a way easily understood by the readers (Tammiksaar, 2017).

In 1940, Estonia was occupied by the Soviet Union and lost much of its elite during World War II: they were either killed, arrested and deported, or fled to the Western world. Science was rebuilt to Soviet standards that had a much stronger focus on fundamental sciences and saw science and technology as an instrument to demonstrate the superiority of the Soviet model of socioeconomic organisation. The scientists had to adapt their work to the official philosophy of dialectical materialism[1] and operate within a system of strong political control that also included censorship and difficulty of access to scientific information published in the West (Medvedev, 1979).

These pressures somewhat eased during the Khrushchev era at the end of the 1950s and beginning of 1960s. This allowed the scientific societies to become more active again, to use more Estonian language in science, and to restore some magazines closed at the beginning of the Soviet occupation (Tammiksaar, 2018). Good-quality Russian-language books and magazines were available, but soon an Estonian magazine was founded that was to become the most influential of its kind: the popular science magazine *Horisont* [Horizon], first published in 1967.

The monthly magazine *Horisont* has often been cited by current Estonian scientists as one of the reasons they chose a scientific career. The magazine offered articles written by Estonian scientists in an accessible language and sometimes provided adaptions from Western popular science magazines. During the first years, the topics mostly covered global science and the latest scientific advances, while later coverage became more timeless and focused on local and Soviet topics (Olesk, 2017). The magazine enjoyed its greatest popularity in the 1970s with a top circulation of 54,000 in 1971.

The magazine was the official publication of the Teadus [Science] society. Founded in 1947 as a branch of the similar all-union society (Znaniye), its aim was to spread political and scientific knowledge among the population. Its main activities—lectures and brochures—did provide a venue for science communication but were foremost of ideological nature (officially, communism was considered a scientific discipline in the Soviet Union). As a result, we can distinguish two types of public science communication: one whose aim was not to introduce or explain science but to use science examples or scientists to reinforce ideological discourses such as legitimisation of the Soviet system; and the other that sought to popularise and explain science. A study on the science coverage in Soviet Estonian media (Olesk, 2017) showed that although the first type was predominant, the second type was also present, especially in *Horisont*, but also in daily newspapers, probably more due to the personal initiative of some scientists and journalists than to editorial policy.

The educational science communication paradigm that was to become the dominant one in 21st-century Estonia traces its beginnings to 1980, to the founding of the National Student Research Society. The society facilitated the mentoring of gifted school children by senior researchers and organised student research conferences.

1 Dialectical materialism is a philosophy of science that serves as the philosophical basis of orthodox communism.

Several scientists played prominent public roles in the popular movement in the late 1980s that finally led to the restoration of independence in 1991. The movement was sparked by scientists highlighting environmental concerns related to Moscow's plan to start mining phosphorite in the ecologically delicate Virumaa county.

The harsh reforms Estonia undertook after the restoration of independence and the abrupt transition to a market economy hit society hard. Scientists had to re-orientate themselves to the Western model, including publishing in international peer-reviewed journals and competing for funding. Under these circumstances, science communication was not a priority—neither for the scientists, nor for the media or the universities. The circulations of *Horisont* and *Eesti Loodus* declined dramatically. The field suffered from lack of resources and support and was only sustained by devoted enthusiasts.

2. Emergence of modern science communication in Estonia

When preparing this chapter and talking to people with a long history in the Estonian science communication landscape, the developments witnessed during the last 20 years were often explained by one key factor: the crucial role of individuals. In a small country, the right person in the right place or with enough determination could trigger long-term processes and have a remarkable impact. Several initiatives that laid the foundation to the period of rapid development and institutionalisation of science communication were attributed to such individuals or small groups.

To give two prominent examples: in the national media, science was kept visible by Tiit Kändler, a former physicist who started editing a weekly science page in 1995 and became eponymous with science journalism in the following decade. The interactive science centre AHHAA that became a crowd magnet in 2011 was established in 1997 and built from scratch by the former chemist Tiiu Sild.

The leap from those endeavours to a modern science communication system in Estonia required several supportive factors to come together in the beginning of the 2000s. First, society was recovering from the ruptures caused by the transition, and now had more resources to focus on issues beyond mere survival. The scientific community started to discuss the same set of perceived problems that helped to launch the science communication movement in Western European countries in the 1980s and 1990s: lack of students in STEM-fields (science, technology, engineering and mathematics), little or inaccurate media coverage of science, and the diminishing role of science and scientists in society.

Second, new resources became available to support science communication activities—most importantly, funds from the European Union (EU). Outside support had been there before as well—in the 1990s the Scandinavian countries provided financial support and know-how for the establishment and growth of several science communication initiatives. For example, the Estonian Association of Science Journalists was founded in 1990 (then also including environmental journalists) with the support of the Finnish association; and science centre AHHAA was directly inspired by the Heureka centre in Helsinki with whom they have had a close collaboration over the years. However, the financial means that became available with Estonia's accession to EU in 2004 moved activities to a whole new scale. By around 2005, the stage was ready for a quick expansion of science communication activities in Estonia. There were dedicated individuals who worked on limited resources and with little or no institutional support, meaning that they were usually not capable of reaching beyond a niche audience. At the same time, there was a growing understanding among the scientific community that science communication can be a tool to solve problems that science is facing in society. Finally, access to EU funds made decision-makers look for fields that needed a development boost.

One case that illustrates how these factors co-contributed to a quick shift in the nature of science communication activities in Estonia was the celebration of the International Year of Physics in 2005. The Estonian Physical Society had been concerned for some time about the sustainability of the field in Estonia, considering low student interest and little public visibility of physics. While they had been doing small events before, the international year prompted them to design a comprehensive program to increase the visibility of physics in the public and among potential students. The activities included a new web portal for physics news, a weekly science experiment presentation on the national broadcaster's morning show, a public event for families and the Science Bus—a science theatre that toured schools (Eesti Füüsika Selts, 2005). Most of the funding for the program came from an EU framework project related to the Year of Physics, but the extent of the activities was also supported by a substantial amount of work done on a voluntary basis by students and university staff.

The impact of the Year of Physics activities extended well beyond the one-year project. The Estonian Physical Society continued and extended many of the activities in the following years. Several people who got their first science communication experience during the Year of Physics are now prominent communicators. The idea of the TV show *Rakett 69* [Rocket 69], one of the biggest science communication success stories in Estonia, was born in preparation for the Year of Physics, although it finally aired only in 2011.

Perhaps the most influential activity was the creation of the Science Bus. It has visited hundreds of schools over the years with its science theatre performances and served as an inspiration for other fields to launch similar initiatives. Today, organised school visits with interactive workshops and mobile laboratories have become a widely used science communication format in Estonia.

In a way, the Science Bus accelerated the formation of the current state-managed science communication system. The wish to officially recognise the Science Bus prompted the creation of annual National Science Communication Awards in 2006, which allowed Estonia to submit the winners to a similar (but now defunct) European Union science communication competition. Further, the awards led to the establishment of the annual science communication conference. Terje Tuisk, the head of the Department of Science Communication in the Estonian Research Council, recollects that in the first years the award ceremony was rather unattractive to anyone but the people immediately involved.[2] Hence, in a discussion about how to increase the visibility of the awards the idea of a science communication conference was born. First held in 2008, it has since then annually brought together science communication practitioners, researchers, journalists, administrators, decision-makers and others. The conference has allowed the sharing of best practices and can be credited as a key component in the creation of a sense of community among science communicators in Estonia.

These stories already highlight the important role of various EU influences. Furthermore, a nudge from the EU can even be considered the beginning of the national science communication program. In 2002, the European Commission approached Estonia, then still a non-member, to submit entries to the pan-European contest of young scientists. For this, a similar competition in Estonia had to be organised. The task was given to Tuisk, a biologist by training who had some previous experience with student research. 'Essentially the Ministry [of Science and Education] was saying that you can do [the contest] if you wish but we have no money. So we did it with no funds,' she recalls.[3]

Later, the ministry handed over the coordination of university-level student research contests to Tuisk, and she began to hire people to manage all the tasks. In 2006, this group was officially named the Department of Science

2 Personal communication, 29 August 2018.

3 Personal communication, 29 August 2018.

Communication (then located at Archimedes Foundation that coordinated the use of EU funds; today it is in the Estonian Research Council, the main funding institution of science in Estonia) and started to gain more functions such as managing the science communication awards and organising the annual conference. In 2008, the Ministry of Science and Education began funding various science communication projects in an annual open call, again coordinated by the Department of Science Communication. In 2018, the budget for the call was €150,000 and a total of 30 projects were financed (Eesti Teadusagentuur, n.d.-c). In 2010 and 2016 there were two additional open calls with a budget of more than €1 million each for systematic long-term science communication projects (including extracurricular activities) aimed at young people.

'When the preparations for the new period of the [EU] structural funds started, the Ministry had already realised that this field needs more resources,' Tuisk says.[4] Hence, a national science communication program was crafted. The TeaMe program (short for *Teadus, meedia ja meie* or Science, Media and Us; also translates as 'we know') had a budget of €3.34 million over the period 2009–15, 85 per cent of which was provided by the European Social Fund. The program had three general aims:

> To increase the interest of young people in science and technology, and for careers in these fields;
>
> To expand the scope of Estonian science media and journalism; and
>
> To spread the scientific way of thinking, bring science closer to people and make it more visible in the media. (Eesti Teadusagentuur, n.d.-a)

The biggest part of the budget was used to commission two TV shows: one for general audiences, introducing Estonian scientists; and the other aimed at young people. The latter was the aforementioned *Rakett 69*, a show where youngsters aged 15–24 compete in solving science-related puzzles. The show has been running on prime-time since 2011 and was declared the best European educational format by the European Broadcasters Union in 2012.

Other activities of the TeaMe program included communication training for scientists and skills training for journalists. For schools, the program commissioned new study materials for science-related elective courses as the possibility for such elective courses was recently introduced in secondary education, and science and technology subjects in particular lacked suitable materials.

4 Personal communication, 29 August 2018.

Figure 12.1: The TV show *Rakett 69* features young people solving science-related tasks.

Source: *Rakett 69*.

The program also supported the Year of Science in Estonia. In the tradition of having theme years, 2011/12 was declared the Year of Science to make science more visible to the public. The execution of the year was designed to get more attention to existing science communication activities rather than to create new ones. For this purpose, a specialised portal was created (miks.ee) and PR support for various activities was provided.

The program did have its critics. A study commissioned towards the end of the TeaMe program concluded that the field of science communication in Estonia is characterised by a lack of strategic guidance or vision by the funding bodies, lack of focus on effectiveness and desired outcomes, and too much emphasis on attracting pupils' attention rather than long-term activities to maintain interest in science and technology (Kirss, Haaristo, Nestor, and Mikko, 2013). This input, along with the comments from stakeholder representatives on the TeaMe advisory board who strongly recommended focusing the activities on young people, was used to design the follow-up program TeaMe+ (2015–20, total budget €3.2 million).

As a result, the current program introduced new measures that support long-term activities at the primary and secondary levels of education. These include networking and training of teachers and supervisors in extra-curricular education and supplying them with methodical materials for teaching. At the same time, the program continued with the TV show *Rakett 69*, which has in the last seasons paid special attention to gender equality (and has produced two female winners to date). The annual competition for young scientists was developed into a full-weekend science fair open to the public and an initiative was launched to involve companies in teaching STEM subjects.

The activities of the Department of Science Communication, including the two TeaMe programs, have been tone-setting in Estonian science communication, both because its focuses define the national priorities and because it is the biggest funder of science communication activities. Tuisk also sees a clear impact of the TeaMe programs and the project calls they organise: 'The fact that public money was given to the field [of science communication] has brought more actors to the field. Since then activities have gained a much wider base'.[5]

It must be noted, however, that science communication has never featured in policy documents of the Estonian government or been the focus of special government initiatives. The national programs and activities mentioned above have been mostly initiated and managed at the ministry level or below. This could be contrasted with the topic that has made Estonia most prominent internationally—the advanced information society characterised by the widespread usage of public e-services, digital society innovations and an active and successful ICT start-up scene (Heller, 2017). This digital transformation was pushed strongly by policymakers (Kattel and Mergel, 2018), with policy documents describing the benefits of ICT adoption as being the improvement of Estonia's competitiveness, democracy and educational system (Runnel, Pruulmann-Vengerfeldt and Reinsalu, 2009).

It might be surprising that science communication has been rarely discussed in the context of the digital transformation. Digital innovations have not been seen as a potential tool for science communication, and nor has science communication been focusing specifically on popularising ICT (although robotics is one of the best-established fields in informal science education). There are some connections on the rhetorical level: the focus on education and young people of Estonian science communication activities has been justified by its potential economic benefits to the country. The program document for the original TeaMe program stated that Estonia lacked enough researchers and engineers to move to a knowledge-based economy, therefore it was necessary to attract more young people to STEM fields:

> To better understand the connections between the society and science and technology we need to increase the awareness of young people and the whole population about the impact that research and development and innovation have on national competitiveness and productivity and thereby on social well-being (Haridus- ja Teadusministeerium, 2013).

5 Personal communication, 29 August 2018.

The follow-up program emphasised a similar motivation: 'Young people's willingness and motivation to acquire a higher education in STEM fields must be given a firm basis already in comprehensive school and secondary school levels' (Eesti Teadusagentuur, n.d.-b). This economic discourse aligns well with the political liberalism that has been prevalent in Estonian politics throughout the current independence period.

Thus, the core goal of Estonian science communication is getting the attention of young people with attractive presentations of science and then guiding their interest towards choosing a career in STEM fields. As well as the already mentioned activities, several other successful examples follow the same discourse—for example, science centres or the robotics contest Robotex, one of the largest in Europe.

In terms of public visibility of science, a major turning point was the opening of the AHHAA science centre in 2011. AHHAA had been founded in 1997 as a project of the University of Tartu. The story goes that the Estonian president Lennart Meri, a person with much symbolic power, visited the Heureka centre in Helsinki and was so impressed that he immediately faxed the Estonian Minister of Education to recommend establishing a similar centre in Estonia.[6] The University of Tartu took on the task and appointed the young chemist Tiiu Sild to run the centre. However, as Jaak Kikas pointed out, the establishment of the centre was not a top-down order but matched the interests of some researchers who were keen to communicate science but had had few opportunities for this.[7]

The determination of Tiiu Sild allowed the centre to develop from modest beginnings and from no permanent exhibition space into one of the most modern science centres in the Eastern and Northern part of Europe. In 2004, the centre was reorganised into a foundation by the University of Tartu, the city of Tartu and Ministry of Science and Education, and applied for EU funds to build its own permanent house. When it was opened in Tartu in 2011, the centre became an immediate public success and still attracts a steady 200,000 visitors per year. The centre also coordinates the annual Night of Researchers, which grew in 2012 into a week-long national science festival.

6 Jaak Kikas, chairman of the board of AHHAA, personal communication, 24 August 2018.

7 Personal communication, 24 August 2018.

Figure 12.2: The new building of science centre AHHAA was opened in 2011.

Source: AHHAA.

Figure 12.3: First director Tiiu Sild (1958–2012).

Source:Lauri Kulpsoo/Tarkade Klubi.

As the examples of AHHAA and the Science Bus show, the support of universities has been crucial for the start of initiatives that have later grown into something bigger. The universities are relevant actors in the field and many of their initiatives are directed towards young people, i.e. potential future students. Their efforts to communicate to the media and the wider public include hiring of communication specialists and providing communication training to scientists. However, these trainings remain on a small scale, are not systematic and have not been integrated into official curricula. Often, they take place within EU-funded doctoral schools. The same applies to the training of science communicators—these are based on ad hoc activities instead of designated programs. Academically, science communication has been researched in the communication departments at the universities of Tartu and Tallinn, but currently there is no designated research group or study program.

The universities have also made efforts to communicate with industry as the lack of knowledge transfer to business has often been cited as one of the major problems in Estonian science. The initiatives sometimes have a dual purpose, also serving to communicate science to the public—for example, TalTech University's innovation centre Mektory. The online science news site *Novaator*, which is now a part of the public broadcaster's online service, was founded by University of Tartu as a public channel to provide university-related news to entrepreneurs.

Recent years have seen the emergence of another major driver for public communication of science. While the rapid developments in the mid-2000s can be attributed to a common concern of the stakeholders about the science interest and career choices of young people, the 2010s brought a new focus on the funding of science. Because state funding to science has been dwindling, the universities, individual scientists, National Academy of Science, and Estonian Research Council started to consider public visibility as a valuable tool to influence the situation and undertook efforts to increase the profile of science. By presenting success stories and increasing the quantity of science coverage in the media, they hope to increase public support to science, which they expect will then lead decision-makers to increase funding (Scheu and Olesk, 2018). In late 2018, the pressure by the scientific community led to the signing of a political agreement to increase research funding to 1 per cent of GDP.

Generally, the Estonian media is a good partner for the scientific community. Science has become a permanent part of the menu that media houses offer and (science) journalists are generally characterised by a favourable attitude towards scientists. This development has taken place within the last 15 years, more as an evolutionary process rather than due to outside influences.

After the collapse of the early 1990s, science was hardly present in the media. The magazines *Horisont* and *Eesti Loodus* survived but became marginalised. For a long time, the only regular occurrence of science in mainstream media was the weekly science page in one of the daily newspapers that the former physicist Tiit Kändler started writing in 1995. The beginning was hard, he recalls, especially due to lack of information and researchers' mistrust towards journalists since the concept of science journalism was almost unknown at the time.[8] Over the years he established himself among scientists, editors and readers as a respected writer. Another few enthusiastic journalists emerged to cover science in addition to their main reporting tasks. Their efforts convinced media managers that it was possible to cover science in an engaging way and that there was interest from the readers, so media channels began to consciously

8 Personal communication, 20 October 2018.

look for science coverage. Today, most main national media channels employ a science journalist or have a science section. The strongest online channel is *ERR Novaator* (err.novaator.ee), co-managed by the University of Tartu and Estonian National Broadcasting. In print, franchise magazines such as the local version of Danish *Illustrerad Vetenskap* enjoy good circulation numbers.

There is no general association for science communicators, only two more specialised umbrella organisations. The Estonian Association of Science Journalists, originally founded in 1990, experienced a long hiatus soon afterwards. It was revitalised in 2007 and has since become an active stakeholder in science communication discussions. As of 2018, it has 23 individual members. The Estonian STEM Education Union, an umbrella organisation for everyone in non-formal STEM education in Estonia, was founded in 2016 and, as of 2018, has 119 members, both individuals and organisations. The main informal network, the Facebook page *Teaduse populariseerijad*, has more than 800 members.

Ten years ago, the initial discussions at the first science communication conferences mostly focused on issues related to media. The poor nature or lack of science coverage was seen as the central problem and the cause for the perceived lack of public appreciation for science. The understanding of the goals and methods of science communication were very much similar to these now described as the deficit model of science communication (Miller, 2001). Improving the quality of science media was also one of the aims of the original TeaMe program. However, actions quickly revealed that the impact of interventions is limited. This, along with the shifting focus to education and the gradual improvement of science media triggered by media houses themselves, contributed to science communication being understood as more like an ecosystem of actors with varying possibilities and roles. Additionally, an increasing number of activities claim to focus not just on presenting science attractively but also on creating a deeper understanding of the scientific process (i.e. increasing the scientific literacy of people).

A notable gap in the Estonian science communication field is the lack of public engagement activities. While many activities with young people can be considered some form of engagement (they are involved in hands-on activities or even forms of citizen science), these mostly serve educational purposes, not democratic participation. One possible explanation is the lack of a perceived need for such formats. The fields that are generally the subject of engagement formats in Western European countries, e.g. synthetic biology, are not being developed here, or there is no appropriate local political process that could be influenced via such mechanisms. Surveys show a strong public trust of scientists (European Commission, 2010) and PISA tests indicate that Estonian students are among the best in the world in natural sciences

(Haridus- ja Teadusministeerium, 2017). Therefore, other formats of science communication might be perceived as more suitable to approach the issues that have been defined as core problems (such as lack of students in STEM fields or insufficient knowledge transfer from academia to industry).

Regarding terminology, 'popularisation of science' is still the most common expression to describe the presentation of science to the public. The expression was widely used during the Soviet period and its continued use can be both attributed to a habit and to a linear one-directional understanding of science communication as transfer of knowledge from scientists to the public. One can see, however, that 'science communication' has started to gradually replace 'popularisation'. The change is not uncontested as it has been argued that in Estonian, 'communication' has a more verbal connotation than in English, and therefore is not the most appropriate term to describe educational activities such as hands-on experiments. Currently, the Estonian Research Council uses 'popularisation' to describe educational activities and 'science communication' for media-related activities. However, the discussion in the Estonian science communication community about the most appropriate term continues.

The science communication system and its development in Estonia is greatly influenced by the fact that Estonia is a small country. This enables some processes to take place quickly and one person can have a great impact on the outcome, as most of the stories in the chapter demonstrate. At the same time, there is a constant lack of resources that leads to questions of sustainability. In a small market, activities are often not able to operate on a fully commercial basis and depend on institutional support. This again depends on the priorities of the individuals currently in the system.

Tuisk attributes the rapid development of Estonian science communication in the mid-2000s to favourable attitudes among the decision-makers: 'There were people [in the Ministry and the science funding body] who saw perspective [in science communication], were supportive of new initiatives and found resources to start things'.[9] The ideas that had been devised by the early enthusiasts and received initial institutional support were then catalysed into major projects once EU funding became available. Many major projects (museums, TV programs, the national science festival) have been or are still partly funded with EU money.

This arrangement means that the science communication system in Estonia is still fragile. Major reduction of EU funding is expected in 2020 and the future funding of many current activities, including those in the TeaMe+ program,

9 Personal communication, 29 August 2018.

is uncertain.[10] A significant section of various science communication activities is dependent on project-based funding. Such an unsteady environment again amplifies the role of individuals: if there is institutional support, the system fulfils its purposes; but once the tide turns towards other priorities, enthusiasm may not be enough to sustain the achievements.

Despite the somewhat uncertain future and some gaps in the science communication landscape, Estonia can be considered as having completed the transition to a modern science communication system. The system is not yet consolidated, and in their cross-Europe analysis, Mejlgaard et al. (2012) place Estonia in the group of countries with a 'developing' science communication culture along with several other Eastern and Central European countries. A clear difference with other post-socialist countries emerges, however, when looking at the science–society relationship, according to Mejlgaard (2017). Taking into account not only the state of science communication but also the use of science in policymaking, public participation in science governance and innovation performance, this analysis places Estonia in the 'science central' cluster. In contrast, the position of science in other Eastern European countries can be considered as 'disregarded'.

Unfortunately, data about the state of science communication in the European post-socialist countries are too scarce to make generalisations. Available literature from individual countries mostly discuss various problems: for example, low level (Lehmkuhl et al., 2012) and low quality of media coverage (Šuljok and Brajdić Vuković, 2013), lack of domestic sources in media coverage (Łach, 2014), dominance of the 'deficit model' in science communication activities (Adamsone-Fiskovica et al., 2009), and fragmented academic research in the field (Valinciute, 2017).

Messages received via personal contacts or anecdotal stories indicate that although science communication activities exist in post-socialist countries, they have a low profile and often remain under the national or international radar. These countries are still struggling to build up a strong and sustainable national science communication system, the weakness of their science media being the greatest concern. While many post-socialist countries have their own associations of science journalists, Estonia and Russia are the only countries from the former Soviet Union to be members of the European Union of Science Journalists' Associations (EUSJA).

10 In 2019, the Ministry of Education and Research launched a process to devise a national strategy of science communication. The strategy aims to define the national priorities, main activities and the actors involved, and address the issues of funding.

References

Adamsone-Fiskovica, A., Kristapsons, J., Tjunina, E. and Ulnicane-Ozolina, I. (2009). Moving beyond teaching and research: Economic and social tasks of Universities in Latvia. *Science and Public Policy, 36*(2), 133–137. doi.org/10.3152/030234209X406836.

Eesti Füüsika Selts. (2005). Aastaraamat 2005. Retrieved from opik.fyysika.ee/index.php/book/section/4025.

Eesti Teadusagentuur. (n.d.-a). Science Communication Programme TeaMe. Retrieved from www.etag.ee/rahastamine/programmid/loppenud-programmid/teame-programm/.

Eesti Teadusagentuur. (n.d.-b). Science Communication Programme TeaMe. Retrieved from www.etag.ee/en/funding/programmes/closed-programmes/science-communication-programme-teame/.

Eesti Teadusagentuur. (n.d.-c). Teaduse populariseerimise projektikonkurss. Retrieved from www.etag.ee/tegevused/konkursid/projektikonkurss/.

European Commission. (2010). *Special Eurobarometer 340. Science and Technology.* Brussels: European Commission.

Haridus- ja Teadusministeerium. (2013). Programm TeaMe. Retrieved from www.etag.ee/wp-content/uploads/2012/05/TeaMe-15.11.2013.pdf.

Haridus- ja Teadusministeerium. (2017). PISA. Retrieved from www.hm.ee/en/activities/statistics-and-analysis/pisa.

Heller, N. (2017, 17 and 25 December). The Digital Republic. *The New Yorker.* Retrieved from www.newyorker.com/magazine/2017/12/18/estonia-the-digital-republic.

Kalling, K. and Tammiksaar, E. (2008). *Eesti Teaduste Akadeemia: Ajalugu. Arenguid ja järeldusi.* Tallinn: Eesti Teaduste Akadeemia.

Kattel, R. and Mergel, I. (2018). Estonia's digital transformation: Mission mystique and the hiding hand. *UCL Institute for Innovation and Public Purpose Working Paper Series (IIPP WP 2018–09).* Retrieved from www.ucl.ac.uk/bartlett/public-purpose/wp2018sites/public-purpose/files/iipp-wp-2018-09_estonias_digital_transformation.pdf.

Kirss, L., Haaristo, H.-S., Nestor, M. and Mikko, E. (2013). *Teadust ja tehnoloogiat populariseerivad tegevused Eestis.* Tallinn: Poliitikauuringute Keskus Praxis.

Łach, A. (2014). Science in the Polish popular press – between eastern and western colonialism? *Revista de Stiinte Politice,* (42), 1–15.

Lauk, E., Brikše, I., Bulota, J., Dimants, A., Kruks, S., Raguotis, B. and Trofimovs, R. (1993). The Beginning of Publishing in the Baltics. In S. Høyer, E. Lauk and P. Vihalemm (eds), *Towards a Civic Society: The Baltic Media's Long Road to Freedom* (pp. 43–66). Tartu: Baltic Association for Media Research/Nota Baltica.

Lehmkuhl, M., Karamanidou, C., Mora, T., Petkova, K. and Trench, B. (2012). Scheduling science on television: A comparative analysis of the representations of science in 11 European countries. *Public Understanding of Science*, *21*(8), 1002–18. doi.org/10.1177/0963662511436070.

Medvedev, Z. A. (1979). *Soviet Science*. Oxford: Oxford University Press.

Mejlgaard, N. (2017). Science's disparate responsibilities: Patterns across European countries. *Public Understanding of Science*, *27*(3), 262–75. doi.org/10.1177/0963662517724645.

Mejlgaard, N., Bloch, C., Degn, L., Nielsen, M. W. and Ravn, T. (2012). Locating science in society across Europe: Clusters and consequences. *Science and Public Policy*, *39*(6), 741–50. doi.org/10.1093/scipol/scs092.

Miller, S. (2001). Public understanding of science at the crossroads. *Public Understanding of Science*, *10*(1), 115–20.

Olesk, A. (2017). Beyond propaganda: science coverage in Soviet Estonian media. *Journal of Science Communication*, *16*(03). doi.org/10.22323/2.16030206.

Peegel, J. (1994). *Eesti ajakirjanduse teed ja ristteed*. Tartu: Olion.

Runnel, P., Pruulmann-Vengerfeldt, P. and Reinsalu, K. (2009). The Estonian Tiger Leap from Post-Communism to the Information Society: From Policy to Practice. *Journal of Baltic Studies*, *40*(1), 29–51. doi.org/10.1080/01629770902722245.

Scheu, A. M. and Olesk, A. (2018). National Contextual Influences on Mediatisation: The Comparison of Science Decision Makers in Estonia and Germany. *Science Communication*, *40*(3), 366–92. doi.org/10.1177/1075547018766917.

Šuljok, A. and Brajdić Vuković, M. (2013). How the Croatian Daily Press Presents Science News. *Science & Technology Studies*, *26*(1), 92–112.

Tammiksaar, E. (2017, March). Stiihiline periood Eesti teaduse arengus (1919–1936). *Horisont*, 30–31.

Tammiksaar, E. (2018, February). Eestist saab moodne teadusriik. *Horisont*, 44–45.

Valinciute, A. (2017). Science communication scholarship in Lithuania: A scoping study. *Studies in Communication Sciences*, *17*(2), 149–164.

Timeline

Event	Name	Date	Comment
First interactive science centre established.	AHHAA Science Centre in Tartu	1997/2011	Permanent exhibition opened in 2011
First national (or large regional) science festival.	European Night of Researchers	2006	
An association of science writers or journalists or communicators established.	Estonian Association of Science Journalists	1990/2007	EASJ founded in 1990, revitalised in 2007 after a hiatus

Event	Name	Date	Comment
First university courses to train science communicators.	Various short training courses	2010	No full-length university course available
First PhD students in science communication graduate.		2020	
First national conference in science communication.		2008	
National government program to support science communication established.	EU-funded TeaMe program	2009	
First significant initiative or report on science communication.	EU-funded TeaMe program	2009	
National Science Week founded.		2006	Organised around the European Night of Researchers
First significant radio programs on science.	*Kristall* [Crystal]	1964	From 1964 to 1985
First significant TV programs on science.	*Rakett 69* [Rocket 69]	2011	A competitive science show for young people
First awards for scientists or journalists or others for science communication.		2006	
Other significant events.	National society Teadus [Science] founded	1966	Lectures and brochures and a venue for science communication
	Popular science magazine *Horisont* [Horizon] founded	1967	
	National Year of Science declared	2011–2012	September 2011–September 2012

Contributor

Arko Olesk is a lecturer in science communication at the Baltic Film, Media, Arts and Communication School at Tallinn University.

13

FRANCE

'The Republic needs scholars!' A rapid history of making science public in 20th-century France

Andrée Bergeron

Reading *Ouest-Eclair* on the 31 May 1937, the readers of the Breton daily newspaper learned about a 'crowd of young folks gaping at the mysterious machines used by modern scholars and at the tremendous machine generating stunning lightnings and artificial thunderstorms' and about their presumed fear in front of such powerful forces that 'meekly obey to the order of a human being' comparatively so small. This lyrical description related to the Paris international exposition *Arts et techniques dans la vie moderne* [Arts and technics in modern life] and described the first visitors to the Palais de la découverte.

The creation of the Palais de la découverte was the first step toward *modern* (Kargon et al., 2015) science communication in France, in both form and content. Designed within the framework of the international exposition (Ory, 1991) by prominent scientists led by the Nobel laureate Jean Perrin, the Palais became permanent in 1938 after the rest of the international exposition was dismantled.

1. A genuine founding act

The preparation of the 1937 international exposition has a long and eventful history (see e.g. Kargon et al., 2015; Ory, 1994). It was prepared during a period of international tension and, in France, of serious political crises of the late Third Republic. Increasing threats of war and growing unemployment was the background on which the exhibition was built.

The theme eventually selected[1] by the government—International Exposition (modern decorative and industrial arts, working and peasant life, intellectual cooperation)—allowed for intellectual cooperation and intellectual work. This choice fell in line with the interests of the International Institute of Intellectual Cooperation (then considered the arm of French diplomacy) and of the powerful Confederation of Intellectual Workers (Confédération des Travailleurs Intellectuels, CTI).[2] The first general commissioner of the future international exhibition, Aimé Berthod, was a member of the Intellectual Workers' Defence Group at the Chamber of Deputies. He put the vice-president of the CTI, the painter André Léveillé (later the first director of the Palais de la découverte), in charge of the section devoted to intellectual cooperation. Léveillé proposed to split this into 'three groups on the manifestation of thought': Expression, Training and Diffusion. Expression of Thought aimed at the promotion of diverse facets of the work of mind: science, literature, museums, theatre, music and dance, cinema, and congresses. The idea of 'reserving a place for science' in an exhibition in part dedicated to thought very much fitted with the conceptions of the CTI (Bergeron and Bigg, 2015), and this first proposal explicitly included a class dedicated to science as early as 1933. By the time Perrin and his colleagues[3] joined the project in 1934, 'science' was already constituted as the first class of the first group 'Expression of Thought' (including literature as the second class, the museum as the third, etc.).

The dedication of a significant section to a subject that was *a priori* difficult to exhibit, the activity of the mind—or in the exhibition's terminology, the Expression of Thought—was considered by many (such as Paul Valéry, who became in charge of the first group) as a 'great and paradoxical novelty'. What was supposed to be shown was not the product of thought but intellectual labour as such that had then to be made materially observable. The challenge was to represent (in an exhibition, with objects, visuals, etc.) this very immaterial thing (thought at work) so that a visitor could observe it. For science that was done by means of experiments; for literature, by means of manuscripts. For Valéry, this fundamental challenge was the very same for literature and for mathematics:

1 Décret relatif à la création à Paris d'une exposition générale internationale (arts décoratifs et industriels modernes, vie ouvrière et paysanne, coopération intellectuelle), *Journal Officiel de la République Française. Lois et décrets,* 17 January 1933, p. 491. The 1937 world fair was the first one to be organised according to the new regulation established by the Convention Relating to International Exhibitions, signed in Paris on 22 November 1928. Among other things, the convention defined new rules for exhibition frequencies: facing three proposals, the government went for the compromise.

2 On the CTI see Chatriot (2006).

3 Let us just mention the two vice-presidents: the physicist Paul Langevin and the mathematician Emile Borel. It is worth noting that the board included a significant number of CTI officers: Emile Borel (founding vice-president); André de Sainte-Laguë (president); Charles Marie (treasurer). Many scientists involved in the creation of the Palais had clear political commitment (Bensaude-Vincent, 2013).

how was it possible to 'put before the eyes what exists only through the mind and in the mind' (Valery, 1960). The option selected was, for any domain of the Expression of Thought, to stress 'the immediate creations of the mind', the *act of creation* rather than the *product*, on *modern* thinking rather than the *repertoire* (Bergeron and Bigg, 2019).

For science, in the first place this meant to display modern science, contemporary rather than historical, and pure rather than applied. The philosophy of the Expression of Thought exhibit had consequences on the museological choices: creation had to be made visible, spectacular, lively. For the Palais' founders, the obvious model was the Hall of Science at the Chicago exposition in 1933, 'A Century of Progress'. This was considered a success in its combination of spectacle and pure science. In the end, as a comparison with the ephemeral Museum of Literature suggests (Bergeron and Bigg, 2019), the Palais should not so much be regarded as a unique museum—neither with respect to the history of science museums nor within the Paris 1937 exhibition—but as part of a recent line of exhibitions and the concrete expression for science of the principles of the Expression of Thought exhibit.

The Expression of Thought exhibit was intended to promote the cultural and social importance of intellectual labour, with the immediate benefit of remedying unemployment by engaging artists and scientists working on the exhibition. For the scientists involved in the preparation of the Palais, this held a particular relevance: it took place precisely when the same group had been wrestling with issues about the organisation of research (Picard, 1990), eventually leading to the creation of the National Centre of Scientific Research (CNRS). The parallels between the two processes are so convincing that in her pioneer work Eidelman (1985) suggests that the Palais could be regarded as a museum transposition of the CNRS.

The Palais de la découverte was a true success with more than 2,250,000 visitors. It became permanent after the end of the International Exhibition and reopened in April 1938, while the biology section reopened in August after being completely reconsidered. The exhibit included the famous 'Lottery of heredity' designed by Jean Rostand and Jean Painlevé. The first temporary exhibit at the Palais in 1938 was organised by the Carnegie Institute in Washington and focused on recent US advancements in biology and astronomy, presented with the support of the New York Museum of Science and Industry. The Palais continued its activities during the war. A commemorative exhibition on Lavoisier organised in 1943 was considered then and today (Beretta, 2004) as a genuine act of resistance.

2. Construction and reconstruction

The Palais was severely damaged during the Battle of Paris in 1944. Nonetheless, it resumed its activities and developed various initiatives very rapidly. As early as 1944, Léveillé began to work to develop international contacts[4] that led to exhibitions (like the exhibition on penicillin organised with the support of the British Council and the US Information Service and with the collaboration of Alexander Fleming) or to lectures (like the series of lectures by English scientists including Dirac, Huxley, Florey and Bernal). Curiously enough for a place affirming its commitment to 'living science', a series of commemorative exhibitions was organised, beginning with Pasteur in 1946. Predictably, the Palais became a natural place for science communication in Paris, and can probably be considered as the foundation upon which the post-war science communication edifice in France would be built.

It was not the only science museum in France at the time. In Paris, the Muséum National d'Histoire Naturelle (Blanckaert et al., 1997) and the technical museum of the Conservatoire National des Arts et Métiers (CNAM) (Dufaux, 2013) had long histories. Paul Rivet's Musée de l'Homme, designed as a museum/laboratory presenting ethnology, opened in 1938 (Conklin, 2013). And, of course, natural history museums existed in the main cities.

Science museums played their part in the first efforts to develop science clubs for youngsters in the early 1960s: the Palais had the Club Jean-Perrin, while the museum had its Clubs Naturalistes and the CNAM its Club des Jeunes Techniciens. Another association named Jeunes Sciences already existed, created in 1958 by the surrealist writer André Thirion. In 1962, the ANCS (Association nationale des clubs scientifiques [National association of scientific clubs]) was created, and subsequently became the Association nationale sciences techniques jeunesse (ANSTJ) in 1977 and Planète sciences in 2002 (Gautier and Las Vergnas, 2010). The keen interest of youngsters for space and astronomy impelled the Centre national d'études spatiales to draw on these associations to channel the activities of young people interested in rocket construction and propulsion (Radtka, 2016).

During the 1950s and 1960s, science communication was also achieved by other means: in the press, on radio and on the nascent television. Specialised magazines (*de vulgarisation*) had existed a long time (like *La Nature*, created in 1873) or had been created after World War II. *Science et vie* had been

4 Léveillé also played a very active role in the creation of the International Council of Museums (ICOM). He was a long-lasting member of its board and chaired its 'Science museums and planetaria section' that he strongly contributed to set.

created in 1913 on the model of the American *Popular science* under the title *La science et la vie*. One could add *La Revue scientifique* (a magazine *de haute-vulgarisation*, created in 1863). Two new magazines were created in the aftermath of World War II: in 1946, *Atomes*, a 'high-level' monthly magazine that would have a promising future under another title (*La Recherche*) and, in 1947, *Sciences et Avenir* that, like *Science et vie,* is still published today.

Science on radio has a long tradition in France. Before World War II, renowned scientists used to give 'lectures' on various topics and a *Conseil supérieur de la Radiodiffusion* had been created in 1937, bringing together personalities close to the Union Rationaliste and the Popular Front. The Conseil included a scientific section: Emile Borel, Henri Laugier, Jean Perrin, Paul Langevin (all involved in the Palais' foundation), Irène Joliot-Curie and Georges-Henri Rivière were members (Morelle and Jakob, 1997). In the 1950s and 1960s the main science broadcast on radio was *La science en marche* (1958–83) hosted by François le Lionnais, who was incidentally head (1950–58) of the Division of Science Teaching and Science Popularisation (Salon, 2016) at the Natural Sciences Department of UNESCO (then directed by the physicist Pierre Auger, who would a few years later host *Les Grandes Avenues de la science moderne* (1969–86), a program on the State radio channel *France-Culture*).

In this post-war period, television[5] was still in its infancy. Science had nonetheless its place, either through specific programs such as *Les bâtisseurs du monde*, a series of biographies of famous scientists (Galiléo, Copernicus, Galois, Curie) directed by André Labarthe (1956–57); or, as for *La page des sciences* (1961–62), as part of the newsreels. *Les médicales* (1954–84) hosted by journalist (and former physician) Igor Barrère and the journalist Etienne Lalou is considered a pioneer program on medical communication (Mansier, 2014). It owed its longevity in part to the excellent relations maintained by its directors with the medical establishment (Marchetti, 2007). As concerns science, one of the most interesting programs was *Visa pour l'avenir*, a one-hour program broadcast once a month from May 1962 to July 1967. The producers were at first two journalists from the legendary news TV magazine *Cinq colonnes à la une*, Jean Lallier and Roger Louis; they were soon joined by two renowned print media science journalists: Nicolas Skrotzky (Agence France-Presse) and Robert Clarke (France-Soir). *Visa pour l'avenir* was

5 Until 1974, television in France was a public monopoly (Radiodiffusion-Télévision Française [RTF: French Radio and Television Broadcasting], and from 1964 Office de Radiodiffusion-Télévision Française (ORTF), the national agency charged, between 1964 and 1974, with providing public radio and television in France). A law voted on the 8 July 1974 dismantled the ORTF. The first private channel was a pay TV channel Canal+ in 1984 and the first free generalist private broadcast was Silvio Berlusconi's *La Cinq* in 1986.

interested in technical progress and in the advancement of science, but it was foremost the result of journalism focused on science. Subjects like the thalidomide crisis (*Après la Thalidomide*: broadcast 19 October 1962) were quite controversial and the social responsibility of science and scientists could be openly questioned (*Faut-il tuer les savants*: broadcast 26 September 1963). In this period science on French television was mostly a matter of generalist journalists who might later specialise in scientific topics.

In March 1955, some journalists (primarily André Labarthe and Robert Clarke) created the Association des journalistes scientifiques de la presse d'information (AJSPI) (Clarke, n.d.), gathering together the nascent group of journalists specialising in science. Their objective was to gain legitimacy (Marchetti, 2007), externally among the scientific communities and internally to convince media owners to make more space for science. Later, in 1967, the AJSPI established 'the Club', bringing together the major research institutions and industries in order to encourage contacts between institutions and journalists (Chavot and Masseran, 2003).

Almost at the same time (June 1956), another association was created, this time under the aegis of UNESCO: the Association des écrivains scientifiques de France (AESF). The AESF brought together writers, journalists, popularisers and scientists. Its president was François le Lionnais. Members shared the view of a 'triumphant science, able to technologically catch up and to solve the underdevelopment of third world countries' (Laigneau, 1989). The members of AESF, Laigneau writes, were very committed to the literary quality of their writings and aimed more at the legitimation of scientific literature within the sphere of literature than at popularisation.

It is worth noting that the location of UNESCO's premises in Paris as well as the personalities of the first officials of UNESCO and of its Natural Sciences Section (Nielsen, 2018) may have played a role in the active development of science communication in post-war France. Léveillé had very good relationships with Huxley and Needham,[6] both of whom were invited to give lectures in the Palais before the creation of UNESCO. The Palais played a leading role in UNESCO in the late 1940s to early 1950s: organisation of the *Month of UNESCO* in 1946; participation in expert panels on the popularisation of science in 1947; commissioning of a report on scientific museums in 1949 (Bergeron and Bigg, 2015). After Needham returned to Cambridge, Pierre Auger, headed the Science Department; shortly after, Le Lionnais joined the division of Science Teaching and Science Popularisation. Both were members of the science committee at the ORTF and hosted a scientific broadcast.

6 Respectively the first director-general and the first director of the Natural Sciences Section of UNESCO (1946–1948).

What has been described testifies that, in late 1950s France, science communication was active in multiple forms and was already quite well organised. From the early 1950s until the mid-1970s, science communication repeatedly appeared as a matter of public concern to politicians, industrialists and scientists. To sum up a long history, let us begin in 1953 when the prime minister-to-be Pierre Mendès-France, probably persuaded that a strong and ambitious science policy was a prerequisite to the reconstruction of the country (Crémieux-Brilhac, 1986) by his many discussions with the scientist, diplomat and former director of the CNRS Henri Laugier, declared in a programmatic discourse in 1953: 'The Republic needs scholars!' Mendès-France became prime minister in 1954. He created the position of Secretary of State for Scientific Research and appointed Henri Longchambon (who, like Laugier, was one of the Palais' founders). Mendès-France only remained in the position for eight months, but a new momentum had been imparted to the field. Soon after, with his support, a group of scientists mobilised to promote the 'expansion of scientific research'. Their efforts led in 1956 to a major event: a National Conference on Scientific Research in Caen, which brought together the *élite* of French public and private research, politicians, industrialists and journalists. Known today as the Caen Colloquium, it is considered a milestone in French public research during the post-war years (Chatriot and Duclert, 2006). Similar conferences followed, like the colloquium entitled 'The Relations between University and Industry' in 1957 in Grenoble, at the precise moment when the success of Sputnik was made public. Discussions at both colloquia emphasised the assertion that the development of research was the indispensable condition for the recovery of the standard of living of the French people and the place of France in the world. This would only be possible, so the theory went, through a radical increase in the number of technicians, engineers and scientists. Therefore, the massive development of science communication (or *propaganda* to use the term employed at that time) using modern techniques (such as cinema, television and records) was fundamental: it was considered as the most powerful means to encourage young boys and girls to enter the scientific and technical professions. These were considered as the best methods for ongoing professional training, allowing technicians and engineers to acquire or maintain scientific and technical knowledge (Bergeron, 2020).

The French public research system was implemented gradually years (Chatriot and Duclert, 2006). Science communication remained a matter of interest to ministers and heads of the research administration. Important reports, like the Boutry report (1963), argued for more efficient science communication, to maintain public attitudes in favour of science and to keep professionals informed of current scientific developments. But this official support did not translate into practical action or achievement, even though interest persisted through the 1960s and the early 1970s. One of the main reasons is that science communication policy depended on the Bureau National de l'Information Scientifique et Technique (BNIST), the state service given the job of dealing with 'scientific information', i.e. information flows and infrastructures, documentation, computer development, implementation of databases and training of documentation professionals. While an urgent issue for science, for industry, for the army and the country, science communication was such a low priority that it worked as an adjustment factor in the face of tight budgets and a limited number of staff (Bergeron, 2020). Better times would soon come.

3. Protests and state affairs

The late 1960s were a time of protest in France. May 1968 was the most conspicuous event (Gobille, 2009). The movement born in the universities was joined by workers and then the whole country, and France was paralysed by weeks of strikes. There were days of intense contestation, but also days of multiple reflections and speeches (Certeau, 1997), and these contests, reflections and speeches pertained to science as well. During the 1960s and 1970s, groups of 'critic scientists' (Debailly, 2015) organised (often in connection with similar groups in Britain, Belgium, etc.), convened meetings and published bulletins like *Impascience* or *Labo-Contestation*. The functioning of scientific communities and institutions was questioned: issues like the low status of women and technicians and social hierarchies as well as the connection of science with the military sector and financial interests were raised. Popularisation of science came under scrutiny. Some social scientists (Jurdant, 1973; Roqueplo, 1974) highlighted its paradoxical effects: popularisation created the gap it was supposed to fill but stood a long way from the empowerment it was supposed to enable. Popularisation was indeed a major problem for those radical scientists looking for new means to present science in a less authoritative way, showing that the knowledge produced by science was less categorical than contingent.

One new approach was proposed: to 'put science into culture'. Jean-Marc Lévy-Leblond, a major figure in the 'science critic' movement and main theorist of the *mise en culture de la science*, took over leadership of the series of books *Science Ouverte*[7] at the Editions du Seuil. It had just published *(Auto)critique de la science*, the book he co-authored with Alain Jaubert in 1973. What was then called *action culturelle scientifique* [scientific cultural

7 The series still exists today and is still directed by Jean-Marc Lévy-Leblond. More than 220 titles have been published in this collection.

action] offered another means. Since the mid-1960s, science was one of the many domains treated in the Maisons de la culture (Bergeron, 2009), in the same way as theatre, music and literature. The Maisons de la culture was the cornerstone of André Malraux's policy at the Ministry for Culture. The idea was to create reception structures across France to 'give access to the major cultural works of humanity' not only in Paris, but to the widest public. Since the first one opened in Le Havre in 1961, they developed polyvalent cultural projects that often dealt with scientific themes (as in Le Havre, Bourges, Reims, Nanterre, Chalon-sur-Saône, Grenoble, Saint-Étienne, etc.). Science was presented in the Maisons de la Culture by means of exhibitions and conferences, etc. Its presence in such cultural institutions seemed then self-evident, fitting very well with Malraux's doctrine.

Unexpectedly, science in these cultural institutions attracted new publics that would probably not have come for the theatre or music. The Ministry for Culture doctrine was 'cultural action' (Urfalino, 1996), and what concerned science was named 'scientific cultural action'. In 1974, the main actors (scientists or sociocultural coordinators) met in Grenoble and created the Groupe de liaison pour l'animation culturelle scientifique (GLACS) [Liaison group for scientific cultural animation] to facilitate mutual information. GLACS was headed by the physicist Michel Crozon and its main activity at the time was Sciences dans la rue [Sciences in the street]. The principle was simple: at scientific conferences (e.g. in Aix 1973, Poitiers 1974, etc.), events were organised for the general public, and scientists engaged in conferences or public demonstrations (often 'in the street'). The Science dans la rue series was a big success and received financial support from the State that (as we will see below) began to engage concretely in the domain.

Contestation was also about ecological matters, technological choices and, above all, a highly strategic matter for France: nuclear energy (Hecht, 1998). Some scientists played an active role in the anti-nuclear movement, particularly the Group of Scientists for Information on Nuclear Energy (Groupement des Scientifiques pour l'Information sur l'Energie Nucléaire, GSIEN). Against what appeared to them as a strong tendency of nuclear officials to turn information into propaganda and to maintain secrecy, they published in *La Gazette Nucléaire* arguments informed by academic literature to counter the official information of the electrical public agency Électricité de France (EDF) (Topçu, 2008, 2013). The GSIEN was created at the end of 1975 as a reaction to a program known as 'Messmer's Plan'. The plan was made public in 1974 and generalised the use of nuclear power through the construction of 13 nuclear plants. 'Information' was a crucial issue: Messmer's Plan had never been discussed in parliament and political opponents declared it was presented as a *fait accompli.* In May 1975, in a declaration on energy policy delivered in parliament, Michel d'Ornano, Minister for Industry and Research, announced a new strategy of the government: to publish scientific information and leave nothing in the shadows. To inform 'all citizens' became, in this respect, as important an issue as developing computers or databases: the time had come for the State service devoted to science information at large (BNIST) to take seriously this part of its mandate, a job that until then had been on standby (Bergeron, 2020). To this end, a 'scientific and technical cultural action' sector was created with Lucie Degail recruited to head it. From 1976, actions were developed including the participation in the youth fairs, support for the program Science in the street, regular support to the monthly *La Recherche*, and commissioning studies on French attitudes to science.

4. Institutionalisation and professionalisation

A few years later, in 1981, another political event had major repercussions on science communication matters in France.

In May 1981, François Mitterrand was elected president of the Republic. For the first time since the immediate post-war period, a left-wing government came to power. From the outset, the new government showed a keen interest in science and technology, considering it the most useful mechanism to help France find a way out its economic crisis. The new Minister for Research and Technology, Jean-Pierre Chevénement, rapidly organised a series of regional conferences for 'Research and Technology', bringing together scientists, industrialists and politicians. The regional conferences culminated in the national conference 'Research and Technology' in Paris on 13–16 January 1982. The general idea was to involve the nation's scientific communities in the definition of a future science policy. The conference was a national event and received wide media exposure. As noted by a TV report: 'Almost all members of the government were there […] Science truly became a political object!' (Journal télévisé d'Antenne 2, 1982). In a closing statement, Chevènement emphasised the necessity to 'integrate a scientific dimension into the culture by means of the development of (scientific) formation, information and animation', and called for more actions in the domain of 'scientific and technologic culture'. As noted by Petitjean (1998), one of the minister's personal priorities was to fight anti-science, though the theme had never been raised in any working group of the conference. The consequences for science communication were concrete: it was from then on a State priority and funds were made available.

Centres for scientific, technical and industrial culture (Centre de culture scientifique, technique et industrielle, CCSTI) were then considered as the best way to communicate science, and their development became a priority. These were based on the model of the 'science cultural centre' opened in Grenoble in 1979 by a group of scientists who wanted to continue scientific cultural animation independently from the local Maison de la culture. Two months after the national conference, a second CCSTI, the Fondation 93, opened in Montreuil. It was directed by Alain Berestetsky, formerly responsible for cultural action in the nearby town of Bagnolet. In 1984, more than 16 CCSTI projects were running in the main cities of the country. Since the very beginning, the CCSTI could take rather different forms: some (such as La Casemate in Grenoble) were true science centres, with dedicated spaces for exhibitions and public events, while others (such as Fondation 93) preferred not to have any space opened to the public and worked as resource centres for cultural institutions, associations, schools, etc. in their geographic areas. Their diversity of forms, sizes and styles was linked as much to the importance of the local context in which they originated (and in which they would later strongly integrate) as to the different philosophies of the founding teams (Bergeron, 2016b). Very little detailed historical analysis of those local science centres exists in the literature, with the remarkable exception of the Lorraine Region studied by Choffel-Mailfert (2002). Her analysis shows that the simultaneous creation of science centres and of heritage and craft-oriented structures in Lorraine can be understood in the context of the severe deindustrialisation of the region. The diversity in the styles of museumification can be regarded as the results of different ways of thinking of the (local/regional/national) stakeholders. Her conclusion is that in the Lorraine region, the development of 'scientific, technical and industrial culture (*la culture scientifique, technique et industrielle*) allowed a shift in social struggle, replacing social struggle by cultural struggle'. This conclusion is consistent with other analyses, such as Debary's (2003, 2004), interpreting the creation of the Creusot ecomuseum as a (cultural) art to use up the (industrial) leftovers. This coincides with a growing interest in France toward its industrial heritage (Gasnier, 2016).

When all this was happening, what was to become in 1986 the main institution for scientific and technical culture was under construction in Paris. The project of a 'national museum for science, technology and industry' dates back to 1977, when the architect Roger Taillibert suggested converting the site of the unfinished abattoir at La Villette into a major museum. Two years later, the physicist Maurice Lévy, former president of the National Centre for Spatial Studies, issued a report outlining the contours of what would later be the Cité des sciences et de l'industrie [City of sciences and industry].[8]

8 With more than 2 million visitors in 2016, the Cité is still today the largest scientific and technical centre in France. In 2010, it merged with the Palais de la découverte to form Universcience.

The prospect of opening such a new and large museum, together with the emergence of scientific and technical centres in the country, motivated scholars to begin specific training for science communication (Devèze-Berthet, 1989). In 1984, Denise Devèze-Berthet set up the first training courses (bachelor's and graduate) focused on 'communication and scientific, technical and medical information'. One year later, in Tours, Jean Lagoutte and colleagues created bachelor-level training at the Institute of Technology (IUT) in Tours, focused more on scientific mediation. Other programs followed in the 1990s: in Strasbourg and Grenoble. During the same decade, the renovation of two major science museums (Muséum National d'Histoire Naturelle, 1993; Musée des Arts et Métiers, 2000) and a national project for the development of research in museology (REMUS, 1993) led to the development of new courses specialising in science museology in Lyon, Paris, St Etienne and other locations.

Science communication at this point had its specific places, trainings and scholars. It also had its own networks. Journalists and writers had become organised in the 1950s; in 1982, professionals and supporters of science centres and of scientific culture created the Association des musées et centres pour la promotion de la culture scientifique, technique et industrielle (AMCSTI) [Association of museums and centres for the development of the scientific, technical and industrial culture]. Its first general meeting was presided over by Hubert Curien, then president of the Centre national d'études spatiales (CNES). In 1989, Pierre Fayard, associate professor in information sciences at Poitiers, took part in the creation of another network: the Public Communication of Science and Technology (PCST) Network. The first meeting was held in Poitiers in 1989 and attracted international participants. The Nuit des étoiles [Night of Stars] was created in 1991 by a group of amateur and professional astronomers (Las Vergnas et al., 2010); of La main à la pâte [Hands-on], created in 1995 by the physicist and Nobel laureate Georges Charpak and other academicians to promote the use of a hands-on approach in science education and immediately supported by the Ministry for Education; and of the emergence of new publishers specialising in science, such as the Editions Odile Jacob created in 1986 by Nobel laureate François Jacob's daughter (Lemerle, 2007).

Between the early 1980s and the late 1990s, numerous initiatives[9] contributed to the development of science communication in France in different forms: print, journalism, museums, cultural animation and heritage. Throughout this period two trends could be clearly observed:

9 In 2010 and 2012, two conferences were organised in Orleans in order to document the memories of 30 years (1980–2010) of 'scientific and technical culture' in France. The interested reader can find numerous stakeholder testimonies in the conference proceedings (Caillet et al., 2014; Bergeron et al., 2014).

- an increasing professionalisation (specialised trainings, specific jobs)
- an increasing institutionalisation with more specific structures, better organised at the national level, growing funding and recognition.

These efforts transformed 'science communication', which was then better known as 'culture scientifique et technique', as a true *dispositif* that was considered with a great deal of interest by the public authorities, particularly the Ministry for Research, as it performed various complementary functions: disseminating scientific knowledge; encouraging scientific vocations; creating areas for and engaging scientists in debates; and promoting local industries' know-how. In 1992, Hubert Curien (Minister for Research and Technology, 1984–86 and 1988–93), really 'invented' the Fête de la science [Science Festival]. It was built on the model of the Fête de la musique [Music Day] established by the Minister for Culture, Jack Lang, in 1982; and its first occurrence took place in the ministry gardens, which were opened to the public for this occasion. This event was a success and the idea was soon adopted in several countries (Ledur, 2016).

In the mid-1990s, science communication in France can be regarded as consolidated in a form close to what we know today. Of course, nothing is set in stone and new modalities (science blogs, YouTubers) emerged in France as in other countries. One of the most conspicuous is the rapid implementation of FabLabs (Lhoste and Barbier, 2016) often hosted in the CCSTIs. Another major trend is undoubtedly the growing interest of universities in their heritage and its promotion (Boudia et al., 2009) that has transformed universities today in a major actor of scientific communication.

5. Meaningful words

All through this chapter, for a matter of homogeneity with the book, I strove to use the expression 'science communication' as a generic name for our topic. In actual fact, the wording 'science communication' (*communication scientifique*) was in use for a rather short time: around the 1980s, coexisting for a while with the broader 'scientific (technical and industrial) culture' (*culture scientifique [technique et industrielle]).*

How to name 'popularisation of science' is an old and seemingly unsolved problem in French. Jeanneret (1994) made this clear in the introduction of his now classical book: by choosing *vulgarisation*, the French language (or rather its speakers) created persistent discomfort due to the predictable links toward 'vulgar' and its derivatives. This partly explains the numerous attempts to introduce alternative designations all through the past century.

I have discussed elsewhere (Bergeron, 2016a) how none of these wordings is neutral. On the contrary, each of them is linked to a specific historical, social and epistemic configuration.

In the post-war period, scientists involved in the French public research system pleaded for more attention to be given to science, more scientists, engineers and technicians, more boys and girls embarking on scientific careers. 'Propaganda' (*propagande*[10]), using modern tools (television, records, etc.), was what they called for. In the 1960s, with the beginnings of the informatisation of society, access to information became a crucial issue. It was then referred to as 'scientific (and technical) information' (*information scientifique*) for the general public. In the late 1970s, the overall increase of 'communication' in society, together with growing contestation about technical choices such as nuclear, put 'scientific communication' (*communication scientifique*) at the forefront. In the 1980s, as a result of the 'cultural turn' of society and of the quest of critic scientists for new forms to communicate science, came the moment of the 'scientific and technical culture' (*culture scientifique et technique*) concept that would soon be adopted in several other countries, mostly (but not exclusively) Francophone (Delicado, 2010).

Presently, the wording 'scientific mediation' (*médiation scientifique*) is the most widely used. In the above-mentioned paper I identified three complementary movements to explain its genesis. First, when considered as the implementation of science for 'cultural mediation', it is linked to professionalisation issues. Second, considering its early emergence in the brand new Cité des Sciences et de l'Industrie, the genesis of which involved working groups in which researchers such as Bruno Latour or Isabelle Stengers participated, this wording indicates a renewed way to communicate sciences within a new framework of analysis. Third, considering the general rise of the notion of mediation at the time interpreted by many scholars as a new sort of soft-law device (Briant and Palau, 1999), this idea of 'mediation' should be set against the background of a new governmentality of sciences that promote participation.

This history, roughly sketched, led us from the interwar period to the late 20th century. All along this period, concerns for science communication remained active (Bensaude-Vincent, 2010). The term was reformulated, reframed and taken over by different groups depending on the prevalent context and concerns of the moment, to which its changing designation bears witness.

10 In French at that time the word commonly designated advertising.

References

Bensaude-Vincent, B. (2010). Splendeur et décadence de la vulgarisation scientifique. *Questions de communication, 17*, 19–32. doi.org/10.4000/questionsdecommunication.368.

Bensaude-Vincent, B. (2013). Popular Science and Politics in Interwar France. *Science in Context, 26*(3), 459–71, doi.org/10.1017/S0269889713000148.

Beretta, M. (2004). Sur les traces du laboratoire de Lavoisier. *La revue du musée des arts et Métiers, 41*, 4–23.

Bergeron, A. (2009). Patrimoine et culture scientifique: sur l'inscription culturelle des savoirs. In S. Boudia, A. Rasmussen and S. Soubiran (eds), *Patrimoine, savoirs et communautés savantes*. Rennes: Presses universitaires de Rennes.

Bergeron, A. (2016a). Médiation scientifique. *Arts et Savoirs* [En ligne], *7*. journals.openedition.org/aes/876, doi.org/10.4000/aes.876.

Bergeron, A. (2016b). Pour une histoire des centres de culture scientifique, technique et industrielle. In P. Poirrier (ed.), *Histoire de la culture scientifique en France: institutions et acteurs*. Dijon: Editions universitaires de Dijon.

Bergeron, A. (2020). From databases to 'information for the general public': the long path toward the emergence of a public action for scientific popularisation in France at the turn of the 70s. In M. Le Roux (ed.), *Communicating science*. Bruxelles: PIE Peter Lang.

Bergeron, A. and Bigg, C. (2015). D'ombres et de lumières. L'exposition de 1937 et les premières années du Palais de la découverte au prisme du transnational. *Revue germanique internationale, 21*, 187–206. doi.org/10.4000/rgi.1529.

Bergeron, A. and Bigg, C. (2019). Built in Thoughts rather than Stone. The Palais de la Découverte and the 1937 International exhibition in Paris. In E. Canadelli, M. Beretta and L. Ronzon (eds), *Behind the Exhibit. Displaying Science and Technology at World's Fairs and Museums in the Twentieth Century*. Artefacts Studies in the History of Science and Technology, Volume 12 (pp. 108–31). Washington DC: Smithsonian Institution Scholarly Press. doi.org/10.5479/si.9781944466237.

Bergeron, A., Caillet, E., Guyon, E., Guillet, P., Guiraudon, J.-C., Maitte, B., Morand, O. and Van-Praët, M. (2014). *Hier pour demain: pour demain: une mémoire de la culture scientifique, technique et industrielle*. Paris: L'Harmattan.

Blanckaert, C., Cohen, C., Corsi, P. and Fischer, J.-L. (1997). *Le Muséum au premier siècle de son histoire*. Paris: Muséum national d'histoire naturelle.

Boudia S., Rasmussen, A. and Soubiran, S. (eds) (2009). *Patrimoine, savoirs et communautés savantes*. Rennes: Presses universitaires de Rennes.

Boutry, G.-A. (1963). *Rapport à M. le ministre d'État chargé de la Recherche scientifique et des Questions atomiques et spatiales.* Paris: Archives de la DGRST, Centre des archives contemporaines.

Briant (de), V. and Palau, Y. (1999). *La médiation. Définition, pratiques et perspectives.* Paris: Nathan.

Caillet, E., Guyon, E., Guillet, P., Guiraudon, J.-C., Maitte, B., Morand, O. and Van-Praët, M. (2014). *Une mémoire pour demain : 30 ans de culture scientifique, technique et industrielle en France.* Paris: L'Harmattan.

Certeau (de), M. (1997). *Capture of Speech and Other Political Writings.* Minneapolis: University of Minnesota Press.

Chatriot, A. (2006). La lutte contre le 'chômage' intellectuel: l'action de la Confédération des Travailleurs Intellectuels (CTI) face à la crise des années trente. *Le mouvement social, 214*(January–March), 77–91. doi.org/10.3917/lms.214.0077.

Chatriot, A. and Duclert, V. (2006). *Le gouvernement de la recherche. Histoire d'un engagement politique, de Pierre Mendès-France à Charles de Gaulle (1953–1969).* Paris: La Découverte.

Chavot, P., and Masseran, A. (2003). French media: Introducing science to everyday-life. In U. Felt (ed.), *O.P.U.S. Final report.* Retrieved from sts.univie.ac.at/fileadmin/user_upload/i_sts/Forschung/Projekte_abgeschlossen/final_report_opus.pdf.

Choffel-Mailfert, M.-J. (2002). La médiation culturelle, territoires d'enjeux et enjeu de territoire. In B. Schiele (ed.), *Patrimoines et identités.* Québec, Canada: Editions Multimondes, Musée de la civilisation.

Clarke, R. (n.d.). Historique. Retrieved from www.ajspi.com/lassociation.

Conklin, A. (2013). *In the Museum of Man: Anthropology, Race, and Empire in France, 1850–1950.* Ithaca, NY and London: Cornell University Press.

Crémieux-Brilhac, J.-L. (1986). Conférence avec Jean-Louis Crémieux-Brilhac au séminaire du professeur Antoine Prost. Centre Malher, Paris I, le 26 juin 1986.

Debailly, R. (2015). *La critique des sciences en France depuis 1968. Critique des sciences et études des sciences en France après mai 1968.* Paris: Hermann.

Debary, O. (2003). *La Fin du Creusot ou l'Art d'accommoder les restes.* Paris: Comité des travaux historiques et scientifiques.

Debary, O. (2004). Deindustrialisation and Museumification: From Exhibited Memory to Forgotten History. *The ANNALS of the American Academy of Political and Social Science, 595*(1), 122–33. doi.org/10.1177/0002716204266630.

Delicado, A. (2010). Exhibiting science in Portugal: Practices and representations in museums. *Portuguese Journal of Social Sciences, 9*(1), 19–32. doi.org/10.1386/pjss.9.1.19_1.

Devèze-Berthet, D. (1989). Pourquoi et comment sont nées les formations à la médiation scientifique. *Brise, Bulletin de recherche sur l'information en sciences économiques et sociales, 14*, 115–20.

Dufaux L. (2013). *Le Musée des arts et métiers. Guide des collections.* Paris: Musée des arts et métiers/Artlys.

Eidelman, J. (1985). The Cathedral of French Science. The early years of the Palais de la découverte. In T. Shinn and R. Whitley (eds), *Expository science. Forms and Functions of Popularisation.* Dortrecht-Boston-Lancaster: D. Reidel Publishing Compagny.

Gasnier, M. (2016). Patrimoine industriel et technique: entre héritage culturel et institutionnalisation. In P. Poirrier (ed.), *Histoire de la culture scientifique en France : institutions et acteurs.* Dijon: Editions universitaires de Dijon.

Gautier, G. and Las Vergnas, O. (2010). Loisir aérospatial et astronomique: la saga de l'Association nationale des clubs scientifiques. I. de 1962 à 1977. Entretiens avec Pierre-Julien Dubost et Jean-Claude Guiraudon. *La revue pour l'histoire du CNRS, 27–28*, 42–47.

Gobille, B. (2009). *Mai 1968.* Paris: La Découverte.

Hecht, G. (1998). *The Radiance* of *France. Nuclear Power and National Identity after World War II.* Cambridge: MIT Press.

Jeanneret, Y. (1994). *Ecrire la science.* Paris: Presses Universitaires de France.

Journal télévisé d'Antenne 2. (1982, 13 January). *Edition de 20h.*

Jurdant, B. (1973). *Les Problèmes théoriques de la vulgarisation scientifique*, Thèse de doctorat, Université Louis Pasteur, Strasbourg. Reedited, Editions des Archives Contemporaines, Paris, 2009.

Kargon, R., Fiss, K., Low, M. and Molella, A. (2015). Modernity à la Française: The 1937 Paris Exposition. In R. Kargon, K. Fiss, M. Low and A. Molella (eds), *World's Fairs on the Eve of War: Science, Technology, and Modernity, 1937–1942.* Pittsburgh, PA: University of Pittsburgh Press. doi.org/10.2307/j.ctt18d83c0.5.

Laigneau, M. (1989). 'Contextes', *Culture Technique.* Centre de recherche sur la culture technique, Neuilly-sur-Seine (FRA), *20*, 10–27. Retrieved from hdl.handle.net/2042/31938.

Las Vergnas, O., Piednoël, E. and Gautier, G. (2010). Loisir aérospatial et astronomique : la saga de l'Association nationale des clubs scientifiques. II. de 1975 à 2000. *La revue pour l'histoire du CNRS, 27–28*, 48–57.

Ledur, M.-C. (2016). La science en fête. *Histoire de la recherche contemporaine, 5*(2). doi.org/10.4000/hrc.1356.

Lemerle, S. (2007). Le biologisme comme griffe éditoriale. Sociologie de la production des éditions Odile Jacob 1986–2001. *Sociétés contemporaines, 64*, 21–40.

Lévy-Leblond, J.-M. and Jaubert, A. (1973). *(Auto)critique de la science.* Paris: Le Seuil.

Lhoste, É. and Barbier, M. (2016). FabLabs. L'institutionnalisation de Tiers-Lieux du 'soft hacking', *Revue d'anthropologie des connaissances, 10*(1), 43–69.

Mansier, P. (2014). Les magazines de santé à la télévision depuis les années 1950. *Le Temps des médias, 23*, 241–45. doi.org/10.3917/tdm.023.0241.

Marchetti, D. (2007). Une place à part. L'information médicale de l'après-guerre au début des années 80. *Questions de communication, 11*, 71–90. doi.org/10.4000/questionsdecommunication.7330.

Morelle, C. and Jakob, P. (1997). *Henri Laugier.* Bruxelles/Paris: Un esprit sans frontières. Bruylant/L.G.D.J.

Nielsen, K. H. (2018). 1947–1952: UNESCO's division for science & its popularisation. *Public Understanding of Science, 28*(2), 246–51. doi.org/10.1177/0963662518787572.

Ory, P. (1991). Une 'cathédrale pour les temps nouveaux'? Le Palais de la découverte (1934–1940). In R. Robin (ed.), *Masses et culture de masse dans les années 30.* Paris: Editions Ouvières.

Ory, P. (1994). *La belle illusion : culture et politique sous le signe du Front populaire, 1935–1938.* Paris: Plon.

Ouest-Eclair. (1937, 31 May). 85000 entrées ont été enregistrées hier à l'exposition, *Ouest-Eclair*, p. 3.

Petitjean, P. (1998). La critique des sciences en France. In B. Jurdant (ed.), *Impostures scientifiques. Les malentendus de l'affaire Sokal*, Paris and Nice: La Découverte/Alliage.

Picard, J.-F. (1990). *La République des savants, La Recherche française et le CNRS.* Paris: Flammarion.

Radtka, C. (2016). De l'enthousiasme aux réalisations pratiques : la fusée, objet d'une culture 'jeunes'. Retrieved from humanites-spatiales.fr/de-lenthousiasme-aux-realisations-pratiques/.

REMUS (1993). *1er colloque sur la muséologie des sciences et des techniques, 12–13 décembre 1991, Paris.* Dijon: Éditions de l'OCIM.

Roqueplo, P. (1974). *Le partage du Savoir: Science, culture, vulgarisation.* Paris: Éditions du Seuil.

Salon, O. (2016). *Le disparate François Le Lionnais.* Paris: Othello.

Topçu, S. (2008). Confronting Nuclear Risks: Counter-Expertise as Politics Within the French Nuclear Energy Debate. *Nature and Culture, 3*(2), 225–45. doi.org/10.3167/nc.2008.030205.

Topçu, S. (2013). *La France nucléaire. L'art de gouverner une technologie contestée.* Paris: Editions du Seuil.

Urfalino, P. (1996). *L'invention de la politique culturelle.* Paris: La Documentation française.

Valéry, P. (1960). Un problème d'exposition. Paris 1937, Revue mensuelle de l'exposition. Reprinted in Œuvres, Vol. 2, pp. 1150–1156. Paris: Gallimard.

Timeline

Event	Name	Date	Comment
First interactive science centre established.	La Casemate, Centre culturel scientifique, Grenoble	1979	Also arguably the Palais de la Découverte 1937
First national (or large regional) science festival.	Aix pop	1973	
An association of science writers or journalists or communicators established.	Association des journalistes scientifiques de la presse d'information	1955	1956: Association des écrivains scientifiques de France (AESF) formed
First university courses to train science communicators.	Université Paris 7	1984	
First master's students in science communication graduate.	Université Paris 7	1984	
First PhD students in science communication graduate.	Perhaps B. Jurdant	1973	
First national conference in science communication.	Colloque de Nice (organised by AJSPI)	1969	1974: National meeting on scientific cultural animation, Grenoble
National government program to support science communication established.		1975	A new strategy to publish scientific information
First significant initiative or report on science communication.	Creation of Palais de la découverte	1937	
National Science Week founded.	Fête de la science (one weekend)	1992	
A journal completely or substantially devoted to science communication established.		19th century	Popular science magazines exist since the 19th century

Event	Name	Date	Comment
First significant radio programs on science.	Regular conferences of scientists	1930s	The *Conseil supérieur de la Radiodiffusion*, 1937, had a scientific section
First significant TV programs on science.	*Visa pour l'avenir* [Visa for the Future] was the most significant	1962	Others already existed (e.g. *Les bâtisseurs du monde*, 1956, or *Les médicales*, 1954)
First awards for scientists or journalists or others for science communication.	Prix Roberval	1987	
Date hosted a PCST conference.	The very first PCST conference was organised in Poitiers	1989	Organiser was Pierre Fayard
Other significant events.	The creation of AMCSTI	1982	This was the Association of Museums and Centres for the Development of the Scientific, Technical and Industrial Culture

Contributor

Dr Andrée Bergeron is an associate professor in epistemology and history of science at Universcience, and a researcher at the Centre Alexandre-Koyré d'Histoire des Sciences et des Techniques (EHESS-CNRS-MNHN) in Paris, France.

14

GERMANY

Continuity and change marked by a turbulent history

Hans Peter Peters, Markus Lehmkuhl and Birte Fähnrich

In Germany, science popularisation through public talks of scientists, public shows, museums, popular science magazines and even newspapers dates back to the 19th century. The post-war development of science communication in West Germany was characterised by the (modified) adoption of two Anglo-Saxon models: US science journalism leading to professionalisation of science journalism in the 1980s and the British public engagement approach in the second half of the 1990s. In both cases, charitable foundations played a key role. Science journalism in communist East Germany had an ideological mission and was therefore supported by the state but adopted the Western approach after German reunification in 1990. Since the 1960s, universities and other science organisations had already built a capacity for public relations that at first mainly targeted journalistic media but subsequently expanded towards events, face-to-face interactions and citizen science in the context of the new 'engagement' paradigm. Since the late 1990s, several interactive science centres that complemented traditional science museums were established. Science organisations, charitable foundations and the government contributed to the creation of a national science communication agency Wissenschaft im Dialog (WiD) that initiates and coordinates many activities. Scientific-technical controversies beginning in the 1970s provided a challenge for science communication. The current government supports public participation in science policy and citizen science.

1. Introduction

Germany has a rich history, culture and infrastructure of science communication that dates back to the 19th century. After World War II, science communication re-emerged in the late 1950s, initially in the form of science journalism and media-focused science PR and later by adoption of the British model of public engagement with science. Today, science communication is a vital and diverse field. Although affected by the economic crisis of traditional media, science journalism still plays an important role in Germany. Quality newspapers (print and online) offer science reporting; popular science magazines serve those with a stronger interest in science; public TV and radio programs broadcast science documentaries and magazines. German scientists are as motivated for public communication as those in other major science nations and involved in a variety of science communication activities (e.g. Peters et al., 2009). Organisational communication by universities and other publicly funded research organisations such as the Max Planck Society, Fraunhofer Society and the Helmholtz Association is diversified and professional (for an overview, see Fähnrich et al., 2019). At the national level there are different forms of science communication such as the German Science-Media Center, a science-in-dialogue agency (WiD) and traditional science museums, as well as several interactive science centres. A report prepared for the Commission of the European Union concluded that Germany has a 'consolidated science communication culture' (Mejlgaard et al., 2012), in the same way as the United Kingdom, France, Italy, Spain and other major science countries.

However, it is important to note that while we use the modern term 'science communication' as an inclusive label for all forms of public communication of science in this chapter, these activities were hardly discussed as 'science communication' (or *Wissenschaftskommunikation* in German) before the mid-1990s. In Germany, the term came into use only when the spectrum of science communication activities broadened with the German adoption of the 'public engagement with science' paradigm. Until then, the terms 'science popularisation', 'science and the media' and 'science journalism' were mostly used. There is still no consensus about the meaning of the term 'science communication'. While we use it as an umbrella term, in line with the definition adopted by the Science Communication Section of the German Communication Association (DGPuK) as well as its international use, some German scholars restrict it to communication activities other than those related to mass mediated science.

While Germany has its own tradition of science popularisation, major transformations in the post-war era were based on the reception and adoption of two Anglo-Saxon models: professional US science journalism in the late

1970s and 1980s (along with the academic study of science journalism and its institutionalisation as an academic sub-discipline) in former West Germany; and the British 'Public Engagement with Science and Technology' approach with its focus on dialogic forms of science communication in the second half of the 1990s.

After World War II, Germany was divided into the (communist) East Germany and the (democratic) West Germany for almost 45 years until it was reunited in 1990. During the period of separation, science communication developed in both parts of Germany, but the approaches and functions differed in each part. In this chapter, while briefly describing the East German situation, we focus on the post-war development of science communication in West Germany during the phase of separation because developments after reunification in 1990 connected to the West German tradition.

2. Roots

Historically, Germany has been a vital place for the development of modern science and its communication. The origins of modern science lay—as in other European countries—in the foundation of science academies and scientific societies (Ausejo, 1994). Whereas universities date back to medieval times (the oldest German university in Heidelberg was founded in 1386), they took their role as central science organisations only in the 19th century in the context of the Prussian university reforms. Under the then minister Wilhelm von Humboldt, the ideals of a humanitarian academic education, the unity of research and teaching and the freedom of science from political intervention were introduced. Moreover, Humboldt (1964) stressed the necessity of the society-orientation of science. The current international discussion of the universities' 'third mission' of interacting with citizens and society (e.g. Laredo, 2007), also taken up in Germany (Himpsl, 2017), seems very much in line with Humboldt's early concept of the modern university.

Other precursors of German science communication were popular presentations of science in the 19th century, especially in the form of scientific experiments conducted in parlours as well as at folk festivals (Weitze and Heckl, 2016, p. 4). Prominent scholars such as Alexander von Humboldt and Ernst Haeckel gave public lectures on science and wrote books that addressed not only academic scholars but also a science-interested public. Since about 1830, the newspaper *Kölnische Zeitung* employed journalists specialising in the reporting of science and technology (Krüger and Ruß-Mohl, 1989). The popular science magazine *Kosmos* was launched in 1903, claiming to follow the tradition of Alexander von Humboldt's *Kosmos* book series, which were published 1845–58 (Faulstich, 2006).

In this spirit, science museums were founded to make science and technological developments open to the public (Weber, 1993). For example, the natural history museum in Berlin was opened in 1889. Another outstanding historical example of the late 19th century was Urania in Berlin (Wilke, 2013), the first public astronomical observatory and science centre in Germany.

At the beginning of the 20th century, universities recognised the importance of public communication and started to institutionalise press relations (Koenen and Meißner, 2019). Further, museums of science and technology were founded, among them the Deutsches Museum in Munich (1903) and Hygiene-Museum in Dresden (1912). With the establishment of the Nazi Regime, however, science and its public communication were under full political control and censorship and exploited in the regime's propaganda machinery (Nagel, 2012).

3. Professionalisation of science journalism

We locate the origin of 'modern' post-war science communication—mostly science journalism at that time—in the mid-1950s. The space flight programs of Russia and the United States led to extensive coverage in print and broadcast media. In West Germany, the newspaper *Frankfurter Allgemeine Zeitung* started publishing the first regular science section in 1959. The two public TV channels ARD and ZDF established editorial offices for science reporting and launched science programs in 1964. New popular science magazines such as *Bild der Wissenschaft* (1964) were founded and, in addition, two very successful TV magazines on nature and environment (*Ein Platz für wilde Tiere*, ARD) and health (*Gesundheitsmagazin Praxis*, ZDF) were launched in 1956 and 1964 respectively.

The 1970s were characterised by a sudden rise of interest in science journalism as well as a growing interest in its academic study. The quality and quantity of science journalism, the training of science journalists and the collaboration between science and the media were widely considered deficient in West Germany (e.g. Lohmar, 1972; Fischer, 1976; Hömberg, 1980). Hömberg (1990) titled his book on German science journalism *The Delayed [Science] Beat* and Grabowski (1982) diagnosed *Structural Problems of Science Journalism in News Media*.[1]

1 See Kohring (2005) for a detailed description of the German discussion of science popularisation up to 1995.

In 1980, the Robert-Bosch Foundation started a large Promotional Program for Science Journalism (Göpfert and Schanne, 1998) that targeted science journalism practice and training in the first place, but also included communication scholars in the discourse. The program awarded about 200 scholarships to journalists, many of whom went to the United States to learn from science journalists there. The program provided the initial funding for the first German Chair for Science Journalism at the Free University of Berlin. Winfried Göpfert, then senior science editor at the public TV channel SFB, took this position to become the first German professor of science journalism. The foundation also funded a series of workshops (1982–92) in which scientists, journalists and communication scholars discussed the status of science journalism in Germany. A number of well-known US scholars were invited to these workshops, illustrating the role of US science journalism and science communication research as a model for Germany (Robert Bosch Stiftung, 1985; Göpfert and Bader, 1998). Referring to the United States, Stephan Ruß-Mohl, the main protagonist of the Robert Bosch program, titled his introduction to one of the workshops 'Science Journalism in the Promised Land' (Ruß-Mohl, 1985). The German orientation towards US science journalism was in line with the general influence of the United States on the reconstruction of the German mass media system after World War II by the United States and Great Britain. Two further initiatives followed in 2002–16, funded by Bertelsmann Foundation, Volkswagen Foundation, Stifterverband[2] and the pharmaceutical company BASF (Lehmkuhl, 2012).

Since the 1980s, science journalism in Germany has broadened, diversified and professionalised in terms of training, associations and infrastructures (e.g. Göpfert, 2006; Wormer, 2006; Hettwer et al., 2008). While there were occasional academic courses on science journalism before 1988 (Krüger and Flöhl, 1982; Robert Bosch Stiftung, 1983), systematic academic training started with a postgraduate science journalism program at the Free University of Berlin in 1988 (Krüger, 1990). Ten years later, in 1998, the Berlin Chair for Science Journalism organised the 5th international conference of the Network for the Public Communication of Science and Technology (PCST) in Berlin.

German associations of communicators of science, technology and medicine have existed since the 19th century. Many science journalists were members of the Technisch Literarische Gesellschaft (Technical Literary Association, TELI) established in 1929. In 1986, the German science journalism

2 Stifterverband describes itself as 'a joint initiative started by companies and foundations', focusing on 'consulting, networking and promoting improvements in the fields of education, science and innovation'. Although formally not a foundation, it operates in a similar manner.

association (WPK) was founded; unlike TELI, its membership is strictly confined to professional science journalists (Pütz, 2006). WPK advocates a clearly 'journalistic' approach as critical observer of science in contrast to the affirmative popularisation of science. It tries to keep science journalism and public relations of science clearly separated (Koch and Stollorz, 2006). Since 2004, WPK has been co-organiser of the annual conferences Wissenswerte [Worth knowing] that are particularly addressed at science journalists. WPK also played an important role in the institutionalisation of the German Science-Media Center in 2015, which, unlike the British model (Fox, 2012), can best be described as a journalistic infrastructure for science reporting rather than as an interface of science with the media (Stollorz and Meurer, 2015).

4. Post-war science communication in East Germany

In the German Democratic Republic (GDR) science communication was considered an instrument to promote a scientific world view (*Weltanschauung*). In 1954, at about the same time when the pre-war Urania was re-established in West Berlin, the Society for the Dissemination of Scientific Knowledge (Gesellschaft zur Verbreitung wissenschaftlicher Kenntnisse and later also named Urania) (Gruhn, 1979) was constituted in East Berlin. In contrast to the Western Urania, which put itself in the tradition of the 19th century version, the Eastern Urania connected to the working-class movement tradition of the Open Educational Institute Urania (Freies Bildungsinstitut Urania). That institute had been founded 1924 in Jena during the Weimar Republic and had published the popular science journal *Urania* until it was banned 1933 when the Nazis took power in Germany (Schmidt-Lux, 2008).

In 1947 the journal was relaunched in East Germany with the same title but a different mission and aiming at a broader public impact (Gruhn, 1979). From the very beginning the dissemination of scientific knowledge was closely connected to the ideological mission of propagating ideas of the biologist Ernst Haeckel (1834–1919) who promoted monism, a kind of anti-religious scientism (Schmidt-Lux, 2008). The general idea was to influence people's worldview through political exploitation of science communication. Taking up scientistic, anti-religious traditions from the Weimar period in Germany, the journal was supposed to support the dominant political doctrine of Marxism-Leninism, which presented itself as a scientific world-view. Scientism became a guiding principle of the state, the socialist party and the education system. It was institutionalised in various organisations, of which Urania was the most important. The role of the Society for the Dissemination of Scientific Knowledge (Urania) was 'to engage in the fight for raising the atheistic consciousness' (Schmidt-Lux, 2008, p. 56).

Given the ideological focus on scientism, science communication was considered important in several fields, including the education of journalists. Training and licensing for journalism was strictly regulated and Leipzig University was the only place where journalists were educated academically in East Germany (Schemmert and Siemens, 2013). The importance of science as a field of journalism is obvious from the available specialisations in the 1970s: 'science' was one of only four possible specialisations besides 'politics', 'foreign affairs' and 'economy' (Sager, 1975). In contrast, there was no academic offering for journalists to specialise on science reporting at that time in West Germany.

The propagation of scientism in East Germany seems to have been effective: Eurobarometer surveys conducted after the reunion revealed a higher knowledge level and more positive attitudes towards science and technology of respondents from the former East compared to those from West Germany (Noelle-Neumann and Hansen, 1991; Commission of the European Community, 1993). More specifically, differences in responses to the item 'We depend too much on science and not enough on faith' were obvious even 15 years after the reunion, when 42 per cent of the West German respondents agreed with the item compared to only 31 per cent of East German respondents (European Commission, 2005). A small but consistent East–West difference is still present in the German science barometer 2018 across several questions, indicating a somewhat more positive image of science and technology in the eastern part of the country (Wissenschaft im Dialog, 2018). Other factors besides ideological indoctrination may have contributed to the differences: citizens of the East were not exposed to the criticism of technology that in the West originated in the civil society and led to protest movements against nuclear power, biotechnology and the chemical industry.

5. Goodbye 'ivory tower': Scientists and science organisations

In the 1960s, journalists, politicians and communication scholars in West Germany diagnosed a societal need for public communication of science and also identified an apparent reluctance of scientists to leave their 'ivory tower'. However, the first German survey of scientists' attitudes towards the mass media and their interactions with journalists by Jens Krüger in 1984 showed instead that a majority of professors at the University of Mainz were actually talking to journalists. They reported positively or neutrally about their experiences with the media and had a positive attitude towards public communication (Krüger, 1985, 1987). Later surveys among a more diverse sample of German scientists confirmed this initial finding (Strömer, 1999; Peters et al., 2009; Pansegrau et al., 2011; Peters et al., 2012). There are hardly formal incentives for German scientists to go to the media, but many perceive the professional benefits of public visibility. The public relations departments of universities and other research centres generally motivate scientists for public communication, and scientists recognise that their organisations want them to communicate with the public (Marcinkowski et al., 2014; Peters, 2019). Since 2000, the German Research Foundation (Deutsche Forschungsgemeinschaft, DFG) has awarded an annual 'communicator prize' to a scientist for excellence in public communication.

To improve the public communication skills of scientists, Walter Hömberg organised the first German media training seminars for scientists at the Catholic University of Eichstätt in 1990 (Hömberg, 1998) and other organisations such as the Forschungszentrum Jülich [Research Centre Jülich] followed (Göpfert and Peters, 1992). Today, media training for scientists is offered on a regular basis by several organisations in Germany such as the Nationales Institut für Wissenschaftskommunikation [National Institute for Science Communication] in Karlsruhe. In a recent survey, 21 per cent of a sample of 240 German researchers from natural sciences and engineering said that they had received some form of communication training (Lo, 2016, p. 133).

Parallel to the individual media contacts of scientists, mostly initiated by journalists, several science organisations had established policies and units for public communication in the 1960s or even before. These included universities as well as the large non-university research institutions such as the Max-Planck Society, Fraunhofer Society, Helmholtz Association and Leibniz Society. The association of German universities (Hochschulrektorenkonferenz, HRK) recommended the establishment of press offices at German universities in the 1950s and 1960s (Bühler et al., 2007) and, in 1969, the association of university press officers (Bundesverband Hochschulkommunikation) was founded. By the 1970s, the majority of German universities had installed press offices or at least assigned staff responsible for public communication. The rising public awareness of the risks of science and technology and the association of universities with student riots, a steady growth of the academic education sector and the high unemployment rate of academics (Tonnemacher, 1982) forced universities to strengthen their communication activities. In addition, there was growing competition between science organisations for public funding. Consequently, the communication functions of universities and other research organisations were professionalised and expanded in the 1980s.

According to the rectors' conference, the main objective of university communication was to support communication within the university and to be in dialogue with the public sphere through media relations. Further activities such as science festivals and other events were only apparent from the 1990s. In 1995, the HRK released new recommendations for public communication that emphasised it as a core responsibility of university management. The recommendations emphasised that public communication should pursue the universities' interests. Most important objectives were to sustain autonomy for research and teaching but also to secure research funding in an unfavourably changing social environment (*Hochschulrektorenkonferenz*, 1995, p. 2).

6. Public controversies over techno-sciences

Since the beginning of the 1970s, big public controversies over nuclear power, biotechnology and environmental issues played out in West Germany as in many other European countries (e.g. Kepplinger et al., 1991; Kliment, 1994; Bauer, 2014). Science was involved in two roles: as technology developer and as provider of scientific expertise for the assessment of risks and benefits. Scientific expertise and counter-expertise were used as political resources in these controversies (e.g. Peters, 1994).

Acceptance of modern technologies was one of the strong motivators for government and industry to engage in or sponsor science communication activities—for example, nuclear power and biotechnology. Industry and nuclear research institutes cooperated in the Informationskreis Kernenergie [Nuclear Energy Information Group]. It was founded in 1975 to provide public information about nuclear power, energy policy and safety aspects. Public information groups were established at research institutes dealing with nuclear power: Kernforschungsanlage Jülich (KFA), for example, tried to raise the knowledge level of citizens through activities including popular books, public talks, information seminars for journalists and teachers and the provision of scientific experts for lectures at schools. These communication initiatives generally relied on 'deficit model' assumptions: that acceptance problems resulted from a lack of knowledge or from misinformation about a technology, and that overcoming acceptance problems was possible by replacing public ignorance with scientifically sound information.

Recognising that information alone was ineffective for overcoming acceptance problems, in 1975 the Ministry for Research and Technology (BMFT) offered an open citizen dialogue on nuclear energy (Bürgerdialog Kernenergie) (Hauff, 1977). Anticipating the later development of the consensus conference model, experiments with citizen participation in energy planning explored citizens' energy-related preferences. So-called 'planning cells' were used to summarise citizens' informed opinions for policymakers in a 'citizen expert report' (Dienel and Garbe, 1985; Renn et al., 1984). But these efforts had no significant consequences for public policy or public acceptance. In the end, the German phasing-out policy on nuclear energy resulted from long-lasting public protests, often involving violence and civil disobedience, and 'supported' by increasingly professionalised counter-expertise and the shockwaves of the Chernobyl disaster in 1986 and the Fukushima disaster in 2011. Both de-legitimised pro-nuclear positions in the public discourse (Arlt and Wolling, 2014).

Concerns about a specific German hostility towards technology (*Technikfeindlichkeit*) spread in the 1970 and 1980s, underpinned by Elisabeth Noelle-Neumann and her influential polling institute Institut für Demoskopie Allensbach, whose time-series data showed a decay of positive attitudes towards technology among the German population (e.g. Noelle-Neumann and Hansen, 1988). Biased and negative media reporting of science and technology was often assumed as the cause for this attitude change. In 1986, BMFT created a research program on public acceptance and media impact (Technikakzeptanz und Medienwirkungen) that led to a number of studies about public opinion of science and technology. In his famous (and controversial) study 'Artificial Horizons', Kepplinger (1989, 1992) presented empirical evidence for the hypothesis that biased media reporting on the risks and benefits of technologies caused changes in public opinion towards the negative. However, other researchers in the program disagreed with the idea of a particular German hostility towards science and technology and concluded rather that the German public had ambivalent attitudes, with clearly negative attitudes restricted to a few specific techno-sciences such as nuclear power or food biotechnology (Jaufmann and Kistler, 1988; Jaufmann et al., 1989; Jaufmann and Kistler, 1991).

7. The German PUSH in science communication

In the years before the EXPO 2000 in Hanover, the big German science associations together with the BMFT (now the Ministry of Education and Research, BMBF) planned the German contribution to the exhibition. Probably for the first time, high-level talks about the public communication of science took place between top research managers and representatives of the government. In the context of these talks the idea of an innovative 'relaunch' of science communication was born, different from communication via the mass media that so far had dominated the German discussion about science communication. The action program Public Understanding of Science and Humanities (PUSH) was launched by a memorandum on the dialogue between science and society that was signed by the head of Stifterverband (the organisation that provided money for the program) and the heads of the large German science associations (Stifterverband, 1999). It called for increased efforts in science communication and the provision of incentives and support for scientists wanting to interact with the public. A call for science communication initiatives was issued and grants for model projects were awarded through the action program.

For a second time (after the adoption of US science journalism in the 1980s), an Anglo-Saxon model set an example for science communication efforts in Germany.[3] While clearly adopting the general ideas of the public engagement approach with its focus on dialogue, involvement, participation and direct experience of science and scientists, the initiators decided to stick to the traditional 'Public Understanding of Science' label—not even translating it into German—but supplementing it with 'Humanities' to account for the German insistence on the unity of the academic world (obvious in the German term *Wissenschaft* that covers the sciences, social sciences and humanities).

With respect to scientific actors in public communication, the consequences of the relaunch of science communication marked by the PUSH initiative and the availability of new online channels of addressing larger public audiences without journalistic mediation were fourfold:

(1) The founding of a quasi-agency Wissenschaft im Dialog (WiD) in charge of the national coordination of science communication initiatives and initiates, and which encourages and coordinates many activities.[4] Controlled by the big science associations and some foundations, it also collaborates with BMBF. Communication activities of WiD include mobile science exhibitions, science cafés and debates. From 2000–12 WiD oversaw the 'science summers', the German version of science festivals taking place in different cities. Since 2003 it has organised the 'science years' on behalf of the BMBF. While the German

3 The importance of the Anglo-Saxon dialogue model for Germany is obvious from the schedule of the initial PUSH symposium. The program started with sessions 'Example Great Britain' and 'Example USA', each country being represented by two keynote speakers, and then moved on to 'Plans for Dialogue in Germany' (Stifterverband, 1999). German science communication scholars, with their almost exclusive focus on science journalism, were largely ousted from the relaunch of science communication.

4 See www.wissenschaft-im-dialog.de/en/about-us/.

science years resemble national science weeks in other countries, their activities are not limited to a particular week. Each science year is devoted to a particular discipline or topic, e.g. chemistry (2003), humanities (2007), digital society (2014) and artificial intelligence (2019). WiD also operates online platforms for citizen science and crowd-funding projects; offers science communication summer schools for scientists; organises annual conferences for science communication practitioners (Forum Wissenschaftskommunikation); and conducts the Science Barometer (Wissenschaftsbarometer), an annual survey of the public image of science in Germany. One focal instrument is the online platform *Wissenschaftskommunikation.de*, which offers opinion pieces, interviews, debates and summaries for actual trends and issues from the field of science communication.[5] Within the almost 20 years of its existence, WiD has developed from a mere event organisation into a major player in the field and considers itself a driving force and think tank for science communication in Germany and beyond (Wissenschaft im Dialog, 2017).

(2) In general, organisational science communication has become tremendously professionalised within the last 20 years after the PUSH memorandum. A survey among university public information officers by Leßmöllmann et al. (n.d. [2017]) showed that university PR is an expanding field. Respondents to the survey believe it will become even more important in the next years and expect budgets and staffing to grow. While this relates to all fields of science communication from online events to media relations, the latter is still considered one of the most important activities of university science communication (Lehmkuhl, 2019). Media relations activities (e.g. sending out press releases) are still very important and even on the rise (Serong et al., 2017). Parallel to relations with traditional media, the importance of online media is rising in organisational PR. All universities and research organisations use a website and further online channels, such as blogs, Facebook, Twitter and Instagram to address their stakeholders (Metag and Schäfer, 2019).

(3) The media-targeted efforts of science organisations have been supplemented with a portfolio of activities aiming at different target groups and serving different goals from science education to participation in research and science policy. Science organisations participate in activities coordinated by WiD such as the science summers and science years mentioned above. They also organise local activities on their own or in cooperation with others such as 'long nights' of science, open days, 'science slams', citizen science, children's universities and school labs at research institutions (Weingart et al., 2007; Schneider, 2012; Fähnrich, 2017).

5 This platform is a joint activity of WiD, the German Nationales Institut für Wissenschaftskommunikation [National Centre for Science Communication], and the science communication unit at the Karlsruhe Institute of Technology.

Figure 14.1: A scene from a so-called 'science slam'; this one took place in 2014 in the Theatre of Marburg. Scientists have a few minutes to talk about a subject of their choice. A jury from the audience rates their performance and determines a winner. Science slams are usually organised as humorous events.

Source: H. P. Peters (used with permission).

Yet there are some concerns that the engagement activities implemented after the PUSH memorandum lack concrete aims and target groups; are largely manifestations of one-way knowledge dissemination by scientists; only preach to the converted; and avoid direct dialogue between scientists and citizens about controversial issues (Schnabel, 2008). In their analysis of science communication in Germany for the German Ministry of Education and Research, Weingart et al. (2007) similarly conclude that while the efforts successfully initiated science communication and established it as a task area for science and politics, they did not succeed in starting a societal dialogue, and reached only those already interested in science.

(4) For a long time, individual scientists communicated with the public mainly through their interactions with journalists. The development of a rich infrastructure of event-based interactions with the public, expansion of organisational public relations and the more widespread access of the general public to websites, blogs and social media, increased their options. The number of opportunities for them to meet the public grew as organisers of science communication events, in particular the PR departments of their own universities or other science organisations, offered more invitations to scientists to participate. Scientists motivated for proactive communication could easily use personal websites, blogs, Facebook, Twitter, Wikipedia or YouTube to address the public directly, and a significant proportion of them have begun to use these online channels (Lo, 2016).

8. Online communication

The paradigm shift towards more dialogical forms of science communication coincided with the development of the internet from a specialised communication infrastructure to a general one. Between 1998 and 2003 the proportion of the German population at least occasionally using the internet rose from about 10 per cent to 54 per cent (Projektgruppe ARD/ZDF-Multimedia, 2017). This development created the principal opportunity for dialogic communication between scientific communicators and citizens without journalistic mediation, thus providing science a means to implement the new public engagement paradigm not only through events and face-to-face interactions but also online.

Yet compared with other countries, online social media are generally less important than other information media in Germany, although the percentage of the population with access to the internet is similar to that of the United States, for example (Hasebrink and Hölig, 2013; Pew Research Center, 2016; Newman et al., 2017). Germans would routinely use the internet when seeking specific information, e.g. about a health problem. For news in general and for science news in particular they still prefer journalistic media, including their online sites. This has an impact on scientists' beliefs, preferences and behaviour regarding the use of online communication with the public. Allgaier et al. (2013) have shown that German neuroscientists have assessed the impact of blogs and social media on science policymaking and public opinion as being lower than their US colleagues. In a survey of scientists in Germany, Taiwan and the United States, Lo (2016) found that Germany ranked lowest among these countries with respect to scientists' actual use of personal websites, blogs and social networks for communication with the public.

The German science academies have started to deal with the challenges of science communication in the digital age and its consequences for the science–society interface. They established a joint working group Relationship of Science, Public Sphere and Media (2015–17) to study the challenges arising from the digitalisation of communication. The working group has published a number of recommendations (acatech, 2017; Weingart et al., 2017) and the topic remains on the agenda. In 2018, the Berlin-Brandenburg Academy of Sciences and Humanities created an interdisciplinary research group to explore the 'Implications of Digitisation for the Quality of Science Communication'.[6]

6 See www.bbaw.de/en/research/wissenschaftskommunikation.

9. Science museums and science centres

Germany has a long tradition and rich landscape of institutions of informal learning related to science and technology including museums that exhibit artefacts from science, technology, medicine and nature (including their history); and hands-on science centres that focus on interactivity with the visitors.[7] The latter are of more recent origin. Planetariums and public astronomical observatories complement the list of institutions for science education. The German report of the *Monitoring Policy and Research Activities on Science in Society in Europe* (MASIS) initiative estimates that 'over 1,000 science centres and museums on science and technology, natural history and natural science provide access to their collections and attracted over 24 million visitors in 2009' (Grunwald, 2011, p. 63).[8]

Since about the mid-1990s approximately 25 science centres in the modern sense—i.e. educational, entertaining and hands-on—have opened their doors in Germany.[9] Among these science centres are very small institutions with only a few thousand visitors per year, but also large centres that count several hundred thousand visitors each year. The Universum Bremen, founded in 2000, is the older of the two largest science centres in Germany (about 215,000 visitors in 2017). The permanent display is supplemented by special exhibitions and events such as science shows, science theatres and talks. The centre engages in collaborations with schools. In 2005 the Phaeno, another large science centre (about 245,000 visitors in 2016), opened in Wolfsburg in a building that received much attention because of its spectacular architecture. The Mathematikum Giessen is a mid-size science centre (about 150,000 visitors per year) with a clear focus on mathematics and addressing a broad audience. It opened in 2002.

Final examples of German hands-on science centres are the Phänomenta centres, small to mid-size centres in four German cities: Lüdenscheid, Bremerhaven, Flensburg, Peenemünde. The first of these centres opened in 1993 in Flensburg. It was based on interactive exhibits that had been displayed in the University of Flensburg since 1986 (Kiupel, 1998). While these four centres have developed locally and independently of each other, they have built a network and created a common brand.

7 This section on science museums and science centres is adopted from an unpublished expert opinion written for the PLACES project (Peters, 2013).

8 Note that in accordance with the broad German definition of *Wissenschaft*, these figures include museums on history and archaeology.

9 A list of German science centres can be found at www.science-museum.de.

Figure 14.2: Universum Bremen, the first 'large' modern science centre in Germany.

Source: H. P. Peters (used with permission).

10. The role of politics in the development of science communication

In a keynote at the 1977 annual meeting of the DFG, the then German Chancellor Helmut Schmidt emphasised science's responsibility towards a democratic society and urged German scientists to become publicly visible. He coined the widely quoted term of a *Bringschuld*, i.e. a duty of individual scientists and researchers to actively engage with the public, and not to delegate that task to public information departments or science journalists (Schmidt, 1977, 2005). He also stated that public visibility of science helps to defend public funding of science against competing demands of other stakeholders.

Schmidt's call was preceded by earlier discussions of the relationship of science and democratic society. For example, the political scientist and social-democratic politician Ulrich Lohmar had pointed to the challenges of the 'productive force science' for a democratic society. Lohmar emphasised the responsibility of the media for making science transparent for citizens and enabling a critical public discourse about science. He diagnosed deficits of journalism in this respect (Lohmar, 1972).

German governments were and certainly are interested in a science-attentive, science-literate and science-friendly German public. However, they have largely remained in the background. Whereas there has been encouragement and funding for science communication and research on it, the leading role has been left to scientific organisations and private foundations. One of the important exceptions is the (failed) citizen dialogue on nuclear power mentioned above.

The core interest of the German government in PCST (probably like that of most governments worldwide) is based on the beliefs that good science communication and public engagement with science will foster technological innovation and thus contribute to the economic competitiveness and social welfare of the country. It is thus no surprise that the German High-Tech Strategy includes a chapter on science communication and participation (BMBF, 2018).

Another motivation appears in the discussion of science communication by politicians: the idea of a democratisation of science and science policy by citizen participation, quite similar to the goals stated in the UK House of Lords Select Committee on Science and Technology (2000) report and probably influenced by it. The goal to enable public participation in science and science policy is evident—for example, in the coalition contract between two major political parties in Germany (the conservative CDU/CSU and the liberal SPD) beginning in 2013 (Koalitionsvertrag, 2013, p. 151). The parties agreed to involve citizens in the discussion of future projects and the shaping of the science agenda and to develop new forms of citizen participation and science communication. In 2016, BMBF issued a position paper on the participation of civil society in science and science policy (BMBF, n.d. [2016]). The benefits of citizen participation are seen in increased interest and trust in science and improved legitimacy of decisions about science policy. This may be seen as just another strategy to create 'acceptance' for science and technology. But the position paper also emphasises the role of civil society in shaping science and technology policy, and points to the large potential for citizens' knowledge to be utilised in citizen science projects and by crowd-sourcing.

In 2015–17, the BMBF experimented with citizen participation in a series of so-called ZukunftsForen [Future Forums] dealing with innovation in health care, the sharing economy, digitalisation and technologies transforming the working environment. Each forum combined a public opinion survey, workshops and a discussion of citizens and representatives from science and politics with the responsible minister. However, it is not obvious whether and how the results of the citizen consultation had an impact on research and innovation policy.

As mentioned above, foundations have played an important role in the development of science communication in Germany as funders and also initiators of programs and activities. Although politicians assumed that public communication of science is both important and deficient, for a long time they did not consider 'promotion' of science communication as a legitimate field of political activity but saw it rather as a genuine responsibility of scientists and of the media. We suspect that the reluctance of the government to intervene in the public communication of science may be a consequence of the experience with Nazi propaganda, which delegitimised direct government influences on media and public communication in the post-war era. We believe that this reservation of the German government created an empty space that was filled by activities of charitable foundations.

11. Academic institutionalisation of science communication

Although interdisciplinary perspectives from sociology (e.g. Weingart, 2001), psychology (cf. Bromme and Kienhues, 2017) and even philosophy (Spinner, 1985) play a significant role in German science communication research, the academic institutionalisation of the field has taken place predominantly in communication and media studies (Bonfadelli et al., 2017). The field of science and technology studies, an important context for science communication in other European countries (e.g. Horst et al., 2017), is less significant in Germany. The development of science communication research in Germany can be regarded as a response to the perceived deficient status of science communication practice. Since the 1970s, several workshops on science and the media have enabled a discourse between scientists, science journalists, public information officers and communication scholars. The Center for Interdisciplinary Research (Zentrum für interdisziplinäre Forschung, ZiF) at the University of Bielefeld established an interdisciplinary research group, Science and Journalism in 1974 (Hömberg, 1974), which led to a number of early studies on science and the media (e.g. Roloff and Hömberg, 1975; Depenbrock, 1976; Hellmann, 1976). These initiatives and activities are glimpses of a comprehensive landscape of workshops, roundtables, reports and declarations addressing the issue of science and the media, starting in the 1970s or even before.

It is interesting to note that the public communication of the social sciences was included in this discussion early on (e.g. Wissenschaftszentrum Berlin, 1976; Hömberg, 1978; Peters, 1982). However, with some exceptions (e.g. Weßler, 1995; Fähnrich and Lüthje, 2017), the focus of science communication research in Germany was and still is on the natural sciences and related topics such as climate change.

Academic science communication programs in Germany are still rare and only four universities have degree programs specifically related to science communication in the broader sense. The Technical University of Dortmund offers bachelor and master's programs in 'science journalism' aimed at training future science journalists; the Karlsruhe Institute of Technology offers bachelor and master's programs in 'science – media – communication', preparing graduates for a more diverse spectrum of functions in science communication. Two bachelor programs on 'science communication & bionics' (English-speaking) and 'technical journalism/PR' exist at the Rhine-Waal University of Applied sciences in Kleve and the Bonn-Rhein-Sieg University of Applied Sciences in St Augustin, respectively. Ten full professors with explicit appointments for science communication are currently teaching at nine universities, and several more are interested in science communication although their chairs do not denote such a specialisation. Science communication can be part of the curriculum of some general communication or interdisciplinary degree programs.

While academic institutes for science communication only evolved after 1990, the topic itself had previously been the subject of many doctoral dissertations and other academic theses (cf. the documentation by Krüger and Flöhl, 1982). The first and (to date) only relevant temporary postgraduate research program leading to doctoral theses on science communication was the DFG graduate program 'On the Way into the Knowledge Society' at the University of Bielefeld in 2002–11. However, the strong majority of doctoral dissertations are done outside formal postgraduate programs. In the past 15 years the DFG, BMBF and the Volkswagen Foundation have funded several research programs on science communication, leading to dozens of projects in which doctoral students were employed as junior researchers and given the opportunity to conduct their PhD research related to the projects. There is a growing number of young researchers who consider themselves 'science communication scholars', contributing to national and international science communication conferences such as the PCST conferences and publishing in international science communication journals. While several general communication journals are published in Germany that also accept articles on science communication, there is no dedicated science communication journal.

An important step in the academic institutionalisation of science communication research in Germany was the creation of a science communication branch (Fachgruppe Wissenschaftskommunikation) of the German communication association (DGPuK) in 2012.[10] It organises annual conferences (Schäfer et al., 2015; Ruhrmann et al., 2017; Hagen et al., 2018)

10 See www.dgpuk.de/en/science-communication.html.

and supports exchanges among scholars in the field. In 2019, the branch had about 120 members from universities and research organisations in Germany and also from Switzerland and Austria.

12. Current issues

While for decades German scientists and science organisations were called to increase their public communication efforts, there is now growing concern about possible negative implications of scientists' and science organisations' increasing interest of being visible in the public sphere (Weingart, 2012; Marcinkowski and Kohring, 2014). It is feared that the 'medialisation of science' (Weingart, 2012) may have negative impacts on the quality of public communication. In order to get the attention of journalists and/or a public audience, scientists and public information officers may be tempted to hype their messages. A second possibility is that a strong media orientation of scientists and science organisations may result in the partial replacement of genuine scientific criteria with criteria of media attention that are then not only used in public communication but also within science. According to the medialisation thesis, decisions about research, the distribution of resources, the interpretation of findings and scholarly publication may be influenced by anticipation of public responses, thereby subtly compromising the autonomy of science, which is highly valued in Germany. In the end the quality of scientific knowledge itself may suffer. The medialisation of science hypothesis has received a lot of interest among science communication scholars and has stimulated several PhD theses and research projects.

A working group of the German science academies has responded to the temptations of public visibility and other challenges of science communication with the development of guidelines urging scientists and science organisations to communicate with the public in line with scientific responsibility. The guidelines also emphasise the importance of independent quality science journalism (acatech, 2014). Similarly, an influential group of university press officers (Siggener Kreis) has issued guidelines for ethical science PR (*Bundesverband Hochschulkommunikation/Wissenschaft im Dialog*, 2016), calling for self-restriction in using the increased power of science PR.

As Davies and Horst (2016) have noted, the 'ecosystem of science communication' is changing and diversifying. The above-mentioned German meta-discourses on science communication reflects the empirical and normative uncertainties about the consequences of ongoing changes caused by the increasingly strategic orientation of communicators, the growing importance of online communication and a focus on public engagement as a model for science communication.

Germany has its own history of science popularisation by scientists and the mass media, but it has also been open to adopting approaches from the Anglo-Saxon countries. The first of these adoptions—learning from science journalism in the United States—had the goal of strengthening the role of journalism as mediator between science and the public. The second adoption—the public engagement approach with its focus on direct interaction and dialogue with citizens—created an alternative channel for public communication and thus reduced science communicators' dependency on science journalism. We are quite convinced that in Germany journalism is still important in the public communication of scientific knowledge and the critical surveillance of science as a social system, but that communicators from science and government believe in the advantage of public engagement as a means to increase public appreciation and trust. An open question is whether public engagement as it is currently implemented in Germany truly can contribute to a re-negotiation of the science-in-society relationship or serves, in effect, as a surrogate in that respect.

The options of science communicators multiplied, new niches arose and were colonised, priorities and resources of science communicators were re-allocated. Science communicators are currently adapting public communication to the new ecosystem. Changes in the ecosystem of science communication also imply changes in the roles of communicators and shifts in the distribution of power between different actors. Science communicators and scholars in Germany pursue a critical discourse not only about the 'efficiency' of communication in terms of impact, but also about the responsible use of options and power in the changed ecosystem of science communication.

Acknowledgements

Peter Weingart, Jens Krüger, Toss Gascoigne and two anonymous reviewers provided valuable comments on a previous version of our chapter. We gratefully acknowledge their criticism and suggestions that guided us in our revision.

References

acatech. (2014). *On Designing Communication between the Scientific Community, the Public and the Media.* München: National Academy of Science and Engineering.

acatech. (2017). *Social Media und digitale Wissenschaftskommunikation. Analyse und Empfehlungen zum Umgang mit Chancen und Risiken in der Demokratie.* München: National Academy of Science and Engineering.

Allgaier, J., Dunwoody, S., Brossard, D., Lo, Y.-Y. and Peters, H. P. (2013). Journalism and social media as means of observing the contexts of science. *BioScience, 63*(4), 284–7. doi.org/10.1525/bio.2013.63.4.8.

Arlt, D. and Wolling, J. (2014). Fukushima-Effekte in Deutschland? Die Reaktorkatastrophe als Ursache von Veränderungen in der Berichterstattung und in der öffentlichen Meinung über die Atomenergie. In J. Wolling and D. Arlt (eds), *Fukushima und die Folgen. Medienberichterstattung, Öffentliche Meinung, Politische Konsequenzen* (pp. 269–96). Universitätsverlag Ilmenau, Ilmenau.

Ausejo, E. (1994). The window case of science: the associations for the advancement of science and the birth of scientific congresses in Western Europe. *Archives Internationales d'Histoire des Sciences, 44*(133), 338–71.

Bauer, M. W. (2014). *Atoms, Bytes and Genes. Public Resistance and Techno-Scientific Responses*. London: Routledge. doi.org/10.4324/9781315775999.

BMBF. (n.d. [2016]). *Grundsatzpapier des Bundesministeriums für Bildung und Forschung zur Partisipation*. Berlin: Ministry of Education and Research. Retrieved from www.zukunft-verstehen.de/download_file/view/379/437.

BMBF. (2018). *Forschung und Innovation für die Menschen. Die Hightech-Strategie 2025*. Berlin: Ministry of Education and Research.

Bonfadelli, H., Fähnrich, B., Lüthje, C., Milde, J., Rhomberg, M. and Schäfer, M. S. (2017). Das Forschungsfeld Wissenschaftskommunikation. In H. Bonfadelli, B. Fähnrich, C. Lüthje, J. Milde, M. Rhomberg and M. S. Schäfer (eds), *Forschungsfeld Wissenschaftskommunikation*. Wiesbaden: Springer VS. doi.org/10.1007/978-3-658-12898-2.

Bromme, R. and Kienhues, D. (2017). Gewissheit und Skepsis: Wissenschaftskommunikation als Forschungsthema der Psychologie (Editorial). *Psychologische Rundschau, 68*(3), 167–71. doi.org/10.1026/0033-3042/a000359.

Bühler, H., Naderer, G., Koch, R. and Schuster, C. (2007). *Hochschul-PR in Deutschland. Ziele, Strategien und Perspektiven*. Wiesbaden: Deutscher Universitäts-Verlag.

Bundesverband Hochschulkommunikation/Wissenschaft im Dialog. (2016). Leitlinien zur guten Wissenschafts-PR. Retrieved from www.wissenschaft-im-dialog.de/fileadmin/user_upload/Trends_und_Themen/Dokumente/Leitlinien-gute-Wissenschafts-PR_final.pdf.

Commission of the European Community. (1993). Biotechnology and genetic engineering. What Europeans think about it in 1993. Survey conducted in the context of Eurobarometer 39.1. Retrieved from ec.europa.eu/commfrontoffice/publicopinion/archives/ebs/ebs_080_en.pdf.

Davies, S. and Horst, M. (2016). *Science Communication. Culture, Identity and Citizenship*. London: Palgrave Macmillan.

Depenbrock, G. (1976). *Journalismus, Wissenschaft und Hochschule*. Bochum: Brockmeyer.

Dienel, P. C. and Garbe, D. (1985). *Zukünftige Energiepolitik. Ein Bürgergutachten*. München: High Tech Verlag.

European Commission. (2005). *Europeans, Science and Technology. Eurobarometer 2005*. Brussels: European Commission.

Fähnrich, B. (2017). Wissenschaftsevents zwischen Popularisierung, Engagement und Partisipation. In H. Bonfadelli, B. Fähnrich, C. Lüthje, J. Milde, M. Rhomberg and M. S. Schäfer (eds), *Forschungsfeld Wissenschaftskommunikation* (pp. 165–82). Wiesbaden: Springer. doi.org/10.1007/978-3-658-12898-2_9.

Fähnrich, B. and Lüthje, C. (2017). Roles of social scientists in crisis media reporting: The case of the German populist radical right movement PEGIDA. *Science Communication, 39*(4), 415–42. doi.org/10.1177/1075547017715472.

Fähnrich, B., Metag, J., Post, S. and Schäfer, M. S. (eds) (2019). *Forschungsfeld Hochschulkommunikation*. Wiesbaden: Springer VS. doi.org/10.1007/978-3-658-22409-7.

Faulstich, P. (2006). Öffentliche Wissenschaft. In P. Faulstich (ed.), *Öffentliche Wissenschaft. Neue Perspektiven der Vermittlung in der wissenschaftlichen Weiterbildung* (pp. 11–32). Bielefeld: Verlag. doi.org/10.14361/9783839404553.

Fischer, H.-D. (1976). Probleme der 'Vermarktung' von Wissenschaft durch Massenmedien. *Aus Politik und Zeitgeschichte*, B44, 9–34.

Fox, F. (2012). Practitioner's perspective: The role and function of the Science Media Centre. In S. Rödder, M. Franzen and P. Weingart (eds), *The Sciences' Media Connection – Public Communication and its Repercussions* (pp. 257–70). Dordrecht, NL: Springer. doi.org/10.1007/978-94-007-2085-5_13.

Göpfert, W. (ed.) (2006). *Wissenschaftsjournalismus. Ein Handbuch für Ausbildung und Praxis*. Berlin: Econ.

Göpfert, W. and Bader, R. (eds) (1998). *Risikoberichterstattung und Wissenschaftsjournalismus – Risk Communication and Science Reporting*. Tagungsbericht zum 4. Colloquium Wissenschaftsjournalismus. Stuttgart: Schattauer Verlag.

Göpfert, W. and Peters, H. P. (1992). Konzept einer 'Sommerschule Wissenschaftsjournalismus' und Erfahrungen bei der Realisierung. *Publizistik, 37*(1), 118–20.

Göpfert, W. and Schanne, M. (1998). *Das Förderprogramm Wissenschaftsjournalismus der Robert Bosch Stiftung GmbH: Evaluation*. Berlin and Zurich: Zusammenfassender Bericht. Retrieved from www.polsoz.fu-berlin.de/kommwiss/arbeitsstellen/wissenskommunikation/media/bericht.pdf.

Grabowski, K. H. (1982). *Strukturelle Probleme des Wissenschaftsjournalismus in aktuellen Massenmedien. Eine soziologisch-kommunikationswissenschaftliche Untersuchung*. Bochum: Brockmeyer.

Gruhn, W. (1979). *Wissenschaft und Technik in deutschen Massenmedien: Ein Vergleich zwischen der Bundesrepublik Deutschland und der DDR*. Erlangen: Deutsche Gesellschaft für zeitgeschichtliche Fragen e.V.

Grunwald, A. (2011). *Monitoring Policy and Research Activities on Science in Society in Europe (MASIS)*. National Report, Germany. Retrieved from rritrends.res-agora.eu/uploads/1/MASIS_Germany.pdf.

Hagen, L., Lüthje, C., Ohser, F. M. and Seifert, C. (eds) (2018). *Wissenschaftskommunikation: Die Rolle der Disziplinen*. Baden-Baden: Nomos. doi.org/10.5771/9783845276427.

Hasebrink, U. and Hölig, S. (2013). Lagging behind or choosing a different path? Information behaviour in Germany. In N. Newman and D. A. L. Levi (eds), *Reuters Institute Digital News Report 2013: Tracking the Future of News* (pp. 81–83). Oxford: University of Oxford, Reuters Institute for the Study of Journalism.

Hauff, V. (1977). Zwischenbilanz zum Bürgerdialog Kernenergie. Sozialdemokratischer Pressedienst, 28. Oktober 1977, pp. 1–2. Retrieved from gateway-bayern.de/BV027894691.

Hellmann, M. (1976). Medisinberichterstattung in der Presse. *Deutsches Ärzteblatt – Ärztliche Mitteilungen, 73*(45), 1–7.

Hettwer, H., Lehmkuhl, M., Wormer, H. and Zotta, F. (eds) (2008). *WissensWelten. Wissenschaftsjournalismus in Theorie und Praxis*. Gütersloh: Verlag Bertelsmann Stiftung.

Himpsl, F. (2017). 'Third Mission'. Forschung, Lehre – und was noch? *Deutsche Universitätszeitung*, no. 5/2017. Retrieved from www.duz.de/beitrag/!/id/428/forschung-lehre-und-was-noch.

Hochschulrektorenkonferenz. (1995). Zur Öffentlichkeitsarbeit der Hochschulen. Empfehlungen des 176. Plenums. Bonn.

Hömberg, W. (1974). Wissenschaft und Journalismus: Interdisziplinäres Forschungsprojekt an der Universität Bielefeld. *Information und Meinung, 5*, 30–31.

Hömberg, W. (1978). Soziologie und Öffentlichkeit. *Soziologie, 1*, 5–23.

Hömberg, W. (1980). Glashaus oder Elfenbeinturm? Zur Entwicklung und zur Lage der Wissenschaftskommunikation. *Aus Politik und Zeitgeschichte*, B28, 37–46.

Hömberg, W. (1990). *Das verspätete Ressort. Die Situation des Wissenschaftsjournalismus*. Konstanz: Universitätsverlag.

Hömberg, W. (1998). Keine Angst vor Journalisten – Medientraining für Wissenschaftler. In W. Göpfert and R. Bader (eds), *Risikoberichterstattung und Wissenschaftsjournalismus – Risk Communication and Science Reporting*. Tagungsbericht zum 4. Colloquium Wissenschaftsjournalismus (pp. 145–54). Stuttgart: Schattauer Verlag.

Horst, M., Davies, S. R. and Irwin, A. (2017). Reframing Science Communication. In U. Felt, R. Fouché, C. A. Miller and L. Smith-Doerr (eds), *The Handbook of Science and Technology Studies* (pp. 881–907). Cambridge, MA: MIT Press.

Humboldt, W. v. (1964). *Werke in fünf Bänden: Band IV: Schriften zur Politik und zum Bildungswesen*. Darmstadt: Wissenschaftliche Buchgesellschaft.

Jaufmann, D. and Kistler, E. (eds) (1988). *Sind die Deutschen technikfeindlich?* Opladen: Leske + Budrich. doi.org/10.1007/978-3-322-95585-2.

Jaufmann, D. and Kistler, E. (eds) (1991). *Einstellungen zum technischen Fortschritt. Technikakzeptanz im nationalen und internationalen Vergleich*. Frankfurt-Main, New York: Campus Verlag.

Jaufmann, D., Kistler, E. and Jänsch, G. (1989). *Jugend und Technik: Wandel der Einstellungen im internationalen Vergleich*. Frankfurt-Main, New York: Campus Verlag.

Kepplinger, H. M. (1989). *Künstliche Horizonte: Folgen, Darstellung und Akzeptanz von Technik in der Bundesrepublik*. Frankfurt-Main, New York: Campus Verlag.

Kepplinger, H. M. (1992). Artificial horizons. How the press presented and how the population received technology in Germany from 1965–1986. In S. Rothman (ed.), *The Mass Media In Liberal Democratic Societies* (pp. 147–76). New York: Paragon House.

Kepplinger, H. M., Ehmig, S. C. and Ahlheim, C. (1991). Gentechnik Im Widerstreit: Zum Verhältnis Von Wissenschaft Und Journalismus. Frankfurt-Main, New York: Campus Verlag.

Kiupel, M. (1998). Phänomenta – where physics is fun! Paper presented at the 5th International Conference on Public Communication of Science and Technology (PCST), Berlin, Germany, 17–19 September.

Kliment, T. (1994). *Kernkraftprotest und Medienreaktionen. Deutungsmuster einer Widerstandsbewegung und öffentliche Rezeption*. Wiesbaden: Deutscher Universitätsverlag. doi.org/10.1007/978-3-663-14607-0.

Koalitionsvertrag. (2013). Deutschlands Zukunft gestalten. Koalitionsvertrag zwischen CDU, CSU und SPD. 18. Legislaturperiode. Retrieved from www.cdu.de/sites/default/files/media/dokumente/koalitionsvertrag.pdf.

Koch, K. and Stollorz, V. (2006). PR-Arbeiter und Journalist: Geht beides? In C. Götz-Sobel and W. Mock (eds), *Wissenschaftsjournalismus heute. Ein Blick auf 20 Jahre WPK* (pp. 103–10). Düsseldorf: VDI Verlag.

Koenen, E. and Meißner, M. (2019). Historische Perspektiven der Hochschulkommunikation. In B. Fähnrich, J. Metag, S. Post and M. S. Schäfer (eds), *Forschungsfeld Hochschulkommunikation* (pp. 39–59). Wiesbaden: Springer VS. doi.org/10.1007/978-3-658-22409-7_3.

Kohring, M. (2005). *Wissenschaftsjournalismus: Forschungsüberblick und Theorieentwurf.* Konstanz: UVK.

Krüger, J. (1985). Wissenschaftsberichterstattung in aktuellen Massenmedien aus der Sicht der Wissenschaftler. Ergebnisse einer Befragung der Professoren der Johannes Gutenberg-Universität. Unpublished magister thesis, Johannes Gutenberg-Universität Mainz.

Krüger, J. (1987). Wissenschaftsberichterstattung in aktuellen Massenmedien aus der Sicht der Wissenschaftler. In R. Flöhl and J. Fricke (eds), *Moral und Verantwortung in der Wissenschaftsvermittlung: Die Aufgabe von Wissenschaftler und Journalist* (pp. 39–51). Mainz: v. Hase & Koehler.

Krüger, J. (1990). In der Aufbauphase. Berliner Modell I: Wissenschaftsjournalismus als Lehrangebot. In S. Ruß-Mohl (ed.), *Wissenschaftsjournalismus und Öffentlichkeitsarbeit. Tagungsbericht zum 3* (pp. 133–36). Colloquium Wissenschaftsjournalismus vom 4./5. November 1988 in Berlin. Gerlingen: Bleicher Verlag.

Krüger, J. and Flöhl, R. (1982). Kein Platz auf dem Stundenplan. Wissenschaftsjournalismus in Forschung und Lehre. *Deutsche Universitätszeitung, 38*(6), 18–21.

Krüger, J. and Ruß-Mohl, S. (1989). Popularisierung der Technik durch Massenmedien. In L. Boehm and C. Schönbeck (eds), *Technik und Kultur, Bd. 5: Technik und Bildung* (pp. 385–415). Düsseldorf: VDI-Verlag. doi.org/10.1007/978-3-642-95790-1_6.

Laredo, P. (2007). Revisiting the third mission of universities: Toward a renewed categorisation of university activities? *Higher Education Policy, 20*(4), 441–56. doi.org/10.1057/palgrave.hep.8300169.

Lehmkuhl, M. (2012). The recent public understanding of science movement in Germany. In B. Schiele, M. Claessens and S. Shi (eds), *Science Communication in the World* (pp. 125–38). Dordrecht, NL: Springer. doi.org/10.1007/978-94-007-4279-6_8.

Lehmkuhl, M. (2019). Journalismus als Adressat von Hochschulkommunikation. In B. Fähnrich, J. Metag, S. Post and M. S. Schäfer (eds), *Forschungsfeld Hochschulkommunikation* (pp. 299–318). Wiesbaden: Springer VS. doi.org/10.1007/978-3-658-22409-7_14.

Leßmöllmann, A., Hauser, C. and Schwetje, T. (n.d. [2017]). Zwischenbericht. Hochschulkommunikation erforschen. Hochschulkommunikatoren als Akteure: Ergebnisse einer Online-Befragung – 1. Welle. Retrieved from www.geistsoz.kit.edu/germanistik/downloads/Zwischenbericht%20Hochschulkommunikation%20erforschen%201.%20Welle%20Le%C3%9Fm%C3%B6llmann%20Hauser%20Schwetje.pdf.

Lo, Y.-Y. (2016). Online communication beyond the scientific community. Scientists' use of new media in Germany, Taiwan and the United States to address the public (PhD thesis). Freie Universität Berlin, Germany. doi.org/10.17169/refubium-11625.

Lohmar, U. (1972). Die Produktivkraft Wissenschaft als publisistisches Problem. *Aus Politik und Zeitgeschichte*, *22*(B21), 3–15.

Marcinkowski, F. and Kohring, M. (2014). The changing rationale of science communication: a challenge to scientific autonomy. *Journal of Science Communication*, *13*(3), C04. doi.org/10.22323/2.13030304.

Marcinkowski, F., Kohring, M., Fürst, S. and Friedrichsmeier, A. (2014). Organisational influence on scientists' efforts to go public: An empirical investigation. *Science Communication*, *36*(1), 56–80. doi.org/10.1177/1075547013494022.

Mejlgaard, N., Bloch, C., Degn, L., Ravn, T. and Nielsen, M. W. (2012). *Monitoring Policy and Research Activities on Science in Society in Europe (MASIS)*. Final synthesis report. Retrieved from mitrends.res-agora.eu/uploads/MASIS%20final%20synthesis%20report.pdf.

Metag, J. and Schäfer, M. S. (2019). Hochschulkommunikation in Online-Medien und Social Media. In B. Fähnrich, J. Metag, S. Post and M. S. Schäfer (eds), *Forschungsfeld Hochschulkommunikation* (pp. 363–92). Wiesbaden: Springer. doi.org/10.1007/978-3-658-22409-7_17.

Nagel, A. C. (2012). *Hitlers Bildungsreformer. Das Reichsministerium für Wissenschaft, Erziehung und Volksbildung 1934–1945*. Frankfurt: Fischer Verlag.

Newman, N., Fletcher, R., Kalogeropoulos, A., Levy, D. A. L. and Nielsen, R. K. (2017). *Reuters Institute Digital News Report 2017*. Retrieved from reutersinstitute.politics.ox.ac.uk/sites/default/files/Digital%20News%20Report%202017%20web_0.pdf.

Noelle-Neumann, E. and Hansen, J. (1988). Medienwirkung und Technikakzeptanz: Kurzfassung des Allensbacher Berichts über ein Forschungsprojekt für das BMFT. In J. Scharioth and H. Uhl (eds), *Medien und Technikakzeptanz* (pp. 33–76). München: R. Oldenbourg Verlag. doi.org/10.1007/978-3-322-83727-1_22.

Noelle-Neumann, E. and Hansen, J. (1991). Technikakzeptanz in drei Jahrzehnten – in der Bevölkerung und in den Medien. Ein Beitrag zur Medienwirkungsforschung. In J. Krüger and S. Ruß-Mohl (eds), *Risikokommunikation. Technikakzeptanz, Medien und Kommunikationsrisiken* (pp. 91–143). Berlin: Edition Sigma.

Pansegrau, P., Taubert, N. and Weingart, P. (2011). Wissenschaftskommunikation in Deutschland: Ergebnisse einer Onlinebefragung. Unpublished report.

Peters, H. P. (1982). Vergleich physikalischer und soziologischer Wissenschaftsberichterstattung und Darstellung einiger Veränderungen auf den Wissenschaftsseiten von Zeitungen seit 1959. *Soziologie, 11*(1), 37–45.

Peters, H. P. (1994). Wissenschaftliche Experten in der öffentlichen Kommunikation über Technik, Umwelt und Risiken. *Kölner Zeitschrift für Soziologie und Sozialpsychologie, 46*(special issue 34), 162–90.

Peters, H. P. (2013). *National Overview on Scientific Culture: GERMANY.* Unpublished report prepared for the PLACES Project.

Peters, H. P. (2019). WissenschaftlerInnen als Kommunikatoren. In B. Fähnrich, J. Metag, S. Post, and M. S. Schäfer (eds), *Forschungsfeld Hochschulkommunikation* (pp. 209–25). Wiesbaden: Springer VS. doi.org/10.1007/978-3-658-22409-7_10.

Peters, H. P., Brossard, D., De Cheveigné, S., Dunwoody, S., Kallfass, M., Miller, S. and Paquez, A.-S. (2009). Kontakte biomedisinischer Forscher mit Journalisten und Öffentlichkeit: internationaler Vergleich von Erfahrungen und Einstellungen in Deutschland, Frankreich, Großbritannien, Japan und den USA. In H. P. Peters (ed.), *Medienorientierung biomedisinischer Forscher im internationalen Vergleich. Die Schnittstelle von Wissenschaft & Journalismus und ihre politische Relevanz* (pp. 45–99). Jülich: Forschungszentrum Jülich.

Peters, H. P., Spangenberg, A. and Lo, Y.-Y. (2012). Variations of scientist-journalist interactions across academic fields: Results of a survey of 1600 German researchers from the humanities, social sciences and hard sciences. In M. Bucchi and B. Trench (eds), *Quality, Honesty and Beauty in Science and Technology Communication: PCST 2012 Book of Papers* (pp. 257–63). Vincenza: Observa Science in Society.

Pew Research Center. (2016). *Smartphone Ownership and Internet Usage Continues to Climb in Emerging Economies.* Retrieved from www.pewresearch.org/global/wp-content/uploads/sites/2/2016/02/pew_research_center_global_technology_report_final_february_22__2016.pdf.

Projektgruppe ARD/ZDF-Multimedia. (2017). Kernergebnisse der ARD/ZDF-Onlinestudie 2017. Retrieved from www.ard-zdf-onlinestudie.de/files/2017/Artikel/Kern-Ergebnisse_ARDZDF-Onlinestudie_2017.pdf.

Pütz, J. (2006). Die WPK profiliert sich. Gründungsgeschichte. In C. Götz-Sobel and W. Mock (eds), *Wissenschaftsjournalismus heute. Ein Blick auf 20 Jahre WPK* (pp. 117–21). Düsseldorf: VDI Verlag.

Renn, O., Stegelmann, H. U., Albrecht, G., Kotte, U. and Peters, H. P. (1984). An empirical investigation of citizens' preferences among four energy scenarios. *Technological Forecasting and Social Change*, *26*(1), 11–46. doi.org/10.1016/0040-1625(84)90042-8.

Robert Bosch Stiftung. (ed.) (1983). *Wissenschaftsjournalismus und Journalistenausbildung. Eine Bestandsaufnahme. Tagungsbericht zum 1.* Colloquium Wissenschaftsjournalismus vom 10./11. Dezember 1982 in Stuttgart-Hohenheim. Stuttgart: Robert Bosch Stiftung.

Robert Bosch Stiftung. (ed.) (1985). *Wissenschaftsjournalismus in den USA: Infrastrukturen, Ausbildungsangebote, Erfolgsgeheimnisse. Tagungsbericht zum 2.* Colloquium Wissenschaftsjournalismus vom 10./11. Februar 1984 in Stuttgart-Degerloch. Stuttgart: Robert Bosch Stiftung.

Roloff, E. K. and Hömberg, W. (1975). Wissenschaftsjournalisten. Dolmetscher zwischen Forschung und Öffentlichkeit. *Bild der Wissenschaft*, *12*(9), 56–60.

Ruhrmann, G., Kessler, S. H. and Guenther, L. (eds) (2017). *Wissenschaftskommunikation zwischen Risiko und (Un-)Sicherheit.* Köln: von Halem.

Ruß-Mohl, S. (1985). Wissenschaftsjournalismus im gelobten Land – Anmerkungen und Fragen aus der Diaspora. In Robert Bosch Stiftung (ed.), *Wissenschaftsjournalismus in den USA. Infrastrukturen, Ausbildungsangebote, Erfolgsgeheimnisse. Tagungsbericht zum 2.* Colloquium Wissenschaftsjournalismus vom 10./11. Februar 1984 in Stuttgart-Degerloch. Stuttgart: Robert Bosch Stiftung.

Sager, D. (1975). Journalistenausbildung in der DDR. In J. Aufermann and E. Elitz (eds), *Ausbildungswege zum Journalismus. Bestandsaufnahmen, Kritik und Alternativen der Journalistenausbildung* (pp. 108–16). Wiesbaden: VS Verlag für Sozialwissenschaften. doi.org/10.1007/978-3-322-85398-1_7.

Schäfer, M. S., Kristiansen, S. and Bonfadelli, H. (eds) (2015). *Wissenschaftskommunikation im Wandel.* Köln: von Halem.

Schemmert, C. and Siemens, D. (2013). Die Leipziger Journalistenausbildung in der Ära Ulbricht. *Vierteljahrshefte für Zeitgeschichte*, *61*(2), 201–37. doi.org/10.1524/vfzg.2013.0010.

Schmidt, H. (1977). Verantwortung der Forschung für die Zukunft der Gesellschaft. Ansprache des Bundeskanzlers vor der Deutschen Forschungsgemeinschaft. In *Bulletin. Presse-und Informationsamt der Bundesregierung*, no. 72/1977, pp. 673–79.

Schmidt, H. (2005). Das vornehme Schweigen: Die Niederungen der Politik und die Wissenschaft. In *WZB Mitteilungen*, no. 107, pp. 11–12.

Schmidt-Lux, T. (2008). Das helle Licht der Wissenschaft. *Geschichte und Gesellschaft*, *34*(1), 41–72. doi.org/10.13109/gege.2008.34.1.41.

Schnabel, U. (2008). Zeit für einen neuen Push. *Gegenworte*, *19*, 16–20.

Schneider, A. (2012). Events als Kommunikationsinstrument im Hochschulmarketing. In C. Zanger (ed.), *Erfolg mit nachhaltigen Eventkonzepte* (pp. 35–58). Wiesbaden: Springer. doi.org/10.1007/978-3-8349-6885-2_4.

Serong, J., Koppers, L., Luschmann, E., Molina Ramirez, A., Kersting, K., Rahnenführer, J. and Wormer, H. (2017). Öffentlichkeitsorientierung von Wissenschaftsinstitutionen und Wissenschaftsdisziplinen – Eine Längsschnittsanalyse des 'Informationsdienstes Wissenschaft' (idw) 1995–2015. *Publizistik*, *62*(2), 153–78. doi.org/10.1007/s11616-017-0336-6.

Spinner, H. F. (1985). *Das 'wissenschaftliche Ethos' als Sonderethik des Wissens: über das Zusammenwirken von Wissenschaft und Journalismus im gesellschaftlichen Problemlösungsprozes*. Tübingen: Mohr.

Stifterverband. (1999). Dialog Wissenschaft und Gesellschaft. Symposium 'Public Understanding of the Sciences and Humanities – International and German Perspectives', 27 Mai, Wissenschaftszentrum Bonn. Stifterverband, Essen.

Stollorz, V. and Meurer, M. (2015). Deutsches Science Media Center kommt. *Meta. Das Magazin über Journalismus und Wissenschaft*. Retrieved from www.meta-magazin.org/2015/06/04/faq-deutsches-science-media-center-kommt-2016/.

Strömer, A. (1999). Wissenschaft und Journalismus 1984–1997: Ergebnisse einer Befragung von Berliner Professoren sowie wissenschaftlichen Mitarbeitern des Forschungszentrums Jülich und Vergleiche mit einer früheren Studie aus Mainz und Jülich. Unpublished magister thesis, Freie Universität Berlin, Germany.

Tonnemacher, J. (1982). Fallbeispiel XIII: Öffentlichkeitsarbeit an Hochschulen – Ein Fallbeispiel über die Vermittlung von Hochschulpolitik und Universitätsforschung an die Öffentlichkeit. In G. Haedrich, G. Barthenheier and H. Kleinert (eds), *Öffentlichkeitsarbeit. Dialog zwischen Institutionen und Gesellschaft. Ein Handbuch* (pp. 367–80). Berlin: Walter de Gruyter.

UK House of Lords Select Committee on Science and Technology. (2000). *Science and Society: 3rd Report of the Session 1999–2000*. London: HM Stationery Office.

Weber, W. (1993). The political history of museums of technology in Germany since the nineteenth century. *History and Technology, 10*(1), 13–25. doi.org/10.1080/07341519308581832.

Weingart, P. (2001). *Die Stunde der Wahrheit? Zum Verhältnis der Wissenschaft zu Politik, Wirtschaft und Medien in der Wissensgesellschaft*. Weilerswist: Velbrück Wissenschaft.

Weingart, P. (2012). The lure of the mass media and its repercussions on science. In S. Rödder, M. Franzen and P. Weingart (eds), *The Sciences' Media Connection – Public Communication and its Repercussions* (pp. 17–32). Dordrecht, NL: Springer. doi.org/10.1007/978-94-007-2085-5_2.

Weingart, P., Pansegrau, P., Rödder, S. and Voß, M. (2007). *Vergleichende Analyse Wissenschaftskommunikation: Abschlussbericht für das BMBF*. Universität Bielefeld, Institut für Wissenschafts-und Technikforschung (IWT), Bielefeld.

Weingart, P., Wormer, H., Wenninger, A. and Hüttl, R. F. (eds) (2017). *Perspektiven der Wissenschaftskommunikation im digitalen Zeitalter*. Weilerswist: Velbrück Wissenschaft.

Weitze, M.-D. and Heckl, W. M. (2016). *Wissenschaftskommunikation – Schlüsselideen, Akteure, Fallbeispiele*. Berlin: Springer Spektrum. doi.org/10.1007/978-3-662-47843-1.

Weßler, H. (1995). Die journalistische Verwendung sozialwissenschaftlichen Wissens und ihre Bedeutung für gesellschaftliche Diskurse. *Publizistik, 40*(1), 20–38.

Wilke, T. (2013). Die Urania Gesellschaft und die Popularisierung naturwissenschaftlicher Erkenntnisse am Ende des 19. Jahrhunderts. *Medien & Zeit, 28*(4), 5–13.

Wissenschaft im Dialog. (2017). *Debatten anstoßen, Erkenntnisprozesse kommunisieren und neue Zielgruppen ansprechen*. Strategiepapier 2017 der Wissenschaft im Dialog gGmbH. Retrieved from www.wissenschaft-im-dialog.de/fileadmin/user_upload/Ueber_uns/WiD_dokumente/Strategiepapier_WiD_2017.pdf.

Wissenschaft im Dialog. (2018). *Tabellenband Wissenschaftsbarometer 2018*. Retrieved from www.wissenschaft-im-dialog.de/fileadmin/user_upload/Projekte/Wissenschaftsbarometer/Dokumente_18/Downloads_allgemein/Tabellenband_Wissenschaftsbarometer2018_final.pdf.

Wissenschaftszentrum Berlin. (1976). *Die Vermittlung von sozialwissenschaftlichen Ergebnissen an die Öffentlichkeit*. Konferenz Wissenschaftszentrum Berlin (WZB) 5./6. Dezember 1975. WZB-paper GS 1976-1. Wissenschaftszentrum Berlin für Sozialforschung, Berlin.

Wormer, H. (ed.) (2006). *Die Wissensmacher: Profile und Arbeitsfelder von Wissenschaftsredaktionen in Deutschland*. Wiesbaden: VS Verlag für Sozialwissenschaften.

Timeline

Event	Name	Date	Comment
First interactive science centre established.	Phänomenta Flensburg opened (Kiupel 1993)	1993	Phänomenta was based on hands-on exhibits presented in the University of Flensburg since 1985 2000: Opening of the larger science centre Universum Bremen
First national (or large regional) science festival.	Launch of 'science summers' organised by Wissenschaft in Dialog	2000–2012	Funded by the Federal Ministry of Education and Research
An association of science writers or journalists or communicators established.	Foundation of Technisch Literarische Gesellschaft (TELI), an association of a diverse spectrum of people interested in science popularisation (based on precursor associations)	1929	1969: Foundation of the Association of University Press Officers (Bundesverband Hochschulkommunikation) 1986: German Association of Science Journalists (WPK) 2012: Science Communication Section of the German Communication Association (DGPuK)
First university courses to train science communicators.	Beginning of sporadic science journalism courses in general communication programs (U Mainz)	1980	1988: Launch of a regular postgraduate program (Aufbaustudiengang) in science journalism (FU Berlin), aimed at science graduates wanting to become science journalists
First master's students in science communication graduate.	Technical University of Dortmund	2009	Master's program on science journalism launched in the winter term 2007/08
First PhD students in science communication graduate.	The first PhD theses on science in the media date back to the early 1970s	Early 1970s	First postgraduate research program: DFG-Graduiertenkolleg, Bielefeld 2002–11. Most Germany doctoral dissertations are outside formal postgraduate programs by individual agreements between a PhD student and a supervising professor
First national conference in science communication.	Many workshops on science and the media with scientists, science journalists, public information officers and communication scholars	1960s to 1970s	2004: Launch of the annual conference series Wissenswerte, for science journalists 2008: Launch of general science communication conference Forum Wissenschaftskommunikation 2012: Launch of annual conferences of the Science Communication Section of the German Communication Association (DGPuK)

Event	Name	Date	Comment
National government program to support science communication established.	Memorandum Public Understanding of Science and Humanities (PUSH). Not a government report but a functional equivalent, issued by the major German research organisations in collaboration with Stifterverband	1999	2016: Position paper on participation of civil society and public in science and science policy by the Federal Ministry of Education and Research
First significant initiative or report on science communication.	Memorandum Public Understanding of Science and Humanities (PUSH)	1999	2016: Position paper on participation of civil society and public in science and science policy by the Federal Ministry of Education and Research
National Science Week founded.	Launch of the series of Wissenschaftsjahre [Science Years]	2003	Science Years (where each year is devoted to a particular discipline or topic, e.g. chemistry (2003), humanities (2007) etc.) are the German equivalent of national science weeks
First significant TV programs on science.	Launch of the popular TV series *Ein Platz für wilde Tiere* [A place for wild animals] with Professor Bernhard Grzimek. It focused on nature, yet scientific expertise was used to describe and explain animals	1956	1960s: Documentaries about the US spaceflight program (journalists: Heinrich Schiemann, ZDF, and Günther Siefarth, WDR) 1964: Launch of the popular TV health magazine *Gesundheitsmagazin Praxis* (ZDF) with Hans Mohl 1964: Creation of editorial offices for science in TV channels ARD and ZDF; two regular TV science programs were launched
Date hosted a PCST conference.	5th International PCST Conference, Berlin	1998	
Other significant events.			1830: Journalists specialising in the reporting of science and technology at the newspaper *Kölnische Zeitung* (Krüger and Ruß-Mohl, 1989) 1903: Launch of the popular science magazine *Kosmos*

Contributors

Dr Hans Peter Peters is a social scientist at the Research Center Juelich, Germany, and adjunct professor of Science Journalism at the Free University of Berlin. He has been a member of the PCST Scientific Committee since 2002, and is editor of *Public Understanding of Science* and editor of the book series *Science Communication*, of World Scientific Publishing (WSPC).

Dr Markus Lehmkuhl is a full professor of Science Communication in Digital Media at Karlsruhe Institute of Technology (KIT). Since 2000 he has done research on the public perception of risk, media coverage of science in Germany and Europe, and on the journalistic construction of scientific uncertainty.

Dr Birte Fähnrich is senior researcher at the Berlin-Brandenburg Academy of Sciences and Humanities and principle investigator in the EU-funded project RETHINK for Zeppelin University. She is a member of the PCST committee.

15

GHANA

When individuals refuse to let science communication die

Hephzi Angela Tagoe and
Thomas Amatey Tagoe

Ghana prides itself in being identified as the gateway to West Africa. This tag line applies because Ghana was the first African country to gain independence, in 1957, paving the way for others to follow. Within a decade of Ghana's independence, over 30 other African countries were inspired to follow suit (Bourret, 1960). Since then, Ghana has continued to exhibit one of the strongest and longest-lasting democracies on the continent, making room for the country to explore and grow in many ways.

Prior to independence, Ghana was known as the Gold Coast, a name coined by the Europeans as a nod to the country's rich natural gold resources. The Portuguese were the first Europeans to arrive on the Gold Coast in 1471, and by 1492 they had built a fortress along the coastal town of Elmina. The Dutch, Danes and British subsequently became the main traders in the Gold Coast, with the trading commodities shifting from gold to human slavery by the 17th century. By 1874, the country had become solely a British colony (Reynolds, 1984). Gold remains one of Ghana's richest resources, along with cocoa.

Geographically, Ghana is positioned to allow easy access to all other West African countries by land, sea or air. It is home to one of the largest artificial lakes in the world, Lake Volta, which has a surface area of 8,500 square kilometres. Ghana has chalked up many firsts in areas ranging from democratic governance to sports and innovation. For example, Ghana was the first African country to win both the FIFA under-17 and under-20 world cup tournaments. The country's most famous footballer, Abedi Pele, is regarded as one of Africa's best footballers of all time, paving the way for

African footballers to play in Europe (BBC, 2019). The country also prides itself in giving the world Kofi Annan, the first black African to hold the office of Secretary General of the United Nations.

Sixty years after independence, Ghana has once again set a precedent, this time in science communication. Ghana is home to the only planetarium in West Africa, as well as the first organisation focused on science communication as a means of building capacity in science, technology, engineering and maths (STEM) within the region. The country is also home to a STEM communication network with over more than 100 active members, a STEM-based show on morning radio, a national science and mathematics quiz that garners as much attention as the FIFA world cup and a hot air balloon as a tool for scientific outreach. Interestingly, many of the individuals spearheading these projects spent a significant part of their academic or professional lives in the UK and draw from best practices in the UK. Dr Jacob Ashong worked in the UK for many years before returning to Ghana to set up the planetarium. GhScientific was founded by a sibling duo (Drs Thomas Amatey Tagoe and Hephzi Angela Tagoe), who both studied and worked extensively in the UK before returning to lead the work of science communication in Ghana.

The question to ask is: How did Ghana get to this position? Science communication in Ghana has historically experienced a bias towards health and technology. Despite this approach, science communication has continued to grow. To fully appreciate how the field has evolved within an unbalanced environment, it is important to go back to pre-colonial Ghana. The foundations of who Ghanaians are as a people is key to appreciate how we arrived at the current state of modern science communication.

Figure 15.1: First meeting of the Ghana STEM Network.

Source: GhScientific.

1. Evidence of science communication in Ghana pre-independence

What was to become known as Ghana is an area home to 100 different ethnic groups with over 44 distinct languages, not counting dialects. Many ethnic groups belong to larger groups and as such share some common cultural practices and traditions. One commonality is the use of folklore as a means of science communication. Storytelling over burning firewood at sunset was common practice for imparting knowledge (Gyekye, 1996). Although the information that came through the stories has been passed on as myths, modern research has been known to demonstrate the scientific basis of these supposed myths. For example, natural conservation has been historically practised in the country through 'taboos', imposing closed seasons for hunters and fishermen as well as maintaining sections of forests as off-limits to the public (Acheampong, 2010). The benefits of these 'taboos' to the ecosystem are only now being appreciated and reinstated.

Folklore as a form of science communication told stories on innovation and problem-solving often using a cunning character by the name of 'Kweku Ananse'. Today, Kweku Ananse is an educative tool for communicating science, and his famous stories have been incorporated into a gamified learning app called 'Ananse the Teacher', downloadable in the iTunes app store. It is also available in the Google Playstore, where it has had over 1,000 downloads as at January 2019. The app infuses culture with technology, delivering lessons relevant to STEM and reaching younger audiences through new tools based on a familiar concept rooted in culture. Since its introduction, 'Ananse the Teacher' has proven to be a popular outreach tool for engaging younger audiences.

With a rich oral tradition, it is no surprise that radio remains one of the most penetrative modes of communication since its introduction in the Gold Coast in 1935. Today, over 46 radio stations can be heard in the capital city Accra, a significant proportion of which have a program relating to health, sharing information related to healthy living and disease control (NCA, 2016). The first station to have a science communication show as we now understand it was one of the state-owned radio stations. In 2014, Uniq FM launched a show touching on all manner of science and the way it relates to everyday life. It was only in 2018 that this type of programming hit primetime morning radio, with the Ghana Science Association launching *The Horizon*, an hour-long segment on one of the nation's leading radio stations, StarrFM.

The introduction of radio in 1935 was followed by television in 1965. It was not until the late 1990s that Ghana's airwaves were liberalised from the control of the post-colonial state with print media also being liberalised in 1992 (Anokwa, 1997). Now, just as with radio, every TV station and print house has a dedicated segment focused on communicating health-related matters, often at the expense of other sciences.

In anticipation of this liberalisation, Dr Kwame Nkrumah, the first president of Ghana, laid down the foundation for training communicators through the establishment of the Ghana Institute of Journalism (GIJ) in 1962. The GIJ is the first institution for training journalists in Africa and is the leading institution of its kind in Ghana.

Over the last few years, it has become evident that there is an increased demand for journalists with the ability to work collaboratively with scientists for the purposes of science communication. As such the GIJ is collaborating with the Department of Communication Studies at the University of Ghana and the Science and Technology Communicators of Ghana to introduce a short course in science journalism. This move is undoubtedly in line with the ideals of Dr Nkrumah, the visionary who set up GIJ in the first place (Reporters without Borders, 2019).

1.1. Case study: *The Horizon*

The Horizon is an hour-long segment on morning radio where professionals and enthusiasts discuss their areas of expertise within STEM. Hosted by Francis Abban on Wednesdays, the show has been airing since November 2017 on StarrFM (103.5 MHz), which boasts the 7th most popular breakfast show (6am – 10am) on morning radio (Botchway, 2019). *The Horizon* launched when the producer of the breakfast show, Alex Mensah, approached the Ghana Science Association (GSA) to discuss the possibility of a partnership to host a STEM-themed segment as part of the morning show. This meeting took place at a time when GSA was under new leadership and on the lookout for alternative avenues to engage with the public. A series of fortunate coincidences led to the first episode of *The Horizon* on 1 November 2017.

Figure 15.2: Advertising sample on *The Horizon* radio show.

Source: GhScientific.

The show is divided into three segments: top science news, fun science facts and the STEM conversation. The first segment, Top Science News, shares three of the most significant findings in the world of science during the prior week, both nationally and internationally. The Fun Science Fact shares unique scientific facts that may not be common knowledge and leads into the main STEM conversation segment. In the STEM conversation segment, the host of the morning radio show interviews invited guests to shed light on their life and work. Guests on the show range from young high school innovators to established professors who are world renowned as experts in their field. Listeners of the show can phone in and ask questions. In addition, the show is broadcast live on Facebook with regular updates on Twitter, ensuring that a social media audience is engaged. Indeed, *The Horizon* is working to democratise scientific knowledge by taking advantage of technology to tap into the oral tradition of communication so pervasive across the country.

2. Dr Kwame Nkrumah the visionary

In the 1940s and early 1950s, in the Gold Coast, as Ghana was known then, a movement for decolonisation was building, triggered by increasing demand by the natives for more autonomy. The country saw the first attempt at a nationalist party, the leadership of which included Dr Kwame Nkrumah, who subsequently became Ghana's first president in 1957 (Lupalo, 2016).

Nkrumah was a visionary who saw value in the application of science and engineering for national development. From extensive plans for industrialisation to building one of the largest man-made lakes at the time for the purpose of generating hydroelectric power, he considered science and technology as critical. His position on the matter was clearly communicated in his speech to the first conference of the Organisation of African Unity (OAU) when the OAU was formed in Addis Ababa, Ethiopia, 25 May 1963 (Lupalo, 2016):

> We shall accumulate machinery and establish steel works, iron foundries and factories; we shall link the various States of our continent with communications; we shall astound the world with our hydroelectric power; we shall drain marshes and swamps, clear infested areas, feed the under-nourished, and rid our people of parasites and disease. It is within the possibility of science and technology to make even the Sahara bloom into a vast field with verdant vegetation for agricultural and industrial developments. We shall harness the radio, television, giant printing presses to lift our people from the dark recesses of illiteracy.
>
> A decade ago, these would have been visionary words, the fantasies of an idle dreamer. But this is the age in which science has transcended the limits of the material world, and technology has invaded the silences of nature. Time and space have been reduced to unimportant abstractions. Giant machines make roads, clear forests, dig dams, lay out aerodromes; monster trucks and planes distribute goods; huge laboratories manufacture drugs; complicated geological surveys are made; mighty power stations are built; colossal factories erected—all at an incredible speed. The world is no longer moving through bush paths or on camels and donkeys.

Nkrumah set the precedent for science communication across the country. During his presidency, he oversaw the inauguration of two scientific bodies with science communication at the heart of their mission: the Ghana Academy of Arts and Sciences (GAAS) as the highest body of scientists influencing policy; and the GSA to be an umbrella body of scientists promoting and popularising science for national development.

GAAS was founded in 1959 with the aim of promoting the pursuit, advancement and dissemination of knowledge in all branches of the sciences and the humanities. It hosted regular public lectures where it facilitated the attendance of young students. GAAS is represented on presidential advisory boards. Recently it established relationships with the Ghana Young Academy, part of the Global Young Academy linking young scientists with promising futures.

GSA is a large national, multidisciplinary association of scientists, technologists and mathematicians that provides the scientific community the broad opportunity to share their knowledge of science. The work of GSA has inspired the formation of other professional associations of scientists to better meet the needs of professionals and citizens. The inauguration of GSA in 1959 offered Ghana a broad scope of activities, from reading of scientific papers to involvement in national and international affairs. The very first international conference hosted by GSA was held in Accra in 1961 under the theme 'The World Without the Bomb'. Since then, GSA has been mandated to promote, popularise and demystify science to create a scientific culture in the country.

2.1. Case study: Professor Frederick Addai

Professor Frederick Addai is the immediate past head of the Anatomy Department at the University of Ghana. He has held many prestigious positions within and outside of academia, including national president of the GSA. Throughout his career, there have been many accomplishments, with the most noteworthy being his work highlighting the benefits of natural cocoa for healthy living.

It all started in 1991, during a Wellcome Trust Fellowship in the UK, when his curiosity was triggered by a Wrigley's Gum advertisement about its effect on teeth mineralisation. Upon his return to Ghana, Professor Addai explored how chocolate could elicit similar responses, and his research came to a head in a 2002 publication 'Responses of saliva pH to ingestion of Golden Tree chocolate and the effect of stick chewing' (Addai et al., 2002). The contents of this paper formed the basis of a public lecture in 2004 at the British Council Auditorium in Accra. In attendance at this public lecture were senior members of government who were intrigued at the proposed health benefits of natural cocoa. They were perplexed why such research had never been explored, particularly because Ghana is the world leading producer of cocoa.

Over the next year, what started out as a curiosity-driven question led to presentations at high-level cabinet meetings and the launch of an advocacy program to enhance cocoa consumption in Ghana. Addai's research kept revealing health benefits of natural cocoa consumption and a series of consistent science communication activities finally led to the declaration of a National Chocolate Day in 2007, which is celebrated yearly on 14 February.

Addai is a perfect example of how science communication can yield results, influencing policy and changing lives.

3. Universities and science communication

Ghana has 212 tertiary educational institutions, of which 10 are recognised as traditional public universities offering the full complement of STEM courses and active research. In addition to the universities, there is the Council for Scientific and Industrial Research (CSIR), which has eight arms with varying areas of research focus including the Science and Technology Policy Research Institute, Food Research Institute and Water Research Institute. Despite what is a strong culture of research, the communication of findings has struggled to go beyond academic journals and conferences. Media engagements are sparse, possibly because of the unbalanced demand for stories related to health and technology. Nonetheless, many professionals make themselves available for interviews on current matters of public interest. Indeed, there is no shortage of these scientists because such activities are considered 'public service', which is taken into consideration when academics apply for promotion in the public universities (University of Ghana, 2019). Incentivising scientists has always been a great way to promote science communication all over the world, and in that respect Ghana is no different.

Dr Patrick Kobina Arthur, a senior lecturer at the University of Ghana recalls how his colleagues over the years have taken a passive approach to science communication. All the institutions he has worked for have had an office of public affairs, but these offices typically do not encourage faculty to engage in science communication. To him, science communication within the universities has been dependent on an intrinsic desire of the scientists in question, independent of what the university policy may be on communication.

Recently the University of Ghana has taken the lead in promoting science communication by setting up an Office of Research Innovation and Development (ORID). Two of the working teams of ORID are engaged in science communication: the Technology Transfer Team, which communicates with industry; and the Publication, Dissemination, and Translation Team, which is exploring how to encourage science communication to the public. The model employed by ORID is being replicated at other public universities including the Kwame Nkrumah University of Science and Technology, which has set up the Office of Grants and Research. In addition to ORID, the University of Ghana is also home to a unit that has become the bastion of modern science communication within the Ghanaian setting of academia, the West African Center for Cell Biology of Infectious Pathogens (WACCBIP).

3.1. Case study: WACCBIP

WACCBIP is an Academic Centre of Excellence created by the Biochemistry Department of the University of Ghana. Set up in 2014 with funding from the World Bank and under the leadership of Professor Gordon Awandare, the centre has over the years distinguished itself in the area of research and science communication.

Fellows and postgraduate students at the centre are actively encouraged to engage in science communication activities, and it is even required of MPhil students who are on a sponsored program. It helps that the director, Professor Awandare, can often be seen on media platforms communicating the vital work that goes on at the WACCBIP. This has fuelled a culture of science communication within the centre.

WACCBIP is currently the only research unit in the country with a dedicated public engagement officer. Her name is Kyerewaa Boateng and she works to identify and create opportunities for the researchers and students within WACCBIP to communicate their work. During her time with WACCBIP, she has observed an increase in the interest of faculty and students to partake in science communication activities:

> Organising public engagement activities can be difficult and sometimes finances are limiting but faculty always find it exciting to share their work. I enjoy what I do and I think that every department should have a public engagement officer, the scientists and the public appreciate my work.[1]

Kyerewaa explains that on average the unit undertakes two public engagement activities every week, about 60 per cent involving media engagements and 40 per cent in-person community engagements. By making science communication an active part of its postgraduate training, WACCBIP is creating a future generation of researchers who have the skills and experience to engage in science communication. This is something that will undoubtedly influence the culture at institutions they will join in the future.

1 Kyerewaa Boateng, personal communication to Thomas Tagoe, 12 March 2018.

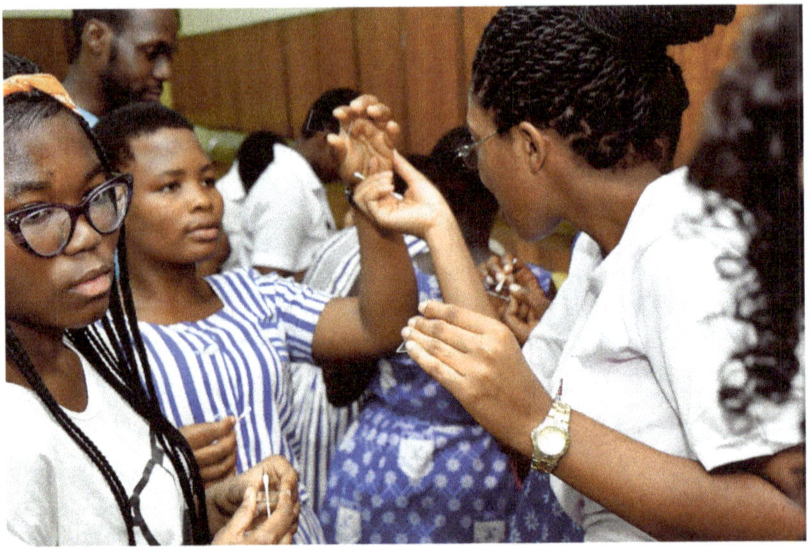

Figure 15.3: Students of WACCBIP engaging with a school audience.
Source: GhScientific.

4. Media and science

The skew of science communication in media (print, TV, and radio) leans heavily towards healthy living and neglected tropical diseases. One can theorise that the extensive intervention programs financed by donor agencies during the 20th century to educate the population would have created a bias that persists to this day. The large amount of funds made available for these programs and the need to use mass communication have made health-related science communication prominent across all media channels in comparison to other aspects of science.

Two aspects of journalism demonstrate this skew in science communication: training and recognition. In respect to training, journalists at the GIJ undertake many courses as part of their syllabus. Although there is no course on science communication per se, there are two related courses: health communication and environmental journalism. Both courses reflect the skew towards health. Then there are the Ghana Journalism Awards, which recognise the achievements of journalists over the course of a year. Among the winning categories are 'Environment', 'Health' and 'Sanitation and Hygiene'. It was not until 2017 that a new category 'Science' was added. The growing need for journalists skilled in science communication is clearly not lost on journalists.

4.1. Case study: GhScientific

In January 2014, shortly after graduating with a PhD in neuroscience from the University of Leicester, Dr Thomas Amatey Tagoe was preparing to return to his home country Ghana. He found himself lamenting the lack of scientific information on the internet that could give insight into what the scientific community in Ghana looked like. His sister, Hephzi Angela Tagoe, then a PhD candidate at University College London, mooted the idea to create an organisation that would fill the gap identified.

In September 2014, the duo started GhScientific as a science communication hub in Ghana, the first of its kind. Over the next four years, GhScientific executed 21 science communication and public engagement projects directly engaging with people ranging from primary school students to professionals and members of the public. The most significant of these projects has been a three-day science communication workshop that brought together 24 professionals from 13 institutions across Ghana and Nigeria. In September 2018, GhScientific announced a project to document the evolution of medical research over a 60-year period starting from the date Ghana gained independence (1957). The findings are to be captured using various forms of art and the final collection will be taken on a tour around the country in late 2020 (Daily Guide Africa, 2018).

Figure 15.4: Early career researchers out on community engagement as part of a three-day science communication training workshop.
Source: GhScientific.

Since its inception, GhScientific has maintained a steady online presence with a growing following across all social media channels. It continues to serve its audience with STEM-related news, events, opportunities and informative blogs. One of the video features produced by GhScientific in support of the Planetarium Science Centre Ghana caught the attention of Dev Varyani, founder and chairman of Resources For Africa (YouTube, 2017). This led to a significant donation in support of the Planetarium Science Centre Ghana, allowing it to keep its doors open (My Joy Online, 2018).

4.2. Science and technology communicators of Ghana

Back in 2013, five experienced journalists from different media houses got together to form what is now the Science and Technology Communicators of Ghana (STCG). The founding members had been involved in reporting science news for large parts of their careers, and have seen the challenges of presenting science accurately. Coming together to start an association was the next logical step to address this challenge and build capacity in other journalists to effectively communicate science. One of the founding members, Mrs Linda Asante-Adjei reflected on the success and challenges faced:

> We started off with five seasoned professionals and now we have 20 members and counting. We have been able to run a series of workshops as well as join the African Federation of Science Journalists. Clearly there is work to be done and now we are looking to roll out quarterly newsletters updating members of parliament about matters of scientific relevance in their various constituencies.[2]

All this is against a backdrop of a need recognised by experienced journalists. Now the STCG is regularly called upon by the CSIR to assist in the dissemination of research findings. This approach removes the apprehension faced by both parties about inaccuracies that often plague media reports of science when inexperienced journalists are involved.

With the proliferation of media houses across the country and the widening influence of social media, the need for more capacity building has been recognised. The STCG is therefore working with Department of Communication Studies at the University of Ghana and the GIJ to introduce a science communication course to be run by the GIJ. It was to start in the 2019–20 academic year but has been postponed until 2020–21. They are also in partnership with the Medical Communicators Society of Ghana, a collection of health professionals with an interest in engaging in science communication. As the African proverb goes: 'If you want to go fast, go alone. If you want to go far, go together.'

2 Linda Asante-Adjei, personal communication to Thomas Tagoe, 8 October 2018.

5. Ghana STEM Centres

At an international mathematics conference in January 2018, the Minister of Education announced plans for the government to open 10 STEM centres across the country. The first STEM centres in the country were established in 1990 as a hub where secondary schools could undertake practical science to supplement the theoretical curriculum. The facilities were referred to as 'science resource centres'. Unfortunately, these were not maintained and soon become empty buildings and remain out of commission. Almost 20 years later, the government has announced plans to bring the abandoned science resource centres back to life. The recent announcement to open STEM centres across the country is laudable, but only time will tell what impact these will have on the STEM education landscape if the promise is fulfilled. Until then the Ghana Planetarium remains the main public STEM centre in the country.

6. Individuals with a passion for science

Having discussed the evolution of science communication in structured institutions (government, universities, associations and media), we will now explore the growth of individuals spearheading a range of activities from community-based initiatives to self-financing the only planetarium in West Africa. Individuals have become integral to science communication within Ghana.

The current state of science communication in Ghana can be traced back to institutional interventions initiated by Nkrumah as the first president of the country. From universities to associations and training institutes for journalism, his contributions cannot be ignored. The most important of all his contributions must be his unwavering belief that science and technology held the key to rapid national development to raise the quality of life for all citizens. Generations later, this passion for science is one that is shared by many.

6.1. Case study: The Ghana STEM Network and Africa Science Week

In 2015, four science enthusiasts working independently in science communication and capacity building got together to discuss the need to create a collaborative platform to support each other's work. The four were Dr Connie Chow, Dr Thomas Amatey Tagoe, Miracule Gavor and Gameli Adzaho, and together they created what came to be known as the Ghana

STEM Network. The size and influence of the network has slowly grown to include more than 100 individuals and organisations working within the STEM space to collaborate, share, lobby and influence policy.

The true benefit of the Ghana STEM Network would be realised in September 2018 during the maiden Africa Science Week – Ghana celebrations. The Africa Science Week – Ghana is an initiative of the African Institute of Mathematical Sciences (AIMS) and the Next Einstein Forum (NEF). This celebration across 33 African countries was the second Africa Science Week and the first time that Ghana had participated. At the time, Lucy Quist, the President of AIMS Ghana said: 'This celebration presents our country with a unique opportunity to elevate the conversation, drive greater participation and celebrate individuals and organisations in STEM in Ghana' (Quist, 2018).

The week-long celebration from 25–29 September leveraged the members of the Ghana STEM Network to ensure that public engagement and science communication activities took place all over the country. There were radio and TV shows, a STEM activity day in schools, a hackathon, a coding camp and a public lecture with many exhibitions. The week was considered a huge success by all involved, and it was evident that this collaborative approach to celebrating Africa Science Week in Ghana was here to stay.

Over the last decade, many individuals have played leading roles in promoting science communication at both the local and national levels. These individuals are convinced that science and technology holds the key to sustainable national development and that a scientifically literate population will raise the quality of life for all. Notable individuals include:

- **Freda Yawson**, founder of Innovate Ghana. Innovate Ghana runs an annual high school innovation challenge where teams develop products and solutions towards national development. Teams communicate their designs to an audience and a judging panel. The process is heavy on people-centred design, and teams are required to engage with the public for whom their designed solutions are targeted.
- **Gameli Adzaho** is founder of Global Lab Ghana, Ghana's main STEM organisation with a focus on citizen science and building a community around its activities. Global Lab particularly focuses on air quality and open science, spearheading the organisation of the first Africa Open Science and Hardware summit, which was held in Kumasi, Ghana, in 2018. Since then, Global Lab has been organising science cafés, an informal gathering of interested persons to discuss various STEM issues.
- **Dr Hephzi Angela Tagoe and Dr Thomas Tagoe** are founders of GhScientific, currently Ghana's premier science communication network, running with the tag 'From the bench to the community'. GhScientific

works with schools and the public to communicate various aspects of science and provides training for researchers on science communication skills. With a pool of freelance journalists, their online portal is able to focus on STEM-related news and opportunities.

- **Kelvin Odonkor** is founder of Ghana Health Nest. A nurse and photographer, he has changed the face of nursing and health care across the country by communicating news from the health sectors in real time. He documents health matters through photography and each image is accompanied by a story that lay audiences can comprehend.
- **Issac Sesi** is founder of Nsesa Foundation, which engages high school students with science through an intensive three-week program after which they pitch final projects for a prize. They also run a campaign—STEM WOW—celebrating women in STEM and spotlighting their career accomplishments via social media channels. This collection of stories was launched as a book series (STEM WOW Chronicles) in January 2019 to increase accessibility and reach parts of the country where internet penetration is poor (STEM WOW Chronicles, 2018).
- **Triump Tetteh** is founder of Starters Tech, a technology company that provides STEM education services for schools, communities and homes. Starters Tech is behind a new series of STEM-based story books, weaving scientific knowledge into gripping stories for young audiences. The aptly named 'Next Gen Stories' was launched in March 2019 to inspire the next generation of scientists.
- **Larisa Akrofie** is founder of Levers in Heels, which promotes women in STEM by highlighting personalities across all levels of their career via social media channels. Levers in Heels runs features on women in STEM and provides an online platform for scientists to share their work with a non-academic audience.
- **Charles Amegamashie** is founder of WeGoInnovate, which seeks to use the power of media to capture the nation's imagination in STEM. WeGoInnovate has collaborated with scientists to release animated one-minute videos sharing fun science facts for TV. In September 2018, they launched a national competition on TV that relies on sharing videos of high school students designing and executing exciting science experiments.
- **Philip Ashon and *CitiTrends*:** Philip is the host of *CitiTrends* on CitiFM, the longest running technology-focused radio show. With news, reviews and engaging interviews, Philip ensures that tech enthusiasts across the country are kept up to date on happenings both at home and abroad. The show also provides handy technology tips for the less technology-savvy audiences.

- **Dr Jacob Ashong and the Planetarium Science Centre** are worthy of the most recognition. The planetarium is the dream project of Ashong and his wife Jane Ashong, who used their life savings to build the planetarium in 2008. It continues to be the only one in West Africa. Their focus is to bring astronomy closer to the masses and promote practical STEM learning through initiatives such as planetarium shows, live astronomy telescope viewings and STEM learning workshops for school students.

Modern science communication in Ghana can be described as a young but integrated space where all players are working collaboratively to increase capacity while reaching a large audience to increase scientific literacy across the country. The tools are diverse, taking advantage of existing cultural traditions and applying technology where it best serves its purpose.

As science communication increasingly becomes a global requirement of researchers, it is encouraging to see science communication also growing among PhD students and postgraduates. This pool of young people share the desire and commitment to communicate science to a wider non-expert audience. Although the current government has pledged to commit 1 per cent of the country's GDP towards promoting all aspects of scientific research, including science communication, this is yet to be realised and it may take another decade for the government to put in place systems and structures that reflect this level of verbal commitment. To accelerate this process, collaborations among this new generation of science communicators and established organisations/associations must create the critical mass needed to effectively lobby for such changes. Indeed, such steps are necessary to avoid the dangers of 'burnout' typical when talented individuals pursue their ambitions in silos.

One of the major hurdles faced by organisations within the science communication space continues to be funding, and many have to rely on international grants, personal funds or charging a small fee for their projects. The latter limits the reach of the science due to economic hardships while the former caps the potential for growth. This is the same challenge faced by young innovators who develop prototypes of clever practical inventions but are unable to transition into production, scale and market. In the past, their achievements were only celebrated by peers, but now the media gets involved in communicating their feats of brilliance. Nonetheless, the financial and structural support to scale it up and, by so doing, to provide more jobs in the science sector remains a challenge. Publications and interviews that follow these inventions and discoveries do serve as a form of science communication and contribute to raising the level of science literacy across the country. Such actions will, in the long run, encourage the willingness of individuals and corporations to invest in scientific innovation, increasing the success rates.

The future of science communication is a promising one as Ghana collectively moves towards becoming a country beyond aid. The vision of the first president, Dr Kwame Nkrumah, has never been more important. Science and technology holds the key to national development, but all this is meaningless if it is never communicated to the masses.

References

Acheampong, E. (2010). *The role of Ghanaian culture and tradition in environmental sustainability*. Elvis Akwasi Acheampong, Green Ghana Foundation. Retrieved from www.ghanaweb.com/GhanaHomePage/features/The-Role-Of-Ghanaian-Culture-And-Tradition-In-Environmental-Sustainability-198388.

Addai, I., Nuamah, K. and Addo, J. (2002). Responses of saliva pH to ingestion of Golden Tree chocolate and the effect of stick chewing. *Journal of Ghana Science Association*, *4*(1), 8–13.

Anokwa, K. (1997). Press Performance under Civilian and Military Regimes in Ghana: A Reassessment of Past and Present Knowledge. In F. Eribo and W. Jong-Ebot (eds), *Press Freedom and Communication in Africa*. Asmara: Africa World Press, Inc.

BBC. (2019). *9 things Ghana brought the world*. Retrieved from www.bbc.co.uk/programmes/articles/48VTLmTVZJtxVKgpKWwvpgT/9-things-ghana-brought-the-world.

Botchway, K. (2019). *Ghana Media Measurement: Top TV And Radio Stations 2018*. GeoPoll. Retrieved from www.geopoll.com/blog/ghana-media-measurement-top-tv-radio-2018/.

Bourret, F. M. (1960). *Ghana, the road to independence, 1919–1957*. London: Oxford University Press.

Daily Guide Africa. (2018, 24 October). *GhScientific received wellcome fund*. Retrieved from dailyguideafrica.com/ghscientific-receives-wellcome-fund.

Gyekye, K. (1996). *African Cultural Values*. Accra: Sankofa Publishing Company.

Lupalo, L. E. K. (2016). *Nyerere and Nkrumah: Shared Vision*. CreateSpace Independent Publishing Platform.

My Joy Online. (2018, 15 March). Academic City, Vista Ghana and NIIT Ghana adopt Planetarium Science Center. *MyjoyOnline*. Retrieved from www.myjoyonline.com/news/academic-city-vista-ghana-and-niit-ghana-adopt-planetarium-science-center/.

National Communications Authority (NCA). (2016). *List of authorised VHF – FM Radio stations in Ghana*. Retrieved from www.nca.org.gh/assets/Uploads/NCA-FM-Radio-Authorisations-2016-Quarter-3.pdf.

Quist, L. (2018, 10 September). Maiden Africa science week Ghana celebrations slated for 25th – 29th September. *B&FT Online*. Retrieved from thebftonline.com/2018/business/education/maiden-africa-science-week-ghana-celebrations-slated-for-25th-29h-september/.

Reporters without Borders. (2019). *Media Ownership Monitor: Ghana*. Retrieved from ghana.mom-rsf.org/en/context/history/.

Reynolds, E. (1984). The Rise and Fall of an African Merchant Class on the Gold Coast 1830–1874. *Cahiers d'études africaines*, *14*(54), 253–64. doi.org/10.3406/cea.1974.2644.

STEM WOW Chronicles. (2018). Nsesa Foundation, published by Heturo Company. Retrieved from online.fliphtml5.com/hoawo/acck/.

University of Ghana. (2019). *Amendments to the statutes of the University of Ghana/*. Retrieved from www.ug.edu.gh/sites/default/files/documents/UG_AMENDMENTS_0.pdf.

YouTube. (2017). *GhScientific at the Ghana Planetarium*. Retrieved from www.youtube.com/watch?v=Ih2Nz2HpZgk.

Timeline

Event	Name	Date	Comment
First interactive science centre established.	The Ghana Planetarium	2008	Still the only planetarium in West Africa
First national science festival.	African Science Week	2018	An initiative of the Next Einstein Forum
First association of science writers, journalists or communicators established.	Science and Technology Communicators of Ghana (STCG)	2013	Formed by five journalists from different media houses
First university courses to train science communicators.	Ghana Institute of Journalism (GIJ)	2019	GIJ has included a science communication module since September 2019
National government program to support science communication established.		1959	The Ghana Science Association has science communication as part of its mandate
First significant initiative in science communication.		2014	GhScientific is Ghana's first dedicated science communication organisation
First collaborative network of science communicators.	Science and Technology Communicators of Ghana. STCG is made up purely of journalists	2013	2015: The Ghana STEM network formed. This first collaborative network of science communicators is broader and more encompassing than STCG

Event	Name	Date	Comment
National Science Week founded.	African Science Week	2018	
First significant radio programs on science.	Uniq FM	2014	Hosts a science radio show focused on STEM
First awards for scientists or journalists or others for science communication.	Ghana Journalism Awards	2017	The first category for science reporting was included in 2017
Other significant events.		1959	The first professional association for scientists established
	Museum of Science and Technology	1965	Established 1965
	National Science and Maths Quiz	1993	Ongoing and embraced by the nation
	Public Engagement Officer appointed at WACCBIP	2014	The West African Cell Centre for Biology of Infectious Pathogens (WACCBIP)
	The National Science and Math Quiz introduces a science fair component	2015	Includes mentoring from scientists
	Global Lab Café Scientifique, a first	2018	Organised by Global Lab Ghana on artificial intelligence
	Science Slam Ghana	2019	Hosted by the African Institute of Mathematical Sciences

Contributors

Dr Hephzi Angela Tagoe is the founding director of GhScientific, a non-government organisation in Ghana focusing on science communication and public engagement and aiming to build capacity in STEM. She is a freelance science writer and runs an education consultancy.

Dr Thomas Amatey Tagoe is a neuroscientist at the University of Ghana and co-founder of GhScientific, an organisation building capacity in science, technology, engineering and math (STEM) through public engagement and outreach activities.

16

INDIA

Tracing science communication in independent India

Anwesha Chakraborty, Usha Raman and Poojraj Thirumal

The shadow of imperialism and European colonisation of the subcontinent looms large in the narration of the history of modern science in India (Phalkey, 2013). So where the role of science and technology in independent India is concerned, the lens used by political and social scientists is that of post-colonial nation-building (Visvanathan, 1997; Abraham, 2006; Roy, 2007). Phalkey further observes, in her introduction to the Focus section on 'Science, History and Modern India' in *Isis*, the premier journal in the field of history of science, that historians have claimed scientific practice in India to be derivative, and that these claims have happened in the absence of national histories of ideas, pedagogy, policies and practices of science. To address this lacuna, she suggests the study of institutional, social, political, economic and cultural contexts with a focus on the experiences of practitioners so that a practice-oriented understanding of science in India can emerge. This short essay, ambitious as it is in its scope, attempts to fill the void by working through science policies; dissemination and communication practices at institutions such as museums and universities; journalistic writings on science and technology; and people's science movements (critical science literacy movements organised by civil society and community-level groups). By doing so, the essay purports to narrate a history of science communication practices in independent India (that is, post-1947).

However, before delving into the practices, it is important to review the sociocultural context of contemporary India and examine its specific communication politics and policies.

1. Science, technology, innovation and the Indian state

In a 2015 news feature in *Nature*, Indian science journalist T. V. Padma listed some of the highlights in India's road to becoming a science superpower and some of the structural and societal challenges impeding the country from fulfilling its technological, industrial and economic potential (Padma, 2015). Several major achievements have emerged: in the fields of space research (the Indian Space Research Organisation (ISRO) completed its Mars mission in 2014 by successfully launching its Mars Orbiter on its first attempt at a fraction of the cost incurred by other space research agencies in the developed world); in the pharmaceutical industry (India produces a large volume of low-cost medication and vaccines); and in renewable energy (there is an increased focus on making India a major solar power; it already is a world leader in wind power).

However, the list of challenges for a country with a population of 1.3 billion people, with many living below the poverty line, is immense and often seems to outweigh the positives. Padma notes that the problem starts at the very top and is a result of a lack of political will. Successive governments, while pledging financial support for the Indian scientific community, have not increased the budget for research and development from about 0.9 per cent of GDP (with conspicuously less investment from the private sector), a figure significantly lower than other BRICS[1] economies. The quality of education at the universities, save for a few that receive central government patronage, has been inadequate to produce world-class research or build the competencies required for its successful dissemination. India has one of the lowest densities of scientists and engineers in the world, which is surprising as the country produces many scientists and engineers who then move on to work in foreign countries, especially the US. Brain drain is a cause for concern in Indian society, as some of the most qualified among Indian students migrate in search of better opportunities. Then there are continuing societal challenges such as concerns over public health: maternal deaths, malnutrition, and high incidence of tuberculosis and malaria. And yet Mèdecins sans Frontières regards India as the 'pharmacy of the developing world' (Padma, 2015) because of its significant output in low-cost drugs. Scientists and entrepreneurs have recommended enhanced ties between universities, research laboratories and industry, and the government has responded by setting up incubators and by supporting start-ups to transfer knowledge from research facilities to industries.

1 BRICS is the acronym for an association of five major emerging national economies: Brazil, Russia, India, China and South Africa.

1.1. Which science for which people: Discussing 'public' and 'policy'

The question of which science and technology should be communicated to the public merits a deeper interrogation of the 'public' itself. For a country as vast as India, with many sociocultural/economic identities that form and inform individual (and collective) choices, the question is especially complex. While solutions cannot only come from government policymakers and experts, it is worth discussing how experts have tackled this debate.

The issue of what science should take precedence in the national imagination to ensure sustainable, equitable development of the population (a majority of which lives in material poverty) has been present in the rhetoric of politicians for a long time. One of the strongest articulations was Indira Gandhi's speech at the 63rd Indian Science Congress at Waltair (now Vishakapatnam) in 1976, when she emphasised the need for science and technology in India to take a rural turn.

> There should be greater attention to rural engineering ... Rural electrification has made rapid progress but we have not yet succeeded in teaching villagers how to use power to advantage. There is a woeful lack of rural technicians and of innovative work. The village and the home should become laboratories for inter-disciplinary scientific and technological investigation (Gandhi, 1976).

That science and technology in India is an elitist enterprise has been argued by critics, scholars and practitioners. Abha Sur, who studies science in India in the interstices of caste and gender, points out that scientific practice and establishments are embedded in a male upper-caste ethos that refuse to translate concerns of ordinary people into agendas of scientific research (Sur, 2011). She argues that the formal laboratory remains an upper caste, elite space of learning and doing, making a strong case that the former prime minister's concerns remain valid to the present day (Sur, 2012).

A study of science policy documents in independent India tells us that in rhetorical terms, the focus has been to harness the capabilities of the population and reach the poorest sections of society. The *Scientific Policy Resolution* (Government of India, 1958), the first national policy, was released in 1958 and placed great emphasis on industrialisation for the creation of national wealth and prosperity, with a focus on science education and training in technical skills. The Technology Policy Statement of 1983 discussed the need to become self-reliant and technologically independent and shifted the emphasis to technological development (Government of India, 1993). In 2003, the government proposed a science and technology policy where the two were brought together and the need for investment in research and development was highlighted. This document is of particular interest to scholars of public communication of science and technology as it has a clearly defined section on the public awareness of science:

> There is growing need to enhance public awareness of the importance of science and technology in everyday life, and the directions where science and technology is taking us. People must be able to consider the implications of emerging science and technology options in areas which impinge directly upon their lives, including the ethical and moral, legal, social and economic aspects. … Support for wide dissemination of scientific knowledge, through the support of science museums, planetaria, botanical gardens and the like, will be enhanced. Every effort will be made to convey to the young the excitement in scientific and technological advances and to instil scientific temper in the population at large (Government of India, 2003).

The most recent government science policy document, *Science, Technology and Innovation Policy* (Government of India, 2013), discusses the aspirations of India for 'faster, sustainable and inclusive growth' (p. 1) and the role of the huge talent pool that India's largely young population offers. The 2013 document makes it clear that the focus will be on people, and the national science, technology and innovation (STI) system must recognise society as its major stakeholder. Thus, the 'emphasis will be to bridge the gaps between the STI system and socio-economic sectors by developing a symbiotic relationship with economic and other policies' (p. 3). To empower people and incorporate them into the STI framework of the country, one of the suggestions in the policy document is the promotion of scientific temper among all sections of society.

It is interesting that over the years the rhetoric of the documents grants the public a greater level of agency. If the 1958 resolution was all about the scientific and technical education of the masses, the 2003 policy shows awareness of the growing interest of the public in science and technology and how they have to be included in debates of the ethical, social and economic dimensions of science. The 2013 document observes that people should not be mere recipients of scientific knowledge but must be made an active part of the scientific innovation framework. This increase in agency of the public in policy formulation goes along with the similar emphasis on the building of 'scientific temper' through school education—the National Curriculum Framework (NCF) of 2005 spends a lot of time discussing science education in this manner.[2] The position paper on science education that formed part of the NCF 2005 formulation process notes that school education in India 'develops competence but does not encourage inventiveness and creativity' (NCERT, 2005, p. 3). It says that 'schools promote a regime of thought that discourages thinking and precludes new and surprising insights' (p. 22). Clearly, there is awareness of the failure of school education to build this scientific temper (NCERT, 2005).

1.2. The journey of 'scientific temper' to the Constitution of India

'Scientific temper' is a phrase attributed to Jawaharlal Nehru's *The Discovery of India* (1946), a monograph he wrote while imprisoned with other leaders agitating against British rule. Presented as a part-autobiography, part-civilisational history of India, the patriotic overtones are evident. While recounting India's many existing social problems, like poverty, overt religiosity, superstition and the caste system, Nehru (1889–1964), a science graduate of Cambridge, emphasised the need to cultivate scientific thinking in order to approach life and its challenges:

> The applications of science are inevitable and unavoidable for all countries and peoples to-day. But something more than its application is necessary. It is the scientific approach, the adventurous and yet critical temper of science, the search for truth and new knowledge, the refusal to accept anything without testing and trial, the capacity to change previous conclusions in the face of new evidence, the reliance on observed fact and not on pre-conceived theory, the hard discipline of the mind all this is necessary, not merely for the application of science but for life itself and the solution of its many problems (Nehru, 1946, p. 512).

It is important to pause here and reconsider the phrases *scientific approach* and *critical temper of science*. Not only did Nehru recognise the material and practical benefits of foregrounding science and technology for national development, he also strongly argued for science (scientific method and approach) as a 'philosophical approach' (Arnold, 2013). This is the enduring legacy of Nehru and his contribution to post-colonial scientific debates: the shift of understanding from science and technology as an imposition of Western authority, to science and technology as answerable to the state and the public for its capability of delivering a better, more inclusive and humane society.

2 Also in the 1990s major policy decisions, such as the overhaul of the electricity generation and transmission systems, were preceded by a series of public consultations—see prayaspune.org, which is one of the advocacy organisations that facilitated these public conversations.

The post-Nehruvian period in Indian politics and policymaking saw a growing interest in the concept and its increasing importance in the mandate of several public institutions working at the interface of science and society (including those in charge of promoting scientific literacy and popularising science). The inculcation of scientific temper was added to the Indian Constitution in 1976 as one of the 10 fundamental duties of every citizen under Article 51(A)(H) by the 42nd constitutional amendment: 'to develop the scientific temper, humanism and spirit of inquiry and reforms' (Raza et al., 2014; Ministry of Law and Justice, 2019). India became the first country to include such a clause in its constitution, and the change occurred while Nehru's daughter Indira Gandhi was at the helm and during the period of political Emergency. The purpose was to fight religious obscurantism: scientific temper was interpreted as a rejection of unscientific, irreligious or superstitious beliefs often fostered by organised religion(s).

It is important to recollect that scientific temper and its promotion appear in the two latest science and technology policy documents of 2003 and 2013 (in the latter, it is even accorded the position of the primary objective of the policy formulation).

In this chapter, one key aim is to engage critically with the rhetoric of scientific temper and examine how it has become a part of institutional and public narratives on science and technology in India. We will examine different institutions that are formally and informally engaged with science communication, and comment on how they have promoted scientific temper and scientific attitude among the public. Information has been sourced from government documents and from personal research and experience in the field.

2. Science communication: Institutions and movements

2.1. Indian science museums

With the sustained importance of nationalism in the period of decolonisation after World War II, museums were recognised as powerful tools for radical socioeconomic transformation (Ghose, 1992; Venugopal, 1995). The first government attempt at defining India's scientific heritage and promoting science education was the establishment of Birla Industrial and Technological Museum (BITM) in Calcutta in 1959. Saroj Ghose, erstwhile president of ICOM, explained the need was felt by the central and state governments, and especially by the then chief minister of West Bengal, Dr Bidhan Chandra

Roy, to preserve artefacts of historical significance in the newly formed and diverse state. Roy had visited the Deutsches Museum in Munich and was influenced and inspired to form a similar institution in India.[3]

The opening of the Exploratorium in 1969 in San Francisco was a challenge to museums around the world. The Exploratorium's hands-on approach to science communication strongly favoured science education and active participation in the understanding of science. As Ghose said, a young country with its policies firmly grounded in the need to become self-sufficient and educate its large rural masses, science communication had to be based on a model where education was foregrounded rather than science appreciation. The narrative of attaining self-reliance and self-sufficiency through the acquisition of technical skills in the Fifth Five Year Plan (1974–1979) (Planning Commission, 1974) spilled over to the creation of scientific and technical institutions of national importance. It is around this time that a task force was set up to evaluate science popularisation efforts, and the Exploratorium's model of hands-on science training gained currency among policymakers and science museum professionals, resulting in the formation of the National Council of Science Museums (NCSM) in 1978.

The promotion of scientific temper, a crucial clause frequently mentioned in the rhetoric of science popularisation in India, was the NCSM's primary activity to ensure that demographic dividends could be reaped in the future. The phrase finds mention in the outcome budget of 2016/17 of the Ministry of Culture, with reference to NCSM:

> National Council of Science Museums (NCSM) has been engaged in creating awareness on Science & Technology, developing scientific temper in society and promoting science literacy throughout the length and breadth of the country and engaging young students in creative and innovative activities. For last 35 years, the Council has developed a nationwide infrastructure of 48 science museums & centres to achieve these goals. Its outreach activities throughout the year aspire to develop a culture of science and innovation by engaging people from all segments of the society in the process of science & technology (Ministry of Culture, 2016, p. 369).

3 Personal communication, August 2015.

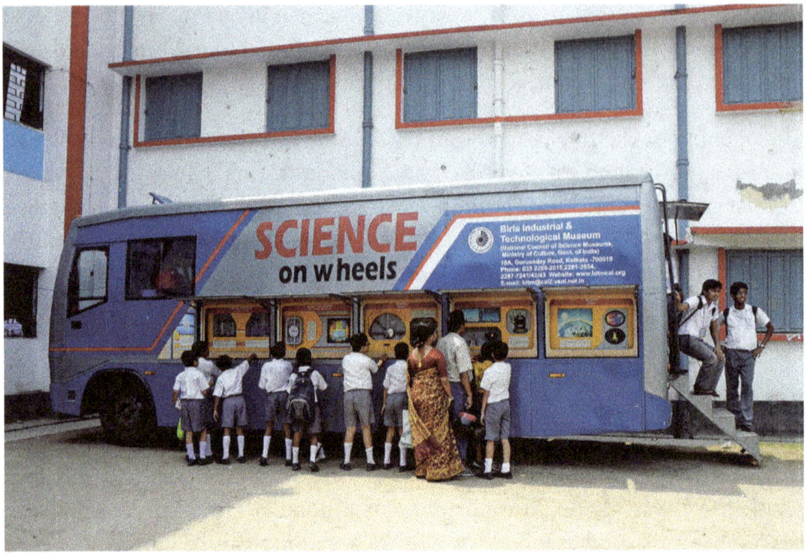

Figure 16.1: Mobile science exhibition.
Source: Birla Industrial and Technological Museum, Kolkata.

Since its inception the NCSM has expanded, developing a nationwide infrastructure for science communication and informal science learning. The number of visitors to the 25 museums and centres of NCSM has grown progressively, with the 2013/14 Activity Report of NCSM recording 9.1 million visitors to the various centres (Ministry of Culture, n.d.). In 2016, the director of NCSM Headquarters revealed in a personal interview that another 15 million people participate in the NCSM's engagement and outreach programs annually. As new centres are built and engagement and outreach programs expanded, these numbers will grow. Attendance figures may appear very low for such a populous country, but India has a disproportionately small number of science museums and centres for its large population. As one of the top officials of NCSM said in a peronsal interview in 2015: 'We have a lot to achieve in the field of public engagement with science. As of now we have about 95 science museums (including centres, planetaria and zoos) for a population of 1.25 billion people.'

Any narrative on scientific temper and how it is being promoted should include the history of mobile science exhibitions (or MSEs) and how they became a huge success with the rural population. In the 1960s and early 1970s, two major science museums, BITM and Bangalore's Visveswaraya Industrial and Technological Museum (VITM), started the first MSEs. These travelling exhibitions ventured out into small towns and villages to create scientific awareness. Each exhibition comprised the following resources and persons: a bus, 24–28 simple exhibits focusing on everyday scientific phenomena and the uses of science and technology in daily life, a technician, an explainer and a driver. The Museobus proved to be immensely popular and created a target group of visitors distinct from school students (Mukherjee, 2003). NCSM today is quick to point out the sheer ambition of the MSEs, given that they manage to reach 2 per cent of the entire population. Their aim is to ensure universal awareness of science and technology, percolating down to the common individual. As the director of one national centre noted in a personal interview in 2015: 'What we are definitely interested in promoting is science as culture through the concept of scientific temper. We want to communicate the idea that science is not just a set of rules and knowledge, but it is a way of thinking, of doing things. A person with scientific temper would be willing to receive inputs from everywhere without perception bias.'

Science Express is a similar travelling science exhibition, but travels through India by train. This collaborative project was started in 2007 by two institutions dedicated to science popularisation and education: the National Council for Science and Technology Communication (part of the Department of Science and Technology, or DST) and the Vikram Sarabhai Community Science Centre (whose partners include the NCSM and the DST).

Another crucial objective of the NCSM is to provide assistance in science communication activities to other institutions of learning, both public and private. This reflects NCSM's important role as a broker of science communication activities, connecting schools, universities, technical institutions, other museums, think tanks, government and the public. Activity reports through the years show that apart from creating and maintaining galleries, the constituent members of NCSM are also in charge of conducting a vast array of public engagement programs. These include science seminars, drama competitions and science quizzes conducted at the local, zonal and national levels. For visitors to the centres in cities and small towns, there are programs that highlight health and environmental issues, such as water scarcity or the need to prevent child marriage. There are also special events for children living below the poverty line. Teacher-training programs help teachers working with students in Indian classrooms. The NCSM has offered a two-year Master of Technology program to science communication aspirants since 2005, in conjunction with the Birla Institute of Technology and Science, Pilani. The degree trains students to work in the sector of science museums and centres. The range of activities may seem overwhelming, but from the perspective of promotion of scientific temper, it is understandable that the NCSM has engaged in a multi-pronged approach to ensure that a strong narrative about science and its role in society is communicated to the people.

One critique of NCSM is that it derives almost all its funding from the national government exchequer. The issue of government control over the rhetoric and activities of NCSM is one of great significance, because the political and ideological affiliation of governments can influence activities. Recent updates on the council's social media pages suggest it is promoting a number of activities unrelated to science: celebrating a special day to commemorate Hindi as the major national language, or the birth anniversary of a nation builder. These activities seem innocuous but clearly do not belong to the resume of a science museum. And while scientific temper is fundamentally about promoting anti-superstitious thoughts, spreading public awareness and understanding of science, NCSM activities also promote careers in STEM without critiquing institutionalised big science in the exhibits and displays.

2.2. Science communication courses in universities

While developing a scientific temper through general education is one strategy for building a critically engaged and scientifically literate citizenry, the media and other communication channels also disseminate information about science. As a corollary to this, creating a pool of professionals able to convincingly and engagingly speak about science to lay audience becomes essential. While there is little formal documentation or scholarship examining science communication/journalism courses in India, we attempt to fill out the picture by drawing on our own experience as journalism educators and science communication professionals. The history and development of science communication education in India may be broadly traced along two paths: one, within the institutional framework of universities; the other, in private and quasi-governmental organisations. The earliest efforts to build capacity to communicate research to lay audiences had their origins in departments of agriculture, in courses known as 'agricultural extension'. It is no coincidence that the first journalism course (a diploma program) within a university (Punjab University) had a strong emphasis on agriculture education as far back as 1941 (Bharthur, 2017).

The focus has been, largely, to equip communicators to convey advances in agricultural technology and farming practices to farmers in a manner in keeping with the 'development communication' paradigm, premised on the belief that an efficient flow of appropriate information will bring about positive social and economic change (or 'development'). This tradition of science communication education to serve the needs of the research establishment, and to inform both the general public and specific 'end-user audiences', dominated through the 1960s and 1970s. 'Science' was by and large interpreted as 'institutional science', and many communication graduates with an interest in 'science' became public relations officers in public sector laboratories established under the Council for Scientific and Industrial Research (CSIR) or Indian Council of Medical Research (ICMR). These science communicators popularised the research output of these laboratories, putting out press releases, organising public events and managing in-house publications.

As journalism and mass communication education programs proliferated within the Indian university system, agricultural extension and development communication remained an important component, particularly in graduate-level education. This took the form of an optional course (or, rarely, a series of courses) in science communication, health communication or science writing. These courses combined institutional science communication with science journalism. A few universities offered graduate degrees specialising in development communication or health communication (Jamia Milia Islamia in New Delhi, Tezpur University in Assam, for instance), and graduates from these programs often secured jobs in the emergent social sector. This tradition of science communication education saw the professional communicator as being in service to the cause of science—his or her job was to clarify and explain and show applicability of science to life. Clearly, there were two career options for the professional science communicator: one was to serve the science establishment, while the other was to report science for the mass media, as a science journalist. Most science communication and journalism courses, however, focused on the 'how' of writing about science rather than on understanding its structure, dynamics or the process of science, much less looking at it critically as another area of human endeavour (for instance, in non-government and multilateral organisations engaged in science popularisation, eduation and advocacy).

In the aftermath of the Bhopal Gas tragedy in 1984 and the growing disenchantment with Big Science and its close links with the military-industrial complex and apparent disregard for ethical and environmental issues, a large number of civil society organisations joined with conscientious scientists to articulate the need for better informed and more critical engagement with publicly funded science as well as corporate research and development. This, along with the unrelated but concomitant effort to set up short-term training courses in science writing for new and practising journalists, led to the growth of the second set of programs: private and quasi-governmental efforts in science journalism training. Brian Trench reports that, by the early 2000s, there were several government-supported science journalism programs in India (Trench, 2012). These range from one-year programs such as those offered by the Indian Science Communication Society (offered online), to the two-week intensive course run annually by the National Centre for Biological

Sciences. The National Council for Science and Technology Communication (NCSTC), a body set up under the CSIR to promote science outreach and foster public understanding of science, offers short-term capacity-building programs to professional journalists and science communicators but does not train new entrants to the field.

Universities have found it difficult to sustain standalone science communication or science journalism diploma or degree programs, although the national higher education governing body, the University Grants Commission, has provided limited funding for new 'innovative' programs. These aim to build capacity in niche areas of journalism perceived to have market demand or that serve certain social or development goals. Apart from science communication, they have included postgraduate diploma or degree programs in areas like community media, agricultural journalism and development communication. The funding is normally provided for a specified period of three to five years, with the host university committing to continuing funding and faculty positions beyond that. These programs have not been able to attract students, who seem to prefer a broad-based program that prepares them for multiple roles in media/institutional communications.

Unlike some of the strongest programs in science journalism/writing in the West (such as MIT's graduate program in science writing, or science journalism programs at Cornell University, University of California, and others) the university-based programs are generally not run by faculty with a science background, although scientists may be invited for guest lectures. The emphasis is on the journalistic process—writing, production, interviewing—with the critique of science taking a backseat. Some courses in recent times—such as the module on health journalism at the Delhi-based Indian Institute of Mass Communication (IIMC)—have included a critical appraisal skills component, introducing students to the process, pitfalls and politics of doing science, and building competence in assessing the claims of science publicists. But in this regard, science journalism is no different from any other specialisation: students learn the nitty-gritty of how to cover a particular field when they are on the job; and acquire with experience the critical temperament to write about science dispassionately, yet with an eye for detail.

2.3. Science in the Indian media

It is curious that despite the huge focus on science and technology (in education, industry and institutional research) in India over the past half century, and the unrelated development of media in the country, there is hardly any scholarship on science journalism or media-based science communication. After an extensive literature search, Dutt and Garg could find only two studies that looked at science and technology content in Indian news media (Dutt and Garg, 2000). This relative lack of scholarly interest is not reflected in the actual presence of science in the mainstream media; even a cursory examination of newspapers and magazines will yield a fair amount of science news, much of it relating to the fields of health, environment, agriculture and, overwhelmingly, technology (computers, consumer electronics, and automobiles). Manoj Patairiya, a scientist at the NCSTC and honorary secretary of the Indian Science Writers' Association, pointed to the demise of magazines like *Science Today* and *Bulletin of Sciences* as indicating there was not much interest in science communication (Patairiya, 2002). However, Dinesh Sharma, editor of the *India Science Wire*, a daily science news service funded by Vigyan Prasar (an autonomous organisation under the Department of Science and Technology, whose name translates as 'Science Dissemination'), says that the enthusiastic uptake of stories from *India Science Wire* by mainstream media suggests a definite interest in such content.[4] Also, the continuing (though limited) popularity of an environment-focused magazine like *Down To Earth*, published by the Centre for Science and Environment, New Delhi, suggests that there is an audience for such material, however niche.

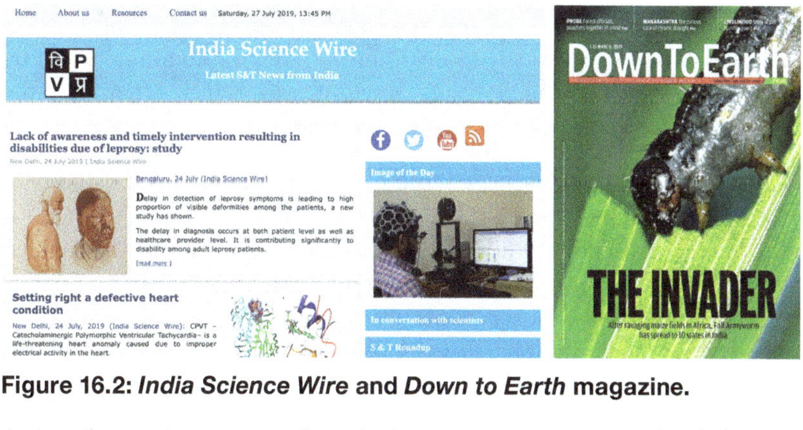

Figure 16.2: *India Science Wire* and *Down to Earth* magazine.

As in other mainstream media vehicles, science coverage in the daily press may appear either in the main body of the newspaper or as part of a news bulletin, or in special sections or programs that focus on science. Until the late 1990s, most large English and regional language dailies had multiple-page sections devoted to science, but by the early 2000s these sections had been incorporated into the main paper and reduced to one or two pages.

4 Personal communication, 8 January 2019.

The Hindu, a major national daily published from Chennai, had for many years a 6–8-page weekly supplement, but this was reduced to a one-page weekly section in about 2005, and subsequently (in 2017) increased to a two-page section in the Sunday magazine. Dinesh Sharma observes that despite the disappearance of exclusive science sections, the daily newspaper has more science and technology news today than in the past. 'It's no longer ghettoised in a separate section,' he notes.[5] Most media organisations have designated journalists covering science—again, *The Hindu* has for many years had reporters specialising in science, health, environment and technology.

As Patairiya (2002) observes, the popular science magazine is practically absent from the Indian media scene. *Science Today*, a publication from the *Times of India* group, was perhaps the only magazine that combined an overt mission of science popularisation with a degree of critical examination of the process and culture of science. The magazine had a line of distinguished public scientists at its helm and a committed though small readership; but it was clearly not a money spinner and was laid to rest in 1992. The National Institute of Science Communication and Information Resources (NISCAIR, under the CSIR) now brings out three lay publications—*Science Reporter* (English), *Vigyan Pragati* (Hindi: Progress in Science), and *Science-ki-Duniya* (Urdu: The World of Science)—that position themselves as serving the goal of 'science education'. The Bangalore-based Indian Academy of Sciences publishes two journals widely read within the scientific community and by science aficionados: *Resonance* and *Current Science*, which often form the basis for science stories in the mainstream media.

With the growth of online media, science writing has gained something of a resurgence. Online news sites like *thewire.in* and *scroll.in* have sections for science and technology, while *indiaspend.com*, a data journalism website, produces critical, long-form pieces on science, the environment and public health. Some (*thewire.in*) have trained the science journalists who manage the section. The wire service *Press Trust of India* is a leader in science reporting, and its first designated science editor K. S. Jayaraman mentored many young journalists who went on to write early critical and investigative reports on science.

While there is fair amount of coverage of science in the mainstream media as news, opinion or features, the scope and quality of this reporting is another matter. Much of what might be classified as 'science writing' is generated from press reports and institutional press releases, mostly from public

5 Personal communication, 8 January 2019.

laboratories. In recent years, public concerns around areas such as genetically modified crops and multi-drug-resistant pathogens has led to a journalistic interest in such stories, fuelled partly by activism. The same may be said for environmental and health areas. Dutt and Garg in their 1996 analysis of 27 major metropolitan dailies found that nuclear science, defence technology and space research were the most common subjects (Dutt and Garg, 2000). A special issue of the Indian edition of the *Global Media Journal*, published by Calcutta University, published a set of papers examining science coverage in the media in India (University of Calcutta, 2013). While most were broad commentaries on the possible impact and quality of science and technology communication, there were a few empirical studies based on small data-sets. A content analysis of 10 major Hindi and English-language dailies found: a) more science news in the English newspapers; and b) that health and environment were the most covered topics (Kumar, 2013).

Another aspect of science coverage is its framing. An analysis of the English-language press in India by Samuel Billet (2010) found that environmental journalists situate 'climate change' as a socioeconomic and political issue, but that coverage tends to place the issue either within a nationalistic frame or along global north–south lines, thus losing much of the nuance that the topic requires.

This 'nationalistic' frame was certainly a feature of science reporting in the first five decades after Indian independence, but one may argue that critical reporting has increased in the last two decades, and multiple viewpoints are available to news consumers who peruse more than one news source. The rise of data based journalism has given science writers another tool to examine scientific claims, and there is a greater pressure on scientists to better communicate the results of their research with an eye to both the public interest and the demand for accountability from funders. But like elsewhere in the world, Indian academics and scientists are focused almost exclusively on scholarly publishing and are not very interested in popularising their work through the mainstream media. There are exceptions, and occasional pieces by scientists do appear in magazines and newspapers such as *The Hindu*'s Sunday edition, which routinely carries a policy-related or perspective piece by a senior scientist.

2.4. Popular science movements

The 1950s were a decade of euphoria and unbounded optimism for the nascent nation. Its flavour and distinctness from the colonial past rose from the extraordinary energy that accompanied the birth of the new nation. But the trauma experienced by partition left deep scars in its psyche, and it was against this violent background that science became an expression of desire for the newly independent nation-state. Science was called upon to provide the moral, material and emotional basis of the largest post-colonial democratic experiment in the world. The bestowing of modern apparatus—in the form of the bureaucratic state —onto the nation required a public/policy imagination that rested on universal algorithms rather than culturally specific formulations. The technocratic elite had to circumvent the historical and cultural logic of the anti-colonial struggle and cast the theme of economic development and growth as a foregrounding principle of building a new India. It was a decade when modern institutional India emerged, laying the foundations for economic, political, industrial, scientific and cultural institutions to be built. It is in this context that efforts to produce an enlightened public was considered as a noble enterprise by the ruling political and cultural elite. In order to fashion such a cultural horizon, a variety of interventions were undertaken by both civil society and the state.

The establishment of the Kerala Sastra Sahitya Parishat (KSSP: Kerala Forum for Science Literature), in the southern tip of the country, was a classic example: scientific literacy was part of a cultural literacy, to be disseminated to an allegedly backward, scientifically distanced population. In one sense, science communication and science literacy alluded to rationalities of a social, political and economic nature. KSSP took up issues that informed the transformation of the cultural and social habitus that had seemingly obstructed the production of modern citizen subjects. The spirit of KSSP spread to neighbouring regions and science activists took up the evangelical role of spreading a secular, scientific form of rationality to engage with an everyday life still mired in deep inequalities. This led to the formation of People's Science Movement (PSM) later re-designated as All India Peoples Science Network (AIPSN).

Multiple scholars have commented on the KSSP, which was primarily interested in communicating scientific information to the masses on wide-ranging topics from environment to gender sensitisation (see, for example, Parameswaran, 1979; Devika, 2005). Others have pointed at the close linkages with leftist parties in Kerala, where the central idea of KSSP was to transform society into a less hierarchical and more egalitarian one (Kannan, 1990). In this kind of thinking, science communication has the responsibility of addressing diverse social inequities. It can be said that KSSP, along with other left-leaning policies, has effectively intervened in public life. The federal state of Kerala has more positive indicators in the realm of primary health, education and sex ratio (i.e. birth ratio of male children to female) than many other states in the country, but there still exist deep gender, caste, class and religious fissures.

Other important organisations related to AIPSN included Delhi Science Forum, Eklavya (Madhya Pradesh), Madhya Pradesh Vigyan Sabha [Madhya Pradesh Science Forum], Pondicherry Science Forum (PSF), Lok Vigyan Sangathana [People's Science Organisation] (Maharashtra), Tamil Nadu Science Forum (TNSF), and Karnataka Rajya Vignana Parishath (KRVP) [Karnataka State Science Council]. AIPSN expanded its base rapidly by chalking out and executing a national-level action plan known as Jan Vigyan Jatha [Peoples' Science Campaign] (Raza et al., 2008). *Jatha* refers to a journey by foot, and scientists and science activists organised long *jathas* in different parts of the country to give talks, demonstrations, stage plays and utilise other indigenous media to convey the spirit and benefit of scientific thought including its technological prowess for the benefit of the masses.

3. Science communication today: An appraisal

The spirit of the Nehruvian state's stance towards science contains within it civilisational and historical dimensions. In that sense, the science policy documents should be seen as a particular form of historical inscription and a distinct variety of civilisational aspiration. It appears that ambitious gestures of such kind take a long time to influence a religiously charged society. Some policies towards building of scientific institutions premised on a secular ethos still seem to be working on the ground, but with little formal power. The aura generated around big dams, commercial agriculture, capital intensive industries, educational institutions like the Indian Institutes of Technology, space sciences and large national laboratories was to dwarf the solidarities built around charismatic, religious and mystical identities, and replace them with industrial productive capacities. The coalescing of science with the building of human capital required for a modern industrial country together with the erasure of loyalties and solidarities attached to primordial identities like religion, caste and region becomes the agenda and the reason for the post-independent nation-state. It is almost as though 'science' becomes another name for the Indian state (Nandy, 1988). While the positive narrative of the fruit of modern science and technology continues to hold sway in policy documents, political parties (especially of those of a nationalist hue) tend to appropriate ancient Indian mythologies and scriptures to posit a continuum of Indian scientific achievements for several millennia.[6] This tendency does actual disservice to the history of ancient Indian contributions in mathematics, astronomy, medicine and architecture.

6 Prabir Purkayastha, noted science activist, has written about this appropriation of mythologies by the government in power (Purkayastha, 2019).

In the rarest of rare moves, the phrase 'to develop the scientific temper, humanism and a spirit of enquiry' was added to the Constitution of India (largely inspired by Nehruvian ideas) in 1976. This provided the modern cultural context for the Indian State to establish institutions to achieve this ambition and became a crucial focus for NGOs to carry out and realise the cultural and civilisational ambition of the nationalist leaders at a post-independence historical juncture. A number of centrally funded institutions were formed, including the National Institute of Science Communication (NISCOM), which started publishing several popular science journals such as the Hindi journal *Vigyan Pragati* in 1952. The National Council of Science and Technology Communication was established in 1982 under Narendra Sahegal and became involved in both science communication and the popularisation of science. *Vigyan Prasar* was established in 1989 as part of the Department of Science and Technology and its mandate included communicating science from laboratories to the field.

The emphasis of the Indian state to build a self-reliant economy around agriculture, industry and commerce presupposed scientific knowledge and scientific practices amongst its population. It was precisely to address these issues that science communication was foregrounded in areas including agricultural extension work, health delivery systems and modern school education. There were two aspects to federally sponsored science communication institutions and programs. University departments and allied educational institutions were sponsored by organisations like NCSTC to train science students to write and make programs for newspapers, radio and television. The second aspect relates to popularisation of science through a variety of indigenous cultural forms like folk theatre and music, *jathas*, exhibitions and storytelling traditions. This was primarily done through liaisons with NGOs and other civil society organisations. Major surveys were carried out by institutions like National Institute of Science, Technology and Development Studies (NISTADS) to study the efficacy of public understanding of science among the general population.

While a number of institutions were created by the central government after independence, and especially after the Scientific Policy Resolution of 1958, 'the Resolution did not succeed in generating "the scientific cultural revolution", so badly needed by the Indian society' (Sharma, 1976, p. 1969). Unscientific religious mores and practices continue to play a major role in civil society, kept alive by self-styled godmen and godwomen, and attract significantly larger audiences than science communication networks. The murder of Narendra Dabholkar in 2013, a rationalist and social activist who espoused the cause of scientific temper by establishing a committee in Maharashtra to fight superstition, is possibly the most lethal instance of the clash of rational and irrational beliefs.

The major issue characterising much of the science communication activity carried out by institutions is that science is too often framed in a nationalistic and often jingoistic way, with government-led science communication efforts aiming to create a sense of pride in the products of Indian scientific activity rather than to build critical appreciation for the process of science (which arguably is the essence of scientific temper). This is seen not only in the framing of science communication education but also in popular science initiatives, including museums and science fairs. While the economic potentials of science have been rhetorically appropriated, the power of science to build an equitable society, non-coercive solidarities and humanise relations in a deeply divided society has yet to be realised. The Indian state continues to battle a plethora of socioeconomic challenges, many of which do not have science and technology-based solutions.

However, one cannot deny the power of effective science communication in a young post-colonial nation, young both in terms of years of existence and the age of the majority of the population, and its ability to create aspirational value for people. Particularly illuminating in this regard was the *India Science Report*, a first-of-its-kind nationally conducted survey with a sample size of over 100,000 people (Shukla, 2005). It was published by the National Council of Applied Economic Research (NCAER) in 2005 and discussed public awareness of science and technology as well as participation of the population in science education and in scientific jobs. Two issues stand out from the report: the first is that, even with a relatively low percentage of literate people (about 64 per cent according to the 2001 census), interest in issues of science and technology and awareness of basic science is very high. A second, more telling figure is the 60 per cent of middle school students who want to pursue a career in sciences, technology, engineering and medicine. This percentage is sustained through all school years with about 57 per cent in high school recording a similar response. About 40 per cent of all middle school students said that they wanted to become 'an engineer or a doctor' (Shukla, 2005, p. 16), thereby indicating the perceived importance of science in the society. However, the number of students aspiring for a STEM career drops significantly in the rural areas. This is where the question of reaping demographic dividends of India's vast young population becomes extremely challenging, and Indian science communicators need to address the urban–rural divide on a priority basis, while also bringing a more critical and interrogative lens to the process and outcomes of science in general, and Indian science in particular.

References

Abraham, I. (2006). The contradictory spaces of postcolonial techno-science. *Economic and Political Weekly*, *41*(3), 210–17.

Arnold, D. (2013). Nehruvian science and postcolonial India. *Isis, 104*(2), 360–70. doi.org/10.1086/670954.

Bharthur, S. P. (2017). Journalism education in India: Quest for professionalism or incremental responses. *Journalism & Mass Communication Educator, 72*(3), 285–96. doi.org/10.1177/1077695817719136.

Billett, S. (2010). Dividing climate change: global warming in the Indian news media. *Climatic Change 99*(1–2), 1–16. doi.org/10.1007/s10584-009-9605-3.

Devika, J. (2005). Modernity with democracy: Gender and governance in the people's planning campaign, Keralam. Working Paper 368, Centre for Development Studies.

Dutt, B. and Garg, K. C. (2000). An overview of science and technology coverage in Indian English-language dailies. *Public Understanding of Science 9*(2), 123–40. doi.org/10.1088/0963-6625/9/2/303.

Gandhi, I. (1976, 31 January). Indira Gandhi strongly emphasizes need for science to have a rural bias in India. *India Today*. Retrieved from www.indiatoday.in/magazine/cover-story/story/19760131-indira-gandhi-strongly-emphasizes-need-for-science-to-have-a-rural-bias-in-india-819563-2015-03-27.

Ghose, S. (1992). People's participation in science museums. In P. Boylan (ed.), *Museums 2000: Politics, People, Professionals and Profit* (pp. 81–96). London: Routledge.

Government of India. (1958). *Scientific Policy Resolution*. Government of India, New Delhi.

Government of India. (1993). Technology Policy Statement – 1983. *Current Science, 65*(5), 374–78.

Government of India. (2003). Science and Technology Policy 2003. *Science, Technology and Society*, *8*(1), 101–12. doi.org/10.1177/097172180300800105.

Government of India. (2013). *Science, Technology and Innovation Policy*. Ministry of Science and Technology, New Delhi.

Kannan, K. P. (1990). Secularism and people's science movement in India. *Economic and Political Weekly*. *25*(6), 311–13.

Kumar, M. (2013). Comparison of science coverage in Hindi and English newspapers of India: a content analysis approach. *Global Media Journal* (Indian edition), *4*(1). Retrieved from pdfs.semanticscholar.org/db1b/5f66b5299cf20ddc0f8ba06ebd4eafa78c62.pdf.

Ministry of Culture. (n.d.). *Activity Report 2013–14*. National Council of Science Museums, Kolkata. Retrieved from drive.google.com/file/d/0B2sxx9FjouliNndNbjMzMkVJV1czdElRUlZMQS0wcUV4VV9n/view.

Ministry of Culture. (2016). *Outcome budget 2016/17*. Government of India, New Delhi.

Ministry of Law and Justice. (2019). *The Constitution of India* (as on 1st April 2019). Government of India, New Delhi. Retrieved from legislative.gov.in/sites/default/files/COI-updated.pdf.

Mukherjee, I. (2003). *National Council of Science Museums: 25 years of service to the nation*. Special Issue, NCSM, Kolkata.

Nandy, A. (1988). *Science, Hegemony and Violence: A Requiem for Modernity*. Tokyo: United Nations University.

National Council of Educational Research and Training (NCERT). (2005). *National Curriculum Framework*. Retrieved from www.ncert.nic.in/rightside/links/pdf/framework/english/nf2005.pdf.

Nehru, J. (1946). *The Discovery of India*. New Delhi: Oxford University Press.

Padma, T. V. (2015). India: the fight to become a science superpower. *Nature*, *521*(7551), 144–47. doi.org/10.1038/521144a.

Parameswaran, M. P. (1979). Significance of Silent Valley. *Economic and Political Weekly*, *14*(7), 117–19.

Patairiya, M. (2002, 20 March). Science communication in India: perspectives and challenges. *SciDev.Net*. Retrieved from www.scidev.net/global/communication/opinion/science-communication-in-india-perspectives-and-c.html.

Patairiya, M. (2002). Emerging scenario of science and technology communication. *Indian Journal of Science Communication*, *1*(1). Retrieved from www.iscos.org/ijsc/Vol1-No1/rp2.htm.

Phalkey, J. (2013). Introduction. *Isis*, *104*(2), 330–36. doi.org/10.1086/670950.

Planning Commission. (1974). *5th Five Year Plan 1974-79*. Government of India. Retreived from niti.gov.in/planningcommission.gov.in/docs/plans/planrel/fiveyr/5th/5th_fyp.pdf.

Purkayastha, P. (2019, 10 January). Science will survive the Modi regime, will science institutions? *Newsclick*. Retrieved from www.newsclick.in/node/14961?fbclid=IwAR1xXsOK-aBFZvf7Lps5Apng3_O50yxOL-y56sTNMGfCVTdhIHw8HG9MHw8.

Raza, G., Singh, S., Kumar V. S. and Nayak, D. (2008). *Public Understanding of Science in India*. New Delhi: National Institute of Science, Technology and Development Studies. Retrieved from www.nistads.res.in/indiasnt2008/t3infrastructure/t3inf14.htm.

Raza, G., Singh, B. P., Venkateswaran, T. V., Gopichandran, R. and Rautela, G. S. (2014). Scientific temper: an arena of contestation in a globalised world. Paper presented at the 13th International Public Communication of Science and Technology, Conference (PCST), Salvador, Brazil. Retrieved from pcst.co/archive/pdf/Raza_et_al_PCST2014.pdf.

Roy, S. (2007). *Beyond Belief: India and the Politics of Postcolonial Nationalism*. Durham: Duke University Press.

Sharma, D. (1976). Growth and failures of India's science policy. *Economic and Political Weekly*, *11*(51), 1969–71.

Shukla, R. (2005). *India science report: Science education, human resources and public attitudes towards science and technology*. Delhi: National Council of Applied Economic Research.

Sur, A. (2011). *Dispersed Radiance: Caste, Gender and Modern Science in India*. New Delhi: Navayana.

Sur, A. (2012). *Women, Caste and Modern Science in India*. Women's and Gender Studies Program, MIT. Retrieved from www.youtube.com/watch?v=LZZVsJ9_Abg.

Trench, Brian. (2012). Vital and vulnerable: science communication as a university subject. In B. Schiele, M. Claessens and S. Shi (eds), *Science Communication in the World*. Springer. doi.org/10.1007/978-94-007-4279-6.

Venugopal, B. (1995). Family groups in museums: an Indian experience. In E. Hooper-Greenhill (ed.), *Museum, Media, Message* (pp. 276–80). London and New York: Routledge.

Visvanathan, S. (1997). *A Carnival for Science*. New Delhi: Oxford University Press.

Timeline

Event	Name	Date	Comments
First interactive science centre established.	Birla Industrial and Technological Museum, Kolkata	1959	1978: National Council of Science Museums established to oversee the administration of science centres in India
An association of science writers or journalists or communicators established.	Kerala Sastra Sahitya Parishat (KSSP) started as a forum of science writers	1962	1972: KSSP become a People's Science Movement. Their motto is 'Science for Social Revolution' 1985: Indian Science Writers' Association (ISWA) established 1988: All India People's Science Network formed
First university courses to train science communicators.	Dept. of Science Communication and Journalism, Madurai Kamaraj University	1988	
National government program to support science communication established.	National Institute of Science Communication (NISCOM) was set up to publish scientific journals and periodicals	1951	1952: Indian National Scientific Documentation Centre (INSDOC) set up with UNESCO help 2002: NISCOM and INSDOC merged to form National Institute of Science Communication and Information Resources (NISCAIR)
National Science Week founded.	National Science Week	1987	Held around National Science Day, on 28 February
A journal completely or substantially devoted to science communication established	*Vigyan Pragati* [Progress in Science] a Hindi popular science journal published by NISCOM	1952	2002: First issue of *Indian Journal of Science Communication* 2013: The quarterly *Journal of Scientific Temper* published by NISCAIR
First significant radio programs on science.	*Maanav ka Vikaas* [Human Evolution]	Early 1990s	144-part radio series
First significant TV programs on science.	*Bharat ki Chaap* [India Key Impressions]	1989	Broadcast on Doordarshan (now DD India), an Indian English news and current affairs channel
First awards for scientists or journalists or others for science communication.	National Science Popularisation Awards	1987	Instituted by National Council of Science and Technology Communication
Date hosted a PCST conference.	11th PCST Conference, New Delhi	December 2010	

Event	Name	Date	Comments
Other significant events.	Scientific Policy Resolution (SPR) to 'foster, promote and sustain' the 'cultivation of science and scientific research in all its aspects'	1958	1983: Technology Policy Statement 1976: Inclusion of 'scientific temper' in the Indian Constitution 2003: Declaration of National Science and Technology Policy 2010: Declaration of Decade of Innovation 2013: Declaration of Science, Technology and Innovation Policy
	India Science Wire, a daily news and features service in English and Hindi	2017	Organised by from Vigyan Prasar, a unit of the Department of Science and Technology, New Delhi

Contributors

Dr Anwesha Chakraborty is a postdoctoral fellow at the Department of Political and Social Sciences, University of Bologna, and previously worked as a postdoctoral researcher at Chalmers University of Technology, Gothenburg.

Professor Usha Raman is at the Department of Communication, University of Hyderabad. Previously, she headed the communications department at L. V. Prasad Eye Institute.

Professor Poojraj Thirumal is a senior faculty member of the Department of Communication, University of Hyderabad, Telangana, India.

17

IRAN

From the ancient world of Elam to modern science communication

Seyede Zahra Ojagh and Zarrin Zardar

1. Introduction

As the inheritor of Elam's civilisation (2700–539 BCE), Iran has a long history in science. It is located in the Middle East, and has 81 million inhabitants. Gondishapur, the intellectual centre of the Sassanid Empire, dates back to the late third century CE, was a remarkable achievement of the Iranian civilisation. Philosophy, astronomy, pharmacy and medicine were some of the scientific fields studied in Gondishapur. It was an international science centre that brought together scientists from around the world. The Greek, Indian and Iranian scholars communicated in Gondishapur in a way that made it a big multilingual university. Islam first came to Iran between 637–651 CE, and subsequently it experienced further brilliant scientific development for more than five centuries. Gondishapur, along with other scientific centres such as Baghdad and Al-azahar, were centres for knowledge production, communication and education. After the Moguls' invasion about 1219 CE, the Islamic civilisation in Iran started to decline and this continued until the country faced modern Western science in the 19th century.

The acquaintance of Iranians with modern science dates back to the Qajar era in the early 19th century when the Iranian army was defeated by the modern weapons of the Russians. The government responded to this failure and its subsequent misery by dispatching students overseas to study in 1811. Then followed the development of the printing press in 1837 and the establishment of the first Iranian modern high school, Dar-ul Fonoun in 1852, which indicated the initiation of the first period of modernisation in Iran (Mohsenian-Rad, 2013; Abrahamian, 2008). Subsequently, translated texts and the graduates of the Western universities paved the way for the progress of modern science in Iran. Among these people, Abdul Rahim Tabrisi, known as Talebov, can be considered as the first writer addressing public knowledge and the first science communicator in Iran. In 1892, influenced by the book *Emile* by Jean-Jacques Rousseau, he wrote the book *Ahmed*, explaining modern science in plain language to the general public.

The advancement of modern science was accelerated by the establishment of the first Iranian university in 1934, and soon diverse fields of science and technology were developed in Iran. Today, over 4 million students are studying in 2,569 academic centres in Iran (Iranian Higher Education Research Institution, 2018).

The first signs of attention to science communication can be found in a few works of journalists who wrote about the advancements of the Western community for the literate urban audience in Iran in the second half of the 19th century (Ojagh, 2012). However, in planning for the nation's scientific development, policymakers rarely considered a role for the public and the participation of citizens in science.

Overall, more science communication activities in Iran were carried out after the Islamic Revolution. At this point, the government's approach to modern science changed and applied science took priority over the humanities, critical and basic science. Islamisation and the application of science in Iran became the government's agenda. Since 1979, the government's policy on science has been focused on education and on encouraging research rather than the public communication of science. Science communication was first recognised in the government policy in 2010 (Supreme Council of the Cultural Revolution, 2010). Scientific societies and individual scientists led the way, setting up foundations, museums and activities to encourage public interest in science. Interest in science communication has grown enormously in recent years thanks to widespread access to international events through the internet, but it is still a new and developing field.

The government thought that the development of science communication was indispensable for both the scientific community and the public because science and technology can help the development of the society only when they are turned to address society's concerns. These subjects need to be discussed at all levels among the stakeholders to reduce the costs and risks of the development of science and technology. Society's needs and expectations should be considered when policymakers are determining the priorities of the development of science and technology, as such an environment will let new scientific and technological issues enter the public sphere and become part of social debates.

There have been efforts from both state and civil institutions to link science and society. However, these experiences have never been systematically analysed and documented, which is why there is not much knowledge of the status of science and technology communication in Iran.

In this chapter, we have attempted to present a relatively comprehensive summary of modern science and technology communication in Iran by collecting evidence from a variety of sources, including academic papers, books, interviews with experts and stakeholders as well as online databases. For this purpose, the experience of science communication in Iran has been considered from three perspectives: policymaking, practical experiences of science communication, and science communication at universities. This exploration is followed by a discussion of the status of science communication in Iran.

2. Science and technology policymaking in Iran

Science and technology policymaking in Iran can be discerned from two categories of documents: a) medium-term five-year development plans in which science and technology have been considered as one of the sections or sub-sections; and b) upstream documents that include the three major legislative documents in this area and outline the country's major strategies.

The first basis for science and technology policymaking can be seen in Iran's medium-term plans, with the starting point dating back to the country's first development plan in 1948. Science and technology were implicitly considered by policymakers in each of the six development plans in the years 1948–79 leading up to the Revolution. All plans had a chapter on education but none ever mentioned communicating science to the public. The country wanted to develop infrastructure and industry and needed educated and skilled people. Thus, many programs were established to increase functional literacy.

Following the eight-year war between Iran and Iraq (1980–88), science and technology were considered in limited and dispersed legislative articles or notes in the first and second development plans in the post-Revolution period. In terms of their content, the plans focused mostly on the development of education (State Management and Planning Organisation, 1990; State Management and Planning Organisation, 1994). However, the turning point of attention to science and technology in policymaking was in the third post-Revolution development plan, where the development of research, innovation and technology was one of its main goals (State Management and Planning Organisation, 2000). This change in policymaking seems to be due to a change of personnel, a rotation of the political elite and the coming to power of the so-called reformist government. Eight years after the war, under stable economic conditions, knowledge-based development was now on the agenda of the government. It demonstrated its intentions by increasing the number of research institutions and developing new academic disciplines and knowledge enterprises. The aim was to acquire leadership status in political and economic competitions in the Middle East.

Following this trend, the fourth development plan (2005–09) focused on increasing scientific production and the promotion of Iran's ranking in knowledge production, and making progress towards a knowledge-based economy. It also emphasised the quantitative and qualitative development of education, especially academic education (State Management and Planning Organisation, 2004). Then the conservatives returned to power and dissolved the Management and Planning Organisation. While the fifth development plan (2011–15) pursued the same goals as the fourth, its policies were too ambiguous to be practicable (Vice President for Strategic Planning and Oversight, 2010). For instance, it introduced the evolution of the humanities with regard to the four criteria of Islamisation, localisation, modernisation and efficiency, but it did not specify how this was to be carried out or the mechanisms of implementation.

Then the Rohani administration took office and the Management and Planning Organisation was revived. The sixth development plan (2016–21) focused on specific scientific and technological priorities and sought to return to the path of the fourth plan. The aims were now the promotion of the scientific ranking of Iran by the development of research, the production of goods and services based on science and technology, and the enhancement of international scientific-research interactions (State Management and Planning Organisation, 2017). In general, the approach of the development plans to science over 32 years has shifted from a focus on the development of general education and higher education, to a new prioritisation of research, the creation of science and its application in the knowledge-based economy. These policies encountered serious problems on the path to implementation, an issue that greatly affected the quality of knowledge-based development. Additionally, although increased international scientific interactions and joint scientific and research activities with other countries of the world were mentioned in the plans, they have always been marginalised.

The government also released larger-scale and more significant documents to explain its plans. The use of large-scale policymaking in Iran is more recent than the medium-term planning. The oldest large-scale policymaking document in science and technology is Expediency Discernment Council's

The future outlook of the Islamic Republic of Iran in the horizon of the next two decades, approved in 2003. This defines the 20-year horizon of the country's development.

The main goal of this vision document in science and technology was to achieve leadership for the economic, scientific and technological policies in southwest Asia, emphasising software development and science production (Expediency Discernment Council, 2003). This document formed the basis of the government's policy in 2003.

However, the most comprehensive reference for science and technology policymaking is the Supreme Council of the Cultural Revolution's *Iran's scientific comprehensive roadmap*, approved in 2010. This map has specified the major strategies of the country's science policy and the priorities of Iran's science and technology in different areas of science and technology. Some of these priorities include aerospace, nuclear technology, biotechnology, nanotechnology and information technology. 'The promotion and simplification of scientific concepts' and the use of the media as a tool for 'transforming science into one of the main discourses of the society' were considered for the first time in this document. The media are proposed as a platform for knowledge transfer, to make up for the lack of public communication of science and technology by government and research organisations.

The latest document formulated by the Office of the Supreme Leader of the Islamic Revolution, in consultation with the Expediency Discernment Council, is *General Policies of Science and Technology*, which was announced to the heads of three judicial, executive and legislative powers in 2014. This brief and general document illustrates the major priorities of the government in the field of science and technology and is based on the idea of 'the Islamic Iranian progress model'.

The common feature of all the development documents and plans in Iran is that the government has always been the main executor (Godarzi et al., 2015): the part played by the private sector, industry and civil institutions in developing science and technology is negligible. These policies focus mainly on the development of tangible achievements of science and technology rather than providing the cultural prerequisites for development. The application of these policies has increasingly accelerated the publication of scientific papers in Iran, resulting in

> the growth of scientific movements in the country, regardless of the needs of society, and sometimes as an independent entity disregarding the requirements of its peripheral environment. This has caused the science to fail to establish communication with decision-making, service and production systems. (Ghaneirad, 2003)

The relationship between science and society in Iran has historically manifested four epistemological approaches: scientism, ideologisation and ritualisation of science, pseudo-positivist phobia towards the ideologisation of science and technology, and post-positivism (Farasatkhah, 2008). The scientistic approach believed in a stable place for science and excluded any influence of social and political conditions in scientific fields, and was dominant until the Islamic revolution. The ideological and ritualistic approach to science paid attention to the connection of social, political and cultural elements with science, showing its effects in the following decades in the projects of Islamicisation and localisation of science (ibid.). This approach was dominant in the post-Revolution period. The reaction to the second approach (ideologisation and ritualisation of science) was that any discussion of local and indigenous knowledge was susceptible to suspicion and distrust (ibid.) within Iranian scientific communities. The post-positivist approach considers science as a sociocultural process linked with the interests of groups, loyalties and social and cultural factors within and outside the scientific community (ibid.). This approach was developed recently as a result of the globalisation.

Of these four approaches, the ideological and ritualistic approach to science has had a strong presence in the political and ideological arena. Specifically, the Leader of the Revolution has repeatedly criticised the extraction of science and technology, and in particular scientific and technological culture, from the Western world (Hasirchi and Niawand, 2011). According to him, the attention to development must be accompanied by the attention to spirituality (Supreme Leader, 2008) and the ultimate goal of the development of science and technology should be human salvation (Supreme Leader, 2010). Iran's scientific comprehensive road map has also been drafted with an emphasis on the Islamicisation of science and technology. In the *General Policies of Science and Technology* document (Supreme Leader, 2014), scientific and spiritual growth are always on the agenda of the policymakers as well as science issues. With the suggestion of the Leader of the Revolution, the Western concept of 'development' has been replaced by the term 'progress', and the concept of 'the Islamic-Iranian progress model' has been formulated. This model incorporates the perceptual foundations and the Islamic worldview. It puts forward the aspirations and desires to achieve progress based on the Islamic-Iranian identity. However, it has not provided a robust theoretical model for the realisation of these aspirations (Radmanesh and Taghavi, 2015).

Its effect has been an unbalanced attention to scientific fields, with some considered as strategic sciences (including tech-sciences such as ICT and nanotechnology), assuming that their content cannot be influenced by ideology or religion. The biggest influences of Islamicisation can be seen in humanities, which are more expected to promote Islamic attitudes and beliefs. In this situation, science communication can be considered closer to the strategic sciences.

A review of the process of science and technology policymaking in Iran shows that three groups of actors play a role with different levels of influence:

a. At the highest level, government and religious leaders (including the most influential legal entities or policymaking organisations) are responsible for formulating the upstream documents.
b. The intermediate level (senior officers in government departments) contributes to the formulation of the upstream documents and is responsible for formulating downstream documents, such as mid-term development plans or executive by-laws resulting from upstream documents.
c. Target groups (researchers, journalists, institutions) that are directly affected by the policies but rarely have a role in the policymaking process. However, they influence the implementation, realisation and formation of policies.

The institutions that carry out the activities of science communication and science promotion are not part of the process of science and technology policymaking, except for brief indications in the comprehensive scientific map of the country (Table 17.1).

Table 17.1: Key actors of science and technology policymaking in Iran

Highest level policymakers
Supreme Council of the Cultural Revolution; Expediency Council; Supreme Leader; Supreme Council of Science, Research and Technology
Medium level policymakers
Science and Technology Deputy of Presidency; Science, Research and Technology Ministry; State Management and Planning Organisation; Strategic Research Centre; Ministry of Health, Treatment and Medical Training; Ministry of Education; Ministry of Commerce, Mining and Industry; Ministry of Agriculture
Target groups
Universities, research institutes, science and technology parks, state organisations, scientific associations, Islamic Republic of Iran Broadcasting, private science promotion institutes, science clubs, knowledge enterprises, centres of excellence, media

3. The practice of science communication in Iran

Despite the lack of specific government policies for the development of science communication in Iran, science enthusiasts and experts have attempted to communicate with the community and have encouraged governmental institutions to build a relationship with the public. That is why the experience of science communication in Iran can be classified into the civil and governmental areas.

3.1. Civil institutions, public sphere and science and technology communication

The scientific community became interested sooner than any other institution in attracting the attention of the public to science. Specifically, the first scientific society of Iran, Physics and Chemistry Society of Iran, was formed in 1931 by a small group of physicists. The evolution of this society played an important role in public science communication.

Over time, promoting science and making the public aware of scientific achievements became the priority of the scientific community and science lovers. These pioneers established centres and foundations whose main mission was the establishment of the relationship between science and society. The most important examples are:

3.1.1. Zirakzadeh Science Foundation

Zirakzadeh Science Foundation was founded as a non-government and non-profit institution in 1993 in order to promote science and technology. Its founder, Ahmad Zirakzadeh, sold his house before his death to provide the capital needed for this project. He believed that purely theoretical education was the most important weakness in the Iranian educational system (Dadar, 2010). In cooperation with municipalities, charitable institutions and people, as well as domestic and foreign experts, the foundation established exhibitions and halls of science and technology in Tehran and other cities. It also participated in setting up scientific centres in order to provide the opportunity for practical experiments and interactive learning of science for children and adolescents.

3.1.2. The Home of Mathematics

The first 'home of mathematics' in Iran was indebted to more than 20 years' experience of mathematics teachers' meetings on Monday afternoons. It was established in the form of a non-government institution in Isfahan in 1998.

Its purpose was to promote mathematical knowledge and to generalise mathematical science. This venture was quickly expanded in other cities of the country (Scientific Associations' Reference, 2016).

3.1.3. The City of Mathematics

One of the Iranian innovations in promoting mathematics is the City of Mathematics. It was established in 1998 by Majid Mirzavaziri, professor at Ferdowsi University of Mashhad, to conceptualise mathematics with simple and attractive games for adolescents. The Mathballmatch (a game like football) and the Math Clinic (the detection of behavioural disorders through math games) are initiatives at the City of Mathematics. Other groups involved in similar activities include the Knowledge and Art Foundation, Amateur Branch of Astronomical Society of Iran, and the Iranian Association for Popularisation of Science and Technology Training Park.

3.2. Science and technology activities supported by government and other groups

Encouraging public interest and support for scientific policies is regarded by the government as one of the most effective measures to encourage scientific and technological development. That is why the prominent role of the government in many areas of science communication can be seen in Iran. Its most important activities are explained below (as well as those of other organisations).

3.2.1. Academies

Given that the non-Persian language of science was considered as an obstacle to its expansion into society, the First Academy was established in 1934. Its main task was to define the Persian equivalents for non-Persian scientific and technical terms. It helped people understand scientific texts. Today, the number of academies has increased in Iran, creating a space for developing new ideas for elites. However, the academies offer suggestions for major scientific issues that are highly complex as they are more in touch with elites than common people (Tayyebi, 2000; Roustaei, 2006).

3.2.2. Institute for the Intellectual Development of Children and Young Adults

The Institute for the Intellectual Development of Children and Young Adults was established in 1965 to develop the abilities of children and adolescents. It focuses on three activities: libraries, publishing organisations, and a cinema centre. With the establishment of mobile and fixed libraries, the institute tried to promote reading among children and adolescents in the early years following its founding. Novelty, attractiveness, simplicity and raising awareness were the main features of books written at the institute as they tried to persuade audiences to look more closely at world issues and get familiar with scientifically analysing them (Sharifi, 1998). The circulation of these books was high: 30–40,000 in 1969 and the minimum circulation was 10,000 volumes (ibid.). In 1970, a cinema centre was established at the institute to provide educational and entertainment films for children and young people (ibid.).

3.2.3. Vice-Presidency for Science and Technology and Promotion of New Technologies

According to the Iranian comprehensive scientific map, the promotion of new technologies such as nanotechnology, biotechnology and stem cells has become a priority. To develop and promote these technologies, to enhance the participation of interested groups, and to foster young talents, the Vice-Presidency for Science and Technology has launched a task force for their promotion and education. The jobs for this task force include holding seminars, workshops, festivals and national competitions aimed at providing appropriate cultural context and raising the awareness of the community and interested groups in these areas (Iran Biotechnology News Agency, 2018; Iran Nanotechnology Innovation Council, 2018).

To introduce nanotechnology, the Nano Task Force publishes a monthly journal, shows films, holds exhibitions and organises a nationwide nanotechnology week with the establishment of nano clubs in collaboration with education departments. The Biotechnology Task Force aims to establish a connection with five groups of audiences: students, scholars, professionals, managers and the general public who are consumers or members of the cycle of knowledge production, technology and industry. Its main purpose is to increase knowledge and technology acceptance among managers at various levels of government, investors and entrepreneurs, and to strengthen national pride. In the field of stem cells, the most important research institute, Royan, has called a part of its activities 'Stem Cells for Everyone'. By establishing mobile and fixed laboratories for the public, it has tried since 2011 to increase citizens' interest (especially students) in this field of study (Iran Biotechnology News Agency, 2018). Other groups involved in similar activities include the Ministry of Education, research centres, municipalities, science parks and science and technology museums.

3.2.4. Awards and festivals of science promotion

The Iranian Science Popularisation Award is the oldest award in the field of Iranian science promotion. It was founded in 1998 and has been the basis for the formation of the Iranian Association for Popularisation of Science.

The association presents this award each year on World Science Day for Peace and Development. The award is given to individuals or organisations that have performed the best and most effective activities for the popularisation of science, the spread of scientific thought and culture in Iran, and the application of science at each level of social life (Iranian Association for Popularisation of Science, 2018).There are similar awards in science fiction, astronomy, etc.

3.3. Media

In general, science journalism from its beginning around 1900 has experienced an upward trend in quantity and quality, but there are challenges that have retarded its pace. First, economic fluctuations have caused difficulties for non-government scientific media. This has decreased the number of the pages and circulation rate of print media and has encouraged them to publish online. The second challenge stems from the lack of trust and legitimacy in the relationship between scientists and journalists. On one hand, scientists expect journalists just to reflect their voices, but on the other hand, journalists do not believe in scientists' ability to effectively communicate science to the public. The third challenge is related to the deficiency of professional science journalism education, which leads to a restricted science journalists' community who usually are self-taught or amateurs. This isolates the Iranian science journalists from global trends and developments arising in their careers. In the following paragraphs we will review the Iranian science journalism experiences on different media platforms.

3.3.1. Radio and television

The production of science TV shows began with the show *Knowledge* in 1968, with the aim of making people interested in and aware of scientific achievements. Iran's first science television channel (Channel 4) was launched in 1995 with the goal of reaching an elite audience and popularising science and technology. The Health Radio and the Health TV Channel were launched in 2003 and 2013 respectively, aiming to improve individual and social health and to popularise health issues.

The process of producing science shows has been slowly evolving in Iran. These shows, with the exception of TV shows called *Night Sky* and *Pangan* and a radio program called *Scientific Talk*, have not succeeded in attracting large audiences.

3.3.2. Documentary TV shows

The science documentary TV shows produced in Iran are only related to the environment, wildlife and archaeology, and there are few spectacular, impressive or even good examples of these shows. The Growth International Film Festival, Scientific and Industrial Film Festival and Cinema Verite Festival are the only arenas available for the supply of science documentaries in Iran.[1]

3.3.3. Journals

The first Iranian public scientific journal is called *Falahat Mozaffari*, published by the Administration of Falahat (agriculture administration) since 1900. The purpose of this journal is to disseminate science, to modify methods of agriculture and cultivation, and to inform and raise public awareness. Two other public scientific journals are related to health and aim to promote applied sciences. Over the 119 years since the publication of public scientific journals in Iran, about 50 diverse journals have been published in various fields of science and technology (Ojagh, 2012). An analysis of the content of 144 issues of 12 series of the Iranian scientific journals demonstrates that medicine, psychology, animal sciences and technology, astronomy and nutrition sciences have enjoyed more coverage and have been often established and operated by specialists in the same disciplines (ibid.). Of course, the circulation of public scientific journals has not been high in Iran and their publication has been continued largely with the support of governments. More than anything else, the relationship between these journals and the public has been defined as educating good citizens, indicating the ability to use science to improve health, extend modern lifestyle, develop industry and solve socioeconomic problems. The journals have accordingly been dedicated to the celebration of science (Ojagh, 2012).

Interestingly, the first Iranian newspaper with a scientific identity was the *Rooznameh Elmiye Dolat Ellie Iran* (scientific newspaper of the Iranian supreme government), which began publication on 11 January 1864 (Hosseini Pakdehi, 1996). The founder of this newspaper, the Minister of Science and Education during the reign of Nasser-al-Din Shah Qajar, was himself a mathematician and an astronomer (Mohit-Tabataba'i, 1987). That probably is why the newspaper had a scientific character. With the proliferation of journalism, scientific pages became a regular part of many national newspapers and included information about new advances and

1 An interview with the producer of documentary shows and the researcher of science communication, Hossein Forotan, by Zarrin Zardar on 28 April 2018.

discoveries, environmental crises, education and pseudoscience. With the development of communication with the Western world, the ratio of translated materials to produced ones has increased in newspapers.

3.3.4. Public scientific books

So far, a systematic study has not been conducted of these books as most of them have been translated from other languages, especially English. The history of science, the biography of scientists, and the concepts of basic science are common topics in public science books. In the field of story-writing, the most significant science fiction example, entitled *S.G.L.L.*, was written by Sadegh Hedayat as a short story and published in 1932. Here the writer portrays the future of the world using his imagination and science plays an important role. The oldest novel in science fiction is *Rustam dar Garne Bist-o-Dovom* [Rustam in the Twenty-Second Century], written by Abdol Hossein Sanatisadehin 1933, and it refers to future scientific advances.

3.4. The importance of cyberspace in public science communication

The internet, cyberspace and social networks have caused many changes in science communication. Their effects can be seen in three areas: first, the possibility of producing and publishing online news has led to the formation of news agencies and specialised channels of science and technology: the Science and Culture News Agency and the Telegram channel of Iran Science Watch are two examples. Second, the existence of the internet has led to the creation of virtual editorials and the publication of online public science journals such as *Saros Astronomy Monthly* (Saros, 2018). Third, science journalists have created scientific and technological events in the audiences' daily lives through blogging, video casts or podcasts.

4. Science communication in the Iranian academic world: Research

At the academic level, research in science communication is highly stressed. There is no independent education program in the field of science communication and it is only taught as a part of journalism or media studies programs at Iranian universities and research institutes. Under such circumstances, it is not unreasonable to say that education and research on science communication in Iran are not connected systematically to each other; therefore, improvements in research studies are dependent more on interested researchers than the pre-arranged academic research agendas. This issue leads to an unpredictable environment for research and education in the field of science communication.

In Iran, science communication was initially introduced by sociologists, and the first systematic study dates back to research conducted by Ghaneirad (2003). He analysed the relationship between science and social, political and economic systems. Today, four main fields of research can be identified:

4.1. The social function of science communication

This field of research examines the importance of science communication for society (Paya, 2008), and analyses the concept of literacy and ways to increase civic education in Iran (Ojagh and Vakil, 2013).

4.2. Science and technology communication

Studies conducted in this field can be divided into four sub-groups:

i. Measuring public understanding of science (Ghaneirad and Morshedi, 2011; Maher and Madaniyan, 2016; Shahriari, 2017). These studies, carried out on the citizens of Tehran and Isfahan, showed a low level of scientific knowledge of the public but also indicated that the public attitudes toward science are positive in Iran.

ii. Representing and covering science and technology on various media platforms. Examples of these studies include the identification of the criteria for the simplification of science and technology (Ojagh, 2012); the television representation of science and technology (Zardar, 2014; Tayyeb-Taher, 2016); the representation of science in the scientific pages of news agencies (Torabi, 2013); and the representation of various scientific topics in social networks (Bahram-Mirzai, 2017).

iii. The nature of the science communication process in Iran, within which one can refer to the process of science communication in Iran (Khaniki and Zardar, 2014); the study and identification of the model governing science communication in the Iranian public scientific journals (Ojagh, 2012); the description of the nature of the public science communication and its common theories (Ojagh, 2011); conceptual issues, methods and the importance of science promotion (Vesali and Ojagh, 2009); the structure of science communication in Iranian society (Sheikh-Jabbari and Ojagh, 2012); the explanation of the nature and necessity of public understanding of science (Vahidi, 2009; Frutan, 2011; Ojagh et al., 2013); the presentation of a theoretical concept for the establishment of effective science communication (Abdollahyan and Ojagh, 2014); the historical analysis of public science communication in Iran (Abdollahyan and Ojagh, 2013); the manner of translating and delivering public science texts (Ojagh, 2013); the study of the logic of content production in Iranian public scientific journals (Ojagh and Abdollahyan, 2014); and the study of the role of Iranian public scientific journals in increasing the public scientific understanding (Ojagh, 2012).

iv. Challenges and barriers facing science journalism (Bonyadi, 2015). Both the role of science journalism in simplifying and translating scientific concepts to society (Ojagh, 2019) and the functions of science journalists (Bonyadi and Borojerdi Alavi, 2016) are among the main research fields of science and technology journalism in Iran.

4.3. Health communication

A large number of researchers in the fields of social sciences, medicine and health have carried out studies in the health area. They have focused on physician–patient relationships (Shafati, 2012), health plans coverage (Ahmadi, 2012), and the status of health communication in Iran (Rasi-Tehrani and Atefimanesh, 2011).

4.4. Policymaking in science and technology communication

Reviewing the experiences of science and technology policymaking in the G8[2] and D-8[3] countries and comparing them with Iran (Vesali, Ojagh, Attari, 2007a, 2007b) is the first and most detailed work in this field. Some of these studies have also considered the network of relations between academic institutions and policymakers (Ghaneirad, 2004; Miremadi, 2013).

5. Education in science and technology communication

The first departments of science and technology communication, Science Promotion, Science and Society, and Politics of Science, Technology and Innovation, are active in the National Research Institute for Science Policy of Iran, educating students in a limited form in addition to training them to conduct research.

2 Group of Eight highly industrialised nations: France, Canada, Russia, Germany, Italy, United Kingdom, Japan and United States.

3 The Developing Eight is an organisation for development cooperation among Indonesia, Bangladesh, Egypt, Nigeria, Iran, Malaysia, Pakistan and Turkey.

The science communication courses can be seen to some extent as lying within the syllabus of the academic discipline of communication. The science communication course at the undergraduate level at Tehran University and the science journalism course at the graduate level of the journalism major at Allameh Tabataba'i University are the only examples. However, the agricultural sector as a precursor cannot be ignored. Since the establishment of the Agricultural Research, Education and Extension Organisation under the supervision of the Ministry of Agriculture Jihad in 1974, all major universities in Iran have established the major of Agricultural Extension and Education in their faculties of Agricultural Sciences and Natural Resources. Tarbiat Modares University Press publishes a research journal with the same title.

Additionally, the establishment of new disciplines may indicate the interest of higher education institutions in this field. The major of Cognitive Science and Media, which has been approved at the Research Centre for Cognitive Science, is one of the first attempts to educate students in the field of science and technology communication. The UNESCO Science and Technology Communication Chair, launched at Allameh Tabataba'i University in 2017, pursues simultaneous research and educational goals (the establishment of new disciplines and short-term education courses) in this field.

5.1. Science and technology communication journals

The journals related to this field can be divided into three categories:

i. The journals relevant to communication and social sciences, which publish the results of studies of science communication scholars. *Popularisation of Science* is the only specialised journal in this field.

ii. In relation to the agricultural sciences, the journals *Extension and Development of Watershed Management*, *Agricultural Extension and Education*, and *Agricultural Extension and Education Research* publish specialised papers in this field.

iii. In relation to medical sciences, *Health Literacy* is interested in the publication of studies by researchers in medical sciences. In addition, there are a number of journals in health education and health promotion in Iran. Interestingly, there is no cooperation between these three categories, even in the publication of papers, indicating the lack of an interdisciplinary perspective on research, education and promotion of science communication in Iran.

5.2. Scientific conferences on science communication

In recent years, the relationship between science and society has been considered in conferences. Two conferences were held in 2018: the National Conference on the Legacy of Science and Society organised by the Iranian National Science and Technology Museum, and the National Conference on the Academic System and Society organised by the Ministry of Science, Research and Technology and the National Research Centre for Science Policy.

6. Discussion and conclusion

This chapter presents a concise but comprehensive overview of the status of science communication in Iran at the three levels of policymaking, civil society and academia. As a review of the development documents show, science and technology communication was first noted in major policymaking documents in 2010. The most direct reference to public science and technology communication has been made in a comprehensive scientific map of the country. Making the language of science simple and intelligible has been taken into account more than anything else. The structure of science and technology policymaking in Iran focuses on the growth of science, regardless of providing necessary prerequisites of effective science communication (including public awareness of science and technology, public participation and engagement) in the public domain and civil institutions. There is almost no attention to the relationship between science and society at this time.

However, reviewing the experiences of science communication in Iran indicates that the government institutions at a micro level have some experience in science communication, and their activities are broadly diverse regarding subject as they cover a wide range in terms of geography. Despite the role of these institutions, the first practical efforts to develop science communication in Iran were made by civil institutions and interested people who sometimes engage in the institutionalisation of science communication with their personal funds. Mathematics, astronomy, information technology and physics are successful examples. However, what distinguishes the scientific communication experience in civil institutions from the government sector is the prospect of the activities of each of these sectors.

On one hand, civil institutions have shown a greater willingness to communicate with the outside world and have not confined the scope of their activities to the borders of the country. The idea of 'the sky without borders' hidden behind the Stars Peace Project or the idea of dialogue in the Good

Wishes Project indicate this fact. On the other hand, they are making more effort to bring scientific subjects to the social arenas, and hints of the public involvement in science can be identified in their activities.

In contrast, despite the extensive activities and the availability of a greater budget, the government has focused on promoting scientific literacy through increasing awareness and education. Science communication has had less chance of engaging the audience in the realm of the government, thus much of the latter's activities can be explained by the transition model and the lack of public understanding of science. Given this evidence, it seems that in the field of science communication, civil institutions appear to be moving one step further than government structures even without supportive policies.

Reviewing media activities in Iran also indicates that the public scientific content of the media is quantitatively increasing. However, the translation and modelling of successful foreign samples still dominate the domestic creative products in all media platforms. The existence of the internet, of course, has led to the development of decentralised activities in the field of science communication, resulting in an unprecedented increase in content producers in the field of science and technology.

The review of academic activities also demonstrates that attention to science communication as a research area has a history of less than three decades in Iran. However, the interest in it has grown enormously in recent years, and numerous dissertations and research papers have been presented in this field. Science communication has also been considered in the arena of academic education and relevant academic disciplines have been established. In general, the development of science communication in Iran has been marked by a relatively long delay compared with the entry of modern science into Iran. However, it is now experiencing a rapid growth.

This widespread attention can be analysed from the perspective of developments in new technologies and policymaking insights. Indeed, despite the unique characteristics of Iran's media or structural status compared with many other countries in the world, Iranian society is part of the international community and is affected by global developments. The relatively broad and open space provided by the internet has strengthened the connection of Iran with the world. This is important in two respects: on the one hand, the Iranian public has access to an extensive source of information outside its national borders, causing their expectations of public scientific content to be defined based on global standards; and on the other hand, the providers of public scientific content can compare their priorities and practices with successful global models. Closer contact with the audience and moving to information-entertainment is one of the most important changes that the professional activity space has experienced in recent years. Therefore, the increased priority of science communication, particularly in such areas as environment, climate change and health have strengthened the status of science communication in Iranian society.

Policymaking affects the function of science communication for the state and government. Scientific fields, and in particular new technologies, are a competitive arena for countries around the world, and thus science and scientific progress have become an important part of the power discourse. The presentation of awards for scientific achievements has become a part of representing the image of governments to both the domestic and international public opinion. As a result, the growth of science communication to represent the image of scientific authority inside and outside the boundaries of the nation-state has become a necessity and has entered the area of science and technology policymaking and educational academic programs.

Therefore, a combination of global developments that increase civil society's capacities for the expansion of science communication and the need and willingness of governments to expand it indicate the clear potential for the expansion of science communication in Iran and its scientific community. Of course, the imbalance of forces and the lack of necessary links among the stakeholders of science communication in the country point to the danger that the expansion of quantitative and accelerated science communication may not necessarily lead to an increase in public participation and involvement in science. This is a concern that can be the focus of scholars' and experts' discussions in this field in Iran and beyond.

References

Abdollahyan, H. and Ojagh, S. Z. (2013). Calling to Audience By Science Communication Magazines. *Culture – Communication Studies, 4*(24), 7–24.

Abdollahyan, H. and Ojagh, S. Z. (2014). A Comparative Analysis of the Role of Iranian Public Science Magazines in Upgrading Scientific Knowledge of their Audience. *Journal of Cultural Studies and Communication, 9*(31), 43–62.

Abrahamian, E. (2008). *A history of modern Iran.* Cambridge University Press. doi.org/10.1017/CBO9780511984402.

Ahmadi, N. (2012). Picture of health in Iranian press: Comparative study of Health weekly and Tandorosti weekly (MA thesis). Allameh Tabataba'i University at Tehran, Iran.

Bahram-Mirzai, M. (2017). Communication and Environment in Social Networks (MA thesis). Allameh Tabataba'i University at Tehran, Iran.

Bonyadi, F. (2015). Obstacles and challenges among science journalism in Iran (MA thesis). Allameh Tabataba'i University at Tehran, Iran.

Bonyadi, F. and Borojerdi Alavi, M. (2016). Function of science journalists in the press. *Quarterly Journal of Iranian Association for Cultural Studies & Communication, 12*(43), 121–53.

Dadar, N. (2010). Khwaja Nasiruddin Tusi Hall of Science and Technology. *Educational Technology, 26*(1), 24–27.

Expediency Discernment Council. (2003). *The future outlook of the Islamic Republic of Iran in the horizon of the next two decades.*

Farasatkhah, M. (2008). Comparing three cognitive paths on 'domestic science' in Iran with emphasis on its conceptual, structural and functional change. *Iranian Journal of Social Studies, 5*(1), 97–110.

Frutan, H. (2011). A Study of the Concept of Science for Public (MA thesis). University of Amirkabir at Tehran, Iran.

Ghaneirad, M. (2003). *Knowledge anachronism: Science relations with social-economic systems in Iran.* Tehran: National Research Institute for Science Policy.

Ghaneirad, M. A. (2004). *Science anachronism: Relationships between science and social systems in Iran.* Tehran: National Research Institute for Science Policy.

Ghaneirad, M. A. and Morshedi, A. (2011). Survey of Public Understanding of Science and Technology, Pilot Study in Tehran. *Journal of Science and Technology Policy, 3*(3), 93–110.

Godarzi, M., Alizadeh, J., Gharibi, H. and Mohseni, M. (2015). Pathology of Science and Technology Policies of Iran: An Analysis of the five-year development plans. *Journal of Technology Development Management, 2*(2), 137–61. doi.org/10.22104/jtdm.2015.184.

Hasirchi, A. and Niawand, A. (2011). Analysis of Islamic-Iranian Model of Progress from Supreme Leader's point of view. *Journal of Basij Studies, 14*(50), 51–82.

Hosseini Pakdehi, A. (1996). Take a look at the first Iranian scientific journal. *Journal of System Management, 8*(5), 30–37.

Iran Biotechnology News Agency. (2018). About us. Retrieved from biotechnews.ir/.

Iran Nanotechnology Innovation Council. (2018). About us. Retrieved from www.nano.ir.

Iranian Association for Popularisation of Science. (2018). Iran Science Promotion Award. Retrieved from www.popscience.ir.

Iranian Higher Education Research Institution. (2018). *Iran's 2018 Higher education report.* Institute for Research & Planning in Higher Education.

Iranian Physical Society. (2009). About us. Retrieved from www.psi.ir/.

Khaniki, H. and Zardar, Z. (2014). Science in the Media; Reconstruction of Science in Media Frames. *Iranian Journal of Information Processing and Management, 29*(4), 875–902. jipm.irandoc.ac.ir/article-1-2514-fa.html.

Maher, Z. and Madanian, S. (2016). Survey of Public Understanding of Science and Technology; Pilot Study in Isfahan, the First International Conference of Social Science and Sociology, Shiraz, 10 February.

Miremadi, T. (2013). The Role of the Analytical and Reflective Capacity in the Twin Processes of Technology Policy and Diplomacy: A Review of the Recent Development in the Process of Policy Making of Uranium Enrichment Technology in Iran. *Journal of Science and Technology Policy, 6*(2), 51–76.

Mohit-Tabataba'i, S. M. (1987). *Iranian Press History* (Persian edition). Be'sat, Tehran.

Mohsenian-Rad, M. (2013). *Yearning for Proper Comprehension: Dar Hasrat Fahm-e Dorost* (Persian edition). Simaye Shargh Tehran.

Office of the Supreme Leader of the Islamic Revolution. (2014). *General policies of science and technology.*

Ojagh, S. Z. (2011). An Overview of the Nature and Theories of Public Communication of Science. *Popularisation of Science, 1*(1), 24–31. Retrieved from www.popscijournal.ir/article_93351.html.

Ojagh, S. Z. (2012). Public Communication of Science and Technology in Iran: Mixed Method Analysis of the Role of Popular Science Magazines for Improving Public Understanding of Science (PhD thesis). Tehran University at Tehran, Iran.

Ojagh, S. Z. (2013). Translation to improve science literacy of public: Ignoring the Functionality in Popular science magazines. *Translation Studies Quarterly, 11*(41), 46–59.

Ojagh, S. Z. (2019). The changing role of science journalism in Iran. *Journal of Culture-Communication Studies, 20*(45), 103–26.

Ojagh, S. Z. and Abdollahyan, H. (2014). Understanding the Managerial Logic Used by the Iranian Public Science Magazines for the Production of Content. *Journal of Iranian Cultural Research, 7*(3), 25–43. doi.org/10.7508/ijcr.2014.27.002.

Ojagh, S. Z. and Vakil, H. (2013). Environmental Impact Assessment: Achieving Sustainable Development through Increased Civic Literacy. *Popularisation of Science, 4*(4), 19–27. Retrieved from www.popscijournal.ir/article_93157.html.

Ojagh, S. Z., Sheikh-Jabbari, M. M., Vesali, M., Zare, M. and Dorostian, A. (2013). On the essence and necessity of public understanding of science. *Iranian Journal of Engineering Education, 4*(56), 117–32. doi.org/10.22047/ijee.2013.3013.

Paya, A. (2008). Public Understanding of Science: A Philosophical Appraisal. *Journal of Science and Technology Policy, 1*(1), 25–38. Retrieved from jstp.nrisp.ac.ir/article_12742_en.html.

Radmanesh, S. A. and Taghavi, M. (2015). Analysis of S&T Section of the Bill of 'Islamic-Iranian Model of Progress' on the Basis of Islamic Requirements of Progress,. *Journal of Science and Technology Policy, 7*(3), 57–70. Retrieved from jstp.nrisp.ac.ir/article_12933_en.html.

Rasi-Tehrani, H. and Atefimanesh, R. (2011). Educational policy making on television. *Social science monthly, 138*(30), 126–36.

Roustaei, M. (2006). *The History of The First Academy of Iran: Tarikh-e NakhostinFarhangestan-e Iran* (Persian edition). Tehran: Nashr-e Ney.

Saros. (2018). About us. Retrieved from saros.ir/.

Scientific Associations' Reference. (2016). The history of Iranian Mathematical Houses Association. Retrieved from www.saref.ir/content-16-30=.

Shafati, M. (2012). A Sociological Analysis of the Doctor-Patient Interactive Relationship (A qualitative study in the city of Ahvaz, Iran) (MA thesis). Payame Noor University at Tehran, Iran.

Shahriari, P. (2017). Mediatisation of Science: The Role of media in Popularisation of Science (PhD thesis). Allameh Tabataba'i University at Tehran, Iran.

Sharifi, S. (1998). 'Institute for the Intellectual Development of Children and Young Adults 1965–1979'. *Goftogu Monthly*, (19), 2–37.

Sheikh-Jabbari, M. M. and Ojagh, S. Z. (2012). 'Structure of Communication of Science and Society in Iran: Popularising Science'. *Science Cultivation, 2*(2), 64–70. Retrieved from www.sciencecultivation.ir/far/pdf/sc04/4-8.pdf.

State Management and Planning Organisation. (1990). *Islamic Republic of Iran's first economic, Social and cultural development plan.*

State Management and Planning Organisation. (1994). *Islamic Republic of Iran's second economic, social and cultural development plan.*

State Management and Planning Organisation. (2000). *Islamic Republic of Iran's third economic, social and cultural development plan.*

State Management and Planning Organisation. (2004). *Islamic Republic of Iran's fourth economic, social and cultural development plan.*

State Management and Planning Organisation. (2017). *Sixth Five-Year Plan of Economic, Social and Cultural Development of the Islamic Republic of Iran.*

Supreme Council of the Cultural Revolution. (2010). *Iran's scientific comprehensive roadmap.*

Supreme Leader. (2008). Speech in presence of university students and professors. Retrieved from farsi.khamenei.ir/speech-content?id=5979.

Supreme Leader. (2010). Speech at first summit of strategic thoughts. Retrieved from www.leader.ir/fa/content/7626/leader.ir.

Tayyeb-Taher, A. (2016). Comparing the observance of the principles of science journalism in the news section of IRIB and BBC1 (MA thesis). University of Islamic Republic of Iran Broadcasting at Tehran, Iran.

Tayyebi, F. (2000). History of the First Iranian Academy. *Monthly Book of Literature, 38*(2000), 105–7. Retrieved from ensani.ir/file/download/article/20120326165305-3016-489.pdf.

Torabi, M. J. (2013). Content Analysis of Three Popular Iranian News Agencies (Mehr, ISNA and Jam-e-jam Online) in subject of Science and Technology in Six Months of 2011 (MA thesis). Islamic Azad University at Tehran, Iran.

Vahidi, M. (2009). *Science-In-Society*. Institute for Social and Cultural Studies, Tehran.

Vesali, M. and Ojagh, S. Z. (2009). Promoting Science: Conceptual topics, Methods and Importance. *Rahyaft, 9*(44), 62–7. Retrieved from rahyaft.nrisp.ac.ir/article_13481.html.

Vesali, M., Ojagh, S. Z. and Attari, M. (2007a). *Conceptualising of Public Understanding of Science in Iran*. Tehran: National Research Institute for Science Policy.

Vesali, M., Ojagh, S. Z. and Attari, M. (2007b). *Monitoring and Studying of the National Policies in popularisation of science in G8, D8, India and China*. Tehran: National Research Institute for Science Policy.

Vice President for Strategic Planning and Oversight. (2010). *Fifth Five-Year Development Plan of the Islamic Republic of Iran*.

Zardar, Z. (2014). Mediatisation of New Technologies in Iran: Cognitive Gaps of Main Actors in Representation of Biotechnology on TV (PhD thesis). AllameTabataba'i University at Tehran, Iran.

Timeline

Event	Name	Date	Comment
First interactive science centre established.	Zirakzadeh Foundation	1993	
First national (or large regional) science festival.	Popularisation of Science	1998	
First university courses to train science communicators.	The University of Tehran	2017	Bachelor's science communication degree
First master's students in science communication graduate.	Hossein Foroutan	2011	He was an MA student in Philosophy of Science

Event	Name	Date	Comment
First PhD students in science communication graduate.	Zahra Ojagh	2012	From Cornell, as a PhD student in communication sciences
First national conference in science communication.	The Heritage of Science and the Society	2018	The main concept was public communication of science and technology
National Science Week founded.	National Science Week	2012	Established by Iranian Association for Popularisation of Science
First significant radio programs on science.	On geography, by Radio Tehran	1939	1940: Ministry of Culture, the Department of Health and the Department of Agriculture produced scientific texts for radio
First significant TV programs on science.	*Danesh* [Science]	1966	By Esmail Mirfakhrai
First awards for scientists or journalists or others for science communication.	National award for popularisation of science	Since 2011	
Other significant events	UNESCO Chair on Communication of Science and Technology	Established in 2017	

Contributors

Dr Seyede Zahra Ojagh is assistant professor in the Department of Cultural Studies and Communication at the Institute for Humanities and Cultural Studies in Tehran.

Dr Zarrin Zardar is an assistant professor of the Communication Sciences faculty at the Allameh Tabataba'i University, Tehran, Iran. She is the executive director of UNESCO Chair on the Communication of Science and Technology.

18

IRELAND

Science in a land of storytellers

Pádraig Murphy

1. Introduction

Ireland is a small country that punches above its weight in scientific terms. Or rather, it may be more accurate to say that it has tended to proclaim a larger-country status in how it communicates its science. Ireland has always been a land of storytellers. And there is a story to be told about the co-evolution of science, technology and the public communication of science.

This chapter will tell one story of science in Ireland and how it is aligned to public communication, scientific literacy, commercial technologies and, eventually, public involvement. It begins with a whistle-stop historical account of how Irish science co-emerged, as it were, with an institutional self-awareness of that science and how it was disseminated. This particular account is a familiar narrative of Enlightenment science as it pertains to Britain and Europe up to the late 20th century—a 'great man' idea of scientific exploration leading to the mathematical knowledge and technologies we enjoy in late modernity. Then the narrative changes, just as it has across the industrialised world. We see a ramping up of funding, from the Irish Government in collaboration with industry and philanthropic enterprises, that takes us away from a linear arrow of progress of science, and towards what Nowotny et al. (2001) call Mode-2 science—networked, contextualised and driven by strategic interests.

The chapter then makes sense of these changes by focusing on indicators for assessing science communication and taking a critical look at the newer discourse of impact and how it might fit into evolving models of communicating science. The chapter briefly profiles some Irish people who have been ambassadors for science communication in this story.

Through all of this, Ireland's ambitious, rhetorical approach to science and how this impacted on science communication is in evidence. The argument is further developed from the analysis in *Little Country, Big Talk: Science Communication in Ireland* (2017), the edited volume from Trench, Murphy and Fahy that focuses particularly on the scale of this ambition and the underlying realities where science and engaging in science overlap (Trench, 2017a, 2017b).

2. Historical context

We could go back a long way. The astronomer-astrologers of the Stone Age passage tomb at Newgrange, older than the Egyptian pyramids or Stonehenge in England; the metaphysical writings of Irish monks on Greek knowledge taken from antiquity, as told by 'knowledge communicators' of the middle ages such as the 12th century scholar and popular writer known as Honorius of Autun; in the 18th and 19th centuries, the quaternions of William Rowan Hamilton, essential to space travel, and the equations of George Boole, necessary for Google searches. Aedh Buidhe's writings on texts from 8th-century Alexandra from around 1415 sit in the Royal Irish Academy (Mulvihill, 2002). On the cover is a *rotula*, Ireland's oldest scientific instrument, used to demonstrate the movement of heavenly bodies across the constellations of the Zodiac. Irish science has contributed handsomely to information retrieval and astronomy.

However, an examination of any formative period of science and its communication on the island of Ireland cannot avoid a post-colonial analysis. It is within the period of British rule ending in 1921 that we can best assess the context for which modern science communication emerged. While MacLeod may say 'science has no nations; but nations have science' (1997, p. 3), the building of scientific nation-states worldwide has created different characteristics of science and how it is communicated. And it can be argued that we have different nations of science—Jasanoff (2005) and Gottweiss (1998) are excellent examples of studies where state-building is at work. In these two particular instances, state policies around biotechnology dictate the essence of the state itself, as demonstrated in the US and Germany. The emergence of a 'Protestant Ascendency' in science is often attributed to colonialism in Ireland

(Whyte, 1999), and Protestant rule over Catholics. The Ascendancy was a period in Irish history where the professional and elite classes were predominantly Protestant, aligned with large property dispossession from Catholics recorded since the time of Elisabeth I (1533–1603). While this moment has often been characterised as a time of 'no Catholics allowed', Nicholas Whyte (1999), among others, has challenged this straightforward reading although he does describe the exclusionary policy of the Royal Dublin Society towards Catholics.

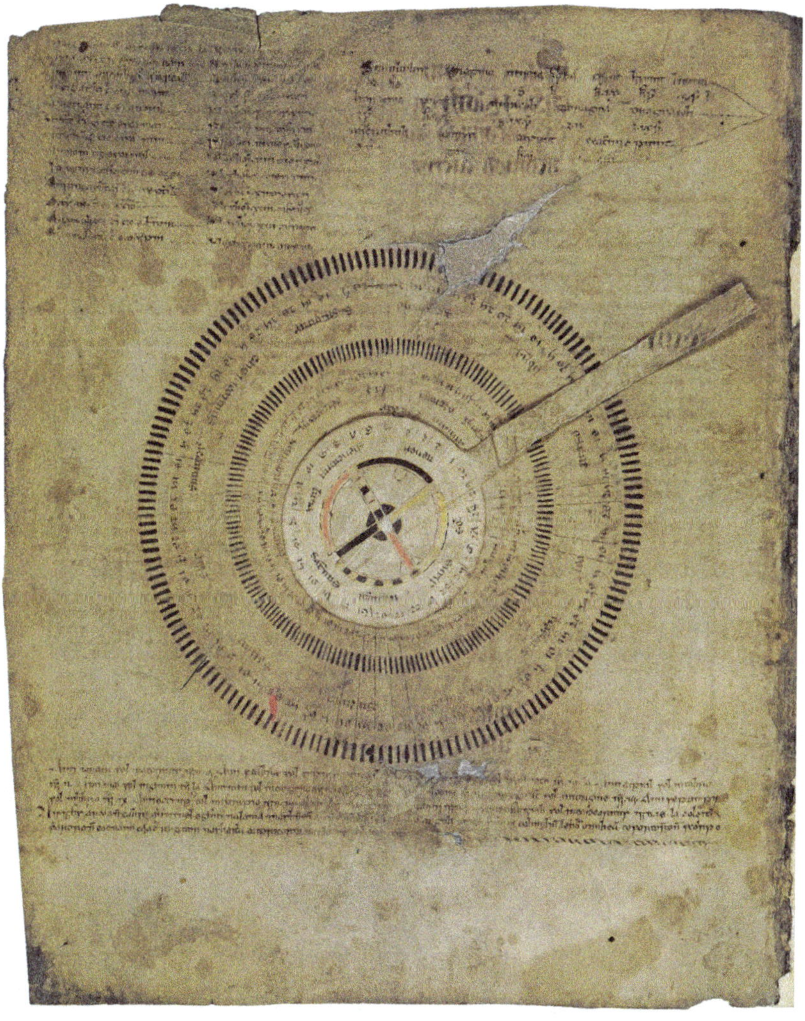

Figure 18.1: The rotula, an ancient astronomical instrument, from Astronomical and Medical Tract a 15th-century text from Aedh Buidhe O'Leighin of Fermoy.

Source: Royal Irish Academy.

There was also open hostility towards Catholics in the Royal Society and British Association for the Advancement of Science (Bennet, 1997). Whyte places greater emphasis on the exclusion of the poor. And the underprivileged were, as is ever thus, the majority in this landscape as we move into the 1800s. The Irish Potato Famine, or Great Famine, devastated the country and almost halved the population of 8 million through death by starvation, or emigration. This was a defining moment for the development of the Irish diaspora worldwide. Jonathan Swift's essay *A Modest Proposal* famously satirised British attitudes to the Irish poor, writing during an earlier period of devastating poverty in Ireland (Swift, 2008). Swift was actually a cogent and compassionate communicator of the horrors that ensued during his time. However, as an elitist science continued, these cataclysmic events did not seem to impact on what could hardly be claimed as an 'Irish science'.

On the other hand, while the Roman Catholic church did renounce Darwinism (Duddy, 2011), there hasn't been sufficient evidence of outright Catholic rejection of science. Whyte also refutes Robert Merton's 'ascetic Protestantism' thesis, which sustains scientific beliefs more readily than the Catholic dogmatic tradition. In fact, Nicholas Callan, a contributor to our knowledge of electromagnetism and batteries, is an example of a Catholic priest and scientist, and there were many more during this Ascendency time.

Wyse Jackson's (2000) edited volume chronicles the science that was taking hold as revolution against the British Crown was happening in Ireland around 1798. Davis reminds us that his tenants in Ireland did the science for Robert Boyle, a founding member of the Royal Society, and that he was their landlord in the heart of the Empire. McNeven, a medical doctor who dabbled in mineral science, associated with Irish Catholicism in 1790s and joined the United Irishman, the rebellious forces against the British in Ireland.

The first recorded group of Irish natural philosophers was the Dublin Philosophical Society (Mulvihill, 2002), founded in the 1680s (Bennet, 2004), a time when ideas from the centre of the British Empire had spread from the tree and taken root in a slightly different way—in this case in Dublin just across the sea. The Royal Dublin Society (RDS), established by the Dublin Philosophical Society members in 1731, has become central to the legacy of Victorian science in Ireland. From this society we have the Botanic Gardens and the Royal College of Science, which once shared buildings with the Irish Government, before evolving into the neighbouring National Museum of Ireland (Whyte, 1999). We also have as its legacy the RDS grounds themselves, and the Boyle Medal, the highest accolade for scientific excellence in Ireland.

In 1785, James Caulfeild, the First Earl of Charlemont—who owned significant property in the Dublin area, including the neoclassical curiosity that is the Casino Marino building, a scientific pleasure house built as part of his 'little Venice' in Dublin—established the Royal Irish Academy (RIA) as an Irish model of the Royal Society, but included in equal parts representations of the humanities and social sciences and the physical sciences. Duddy (2011) recalls how Darwin was elected as honorary member of the RIA. Under its royal charter, which continues to this day, there is a three-year presidency rule where a representative of each of 'the two cultures' alternate the chair. The fauna exhibits were moved to a special site built in 1856—the Natural History Museum. This building is another historical curiosity, taking a cultural-historical approach to science communication. The museum itself is a museum piece, left exactly as it was then, with the same exhibition design preserved from late 19th-century evolutionary taxonomy knowledge of the time.

Northern Ireland's history in some ways parallels the Republic's. The Church of Ireland Primate of Ireland, Richard Robinson, was the founder of the Armagh Observatory in 1789 and a huge contributor to world astronomy, as was Dunsink Observatory in Dublin. The globally famous physicist William Rowan Hamilton worked in Dunsink from 1827 until 1865. This observatory, like many others on these islands, are excellent spaces for communicating the past. Mary Mulvihill (2002) has alerted Irish tourists to look for Broom Bridge in Dublin where Hamilton inscribed his famous quaternion equations.

The giant reflective telescope at Birr Castle was erected by the Earl of Rosse, the first Leviathan in our story of Irish science. For half a century until 1917, Leviathan of Parsonstown was the largest telescope in the world. The interpretive centre currently at the site is testament to its stature, but its fame was recovered by the great astronomer and science communicator Patrick Moore on the TV series *The Sky at Night*. In reflecting on the role of these buildings, structures and instruments of science, Carroll (2006) has brought a material culture view into the history of Irish science, bringing this 'pre-modern' period into a category of 'meters (barometer, hydrometer etc.), scopes (telescopes, stethoscopes, etc.) graphing technologies (cartographic instruments) and chambers (e.g. hydraulic and pneumatic technologies)' (p. 23). These are the objects that contribute to a history, and now to the communication of that history. Indeed, during this period, scientists with Irish blood or Irish addresses enjoyed something of a mini-Renaissance.

And if we go back a couple of centuries there was another, earlier artefact that spurred a new wave of science and presented science scholars with a different look at the enterprise of science—the humble air pump, or vacuum pump. The epistemological nature of Robert Boyle's work in the 1660s, in collaboration with Robert Hooke, and his disputes with Thomas Hobbes, was brilliantly captured in Simon Shapin and Steven Schaeffer's (1985) *Air-Pump and the Leviathan*—the second Leviathan of our story, that of constitutional state and the body politic. Boyle, in this reading, was not only the prime mover for a new empiricism that became modern science by demonstrating against his old foe Hobbes that creating a vacuum in a room was not only possible but necessary for it to be a testable, *witnessed* science.[1] But the act of creating this vacuum (and in the process killing animals!) also marked the beginnings of a type of upstream engagement that characterises our contemporary, more networked, 'Mode-2' science (Gibbons et al., 1994). A select audience needed to witness it, and the gathering (male) natural philosophers needed to replicate it, while women and children reputedly looked on in astonishment, taken by the spectacle of the new magic of science they had seen. However, Boyle was unhappy to be in a 'barbarous country where chemical spirits are so misunderstood and chemical instruments so unprocurable' (Silver, 1998, p. 119).

Irish scientists in early 'science' were influential in disseminating to the elites of Georgian and Victorian Britain. They were part of the 'Big House' scientific endeavour of Britain through the Royal Society, and subsequently through the emergence of Royal Dublin Society and the Royal Irish Academy. These 'influencers' were also orators and writers. John Tyndall, for example, was not only a scientist whose work led to a better understanding of the properties of air, infrared and greenhouse gases, but also wrote a regular newspaper column. Tyndall was also the main Irish pro-Darwinist, a passionate supporter (Duddy, 2011) who took seriously his role as a defender of science, sometimes against religion, and a science communicator. His lectures to the Royal Institution were the stuff of legend, such was their oratorical power, and his US tour generated great interest. It is Tyndall who is credited with finding the answer to the question every child asks: 'Why is the sky blue?' (He was partly right, by today's accumulated knowledge of light and the Earth's atmosphere. He used a special apparatus to conclude that the blue light on the electromagnetic spectrum is most likely to scatter off particles and show visibility. But it is also because blue has the shortest wavelength in the visible spectrum (Royal Institution, 2018).)

1 The vacuum created by these early air pumps was often demonstrated to a live audience by killing a bird, as provocatively captured in Joseph Wright of Derby's painting *An Experiment on a Bird in the Air Pump* (1768).

Then there is a school of thought that states that the Irish embrace of culture and the arts was a way to drive a wedge between this new state called Ireland and the Empire. With the separation came a rejection of science—a separation on cultural and political fronts. For the poet William Butler Yeats, a Protestant, along with the first President of Ireland Douglas Hyde, another Protestant, and others, this new nation needed to define itself as being *not-Britain* as much as *being Ireland*. They spearheaded the Irish Literary Revival, often referred to as the Celtic Twilight. The thinking behind this thesis is that this constructed Ireland resisted the science that represented British Empire—Ireland would instead be a land of artists, singers and poets. Economically, agriculture thrived. The Literary Revival created a hostility to the scientific heritage (Patten, 2003) of William Rowan Hamilton, Robert Boyle, John Tyndall and Joseph Callan. Exclusion cannot be accounted for on grounds of religion or wealth alone: Irish history has regrettably forgotten the great 19th-century women of science, at least until Mary Mulvihill wrote about them; scientists of stature such as Mary Ward, Kathleen Lonsdale, Lilian Bland and Cynthia Longfield. While it cannot be understated how influential this Revival was—it was the imaginary through which banshees, wild spirits and Irish literature flourished (one could almost see the motif woven into the dance costumes of Riverdance)—this is not the full story as the earlier accounts of Irish science advancement demonstrate.

Interestingly, while Ireland became a central point of astronomy during the years of the Earl of Rosse and Leviathan, the Irish state attempted to reclaim leadership again in the mid-20th century, this time as a focal point for physics. In some ways, this was a bizarre occurrence in history. Eamon De Valera, the third Irish president, having earned his legacy from the 1916 Rising, established the Dublin Institute for Advanced Study to attend to his own twin loves of Irish and mathematics (the institute had two schools of study: Celtic studies, and theoretical physics). Suddenly both 'cultures', to brashly reuse the C. P. Snow phrase, were privileged in Irish political thought. De Valera extended an invitation to Erwin Schrödinger to lead the physics part with a Professor's Chair. Others, such as Walter Heitler, continued this great theoretical work on quantum field theory and thermodynamics. The inspirational series of 'What is Life?' lectures Schrödinger delivered in 1943 were hugely significant science communication moments not just for a learned audience, but for the various developments happening within biology, and particularly the new field of molecular genetics. Earlier, Ireland had been home to its only Nobel Laureate, Ernest Walton, for his work on splitting atomic nuclei—again under the proud gaze of De Valera. After that period, Ireland became relatively isolated once again from the discourses of science in the 20th century, although several high-profile Irish scientists contributed, most notably Jocelyn Bell Burnell's discovery of radio pulsars. These pioneers of the past are interwoven into the cultural fabric of Ireland's story.

3. Milestones of science communication in Ireland in the 20th century

This section will outline some of the key moments in Irish science communication. We will see in the next section the changing contexts of science communication, and how conceptually these moments became an 'Irish' version of what international scholarship attempts to categorise as science communication. While some version in Ireland emerged from the late 1990s and early 2000s, in this section the story goes back earlier in the 20th century and the formations of modern science communication in Irish society. They were the rare ways that the public saw, heard and witnessed up close science in action.

The Young Scientist Exhibition was founded in 1963 by Father Tom Burke, a physics teacher and also a priest. It has become the most enduring and arguably the most important public celebration of Irish science on the calendar. It is open to all secondary schools in Ireland. There can be as many as 600 entries and winning is highly prestigious, for both student and school.

When Ben Sherry opened the children's educational TV program *Teilifís Scoile* (1964) [School Television] in native Gaelic on the new national broadcaster Teilifís Eireann (now RTE), he pondered slowly and philosophically to his young audience 'What is Physics?'. He perhaps didn't realise he was starting a new genre in Irish broadcasting, science television. While the answer to his own question perhaps did not take hold on the genre as it might have, there was a little of the future Carl Sagan about it: 'The question is so big and has so many ramifications. It requires a superhuman effort … to answer it. But let's try and at least give an indication of what it's about.' From the 1970s, imported TV shows such as BBC's *Horizon* and *Tomorrow's World*, David Attenborough's *Life on Earth* and Sagan's *Cosmos* appeared on RTE, as more Irish TV receivers picked up international stations. These became influences for the magazine format of Irish-produced shows of the 2000s with Science Foundation Ireland (SFI) support for programs such as *Scope*, *The Science Squad* and *10 Things to Know About*.

A major milestone was the establishment of the Irish Science and Technology Journalists' Association (ISTJA) network in 1985, linked with the European Union of Science Journalist Associations (EUSJA). While ISTJA went into hiatus for a few years, it has become a vibrant organisation again since 2018, and its awards have been running since 1996. Since 2017, the Mary Somerville Medal has been presented by the Institute of Physics in Ireland to the country's top science communicator or spokesperson for the greater public engagement or understanding of science.

The mid-1990s period was a coalescing of areas that saw the move towards an emphasis on science connecting with society. The same year as the White Paper on Science Technology and Innovation (Government of Ireland, 1996) was published, the first science week was launched. The MSc in science communication, also launched in 1996, was a joint program between Dublin City University (DCU) and Queen's University, Belfast. It was the first north–south cross-border course, benefiting from funds from the peace process on building links between the two jurisdictions. Founded by Brian Trench and Ian Hughes, the MSc is still running but at DCU only, and added a health communication strand in 2018.

The next milestone was the 'turn to science' at the beginning of the millennium. The Programme for Research at Third Level Institutions (PRTLI) began the first of five tranches of funding for Irish higher education institutes in 1998, totalling €1.2 billion (HEA, 2004). This was a national government program that also had extra financial support from private industry, EU structural funds and Atlantic Philanthropies to create collaboration potential among different scientific sub-programs based in Irish universities. It represented a significant increase in expenditure, given international recognition that Ireland—despite its historical context—had fallen well short in science spending. Part of the PRTLI remit was support for science communication with dedicated funding for 'education and outreach'. SFI was established in 2003 to oversee policy in science and technology in Ireland, and the Discover program, which funds the majority of Education and Public Engagement (formerly Education and Outreach) programs also comes under its remit.

With PRTLI funding, researchers at DCU linked to the MSc program began to establish themselves as a potential hub for science communication research. They were led by Brian Trench, then chair of the MSc and current president of the international Network for the Public Communication of Science and Technology (PCST). Trench and others were awarded funding from the Framework Programmes in European Commission research. An early success was the European Network of Science Communication Teachers (ENSCOT), training hundreds of researchers in science communication across Europe from 2000 to 2003, with partners including University College London, Pompeu Fabra, Barcelona, as well as DCU and others (Miller et al., 2009). Not long after, the first PhD student in science communication came through.

Dr Fiona Barbagallo successfully defended her thesis 'Public participation and controversy involving science: an Irish perspective' in 2003. There have been six other PhDs in science communication awarded since then, ranging from research on environmental audiences and science TV programming, to celebrity science and young perspectives on biosciences in culture. That first wave of PRTLI funding kick-started the progenitor for the Celsius research group, then called BioSciences and Society, the first research team in Ireland dedicated to science communication research and based in DCU. The Celsius group is now involved with multiple research projects funded nationally and by European Commission Horizon research framework programs in communication, engagement and responsible research and innovation (RRI). Some projects are collaborations with several European and global partners, including the US, South Africa and China.

Perhaps the most significant milestone in terms of Ireland's culture of science, and a demonstration of a new confidence in the intertwining of science and science communication, was the opening of Science Gallery Dublin in 2008. This was a departure for the representation of science in Ireland and arguably the first practical step in UK-style engagement for the country, away from traditional deficit-model marketing. Here was an interstitial public space for the clash of ideas at the edges of science and the arts. Because of private and government funding, as well as its dedication to young people, Science Gallery perhaps cannot venture too far into controversies of science, but the gallery has a unique licence for edginess within the SFI ecosystem. There are now Science Galleries following the same template in London, Bengaluru, Melbourne, Venice and Detroit.

In 2012, we see a return to Schrödinger. Dublin's winning of the significantly competitive bid to host the EuroScience Open Forum (ESOF) City of Science 2012 allowed the country to once more boast what it has to offer. J. Craig Venter, a synthetic biologist venturing into the production of new life-forms, delivered a keynote entitled 'What is Life 2.0?' But there was something else on the minds of the many prominent Irish scientists present, above and beyond this old-school, great-man pioneer replicating Schrödinger from decades earlier at the same institution (Murphy, 2014). The SFI had recently shifted policy even further from 'pure science' to 'applied technology' based on the most recent rounds of funding; and, indeed, the STEM policy had contributed to a greater emphasis in engagement since PRTLI and the establishment of the SFI. Here now, publicly at this grand forum, local scientists were voicing concerns that the days of Boyle, Hamilton, Bell Burnell and Tyndall were numbered if scientists could no longer be allowed to tinker, explore and do creative science alone in the lab. Perhaps the small changes towards a science-for-society had come at a cost to scientific inquiry itself.

There has been a recent demonstration of Irish science communication fitting in with global trends: while the Celsius group has held small conferences and seminars each year since 2008, the launch of the annual SCI:COM Conference in 2015 was a landmark for Ireland as an international player in how it communicated science.

4. A closer look: The Irish turn to science at the turn of the millennium

The period described in this section is a consolidation of science communication activities in Ireland, a new direction based on the changing aims of government for science. We zoom in at the turn of the millennium and see for the first time that policymakers were making a coherent effort to communicate science, albeit initially in a strategic top-down way. This was an improvement, even if these steps into science communication were less about *connecting* and more about *strategic economic interests*: getting an acceptable level of information about the science that the Irish state funded and pleased investors; and simultaneously making efforts to supply a pipeline of future engineers. It was a supply-and-demand type of science communication.

Experts from science and academia were appointed to a government advisory group, the Science, Technology and Innovation Advisory Council (STIAC), established in 1995 (eventually the Irish Council for Science, Technology and Innovation, ICSTI). STIAC's Tierney Report (1995) was the torch that lit the paper, signalling what has often been characterised as 'Ireland's turn to science' and the major changes to science funding that happened with PRTLI, including dedicated science communication funding. Science communication was an add-on, a small but necessary awareness-and-persuasion element of the overall program. STIAC was set up to create a more coherent innovation plan for Ireland. The Tierney Report, in Bodmer Report[2] fashion, highlighted the need to upgrade knowledge and skills in a post-industrial society. Within this text, 'innovation' becomes the new organising phrase. Policy papers and government briefings begin using the acronym STI (science, technology and innovation) to emphasise that 'innovation' was the direction for science. The Tierney Report recommendations included doubling the level of R&D undertaken by the business sector by 1999; increasing funding for basic research from £1.5 million to £6 million (which eventually occurred with PRTLI); looking to universities and the rest of third-level education and encouraging greater interaction between universities and business;

2 The Bodmer Report of 1985 was the genesis of concerns about science literacy in the UK.

establishing a National Task Force to achieve greater awareness of the value of STI for the achievement of national social and economic objectives. All were implemented.

The White Paper (Government of Ireland 1996), influenced by Tierney's recommendations, became the first framing paper for STI in Ireland. Suddenly new buzz phrases appeared, with emphasis on 'contextual' and 'societal' indicators for 'STI'. There were references to Denmark's 'open dialogue model'—a cue to Nordic engagement, seen as being way ahead of Ireland on matters of science communication. Here was a toe dipped in the water of engagement, reflecting discussions that had progressed rapidly in the UK, the one-time colonial ruler. 'There are three simple but profound questions we can ask of the scientific and technological community which seeks public funding,' the White Paper asks. 'What does your project do for jobs? What does it do for society? What are its implications for the environment?' In some ways, this seemed like the beginning of a Rosseau-like contractarian questioning of science's role for society, but the questions were never answered by policy nor asked again. Irish science policy glanced across Europe to note these engagement ideas but never itself engaged. Included in the White Paper also, as if to demonstrate that 'scientists are human too', were snapshots of scientists' lives.

The concept of a modern, 21st-century 'foresight' process enters the conversation three years later with the Technology Foresight report (ICSTI, 1999). The report assessed the best strategic investment for the development of science and technology in Ireland. Eight disciplines were represented by a separate foresight panel, ranging from the life sciences to the construction industry to logistics. A 'stakeholder' analysis was carried out. The foresight exercise concluded that Ireland should be a 'knowledge-based economy', a 1990s buzzword for intensive, post-industrial activities. It could be argued that this was a change from a democratic idea of how knowledge might work for society to one that presented society as analogous to markets (Bell, 1999; Castells, 2000).

Although research support for indigenous companies was mentioned, foreign direct investment was greatly emphasised and this became a theme all the way through: this small country out on the Atlantic moved quickly from post-colonial shock to an agricultural economy, only to move quickly once again to a 'knowledge-based', *open* economy. Again, Ireland wanted the world's scientific leaders to visit, and perhaps stay a while. This report also suggests that working in these new knowledge institutions of universities as researchers should have an attractive career structure, an environment conducive to innovation with more investment in the physical and human infrastructure supporting them. As with other deficit model approaches, the report expects 'citizens [to be] well informed on scientific issues in the context of an innovation culture' (ICSTI, 1999, p. 6). Foresight required a form of prediction, but to control the outcome of this prediction messaging was important. At this point, communicating science was about positive messaging.

During the mid-2000s there were several ICSTI reports on such issues as public awareness of science as well the commercialisation of 'modern biotechnology'. One of these reports appears to attempt to start a 'national conversation' on biotechnology, outlining some ethical issues and risks. While an online forum was set up, it was quickly closed down. This closure was never explained, but if there was no political will for deep engagement with science and technology beyond a deficit approach, then it is clear that such a forum would not last long. In 2004 a statement on nanotechnology by ICSTI was issued, *The Science of Small Things*, coinciding with influential report by the Royal Irish Academy and the Royal Association for Engineering (2004) that progressed the language of dialogue and engagement in the conversation about science policy. However none of the language of the UK report is present in the ICSTI statement.

The ambitious pre-recession *Strategy for Science, Technology and Innovation, 2006–2013* (Government of Ireland, 2006) committed to an R&D spend at 2.5 per cent of GDP, around €2.7 billion at the time, and prioritising food, health, environment, marine and energy as well as biotech and ICTs. It sought to double the PhD output by 2013. The global downturn and particularly the Irish banking crash ensured that these steps did not occur.

Innovation 2020 is the most recent SFI strategy to date. As a crude first-step analysis: when one searches two reports that might be expected to promote public engagement with science and technology (PEST) initiatives in Ireland—*Innovation 2020* (Science Foundation Ireland, 2018) and a recent *SFI Barometer* report (Science Foundation Ireland, 2015)—for the presence of four common words that denote engagement with science and technology, the following is revealed:

- **Communication:** 0 references in the *Barometer*; references only in the context of ICT in *Innovation 2020*.
- **Engagement:** 0 references in the *Barometer*; references only in relation to engagement with IP, industry, global researchers and markets, and funding stakeholders in *Innovation 2020*.
- **Public:** *Barometer* references 'public awareness', 'public trust', a sense of 'public value' of STEM and, promisingly, approaches to 'democratise' science for the public; for *Innovation 2020*, only in relation to public investment and public awareness of this investment and outputs.

- **Participation:** 0 references in the *Barometer*; references in the context of gender and international mobility of researchers in *Innovation 2020*.

Communication, engagement, public (or indeed publics) and participation are common parlance for science policy that is serious about connecting with citizens. Only the word 'public' appears, which may demonstrate a lack of serious consideration for a deeper form of science communication.

While SFI has been slow to catch up with engagement and RRI discourse, it has been proactive in bringing to fruition science communication initiatives under the Discover program, such as the TV series *Scope*, *The Science Squad* and *10 Things to Know About*, the radio series *Future Tense*, co-sponsorship of SCI:COM and Science Gallery Dublin and Smart Futures, Science Week and Discover Primary Science and Maths.

SFI Discover Centres were established in 2005. They include Birr Castel, Armagh Observatory and Planetarium, as well as the Blackrock Observatory in Cork, the National Botanic Gardens, Fota Island Wildlife Park, Dublin Zoo, Airfield Park, the Ailwee Caves and the Imagonisity fun centre for children (the closest Ireland has so far to a traditional science centre). More centres have been added around the country, such as the Arigna Mining Experience (Co. Roscommon), which explores ways in which people use the earth's resources and the themes of energy past, present and future; Bricks 4 Kidz Creativity Centre (Dublin City and Wexford Town), which provides LEGO Technic workshops focused on imaginative and multi-sensory fun; Laois Outdoor Education nature and science walks; Cool Planet Experience (Powerscourt Centre, Co. Wicklow) where students learn the science behind climate change; and the National Reptile Zoo (Co. Kilkenny). Other initiatives include the Festival of Curiosity, Alchemist Café, both funded by SFI, and a science-themed St Patrick's Day parade to mark ESOF City of Science in 2012.

However, there is something else within the Irish science policy system that has not been as successful: the failure, after decades of effort, in establishing a dedicated science centre. The Irish Science Centres Awareness Network (iSCAN) set the groundwork. In 2006, these plans, with support from various commercial interests, crystallised into Exploration Station, with a board chaired by former DCU President Danny O'Hare. Development is finally underway for Exploration Station, the National Children's Science Centre in Dublin. With over 30 years of planning and lobbying, €13 million is still needed to complete the project. It demonstrates the fragility of Irish science culture, and Irish science's inability to engage, that this relatively modest amount of funds (in development terms) cannot easily be raised for what the Irish Government and SFI have declared a priority, namely an 'engaged public … one that understands the role of science' (Science Foundation Ireland, 2020).

The SFI and its current director Mark Ferguson (also the Chief Scientific Advisor to the Irish Government) have expressed satisfaction with public attitudes to science, citing the *SFI Barometer* report results (Science Foundation Ireland, 2015) that demonstrated a healthy respect for how SFI's work is progressing. An earlier Eurobarometer report showed a slightly different Ireland with some ambivalence towards science as well as positive attitudes (European Commission, 2013).

But the discourse of attitude research is a legacy of strategic communication: controlling the thinking around science, while reneging on a public responsibility towards involvement and public ownership as befits a social contract.

Possibilities for greater engagement with science can arise from different initiatives. For example, each higher education institution in the country signed the Campus Engage charter in 2013 (Campus Engage, 2017). Engagement was also a central principle in the Irish Universities Association Charter in 2018, and higher education is still the dominant source of science communication. The Irish University Association Campus Engage initiative launched the *Engaged Research* report (Campus Engage, 2017), a new template for communication and involvement that brings in reflexive techniques and multiple-impact indicators from many disciplines including contemporary science communication (Stilgoe et al., 2014). The two main research funders in Ireland, SFI and the Irish Research Council are now taking on board the concepts of engaged research where publics are invited at various stages of the research and development cycle to be involved and to influence outcomes. The PPI Ignite projects (Public-Patient Involvement) funded by the Health Research Board present models for deep engagement such as user-created diagnostics, where end-users co-design outputs with technologists and co-research with researchers. If using patient groups as co-researchers is a long way from Robert Boyle's air-pump, then maker fayres, hackerspaces, citizen science and DIYbio are actually new ways of getting people in that drawing room to create a vacuum, without killing animals.

From the mid-2010s onwards the term 'education and outreach' was replaced with an official, professional title: Education and Public Engagement (EPE). Here now at last we see—ironically preceding Brexit—an alignment and adoption of an Irish approach with British science communication, re-branded as 'public engagement' as distinct from marketing, persuasion or increasing science literacy.

Box 18.1: Examples of science communication pioneers

Mary Mulvihill

The late Mary Mulvihill was a giant of Irish science communication and her impact continues to grow. This goes far beyond the public communication of science: Mary was a champion for women in STEM a long time before this was part of anyone's agenda in Ireland, and she was an innovator in how she re-imagined Dublin as a city with science at every corner, and Ireland as a country with a cultural heritage of science. Although trained as a geneticist at Trinity College Dublin, Mary went on to study journalism at the institution that became Dublin City University (DCU) after spending some years as an agricultural advisor in a state body. These experiences shaped Mary's craft as a gifted writer and communicator, working as a science journalist and broadcaster. She was editor of *Technology Ireland* and hosted several radio shows on RTE, the national broadcaster, such as *The Goldilocks World*, *The Quantum Leap* and *Left Brain, Right Brain*. She was the go-to science historian for pieces in the *Irish Times*. In 1990, she helped found WITS – Women in Science and Technology. She taught on the MSc in Science Communication at DCU and she developed science communication training programs. Perhaps her most enduring legacy is *Ingenious Ireland*, regarded by many as a definitive book on Irish scientists, which also became a walking tour with the same name of scientific sites of interest in Dublin, demonstrating how science has been an integral part of Irish culture.

Dick Ahlstrom

When Dick Ahlstrom was appointed science editor of the *Irish Times* in 1998, he was the only journalist with such a role in Ireland. In fact, he was the only science journalist employed as staff in an Irish publication. When Dick retired in 2017, this still remained the case. While this may be an indictment of science journalism in Ireland, what it also means is that, during this period, Dick was the principal science writer in Ireland. Over the years, the *Science Today* page has varied with Dick at the helm, organising the main written outlet for science news in Ireland. Dick has honorary life membership of the Royal Dublin Society and has also been appointed honorary fellow of the British Assocation of the Advancement of Science.

Aoibhinn Ní Shuilleabháin

Aoibhinn Ní Shuilleabháin's star has risen quite sharply as one of the faces of science communication in Ireland. She first came to national prominence when she won the international Rose of Tralee Contest in 2005. She is an Assistant Professor in Mathematics Education at University College Dublin (UCD) and is heavily involved in the management and implementation of public engagement projects. Aoibhinn was a champion for the Project Maths initiative, which radically changed the maths curriculum for junior cycle at secondary schools, making the subject more relevant to everyday life. Aoibhinn also co-hosts the annual SCI:COM Conference with Jonathan McCrea.

Leo Enright

For many of a certain generation in Ireland, Leo Enright was the voice of astronomy. As the science correspondent on RTE during the 1980s and 1990s, he covered exciting developments and discoveries of the time: the Space Shuttle program, the Pioneer and Voyager probes and exploration of Mars. Leo continues to be a broadcaster, having appeared on BBC as well as RTE. His later career has concentrated on a greater role for his passion for the public understanding of science: he was Chairman of the Irish Government's Discover Science and Engineering Programme, and was science advisor to centres such as the Blackrock Castle Observatory.

Brian Trench

It is likely that Ireland would only recently have heard of science communication as a practice or a discipline were it not for Brian Trench. Because of him, the country has had a head start in the field. He was founder of one of the earliest science communication master's programs in Europe, as well as board member and campaigner for Exploration Station and, before that, Science Gallery Dublin. He is one of the latter's hallowed 'Leonardos'. Brian started his career in the early 1970s as a journalist, covering cultural and social issues, socialism, politics, jazz and technology. A trip to Carnsore Point, Wexford, the proposed site for Ireland's first nuclear power station, to cover the massive protests there, oriented him towards studying and researching the power and politics of communicating science to publics and publics speaking back. As journalist and editor, he broke some significant stories that are milestones in Irish history. He was the driving force behind the founding of the MSc in Science Communication at DCU, one of the earliest in Europe. The master's degree commenced as a jointly run program between DCU and Queens University Belfast, making it the first cross-border degree program.

5. International indicators for assessing science communication

The story of how science communication emerged as an area of study include debates about whether or not it deserves the status of 'discipline' (Gascoigne et al., 2010; Lewenstein, 2015; Stilgoe et al., 2014; Trench and Bucchi, 2010). These have covered the area of science communication and how it has understood itself internationally through the quite reflective and reflexive works of Lewenstein, Horst, Davies, Trench, Brossard, Irwin and many others. Cultural contextualisation, dialogue, inclusivity and RRI now dominate the discipline. As reported at the 'Big challenges for small countries in science communication' roundtable at PCST 2018 in Dunedin, chaired by Brian Trench, Ireland has found itself being pulled in three directions: the pull of internal national directives to boost the economy and respond to foreign direct investment (FDI), keeping science communication promotional only; the increasing pressures from below from the practice and scholarship of science communication on the grounds that current S&T engagement is not fit for purpose given the positive influences over the last 20 years of upstream engagement in Britain; and finally the related RRI and open science agenda of the largest funder, the European Commission. The first part of the triangle is the most likely to give way.

6. Concluding remarks

Ireland only became a player in this game recently. Brian Trench was a lone voice for a long time, but when funding emerged for engagement initiatives, they tended to be 'deficit model' in approach. The MASIS Report established a framework for assessing science communication across Europe and has become the 'gold standard' for assessing the scientific culture of a country in Europe and its relationships globally (Mejlgaard et al., 2012). RRI (or consideration for ethics, gender, open access, public engagement and good, inclusive governance embedded into technological assessment processes) became the new language; and Ireland measured up as quite 'fragile' in this assessment of the culture of science. For RRI, two-way communication is at the core of its conceptualisation and operation. The opening of Science Gallery Dublin, and the creative 'STEAM' (science, technology, engineering, art and mathematics) movements may change this. Ireland has had a history of culture and science intertwined, and this is the time to make them work together.

Ireland has, at last, staked a claim within the emerging models of science communication and the new paradigm of engagement. Science communication has become embedded in Irish research institutions with the launch of the Programme for Research in Third Level Institutions. Funding developed on an institutional level to avail of infrastructural and personnel support was required to demonstrate a commitment to, and demonstration with validation of, public communication. A percentage of funding needed to be dedicated to what was known as 'education and outreach' (E&O). The latter word has certain ivory tower connotations: borrowing a semiotic idea from the UK again, 'education and outreach' was subsequently changed to 'Education and Public Engagement'.

The familiar 'deficit to PUS to PEST' story came to Ireland later than the UK and the rest of Europe. However, Ireland is now positioning itself within various elements of this mapping out of engagement models and activities. Care needs to be taken that the impact indicators coming with the new concepts of evaluation (and that necessarily capture 'communicating', 'engagement' and 'involvement') do not create extra-strategic communication objectives that ignore the unexpected and, in particular, ignore large-scale public input. The Campus Engage initiative is an example of a cross-sectoral approach, using engaged research as a way that benefits science, universities and communities surrounding those institutions. Ireland has at last reached something like critical mass for science communication research (for example DCU), practice (broadcast radio and TV, Science Gallery Dublin) and the showcasing of best practice (SCI:COM).

Ireland is now positioning itself for engaged research, open science and RRI. The pull of three sides of a triangle—with Irish economic policy itself, UK-inspired public engagement and the European Commission—is getting ever tighter. Ireland's economic policies will continue to progress as an open economy that will also be a driver for how science and technology is both imagined and enacted. This is not necessarily the best for our culture of science. Would that the Royal Irish Academy or the Royal Dublin Society had input into democratic processes aligned with participatory NGOs there might then be challenge-based research as well as ring-fenced blue-sky research. This type of policy would foster the individual 'crazy ideas', but also commit to address real Irish problems such as homelessness, heart disease, cancer, mental illness and local climate action. Although language and emphases are slowly changing, science communication is still equated with STEM education within national policy. Global policies demonstrate significant overlap, but where emphasis is on literacy only, the value of contemporary communication, RRI and engagement theories—participation, inclusivity, dialogue, knowledge exchange—can be neglected.

The hope is that external pressures will lead to a science that fits public policy rather than a policy for Irish science. Impact is important, but blue-sky research and creativity is still a part of Irish science, as our history, so entwined with Britain and British science, shows us. Aligning these objectives of science—the need to address world problems, keep creativity and 'out-there' research, and still include as many non-experts and publics as possible as guides and co-innovators—is not easy in the context of the growing international trend for challenge-based research. But this is how the story should end—a multifaceted science for the doers, the dreamers, the outsiders and all those great women and men of future science.

References

Bell, D. (1999). *The Coming of the Post-Industrial Society*. New York: Basic Books.

Bennet, J. (1997). Science and social policy in Ireland in the mid 19th century. In Bowler and Whyte (eds), *Science and Society in Ireland: The Social Context of Science and Technology in Ireland, 1800–1950* (pp. 37–47). Belfast: Institute of Irish Studies, QUB.

Bennet, J. (2004). Why the history of science matters in Ireland. In D. Attis (ed.), *Science and Irish Culture: Vol. 1* (pp. 1–14). Dublin: RDS.

Campus Engage. (2017). *Engaged Research: Society and Higher Education Working Together to Address Grand Societal Challenges*. Dublin: HEA.

Carroll, P. (2006). *Science, Culture, and Modern State Formation.* Berkley: University of California Press. doi.org/10.1525/california/9780520247536.001.0001.

Castells, M. (2000). *The Rise of the Network Society* (2nd ed.). Blackwell Publishing.

Duddy, T. (2011). *The Irish response to Darwinism.* In J. Adelman and E. Agnew (eds), *Science and Technology in Nineteeth-Century Ireland* (pp. 18–31). Dublin: Four Courts Press.

European Commission. (2013). Special Eurobarometer 401: Responsible Research and Innovation (RRI), *Science and Technology.* Retrieved from ec.europa.eu/commfrontoffice/publicopinion/archives/ebs/ebs_401_en.pdf.

Gascoigne. T., Cheng, D., Claessens, M., Metcalfe, J., Schiele, B. and Shi, S. (2010). Is science communication its own field? *Journal of Science Communication, 9*(3), C04. doi.org/10.22323/2.09030304.

Gibbons, M., Limoges, C., Nowotny, H., Schwartzman, S., Scott, P. and Trow, M. (1994). *The New Production of Knowledge: the Dynamics of Science and Research in Contemporary Society.* London: Sage.

Gottweiss, H. (1998). *Governing Molecules: The Discursive Politics of Genetic Engineering in Europe and the United States.* Cambridge, MA: MIT Press. doi.org/10.7551/mitpress/3334.001.0001.

Government of Ireland. (1996). *White Paper on Science, Technology and Innovation.* Dublin: Stationery Office.

Government of Ireland. (2006). *Strategy for Science, Technology and Innovation, 2006–2013.* Dublin: Stationery Office.

HEA. (2004). *The Programme for Research in Third Level Institutions (PRTLI) Impact Assessment: Volume 1.* Retrieved from hea.ie/assets/uploads/2017/06/PRTLI-Impact-Assessment-Volume-1-.pdf.

Irish Council for Science, Technology and Innovation (ICSTI). (1999). *Technology Foresight Ireland: An ICSTI Overview.* Dublin: Stationery Office.

Irish Council for Science, Technology and Innovation (ICSTI). (2004). *Statement on Nanotechnology: 'The Science of Small Things'.* Dublin: ICSTI Secretariat.

Jasanoff, S. (2005). *Designs on Nature: Science and Democracy in Europe and the United States.* New York: Princeton University Press. doi.org/10.1515/9781400837311.

Lewenstein, B. V. (2015). Identifying What Matters: Science Education, Science Communication, and Democracy. *Journal of Research in Science Teaching, 52*(2), 253–62. doi.org/10.1002/tea.21201.

MacLeod, R. (1997). On science and colonialism. In Bowler and Whyte (eds), *Science and Society in Ireland: The Social Context of Science and Technology in Ireland, 1800–1950.* Belfast: Institute of Irish Studies, QUB.

Mejlgaard, N., Bloch, C., Degn, L., Ravn, T. and Nielsen, M. W. (2012). Monitoring policy and research activities on science and society in Europe (MASIS) – final synthesis report Brussel: European Commission Directorate-General for Research and Innovation.

Miller, S., Fahy, D. and the ESConet Team (2009). Can science communication workshops train scientists for reflexive public engagement?: The ESConet experience. *Science Communication 31*(1), 116–26.

Mulvihill, M. (2002). *Ingenious Ireland: A County-by-county Exploration of Irish Mysteries and Marvels*. Royston, UK: Town House.

Murphy, P. (2014). *Biotechnology, Education and Life Politics: Debating Genetic Futures from School to Society*. London: Routledge.

Nowotny, H., Scott, P. and Gibbons, M. (2001). *Re-thinking Science: Knowledge and the Public in an Age of Uncertainty*. Cambridge, UK: Polity Press.

Patten, E. (2003). Ireland's 'Two Cultures' debate: Victorian science and the Literary Revival. *Irish University Review*, *33*(1), 1–13.

Royal Institution. (2018). *John Tyndall's blue sky apparatus*. Retrieved from www.rigb.org/our-history/iconic-objects/iconic-objects-list/tyndall-blue-sky.

Royal Society and the Royal Academy of Engineering. (2004). *Nanoscience and nanotechnologies: Opportunities and uncertainties*. Retrieved from royalsociety.org/~/media/Royal_Society_Content/policy/publications/2004/9693.pdf.

Science Foundation Ireland. (2015). *Science in Ireland Barometer: an analysis of the Irish public's perceptions and awareness of STEM in society*. Retrieved from www.sfi.ie/resources/SFI-Science-in-Ireland-Barometer.pdf.

Science Foundation Ireland. (2018). *Innovation 2020: Ireland's Strategy for Research and Development, Science and Technology – Excellence, Talent, Impact*. Retrieved from: dbei.gov.ie/en/Publications/Publication-files/Innovation-2020.pdf.

Science Foundation Ireland. (2020). *Engagement*. Retrieved from www.sfi.ie/engagement/.

Science, Technology and Innovation Advisory Council (STIAC). (1995). *Making Knowledge Work for Us, Vols I, II and III (The Tierney Report)*. Dublin: Stationery Office.

Shapin, S. and Schaffer, S. (1985). *Leviathan and the Air-Pump: Hobbes, Boyle and the Experimental Life*. New York: Princeton University Press.

Silver, B. L. (1998). *The Ascent of Science*. New York: Oxford University Press.

Stilgoe, J., Lock, S. J. and Wilsdon, J. (2014). Why Should we Promote Public Engagement with Science? *Public Understanding of Science*, *23*(1), 4–15.

Swift, J. (2008). *A Modest Proposal And Other Short Pieces Including A Tale of a Tub: A Penn State Electronic Classics Series Publication*. Pennsylvania: Pennsylvania State University.

Trench, B. (2017a). The rocky road of science communication. In B. Trench, P. Murphy and D. Fahy (eds), *Little Country, Big Talk: Science Communication in Ireland* (pp. 1–25). Luton, UK: Pantaneto Press.

Trench, B. (2017b). Science in culture, cultures of science. In B. Trench, P. Murphy and D. Fahy (eds), *Little Country, Big Talk: Science Communication in Ireland* (pp. 137–55). Luton, UK: Pantaneto Press.

Trench, B. and Bucchi, M. (2010). Science communication, an emerging discipline. *JCOM: Journal of Science Communication, 9*(3), C03.

Whyte, N. (1999). *Science, Colonialism* and *Ireland*. Cork: Cork University Press.

Wyse Jackson, P. (2000). *Science and Engineering in Ireland in 1798: A Time of Revolution*. Dublin: Royal Irish Academy.

Timeline

Event	Name	Date	Comment
First interactive science centre established.	Science Gallery opens	2009	
First national (or large regional) science festival.	Euroscience Open Forum, Dublin	2012	
An association of science writers or journalists or communicators established.	Irish Science & Technology Journalists' Association (ISTJA)	1985	
First university courses to train science communicators.	MSc Science Communication	1995	Jointly with Dublin City University and Queen's University Belfast
First master's students in science communication graduate.	MSc Science Communication	1996	DCU/QUB
First PhD students in science communication graduate.	Fiona Barbagallo, an Australian	2003	Thesis on public participation and controversy
First national conference in science communication.	SCI:COM 2015	2015	Science communication conference held in Athlone
National government program to support science communication established.	The Programme for Research in Third Level Institutions	2000	
National Science Week founded.	Science Week	1996	

Event	Name	Date	Comment
First significant radio programs on science.	Possibly *Future Tense* or *Spectrum*	2001	
First significant TV programs on science.	*Teilifis Scoile Horizon* (BBC)	1964	
First awards for scientists or journalists or others for science communication.	ISTJA Awards	1996	2017: Institute of Physics Mary Somerville Medal
Other significant events.	Young Scientist Exhibition	1963	

Contributor

Dr Pádraig Murphy is assistant professor in communication at Dublin City University and chair of the MSc in Science and Health Communication program.

19

ISRAEL

Developed Science, developing Science Communication

Ayelet Baram-Tsabari, Daniela Orr, Avital Baer, Erez Garty, Yaela Golumbic, Maya Halevy, Eitan Krein, Adi Levi, Noam Leviatan, Neta Lipman, Ronen Mir and Ettay Nevo

1. Introduction

Israel is a relatively young, small country, with highly developed innovative science and high-tech sectors (Getz, Buchnik and Zatcovetsky, 2020). The Israeli public takes great pride in Israeli science, technology and innovation and is convinced of their importance for national security, prosperity and quality of life (Yaar and Alkalai, 2010). In contrast to the public interest, trust and positive attitude towards science (Israeli Ministry of Science, 2018), the science communication landscape is far from ideal: the Israeli media are rather reserved towards science stories (Barel et al., 2015), infrastructure is slim and based on a few committed individuals, norms within academia do not necessarily support engagement with the public, a culture of public involvement in science-related policymaking is lacking, and government interest in science communication is faint at best. Indeed, a pan-European comparison classified Israel in 2012 as possessing a 'fragile' science communication culture, with weak infrastructure and a lack of science journalists (Mejlgaard et al., 2012).

However, we believe this is changing. In this chapter, we review the history and evolution of science communication in Israel as well as documenting some of the individuals and institutions who are orienting it in exciting new directions. We start with four of the figures who helped shape science communication in Israel in the 20th century; continue with dissemination efforts in print and broadcast media in the 20th century and new media in the 21st; review the historical roots of our science museums and their major role in establishing public engagement with science; introduce the budding participation and citizen science scene; and conclude with the relatively new phenomenon of science communication as a field of research.

2. The founding fathers

Many individuals have shaped the early history of science communication in Israel. Four who played a leading role are Zvi Yanai and Nathan Sharon (popularising science in the media); Peter Hillman (founding science museums); and Azaria Alon (laying the groundwork for environmental mobilisation).

Zvi Yanai (1935–2013) was a unique figure in Israeli science media. After surviving the Holocaust in Italy, he immigrated to Israel, dropped out of high school, volunteered as a paratrooper for his army duty and, after several jobs, became the spokesperson for the Israeli branch of IBM computers in 1970. Yanai was an autodidact and an avid science reader. He transformed IBM's customer magazine into an intellectual science and culture magazine entitled *Machshavot* [Thoughts]. Although it was never sold commercially, the magazine gained popularity and issues were passed from IBM clients to other eager readers. Later Yanai hosted public events, gave talks on scientific issues and published popular science books. In 1993, he was appointed director-general of the Ministry of Science,[1] despite his lack of formal education. He was a popular interviewee on TV and radio and a leading science figure in Israel.

Yanai never practised science, in contrast to Professor Nathan Sharon (1925–2011). Sharon was a highly acclaimed biochemist and biophysicist and was awarded the prestigious Israel Prize in 1994. He immigrated from Poland in 1934 and studied chemistry at the Hebrew University. He was a researcher in the military science corps before earning a PhD from the Weizmann Institute. Sharon was devoted to the public communication of science, as scientific editor of the popular science magazine *Mada* [Science], editor of science features on public radio and science editor for the newspaper *Ha'aretz*.

1 The Ministry of Science has been through 13 changes of name since 1982 (including 'Science and Development', 'Science and the Arts', and most recently 'Science, Technology and Space'). We have used the term 'The Ministry of Science' when referring to any ministry incorporating the word 'science'.

Professor Peter Hillman (1928–2013) immigrated from South Africa to Israel in 1960 after earning his PhD in nuclear physics at Harvard University. In 1964 he was appointed head of the Nuclear Physics Department at the Weizmann Institute of Science in Rehovot and, in 1967, he changed his scientific focus to neurobiology and brain research.

The idea for the Jerusalem Science Museum germinated in Hillman's laboratory in the Hebrew University (HU) in 1980, and it was supported by the Hebrew University and the Mayor of Jerusalem, Teddy Kollek. Subsequently, Hillman tested new approaches and ideas in a pilot museum at the HU science campus. Through donations raised by the Jerusalem Foundation, the first wing of the Bloomfield Science Museum Jerusalem was inaugurated in 1992. Hillman acted as the museum's first director in a voluntary capacity for three years; and after his retirement in 1995 he continued to act as the museum's scientific director until his death. Professor Hillman was acclaimed for making science accessible to the public. In 2002 he was awarded the science minister's prize for his work promoting and advancing science in the community. Hillman was the first and only recipient of this prize.

Azaria Alon (1918–2014) was among the leaders of the nature conservation movement in Israel. Born in the Ukraine, Alon lived his entire adult life in a kibbutz (a collective community). He was a co-founder of the Society for the Protection of Nature in Israel with Amotz Zahavi and awarded the Israel Prize in 2012. In the 1950s Alon started broadcasting about nature and the environment on public radio, which led to a weekly radio program called *Encounters with Animals and Plants*. He continued weekly broadcasts on Kol Israel public radio for more than 50 years. Alon wrote popular books about Israeli nature and wildlife and led campaigns to preserve the environment. His efforts raised generations of nature lovers in Israel, as well as increasing awareness of the importance of public engagement with science and the environment.

3. Disseminating science

3.1. A brief history of science in the Israeli media

Modern Hebrew is a language revived in the 19th century that has had to invent its modern scientific vocabulary. This process involves adapting foreign words and inventing new ones. So 'meiosis', for example, either becomes '*miyosa*' or '*halokat haf-hata*', which means 'divide which decrease'. An example of a new word is '*galai*' for 'detector' (for science glossaries, see the Hebrew Academy site: en.hebrew-academy.org.il/).

Written science communication in Hebrew is rooted in the Haskalah, or Jewish Enlightenment movement. During the 19th century, scientific books using almost biblical terminology were published in central and eastern Europe, mainly with the aim of preparing youngsters for the modern era. A new scientific vocabulary had to be invented.

Science communication in the Hebrew media existed before the establishment of the State of Israel in 1948. Several Hebrew daily and weekly newspapers were published in Palestine under the British mandate (1922–48), and some covered science. For example, the daily *Davar* published a translation in 1935 by British entomologist Evelyn Cheesman about her research voyage to Papua, under the headline 'Large Insects and Small People in the Land of Giant Worms'.

In subsequent decades, although Israeli newspapers reported regularly about science (Golan, 1998), coverage was small compared to newspapers in many other countries. The exception is the quality daily *Ha'aretz*, which still covers more science than any other Israeli newspaper (Barel et al., 2015). None of the major dailies publishes a regular science section, and almost none employs a full-time science reporter. Their online presence usually includes a 'science channel' that relies heavily on external sources (universities' public relations, non-government organisations) for content.

Israeli public radio also came into existence before the State of Israel. Kol Yerushalayim [Voice of Jerusalem] started broadcasting in the 1930s. By 1948 it had changed its name to Kol Israel [Voice of Israel] and had longer broadcast hours (although 24-hour broadcasting only started in 1991). During the 1950s Kol Israel aired brief science features in news programs such as *What's New in Science and Technology?* (Mann, 2008). The first regular science program was *Encounters with Animals and Plants*, aired weekly from 1959 to 1963.

In 1950 the Israeli army (IDF) set up a military radio station Galei Zahal [IDF Waves]. It was popular with the general public and became another public station, although still operated by the army. In 1977, the Israeli military broadcasting service joined Tel Aviv University to broadcast *University on Air*, a daily slot of academic courses running over three months. Many courses were complemented by popular science books (Boas and Baram-Tsabari, 2016). The program is still aired in a different format today and has contributed to the interest and education of many people in Israel. Dozens of lectures were printed in accessible language and gained wide audiences during the last quarter of the 20th century.

Israel public television started in 1968, shortly after educational television (1966). In 1968 public television began broadcasting a monthly prime-time science magazine program, *Mada va'Daat* [Science and Knowledge]. This was followed by a prime-time weekly science magazine, *Tazpit* [Observation] and a science magazine aired by Israeli television in Arabic called *Innovations and Inventions* that ran from 1977 to 1996 (Katz-Kimchi, 2012). All were very popular, but science coverage has been considerably reduced since the introduction of commercial television in Israel in the 1990s. In recent decades, science magazines are confined to educational TV and designed for children and adolescents. One exception is *Science News*, a science magazine put to air over five years (2003–08) on a private cable channel focused on culture, nature and science. In addition, there is generally little science coverage in news programs. An exception was *London et Kirschenbaum*, a daily show on commercial Channel 10, which topped the list of science items per show on Israeli television during its lifetime 2003 to 2019 (Armon et al., 2017; Armon and Baram-Tsabari, 2017; Barel et al., 2015). A bright corner in televised science is the new Israeli Public Broadcasting Corporation (on air since 2017), which puts more emphasis on science and academic knowledge than its commercial counterparts.

Over the years there have been several attempts to publish printed science magazines for adults. The most successful was Nathan Sharon's *Mada*, published six or 12 times a year between 1965 and 1991 by the Weizmann Institute, with many senior science professors contributing. Several other magazines were also published in Hebrew, including *Galileo*, which was published for 20 years starting in 1993. *Galileo*'s science editor was Zvi Atzmon (born 1948), a neurobiologist and a poet who was the former editor of *Mada* and *Ladaat* as well as literary magazines. During the 1990s, *Galileo* was the most successful popular-science magazine in Israel but ceased publication in 2016 as a result of the print media crisis. Another victim of the media crisis was *Odyssey – A journey between ideas*, inspired by Zvi Yanai's *Machshavot*, which was published four times a year 2008–15. Since the 1990s, international science magazines have been published in Hebrew. *Popular Science* (1994) and *National Geographic* are still published but *Scientific American Israel* (2002–17), a bimonthly offering a selection of translated and original Israeli articles, has moved online.

Educational magazines for children and teenagers were popular in the second half of the 20th century, including *The Young Technician* (1945–65), *Ladaat* [To Know] for teenagers (1970–91), the environmental children's magazine *Pashosh* [Warbler] (1976–2004), and *Kimat Alpaim* [Almost 2000] (1994–2000). *Young Galileo*, *Galileo*'s companion magazine for children is published to this day. A new addition is *Frontiers for Young Minds* in Hebrew, an open-access online scientific journal written by scientists and reviewed by a board of adolescents. This was the first time this online magazine has been published in a language other than English.

In the 1950s and 1960s, encyclopedias played an important role in communicating science to the Hebrew reader. Most popular science books published in Israel are translations of bestsellers in English. There were very few popular science Hebrew titles during the first decades of Israel, but this began to change in the 1980s, when a translation by Emanuel Lottem of Stephen Hawking's *A Brief History of Time* was the first popular science book in Hebrew to have commercial success. Other successful books followed, but today the field is weakening and only a handful of new books are published each year. An outstanding exception is historian Professor Yuval Noah Harari's *Sapiens: A Brief History of Humankind* (2011, 2014), based on an academic course at Hebrew University. It became a worldwide bestseller, as did its sequels.

3.2. Various forms of science in the Hebrew new media

New media, including podcasting, Facebook, blogs, YouTube and sites providing science content for media outlets have been a major force in science communication in Israel over the last decade.

In the early 2000s, Israel's science journalism was written by a handful of in-house science journalists, and published by a few popular science magazines and two to three popular science websites such as *Hayadan* [The Erudite]. *Hayadan* was founded in 1997 by reporter Avi Blizovsky and was a milestone in Israel's online popular science publishing. It serves as a database of popular science articles, news items, interviews and reviews over the last 20 years, containing over 20,000 items on 550 subjects in science and technology. Entries are written by Blizovsky and hundreds of contributors, and it is still updated daily.

The rise of Web 2.0 has increased the number of science blogs authored by graduate students, scientists or science enthusiasts on topics such as space, diseases and insects. One example is Roey Tzezana's blog, *Ha Madrich La'Atid* [Guide for the Future]. Figures such as Yaron Assa and Gilad Diamant operate forums and blogs criticising pseudoscience, media scares and inaccurate science reporting in news media. These bloggers and informal science communicators set the stage for more formal online science communication.

In Israel, the main social network for science communication is Facebook, although there is some activity on Twitter, YouTube and Instagram. One Facebook page is run by PhD biology student Yomiran Nissan, *Mada Gadol Baktana* [Little Big Science], the largest popular science Facebook page to date. This page was started in 2013 by Ofer Sadan as a 'jokes and anecdotes' page and began to attract followers when it published longer posts about studies and the science behind everyday life. Today the page has more than 125,000 followers. Their most popular posts are 'debunking posts' about misleading articles on news websites or about health scares. One of *Mada Gadol Baktana*'s posts that had a great impact was a long response to a supposedly investigative report on the side effects of various drugs. The report, broadcast on one of Israel's commercial TV networks, conducted little or no fact-checking and did not consult a science advisor. After scathing criticism on the Facebook page (and other similar Facebook pages), some media outlets started to seek advice from scientists and physicians for their shows. *Mada Gadol Baktana* is today an NGO with 45 scientist volunteers promoting science mostly through digital media.

Another recent online science communication model is the NGO *Midaat* [Informed]. In the summer of 2013, Israel experienced a silent polio outbreak. The Ministry of Health decided to re-vaccinate children, triggering virulent debates in the media and social networks. Science communicators started advocating and answering questions from concerned parents (Orr et al., 2016; Orr and Baram-Tsabari, 2018). Most activity was on Facebook, where long discussions and debunking of anti-vax claims were led by Dr Keren Landsman and Adva Lotan. Along with others, they subsequently founded *Midaat* to deal with responsible communication of medical information, especially concerning vaccinations. *Midaat*'s activity is mostly on digital platforms (Facebook, YouTube and Twitter) and the organisation offers accessible medical information and free advice to journalists. *Midaat* also has a column on other health issues published on the website of *Ha'aretz*.

Science podcasts are relatively new in Israel. The first was in 2007, a science and technology history podcast called *Osim Hisotia* [Making History] produced by Ran Levi, an electronics engineer. This amateur podcast has grown to a company producing 11 podcasts on different topics, with roughly 750,000 episode downloads each month. Since 2017, the broadcasting corporation and the former educational TV channel have reached out to hire podcasters in many fields including science.

Box 19.1: Organised science communication effort

A telling example of the importance of organised science communication took place on 27 June 2018. One of Israel's commercial TV networks published an anti-vaccine video filled with misinformation on their Facebook page. About 24 hours later, the TV network took it down, mostly due to the efforts of *Midaat*, which refuted every part of the video on Twitter and Facebook and contacted the TV network to warn them of the dangers of spreading anti-vaccination propaganda. At first, the producers of the video asked *Midaat* and other science communicators to help create a counter-video to 'show both sides' of the vaccine debate, but *Midaat* refused, explaining that this would only create a 'deceptive balance'. *Midaat* worked with other organisations to make sure that there would be no debate and no new videos showing 'the other side' on that TV network. On other networks, news websites and social networks there were many posts and articles criticising this video. As a result, the TV network published a retraction stating that they had no knowledge or expertise about vaccines and retracting it was the ethical thing to do to protect the public. In addition, *Midaat* recently aided in forming a bill, stating that information on measles should be made accessible to the public, resources should be allocated to public health infrastructure and incentives should be offered to promote vaccination rates against measles.

In the last decade, scientific institutions and projects are slowly associating themselves with individual online actions, and there are growing numbers of collaborations between science communicators and the media. In 2006, the Davidson Institute of Science Education (the educational arm of the Weizmann Institute of Science) established the Davidson Online website for popular science. This website employs and trains dozens of graduate students as popular science writers and has published thousands of articles achieving millions of views each year. This was the first organised online initiative of science journalism based on scientist-writers in Israel, and had 3.65 million unique views in 2018.

In 2014 the Israel Society of Ecology and Environmental Sciences created *Zavit* [Angle], a science news agency offering popular science items in the field of ecology and environment. The aim of *Zavit* is to provide media with accurate, up-to-date stories about environmental issues in Israel, using environmental scientists from all Israeli universities and research institutions in public outreach. *Zavit*'s editorial staff supplies journalists with new stories every day, and journalists can choose stories independently. Following a similar rationale, in 2015 the Davidson Institute opened the Department for Science Communications to promote scientific items to the media based on content published by Davidson Online.

The *Zavit* and Davidson projects aim to promote science in the media, but they differ in the way they address journalists and the topics they cover: *Zavit* covers mainly environmental topics, while Davidson covers all science disciplines. Both initiatives reach print, online, TV and radio in Israel, and interest readers outside the circle of science and environment enthusiasts with stories on sport, the economy, tourism and food. In 2018 the Davidson Institute published 728 science items, 305 of which were re-published on general news websites, on TV (70 items) and radio (150 items).

The year 2018 saw two new popular science websites in Arabic: Davidson aired an Arabic version of their popular science website aiming to reach science teachers and students; and Arab life and exact science professors collaborated to air a new popular science site with original content in Arabic called *Al-Maram*.

In both the Davidson Institute and *Zavit*, articles by scientists are edited by science writers and published in the media. A long trust-building process taught editors and reporters from the mass media that stories from Israel's science and environment news agencies are scientifically accurate and written in a language their readers understand (with similar engagement outcomes to those of items by organic reporters (Barel-Ben David et al., 2018)). But this is a double-edged sword. On the one hand, more high-quality science items are being published in the media and reach a broader public that would not actively seek them. On the other hand, 'feeding the media' with science items at no cost may act as a disincentive to media organisations to employ in-house science reporters.

Box 19.2: Impact on policy

There are many examples of scientific items promoted by the media to influence the public agenda and promote data-based decision-making in health, education and the environment. A striking example is an item published by *Zavit* reporting a study linking drinking of desalinated water and an increased risk of heart attacks. The study was originally presented at a small professional conference and was not meant to be published in the media at all. It caught the attention of *Zavit*'s reporter who attended the conference and understood its significance. A story was written that same day and sent to the health correspondent of Israel's leading news program. The item opened the leading evening news show that day, created a public outcry and eventually attracted the attention of Prime Minister Benjamin Netanyahu, who ordered a special committee to look into these findings and the funding of a pilot program to try solving the problem. This example shows how the work of an independent, science-based news agency can affect the country's environmental agenda and make a real impact on its people's lives.

4. Public engagement activities

4.1. The history of science centres and museums in Israel

The history of interactive science museums in Israel starts with researchers who considered that increasing the numbers of those interested in science was a national priority and one of the duties of academic institutions. Professor Amos de Shalit and Professor Gabi Goldring from the Weizmann Institute of Science started the *Science Camp for Science-Oriented Youth* in 1964, on the grounds that the research community had an obligation to encourage the scientific education of the younger generation.

When Professor Peter Hillman, the founding director of the Bloomfield Science Museum in Jerusalem, visited the San Francisco Exploratorium he heard Exploratorium founder Frank Oppenheimer's statement: 'The whole point is to make it possible for people to believe they can understand the world around them'.[2] In 1981 with a group of scientists, Hillman began the construction of exhibits based on the principles of the Exploratorium. The 'Simply Science' centre was established in a hall allocated by the Hebrew University in the National Library building on the Givat Ram campus. Ten years later, this initiative led to the building of the Bloomfield Science Museum Jerusalem on the National Museums Mall in 1992.

In 1983, Professors Yitzhak Oref and Tzvi Dori[3] from Technion's Chemistry Department began dreaming of a science museum in Haifa. They built interactive exhibits like those in the Exploratorium and in 1983 displayed their efforts in a hall provided by the Technion. The Haifa museum moved to the historical Technion building in 1986 as the MadaTech—National Museum of Science and Technology, Daniel and Matilde Recanati Center (originally called the Technoda).

At the Weizmann Institute in Rehovot, as part of the programs of the Youth Activities Department operating since the mid-1960s, the department's director Dr Moshe Rishpon[4] developed interactive exhibits displayed in the institute's outdoor areas. The Garden of Science, which later became the Clore Garden of Science, opened in 1998. The Science Garden has twice

2 Discussions held by Maya Halevy with Professor Peter Hillman.

3 An interview held by Dr Ronen Mir with Professor Tzvi Dori and Professor Rivka HaShimshoni.

4 Comments by Dr Moshe Rishpon on the origins of the Science Garden, July 2018.

been awarded a prize for innovation from America's Association of Science and Technology Centers (ASTC). The first prize was awarded in 1999 for the concept and the second in 2013 for the Ecosphere in the park.

In 2013, Israel's fourth science museum, the Carasso Science Park, opened in Beer Sheva under the auspices of Ben Gurion University, the Beer Sheva city council and the Rashi Foundation. They saw a science museum as a crucial step for the development of science education in southern Israel. In 1990, all this activity led the Israeli branch of the International Council of Museums (ICOM) to initiate a conference with the museum department of the Ministry of Education and Culture.[5] Complementary areas of activity were presented: Professor Hillman from Bloomfield said the most important goal was establishing a scientific culture in Israel. It should be open to the public and encourage curiosity and inquiry into scientific principles and natural phenomena to connect science to daily life. Professor Tsvi Dori of Haifa noted that the main goal of the Haifa Science Museum was to operate as an educational and learning centre for the formal education system. Dr Neta Maoz from the Weizmann Institute focused on the strongest students to expose them to the latest research and support those the education system had failed.

Thirty years later, these issues continue to challenge the world of science museums and to define their activities. The specific foci of each museum still reflect these early distinctions, although goals and audiences have widened substantially for all.[6]

There are also nature and natural history museums in Israel. The newest is the Steinhardt museum in Tel Aviv, which opened to the public in 2018. Smaller nature museums can be found in Jerusalem, Maayan Baruch, Ein Harod and Kibbutz Dan.

Along with the original development of exhibits and exhibitions and supplementary educational activities, Israeli science museums have gone beyond the 'walls' of museums. The Clore Science Garden initiated science festivals, the Bloomfield Science Museum promotes and develops the Science Theater in Israel, MadaTech opened the first 3D movie theatre in Israel,

5 *Science and Technology Museums*, summary of a one-day seminar, Israeli Association of Museums and the Museums Department of the Cultural Administration of the Ministry of Education and Culture, Tefen Industrial Park, 13–14 March 1990.

6 Israel's first science museum was actually built in Tel Aviv in 1958 in one of the pavilions of the Ha'aretz Museum (today the Eretz Israel Museum). The science museum operated under the management of Ivan Moskovitch who, along with Shabtai Levy, developed and built the first interactive exhibits in Israel. The museum closed in the mid-1960s and parts of its collection moved to the MadaTech archive in Haifa.

and the museum produces and screens unique scientific films. Science cafés and 'Science in Movies' encounters are initiated by the Bloomfield Science Museum and held at public locations. The Maker Faire is a yearly event at the Bloomfield Science Museum, exposing the public to the 'makers' culture; and the Carasso Science Park runs a scientific youth movement. Israeli science museums now operate as a national network, promoting the National Science Day and National Space Day in collaboration with the Ministry of Science. They also develop special programs with the Ministry of Education to provide pedagogical support for STEM in the formal education system.

In addition to their work in Israel, Israeli science museums have helped set up science museums in developing countries. The Bloomfield Science Museum Jerusalem collaborated with the Città della Scienza in Naples to build the first Palestinian Science Museum on the Abu Dis campus of Al-Quds University, funded by the European Union. For several years, the MadaTech in Haifa has helped establish a network of science museums in Ethiopia.

The special connection between Israeli science museums, research and institutions of higher education continues to be one of the outstanding characteristics of the science museum network in Israel. Institutional oversight provides these museums with advice and assistance from active researchers and serves as a basis for exhibitions and public programs. The proximity to research institutions makes it possible to employ science and engineering students as instructors/guides. This has several benefits: they serve as role models for the museum's young visitors, while being trained in science communication and acquiring skills in science-based professions of value in their future professions.

The science museums were established with philanthropic funds and their development depends on philanthropy. They receive some funding from the Ministry of Culture and Sport, like other Israeli museums. The government provides funding for specific educational programs through contracts and calls. Similarly, the Israel Society of Ecology and Environmental Sciences is funded mainly by philanthropic foundations, but also by joint ventures with government and government agencies. A survey in 2011 showed that there were no national funding programs that specifically target science in society issues (Mejlgaard et al., 2012).

4.2. Public engagement and outreach initiatives

From 1949 to 1998 no major institution dealt with public engagement with science or science communication. This included the Israel Academy of Science, the Council for Higher Education, the Ministry of Education, the universities, the Israel Science Foundation (established in 1992) and even the Ministry of Science (only created in 1982). Instead, their efforts regarding public understanding of science were targeted towards science education. But in 1999 senior faculty members from various universities established the Bash'ar [At the Gate] association, with the aim of increasing the involvement of Israeli academia in public discourse and society. The organisation operates in the geographic and social periphery of the State of Israel, and its activities include sending lecturers to schools and providing expert answers to teachers.

A few major figures (including some within official bodies) made pioneering individual efforts to reach out and engage wider publics, including Israel's first president Professor Chaim Weizmann and Professor Aharon Katzir, winner of the Israel Prize, who made social and political efforts to promote science at the national level. Professor Alex Keynan and his wife Malka joined Professor Joshua Jortner, President of the Institute for Advanced Studies in the Humanities (IASH) at the University of Edinburgh to initiate the Batsheva de Rothschild Fund, which hosted two seminars dedicated to encouraging public engagement activities (in 1982 and 2000).

The year 1998 was a turning point. Hemda, the Science Education Center of Tel Aviv-Yaffo funded by the Rothschild Foundation and the Tel Aviv municipality, initiated a season of public lectures. When Hemda turned to Tel Aviv municipality for support it was met with scepticism and doubts about the public interest. Nevertheless, Hemda went on to host hundreds of popular lecturers throughout the 2000s. In 2002, Hemda's director Dr Tehilla Ben Gai was invited by the British Council to visit science media centres in the UK. This led to the Science–Culture project, directed by Dr Eitan Krein. At the same time, the Davidson Institution of Science Education, led by Professor Haim Harari, started the Frontiers in Science meetings and workshops; and Professor Itzchak Parnas in the Hebrew University initiated the 'Why' monthly lecture series for young adults and the general public (2000–17).

Other public engagement events followed. These included Famelab, a contest 'imported' by the British Council Israel's science officer Sonia Feldman from the science festival in Cheltenham and ran in Israel from 2007–12 and in 2015; Researchers' Night, an event involving public activities in research centres and science museums and sponsored by the EU and the Israeli Ministry of Science (2007–present); the annual Science on Tap events led by Weizmann Institute President Professor Daniel Zaifman and PR director Yivsam Azgad (2009–14) (Rehovot), 2011–present (Tel Aviv); Einstein on Trains that later became Professors in Slippers organised by the Hebrew University (2009, 2012–15); the Wolf Prize Public Events initiated by the Wolf Foundation and led by its director Dr Liat Ben-David (2012–16); and the newest addition, Davidson's S-Factor science talent contest.

There are numerous public lectures, youth science festivals and shows initiated by universities, colleges and municipalities. In the last 20 years, Israeli versions of science projects from around the world such as Famelab, Scientific PechaKucha, Science-TED and others have been produced. The private initiative Think & Drink Different was started in 2012 by a former history teacher Tuval Rozenwasser. It offers popular science lectures, along other topics such as politics and art. This initiative has expanded from 16 locations across Israel to New York. The initiative WIZE was established in 2011 as a social entrepreneurship project. It is now an NGO promoting public interest in science, innovation and technology through bar lectures across the country. These lectures, also by other providers (e.g. TalkHouse), have become a popular way to spend a night out.

5. Public participation in science

Much of the early public engagement in Israel can be attributed to the activities of environmental movements active in Israel since the 1950s. The main movement is the Society of Protection of Nature in Israel (SPNI), founded in 1953 by Azaria Alon and Professor Amotz Zahavi, who responded to a deep public need to preserve Israel's flora and fauna (Tal, 2002).

The environmental movement in Israel can be broadly divided into three periods, in line with trends in Israeli civil society: the state-oriented period (until 1990), the civil society–oriented period (1990–2007); and the partnership-oriented period (2007–present) (Greenspan, 2016; Orenstein and Silverman, 2013). The two first stages corresponded to identification and then conflict with government agenda, whereas the third is characterised by coalitions and partnerships between NGOs, government agencies and businesses. The current partnership-oriented period sets the groundwork for increased public engagement and participation in promoting scientific, social and environmental agendas, still in its infancy in Israel (Greenspan et al., 2016).

Research shows that participation mechanisms are not well developed and have not influenced the decision-making structure in urban planning or health care systems (Alfasi, 2003; Efron and Davidovitch, 2011; Miron-Shatz et al., 2012). This is due in large part to the structure of public participation, which is generally based on objections and appeals rather than representation, consultation and deliberation (Alexander, 2008). In fact, public participation in these issues is generally passive in that citizens are typically informed rather than engaged (Efron and Davidovitch, 2011).

Changes in public involvement may come from the new *Strategic Planning for Israel Towards 2048*,[7] which includes public participation components, the increased consideration of social justice issues in local planning, partnerships and joint programs between local authorities and residents, and the emergence of citizen science (Sadan and Churchman 2012; Golumbic et al., 2015). An example of a public participation process revealing many of the underlying tensions is the participatory health impact assessment (HIA) report on the danger of living near a national hazardous industry site in southern Israel (Negev et al., 2013). The assessment helped uncover health issues known to the local community but not addressed by urban planning. It highlighted the difficulty of reaching diverse groups in society, the conflicting views of the stakeholders, the uncertainty and finally the challenge of making the results impact a society that predominantly acknowledges experts' knowledge (Negev et al., 2013).

Other recent public participation initiatives include the SPNI urban community projects for promoting quality of life by developing public spaces and involving residents in planning processes; and citizen science projects such as 'Sensing the Air' to facilitate air-quality research through public involvement in collecting and interpreting meaningful air-quality data (Golumbic et al., 2019). Together, these and similar activities constitute the basis for active and influential community participation, public awareness and public engagement (Greenspan et al., 2016).

6. Researching and teaching science communication

6.1. Terminology

Science communication is a nascent academic field in Israel, so new that until recently it did not have a name in Hebrew. The first and second Israeli conference in science communication in 2009 went under the name of Mada Batikshoret, which translates as 'science in the media'. Following advice by Uri Aviv, director of the Tel Aviv International Festival of Science Fiction and Fantastic Genre Films, Baram-Tsabari and Baer presented 'a translation challenge' to a forum of translators on Inga Michaeli's blog *Translating the Globe* in February 2011, asking them to suggest a Hebrew term for 'science communication'. Many suggestions were made, playing with words such as information, explanation, messages, popular science, science in culture, scientific communication and more. The suggestion that was chosen is *Tikshoret Hamada*, suggested by the

7 *Israel 100: Planning now for 2048 multi-institutional panel* (israel100.org).

translator Ofra Hod and by the editor Ruth Almagor-Rimon. Verbatim it translates as 'communication of science'. The main problem with the term is when using it as a verb and as an infinitive, *Le'taksher*. This unfortunately has been taken over already by the parapsychological sector, which uses it to describe communicating with aliens, the dead and other fantastic beasts. Therefore, the verb being normally used for communicating academic knowledge is *le'hangish*, a word that stems from the word 'accessibility'.

6.2. Academic degrees and courses

Although science education, history, sociology and the philosophy of science are well-established disciplines in Israeli academia and there are many master's and PhD theses carried out in related fields, to date there is only one science communication research group. The Applied Science Communication Research Group at the Technion – Israel Institute of Technology is led by Associate Professor Ayelet Baram-Tsabari and operates within the Faculty of Education in Science and Technology. There is still no specialised department for the field. The group focuses on empirical examinations of how non-scientists navigate science in their everyday lives and supports scientists in learning how to communicate effectively. The group is interested in bridging science education and science communication scholarship, and studies expressions of science literacy in online public engagement with science environments. Its alumni are the first science communication graduates in Israel.

The first academic course to teach STEM students to communicate science began in 2009 at the Technion. Professor Baram-Tsabari still leads the course today (for the syllabus see Baram-Tsabari and Lewenstein, 2017). Additional courses include a graduate seminar in science communication (since 2014) and a practicum course in science communication (since 2017).

6.3. Conferences and workshops

6.3.1. National conferences

The first national science communication conference, 'Science in the Israeli Media: from Apathy to Dialogue', took place on 16 March 2009 at the Technion in Haifa. The event was organised by Professor Baram-Tsabari, and although the potential attendance was a complete unknown prior to the conference, it ultimately attracted 85 scientists, science journalists, formal and informal science educators and university spokespersons. This confirmed an assumption that many people are interested in this topic. An important institutional outcome was the collaboration between Technion's researcher Baram-Tsabari and Avital Baer, (then) Director of Media and Public Relations for the Israel Academy of Sciences and Humanities.

In an effort to establish and consolidate the budding Israeli science communication community, Baram-Tsabari and Baer have organised five more national conferences with various academic collaborators. The second Israeli Science Communication Conference took place immediately after the first, on 24 December 2009 in Jerusalem. It was part of the National Academy's 50th anniversary events and the first conference to involve international speakers.

The third conference was the first to have an academic track where studies were presented as well hearing about science communication initiatives.

The fourth Israeli science communication conference was a two-day event and marked Technion's Cornerstone Centennial. It was the first to include master classes for scientists and science writers. The fifth conference entitled 'Mada, Yeda, o Dea' [Science, Knowledge or Opinion] in 2013 was at Tel Aviv University's Edmond J. Safra Center for Ethics. A dozen international science journalists attended it as part of the first science journalist mission to Israel sponsored by the National Academies and the Jerusalem Press Club.

In parallel, with support from Professor Ruth Arnon and Dr Meir Tzadok, Baram-Tsabari and Baer founded the Academic Forum for Science Communication in Israel at the National Academy of Sciences. This forum allowed key actors to meet and discuss critical science communication–related issues. It led to the sponsorship of a science communication workshop and educational tour in London in February 2015. Fifteen leaders from the science communication community in Israel participated and the meeting was sponsored by IASH and the British Council.

The sixth Israeli science communication conference at the Davidson Institute of Science Education in 2015 built on the success of earlier events, showcasing local science communication initiatives and a second day of master classes for journalists and scientists, and academic sessions presenting peer-reviewed research.

As the field has matured in Israel, specialised conferences have been organised by the Davidson Institute and the Israeli Young Academy. The Davidson Institute ran three events in 2016, 2017 and 2019 called 'How to Connect People with Science', designed to serve as a meaningful bridge between practitioners and the public. 'Pass it Forward' was organised by the Israeli Young Academy and the Davidson Institute in 2018, bringing together humanities, social sciences and STEM communicators to discuss mutual challenges.

6.3.2. International conferences

In 1982, Hillman organised the first international science communication seminar in Israel: the Batsheva Seminar was attended by an invited group of science museum directors from leading science museums in the United States and Europe and researchers from Israel's higher education institutions. It was held under the auspices of the National Academy of Sciences and was the first in a series of meetings that resulted in the construction of science centres in Israel.

In 1998, 10 international science museum directors attended an international seminar to promote the building of a new science museum in the south of Israel. Along with representatives of science museums in Israel, they discussed the challenges facing all the science museums and formulated recommendations. These became the guidelines for the Carasso Science Park, which opened in 2013.

A further meeting in 2000 at the Bloomfield Science Museum set up an Israeli umbrella organisation for all bodies and institutions to create a fruitful dialogue between the public, scientists and authorities around the issue of science and community. The seminar 'Understanding Science' adopted as their model the British Committee on the Public Understanding of Science (COPUS), founded in 1985 by the British Association for the Advancement of Science (BAAS), the Royal Institution and the Royal Society.

6.3.3. Science communication workshops

In 2011, Professor Bruce Lewenstein from Cornell University led the first science communication workshop for scientists in Israel at the Technion. Other international visitors followed. In 2012, Dr Neta Lipman of the Israel Society of Ecology and Environmental Sciences organised science communication workshops for members of Mimshak, a prestigious postdoctoral program training young scientists to apply science in government administrations. Since then, general workshops for environmental scientists focusing mainly on interactions with the media have been held regularly, attracting an audience of hundreds of scientists.

Since 2010 there has been a growing demand for science communication workshops, fuelled by the needs of individual scholars who wish to enrich their research groups, alongside funders, NGOs, the high-tech sector and research institutions. The Israel Young Academy, the Technion applied science communication group, the Alan Alda Center for Science Communication and the Zukerman foundation have all been involved in providing science communication training, as well as private providers.

6.4. National and governmental programs and funding schemes

Currently, there is no funding scheme or program specifically earmarked for science communication research or practice in Israel. However, a 2018 grant call by the Ministry of Science addresses science policy, and specifically included a subsection on science communication funding. The Israel Science Fund (ISF) and the Israeli Ministry of Science fund science communication research under their general schemes. Recently, the Ministry of Science asked all grant recipients to dedicate a small sum of the grant to public outreach activities. Recipients of ISF grants are now asked to write a lay summary of their findings.

6.5. Public understanding of science and attitude surveys

Since the year 2000, public attitudes toward science have been irregularly assessed by the Samuel Neaman Institute at the Technion, and usually compared to the extent of pride and trust in different institutions (Yaar, 2000, 2006; Yaar and Alkalai, 2010). The Samuel Neaman Institute also sponsored the only survey so far of science coverage in the Hebrew media (Barel et al., 2015).

In 2012, the Israeli Ministry of Science began administering an annual survey of public attitudes toward science in a representative sample of Hebrew- and Arabic-speaking Israeli adults (Smith and Faniel, 2014). Beginning in 2014, the ministry's spokesperson Libi Oz consulted Baram-Tsabari on adding questions to the survey, resulting in a new module on public understanding of science based on questions reported by the US National Science Board's Science and Engineering Indicators (Israeli Ministry of Science, 2015, 2016, 2017, 2018). Attitudes and knowledge are analysed according to gender, age and education, but also regarding first language (Hebrew spoken by Jews or Arabic spoken by Muslims, Christians and Druze minorities). The results are also reported as a function of degree of religious observance among the Jewish population. The main findings are published as a press release and as a full report on national science day, Albert Einstein's birthday. Generally speaking, the Israeli public takes great pride in Israeli science, technology and innovation and is convinced of their importance for national security, prosperity and quality of life (Israeli Ministry of Science, 2015, 2016, 2017, 2018; Yaar and Alkalai, 2010). Overall, science content knowledge levels are comparable with US statistics (Table 19.1), with major disparities between secular and religious people as regards the origin of the universe and evolution (Israeli Ministry of Science, 2018).

Table 19.1: Percentage of adult representative sample (n = 500) correctly answering true/false and other questions, by gender, religiosity and sector.

	All	Gender		Religiosity (Jewish sector)			Sector	
		F	M	Secular	Traditional	Religious and Ultra-orthodox	Hebrew speakers	Arabic speakers
Horoscope is an exact science	89	87	92	92	90	91	91	79
The centre of the Earth is very hot (True)	84	84	84	89	84	76	85	78
The continents on which we live have been moving their locations for millions of years and will continue to move in the future (True)	82	84	81	89	85	61	83	80
All radioactivity is man-made (False)	73	67	80	77	73	77	76	59
It is the father's gene that decides whether the baby is a boy or a girl (True)	73	80	65	77	72	65	73	71
According to astronomers, the universe began with a big explosion (True)	70	66	74	81	70	54	73	54
Human beings, as we know them today, developed from earlier species of animals (True)	64	62	67	88	60	21	68	44
Electrons are smaller than atoms (True)	64	63	65	63	60	64	62	73
Lasers work by focusing sound waves (False)	60	55	65	61	62	62	61	53
The universe began with a huge explosion (True)	60	55	65	80	52	21	61	51
Antibiotics kill viruses as well as bacteria (False)	52	49	55	62	42	55	55	35
Does the Earth go around the Sun, or does the Sun go around the Earth? (Earth around Sun)	81	77	86	79	79	80	79	88
Do you agree with the statement 'Earth is getting warmer due to human activity'?				Secular	Traditional	Religious and Ultra-orthodox	Hebrew speakers	Arabic speakers
Yes, the Earth is getting warmer due to human activity (True)	78	NA	NA	84	78	63	78	78
No, the Earth is getting warmer but not due to human activity (False)	4	NA	NA	1	1	5	2	15
No, Earth does not get warmer (False)	12	NA	NA	10	16	23	14	3

Source: Israeli Ministry of Science (2018).

7. Concluding remarks

Science communication in Israel is thriving. Even though it lacks national funding and institutional infrastructure, it is spread widely across the country through various outlets. Science was not a priority in the main media outlets of the 20th century, and the media crisis of the 2010s further diminished the availability of quality science coverage in print newspapers, magazines and on television. However, individuals and institutions are successfully working today to fill in this void with dissemination, engagement and participation-based initiatives, alongside university-based research, research–practice partnerships and training opportunities for scientists. The four science museums in Israel today serve as hubs for science communication for people from all Israeli sectors and are spread from north to south. All operate under the academic auspices of one of the leading research institutions.

Alongside this exciting capacity-building among local science communicators, the next stage should be greater involvement and interest in science communication by funders and leaders of science in Israel, as a means to address social problems and to inform science policy (rather than marketing). These institutions have yet to play a major role in changing norms and practice among Israeli scientists.

There is still a long way to go, but while Israel's science communication culture was classified in 2012 as 'fragile' by Mejlgaard and others, judging today in 2020 by the same criteria we would classify it as 'alive and kicking'.

References

Alexander, E. (2008). Public Participation in Planning – A Multidimensional Model: The Case of Israel. *Planning Theory & Practice*, *9*(1), 57–80. doi.org/10.1080/14649350701843853.

Alfasi, N. (2003). Is public participation making urban planning more democratic? The Israeli experience. *Planning Theory and Practice*, *4*(2), 185–202. doi.org/10.1080/14649350307979.

Armon, R. and Baram-Tsabari, A. (2017). Our findings, my method: Framing science in televised interviews. *Public Understanding of Science*, *26*(8), 986–1002.

Armon, R., Barel Ben-David, Y. and Baram-Tsabari, A. (2017). Ribosomes as current affairs: The science of London and Kirschenbaum [Hebrew]. *Media Frames*, *16*, 85–108.

Baram-Tsabari, A. and Lewenstein, B. V. (2017). Preparing Scientists to be Science Communicators. In P. Patrick (ed.), *Preparing Informal Science Educators: Perspectives from Science Communication and Education* (pp. 437–71). Springer. doi.org/10.1007/978-3-319-50398-1_22.

Barel, Y., Baram-Tsabari, A., Peleg, R., Armon, R., and Rave, A. (2015). *Towards evidence-based policy in science communication in Israel: Science coverage characteristics in Hebrew-language print, broadcast and online media, 2013–2014* [Hebrew]. Shmuel Neeman Institute.

Barel-Ben David, Y., Baram-Tsabari, A. and Garty, E. (2018). Are science stories inherently boring? A comparison of public engagement with science and general items on two popular news sites. Presented as an individual talk at the 15th Public Communication of Science and Technology (PCST) conference, Dunedin, New Zealand.

Boas, H. and Baram-Tsabari, A. (2016). The Puzzle of the University on Air A Story of Media and Academia in Israel, 1977–2013. *Israel Studies Review, 33*, 1–21. doi.org/10.3167/isr.2016.310202.

Efron, N. and Davidovitch, N. (2011). *Monitoring Policy and Research Activities on Science in Society in Europe (MASIS), National Report, Israel.* European Commission DG Research and Innovation.

Getz, D., Buchnik, T. and Zatcovetsky, I. (2020). Science, Technology and Innovation Indicators in Israel: An International Comparison – 2019 – Part A – Key figures. Haifa Israel: Samuel Neaman Institute. Retrieved from www.neaman.org.il/EN/Science-Technology-and-Innovation-Indicators-in-Israel-An-International-Comparison-2019-Part-a-Key-figures.

Golan H. (1998). Enlightened ignorance: Representations of science in the Israeli daily press [Hebrew] (unpublished MA dissertation). Tel Aviv University.

Golumbic, Y. N., Baram-Tsabari, A. and Fishbain, B. (2015). Citizen Science: Public participation in science [Hebrew]. *Ecology and the Environment, 6*(1), 14–23.

Golumbic, Y. N., Baram-Tsabari, A. and Fishbain, B. (2019). User centered design of a citizen science air quality monitoring project. *International Journal of Science Education, Part B, 9*(3), 1–19.

Greenspan, I. (2016). Development and Change in the Israeli Environmental Movement: From State Orientation to Civil Society and Partnerships [Hebrew]. *Social Security, 98*, 157–93.

Greenspan, I., Cohen-Blankshtain, G., Geva, Y. and Naali-Yosef, H. (2016). *Public participation in environmental decision-making: International perspectives.* Society of Protection of Nature in Israel.

Israeli Ministry of Science. (2015). Public Perceptions and Attitudes in Israel: Science, Technology and Space. GeoCatography for the Israeli Ministry of Science, Technology and Space, Tel Aviv, Israel.

Israeli Ministry of Science. (2016). Public Perceptions and Attitudes in Israel: Science, Technology and Space. GeoCatography for the Israeli Ministry of Science, Technology and Space, Tel Aviv, Israel.

Israeli Ministry of Science. (2017). Public Perceptions and Attitudes in Israel: Science, Technology and Space. GeoCatography for the Israeli Ministry of Science, Technology and Space, Tel Aviv, Israel.

Israeli Ministry of Science. (2018). Public Perceptions and Attitudes in Israel: Science, Technology and Space. GeoCatography for the Israeli Ministry of Science, Technology and Space, Tel Aviv, Israel.

Katz-Kimchi, M. (2012). Screening science, producing the nation: Popular science programs on Israeli television (1968–88). *Media Culture Society*, *34*(5), 519–36.

Mann, I. (2008). *This is the voice of Israel broadcasting from Jerusalem ...: A nation behind the microphone*. Jerusalem, Israel: Printiv.

Mejlgaard, N., Bloch, C., Degn, L., Nielsen, M. W. and Ravn, T. (2012). Locating science in society across Europe: Clusters and consequences. *Science and Public Policy*, *39*(6), 741–50.

Miron-Shatz, T., Golan, O., Brezis, M., Siegal, G. and Doniger, G. M. (2012). Shared decision-making in Israel: Status, barriers, and recommendations. *Israel Journal of Health Policy Research*, *1*(1), 5.

Negev, M., Davidovitch, N., Garb, Y. and Tal, A. (2013). Stakeholder participation in health impact assessment: A multicultural approach. *Environmental Impact Assessment Review*, *43*, 112–20.

Orenstein, E. D. and Silverman, E. (2013). The future of the Israel environmental movement: Is a major paradigm shift underway? In E. D. Orenstein, A. Tal, and C. Miller (eds), *Between ruin and restoration: An environmental history of Israel* (pp. 357–82). Pittsburgh: University of Pittsburgh Press.

Orr, D. and Baram-Tsabari, A. (2018). Science and politics in the polio vaccination debate on Facebook: A mixed-methods approach to public engagement in a science-based dialogue. *Journal of Microbiology & Biology Education*, *19*(1). doi.org/10.1128/jmbe.v19i1.1500.

Orr, D., Baram-Tsabari, A. and Landsman, K. (2016). Social media as a platform for health-related public debates and discussions: The polio vaccine on Facebook. *Israel Journal of Health Policy Research*, *5*(1), 34. doi.org/10.1186/s13584-016-0093-4.

Sadan, E. and Churchman, A. (2012). The experience of public participation in Israel: Obstacles and opportunities. *Journal of Architectural and Planning Research*, *29*(1), 65–75.

Smith, R. and Faniel, O. (2014). Public attitudes on science and technology related issues. Prepared for the National Committee for R&D in the Ministry of Science, Technology and Space, Jerusalem, Israel.

Tal, A. (2002). *Pollution in a promised land: An environmental history of Israel.* University of California Press.

Yaar, E. (2000). Education, Science and Beliefs in the Eyes of the Israeli Public. Samuel Neaman Institute for National Policy Research. Retrieved from www.neaman.org.il/EN/Education-Science-Beliefs-in-the-Eyes-of-the-Israeli-Public.

Yaar, E. (2006). Science and Technology in the Israeli Consciousness. Samuel Neaman Institute for National Policy Research. Retrieved from www.neaman.org.il/EN/Science-Technology-in-Israeli-Consciousness.

Yaar, E. and Alkalai, Y. (2010). *Trust in the institutions and pride in Israel's achievements in the first decade of 2000.* Haifa: Neaman Institute.

Timeline

Event	Name	Date	Comments
First interactive science centre founded.	Simply Science – Hebrew University Jerusalem	1982	1986: MadaTech, now Israel National Museum of Science, Technology and Space, Haifa
First national (or large regional) science festival.	Science and Music Festival at the Weizmann Institute	2000	
First university courses to train science communicators.	Technion – Israel Institute of Technology	2009	Taught by Professor Baram-Tsabari
First master's students in science communication to graduate.	Maya Kallir-Meyrav	2014	Topic: the learning of early career scientists in a science communication course
First PhD students in science communication to graduate.	Dr Ran Peleg	2013	Topic: Theater as a venue for science education and communication
First national conference in science communication.	'Science in the Israeli Media: From Apathy to Dialogue'	2009	

Event	Name	Date	Comments
First significant initiative or report on science communication.	'Towards evidence based policy in science communication in Israel'	2015	Report led by Yael Barel on science coverage in Hebrew language print, broadcast and online, 2013–14
National Science Week founded.	Occurs around Science Day on 14 March	2001	The National Science Day was announced by the Knesset (Israeli parliament)
First significant radio programs on science.	*Meetings with Animals and Plants*	1959	Hosted by Azaria Alon. Holds *Guinness Book of Records* as the longest-running host of a radio program
First significant TV programs on science.	*Mada va'Daat* [Science and Knowledge]	1968	
First awards to scientists or journalists or others for science communication.	Professor Peter Hillman (award from the Ministry of Science)	2002	This award was given only once

Contributors

Professor Ayelet Baram-Tsabari is an associate professor at the Faculty of Education in Science and Technology at the Technion – Israel Institute of Technology.

Dr Daniela Orr is a researcher and a lecturer in the fields of science communication, health and risk communication at the Faculty of Education in Science and Technology at the Technion – Israel Institute of Technology.

Avital Baer has been director of media at the Israel Academy of Sciences and Humanities for the last 17 years.

Dr Erez Garty is head of the science communications team at the Davidson Institute of Science Education.

Dr Yaela Golumbic completed her PhD at the Technion – Israel Institute of Technology.

Maya Halevy has been the executive director of the Bloomfield Science Museum, Jerusalem, since 1995 and teaches at Tel Aviv University.

Dr Eitan Krein holds a PhD in chemistry, with a specialisation in organic geochemistry, from the Hebrew University of Jerusalem. He was a co-editor of *Scientific American Israel* until 2017.

Dr Adi Levi is the scientific director of the Israel Society of Ecology and Environmental Sciences (ISEES).

Dr Noam Leviatan is a science journalist for the Davidson Institute website.

Dr Neta Lipman is the director of the Israel Society of Ecology and Environmental Sciences (ISEES).

Dr Ronen Mir is the founding director of the Schwartz/Reisman Science Education Centers at the Weizmann Institute of Science, Rehovot, Israel.

Ettay Nevo is editor-in-chief of the Davidson Institute website.

20

ITALY

The long and winding path of science communication

Giuseppe Pellegrini and Andrea Rubin

1. Setting the scene

The relationship between science and the general public in Italy is longstanding. The publication in the 17th century of *Il Saggiatore* by Galileo Galilei is an early example of popular science communication. Writing in vernacular Italian, Galileo, the father of modern science, became Italy's first important communicator. His work used language that would be accessible to a non-expert readership, but at the same time in *Il Saggiatore* Galileo also focused on topics that were subjects of dispute among scholars.

Nor should we forget the work entitled *Il dialogo sopra i due massimi sistemi del mondo* [The Dialogue on the Two Chief Systems of the World], in which Galileo proposed a debate between two experts and a reader, who in the course of the narration seeks explanations, starting from simple observations on everyday life.

With Galileo, Bacon, Descartes, Harvey and other scientists, the figure of a person dedicated to the sciences was established in Europe in the 17th century, presenting a new method and a new vision of the world. In Italy, in the wake of Galileo, various scientists, such as Torricelli, Redi, Morgagni, Cassini and others, followed the experimental method and asserted themselves not only in Italy but also in the rest of Europe. Many scientists were financially assisted by local patrons at a time when Italy was still divided into several autonomous states.

In this particular cultural context, scientific divulgation was limited to groups of scholars, unlike other countries such as England, where science courses were organised (Golinskj, 2002). In Italy, scientific divulgation occurred mainly in institutions such as the Accademia dei Lincei and the Accademia del Cimento, where communication focused on scientific experiments and philosophical discussion was developed with the support of the State.

In the 18th century, a public space dedicated to science communication developed progressively in Europe and in Italy. Middle-class citizens could participate in experiments organised by the academies and carried out in an open setting. From the mid-18th century, Italian journals publishing scientific articles became more popular and scientific periodicals developed, especially in northern Italy (Delpiano, 1989). The institutionalisation of research practices ensured a degree of continuity in the activity of scientists, helping both the dissemination of scientific knowledge through the press and the development of specialised disciplines (Farinella, 2003).

The early 1800s saw the beginning of two major developments: a second industrial revolution and the *Risorgimento*, the 'resurgence' movement that aimed to unite the individual Italian states and form a national community. Italian scientists contributed in important ways (Ciardi, 2013) including through their organisation of congresses that strengthened the community of scholars. Italian scientists modelled their behaviour on their British counterparts, and this contributed to the outstanding success of Italian popular science in the 1870s and 1880s. The decades following political unification in 1861 represented the period of greatest success for the 'Science for All' movement in Europe (Govoni, 2007).

Interest in the sciences grew at the end of the 19th century with an increase in the number of public and private initiatives aimed at promoting scientific interests. Science was professionalised and became linked to industrial development. These developments established the need to make the general public aware of technological innovations such as electricity and transport.

The Milan International Expo held in 1906 represented one of the most important events related to the public communication of science and technology. Other contributions were conferences, presentations and the publication of a variety of popular magazines (such as the *Nuova Illustrazione Universale*) aimed at disseminating scientific culture in the country. The outbreak of World War I in 1914 caused many magazines stop publishing and only one periodical, *La Scienza per Tutti*, managed to continue during the war period (Battifoglia, 2004).

The end of World War I saw the introduction of research policies aiming to develop laboratories and institutions and foster scientific and technological autonomy in the country. The National Research Council (Consiglio Nazionale delle Ricerche, CNR) was founded in 1923 to promote the application of the results of research, and is still the most important body of its kind in Italy.

The Fascist regime of the 1930s was highly destructive to science, making matters steadily worse. Mussolini's government controlled the dissemination of information and reduced investments in research and innovation to organisations such as the Accademia Nazionale dei Lincei, one of Europe's oldest scientific institutions which was founded in Rome in 1603. At the same time, his fascist government established and financed the short-lived Reale Accademia d'Italia [Royal Academy of Italy] in 1929; and also supported important scientists like Enrico Fermi. The number of graduates in scientific subjects dropped from 15.9 per 100,000 inhabitants in 1925 to 11.6 at the beginning of World War II. There was no freedom of the press and all initiatives involving free expression were controlled by the regime. Scientists could not express themselves freely and scientific activity was strongly conditioned by the choices made by Mussolini's government.

Fascism led to the removal of the best brains from every area of society. Many scientists and scholars were imprisoned and others emigrated. In addition, scientific institutions were heavily damaged in the final phase of World War II by a combination of Allied bombing raids and the systematic requisitioning of laboratory instruments by the retreating German army. By 1945, Italian science was in ruins.

The first post-war government led by Alcide de Gasperi did not consider research as a priority. The United Nations Relief and Rehabilitation Administration showed minimal interest in the resumption of scientific activities. It did propose the construction of a penicillin factory, but at the same time forced a reluctant Italian government, when science was not high on the list of national priorities, to make an extraordinary 200 million lire contribution to the Italian National Research Council.

But science had its champions. The freedom of the press had been restored and some intellectuals, including Ludovico Geymonat, sought to bring science back to the centre of public attention, attempting to promote and draw attention to the value of a new scientific humanism to avoid the traditional contrast between humanistic culture and scientific culture. An important role was played by private businesses: Olivetti in the IT sector and Montecatini in the chemical sector, are two examples of enterprises that engaged in research and development. The former produced what is considered to be the world's first personal computer, and the latter allowed Giulio Natta to win the Nobel Prize for chemistry in 1963 in recognition of his studies of high polymers.

Physicists had a great influence in the post-war reconstruction of Italian science, taking advantage of the international prestige enjoyed by Nobel Prize winner Enrico Fermi and the school in via Panisperna. In the early 1950s, the National Institute of Nuclear Physics (Istituto Nazionale di Fisica Nucleare, INFN) and the National Nuclear Research Committee (Comitato Nazionale per l'Energia Nucleare, CNRN) were created. The INFN managed fundamental research, while the CNRN assumed the main responsibility for the Italian nuclear power sector. Nuclear physics was elevated to a position of supremacy in the Italian scientific sector, not only in terms of funding but also with respect to national and international cultural prestige. Italy became one of the leading countries in European scientific integration for large structures such as the CERN, EURATOM (the conventional name of the European Atomic Energy Community) and the European Space Research Organisation, and the Italian physicist Edoardo Amaldi was appointed as the first Secretary-General of the CERN.

These historical developments in science and research helped shape science communication, which had worked through developments of its own (Govoni, 2002). The Italian word *divulgazione* [divulgation], as 'science communication' is known in Italy, dates back to the 16th century. The term 'popular science' was previously used in the Italian context; however, the use of the adjective 'popular' was excessively reminiscent of the idea of a 'popular culture', from which individuals involved in scientific divulgation would tend to distance themselves. Corresponding to the Italian word *divulgazione*, the English term *popularisation* appeared between 1797 and 1801, and the French word *vulgarisation* came into use between the 1850s and 1870s.

2. Changes from the 1960s onwards: Research policies

The growth of science and technology in Italy after World War II followed a different path to other European countries and the United States, which had grasped the political and economic importance of scientific discoveries for the development of nations. Numerous leading researchers had emigrated following the anti-Jewish measures of the Fascist regime (Israel and Nastasi,

1998) and research policies had slowed down with a reduction in investment. The only exception was nuclear research, in which there was a renewed commitment, both private and public.

In 1962, the government carried out various reforms and, in particular, nationalised the production of electricity. The conflict this policy caused led to crisis, and subsequent scandals (the 'Ippolito affair') paralysed the Italian scientific system. However, the CNR, untouched by the scandals, embarked on a path of renewal that established new committees (including one for the humanities), additional facilities for the organisation of laboratories and research programs, and greater integration with industry. The number of national committees rose from seven to 11, and financing was raised from 4 to 6 billion lire in 1961 (Simili, 2013).

Over a 10-year period, from 1958 to 1968, the share of national wealth allocated to research rose from 0.3 per cent to 0.7 per cent. This addressed a widespread concern about the backwardness of Italy in certain technical sectors and a lack of technical and scientific personnel in particular. Despite these increases, funding was still significantly lower than other European countries such as Belgium or the United Kingdom, or the United States, which was allocating 3 per cent of its national income to research activities (Bucchi, 2001). Nor did the boost to funding spare Italy from severe criticism from the OECD, whose report highlighted the backwardness of Italy's research system in comparison with other developed countries (OECD, 1975a).

Despite changes to the administration of research policies in the 1980s (including the establishment of the Ministry of Education, Universities and Research, which reaffirmed the importance of relating research activities to the content of university courses), the decade ended with another harsh judgement concerning Italian scientific and technological policies by the OECD. They wrote another report that showed that Italy was under-investing in research: 1.29 per cent of GDP against an average of 2.5 per cent in OECD countries. By 2017, this percentage had barely increased and funding for public institutions and universities has actually gone down (OECD, 1975b).

To complete a gloomy picture, the OECD report showed that the number of researchers was below average levels: 27 per 10,000 inhabitants against an OECD average of 49 (OECD, 1991). The distribution of researchers was severely unbalanced, with 24 researchers for every 10,000 inhabitants in the central-northern area of the country and only three in the southern regions (Cannavò, Agnoli and Ciampi, 1989). From the 1960s onwards, the Italian government had chosen to focus its economic policies on public expenditure for goods and services and rather costly welfare programs.

3. Changes from the 1960s onwards: Science in the media

Until the end of the 1950s, the popularisation of the sciences was entrusted to periodicals. This built on a tradition of the post-unification period some 70 years earlier, featuring an authentic flourishing of the production of scientific information for the public, marked by the visibility of scientists who became authors of editorial series and magazines (see Govoni 2002, 2011). Only a few weeks after its first appearance in 1876, the principal national daily newspaper the *Corriere della Sera* ran a science story on its front page (Caprara, 2009). But this was a relatively short phase, and the tradition had begun to wane by the end of the 19th century.

In Italian newspapers the dedication of an entire page to scientific subjects did not occur until 1958, appearing first in the newspaper *Il Giorno*. Subsequently, and inspired by the 'race to conquer space', newspapers such as the *Corriere della Sera* and *La Stampa* began to dedicate pages to science and technology.

Until the 1950s public radio was an important medium for the dissemination of scientific news. The advent of television in 1954 was a significant turning point, allowing the public to see science through television quiz shows, specialist series and documentaries. Science on television increased in the 1960s covering topics such as nuclear energy, astrophysics and medicine. Piero Angela, who produced the most popular television science shows in Italy, directed his first program in 1969: *Il Futuro nello Spazio*. Such programs disseminated information while educating and entertaining the general public (Apollonio, 2002).

Anno I. Domenica-Lunedì, 30 Aprile-1 Maggio 1876 Num. 56.

CORRIERE DELLA SERA

Centesimi 5 in Milano — 7 fuori — numeri arretrati 10.

Ufficii di Direzione ed Amministrazione in Milano, Galleria Vittorio Emanuele, 77

ABBONAMENTO STRAORDINARIO

Un abbonamento straordinario è aperto a tutto l'anno 1876 ai seguenti prezzi:

Per Milano a domicilio.

Dal 1° maggio al 31 dicembre L. 9.

In tutto il Regno franco di porto

Dal 1° maggio al 31 dicembre L. 12.

Estero (Stati dell'Unione Postale)

Dal 1° maggio al 31 dicembre L. 20.

MILANO, 30 APRILE.

DIARIO

GARBUGLI MINISTERIALI

DALLA CAPITALE

SCIENZE

Le fermentazioni.

Figure 20.1: A front page story on science in the *Corriere della Sera* from 1876.

Source: Archivio Storico del Corriere della Sera (Historical Archive of Corriere delle Sera).

4. Moving towards the modern era of science communication

4.1. The media

The period from the 1970s has seen fluctuations in the way the media covered science as the industry responded to changing fashions and economic circumstances. Thanks to the influence of the student movements that arose in 1968, more rapid cultural growth had occurred in the 1970s. Supporting and reinforcing this development, the public was increasingly eager to learn about new fields of knowledge. This was a positive period for scientific information, influenced by the success of the first television programs dedicated to science, the founding of numerous magazines dedicated to the popularisation of the sciences and the emergence of new social concerns (such as that relating to environmental issues), which shifted the interest of the public towards new disciplines. But the competition from other sources had a negative effect on newspaper coverage, and in 1972 the page dedicated to scientific and technical matters in the daily newspaper *Il Giorno* was abandoned.

However, in the 1980s an interest in the 'storytelling of science' was rekindled on the pages of newspapers. The daily newspaper *Il Giorno* once again included a section entitled *L'Era della Tecnica* four times a week. There was an increase in the number of pages dedicated to science and technology in the other main national newspapers (*Corriere della Sera* and *La Stampa*). At the end of the 1980s the creation on a daily basis of a page dedicated to scientific communication had become a clear objective of newspapers.

By 1987 the cycle had changed again, and the only newspaper to publish a daily page dedicated to science was *L'Unità*, originally founded as the official journal of the Italian Communist Party. (At that time in Europe, only the French newspaper *Le Monde* presented on a daily basis a page dedicated to science.) This position may have been influenced by the fact that the Central Committee of the Italian Communist Party included many scientists and the director of the daily *L'Unità* was the engineer Gerardo Chiaramonte. Following an initiative of the latter and, above all, with the support of the journalist Romeo Bassoli, it was decided that the newspaper should present a daily page covering current topics relating to the sciences, technology and the environment. This section of the publication was highly appreciated by the readership, who followed it assiduously, and it became the most popular page of the whole newspaper.[1] A situation such as this had never been experienced by any other newspaper.[2]

Today all the main daily newspapers have a scientific supplement. *La Stampa* publishes *TuttoScienze*, *La Repubblica* publishes *RLab* and the *Corriere della Sera* publishes *Corriere innovazione*. The economic-financial newspaper *Il Sole 24 Ore* also offers a supplement (*Nòva24*) dedicated to technological and scientific development.

Television also ran dedicated programs to scientific subjects. Since it began in Italy in 1954, television has played a fundamental role in increasing the cultural level of the Italian population. It remains as the source of information most frequently used by people who want to learn about science and technology, followed by newspapers, websites, blogs, magazines and the radio (Pellegrini, 2018).

Science programs were included in the programming schedule from the beginning (Schiavini, 1988). *Piccola Enciclopedia Scientifica* (presented in the early evening) lasted about half an hour and was aimed at a very broad and heterogeneous audience. A very different audience, more keen on learning about scientific issues, was catered for in the late evening. Professor Enrico Medi, then professor of terrestrial physics at the University of Rome and director of the National Institute of Geophysics, conducted six episodes of a pioneering program entitled *Avventure nella Scienza*. This program, highly appreciated by the general public, was repeated in 1955 and 1956.

In 1957, *La Macchina per Vivere* focused on the subject of the biological and physiological aspects of the human body and was broadcast in the late evening. Anna Maria Di Giorgio was a communicator who assumed the role of the 'expert' to teach or explain certain topics to a general audience. The model of presentation was that of a university lesson, conducted in a clear and precise manner, and using simple language so the general public could understand.

The role of the conductor/presenter was a typical feature of the communicative modality in other programs broadcast in 1958: *Quarta Dimensione* or *Uomini nello Spazio*. Later, this 'top-down' model of communication tended to be replaced by the 'edutainment' educational model. In the early 1960s, Italy

1 A reconstruction resulting from a conversation on 26 July 2018 between the authors and Pietro Greco, the journalist who was the science specialist of *L'Unità* in that period.

2 An attempt was also made by the newspaper *L'indipendente* in the period 1991–2007.

was engaged in the construction of plants for the production of atomic power and, simultaneously, programs such as *Italia Nucleare*, *Storia della Bomba Atomica* and *Atomo Pratico* were broadcast on television.

Between 1966 and 1973, the program *Orizzonti della Scienza e della Tecnica* was broadcast every Sunday in the late evening. A new range of rather diverse programs flourished, where presenters from different cultural and professional backgrounds presented heterogeneous subjects, using different expressive modalities and communicative intentions. These included *Sapere* (1967), *Planetario* (1968), *Verso il futuro* (1968) and *Dopo Hiroshima* (1969) (Giaccardi, 1988). The medical sector in particular provided a rich source of news (Bauer, 1998), and was communicated to the general public through programs such as *L'altra Medicina* (1970), *Medicina Oggi* (1970) and *Boomerang* (1971).

However, it was only when the program *Quark* was founded in 1981 by Piero Angela (the former presenter of *Il Futuro nello Spazio*, 1969, and *Destinazione Uomo*, 1971) that the Italian public began to follow the development of science and technology on television. It is interesting to note that from the very beginning, the Italian state television company devoted space to science also in programs for children.

Figure 20.2: Piero Angela, founder of the science TV program *Quark*.
Source: Wikiquote.

The media drastically increased their attention to science and technology in the 1990s and the television news program *Leonardo,* produced by RAI (known until 1954 as Radio Audizioni Italiane, now Radiotelevisione italiana), the Italian national broadcasting company, was founded in May 1992. Italy thus proved it was at the forefront of European scientific television broadcasting: *Leonardo* was the first daily scientific news report in Europe.

Radio is the medium least frequently used by Italians to learn about science and technology. Nonetheless, some science communication programs have been broadcast such as *Radio3 Scienza,* the first episode of which went on the air on 6 January 2003. This daily scientific news report is transmitted by Rai Radio 3 five days a week. The program was originally conceived by Rossella Panarese, who is still its editor. This cultural report offers a reflection on scientific issues and the relationship between science and society. It provides the news of the day, explores and offers a reflection on various topics, and presents interviews conducted with researchers. It also provides overviews of books and presents relevant articles that appear in newspapers, discussing their content from a scientific point of view. All *Radio3 Scienza* episodes are available online in the broadcasting archive, offering a form of synergy between different media.

4.2. Science centres

As a member state of the European Union, Italy has benefited from research funds, which to some extent have permitted an investment in public science communication activities. The framework programs have increasingly dedicated resources to foster better communication between the research world and the public. Since 2002, with the adoption of the Science and Society Action Plan, a set of activities has been developed to support communication and dialogue between science and society (European Commission, 2002). These resources have been invested to increase public engagement activities and to modernise old structures. Since the 1990s, significant growth has been achieved in the number of science centres and the modernisation of old science museums.

Since 1995, an educational playground centre for children called the Città dei Bambini e dei Ragazzi has been open to the public in Genoa. Originally inspired by the Cité des Sciences de la Villette in Paris, this was also influenced by the modern approach adopted at the Museo del Balì interactive science centre. The only national museum specifically dedicated to science and technology is in Milan, and Florence has a Museum of the History of Science dedicated to Galileo Galilei. Other Italian science centres include the Città della Scienza [City of Science] in Naples and the MUSE science museum and educational centre in Trento. A branch of the innovative international Science Gallery network will soon be opened in Venice.

4.3. National associations

In Italy, many associations promote the communication of scientific knowledge, not only to experts and scholars, but also to the public. The National Association of Scientific Museums, for example, was founded in 1972, while the Italian Society for the History of Science organises conferences, seminars, lectures, exhibitions and the publication of specialist works to promote research in various fields.

Science and Technology Studies (STS) in general, and studies on the Public Communication of Science and Technology (PCST) in particular, emerge from the convergence of a variety of disciplines and cultural fields.

The Unione Giornalisti Scientifici Italiani (UGIS), the first association of Italian science journalists, was established in 1976 and was one of the promoters of the EUSJA (European Union of Science Journalists' Associations) founded by the presidents of the national associations of seven European countries.

In 2005, a group of Italian academics founded STS Italia—the Italian Society for Social Studies of Science and Technology—to build up an Italian network of researchers oriented towards science and technology and science communication studies, and creating opportunities for the exchange and sharing of research experiences, projects and research activities related to the social dimensions of techno-scientific phenomena.

Regarding scientific journalism, in 2009 the Ettore Majorana Foundation and Centre for Scientific Culture promoted the first course organised by the International School of Science Journalism, the 11th edition of which will be held in 2021. In March 2010, the association Science Writers in Italy (SWIM) was created in Milan and subsequently became associated with the World Federation of Science Journalists (WFSJ). In 2018, it launched the new European Federation for Science Journalism together with corresponding associations in France, Switzerland, the United Kingdom, Germany, the Netherlands and Russia.

Among the initiatives undertaken to promote and support scientific culture is the National Scientific Degree Plan, which has carried out numerous activities to improve the knowledge and perception of scientific disciplines in Italian schools, bringing students closer to the world of research. In 2010, various institutional bodies induced the Ministry of Education, Universities and Research to relaunch the Science Degree Project with the aim of establishing best practices and experimenting with new activities that might further strengthen the relationship existing between scholastic institutions and universities, and also between universities and the professional world.

5. Changing attitudes towards science

On 6 April 2009, a magnitude 6.3 earthquake occurred in the Abruzzo region in central Italy, seriously damaging the city of L'Aquila and destroying some of the nearby villages. There were 309 fatalities. Six years later, the Court of Law at L'Aquila sentenced six experts of the National Major Risks Commission and Bernardo De Bernardinis, all of whom were members of a technical-scientific advisory board of the Italian Civil Protection department, to six years' imprisonment for criminal negligence resulting in multiple homicide. According to the prosecutors' allegations, in the days preceding the earthquake these scientists did not correctly inform the population about the seismic risk. Such an important event drew attention to the role of public communication in the scientific world. In fact, many scientists and the national and international media interpreted the sentence as a condemnation of Italian scientists for not having foreseen the earthquake. In fact the trial was based on the accusation of having provided the local population with 'inaccurate, incomplete and contradictory information'. The defendants were not challenged on account of their inability to predict the earthquake, but for having misinformed the public. The defendants were later acquitted in subsequent sets of proceedings.

This event and other issues involving the judicial system with science have contributed to the idea that in Italy a prevailing 'anti-scientist' attitude is present or, worse still, that a real 'war on science' is under way. Data on the public perception of science, however, would appear to disprove this prejudice.

An analysis of the scientific knowledge of Italians in recent years reveals significant changes. For example, the level of scientific knowledge is one of the most frequently cited indicators in debates on public attitudes towards the sciences through three questions: Is the Sun is a planet? Do antibiotics kill both viruses and bacteria? Are electrons smaller than atoms? Since 2007, the year of the first survey conducted by Observa Science in Society, an Italian Social Research Institute, the level of 'scientific literacy' of Italians has increased. In 2016, it reached a new peak. One third of the respondents answered all questions correctly, and only 13 per cent got them all wrong. In the latest round, 62.5 per cent of Italians know that the Sun is not a planet, while more than half of the population know the function of antibiotics and know that electrons are smaller than atoms.

As in previous years, the survey showed that scientific knowledge decreases in relation to age and increases with the level of education. The highest percentage of respondents unable to answer any of the questions were individuals over 60 years of age and those who have a low level of education. The number able answer all three questions exceeds 50 per cent among graduates.

The use of scientific and technological content in the media is also a relevant indicator of the relationship between science and the general public.

The data for 2017 show an increasing interest in scientific and technological programs or presentations in the press, on television and on the web. Over the last eight years the percentage who enjoy scientific and technological content on television or via the internet has grown by more than 20 points. This is now 58 per cent of the number of people who report reading news relating to science and technology in newspapers at least once a week. The sources of information deemed to be the most credible are public conferences held by researchers and magazines specialising in the divulgation of scientific news. These are judged positively by almost four out of five respondents. Television programs focusing on scientific and technological issues and websites and social media organised by research institutes are not far behind. The levels of reliability attributed to radio programs dedicated to scientific subjects, science pages in newspapers and blogs or the social media presentations of researchers are not quite so high but remain above 65 per cent.

The number of Italians who believe that science and technology change our lifestyle too quickly remains high (seven out of 10), and an equal proportion believe that only science can reveal to us the truth about humans and our place in nature. More than one in two citizens believe that in Italy the freedom of scientists is excessively restricted by religion. In Italy, above all, utilitarian expectations regarding 'relapses' in technological (and scientific) development drive the people's trust in scientific research. Young Italians seem to have great confidence in scientists and in the possibility that science may have a beneficial impact on everyday life. Italian citizens believe that thanks to science and technology there will be more opportunities for the next generation, a consistent result since the mid-1990s. However, this is consistent with all the major international surveys on public attitudes and opinion with respect to science and technology. For a long time, for example, the Eurobarometer recorded the confidence in science shown by Italians, establishing that it is in line with—if not higher than—the European average (European Commission, 2001, 2005). In public debate on science, some people express a hostile attitude towards science and those representing the scientific world. However, empirical studies show this is a minority view:

from 2011 at least 73 per cent of Italian citizens have recognised the benefits of science, and occasionally this proportion has risen to over 80 per cent (European Commission, 2014; Saracino, 2017).

In recent years, exposure of the Italian public to techno-scientific issues has increased. The number attending centres dedicated to the sciences, museums and scientific exhibitions, public participation in festivals or open meetings, and conferences held by scientists and public engagement initiatives has increased (Pellegrini, 2018). In fact, public speaking has been a traditional activity of scientists and researchers for over 25 years. The Italian initiative Settimana della Cultura Scientifica e Tecnologica, first held on the 18–23 May 1991, became the European Week for Scientific Culture in 1993, and in the period 1989–2005 there was a great increase in science events, festivals, conferences and in the publication of books.

Italy is also active in public engagement initiatives: an edition of the EuroScience Open Forum was organised in Turin in 2010, and the next edition is scheduled for Trieste in 2020 (although COVID has caused a suspension to registrations). The Genoa Science Festival, the largest European event dedicated to the sciences, is held in the region of Liguria, but many other initiatives have been organised in other parts of the country including Trento, Iglesias, Mantua (Food&Science Festival), Oristano, Agrigento, Palermo, Cagliari, Frascati, Turin, Naples, Trieste, Bologna, Bergamo, Perugia and Spoleto. Italy has had five European projects funded in relation to the organisation of the European Researchers' Night.

The increasingly important role of the internet in the dissemination of news has modified the science communication process (Trench, 2007) and the use of science news by the public. More importantly, and with respect to the purposes of this sub-section, we may note that transformations occurring in the practices of specialist communication are flanked by evident changes also in the field of the public communication of science. In particular, interactive digital media—the so-called Web 2.0—greatly widens the breadth of possibilities to communicate science at the popular level through blogs, videos, interactive infographics and podcasts. Many programs or publications in Italy exclusively dedicated to online scientific popularisation have flourished, including *Scientificast.it*, *Galileo* or *OggiScienza*.

The 'digital revolution' in science communication began with the advent of 'science blogs' in the early 2000s, when scientists began to set up web pages to inform the public about their work (Tola, 2010). Although their significance tended to be overrated, these resources acquired an excellent reputation for their capacity to communicate (Yo and Peters, 2016), and scientific and popular science journals copied the blog process (Kouper, 2010; Trench, 2008) followed by the digital editions of Italian newspapers. This was an innovation of scientific journalism. In Italy, a differentiation between the content of hardcopy newspapers and their online versions has been a recent development. Multimedia products have given online newspapers new opportunities to communicate with and attract the public.

This profusion of online material has caused an information overload. While the internet represents an opportunity to support and improve scientific communication and contribute more effectively towards public discussion, participation and dialogue (Trench, 2007), it has also caused problems. The validity and reliability of the information available through the internet has been questioned and the terms 'fake news' and 'post-truth' may now refer to matters discussed and presented in the field of science communication. This is a challenge for scientific journalism, in Italy and internationally.

Nor is there a scarcity of private initiatives in the field of science communication. Psiquadro is a science communication company established in 2002 by scientists and science communicators working in the field of science communication since the 1990s. It contended with and filled a gap present in the national scenario, and soon acquired a solid international reputation[3] that allowed it to introduce important initiatives such as FameLab Italia or to participate in European projects for the organisation of the European Researchers' Night. Psiquadro has organised four editions of the Perugia Science Fest and other initiatives including Einstein's Island. Since 2003, it has developed a strong collaboration with important national research institutions and the Ministry of Education, Universities and Research.

6. University research and courses

If, as we have seen, an increasing number of 'popular science' books has been published since the 18th century (Turney, 2007), the development of large scientific and technological exhibitions and magazine articles aimed at satisfying the growing interest of the public (Raichvarg and Jacques, 1991) and research in the field of science communication have occurred only over the last 50 years (Gascoigne et al., 2010; Trench and Bucchi, 2015). Scientific communication has become a dynamic and interdisciplinary field of research that draws on a wide range of disciplines and includes a wide spectrum of scientific approaches (Schiele, Claessens and Shi, 2012). Scholars in this

3 See EUSEA – European Science Events Association: eusea.info/.

field are typically trained in social science disciplines, such as sociology, communication studies, media studies or in related fields of humanistic disciplines, such as philosophy or rhetoric (Hornig Priest, 2007, 2010).

In the Italian academic world, science communication may be seen as a sector that still has to be developed. This is due to a series of historical reasons. One of these reasons is that in the past 'scientific culture' and the 'humanities' were kept separate from a prevailing idealistic culture. The reform of the scholastic system (1923) promoted by Giovanni Gentile and the influence of the thought of the philosopher Benedetto Croce privileged humanistic culture.

It was the philosopher Ludovico Geymonat who first introduced the academic discipline of philosophy of science in Italy at the University of Milan in 1956. This discipline had an important role in the development of the social studies of science. The work of Paolo Rossi, a science historian, is also an important milestone in the academic reflection on social studies of science, particularly his interest in the 'scientific revolution' of the 17th century. More specifically, he identified in this historical period the moment when a revolutionary change occurred in the manner of engaging in scientific activities. This was promoted by a series of factors, such as the new vision of nature, no longer divided between natural and 'artificial' bodies, the continental dimension—or world—of new scientific culture, autonomy from religious thought, the publication and dissemination of results and, above all, the formation of an independent international scientific community.

In the wake of these early works, a multidisciplinary and heterogeneous group of science communication scholars was established, including sociologists specialising in scientific knowledge, historians of technology and philosophers of science.

In 1993, a professional course in science communication was created at the International School for Advanced Studies (Scuola Internazionale Superiore di Studi Avanzati, SISSA) in Trieste. The course, initially a series of seminars for journalists, was referred to as a master's-level degree course in science communication from 1994 onwards, when the master's degree (ISCED – 5A) as such did not exist in Italy.

Another postgraduate course in science communication (Journalism and Institutional Communication of Science) was introduced at the University of Ferrara in the academic year 2000/01. Recently, other postgraduate courses in this field of study have been held at the Milano-Bicocca University, the University of Padua, at the Sapienza University of Rome and, since the beginning of the 2018/19 academic year, at the University of Trento.

Only a few undergraduate courses in science communication are currently available. These include a course on sociology and science communication at the Milano-Bicocca University, a course held at the University of Insubria as part of the syllabus leading to the science of communication degree at the University of Turin (mathematics degree syllabus) and Teaching and Understanding Science, which is part of the degree course in educational sciences at the University of Parma. The fact that only a few courses are currently available may be attributed to two factors in particular: a) research activity relating to science communication is a very recent development in the whole of Europe; and b) in Italy, media scholars and social scientists show little interest in science communication.

In Europe, it was only in 1992 that the scholarly journal *Public Understanding of Science* was founded. In 1994, the oldest peer-reviewed journal in this field, *Knowledge: Creation, Diffusion, Utilisation*, changed its name to *Science Communication*. It was not until 2002 that the first Italian scholarly journal in the field of science communication was published: the *Journal of Science Communication* (*JCOM*), edited by the SISSA in Trieste. At the moment *JCOM* is the only peer-reviewed journal published in Italy specifically dedicated to science communication. This open-access journal was founded when a group of lecturers and former students who had been awarded the master's-level degree in science communication concluded that training should include a commitment to research on science communication issues. For the first time in Italy the favourable educational environment of the master's-level degree course promoted the awareness that the community of professionals—not only academics—and sociologists in particular should identify more specific and systematic analytical instruments in order to comprehend the role and functions of communication in the science–society relationship. The insight and the proposals of the science journalist Pietro Greco were recognised and granted the necessary institutional support at the SISSA (Pitrelli, 2009). Another journal focusing on the relationships between science, technology and society is *Tecnoscienza – Italian Journal of Science & Technology Studies*.[4] It is an open-access journal for academic discussions of religious, gender-based, environmental, ethical and political topics about science and society.

The early years of the 21st century have been a particularly prosperous period for research activities in science communication. In 2002, the first National Conference in Science Communication was held in Forlì, and three years later a group of social scientists gathered at the Observa – Science in Society

4 The journal is available at www.tecnoscienza.net/index.php/tsj/index.

institute, a non-profit, independent, legally recognised research centre that promotes the study and discussion of the interaction of science, technology and society, with a view to stimulating dialogue among researchers, policymakers and citizens. Based in Vicenza, in the North East of Italy, it has published annually since 2005 the *Science Technology and Society Yearbook*, which probably represents the most complete and updated source of data and information on the relationship between the Italian public, science and technology (Pellegrini and Rubin, 2020). This is the most authoritative reference for those who aim at reconstructing the position of Italy on the relationships between public opinion and the principal technical and scientific issues within the national public debate, the image and reputation of science and its producers, and media coverage of the most topical scientific issues. The activities of the Observa research centre are supervised by an international and interdisciplinary scientific committee. Its activities focus on three main areas: science communication, research and innovation policies and science, citizens and technology. Observa was founded because in Italy, unlike other European countries, there were no research centres that dealt with the relationship between science and society and, in particular, the role of science communication.

In 2012, collaborating with the PCST International Network, the Giannino Bassetti Foundation, the Galileo Museum in Florence and the National Institute for Astrophysics (INAF), Observa organised for the first time in Italy the 12th International Public Communication of Science and Technology Conference. The conference attracted 700 registered participants from 50 countries. A total of 368 papers were presented together with over 450 presentations of various kinds.

Through the National Agency for the Evaluation of Universities and Research Institutes (ANVUR), in 2004 the Italian universities introduced a program involving the analysis and assessment of activities of a social, educational and cultural nature that produce public assets not directly linked to the initiation of innovative processes by enterprises. Many of these activities relate to the public communication of science and are commonly referred to as 'third mission' activities of the universities. This process has been adopted by universities across Europe and the Observatory of European Universities (OEU) was called upon to study the various activities of universities, drawing attention to their relations with enterprises, government authorities and society with a view to measuring not only economic effects but also the impact on public policies and on cultural and social life. The OEU examines four economic dimensions and four social dimensions.

In 2006, the University of Turin established the Agorà Scienza Inter-University Centre, and by 2009 all of the universities in the Piedmont region were participating. The centre facilitates the dissemination of science, contributing towards the dissemination of scientific reasoning and knowledge among citizens (cf. 'scientific literacy') with a special focus on schools and, in a symmetrical manner, making university researchers aware of their responsibilities towards society. In other words, Agorà Scienza undertakes the task of promoting public engagement as an objective of the third mission assigned to universities.

7. Latest developments and issues

Public engagement programs have been activated not only to inform but also to actively involve citizens and civil society organisations and develop a public debate that will allow for an appropriate orientation of choices regarding research policies. In this process, the media are mainly involved in expanding the communication offered through broadcasts, events, and television and internet programs. Italy is active in the promotion of scientific topics. For example, five projects have been established to implement the European Researchers' Night.

This process of the dissemination of knowledge and the involvement of the public occurs parallel to a considerable effort made by European universities to meet their third mission obligations for 'the generation, use, application and exploitation of knowledge and other university functions outside academic environments' (Molas-Gallart et al., 2002). This function aims to broaden the spectrum of intervention by universities, alongside the two main research and teaching missions.

The Italian government has recently financed activities in all fields including the promotion of public science communication initiatives, and above all for the dissemination of the results of scientific activities, through the Research Projects of National Interest (PRIN) fund.

Through the National Agency for the Evaluation of Universities and Research Institutes (ANVUR) the Italian government has also established a program for the study of third mission activities dating back to 2004, including the communication of science, with the introduction of a particular monitoring and evaluation method. This institutional process involves the use of an assessment program managed by the ANVUR relating to the period 2004–2010. The aim of this system is to test result indicators and determine a form of stabilisation in view of a periodic assessment that will be performed on an annual or biennial basis.

Over the years, the Ministry of Education, Universities and Research has promoted and developed the Week for Scientific Culture, which provided a model for the European Weeks for Scientific Culture, promoted in 1993 by the European Economic Community (EEC) as an initiative of the European Commissioner for Research, Antonio Ruberti. Numerous public and private entities organise events for the public, offering encouragement and support and coordination at the local and regional levels.

Despite this effort to focus on research communication activities, incentives for scientists in research institutes and universities to communicate their work are rather poor. In the Italian case, but also in many European countries, career advancement is linked to intellectual production and, to a lesser extent, to the quality of teaching offered to students and public communication activities. Currently there are no incentives and no economic or disciplinary mechanisms to reward third mission activities. Preparing public conferences, holding workshops to involve the recipients of experimentation and activating dialogue and discussion with secondary school students to gather opinions on technological innovations are not considered initiatives for which economic support and/or credits may be obtained. These activities cannot be used to achieve an advancement in the ranking of a university or in the national research setting (Pellegrini, 2016).

8. Conclusions

Italian scientists have always enjoyed broad public visibility. In certain periods Italy was a pioneering state in the field of science communication, but these were followed by darker moments. However, in recent years numerous activities, organisations and initiatives in the field of science communication have allowed Italy to make up for lost time and attain the level of other European countries. There is no lack of excellence.

Scientists, institutions, associations and citizens nowadays have to employ new methods of communication. These affect not only content but often involve the use of new instruments, making it possible to produce and adopt scientific information in ways never before imagined. The capacity to manage such great potential and the resulting complexity currently represents the principal challenge to the development of effective and inclusive public science communication.

References

Apollonio, U. (2002). *Scienza e ricerca: conquiste, sfide e dilemmi*. Rubbettino: Soveria Mannelli.

Battifoglia, E. (2004). Modalità di comunicazione divulgativa nelle riviste italiane di scienza popolare (1788–2002). *JCOM*, *1*(3), 11–18.

Bauer, M. W. (1998). The medicalisation of science news – from the 'rocket-scalpel' to the 'gene-meteorite' complex, *Social Science Information*, *37*, 731–51. doi.org/10.1177/053901898037004009.

Bucchi, M. (2001). Ricerca, politica della. In *Enciclopedia delle Scienze Sociali Treccani, IX* (pp. 245–58). Roma: Istituto dell'Enciclopedia Italiana.

Cannavò, L., Agnoli, M. S. and Ciampi, L. (1989). *Professione scienziato: Organizzazione della ricerca pubblica e professionalità in Italia* (edited by L. Cannavò). Milano.

Caprara, G. (2009). *L'avventura della scienza: Sfide, invenzioni e scoperte nelle pagine del Corriere della sera*. Milano: Fondazione Corriere della Sera-Rizzoli.

Ciardi, M. (2013). Scienza e Risorgimento nazionale. In AA. VV., *Enciclopedia Treccani*. Roma: Istituto dell'Enciclopedia Italiana.

Delpiano, P. (1989). I periodici scientifici nel Nord Italia alla fine del Settecento: Studi e ipotesi di ricerca. *Studi Storici*, *30*(2), 457–82. Retrieved from www.jstor.org/stable/20565892.

European Commission. (2001). *Europeans, Science and Technology*, Eurobarometer, 55.2.

European Commission. (2002). *Science and Society Action Plan*. Luxembourg: Office for Official Publications of the European Communities.

European Commission. (2005). *Europeans, Science and Technology*. Luxembourg: Office for Official Publications of the European Communities.

European Commission. (2014). *Special Eurobarometer 419. Public Perceptions of science, research and innovation*. Luxembourg: Office for Official Publications of the European Commission. Retrieved from data.europa.eu/euodp/en/data/dataset/S2047_81_5_419_ENG.

Farinella, C. (2003). Veritas et utilitas. Sull'istituzionalizzazione della scienza nell'età del Settecento. In S. Ferrari (ed.), *Cultura letteraria e sapere scientifico nelle accademie tedesche e italiane del Settecento*. Memorie della Accademia Roveretana degli Agiati', CCLIII, s. II, vol. VII. Rovereto: Accademia Roveretana degli Agiati.

Gascoigne, T., Cheng, D., Claessens, M., Metcalfe, J., Schiele, B. and Shi, S. (2010). Is science communication its own field?, *JCOM*, *9*(3), C04.

Giaccardi, C. (1988). Scienza e divulgazione scientifica in televisione dagli anni '60 a oggi. In Bettetini, G. and Grasso, A. (eds), *Lo specchio sporco della televisione. Divulgazione scientifica e sport nella cultura televisiva* (pp. 91–106). Torino: Fondazione Giovanni Agnelli.

Golinskj, J. (2002). L'Età dei Lumi: la fine della conoscenza naturale 1700–1770. Esperimenti, strumenti e luoghi di lavoro, Storia della Scienza. In AA. VV., *Enciclopedia Treccani*, Roma: Istituto dell'Enciclopedia Italiana. Retrieved from www.treccani.it/enciclopedia/l-eta-dei-lumi-la-fine-della-conoscenza-naturale-1700-1770-esperimenti-strumenti-e-luoghi-di-lavoro_%28Storia-della-Scienza%29/.

Govoni, P. (2002). *Un pubblico per la scienza. La divulgazione scientifica nell'Italia in formazione*. Roma: Carocci.

Govoni, P. (2007). The rise and fall of science communication in late nineteenth-century Italy. In M. Bauer and M. Bucchi (eds), *Journalism, Science and Society*. Routledge.

Hornig Priest, S. (2007). Science Communication: From Your New Editor, *Science Communication, 29*(2), 145–46. doi.org/10.1177/1075547007309214.

Hornig Priest, S. (2010). *Encyclopedia of Science and Technology Communication*. London, UK: SAGE Publications.

Israel, G. and Nastasi, P. (1998). *Scienza e razza nell'Italia fascista*. Bologna: Il Mulino.

Kouper, I. (2010). Science blogs and public engagement with science: practices, challenges, and opportunities, *JCOM, 09*(01), A02. doi.org/10.22323/2.0901 0202.

Molas-Gallart, J., Salter, A., Patel, P., Scott, A. and Duran, X. (2002). *Measuring third stream activities*. Brighton: SPRU.

OECD. (1975a). *Educational, Inequality and Life Chances*, 2 volumes. Paris: OECD.

OECD. (1975b). *Évolution du niveau et des structures de l'activité de R & D dans les pays membres de l'OCDE depuis 1971*. Paris.

OECD. (1991). *Esame della politicascientifica e tecnologica dell'Italia. Rapporto degli esaminatori*. Paris: OECD.

Pellegrini, G. (2016). La terza missione delle università: opportunità e sfide. In Pellegrini G. and Saracino B. (eds), *Annuario Scienza e Società 2016*. Bologna: Il Mulino.

Pellegrini, G. (2018). *Annuario Scienza Tecnologia Società 2018*. Bologna: Il Mulino.

Pellegrini, G. and Rubin, A. (2020). *Annuario Scienza Tecnologia e Società 2020*. Bologna: Il Mulino.

Pitrelli, N. (2009). Filling the gap between theory and practice. *JCOM 08*(03). doi.org/10.22323/2.08030501.

Raichvarg, D. and Jacques, J. (1991). *Savants et ignorants. Une histoire de la vulgarisation des sciences*. Paris: Seuil.

Saracino, B. (ed.) (2017). (a cura di) *Annuario scienza tecnologia e società 2017*. Bologna: Il Mulino.

Schiavini, E. (1988). Scienza e divulgazione scientifica in televisione dalle origini agli anni '60s, In G. Bettetini and A. Grasso (eds), *Lo specchio sporco della televisione. Divulgazione scientifica e sport nella cultura televisiva*. Torino: Fondazione Giovanni Agnelli.

Schiele, B., Claessens, M. and Shi, S. (eds) (2012). *Science Communication in the World*. Springer.

Simili, R. (2013). *L'impresa scientifica 1923–2013*. Roma: CNR Edisioni.

Tola, E. (2010). Una specie in via d'estinzione, Tecnoscienza, *Italian Journal of Science and Technology Studies*, *1*(1), 101–06.

Trench, B. (2007). How the Internet changed science journalism. In M. W. Bauer and M. Bucchi (eds), *Journalism, Science and Society* (pp. 133–41). Routledge.

Trench, B. (2008). Internet: Turning science communication inside-out? In M. Bucchi and B. Trench (eds), *Handbook of Public Communication of Science and Technology*, (pp. 185–98). Routledge.

Trench, B. and Bucchi, M. (eds) (2015). *Handbook of Public Communication of Science and Technology* (Second edition). London; New York: Routledge.

Turney, J. (2007). The latest boom in popular science books. In M. W. Bauer and M. Bucchi (eds), *Journalism, Science and Society*. London; New York: Routledge.

Yo, Y. Y. and Peters, H. P. (2016). *Blogging by scientists: A rare and peripheral activity*, Proceedings of the PCST Conference 2016. Istanbul, Turkey.

Timeline

Event	Name	Date	Comment
First interactive science centre established.	Città della Scienza (Napoli)	1996	Many science museums have opened, including 4 science centres designed for children and teenagers
First national (or large regional) science festival.	Festival della Scienza di Genova	2003	
An association of science writers or journalists or communicators established.	Unione Giornalisti Scientifici Italiani (UGIS)	1966	2010: Science Writers in Italy (SWIM) founded by freelance journalist and communicators 2011: SWIM joins World Federation of Science Journalists (WFSJ)

Event	Name	Date	Comment
First university courses to train science communicators.	Lectures on science communication	1990	There are no degree courses on public communication of science, just lectures
First master's students in science communication graduate.	Master in Comunicazione della Scienza 'Franco Prattico' SISSA Trieste	1993	The first course dedicated to the public communication of science
First PhD students in science communication graduate.	Bicocca-Milano University offered a PhD in science and society	2007	The PhD course was not re-activated after a few years
First national conference in science communication.	National Conference on Science Communication	2002	The conference in Forlì was organised by the SISSA group
National government program to support science communication established.	Law No. 113 on Dissemination of scientific culture	1991	To support and strengthen dissemination of scientific culture and contribute to the protection of the technical-scientific heritage of historical interest preserved in our country
First significant initiative or report on science communication.	European Open Science Forum in Italy, Turin	2010	For the second time in Italy, ESOF will be hosted in Trieste in 2021
National Science Week founded.	National Science Week established	1991	Known as Settimana della cultura scientifica
A journal completely or substantially devoted to science communication established.	*Le Scienze*	1968	*Le Scienze* is the Italian edition of *Scientific American*. Felice Ippolito, created this in 1968, and is still a widely read weekly magazine
First significant radio programs on science.	*Radio3 Scienza*	2003	
First significant TV programs on science.	*Piccola enciclopedia scientifica*	1954	1981: The longest running program is *Quark*. Spin-offs include *Il mondo di Quark* and *Superquark*
First awards for scientists or journalists or others for science communication.	Premio letterario Galileo per la divulgazione scientifica (Galileo Award)	2007	2013: Premio Nazionale di Divulgazione Scientifica [National Science Popularisation Award] was established in 2013
Date hosted a PCST conference.		2012	Florence hosted the 12th PCST Conference

Event	Name	Date	Comment
Other significant events.	European Researchers Night (ERN)	2009	Italy has participated with a high number of projects since ERN started
	Scienza in Parlamento [Science in Parliament]	2019	An independent initiative of a group of researchers, scientists and journalists aiming to put science and technology at the service of democracy
	Patto trasversale per la scienza [Cross-cutting pact for science]	2019	Aims to ensure legislation is based on scientific evidence, to promote the culture of science and to fight hoaxes and pseudoscience

Contributors

Dr Giuseppe Pellegrini is a lecturer in innovation, technology and society at the University of Trento.

Dr Andrea Rubin teaches social sciences theory at the University of Bergamo.

21
JAMAICA
Science communication in the land of wood and water

Zahra H. Oliphant, Cliff K. Riley, Kerry-Ann C. Curtis, Setu N. Monroe, Aisha D. Jones and Charah T. Watson

1. Introduction

The island of Jamaica is located in the Caribbean, with a land area of 10,831 km^2 and a population of 2,728,864 (JIS, 1965; STATIN, 2017). The name Jamaica was derived from the original name *Xaymaca* (land of wood and water) ascribed to the island by the first inhabitants, the Tainos, who noted the abundance of mountains, springs and rivers. The Tainos were wiped out by the Spanish, who were subsequently defeated by the English in 1655. Slavery was instituted under English rule and lasted until full emancipation in 1838 brought freedom to all slaves; from that point, the nation slowly moved towards its eventual independence on 6 August 1962.

Science and technology (S&T) has always been a part of Jamaica's history. Its institutionalisation started during the British era (1655–1962), while the development of S&T infrastructure, promotion of research and development (R&D) facilities to advance agriculture, and the contribution of S&T to national development were features of the 'War Years' (1914–45) (Lowe, Brown and Magnus, 2000). From post-war to independence (1946–62), the development of Jamaica's endogenous capacity began with the establishment of tertiary institutions that trained individuals in S&T and the strengthening of the government policy in this area.

UNESCO reported that Jamaica was one of the earliest developing countries to legislate the use of S&T in exploiting natural resources, being among the first in the western hemisphere to obtain electricity, construct a railway and increase sugar cane production through research (Villavicencio and Ponce, 1990). There have been a number of scientific achievements for this island, especially in the pharmaceutical and nutraceutical industries where patents have been obtained for novel treatments and where research is still ongoing.

The need to communicate science adequately to the populace has been recognised, even if the term 'science communication' is not used in the Jamaican S&T landscape. The efforts made to communicate science fall under the definition of 'popularisation/promotion of science', bringing scientific information to the general public to increase their scientific knowledge and create a more informed society. The following sections will take the reader through how science communication has emerged over time.

2. Planting the seed of science: Pre-independence

2.1. Earliest science institutions

2.1.1. The Natural History Museum of Jamaica

The earliest institution that catered to the dissemination of scientific knowledge was the Institute of Jamaica (IOJ) through its Natural History Division (NHD). The IOJ was enacted in 1879 and the island chemist, J. J. Bowrey was appointed as the first curator (Farr, 1985; *Kingston Gleaner*, 1879). The mission of the museum is to:

> Encourage the study and dissemination of scientific knowledge of Jamaican flora and fauna. Promote the conservation of the Jamaican natural environment. Maintain collections of Jamaican flora, fauna and reference books Assist with the identification of plants and animals found in Jamaica. (Find Glocal, n.d.)

The IOJ is now home to a number of museums; the first one was dedicated to science, underscoring the importance placed on public communication. The NHD stands as one of the oldest science museums in the western hemisphere. Renamed the Natural History Museum of Jamaica (NHMJ), it is open to the public at a minimal cost of J$400 (about US$3). Its permanent exhibitions are generally focused on the island's flora, fauna, geology, geography, ecosystems and ethnobiology. Other disciplines such as physics and chemistry are promoted through the NHMJ's Education and Outreach Department. The science library of the NHMJ consists of over 10,000 volumes of scientific literature and an audio-visual facility; and although the exhibitions may be limited to natural history, they provide access to scientific information for the general public. The museum houses the largest plant and animal collection in the Caribbean. The number of plant specimens in the museum increased from 55,000 in 1965 (JIS, 1965) to over 130,000 in 2016 (Williams, 2016), in addition to several faunal groups and insects.

The museum's location in Kingston means that individuals from more distant parishes within the island face a transportation challenge, especially relevant in early years when infrastructure was not well developed. Additionally, in the years of slavery and those immediately following, popularising science according to the NHMJ's mandate was likely limited to the colonial upper class of the later 1800s to early 1900s.

In 1955, it was reported by C. Bernard Lewis in the newspaper *The Daily Gleaner* that 'the high reputation of our museum [NHMJ] as a scientific institution in the nineties [1890s] declined after Duerden, the marine zoologist, resigned in 1901, for no full-time successor was appointed'. Duerden was the curartor and, as the author noted, although regular citizens did their best voluntarily, lack of support led to a 'deplorable decline'. The decline of the sole museum dedicated to science would have had a negative impact on the efforts being made in the 1950s to popularise science.

2.1.2. Geological Society of Jamaica

The Geological Society of Jamaica (GSJ) was established in 1955 as an organisation of geologists and persons interested in promoting and encouraging the study of geology and its allied sciences (GSJ, 2015). The society promoted geology through regular field tours, public seminars and conferences, as well as publication of their activities in newsletters and journals. Field excursions and public lectures were advertised in *The Daily Gleaner*. On 10 April 1968 it was reported that the GSJ would depart from its normal program of lectures and field excursions and instead open an exhibition to the public on 'Geology in Jamaica' at the IOJ.

In 1958 the GSJ began publication of the journal *Geonotes*, which became the *Journal of the Geological Society of Jamaica* in 1965, and then *The Caribbean Journal of Earth Sciences* in 2000. It is remarkable that the journal has been freely accessible to the public since its inception. With the advent of the internet, the contents of the journal from current and previous issues dating back to 1970 are available for download free of cost.

2.1.3. Scientific Research Council

One of the most significant milestones in Jamaica's scientific history and its attempts to popularise science was the establishment of the Scientific Research Council (SRC) on 6 June 1960 as a statutory body, using the Scientific Research Council Act. Sections 5(1) and (2)(e) of the Act state:

> It shall be the duty of the Council to undertake and foster scientific research in this Island and to encourage the application of the results of such research to the exploitation and development of the resources of the island.
>
> In particular and without prejudice to the generality of the provisions of subsection (1), it shall be the duty of the Council—(e) to establish and maintain a scientific information centre for the collection and dissemination of scientific and technical information.

In noting the vision for the first annual report of the SRC in 1960, Prime Minister Norman Washington Manley stated that 'Jamaica has now taken action to set up a permanent organisation for the promotion of research' (Lowe et al., 2000). The SRC is an important institution that is still relevant today.

2.2. Science communicated via media (1945 – early 1960s)

2.2.1. Print media

The year 1945 marked the end of World War II and, by that year, Jamaica was emerging from the global depression. The way in which science was communicated to the populace and how it was perceived in the immediate post-war era is interesting. The stories told by science as seeds planted in the minds of Jamaicans were examined via the archives of the country's oldest newspaper, *The Daily Gleaner*. Television sets were commercialised globally after World War II, but until they became popular in Jamaica the primary modes of communication were the newspaper and radio.

Agricultural science was an important topic, and a key government institution with responsibility for that area was the Department of Science and Agriculture (DSA). The public was advised via the newspaper on a wide range of topics such as mating and cross-breeding animals, chemistry of soil fertilisation, sugar industry research, use of science in agriculture to improve the country's economy and the R&D tasks of the DSA. One article lamented that scientific inventions were not being embraced in Jamaica to improve agricultural research. The DSA enjoyed high visibility in the print media. By 1948, *The Farmers' Page* was a regular column dedicated to giving information on agricultural science borne out of research. The page served to address a wide range of issues such as successfully harvesting seeds, improving livestock breeding and treatment of diseases. A feature on this page noted that 'science brought to the practice of agriculture such marvels as soil-tests and sucrose-tests, and the vast wonderous field of genetics. Science even taught us to increase our cattle herds through a kind of remote control by distant bulls' (Broderick, 1970).

There was an undercurrent of fear in the scientific articles of 1945 because of the atomic bomb, which ended the war. A particularly scathing article, 'A Balanced Life', denounced the devotion to and worship of science, stating that with the new atomic science the way of life was changing too fast (Simple, 1945). Almost one month later (5 December), another author, as if in direct response, made the argument that the use of scientific discoveries is not for the purpose of advancing war but rather to improve people's daily lives (*The Daily Gleaner*, 1945). The debate was divided: many articles centred on the destructive power of science and its use as an instrument of oppression and enslavement, while other articles sought to assuage fears by providing scientific explanations. By 1948, atomic science was reported less frequently and one article noted that nuclear energy would be a significant power resource in the future: 'the world might have to wait another 10 years before nuclear energy could contribute "anything appreciable" to world power-resources' (*The Daily Gleaner*, 1948).

Although agricultural and atomic science stood out in the mid-1940s, other articles focused on general science and its importance, as well as science regarding drugs, artificial insemination and health (malaria, tuberculosis and vitamin 'P' (now known to be a flavonoid)). In general, the science landscape of Jamaica was undergoing significant changes in this period, and the establishment of the University College of the West Indies (UCWI) in 1948 was a milestone, with the Faculties of Medicine (1948) and Natural Sciences (1949) among the first to be established. In 1951 the *West Indian Medical Journal* (WIMJ) was launched at UCWI as the first peer-reviewed scientific journal to originate in Jamaica. WIMJ is still published today with its target audience limited to scholars and medical professionals.

By the 1950s, scientific themes reported in *The Daily Gleaner* were more varied. The importance of the discovery of penicillin and streptomycin to mothers, a three-part *Triumphs of Science* series in 1950, science as a tool for warfare, geology's importance, discovery of the electroencephalogram (EEG) in medicine, and the death of Einstein were areas covered by the media. In 1954, the newspaper began an irregular series entitled 'Research, Discovery and Invention', and within that inaugural year 12 articles were published. This feature was copied from a British newspaper and highlighted new scientific discoveries, with detailed explanations for how these inventions worked to the benefit of humankind.

Eleven students graduated from the Faculty of Natural Sciences in 1952; these were the first Jamaican-trained scientists to graduate from UCWI. A news article aptly titled 'The First Fruits', published in *The Daily Gleaner* in 7 July 1952, highlighted this significant achievement and signalled to readers that science was becoming more relevant to the country. One of the early graduates was Kenneth Magnus, a Jamaican who went on to become professor emeritus and pioneered teaching programs and initiatives across the Caribbean (NIHERST, 2009). He started the applied chemistry program at UWI, helped to develop the science curriculum for Jamaica's primary and secondary schools, and jointly synthesised the antibiotic Monamycin (named after the Mona campus). Another important milestone in science education occurred towards the end of the 1950s with the establishment of the College of Arts, Science and Technology in 1958.

The science narrative continued into the 1960s with the publication of science articles as well as the continuation of the 'Research Discovery and Invention' feature until its demise in 1964. Thirty-seven articles were published under this feature in 1960–61. It is difficult to assess the effectiveness of science articles in the print media as a way of communicating science because for that era, no readership data are available. For many Jamaicans living in the pre-independence period, copies of the newspaper would be received in their towns only occasionally, on the basis that a community member would be expected to bring the paper back when they returned from a journey. There are no data available on the literacy rate of the population or the extent of the newspaper's reach across the island.

Assuming that readership increased from 1945 to 1961, more Jamaicans would have been in a position to access scientific information via the newspaper. The public at that time would have received scientific information to improve agriculture, health and general information on the new discoveries through research. (Figure 21.1 highlights science headlines.) Access to print media may have been limited to a select number of persons, but radio broadcasts were far more widespread and accessible, particularly after 1950.

Figure 21.1: Snapshots of science articles printed in the newspaper (pre-Independence).

2.2.2. Radio broadcasts

Initially in the earliest years (1939–49), amateur equipment was used for radio broadcasts. These would feature relayed programming from overseas stations (e.g. BBC and NBC Radio). The granting of the first broadcast licence in 1950 expanded radio's reach to rural communities, which meant that more persons were able to access and benefit from any scientific content. The station, time and title of each radio broadcast were printed daily in the newspaper, and programs dedicated to science were aired continually from 1945 to 1961.

Science Notebook was a 15-minute program that aired from 1944 to 1948 on Mondays and Thursdays. The program was broadcast in the evenings anywhere between 6.30 pm and 8.00 pm (depending on the year) and in some instances there was a repeat broadcast at 10.15 pm. For the most part, what was discussed in these broadcasts was not stated in the print media.

However, a few broadcasts highlighted topics such as industrial ophthalmology (1 February 1946) and malaria (25 August 1948). *Science Notebook* was the earliest radio broadcast focusing on science.

Although the last broadcast of *Science Notebook* was in 1948, that year brought three additional science programs. *New Roads in Science and Education* was a 30-minute program that aired at 9.00 pm on Tuesdays from 1948 to 1949. *Science and Everyday Life* aired from 1948 to 1950 as a 15-minute broadcast from Tuesdays to Friday, discussing a range of topics related to everyday living such as water, minerals, the 'Common Leaf'[1] and leather. The longest-running science program of that time was *Science Review*, which aired from 1948 to 1959, receiving both an early and late timeslot: 3.15 pm and 7.15 pm after 1955. After it ended in 1959, the new and sole scientific broadcast was *Talk – Frontier of Science*, which was the first science broadcast to air on a licensed local station, the Jamaica Broadcasting Corporation (JBC) Radio founded in 1959. *Frontier of Science* aired for 15 minutes from Monday to Friday at 2.00 pm (1960–61).

The radio content dedicated to science in the late mid-1940s to 1950s was impressive and showed a clear emphasis on communicating science to the public. The final such broadcast ended immediately before independence in 1962. The next year signalled a different approach to science communication in Jamaica, as science became more institutionalised and communication extended beyond the media.

3. Science growing branches post-independence (1962–89)

Lowe et al. (2000) noted that the first decade of Jamaica's independence (1962–72) saw a rapid growth in public sector institutions for S&T, which was in line with a national strategy for industrialisation and modernisation. Institutions established pre-independence (NHMJ, GSJ and SRC) and continued their individual mandates to promote and educate the public, while new avenues of communicating science were instituted.

3.1. Post-independence science communication

In the years following independence, S&T research focused on agriculture, agroindustry, food technology, nutrition, minerals, energy and the environment. There was no government ministry with responsibility

1 There is no information if the program was referring to a specific type of leaf or the biology of the leaf itself.

for science in this post-independence period, except for the short-lived Ministry of Science, Technology and Environment (January–August 1984). The Department of Science, Technology and Research was formed within the Ministry of Agriculture on 1 January 1985 and combined the functions that previously existed in the Ministry of Science, Technology and Environment. Within this department, the focus was mainly on agricultural research and natural resources conversations, but there is no further record of the department after 1986. At this point, the government's focus on science was primarily to use it as a tool for improving the economy and public lives through improved agricultural techniques.

The actual communication and popularisation of science were activities undertaken by independent groups. Although the SRC had a legal mandate to disseminate scientific knowledge, more than a decade after independence the institution noted in its 1977/78 annual report that the budgetary allocation by the government was insufficient to support its activities (*The Daily Gleaner*, 1979). This suggests that science was not sufficiently high on the government's agenda. On a more positive note, the 1977/78 report noted that the SRC at that time had achieved public acceptance that science and technology is important to national growth through a publicity campaign.

3.1.1. The Jamaican Society for Scientists and Technologists (JSST)

In 1962, the University College of the West Indies achieved independent degree-granting status as the University of the West Indies (UWI). Research scientists from UWI created the Jamaican Association of Scientists in 1966, which evolved into the Jamaican Society for Scientists and Technologists (JSST) in 1978, under the leadership of Dr Henry Lowe. The advancement and application of S&T to the nation's development was the JSST's mandate. Listed in their objectives was 'to promote public appreciation of the roles of Science and Technology in daily life and education', and among their activities was to 'provide for the delivery and holding of lectures, exhibitions, public meetings, classes, examination, seminars and conferences covering a wide variety of topical issues' (Lowe et al., 2000).

The public was invited to attend all meetings of the JSST, allowing them to be informed and also to be part of the conversation. The JSST operated from 1966 to at least 1990 and represented the second institution formed by scientists with an objective of bringing science to the public. However, unlike the first institution (the GSJ, which still operates today), the JSST dissolved in the early 1990s. During its time of operation, however, its work impacted the future science policy (discussed in Section 4).

Jamaica Journal is published Quarterly by the Institute of Jamaica, 12-15 East Street, Kingston, Jamaica, West Indies.

Frank Hill, Chairman.
C. Bernard Lewis, Director.

EDITOR
ALEX GRADUSSOV

Design and Production
RAPHAEL SHEARER

Lithographed in Jamaica
by
STEPHENSON

SUBSCRIPTION

£1 per annum, post paid, for four issues, – 3 years - £2.15.0., 5 years - £4.10.0. Individual copies will be on sale at all Jamaican Book Sellers and Magazine Vendors at 5/- per copy. Bulk purchases will be at special discount rates of 10% for 10 copies, 15% for 100 copies.

Jamaica Journal

QUARTERLY OF THE INSTITUTE OF JAMAICA
DECEMBER '67 VOL.1 NO.1

NEXT ISSUE

1. A Chapter from a Novel in progress by *Orlando Patterson*, illustrated by *Albert Huie*.
2. A short story by *Vic Reid*.
3. The History of the Jamaican Theatre by *Henry Fowler*.
4. The History of Mico College by *A. J. Newman*.
5. The Problem of violence by *Jim Whetton*.

Figure 21.2: Table of contents for the first issue of the *Jamaica Journal* in December 1967.

3.1.2. The *Jamaica Journal*

The *Jamaica Journal* is an academic journal published by the IOJ and its first issue was made available in 1967 (see Figure 21.2). In accordance with the updated IOJ legislation of 1978 to research, study, encourage and develop culture, society and history, the journal features articles related to these three aspects. The first issue included a section entitled 'Science for the Layman', where scientific topics were explained in a format and language that could be easily understood by the general public. The contents of the journal from the first issue to the current one are freely available online to the public.[2]

From 1967 to March 2015, there have been 139 science articles published in the *Jamaica Journal*. It is noteworthy that 100 of these articles were published from 1967 to 1989, but only 39 from 1990 to 2015, representing a significant decline in output. In the immediate post-independence period, most issues contained several science-based articles, but in more recent issues only a few articles are published, and sometimes only one. Since the very beginning, there have always been more articles in the sections dedicated to art and culture.

In the earlier years of the journal up to 1988, Thomas H. Farr contributed eight articles, and from 1984 to 1997, John Rashford contributed seven articles. Throughout the journal's history, there have been contributions from a wide range of scientists in many different fields, and these have been grouped into six categories in Figure 21.3.

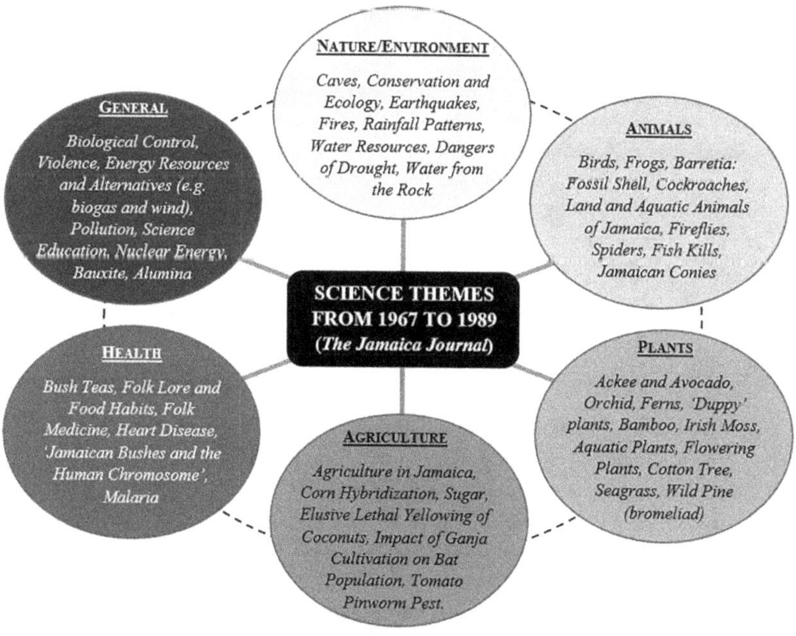

Figure 21.3: Themes covered in articles under six categories under the science sections of the *Jamaica Journal* from 1967 to 1989.

2 Digital Library of the Caribbean – The Jamaica Journal: www.dloc.com/UF00090030/00059/allvolumes?search=jamaica+%3djournal.

3.2. Science in the media post-independence (1962–89)

3.2.1. Post-independence printed media

From the 1960s to 1980s, *The Daily Gleaner* continued to publish science articles that would be both points of information and points of controversy for the public. In 1970, for example, there was an article 'Prepare Yourself for Staggering 70s' that covered the growth of 'test-tube' babies, artificial manipulation of human eggs and the fact that men would have a contraceptive pill by the end of the 1970s. We now know that the latter has only just reached a trial phase, but the author wrote with such conviction that the average reader may have felt that all that was predicted was likely. Other articles of the time covered cancer, contraceptives, nuclear science and pollution.

By the late 1970s, there were more articles in the print media on the use of natural herbs for treating illnesses (e.g. 'Ganja and the Treatment of Glaucoma' and 'Nerves need Calming? Drink Soursop'). 'Science and You' featured every Sunday in *The Daily Gleaner* from 1978 to 1982 and a variety of topics were discussed, such as 'The Issues Beyond' by Dr Henry Lowe, which looked at the future of science where oil would be depleted between 2000 and 2005. An example of the impact that 'Science and You' had on the audience can be found in a letter to the editor (2 April 1980), which responded to 'The Return to Solar' and highlighted the need for solar energy to be used at the National Chest Hospital.

By the 1980s, there was much dialogue on the need for a national science policy and, from the government's perspective, science continued to be an important driver of economic growth, primarily through improved agricultural practices. Martin Henry, who studied both communication and science (discussed further in Section 4.2.1), wrote in an article published 17 January 1982 that the entire society awaited a science policy:

> Nor can we overlook the role of an informed society in the success of any government planning. For many of our people, even among the educated ones, science is a sort of magic word shrouded with mystique. A widespread and basic understanding of science ought rightly to be one of the fundamental goals of national education. The fulfilment of this goal ensures that science-based change cannot be foisted onto an inexperienced population by a paternalistic government. (Henry, 1982)

From the excerpt above, the tenets of science communication are described without the author specifically using that term.

3.2.2. Post-independence broadcast media

Post-independence saw a move towards more local content being broadcast via JBC, with JBC TV airing its first television broadcast one year after Independence Day on 6 August 1963 (Jamaicans.com, 2006). By the 1980s, JBC had television, two radio stations and regional radio stations. From 1965 to 1970, JBC Radio aired a 30-minute feature *Science Corner* and the program was also aired by the Jamaica Information Service (JIS) radio station. As with the pre-independence broadcasts, varied topics were covered.

Science Review was broadcast on JBC Radio in 1972, and it is uncertain if this program was the same as that which aired from 1948 to 1959. According to an article published in *The Daily Gleaner* on 10 September 1972, Alma Mock Yen was responsible for compiling educational broadcasts on JBC, and the *Science Review* series she selected had local and international reports on advances, improvements and problems in S&T. Another science program *JBC Radio Science Magazine* aired for at least one year in 1975.

Television was another means by which science could be communicated to the general public. JBC had an 'Educational TV' component from its inception; and featured from the 1970s to 1980s different science segments such as *Science Grade 6 (to 9)* and *Science I, II* and *III*. The former dealt with science topics being taught in the classroom and was aimed at students, while the latter dealt with general science matters and was aimed at the wider public.

3.3. Increased science focus

At the United Nations Conference on Science and Technology (UNCST) in 1979, Jamaica made a significant contribution and commitments to the Vienna Programme of Action (VPA) that emerged from the conference (Lowe et al., 2000). The Jamaican national paper prepared for the conference highlighted the state of S&T and in particular noted that a state of S&T consciousness must be produced so ordinary citizens can become aware of, and appreciate, modern S&T in their lives. The following is an excerpt from the paper presented at the UNCST:

> At the community level, the mass media, science and technology museums, accessibility to nature and wildlife preserves and annual science exhibitions may be used to popularise S&T, and create a better national awareness of the importance of these assets in every life. These avenues may promote a more deductive approach to national issues and will undoubtedly attract the youth to scientific endeavours … The public should also be educated regarding the benefits of employing scientific methods to the conservation of energy, water and other diminishing resources. (Lowe et al., 2000)

This was the first instance where the government communicated its intent for science on the global stage post-independence. This formally recognised the need for science to be promoted to the general public and the statements communicated this intention to the world by sharing them on the global stage.

Following the Vienna Conference, the roles and functions of the SRC were re-examined in 1979, taking into account its legal responsibility to develop and promote S&T (Section 5 of the Scientific Research Council Act). In the following years, the SRC became more visible to the public and even provided channels for the public to request and receive information from the US Department's National Technical Information Service. The research being carried out by SRC featured in the newspaper under the section 'Scientific Research Council Supplement'.

The SRC had an active public education campaign using roving exhibitions at parish libraries and public exhibitions. These exhibitions featured contributions made by the SRC to national development and covered topics such as agroindustry, microbiology and industrial fermentation (e.g. making fruit wine). Additionally, in its attempt at ensuring that the public was sufficiently informed, the SRC not only disseminated information on science, but also published a directory of all scientists active in Jamaica for the year 1986 (SRC, 1986). The directory listed their qualifications from BSc to PhD, along with the specific field of science and place of employment for each person. Altogether, 674 scientists from 16 fields were listed, with the majority working in civil engineering (182), chemistry (112) and electrical engineering (88). Although agricultural science was important from a national perspective, only 40 agricultural scientists were in the directory. As the first directory of its kind, it is possible that an accurate account of agricultural scientists was not captured, or these scientists may have migrated to other countries.

The first National Science and Technology Conference was held in 1987, an activity that started with the revamped SRC. The conference was held annually up until 2008 (only a few years missed) and then biennially up to the current year (19–20 November 2018). The two-day conference is free to the general public and presents a wealth of information under different themes, most of which address the use of S&T in national development. The main objective of these conferences is to provide a forum for scientists, technologists, entrepreneurs, business persons, policymakers, students and members of the public to share their knowledge and experiences, as well as an opportunity for us to sensitise the nation on the importance of S&T for national growth and development. The conferences are generally well supported by the public and professionals, and 334 persons registered as attendees in 2018. Actual numbers were probably higher. The conference takes the format of a series of public lectures on topics such as health, nutraceuticals and innovation, and conference proceedings are available to the public free of charge at the National Library of Jamaica. The institution of these conferences was an important event in the history of promoting science.

4. Science producing fruit: Policy era (1990–2008)

At the annual awards dinner of the JSST in 1981, the Honourable Dr Ronald Irvine (minister assigned to S&T affairs) spoke of the government recognising the need for a national science policy and challenging scientists and technologists to draft one within six months for presentation to the political directorate (*The Daily Gleaner*, 1981). In response to this challenge, a National Science Policy Committee was formed and at its inaugural meeting on 11 June 1981, Dr Arnoldo K. Ventura, the director of the SRC, was elected chairman. This began the process towards the development and promulgation of the first national science policy in 1990.

The foreword of the National Science and Technology Policy of 1990 (GOJ, 1990) was written by then Prime Minister P. J. Patterson. He highlighted that the scientific community had long advocated for a science policy that 'would guide the development and application of science and technology in the country and form part of the national strategy for development', and this formed the basis for the policy's development. The policy's main goal was to increase the role of S&T in the attainment of economic and social development by bringing about social transformation, removing injustice in the society and improving the quality of life.

The National Science and Technology Policy outlined the importance of the public's awareness of S&T in part of Section 10:

> In the end, the support of science and technology must come from the people. This can only be maintained in the face of competing priorities if the work done is worthwhile and is appreciated as such by the general public as well as by the scientists and technologists. Thus every effort will be made to increase public awareness of S&T and of their social implications, and to involve as much public participation as possible, in decisions which could have a significant and critical influence on the lives of the people of Jamaica (GOJ, 1990).

The mechanism by which awareness was to be increased was the widespread dissemination of information through media, information services, publications, S&T centres and museums, exhibitions and fairs, as well as seminars and other efforts to popularise appropriate aspects of S&T.

Although a national science policy was in place, there was no ministry responsible for science from 1991 to 2003. Up to 1999 there was instead a science unit staffed by three individuals within another ministry (usually the Office of the Prime Minister). Science primarily resided in the divisions with responsibility for planning and development and, by 2002, the staff complement was reduced to a sole S&T director. The scope of the work carried out by the person within this role is uncertain, but for the government responsibility of science to rest solely on the shoulders of one individual would have had negative implications for the growth of science.

4.1. National focus in the policy era

The promulgation of the S&T policy was accompanied by a five-year development plan for S&T, which highlighted, among others, the challenges and inadequate appreciation of the importance of innovating technology in society (PIOJ and Ministry of Finance, 1991). The loss of qualified professionals leading to the deterioration of the S&T sub-sector and minimum budgetary support for R&D. These issues hampered S&T activities needed to increase awareness and appreciation for S&T in society. Section 5.5 of the plan—'Promotion of S&T in Society'—indicated an increase in efforts to raise public awareness on S&T by disseminating information through the media and information services, establishing S&T centres and museums, as well as hosting exhibitions, fairs and seminars, among others, but these events did not happen at the necessary scale.

The plan projected the establishment of a national museum of S&T, as well as the development of science parks, but, up to the present, these tasks have not been accomplished. The absence of a Ministry of Science and accompanying science portfolio is a factor that likely impacted the establishment of these institutions. The S&T policy also gave rise to the establishment of a National Commission on Science and Technology (NCST) in 1993, which has responsibility for fostering and advancing the national policy and strategy for S&T in Jamaica. However, a low staff complement and limited funding have provided challenges to the NCST in attempting to fulfil these objectives.

4.1.1. First science and technology survey

In 1992, a report was published by Trevor Hamilton and Associates entitled *Analysis of the public perceptions of science and technology in Jamaica* as part of a UNDP/Jamaican Project entitled 'Strengthening Endogenous Capacity in Science and Technology through Stakeholders Policy Dialogues'. According to Trevor Hamilton and Associates (as cited in Henry, 1998), the study reported that awareness of S&T was generally low among businesses, media, workers, government officials and women. Unfortunately, the 1992 report is no longer available in the public or private domain to review what the findings of the survey were.

4.2. Key persons involved in science communication

4.2.1. Martin E. D. Henry

Martin Henry, previously referenced in Section 3.2.1, was responsible for the radio program *Science Serving Us* along with the broadcaster Alma Mock Yen. The 15-minute program began airing in Jamaica in 1991 with an aim to promote the dissemination of science to non-scientific audiences. Prior to its broadcast, selected journalists were given exposure and basic training in scientific journalism through a workshop sponsored by the Inter-American Development Bank (IDB) in 1986 entitled 'Science and Technology by Radio for Non-Scientific Audiences'. The workshop was facilitated by Alma Mock Yen.

Science Serving Us was highly successful and aired up until 2003. In his 1998 thesis that analysed the program, Henry noted that science was frequently reported, with matters related to AIDS and the environment being covered by the mass media (Henry, 1998). He also indicated that scientists were rarely called upon to comment on current affairs, give interviews and speak on talk shows; this was probably a result of the media's general disinterest in matters of science. The Caribbean School of Media and Communication (CARIMAC), which graduated 1,000 media professionals from 1974 to 1995, did not produce any graduate who specialised in science journalism. This was the reality of Henry's time even with the new S&T policy being in place. *Science Serving Us* was thus developed as a way of bridging the knowledge gap between scientists and the public.

Science Serving Us was aired on Sundays in time allotted for government broadcasts via JIS on four radio stations. The program adopted an 'edutainment' format, where a representative of the audience had a one-on-one light and upbeat conversation with a credible expert. The broadcasts were based on local events and research conducted by UWI, SRC and others. The feedback was overwhelmingly positive with many persons indicating that they regularly listened. The program received over 2,000 letters from its inception up to 1998, and 90 per cent of letters received requested a transcript of the broadcast, such was the desire for the knowledge given. Although at the time there were more males in S&T disciplines, there was a higher percentage of females (58 per cent) who tuned in to the program. Negative feedback was not received, and one broadcast, 'Flame of Life', won the 1995 Pan American Health Organization radio broadcast award. In response to the death of prisoners from asphyxia in 1992, 'Flame of Life' discussed respiration, gas transportation by the blood, physiological consequences of oxygen deprivation, and other scientific matters.[3]

Science Serving Us was not without its challenges, and a major blow to its efforts was received when the sponsorship of the timeslot ceased. The program experienced resistance from the radio stations and some stations allocated it to 'graveyard' timeslots. The fact that the program had high listenership and was well received did not influence their decision as the media opted for more 'entertaining' content. The program aired on and off with many changes of radio stations and survived until 2003, after which it was fully taken off air.

Martin Henry highlighted the work done on *Science Serving Us* in an article 'Science and the Media' published in *The Daily Gleaner* on 25 November 1998. By then the program was only aired on one station on Sundays at 6.15 pm. He noted the lack of science journalism as an issue and, in response, he along with Professor Aggrey Brown, Director of CARIMAC, developed a course Science, Society and Media (SSM). The course was developed for trainee journalists and was equally applicable to trainee and practising scientists. SSM was an elective course offered by CARIMAC from 1997 to 2000 and was taken by about 50 students, some of whom are active journalists today. The course, however, was discontinued due to a decreasing number of students, and to date there has been no other course that can be considered its equal in the field of science journalism.

Martin Henry was one of very few in the country who had received formal training in both science and communication. At the time of his death on 29 May 2019, he was still writing science-based articles for the newspaper and serving as editor-in-chief for the University of Technology's *Journal of Arts Science and Technology*.

4.2.2. Dr Arnoldo Ventura

Dr Arnoldo Ventura may not have been trained in science and communication but his efforts at promoting science are noteworthy. Dr Ventura is known internationally for promoting S&T as a mechanism to alleviate poverty; his passion developed after his work in under-privileged communities showed the painful reality of the lives of many persons. He believed that science could be leveraged as a tool to solve economic and social problems, and when he became executive chairman of the SRC in 1977, the position provided an avenue to put his ideas into action. In the years leading up to the policy era

3 In October 1992, in a horrible example of overcrowding and police negligence, 19 men were confined in an 8 x 7-foot cell at the Constant Springs lockup in Jamaica. Three men died of asphyxiation (Human Rights Watch, 1993).

in 1990, Dr Ventura used television, newspapers and booklets to promote science and encouraged the media and citizens to develop a greater awareness of S&T.

In 1992, as special advisor to the Prime Minister, he wrote a booklet entitled *An ABC to S&T: An Introduction to Science and Technology*. The introduction noted that 'This publication is written primarily to provide in the simplest way possible, an explanation of the importance of science and technology to society' (Ventura, 1992). The 16-page booklet was written in simple language, provided definitions and explanations of S&T, demonstrated S&T's role in poverty reduction and used large fonts, as well as illustrations (Figure 21.4) to enhance the learning experience. For example, in defining science's importance, Ventura wrote that 'Science allows us to predict happenings and consequences with some degree of accuracy, as in the cases of outbreaks of diseases and the effects of pollution on our lives'.

Selecting the right foreign technology or creating the appropriate methods here, will depend on having more local scientists, engineers and technicians. Also, every worker who takes an interest in his work can suggest ways to improve the technology on his job.

Parents should encourage their children to learn more about science. Encouraging some of them to become scientists is a

Parents should encourage their children to learn more about science.

patriotic duty. It would be ideal, if every Jamaican became "science minded", so that whatever they do, they can do better.

Figure 21.4: An Illustrated Page from *An ABC to S&T* booklet.

4.3. Science activities of the 1990s to early 2000s

Two important events that have become staples for science communication were launched in the late 1990s: November was declared Science and Technology Month in 1997, and UWI Research Days were established in 1998. 'November is Science and Technology Month' was the title of the government bulletin published in the newspapers on 3 November 1997 as it highlighted the proclamation made on Friday, 31 October by the Governor-General Sir Howard Cooke. It was noted that this new annual celebration 'represents a part of the SRC's continuing effort to foster a culture of science in the population and to encourage greater application of technology in our social affairs and development strategies' (*The Gleaner*, 1997). To date, November as S&T month has various activities open to the public and, likewise, the

UWI Research Days showcase, over a three-day period, the research being done at the university. There have also been several activities in various private and public sector organisations that are involved in scientific research and that aim to educate the public on such activities.

5. Growing science in the *Vision 2030* era

In 2009, Jamaica ushered in a new era with the launch of the *National Development Plan: Vision 2030*, a roadmap that shows the steps needed to attain 'developed' status by the year 2030. 'National Outcome #11 – A Technology-Enabled Society' addresses the need for science, technology and innovation (ST&I) to be integrated into all areas of development. One key action given is the implementation of a national public education program on S&T (PIOJ, 2009). The *Vision 2030* plan states that Jamaica should popularise an ST&I culture as a viable agent of 'social and economic transformation' as a key objective. Although a national communication strategy for S&T is lacking, *Vision 2030* provides an avenue through which such a strategy could be launched. *Vision 2030* created a national shift in priorities and highlighted the need for a new S&T policy. This has led to the release of a new *National Science, Technology and Innovation Policy: Catalysing National Development 2019–2029*, which was tabled in parliament as a 'green paper' in November 2019 (Ministry of Science, Energy and Technology, 2019). Public consultations on the policy began in March 2020 and it is anticipated that the policy will be promulgated by the end of 2020.

This period has also brought changes at the government level. By 2013, there was a named science ministry (Ministry of Science, Technology, Energy and Mining). However, while there were established divisions for technology, energy and mining, there was none for science. Therefore, although 'science' was in the name of the ministry, capacity was minimal due to there being no established posts to drive the science mandate.The year 2016 saw a change in administration and the establishment of a Ministry of Science, Energy and Technology and, in 2018, a Science Division was established with three posts. Two of the three posts were filled in 2020 and, at the time of writing, the recruitment is being finalised for the third post. This is indicative of a shift in priorities at the national level, where science is now no longer the sole responsibility of entities such as SRC and NCST. The government will now play a more direct role in leading the national S&T agenda for the country, and activities that have been lagging from 1990 (science museum and scientific parks) are now more likely to be lifted from paper to reality. Additionally, the

Minister of Finance declared that as of the 2019/20 financial year, funds will be made available in the national budget for research and development to drive innovation.

At present, there is still no course offered at the local universities that pair science and communication. In a recent survey of 52 S&T graduates from UWI, 95 per cent of respondents agreed that the communication of science to the public should be taught to postgraduate students, and 65 per cent disagreed that there were sufficient avenues to communicate their science to the public (Oliphant, Mattocks and Monroe, 2018). This shows that scientists have a desire to receive training on methods of effectively communicating their work.

5.1. Recent progress in science popularisation/ promotion

In November 2014, a private sector organisation, GraceKennedy, through its Grace & Staff Community Development Foundation, opened the island's first STEM (science, technology, engineering and mathematics) centre with the intention of disseminating knowledge of STEM to young people in underprivileged communities. The centre offers classes on a weekly basis to dozens of students, and most of the teachers are professionals who volunteer their services. With its brightly coloured walls, murals and high-tech equipment (such as a 3D printer), the centre has become a beacon of light in the downtown Kingston community where it resides.

Additionally, the SRC has continued to offer a wide variety of activities that are geared towards promoting science:

- **National science & technology fair:** this island-wide event brings together primary, secondary and tertiary students to showcase their science-based inventions and innovations under thematic areas. Many excellent ideas and innovations are displayed on this platform, presenting ideal business and investment opportunities.
- **Schools' science & technology societies:** this program occurs in both primary and secondary institutions across the island and serves as an avenue to strengthen science education and teaching strategies in schools.
- **Essay, oratory & poster competitions:** these competitions are open to primary and secondary students and seeks to engage those with an interest in the creative arts on the importance/relevance of science. The essay component has been a useful training ground for students who wish to convey the extent to which science is important.

- **Open day:** this event involves displays from educational, research-based, financial and implementation entities that showcase the application of science in our lives, and is open to the public.
- **SRC in the community:** this showcases the services offered by the SRC in various communities across the island, with specific emphasis on the manufacturing and agro-processing sectors.
- **National innovation awards:** this competition encourages and supports innovations among members of the public who applied scientific or technological approaches in addressing matters of importance.
- **Prime Minister's Medal for Science and Technology:** highest award granted to an individual who played a major role in advancing the ST&I agenda leading to positive impacts on socioeconomic development.
- **SRC Young Scientist and Technologist Award:** this award recognises young scientists/technologists (40 years old or younger) for the excellence of their work and potential contribution to Jamaica's development. This award is presented at the biennial S&T conference.

5.2. Presenting science via media

A lot has changed in the science landscape from the 1940s to current times. Although S&T has progressed far beyond what the writers of that time predicted, the communication of such progress has declined in its representation in the media in Jamaica. Whereas there were dedicated media features such as *Science and You* in the past, science is mainly communicated in the current print media via columnists' contributions or in response to a particular issue (e.g. outbreak of Chik-V and ZIKA viruses). There was, however, a weekly feature published in *The Gleaner* entitled 'Ounce of Prevention' by Dr Tony Vendryes up to the time of his passing on 2 May 2019. In general, most of the science-related articles in the print media are related to health and the environment, as radio broadcasts focus mainly on health and agriculture.

JIS, as a government entity, continues to broadcast on all radio stations in time allotted for government broadcasts. The subjects of these broadcasts were examined from January 2013 to October 2018 to determine how many science topics were covered.[4] Only 346 of 17,502 topics broadcast were science-related (2 per cent). The topics related to energy, agriculture, climate change, natural disasters and different aspects of health.

4 See jis.gov.jm/radio_programs/.

6. Conclusion and future outlook

Since pre-independence, Jamaica has made strides in S&T research and, along the way, the public has been engaged via different media and events. Activities geared towards science popularisation and promotion are carried out primarily by the government through the SRC and secondarily by other entities, both private and public. However, more work is required in the area of science communication that engages not only schools but also the general public. It can be surmised that the concept of 'science communication' is not yet embedded in the psyche of policymakers as evinced by the fact that, despite the country's successes in S&T, there is no targeted approach to communicating science and training science communicators. The lack of a science ministry and/or division in the years following the 1990 policy is indicative of the extent to which science was prioritised, as ministries are formed based on the political priorities of the ruling party.

Examining Jamaica's history in the preceding sections shows a decline in science communication via the media channels, which suggests that part of the general public is not on the receiving end of the information being presented. In the past, by contrast, with radio and newspapers being the main modes of communication, science features were presented weekly and thus would be difficult to ignore.

In this current era of the internet, a number of entities share their 'science' via social media but most, with only a few followers, are limited in their reach. The absence of science journalism as a distinct field of study suggests that most journalists currently reporting do not have adequate capacity to write on science matters, nor are scientists adequately equipped to communicate with the public at large. With the recent changes to the government structure, where a science division has been established and budgetary allocation made for R&D, an important signal has been sent that science is critical. It is expected that these changes will give rise to increased levels of interaction and dialogue that will require effective modes of science communication. In conclusion, Jamaica is not short of scientists or low in its research output, but the modes of communication need to be improved so the dialogue can flow backwards and forwards from scientist to non-scientist for the benefit of society.

References

Broderick, P. (1970, 30 October). JAS to Make Awards at 75th Anniversary Luncheon. *The Daily Gleaner*.

Farr, T. (1985). Early Years of the Natural History Division. *Jamaica Journal, 18*, 19–28.

Find Glocal (n.d.). *The Natural History Museum of Jamaica*. Retrieved from www.findglocal.com/JM/Kingston/136803693019176/Natural-History-Museum-of-Jamaica.

Geological Society of Jamaica (GSJ) (2015). About the GSJ. Retrieved from geologicalsocietyofjamaica.org/index.html.

Government of Jamaica (GOJ) (1990). *National Science and Technology Policy*. Kingston: Government of Jamaica.

Henry, M. (1982, 17 January). The National Science Policy. *The Sunday Gleaner*.

Henry, M. (1998). Audience Response to 'Science Serving Us', a radio programme for the popularisation of science (Master's thesis). University of the West Indies, Mona, Kingston.

Henry, M. (1998, 26 November). Science and the media. *The Gleaner*.

Human Rights Watch. (1993). *The Human Rights Watch Global Report on Prisons*. Human Rights Watch.

Jamaica Information Service (JIS). (1965). *The Handbook of Jamaica for 1965*. Jamaica: Government Printing Office, Kingston.

Jamaicans.com. (2006). This Day In Jamaican History – Jamaica Broadcasting Corporation (JBC) Launches. Jamaicans.com. Retrieved from jamaicans.com/jamaica-broadcasting-corporation-jbc/.

Kingston Gleaner. (1879, 25 March). The Legislative Council: The Institute of Jamaica Law. *Kingston Gleaner*.

Lewis, C. Bernard. (1955, 10 May). The Institute of Jamaica. *The Daily Gleaner*.

Lowe, H., Brown, Y. and Magnus, K. (2000). *Discovering the Future: The Emergence, Development and Future of Science and Technology in Jamaica*. Kingston: Canoe Press.

Ministry of Science, Energy and Technology. (2019). *National Science, Technology and Innovation Policy: Catalysing National Development 2019–2029*. Retrieved from www.mset.gov.jm/wp-content/uploads/2019/07/UPDATED-The-National-STI-Policy-JULY-2019.pdf.

National Institute of Higher Education, Research, Science and Technology (NIHERST). (2009). *Caribbean Icons in STI (Volume 2)*. Trinidad and Tobago: National Institute of Higher Education, Research, Science and Technology.

Oliphant, Z., Mattocks, V. and Monroe, S. (2018). Jamaica: Assessing Science Communication at the Post-Graduate Level. Presented at the 15th International Public Communication of Science and Technology Conference, PCST Conference, Dunedin, New Zealand.

Planning Institute of Jamaica (PIOJ). (2009). *Vision 2030 Jamaica: National Development Plan*. Kingston: Planning Institute of Jamaica.

Planning Institute of Jamaica (PIOJ) and Ministry of Finance. (1991). *Jamaica 5 Year Development Plan for Science and Technology (1990–1995)*. Kingston: Planning Institute of Jamaica.

Simple, P. (1945, 7 November). A Balanced Life. *The Daily Gleaner*.

Scientific Research Council (SRC). (1986). *Skills Bank – A Directory of the Available Expertise in Science and Technology in Jamaica*. Scientific Research Council.

Statistical Institute of Jamaica (STATIN). (2017). Population Statistics. Retrieved from statinja.gov.jm/Demo_SocialStats/PopulationStats.aspx.

The Daily Gleaner. (1945, 5 December). Premier Attlee calls for understanding of world problems in the cause of peace. *The Daily Gleaner*.

The Daily Gleaner. (1948, 31 December). Long time before largescale use of atom in industry. *The Daily Gleaner*.

The Daily Gleaner. (1952, 7 July). The first fruits. *The Daily Gleaner*.

The Daily Gleaner. (1972, 10 September). Educational, cultural programmes on JBC FM. *The Daily Gleaner*.

The Daily Gleaner. (1979, 15 November). SRC gets the squeeze. *The Daily Gleaner*.

The Daily Gleaner. (1981, 13 July). Ventura science committee chairman. *The Daily Gleaner 29*.

The Gleaner. (1997, 3 November). November is Science and Technology Month." *The Gleaner*.

Ventura, A. (1992). *An ABC to S&T: An Introduction to Science and Technology*. Jamaica: Scientific Research Council, Kingston.

Villavicencio, D. and Ponce, L. (1990). *The Science and Technology System in Jamaica*. UNESCO.

Williams, P. (2016, 25 November). Natural History Museum of Jamaica Preserving Our Natural Heritage. *The Gleaner*. Retrieved from jamaica-gleaner.com/article/art-leisure/20161127/natural-history-museum-jamaica-preserving-our-natural-heritage.

Timeline

Event	Name	Date	Comment
First interactive science centre established.	Natural History Museum opened	1879	Primarily focused on science
First national (or large regional) science festival.	First National S&T Conference	1987	
First university courses to train science communicators.	'Science, Society and Media' offered at UWI Mona (CARIMAC)	1997–2000	Short-lived because of decreasing demand
National Science Week founded.	November declared Science and Technology Month	1997	
First significant radio programs on science.	*Science Serving Us*	1991	First local science broadcast
First significant TV programs on science.	*Educational TV* on JBS TV; *Science I, II* and *III*	1970s – 1980s	Aired general science information

Contributors

Dr Zahra H. Oliphant is the chief research officer with the Ministry of Science, Energy and Technology in Kingston, Jamaica.

Dr Cliff K. Riley is the executive director of the Scientific Research Council, Jamaica, and associate professor and associate dean in the College of Health Sciences, University of Technology.

Kerry-Ann C. Curtis is the manager for Information Services at the Scientific Research Council (SRC).

Setu N. Monroe works as the research officer in the Ministry of Science, Energy and Technology, Kingston, Jamaica.

Dr Aisha D. Jones is the director of Research at the National Commission on Science and Technology, Jamaica.

Dr Charah T. Watson is the manager of Product Research and Development at the Scientific Research Council in Kingston, Jamaica.

22

JAPAN

Western science and Japanese culture

Masataka Watanabe and Mitsuru Kudo

1. Introduction

Western science and technology (S&T) were brought to Japan from the 16th century via Portugal and the Netherlands. Through the mid-19th century, Japan was a closed country, permitting commerce only with the Dutch. This precedent was broken by the American naval expedition that came to Japan in 1853 from the east and, in one stroke, the country reluctantly opened its doors to Western countries. The Tokugawa shogunate was overturned by the Meiji Restoration in 1868, and that marked the beginning of Japan's modernisation.

It was only after the Meiji Restoration that education in modern Western S&T could formally be offered, though Japan had its own tradition of S&T before then (Nagahama, 1994). As a consequence, the Japanese people have interacted with modern science for only about 150 years. After taking political steps to introduce Western S&T, Japan as a nation hastened along in the spirit of trying to catch up and then surpass it. During that time, most Japanese people believed that S&T would gradually advance if left in the hands of specialists, and Japan would go on to win fame as an advanced country. This chapter presents a short history of science communication in Japan in this context (see also Watanabe, 2017).

2. Promoting public understanding of science and technology (PUST)

As part of its efforts to recast itself as a country of peace following World War II, Japan sought economic recovery centring on emerging S&T. In 1958, the Science and Technology Agency (founded in 1956 and integrated with Ministry of Education in 2000 to become the Ministry of Education, Culture, Sports, Science and Technology) published its first White Paper on Science and Technology.[1] In Chapter 3 of that White Paper, one can find the following pronouncement:

> In recent times, Japanese people have had many opportunities to build familiarity with S&T. The more scientific and technological issues and successes are reported, the more people of all ages will place their dreams on S&T. If we are able to offer educational opportunities to encourage the sound growth of dreams and aspirations relating to S&T, then various self-motivated activities that make use of S&T will effectively develop in the near future.[2]

The Japanese government followed this policy during the latter half of the century.

The Council for Science and Technology was established in 1959 to advise the prime minister on S&T-related policies. In its first policy proposal in 1960, the council opined that while it was vital for Japan to develop a talented workforce to drive its long-term pursuit of the sciences and technologies needed to grow the economy and improve lives, the Japanese people lacked basic knowledge of and education on these subjects. Moreover, there just wasn't the required political will and public sentiment to provide support for such activities. Therefore, the council said, the government must start by raising awareness of S&T among the populace. It can be said that this policy marked the dawn of public administration aimed at boosting the public understanding of science and technology (PUST).

That same year, National Science and Technology Week was established. It is the week around 18 April each year, the day itself having been known as Invention Day since 1954. It might be one of the earliest attempts of its kind in the world, with even countries like the UK not starting a national science week until 1994 (Briggs, 2001). That same year the Japanese government and industry together established the Japan Science Foundation to contribute to

1 This is an authoritative annual report on the Japanese goverment's science and technology policies. It features special themes as the main topics every year.

2 All citation from Japanese documents are translated by authors.

the improvement of S&T by effectively conducting activities to deepen the general public's understanding of, and interest in, fundamental scientific knowledge and industrial technology. The foundation would later open a science museum in Tokyo and launch a local TV company in 1964. Known as Tokyo Channel 12 'Science TV', the channel was given its broadcast licence on the condition that 60 per cent of its air-time would be dedicated to S&T educational programming. Initially, it met this requirement only in a technical sense: the 'programs' were simply broadcasts of distance-learning classes offered by Kagaku Gijutsu Gakuen High School, a S&T high school also established by the Japan Science Foundation. Only three years after its launch did Tokyo Channel 12 finally begin to air regular programs such as news and dramas.

The 1960 policy proposal by the Council for Science and Technology mentioned above also set long-range targets in various S&T fields. These focused on achieving, within 10 years, a general advance in living standards and proposed the necessary strategies for fostering capable human resources. These were presented as being necessary for economic development. It also highlighted the need to promote PUST on the grounds that 'Public knowledge and literacy regarding S&T are very poor and the political and public bases for the support of S&T are very weak' (from the council's recommendation (Council for Science and Technology, 1960)). Such a top-down policy was sustained during the 1960s and 1970s.

The Japanese government continued to tout 'the dream of S&T' until the 1970s. In Japan, a national opinion survey of public attitudes toward S&T—its main question being whether or not people have an interest in science news—has been conducted almost every five years since 1976. Although the significance of such a survey has been controversial (Durant, Evans and Thomas, 1989), we can recognise an interesting trend by analysing differences among generations. From the results of the survey, it was shown that people in their 20s and 30s reported the highest levels of interest in S&T in 1976, and this generation would maintain its interest in S&T throughout the survey period (Watanabe and Imai, 2003). This may be attributable to the PUST Policy implemented by the government during the 1960s and 1970s.

As the 1980s began, while people's lives had become richer to some degree, the negative aspects of cutting-edge S&T had also become apparent, and society on the whole had grown increasingly indifferent to science. The White Paper on Science and Technology: Trajectory and Prospects on the Development of Science and Technology published in 1980, the year following the Three Mile Island accident in the US, contained a section titled 'Requirements for Promoting Science and Technology'. It claimed that 'Public understanding and cooperation are necessary for promoting S&T'. We should note that this usage of 'understanding' references a viewpoint from the government that expects the public to agree with and accept its policies. The White Paper on Science and Technology in 1982 continued the same tone that was ruled by archetypal phrases about the importance of science education. In a section titled 'Promotional Action Plan to Gain *Public Acknowledgement and Support for Science and Technology*' [emphasis added], it claimed:

> We should carry out the proper evaluation at each step of research and development and increase public awareness so as to advance S&T effectively and to raise creativity in S&T. In particular, it is much more important to gain *public acknowledgement and support by means of enlightenment* to foster a scientific mindset and awareness among the younger generation' [emphasis added].

Thus, the government policy emphasised 'enlightenment of the public' and promoted the construction of science museums and science centres across the nation. At the national level, Japan convened the International Science and Technology Exposition in 1985 at the new science city Tsukuba. Since 1992, Youngsters' Science Festival events have been supported. These festivals collectively offer science shows, booth displays and workshops under one roof. Drawing the engagement of many science volunteers, this series of events was at first held in only three cities. Local governments and various industries offered their support, and the festivals have spread to more than 100 cities around the country with some 420,000 people taking part. However, these actions had only a limited effect. It is a part of the reason why the White Paper on Science and Technology: Young People and Science and Technology published in 1993 has a different tone. In Part 1 of the White Paper, 'Young People's Indifference to Science and Technology', it discussed this apathy and espoused 'fostering an atmosphere for making science issues relevant to young people'.

However, even these kinds of events that convey the pleasures of science to the youth would appear to be insufficient for instilling a recognition of the importance of knowing how to make the most out of science in daily life. Evidence of this comes from the Program for International Student Assessment (PISA) survey of 15-year-olds in 2006. It showed that only 8 per cent of Japanese students expected to have a science-related occupation at the age of 30, the lowest proportion in the world (OECD, 2007). Although Japan's children may get good grades in these subjects, it appears that they do not wish to work in S&T-related jobs.

At the same time, it seems important that people appreciate S&T not just as mere tools but also as a great cultural heritage or property that has been built by humankind. The days when it was thought best to leave matters in the hands of specialists tied to narrow specialised fields are gone. The time has come for each and every citizen to think about the ways in which they interact with science. Achieving this calls for a new goal. This was the situation in Japan at the end of the 20th century when the new concept of science communication was born. But despite the new ways of thinking about science communication in international discussions, the Japanese government still focused on education, understanding and interest with regard to science. The government enacted the Science and Technology Basic Law in 1995 with the aim of raising the standards of S&T in Japan and set out the First Science and Technology Basic Plan, a five-year government plan that included the promotion of PUST (Science and Technology Agency, 1996).

In the UK, the Select Committee on Science & Technology of the House of Lords published its *Third Report: Science and Society* (House of Lords, 2000) and the report *Science and the Public* (Office of Science and Technology and Wellcome Trust, 2000) was also published in the wake of the bovine spongiform encephalopathy (BSE) outbreak. This marked a shift in the S&T policy of the UK government toward promoting public engagement with science and technology (PEST).

In response to such world trends, the Japanese Society for Science and Technology Studies was founded in 2001. Japanese researchers in the field of S&T studies have sparked a new wave of PEST in Japan, holding events such as consensus conferences on the topic of gene therapy. Preceding this, two reports championing science communication were published in Japan, one proposing the establishment of 'science communication plazas' (Nakamura, 1991) and the other proposing the founding of 'S&T communication centres' (Nagahama, Kuwahara and Nishimoto, 1991). The former proposal was realised in 1993 in Osaka, Japan, with the JT Biohistory Research Hall, a unique research centre with exhibitions open to the public. The latter proposal was for facilities such as S&T study (STS) centres. These have yet to be realised despite being the focus of a report published by a government think tank, the National Institute of Science and Technology Policy (NISTEP). An informal meeting held by the Minister of Science and Technology ventured that 'Interpreters who can explain cutting-edge science topics to the layperson are essential' (the detail of the informal meeting is discussed in Watanabe and Imai, 2003).[3] Consequently, the early inroads made by the science communication movement in Japan were driven by government promotion

3 National Museum of Emerging Science and Innovation, Miraikan, was founded in 2001. The concept of Miraikan might be based on this declaration. Miraikan introduced an on-the-job training system for science communication professionals. Science Interpreters—this name was changed to Science Communicators a few years later—are appointed on a fixed-term basis for a maximum of five years. About 40 Science Communicators engage in science communication activities during their terms.

based on the 'deficit model' or people in the academic field of STS within their community. This was one reason why science educators and science centre personnel who had been supporting PUST since the 1960s were unfamiliar with the new concept and practice of science communication.

3. Introduction of the term 'science communication' into the government's S&T policy statement

The situation changed dramatically from 2003 onward. The new concepts of 'science communication' and 'interactive two-way communication about science' spread amongst science communication practitioners. Several things coincided in 2003. First of all, two publications appeared. One was a Japanese edition of *Science Communication in Theory and Practice* (Stocklmayer, Gore and Bryant, 2001). The other—which has been most influential—was a report titled *Research on the Promotion of Public Understanding of Science & Technology and Science Communication* (Watanabe and Imai, 2003) published by NISTEP. It served to change government policy and triggered a cascade effect. In 2004, a new term in the Japanese language, namely 'science communication', first appeared in the White Paper on Science and Technology (Ministry of Education, Culture, Sports, Science and Technology, 2004). Furthermore, the Third Science and Technology Basic Plan from 2005 announced the promotion of science communication (Government of Japan, 2006; see also Watanabe and Imai, 2005). Since then, Japanese government policy for promoting PUST has shifted to PEST to some extent.

In 2005, formal training programs in science communication for postgraduate students supported by five-year-limited government subsidies (each worth about US$1 million per institution per year) began at three universities: University of Tokyo, Hokkaido University and Waseda University. The Science Interpreter Training Program was launched at the University of Tokyo with a goal to

> nurture scientists and engineers who are equipped with social and political literacy and treasure the presence of multiple perspectives, as well as scholars in humanities and social sciences who can identify common grounds between their disciplinary standpoint and visions and values in science and technology (University of Tokyo Science Interpreter Training Program, n.d.).

Hokkaido University launched the Communication in Science & Technology Education & Research Program (known as CoSTEP) to produce 'science and technology communicators, who can enhance two-way information transfer between experts and the public in societal issues related to the two subjects' (Hokkaido University CoSTEP, n.d.). These two programs were offered as certificate programs rather than as full degree programs, so that students specialising in any disciplinary subjects for their degree could participate. In contrast, the Master of Arts Program for Journalist Education in Science and Technology (MAJESTy) at Waseda University was set up as a postgraduate degree program in order to 'train students as journalists who can make a balanced assessment of issues surrounding science and technology both today and in the future' (Waseda University, n.d.). These programs were partly modelled on overseas examples.

Another certificate training program in science communication for postgraduate students was started at the National Museum of Nature and Science in 2006 (Ogawa, Kamei and Shimizu, 2006). The program is run in collaboration with universities.

In 2006, Science Agora was started in Tokyo with the support of the Japan Science and Technology Agency. It is an annual forum that aims to be a pivot for a network linking all kinds of science communication activities together. The event is essentially a miniature version of the AAAS Annual Meeting and similar to Europe's EuroScience Open Forum (ESOF), except that Science Agora features free admission and anyone can attend any session. Science Agora is said to be 'like a big salad bowl' (Umehara and Watanabe, 2012): a wide variety of people including families, students, teachers, researchers, administrators, politicians and science communication practitioners are gathered in one place and mixed together. Science Agora 2016 hosted 213 programs, with roughly 6,000 visitors over the course of four days. Science Agora has fostered network-building among key sectors of science communication. In 2009, a new type of science festival based on the modern concept of science communication was launched in two cities: Hakodate in Hokkaido and Mitaka in Tokyo (these two cities have no science centres). This was an additional side effect of Science Agora. These festivals have built up positive reputations and a number of other cities have launched their own new-type science festivals.

A further example of the rise of science engagement opportunities is the emergence of science cafés, with more than 1,000 being held around the country every year since 2009. They were originally convened in response to an appeal from the Science Council of Japan during the 2005 Science and Technology Week, when such café events were held in more than 20 places across the country. Although they may have begun as a somewhat top-down contrivance, they have subsequently put down firm roots throughout Japan.

It is amazing that science cafés have become so popular in Japan because the country does not have the same level of pre-existing 'café culture' that is found in European countries. Science was thought of as a high-threshold topic before science cafés, but these events are now perceived as being open to all-comers thanks to the relaxed, informal environment where people enjoy talking about science over coffee. It can be compared to the *idobata kaigi*—the 'well-side chat', or, in other words, the neighbourhood gossip session. If these science cafés—which are held in all manner of locations and venues—can be linked up as a network, they will eventually fall into sync and turn into a substantial movement. The critical factor explaining why the new concept of science communication has become popular so quickly in Japan can be attributed to the new key phrase and concept of 'science communication' having been introduced first. There is some truth to the old dictum, 'new wine must be put into new wineskins' about this Japanese context (Watanabe, 2010). The situation resembles that which followed the introduction of Western science about 150 years ago.

National Museum of Emerging Science and Innovation, Miraikan, has played a leading role as one of the flagship science museums based on the science communication concept. There are some exhibits that demonstrate cutting-edge science such as androids (human-like robots). Visitors can meet the most advanced androids and reflect on human existence. Miraikan also holds various kinds of two-way communication events.

4. Critical reflection on the early development of science communication in Japan

While the official introduction of the term 'science communication' led to the establishment of related activities including science cafés and training programs, this official launch of Japanese science communication attracted criticism from Japanese researchers in science communication and STS. While admitting that the 2004 White Paper on Science and Technology and the Third Science and Technology Basic Plan (Government of Japan, 2006) had taken into account theoretical and conceptual frameworks of Western science communication after the 'PUS movement', critical voices pointed out that the official promotion of science communication in Japan had failed to shift its emphasis from its older, conventional understanding-centric approach, to a more engagement-oriented approach. In other words, the overall framework of Japanese science communication still focused on promoting public enlightenment and increasing public interest in S&T (Hirakawa, 2010). This tended to accentuate only the positive aspects of doing and learning science, which would distract public attention away from uncertainty in science and thus possibly lead to uncritical trust in science (Kobayashi, 2007). Those critical views on the early development of official science communication in Japan pointed to the imbalance between understanding-oriented programs and engagement-oriented programs, and they called for more systematic attention to be paid to developing other models for the democratic governance of S&T.

Training programs for practitioners of science communication (see Section 3 for details) were also questioned. For example, their inclination towards *promotion* of public understanding, built on the view that science communication was for the purpose of either dealing with the decreasing popularity of science among youngsters or gaining more social support for basic scientific research, was criticised for paying little attention to nurturing science communication practitioners who would be capable of building bi-directional channels to address issues between science and society (Yagi, 2007).

Criticisms of Japanese science communication were also raised with regard to the lack of attention to previous failures and shortcomings in dealing with science-related social issues. The national institutionalisation of the public communication of science and technology under the name of 'science (and technology) communication' had not taken much account of the previous failures in establishing appropriate science–society relations. For example, lessons learnt from industrial pollution and consequent endemic diseases and issues around nuclear power plant construction in the 1970s did not inform policymaking about contemporary science communication (Fujigaki and Hirono, 2008). More recent prominent science-related social issues, including the Great Hanshin-Awaji Earthquake in Hyogo prefecture and the nuclear power plant accident (a sodium leak and a consequent major fire) at Monju in Fukui prefecture, both of which took place in 1995, had a minimum impact on pushing the early development of a national framework of science communication towards democratic engagement (Hirakawa, 2010). In this sense, Japanese official science communication was perceived as disconnected from previous communication disputes at the science–society nexus, and thus was criticised for lacking the 'pain': the pain that science communication in Europe had gone through during its development when controversies and debates about nuclear power, BSE and genetically modified organisms (GMOs) had taken place among interest groups and stakeholders including the public, established scientists and the government (Fujigaki, 2008, 2009).

It should be noted that these critical views of Japanese official science communication, particularly on its orientation towards promotion of PUST and on its tacit adoption of the deficit model, did not necessarily ignore values for the public to acquire scientific information through understanding-oriented science communication. The STS critics of Japanese science communication cited above were aware of the diversity of science communication essential to achieve truly democratic governance of S&T. Therefore, it was with respect to the *balance* rather than the *choice* between understanding-oriented and engagement-oriented models of science communication that the critics called for more reflective discussion and more resources to spend.

5. The Fukushima disaster and resetting science communication policy

A large-scale earthquake hit Japan on 11 March 2011 and caused a sequence of explosions at the Fukushima Daiichi Nuclear Power Plant. It showed that the government had not considered how it should respond to such an unprecedented, large-scale nuclear accident and what information the public would want and need. The government and scientific community experienced a great loss of public trust as a result of this disaster. The government was found to have concealed information about radiation data because they wished to avoid a resultant panic. Failure to release radiation data during the early stages of the crisis is said to have delayed evacuations of communities located near the plant. At first the government was unable to recognise the meaning of the data, and later pursued an official campaign to play down the scope of the accident and the potential health risks in order to prevent panic as mentioned previously. This policy went counter to the science communication policy of openness and transparency. It revealed a fundamental misunderstanding by the government regarding the idea and concept of science communication despite its previous declarations promoting science communication in the Third Science and Technology Basic Plan (Government of Japan, 2006). The Japanese scientific community did not help: most of them kept silent although they knew the implications of the nuclear power plant accident.

The government intended to publish the Fourth Science and Technology Basic Plan and 2011 White Paper on Science and Technology at the end of March 2011. Ironically, the basic plan would declare that science and technology policy should be created together with society, i.e. through democratic participation in science and technology policymaking. The public announcement of the basic plan was delayed by four months. Another irony was that one of the main topics planned for the White Paper was science communication. The result was that the publication was delayed and the content revised. A trustworthy relationship is the most important consideration for establishing science communication.

The Japanese public learned a great deal after the March 11 earthquake and the Fukushima nuclear power plant accident. Since then people have set up their own local networks to exchange information about radiation risks. For example, many regional communities have procured their own Geiger counters and begun monitoring radiation levels in their local areas with experts' advice for peace of mind. Over 30 science cafés about radiation effects or the earthquake were held all over Japan during the two and a half months immediately following March 11. This unfortunate incident has taught us a major lesson and encouraged people to adopt a bottom-up approach.

Toward the end of 2011, the Japanese Association for Science Communication (JASC) was established. The mission of JASC is to construct a network of science communication practitioners and to propagate and share the concept and methods of science communication across all communities nationwide. The association started out with about 200 members and has since increased to roughly 400. It operates self-sufficiently using just membership fees.

6. Beyond the PUST–PEST dichotomy and towards a complementary relationship

While the policy frameworks of science communication in Japan have been making a gradual shift from understanding-orientation to engagement-orientation, we should be aware that a number of empirical studies in Europe have suggested that the actual practice of science communication would often incline towards PUST. The image of knowledge-deficient publics is still commonly found, and a linear causation between the increase of PUST and public support for science would often be assumed. Difficulties in putting thoughts and theories of dialogical, engaging science communication into practice are also much discussed (e.g. Chilvers and Kearnes, 2015). Japan is no exception, and it is faced with numerous difficulties in putting the blueprint of science communication ideals into actual practice (e.g. Ishihara-Shineha, 2017; Nakamura, 2011; Shineha, 2016). At the same time, we should also note that the understanding-oriented and engagement-oriented models of science communication, which tend to be seen as at the opposing sides on the spectrum of science communication, should be seen more as complementary rather than contradictory, both together aiming at democratising science–society relations (e.g. Tanaka, 2013).

The strong orientation of Japanese science communication towards the promotion of PUST should not be flatly dismissed by employing conventional criticisms of the 'PUS movement' or the 'deficit model'. Simple, clear-cut categorisation of science communication practice can be misleading, and it would potentially turn our attention away from visions, thoughts and broader contexts that are behind such seemingly understanding-oriented approaches to science communication. It would also possibly prevent us from exploring what the practice of science communication—hether it be oriented towards understanding or engagement—actually means to science–society relations. What we need in future science communication research, therefore, should be to map out a wide variety of forms of science communication in our society—some are initiated by the government and/or scientific research institutions and others are more or less bottom-up—and to empirically investigate their meanings from the perspectives of people involved in them on the ground. In such empirical, exploratory and interpretive research practice, we need to go beyond the understanding–engagement dichotomy.

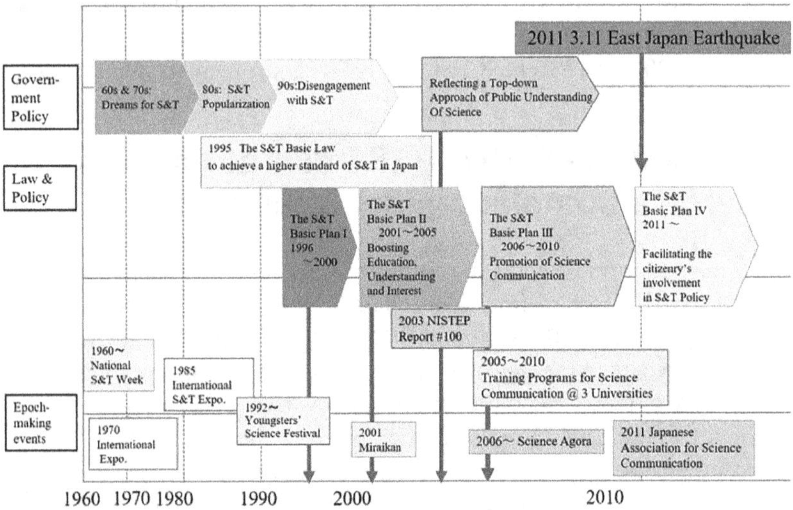

Figure 22.1: A short history of science communication in Japan.

Science communication education, including the training of science communicators, is today offered in various forms. Although there are no full-degree coursework programs in science communication at either undergraduate or postgraduate level in Japan, there are a number of ways to study and learn skills in science communication. Many programs are primarily for students and researchers in the natural sciences (e.g. Mizumachi et al., 2011; Yokoyama, 2009), but there are also courses designed for students studying and researching in the humanities, social sciences and public policy (e.g. Ema, 2015; Yoshisawa and Taniguchi, 2016). It should be mentioned here that these educational activities are not conducted with a specific focus on science communication per se, but they aim to develop learners' skills to work with people from different disciplinary and/or institutional backgrounds to tackle issues about science–society relations that are increasingly trans-disciplinary. Training scientists in such skills—some would refer to as 'transferable skills'—is now becoming an important focus of higher education policy in Japan (Yamanouchi and Nakagawa, 2012) in response to the growing complexity of science–society relations, and education and/or training in science communication is expanding its scope accordingly (e.g. Kamisato and Hosono, 2014; Kudo, Mizumachi and Yagi, 2018; Shineha et al., 2014).

7. Conclusion

Figure 22.1 above represents the development of science communication policy in Japan. Japan's policy of PUST has shifted to PEST since 2003. That year there were a number of simultaneous developments with regard to science communication. The key report that advocated for the promotion of science communication and a textbook on science communication were published. The most important consequence was that the report triggered a change in government policy. Although the shift may have begun as a somewhat top-down contrivance and still PUST-minded, it has gradually put down firm roots across Japan.

Things changed dramatically in the wake of the large-scale earthquake and the Fukushima nuclear power plant accident on 11 March 2011. The Japanese government had to change its science and technology policy, and the public gained the realisation that the government is not necessarily trustworthy and people have to look out for themselves. This would appear to be counter to the principles of science communication. Nevertheless, at a local community level people have shown mutual compassion and established solid links amongst themselves, so a ray of hope can be found there. An updated version of science communication, i.e. 'Science Communication 2.0', must be launched. For this we must look to grassroots science communication.

References

Briggs, P. (2001). New Visions for Associations for the Advancement of Science: A Case Study. In S. M. Stocklmayer, M. M. Gore and C. Bryant (eds), *Science Communication in Theory and Practice* (pp. 191–201). Dordrecht: Kluwer Academic Publishers. doi.org/10.1007/978-94-010-0620-0_13.

Chilvers, J. and Kearnes, M. (eds) (2015). *Remaking Participation: Science, Environment and Emergent Publics*. London and New York: Routledge.

Council for Science and Technology. (1960). *The First Policy Proposal: 'A Comprehensive Basic Policy for the Promotion of Science and Technology for the Next Ten Years'.* Science and Technology Agency, Prime Minister's Office.

Durant, J. R., Evans, G. A. and Thomas, G. P. (1989). The public understanding of science. *Nature, 340*(6228), 11–14. doi.org/10.1038/340011a0.

Ema, A. (2015). 'Science, Technology and Society' class program at a university: a report of teaching technology assessment and research ethics. *Japanese Journal of Science Communication, 18*, 3–16.

Fujigaki, Y. (2008). Citizen participation and science communication. In Y. Fujigaki and Y. Hirono (eds), *Theoretical Perspective for Science Communication.* Tokyo: University of Tokyo Press.

Fujigaki, Y. (2009). STS in Japan and East Asia: Governance of Science and Technology and Public Engagement. *East Asian Science, Technology and Society: an International Journal, 3*(4), 511–18. doi.org/10.1215/s12280-009-9110-9.

Fujigaki, Y. and Hirono, Y. (eds) (2008). *Theoretical Perspective for Science Communication.* Tokyo: University of Tokyo Press.

Government of Japan. (2006). The Third Science and Technology Basic Plan. Tokyo.

Hirakawa, H. (2010). *Who Owns Science?* Tokyo: NHK Publishing.

Hokkaido University CoSTEP. (n.d.). The role of CoSTEP – three fields of activity. Retrieved from costep.open-ed.hokudai.ac.jp/costep/good_design/en/edu_program01.html.

House of Lords. (2000). *Third Report: Science and Society.* London: The Stationery Office.

Ishihara-Shineha, S. (2017). Persistence of the deficit model in Japan's science communication: Analysis of White Papers on Science and Technology. *East Asian Science, Technology and Society, 11*(3), 305–29. doi.org/10.1215/18752160-3819961.

Kamisato, T. and Hosono, M. (2014). Creating a hub for ELSI/TA education, research and implementation in Japan. In T. Michalek, L. Hebáková, L. Hennen, C. Scherz, L. Nierling and J. Hahn (eds), *Technology Assessment and Policy Areas of Great Transitions: Proceedings from the PACITA 2013 Conference in Prague* (pp. 117–22). Prague: Technology Centre ASCR.

Kobayashi, T. (2007). *The Age of Trans-science.* Tokyo: NTT Publishing.

Kudo, M., Mizumachi, E. and Yagi, E. (2018). Towards an integration of practical application, research and education of public engagement with emerging technologies: a case on ethical, legal and social issues of 'Society 5.0', *Co* Design, 3*, 35–53.

Ministry of Education, Culture, Sports, Science and Technology. (1982). *White Paper on Science and Technology: The Relationship between Young People and Science and Technology.* Prime Minister's Office. Retrieved from www.mext.go.jp/en/publication/whitepaper/title01/detail01/1372762.htm.

Ministry of Education, Culture, Sports, Science and Technology. (2011). *White Paper on Science and Technology: Science and Technology Created Together with Society.* Prime Minister's Office. Retrieved from warp.ndl.go.jp/info:ndljp/pid/11293659/www.mext.go.jp/b_menu/hakusho/html/hpaa201101/detail/1308357.htm.

Mizumachi, E., Matsuda, K., Kano, K., Kawakami, M. and Kato, K. (2011). Scientists' attitudes toward a dialogue with the public: A study using 'science cafes'. *Journal of Science Communication, 10*(4), A02. doi.org/10.22323/2.10040202.

Nagahama, H. (1994). Scientific Literacy in Japan: Historical and Cultural Permeation and Current Changes. In B. Schiele (ed.), *When Science becomes Culture.* Boucherville, Quebec, Canada: University of Ottawa Press.

Nagahama, H., Kuwahara, T. and Nishimoto, A. (1991). *Science, Technology, Society and Communication. NISTEP REPORT 017* [in Japanese with English summary]. Retrieved from hdl.handle.net/11035/535.

Nakamura, K. (1991). *Establishing science as culture: proposal for science communication plaza* [in Japanese]. Nippon Institute for Research Advancement.

Nakamura, M. (2011). Policy on promoting science and technology communication. In H. Yoshioka (ed.), *A Social History of Science & Technology in Japan at the Turn of the Century*, Tokyo: Hara Shobo.

Office of Science and Technology and Wellcome Trust. (2000). *Science and the Public: A Review of Science Communication and Public Attitudes to Science in Britain.* London: Department of Trade, Industry and the Trustees of the Wellcome Trust.

Ogawa, Y., Kamei, O. and Shimizu, M. (2006). A training program for science communication in collaboration with universities and museums. In *Proceedings of PCST2006 Conference (Seoul)* (pp. 1359–64). Public Communication of Science Technology Network. Retrieved from www.pcst.co/archive/paper/1350.

Organisation for Economic Co-operation and Development (OECD) (2007). *PISA 2006 Science Competencies for Tomorrow's World.* OECD Publications.

Science and Technology Agency. (1958). *White Paper on Science and Technology. From Foreign Dependence to Independent Development.* Prime Minister's Office. Retrieved from warp.ndl.go.jp/info:ndljp/pid/11293659/www.mext.go.jp/b_menu/hakusho/html/hpaa195801/index.html.

Science and Technology Agency. (1980). *White Paper on Science and Technology: Trajectory and Prospects on the Development of Science and Technology.* Prime Minister's Office. Retrieved from warp.ndl.go.jp/info:ndljp/pid/11293659/www.mext.go.jp/b_menu/hakusho/html/hpaa198001/index.html.

Science and Technology Agency. (1982). *White Paper on Science and Technology: In Search of Creative Science and Technology.* Prime Minister's Office. Retrieved from warp.ndl.go.jp/info:ndljp/pid/11293659/www.mext.go.jp/b_menu/hakusho/html/hpaa198201/index.html.

Science and Technology Agency. (1993). *White Paper on Science and Technology: The Relationship between Young People and Science and Technology.* Prime Minister's Office. Retrieved from warp.ndl.go.jp/info:ndljp/pid/11293659/www.mext.go.jp/b_menu/hakusho/html/hpaa199301/index.html.

Science and Technology Agency. (1996). *First Science and Technology Basic Plan.* Prime Minister's Office. Retrieved from www.mext.go.jp/b_menu/shingi/kagaku/kihonkei/honbun.htm.

Science and Technology Agency. (2011). *Fourth Science and Technology Basic Plan.* Prime Minister's Office. Retrieved from www8.cao.go.jp/cstp/kihonkeikaku/4honbun.pdf.

Shineha, R. (2016). History and issues of science communication: a review of discussions and perspectives in policy-making. *Seijo Communication Studies, 27,* 13–29. doi.org/10.13169/polipers.13.2.0029.

Shineha, R., Iida, K., Nakao, H., Kikuchi, Y. Mikami, K., Ito, K. Hirata, K. and Hasegawa, M. (2014). 'Science and Society' education for PhD candidates: the case of SOKENDAI. *The Journal of Science Policy and Research Management, 29*(2/3), 90–105.

Stocklmayer, S. M., Gore, M. M. and Bryant, C. (eds) (2001). *Science Communication in Theory and Practice.* Dordrecht: Kluwer Academic Publishers. doi.org/10.1007/978-94-010-0620-0.

Tanaka, M. (2013). Discussion on phases of communication about science and technology. In M. Nakamura (ed.), *Science and Politics after the Disaster of March 11 in Japan.* Kyoto: Nakanishiya Shuppan.

Umehara, S. and Watanabe, M. (2012). *Science Agora: The Evolving Science Festival and Its Roles in Japan's Science Communication Policy-Making.* AAAS General Poster Session. Retrieved from aaas.confex.com/aaas/2012/webprogram/Paper7294.html.

University of Tokyo Science Interpreter Training Program. (n.d.). *Concept.* Retrieved from science-interpreter.c.u-tokyo.ac.jp/en/.

Waseda University. (n.d.). *Five requirements.* Retrieved from www.waseda-j.jp/majesty/eng/program/index.html.

Watanabe, M. (2010). New wine must be put into new wineskins: The new key word 'Science Communication' has changed PUST policy in Japan. In *Proceedings of PCST2010 Conference (New Delhi).* Retrieved from www.pcst.co/archive/paper/365.

Watanabe, M. (2017). From top-down to bottom-up: a short history of science communication policy in Japan. *Journal of Science Communication, 16*(03), Y01. doi.org/10.22323/2.16030401.

Watanabe, M. and Imai, K. (2003). *Research on the Promotion of Public Understanding of Science & Technology and Science Communication. NISTEP Research Material 100.* Retrieved from hdl.handle.net/11035/787.

Watanabe, M. and Imai, K. (2005). A Discussion About the Promotion Of Science & Technology Communication. NISTEP Discussion Paper 039. Retrieved from hdl.handle.net/11035/459.

Yagi, E. (2007). A proposal of science communication education as liberal arts in graduate school. *Communication-Design, 10*, 121–43.

Yamanouchi, Y. and Nakagawa, C. (2012). Transferable skills training in the United Kingdom: suggestions for developing human resource in science and technology. *Japanese Journal of Science Communication, 12*, 92–107.

Yokoyama, H. (2009). Science communication, education for graduate students, researchers in science. *Japanese Journal of Science Communication, 5*, 79–85.

Yoshisawa, G. and Taniguchi, T. (2016). Teaching technology assessment. *Proceedings of the 31st Annual Conference of the Japan Society for Research Policy and Innovation Management* (pp. 74–77). Tokyo: JSRPIM.

Timeline

Event	Name	Date	Comment
First interactive science centre established.	Miraikan – The National Museum of Emerging Science and Innovation	2001	
First national (or large regional) science festival.	Youngsters' Science Festival	1992	A generic name of science festivals and a network. First held in Tokyo, Nagoya and Osaka, and now over 100 cities hold their own festivals. 2006: Science Agora is an annual science communication event in Japan held in Tokyo, and the biggest in Japan
An association of science writers or journalists or communicators established.	Japanese Association of Science and Technology Journalists	1994	2011: Japanese Association for Science Communication
First university courses to train science communicators.	University of Tokyo Hokkaido University Waseda University	2005	Courses for postgraduate students started with support of five-year limited government subsidies
First master's students in science communication graduate.	JT Biohistory Research Hall in affiliation with Graduate School of Science, Osaka University	1995	It was not a training course for science communicators but a research course of science communication

Event	Name	Date	Comment
First PhD students in science communication graduate.	JT Biohistory Research Hall in affiliation with Graduate School of Science, Osaka University	1997	
First national conference in science communication.	Seamless Culture through Science Communication	2005	International colloquium organised by National Institute of Science and Technology Policy (NISTEP) 2006: Science Agora
National government program to support science communication established.	Loving Science and Technology Plan. Ministry of Education, Culture, Sports, Science and Technology (MEXT)	2002	2005: Special Coordination Funds for the Promotion of S&T. MEXT funded a program for training postgraduate students
First significant initiative or report on science communication.	*Research on the Promotion of Public Understanding of Science & Technology and Science Communication*	2003	Watanabe and Imai (2003) wrote a report for NISTEP
National Science Week founded.	The week around 18 April each year.	1960	Known as Invention Day since 1954
A journal completely or substantially devoted to science communication established.	*Japanese Journal of Science Communication* organised by CoSTEP	2007	Online journal run by Communicators in Science and Technology (CoSTEP) Education Unit at Hokkaido University
First awards for scientists or journalists or others for science communication.	Public Understanding Promotion Category of the Commendation for Science and Technology by MEXT	Not known	
Date hosted a PCST conference.	Satellite Symposium of PCST 2006 in Tokyo	2006	Organised by NISTEP
Other significant events.	First science café in Japan	2004	An NPO held it in Kyoto

Contributors

Professor Masataka Watanabe is president of the Japanese Association for Science Communication and Specially Appointed Professor at Tohoku University, Sendai, Japan.

Dr Mitsuru Kudo is an associate professor at the Centre for the Study of Co*Design, Osaka University, Japan.

23

MALAYSIA

Science communication in a pluralistic society

Mahaletchumy Arujanan, Noorshamira Shamsuddin and Farahana Nadzri

1. Malaysia in a nutshell

The British ruled Malaysia from the 17th century until 1957; before the British, parts of Malaysia were colonised by the Portuguese and Dutch. Prior to the Westerners' invasion of ancient Malaysia, the region was a hotbed for traders from around the world. Besides traders from Europe, Indians, Chinese and Arabs flocked to the peninsula as its location was a strategic place for traders to meet.

The migration of Indians, Chinese and the Indonesians formed the pluralistic society Malaysia has today. The main religions in Malaysia are Islam, Buddhism, Hinduism, Christianity and Taoism. Malay is the national language, English is the second official language, and Mandarin and Tamil are widely spoken.

Malaysia is the third largest economy in Southeast Asia, with a GDP of $US296 billion (ASEAN UP, 2018). Since the late 1970s, the country has evolved from an agrarian economy to a diversified one with the manufacturing and service sectors comprising a larger proportion of the economic pie (World Bank, 2018). The share of the agricultural sector in GDP terms has declined steadily from 29.9 per cent in 1970 to 8.2 per cent in 2017 (Department of Statistics Malaysia, 2018).

The Malaysian population stands at 31.1 million people, with ethnic Bumiputera (Malay and indigenous) at 68.6 per cent, followed by Chinese (23.4 per cent), Indians (7.0 per cent) and others (1.0 per cent) (Department of Statistics Malaysia, 2018). The assimilation of all these races and the influences of the colonisers make modern Malaysia a melting pot with rich traditions of history, culture, languages and religions. The pluralistic society, diversity and historical background play a role in how science is effectively communicated, the tools and approaches employed, and also the issues and concerns.

2. The start of modern science communication in Malaysia

Science communication, enculturation of science, democratisation of science, and bridging science and society have become buzz phrases in Malaysia in the last 15 years. Today we see science communication initiatives garnering attention from Malaysian researchers and policymakers, although it is still not actively practised by most researchers. It is driven by a few individuals with many limitations in terms of resources.

While researchers, academia, ministries and policymakers are jumping on the bandwagon, policy measures to formalise this field are still lacking. Capacity building, academic programs, public outreach, media engagement, meeting politicians and Science, Technology, Engineering and Mathematics (STEM) promotion in schools are conducted in an ad-hoc manner through a bottom-up approach, with initiatives driven by individuals or organisations and not driven by policies. This should be strengthened through a coordinated effort between ministries, universities and research institutes. Funding and training are hurdles that can be overcome if there is a top-down effort, where the need to engage the public is spelled out in science, technology and innovation (STI) policies and becomes an obligation of the scientific community. But support must be provided so this does not become a burden to them. This is yet to happen.

There are two driving forces contributing to the emergence of this field in Malaysia. The main factor is an emphasis by the government in nurturing the country to be a knowledge-based economy. The momentum of science communication is growing in tandem with the number of STI policies and initiatives. These initiatives are largely attributed to Dr Mahathir Mohamad—the fourth (1981–2003) and seventh (since May 2018) Prime Minister of

Malaysia, and a visionary leader. However, while STI policies are strong in Malaysia, science communication has not been strengthened with dedicated funding, training and human capital.

The other driver is the decline in students wanting to pursue STEM education. In response to this challenge, researchers have started science communication initiatives, mainly with teachers and students as target audiences. Science communication initiatives in Malaysia are in response to the two factors. Most science communication players in Malaysia align their aims, strategies, messages and target audience to these factors, with STEM promotion at a higher priority.

2.1. The driving force from the government

Science-related policies in Malaysia give an impetus to researchers to bring science to the public domain and engage with their key stakeholders.

2.1.1. Policies related to STI policies: The Mahathir factor

Popularly known as Malaysia's Father of Modern Development, Dr Mahathir Mohamad was the man behind many science-anchored national policies in Malaysia. During the first term of his leadership, Malaysia saw the unfolding of many policies and initiatives related to STI. Major ones are shown in the timeline in Figure 23.1. Currently there are 56 national policies related to STI and under the purviews of various ministries.

2.2. STEM education policies in Malaysia

Talent development in STEM is one of the priorities for the country. There is a strong need to draw students into STEM as Malaysia is facing a decline in the number of students pursuing this field, both at schools and universities. This problem was identified in the late 1960s, way before the term STEM was coined. Malaysian students are streamed into either arts or science classes according to their results in the national examination at the age of 15. The Malaysian Higher Education Planning Committee (JPPT) reported in 1967 that out of the 3 per cent of Malaysian secondary school students who continue their education at a tertiary level, 70 per cent enrolled for arts and humanities programs. The low interest in pursuing science prompted JPPT to recommend that 60 per cent of upper secondary school students be enrolled in science programs to meet future needs (Curriculum Development Division, 2016; Ministry of Education, 1967).

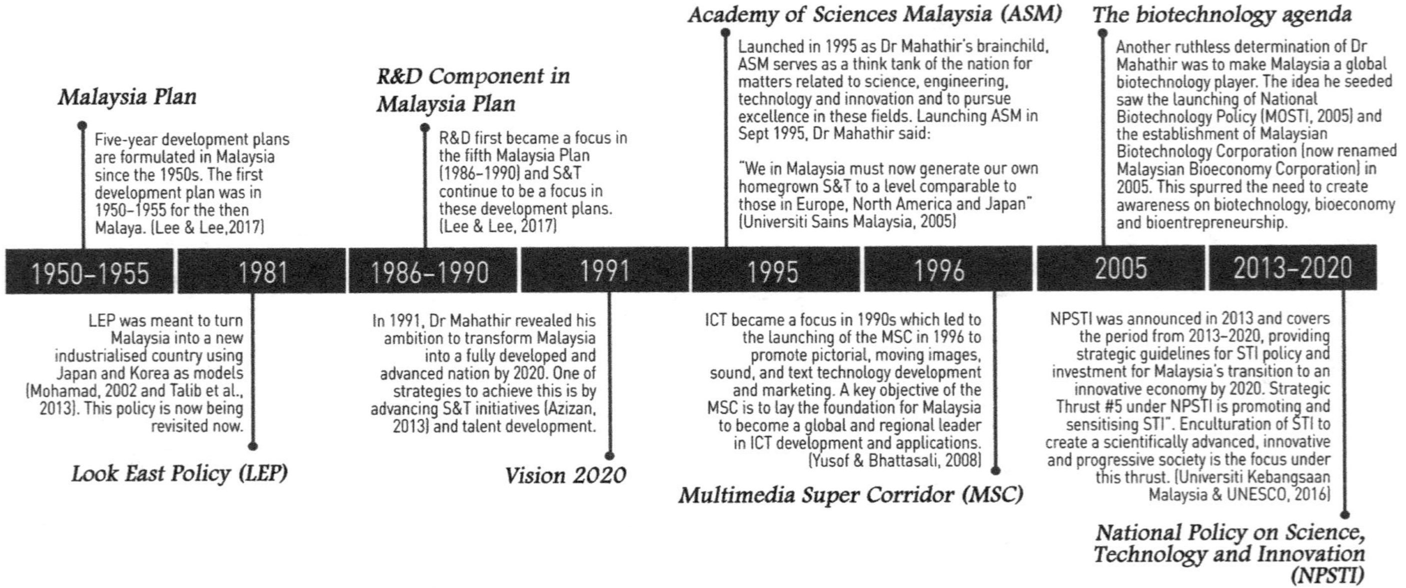

Figure 23.1: Timeline of selected STI policies and milestones in Malaysia.

The last four decades saw Malaysia embarking on various initiatives (discussed below) to achieve the 60:40 ratio of science to arts students. To date this threshold has not been achieved and currently only 42 per cent of students in upper secondary school level opt for science and technical streams (Ministry of Education Malaysia, 2016). The Malaysian Education Blueprint (2013–2025) launched in 2012 further strengthened STEM initiatives and its implementation was outlined in the *Report on Strategies to Achieve the 60:40 Science/Technical: Arts Stream Policy* (Ministry of Education, 2012).

In 2017, the Ministry of Education in collaboration with Ministry of Science, Technology and Innovation (MOSTI), the Ministry of Higher Education and the private sector developed the National STEM Action Plan. This plan looks into many aspects of STEM such as awareness, education, infrastructure, research, career opportunities and information gathering. With the change of government administration in May 2018, the fate of this action plan is not yet known.

The onus of promoting STEM is largely taken up by NGOs and researchers who organise exhibitions, talks, workshops, science competitions, fashion shows and roadshows on a voluntary basis. This is further discussed under Section 6. Most of these activities are funded by universities and NGOs with minimal support from the government. Funding is the main challenge faced by these communicators.

3. Research in science communication

While communicating science is becoming a popular activity, research in this area is still limited in Malaysia. Current research focuses on science and religion, legal and ethics in STI, public understanding of science, indicators in STI, science and gender, and STI policies. The emphasis on ethics and religion is possibly due to the large Muslim population in Malaysia. Areas that could be strengthened are research on science communication strategies, tools and analysis of human cognition.

3.1. The researcher's perspective

One notable researcher producing empirical data on public understanding, awareness, appreciation, perception and psychology, and the factors that drive public attitudes is Professor Latifah Amin from the National University of Malaysia (UKM).[1] Professor Amin is trained in biochemistry and

1 Latifah Amin, personal communication, 21 September 2018.

molecular genetics, but chose to pursue her PhD in consumer behaviour on modern biotechnology due to her curiosity about Malaysians' attitudes and perceptions towards biotechnology. Latifah now teaches bioethics at UKM.

Her research covers consumer behaviour, bioethics, biosafety, communication and education, biotechnology in the religious perspective and technology diffusion. She has published about 200 works including journal articles, book chapters and proceedings; she has been doing research in this area for 17 years. She found that there is a growing interest in pursuing research on public understanding of science among graduates who aspire to be researchers and academic staff but discontinue their research in this field upon graduation. Based on her experience working with them, Amin attributes this to their background in social science (most have bachelor's degree in *Syariah*[2] or media studies) that makes them uncomfortable and not confident enough to tackle issues related to STI.

This was also found by Dr Mahaletchumy Arujanan, the first PhD graduate in science communication in Malaysia (and lead author of this chapter). She had difficulties being accepted as a PhD student by professors in media studies, journalism and social sciences due to her initial qualifications in natural sciences. There is a disconnect between social sciences and natural sciences in Malaysia. This is being addressed by Amin's Centre for Liberal Studies at UKM and the University of Malaysia's Department of Science Studies, with both conducting research in areas where science intersects with the society: ethics, humanity, religion, public acceptance. Despite this activity, research on effective science communication is rare.

Like many other countries, science communication is seen as a social science in Malaysia but due to its hybrid nature, both Professor Amin and Dr Arujanan concur that in the Malaysian context, with the disconnect between social and natural scientists, science graduates are a better fit for research in science communication and the public understanding of science. While researchers from a natural sciences background should be complemented with graduates from media studies, social sciences and journalism, a stronger network and supporting system (mentors, role models and collaborators) made up of senior social and natural scientists is needed to provide confidence and support to those with a background in social sciences.

2 *Syariah* (the Malay spelling of *Sharia*) refers to *Sharia* law in Islamic religious law and deals with exclusively Islamic laws, having jurisdiction upon every Muslim in Malaysia (Wikipedia: en.wikipedia.org/wiki/Syariah_Court (accessed 27 March 2020)).

Professor Amin will be retiring soon and hopes more researchers will show interest in research in the fields of science communication, bioethics, biosafety and religious perspective of STI in Malaysia.

3.2. Public awareness, interest, attitude and acceptance towards STI

The two main sources of data on public perceptions and attitudes towards science and emerging technologies in Malaysia are Latifah Amin and the Malaysian Science and Technology Information Centre (MASTIC) under the Ministry of Energy, Science, Technology, Environment and Climate Change (MESTECC, formerly known as MOSTI). MASTIC has conducted surveys to gauge public interests, attitudes, knowledge, understanding and awareness of the Malaysian public towards STI since 1996, with the last survey being conducted in 2014. The study was conducted biennially until 2004, after which it was conducted once every Malaysia Plan (five years), on the grounds that the trends do not show much change biennially. MASTIC's main objective of monitoring public attitudes and understanding of STI is to provide baseline information for drawing up STI policies.

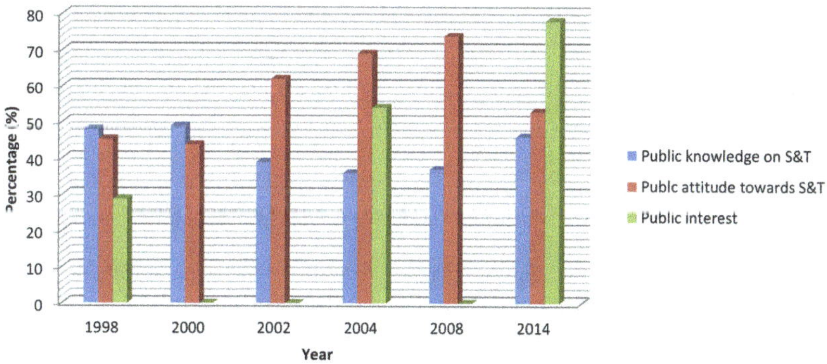

Figure 23.2: Public knowledge, attitude and interest towards STI.
Source: MASTIC, 1998, 2000, 2002, 2004, 2008, 2014.

It is a challenge to do a comparative analysis based on MASTIC's survey as the results are not presented in a consistent manner. 'Mean score' is used in some years, whereas percentages are used in others. The average for each indicator is also not included in the reports. For example, the public knowledge on STI is based on STI topics such as environment, pollution, information technology, etc. The national average is not found in most of the reports, making it very difficult to compare them between years. The definitions used for public knowledge, attitude, awareness and interest are also not in agreement with what most science communicators use internationally. The shortfalls in these reports show that research in science communication is still a relatively new field in Malaysia.

Figure 23.2 shows results extracted from MASTIC reports from 1998–2014 on public knowledge, attitude and interest that could serve as a baseline information to enhance public understanding of science and technology.

4. Islam, culture and ethics: The concerns and motivation

Malaysia is a multiracial and multi-religious country with Islam as the official religion. Out of 31.1 million population, 61.3 per cent are Muslims. Religion is an integral part of Malaysians' lifestyle.

Muslims take the concepts of *halal* (permissibility) and *haram* (forbidden) seriously and products developed through STI must adhere to principles of *Shariah* (Islamic law). Interpretation of Islam is often considered when approving new technologies and in labelling laws. So, ethics and religious consideration and principles play an important role in developing science communication messages. Although there are four main religions in Malaysia (Islam, Buddhism, Hinduism and Christianity), the role of Islam is presented here as more studies are done in this area.

Here is a conclusion from Amin's paper on 'Decision making on agro-biotechnology issues: An Islamic perspective' (Amin et al., 2013):

> Islam is seen as more emphasising on protecting the major *maslahah* which are religion, life and health, progeny, intellect and property. This is to ensure the safe development of agro-biotechnology products for the benefits of mankind, environment and other living organisms in this world. If a certain GMOs or products largely benefit mankind as a whole, without any serious risks to the five purposes of *Maqasid-al-Syariah*, thus the use of the product is allowed. On the other hand, if the GMOs or products are proven to be harmful to the five purposes of *Maqasid-al-Syariah*, thus the use of it is prohibited, even if the product is beneficial to human and society. In cases of uncertainty of the risks, the permission to use the product must be withheld until evidence of either the benefits or risks can be proven.

Amin et al. (2011b) concluded that background variables such as religion, race, age, education level and gender have significant effect on some of the dimensions of Malaysians' ethical perception of modern biotechnology. These findings demonstrate that communication strategies in Malaysia must be customised to include religious concerns. Research on genetically modified (GM) crops, stem cells and cloning do take religious views into account. *Fatwa* (Islamic decrees) are binding on researchers. Malaysia, for example, has a *fatwa* that decrees a genetically modified organism (GMO) made from a gene from forbidden animals as *haram* for Muslims.

Another aspect that affects the way science is communicated in Malaysia is the values and belief of the public. MASTIC (2014) showed that Malaysians have a varied perception of horoscopes, faith healing and fortune telling. About 54.5 per cent of total respondents believed that horoscopes were scientific (36.4 per cent) or sort-of scientific (18.1 per cent). Framing science information and addressing public concerns may have to take these values and beliefs into consideration, as well as cultural sensitivities.

4.1. Playing the mediator between researchers and religious scholars

The Malaysian Biotechnology Information Centre (MABIC),[3] a not-for-profit organisation, is an important player in the area of science communication in Malaysia. MABIC's aim is to build public understanding in science and biotechnology. It is part of an international network of Biotechnology Information Centres (BICs) with the International Service for the Acquisition of Agribiotechnology Applications (ISAAA) as the parent organisation. Concerned that modern biotechnology can be wrongly perceived through the lens of Islam and vice versa, MABIC organised three dialogues between *ulama* (religious scholars) and researchers to bridge the knowledge gap and to help researchers understand Islam's stance on modern biotechnology, especially genetic modification.

Tackling religious issues in Malaysia is a delicate matter and it may lead to serious repercussions if the relevant sensitivities are neglected. Islamic law, education and other matters are under the jurisdiction of Department of Islamic Development (JAKIM) housed in the Prime Minister's Department.

The first attempt at dialogue was in 2010 but it attracted much scepticism, with MABIC's role and agenda questioned. The workshop fell apart and the initial aim to adopt a resolution on Islam's position on agribiotechnology was not achieved. There were useful lessons from this failure. Organisers understood the need for a credible Islamic partner, and that participants should be high-level religious scholars and not middle-level officers.

3 Lead author of this chapter, Dr Mahaletchumy Arujanan, is the Executive Director of MABIC.

Figure 23.3: The highest Islamic scholar from Iran at the dialogue organised by MABIC on agribiotechnology involving GM technology, 2010.

Source: MABIC (used with permission).

The second workshop was organised during the World Halal Forum 2010 in Kuala Lumpur with the International Halal Integrity Alliance (IHIA), a credible organisation with international standing and network. The workshop resulted in a resolution that said GM crops are *halal*, public awareness on biotechnology needs to be strengthened and the role of *ulama* in scientific discussions must be enhanced (World Halal Forum, 2010).

Reinforcement is important for high-concern issues, so MABIC organised another stakeholder engagement in December 2010 with IHIA. The workshop titled 'International Workshop for Islamic Scholars on Agribiotechnology: Shariah Compliance' yielded another set of resolutions (Shaikh Mohd Salleh, 2012). These resolutions are used as references in the Muslim world with regards to GM crops.

While approved GMOs are available in the Malaysian market, researchers of all faiths will not attempt to develop GMOs from non-*halal* genes so as not to exclude Muslim researchers. Further, this step would ensure that Muslim consumers would not be excluded from being able to purchase the end product—as these consumers make up more than 60 per cent of the market, this also makes good business sense. However, biosafety laws do not have any prohibition on imports of GMOs that have been evaluated and approved as safe, even if they have *haram* genes, though this scenario is yet to arise. Clear labelling of such products would be needed to keep consumers informed.

4.2. The motivation

The scientific legacy of Islam during its renaissance is often used to motivate students to undertake STEM education and to create science literacy and public interest.

Petrosains, a major science centre in Kuala Lumpur, organised an exhibition on 'Sultans of Science—Islamic Science Rediscovered' from December 2012 to June 2013. The exhibition showcased the inventions and scientific breakthroughs achieved by Muslim scholars during the peak of the Islamic civilisation (700–1700 CE)[4] in the fields of engineering, medicine, astronomy, geography and agriculture. Notable breakthroughs by ancient Muslim scientists were exhibited, including Al-Jazari, who authored *Book of Knowledge of Ingenious Mechanical Devices* (1206 AD), which describes 50 machines such as animal and humanoid automata, automatic gates and doors and clocks; and Al-Hambra's work on landscaped gardens. The exhibition raised the possibility that the first aviator might have been Abbas ibn Firnas, a Kurdish Arab who took off from a hill near Cordoba, Spain, 1,000 years before Otto Lilienthal in Germany. Part of the exhibition included a rich history of Muslim physicians who pioneered modern medicine.

In 2007, MOSTI held the *Scientific Excellence in Islamic Civilisation* exhibition (IKIM, 2013). There is a need for more programs and exhibitions of this nature, to remind people of the proud Islamic record in science. The ancient Islamic legacy in STI could be used effectively to trigger Malaysian Muslim interest and curiosity in these fields. Communication strategies and messages could be developed, framed and aligned to create the requisite pride and affinity.

5. Democratising science through media

Science news takes a backseat in mainstream media in Malaysia. A dialogue organised by MABIC between researchers and journalists in 2012 revealed a number of reasons for this: reluctance among researchers to engage with the media due to lack of science communication skills; distrust between researchers and journalists; and the knowledge and cultural barriers between journalists and researchers. The other major problem is the priority

4 1700 CE is the same as 1700 AD: 'Common Era' (CE) is a modern alternative for 'Anno Domini' (AD).

given to high-impact journal papers that help push a university's international ranking, but engaging the public does not help in career advancement of researchers. Universities and research institutes do not provide the grants, training and human resources to support researchers' involvement in engaging the public.

5.1. *The Petri Dish*

As the executive director of MABIC, the only organisation in Malaysia with a full-time mandate to create public understanding of biotechnology and science, Dr Arujanan wanted a platform for the scientific community to reach out to the public. Her vision was for a newspaper with empathy for researchers who are not able to translate their research into popular science articles, where science news hits the headline, and where science will be the topic of discussion at home and at coffee shops. She saw several national benefits. Science and research must be in the public domain to bridge the gap between research and the market. Media coverage would facilitate the development of enabling policies and regulations, and inform the public how taxpayer money is spent. It would inspire students to enter research careers and encourage the government to allocate more money for R&D, while helping the public to appreciate and accept emerging technologies.

With the help of a journalist friend, Joseph Masilamany, the newspaper *The Petri Dish* (www.thepetridish.my) was first published in February 2011 as a 12-page monthly English newspaper. The initial business model of a free newspaper supported by advertisements was not feasible and it was later tagged at RM8 per issue (US$2). The newspaper was first circulated to all the universities, research institutes, government agencies, ministries and other relevant organisations. It received very good feedback from the scientific community and the circulation was extended to public places such as shopping malls, private hospitals, airports and Starbucks outlets to reach non-technical readers.

The main challenges to sustain and expand this newspaper are funding and active contributions from researchers. The aim of MABIC is to expand the newspaper beyond biosciences to other science fields and to add content in the national language, Malay, to garner more readers.

5.2. *Majalah Sains*

Majalah Sains [Science Magazine] is a Malay-language science portal, the brainchild of Mohd Fa isal Aziz.[5] Aziz's aim was to inculcate interest in STI among the Malaysian public, especially the younger generation.

5 M. F. Aziz, personal communication, 13 September 2018.

As a researcher at the Institute of Micro and Nanoelectronic Engineering (IMEN) in UKM, Aziz had contributed to popular local science magazine *Dewan Kosmik* since 2004. He then wanted to expand his audience and share the advances in nanotechnology and microelectronics with the general public. Aziz wrote 40 science articles in Malay for *Dewan Kosmik* until 2009; he then looked for a platform to publish articles on the internet. He found hundreds of science portals in English but no Malay-medium portals, and he saw the need for a Malay platform to disseminate information that could reach locals. In his view, readers understand science better when presented in Malay language, especially in rural areas. This saw the birth of *Majalah Sains*.

Other researchers were roped in to support the editorial tasks. Dr Rosdiadee Nordin, Dr Ismayadi Ismail, Ahmad Amryl and Hasfazilah Hassan became part of the team. The portal receives an average of 8–12,000 visitors per day depending on the issues carried—bauxite mining, for instance, was a hot story in Malaysia; special interviews with Nur Adlyka Ainul Annuar about the discovery of the Supermassive Black Holes were also popular. *Majalah Sains* has more than 300 contributors from universities and research institutes, and receives more than 20 articles a week. Dr Nordin said *Majalah Sains* has become a platform for aspiring researchers to communicate their research, although there is one challenge: persuading contributors to write in a manner that laypersons understand. This problem could be solved by offering seminars in how to write popular science articles.

5.3. *Dewan Kosmik*

Dewan Kosmik is a monthly science magazine in the Malay language that aims to raise public awareness of science and how it relates to life and humanity.[6] The first issue was published in January 1993 by Dewan Bahasa dan Pustaka, a department under the Ministry of Education. With 68 pages, the magazine covers a wide range of topics including engineering, technology, geology, medicine, psychology, astronomy, biology, pharmacy, film technology, information technology, botany and science fiction. Target readers are professionals, researchers, industry executive, students, teachers, professors, trainee teachers, high school students and the general public. The magazine accepts articles from the public, and an honorarium is paid to encourage citizen journalism.

6 K. A. Mohd Isa, personal communication, 22 October 2018.

5.4. *Estidotmy*

Estidotmy was Malaysia's favourite science magazine published by *Utusan Malaysia*, a mainstream Malay-language newspaper.[7] With 36 pages, this free magazine was circulated as a pull-out every last Wednesday of the month from 2002 to 2012. MOSTI funded the venture to increase the interest and awareness of STI among communities, especially students, while the Academy of Sciences Malaysia (ASM) provided the content. The content was in both English and Malay. It was an excellent partnership between the ministry, media and the ASM, but had to be halted due to limited funding.

The magazine had a big impact on its readers. Zamir Mohyedin, through his Facebook post,[8] said that he loved science because of *Estidotmy*. Zamir, a Solid-State-Physics researcher at Institute of Science, University of Technology MARA (UiTM), attributes his career to *Estidotmy* as he says science was taught in a boring way in school. Many are hoping for the magazine to make a comeback.

5.5. Mainstream media

STI is not a main feature of mainstream media in Malaysia and the only newspaper to have either a science desk or journalists trained in science is the main Malay daily *Utusan Malaysia*. Health and medical articles form the bulk of science news in Malaysian newspapers as they are the most relevant topics for the general public (Arujanan, 2013). Science coverage is intermittent and typically occurs during times of crisis such as an epidemic or to report a major breakthrough in research. Most science articles are sourced from wire services.

6. STEM promotion

The government is concerned that the shortage of highly skilled talent in the area of STEM will be a handicap for Malaysia in its goal of becoming a knowledge-based economy. Players from the research fraternity are working to create awareness in STEM education, disciplines and careers. These players want to achieve the 60:40 ratio of science to arts students set by the government, and science communication efforts in Malaysia focusing on this objective are targeted at students and teachers.

7 Z. Mohyedin, personal communication, 21 October 2018.

8 www.facebook.com/profile/100000305631702/search/?q=estidotmy, 14 February 2016.

6.1. Government-driven initiatives

With 56 national policies related to STI in Malaysia cutting across various ministries, the government takes promotion of STI and STEM seriously and has promulgated various national-level mega activities.

During these national-level programs, a host of carnival-like activities are organised throughout the country. Some attract more than 10,000 visitors in a week. While there is much rhetoric during these thematic mega-events, the impact of these activities is often not measured (possible measures include measuring an increase in media coverage on science, or the number of STEM students, or public knowledge, awareness and acceptance on science). This is one area for improvement.

Table 23.1: Government-driven initiatives to promote STEM education

Initiative	Aim	Target audience	Ministries
Innovate Malaysia, 2010	To encourage a culture of creativity and innovation among Malaysians	Researchers, students, industry players, NGOs and the general public	MOSTI
Promotion of Science and Mathematics Year, 2011	To foster interest and participation of Malaysians in mathematics and the sciences	Students, general public and NGOs	MOSTI; Education; Youth and Sports; Women, Family and Community Development
Year of Science and National Innovation Movement, 2012 (3GI2012)	To encourage Malaysians to adopt a culture of science and innovation	Malaysian youths	MOSTI
National Science Week, 2018	To increase consciousness of the role of STI and how it affects the country's social economic state	Students, teachers, general public	MOSTI (MESTECC)[1]

[1] In 2018 MOSTI was restructured and became part of the Ministry of Energy, Science, Technology, Environment and Climate Change (MESTECC). It was subsequently restructured in 2020 and reverted to MOSTI.

6.2. Scientist-driven initiatives

Many Malaysian researchers promote science literacy among students and teachers in their own time. Individual researchers have championed this cause and mobilised others to join forces.

The Malaysian High School Biotechnology Awareness Program was initiated in 2001, and involved roadshows to schools in Malaysia. Academics and researchers from universities and research institutes facilitated hands-on sessions in schools and spoke about careers in biotechnology (Firdaus-Raih et al., 2005). Standard modules were prepared with two sessions: one consisting of lectures and series of talks; and the other of three different hands-on sessions (games, wet labs and a multimedia self-exploration fun quiz). In 2005, the program reached 563 schools and 18,000 students. It was funded by MOSTI.

The National STEM Movement is led by Professor Dr Noraini Idris.[9] It partners with universities around Malaysia, the Ministry of Education, MESTECC and industry players to organise activities for schools. A number of private players who offer STEM-based educational programs joined forces with the movement as a STEM Content Providers Network. Science fairs and exhibitions are organised at malls and convention centres, targeting schools, parents and the general public with content and exhibits coming from universities and industry players. The network of academics from universities in all 13 Malaysian states allows events to be organised at the state level. This addresses a big challenge where students from states distant from the capital city are often left out as events are usually organised in cities around the capital.

The STEM Mentor–Mentee Program was launched in 2016. Lecturers, researchers, engineers and mathematicians from universities and industry act as mentors to provide guidance to students to pursue their education and career in STEM. To date, more than 25 universities have introduced the STEM Mentor–Mentee Program, with more than 100 schools and 3,000 students participating in various STEM activities involving hands-on laboratory sessions and field visits. The mentors also train teachers to enhance their pedagogy to make science and mathematics lessons more interesting in classrooms.

6.3. STEM promotion by MABIC

Biotechnology workshops in schools are MABIC's regular events, but the major engagement with schools was nation-wide competitions with biotechnology as the theme. These were organised in collaboration with the Ministry of Education, Malaysian Biotechnology Corporation (BiotechCorp), the National Science Centre and Taylor's University in 2010 and 2011, with more than 2,000 students participating. Public speaking, debates, quizzes, spelling competitions, essay writing and poster drawing were open to schools nationwide.

9 S. Thirugnana, personal communication, 5 October 2018.

Figure 23.4: Science-based board game at MyBio Carnival 2011 organised by Ministry of Science, Technology and Innovation (MOSTI), now known as Ministry of Energy, Science, Technology, Environment and Climate Change (MESTECC).

Source: MABIC (used with permission).

Between 2007 and 2012, MABIC organised a series of teacher workshops to enhance pedagogy and delivery of science in classrooms in several states, in collaboration with the Ministry of Education and universities. The focus was on biotechnology and the topics were based on the expertise of the partnering academics at the universities.

A non-traditional approach from MABIC was a fashion show organised with a fashion school at University Technology Mara (UiTM) in 2010 in collaboration with BiotechCorp. It was later replicated at other outreach events, including one in Nairobi by MABIC's sister organisation, ISAAA AfriCenter. It included a competition for fashion students to design fabrics and outfits based on any biotechnology theme. Students came out with designs inspired by Dolly the Sheep, cloning, stem cells, neuron cells, genetically modified crops, palm oil hybrids, bacteria and viruses. The fashion show was covered by women's magazines and pull-outs in mainstream newspapers, a space often not occupied by biotechnology stories.

Figure 23.5: Outfit designs inspired by biotech and life science motifs during a fashion show organised by MABIC and BiotechCorp in 2010. Clockwise from top left: inspiration from neuron cells, Dolly the Sheep, DNA and palm oil. The outfits were designed by fashion students from University Technology Mara (UiTM).

Source: MABIC (used with permission).

7. Influencing policies and regulations

One aim of science communicators is to influence policymaking and regulations on science-based decisions. Policymakers and politicians need to be well informed to ensure R&D and commercialisation are not stifled by hurdles created by scaremongers and misinformation about science.

7.1. MABIC's role

Science communication efforts in Malaysia hinge heavily on STEM promotion. Science communication is still not well understood in Malaysia, even by those who are involved. Almost all players assume it is about getting the general public and students interested in science, and focus on STEM promotion. Outreach on regulations and policies related to science with target audiences such as policymakers, parliamentarians and religious scholars are not attempted by universities, research institutes or industry players.

MABIC fills in the gap to reach out to policymakers, politicians and regulators through a number of initiatives:

- *The Petri Dish*
- forums/conferences/seminars related to policies, regulations and Islam's position on new technologies
- media engagements to enable balanced coverage of biotechnology in the media
- media–scientist dialogue to bridge the knowledge and cultural barrier so science gets into the public domain to influence policymaking.

7.2. Outreach programs on agribiotechnology regulations

Two key players in this area are MABIC and the Department of Biosafety (DOB), under the Ministry of Water, Land and Natural Resources. Prior to the passing of the Biosafety Act in the parliament in 2007, MABIC regularly organised a number of conferences and seminars to create awareness among researchers, policymakers, politicians, industry players, regulators and traders to enable the development of science-based laws and regulations. These took place between 2006 and 2007. In 2013, MABIC organised a session for members of parliament from both the government and opposition parties to discuss agribiotechnology. MABIC understands the need to have 'Scientists meet MPs' sessions on a regular basis; strong champions and partners are needed to break the firewall for science to enter the corridors of the lawmaking house. In 2010, a study tour to the US was organised for regulators to understand the entire agriculture value chain involving the development of seeds, farming techniques, regulations, transportation and trade.

The DOB enforces the Biosafety Act, and monitors and regulates the activities related to GMOs: R&D, environmental releases, trade and commercialisation (Department of Biosafety, n.d.). It creates awareness in these areas among researchers, regulators, traders, enforcement officers, farmers, media and the general public. Various aspects of GMOs are covered in their outreach and capacity-building programs, ranging from how GMOs are regulated, their safety, the science involved in developing GMOs, their benefits and potential risks, and how risks are assessed and managed.

A number of educational materials developed by the DOB disseminate information on biosafety and agricultural biotechnology to the public, including Q&A kits, biosafety guideline handbooks, model farm and field trials, brochures and replicas of GM crops. MABIC provides technical input and content to the DOB for some of its public awareness material. Two main challenges the DOB faces are limited personnel and the budget to do countrywide outreach programs.

8. Capacity building in science communication: Academic programs

Malaysia lacks academic programs in science communication at all levels, from bachelor's degrees to PhDs. Science communication is not taught as a unit for undergraduates in the STEM fields. Monash University Malaysia makes it compulsory for its undergraduates at the School of Science to take a unit with components in science communication, making it the only university in Malaysia to do so.

The closest to a science communication program is the Master of Communication (Science and Environmental Journalism) offered by Universiti Sains Malaysia (USM), although the focus is on environment. It is offered by USM's School of Communication and is described on the university's website:

> The programme is designed mainly to prepare students for a professional career as science and environmental writers, reporters and editors in the media industry, research institutions and environment-related organisations.[10]

10 web.usm.my/comn/acd.asp?tag=p2 (accessed 27 March 2020).

Dr Arujanan was the first to receive a PhD in science communication in Malaysia. It was a awarded by the University of Malaya (UM) in 2013:

> I tailor-made my PhD in science communication as this was not a research field at any Malaysian university and there were no experts in this field. Knowing that I would not be able to secure a job in Malaysia as a science communicator if I left my position as the Executive Director of MABIC to pursue a PhD overseas, I chose a professor from UM whose work is related to science and society. Science communication is not a mainstream profession in Malaysia and there are few employers in this field except for the National Science Centre and a handful of other science centres. In 2008, science communication was still a relatively new area of research (and it still is today). It was a tough journey without any proper guidance and expertise, and irrelevant questions and comments during defence seminars.
>
> Two years passed by without much progress and groping in the dark, and I was devastated. My previous background on microbiology, biochemistry and biotechnology was not helping me comprehend research methodologies in social science, and there were no colleagues in this field for any intellectual discourses. Finally, after lots of discussion with the dean, a concrete suggestion was made: that I find a consultant to help me. Having worked in this area for five years, I had built my network, so suggested Dr Craig Cormick from Australia as a consultant, and he unofficially stepped up to be my supervisor.
>
> I submitted my thesis in 2012 and received my PhD in 2013, for my study on 'Biotechnology Communications in Malaysia: Understanding the Issues, Influence and Audience towards Developing a Better Communication Matrix.

MABIC hopes to develop a postgraduate diploma in science communication in collaboration with a local university and has been invited by USM to develop online microcredential modules for this field. These will be the first structured academic programs in science communication in Malaysia. Work is in progress in these areas.

MABIC also plays a key role in providing short courses and training workshops in science communication and developed the first home-grown science communication module in 2018. The two-day module was endorsed by Monash University Australia as a School of Science Monash Doctoral Program (MDP) for PhD students in the natural sciences. It is also used to train researchers, policymakers, STEM practitioners and government officers. The module covers translating research into media articles, understanding media culture, creating the 'hook' to attract readers and audiences, developing metaphors and analogies, humanising research, translating research titles to headlines, and common blunders made by researchers when communicating to the public.

Another one-day module was developed on risk communication and the first workshop was carried out for Department of Biosafety for regulators and researchers in the field of genetic engineering research on June 2018. A one-day science communication module was also developed by MABIC for MESTECC officers from various agencies, and training was conducted in October 2018. MABIC's training modules are gaining traction and in the pipeline are modules on media training, risk communication (two-day module), non-traditional approaches and storytelling in communicating science. MABIC also attempts to customise its modules to suit the objectives of the training workshops and its participants.

9. Science and society: Reaching out to the public

The Young Scientists Network (YSN), under the ASM reaches thousands of children and members of the public annually through their activities.[11] Professor Abhimanyu Veerakumarasivam, chairman and founding member of YSN–ASM, says what he found most compelling is that most of the members assert that engaging the public actually helped elevate their research by broadening their network, increasing research collaborations and diversifying their source of research funds.

In collaboration with British Council and Malaysian Industry-Government Group for High Technology (MIGHT), YSN–ASM through its science communication working group organised science communication training for researchers in 2016 and science journalism training for journalists in 2017. The training focused on enabling researchers to communicate to non-technical audiences and journalists to appreciate science and do balanced reporting, as well as bridge the knowledge and cultural barriers between them. YSN–ASM acknowledged that while its members are eager to reach out to the public, their communication skills are still lacking, hence they strive to empower their members with such training.

11 A. Veerakumarasivam, personal communication, 30 August 2018.

Science Café is another initiative. To feed her curiosity about science, Liuyi Yeoh, founder of Science Café KL, frequented Science Café in the United Kingdom where she lived for 15 years.[12] On returning to Kuala Lumpur in 2016, she wanted to replicate it and, being a die-hard science fan, she started Science Café KL at The Bee, a casual café in the upscale Publika shopping mall. Yeoh, a full-time business development manager in a local infrastructure company, said that patrons would be able to discover new, curious and interesting things in the world of science at Science Café KL. She initially made cold calls to researchers asking them to share their research with the public but now she gets her contacts from YSN.

Shawn Keng, co-chair of YSN's science communication working group and a PhD student in science communication, hopes Science Café KL can help overcome public perceptions that science is boring, and also debunk pseudoscience and misinformation. He said the talks contain a higher dose of technical content than the average TV documentary and yet are palatable to laypeople. Science Café KL has tackled a diverse array of topics, from quantum physics to the link between the human genome and cancer, from nanotechnology to how the zebra fish may one day save human lives. The open format allows audience to interrupt and ask questions, and gives it a casual and relaxed air.

A third initiative, the National Science Centre (PSN) program, was a culmination of one of the strategic challenges listed by Prime Minister Dr Mahathir Mohamad in a paper entitled 'Malaysia: The Way Forward' in 1991:

> The sixth is the challenge of establishing a scientific and progressive society, a society that is innovative and forward-looking, one that is not only a consumer of technology but also a contributor to the scientific and technological civilisation of the future.

In November 1996, Malaysia formed its first science centre with a mandate to raise awareness, understanding and appreciation towards science and technology towards the creation of scientific society. The centre is interactive with explorative, hands-on exhibits requiring visitors' active participation. Facilitators conduct demonstrations and often researchers are invited to give talks on scientific phenomena. According to Elena Mazlan, science officer at PSN, the centre receives about 10–18,000 visitors during weekends.[13] In 2006, the PSN established its first branch in the Northern Region, Alor

12 S. Keng, personal communication, 27 August 2018; L. Yeoh, personal communication, 15 September 2018.

13 E. Mazlan, personal communication, 17 September 2018.

Setar, Kedah. In addition to the main exhibition, the centre also organises activities such as 'Special Science Day', 'Meet the Scientist' and 'Education Innovation for Teachers'. Besides the PSN, there are other public and private science centres such as Petrosains, planetaria and Tech Dome Penang.

10. Current challenges and the way forward

The main challenges for effective science communication in Malaysia are:

1. There is no policy that makes science communication an obligation among researchers.
2. Lack of funding and training for researchers to get involved.
3. Lack of understanding and knowledge of science communication among researchers currently involved in engaging with the public.
4. Lack of cooperation and coordination among the players who are involved in science communication.
5. Disproportionate emphasis given to STEM promotion, neglecting communicating science to policymakers, politicians and regulators.

A few recommendations are:

1. Establish a science communication office at all universities and research institutes where trained science communicators can help researchers to develop messages for their outreach programs, translate research into media articles, create social media platforms, provide science communication training, engage with the public through various activities, and promote STEM.
2. All grants received for research must allocate a small portion for public engagement activities. The science communication office could help researchers to develop their programs.
3. Academic programs in science communication at bachelor's and postgraduate levels should be initiated.
4. A science communicator association should be launched given the growing number of players in this space. This will promote collaboration and exchange of experience and knowledge.

The current landscape shows a progressive environment in the area of science communication and Malaysia could be in the forefront in southeast Asia in this field if measures are taken to fill the void.

References

Amin, L., Azlan, N. A. A., Hamdan, M. F., Samian, A. L. and Haron, M. S. (2011a). Awareness and knowledge on modern biotechnology. *African Journal of Biotechnology*, *10*(58), 12448–56.

Amin, L., Azlan, N. A. A., Hashim, H. and Ahmad, J. (2011b). Ethical perception of modern biotechnology. *African Journal of Biotechnology*, *10*(58), 12435–47.

Amin. L., Jahi, J. M. and Nor, A. R. M. (2007). Malaysian Public Awareness and Knowledge on Modern Biotechnology. *J. Pengajian Umum*, *8*, 195–200.

Amin, L., Sujak, S. F., Samian, A. L., Haron, M. S. and Mohamad, N. S. (2013). Decision making on agro-biotechnology issues: An Islamic perspective. *Journal of Food, Agriculture & Environment*, *11*(3&4), 1207–15.

Arujanan, M. (2013). Biotechnology Communications in Malaysia: Understanding the Issues, Influence and Audience towards Developing a Better Communication Matrix (PhD thesis). University of Malaya.

ASEAN UP. (2018, 6 March). Malaysia: 5 Infographics on Population, Wealth and Economy. Retrieved from www.aseanup.com/malaysia-infographics-population-wealth-economy/.

Azizan, S. A. (2013). Strengthening Malaysia's Scientific and Technological Development through Human Capital Development. *Procedia – Social and Behavioral Sciences*, *91*(10), 648–53. doi.org/10.1016/j.sbspro.2013.08.465.

Curriculum Development Division, Ministry of Education Malaysia. (2016). Sharing Malaysian Experience in Participation of Girls in STEM Education. Geneva, Switzerland, UNESCO International Bureau of Education (IBE).

Dardak, R. A. (2015). Transformation of agriculture sector in Malaysia through agriculture policy. FFTC Agriculture Policy Platform. Retrieved from ap.fftc.agnet.org/ap_db.php?id=386.

Department of Biosafety. (n.d.). *Objectives*. Retrieved from www.biosafety.gov.my/en-my/Corporate-Info/Pages/Objectives.aspx.

Department of Statistics Malaysia. (2018). Selected Agricultural Indicators. Retrieved from dosm.gov.my/v1/index.php?r=column/pdfPrev&id=UjYxeDNkZ0xOUjhFeHpna20wUUJOUT09.

Department of Statistics Malaysia. (2016, 22 July). Current Population Estimates, Malaysia, 2014–2016. Retrieved from www.dosm.gov.my/v1/index.php?r=column/cthemeByCat&cat=155&bul_id=OWlxdEVoYlJCS0hUZzJyRUcvZEYxZz09&menu_id=L0pheU43NWJwRWVSZklWdzQ4TlhUUT09.

Firdaus-Raih et al. (2005). A nationwide biotechnology outreach and awareness program for Malaysian high schools. *Electronic Journal of Biotechnology*, *8*(1), 8–15.

IKIM. (2013, 5 March). Insight into Islamic Science. *The Star*. Retrieved from www.thestar.com.my/opinion/columnists/ikim-views/2013/03/05/insight-into-islamic-science/.

ISAAA. (2002). Speech by Former Malaysian Prime Minister, The Hon. Dato Seri Dr. Mahathir Bin Mohamad at the International Convention on Biotechnology. Retrieved from www.isaaa.org/kc/publications/htm/articles/position/mahathir.htm.

Lee, C. and Lee, C. G. (2017). The Evolution of Development Planning in Malaysia. *Journal of Southeast Asian Economies*, *34*(3), 436–61. doi.org/10.1355/ae34-3b.

MASTIC. (1996). *Public Awareness of STI Malaysia 1996*. Ministry of Science, Technology and Innovation. Putrajaya. Retrieved from mastic.mosti.gov.my/publications.

MASTIC. (1998). *Public Awareness of STI Malaysia 1998*. Ministry of Science, Technology and Innovation. Putrajaya. Retrieved from mastic.mosti.gov.my/publications.

MASTIC. (2000). *The Public Awareness of Science and Technology Malaysia 2000*. Ministry of Science, Technology and Innovation. Putrajaya. Retrieved from mastic.mosti.gov.my/publications.

MASTIC. (2002). *Public Awareness of STI Malaysia 2002*. Ministry of Science, Technology and Innovation. Putrajaya. Retrieved from mastic.mosti.gov.my/publications.

MASTIC. (2004). *The Public Awareness of Science and Technology in Malaysia 2004*. Ministry of Science, Technology and Innovation. Putrajaya. Retrieved from mastic.mosti.gov.my/publications.

MASTIC. (2005). National Biotechnology Policy. Ministry of Science, Technology and Innovation. Retrieved from mastic.mosti.gov.my/publications.

MASTIC. (2008). *The Public Awareness of Science and Technology Malaysia 2008*. Ministry of Science, Technology and Innovation. Putrajaya. Retrieved from mastic.mosti.gov.my/publications.

MASTIC. (2014). *The Public Awareness of Science, Technology and Innovation in Malaysia 2014*. Ministry of Science, Technology and Innovation. Putrajaya. Retrieved from mastic.mosti.gov.my/publications.

Ministry of Education. (2012). *Report on the Strategies to achieve 60 per cent Science and Technical Stream: 40 per cent Arts stream Policy*. Ministry of Education.

Ministry of Education Malaysia. (1967). *Educational Statistics of Malaysia, 1938–1967*. Educational Planning and Research Division, Ministry of Education Malaysia. Retrieved from www.fmm.org.my/upload/educational%20statistics%20of%20malaysia%201938-1967.pdf.

Ministry of Education Malaysia. (2016). *Implementation Guide for Science, Technology, Engineering, and Mathematics (STEM) in Teaching and Learning*. Ministry of Education Malaysia.

Mohamad, M. (1991). Malaysia: The Way Forward (Vision 2020). Speech to the Malaysian Business Council, 28 February. Kuala Lumpur.

Mohamad, M. (2002). Look East Policy – The Challenges for Japan in a Globalized World (Speech). Retrieved from www.mofa.go.jp/region/asia-paci/malaysia/pmv0212/speech.html.

MOSTI. (2006). *National Survey of Innovation, 2002–2004*. Malaysia: Malaysian Science and Technology Information Centre (MASTIC).

Shaikh Mohd Salleh, S. M. S. (2012). International Workshop for Islamic Scholars on Agribiotechnology: Shariah Compliance. Georgetown, Penang, Malaysia (1–2 December 2010). Malaysia Biotechnology Information Center: Selangor, Malaysia; International Service for the Acquisition of Agri-biotech Applications: Los Ba-os, Laguna, Philippines.

Talib, K. A., Nidzam, S., Isa, S. M. and Saad, S. (2013). Assessing the Look East Policy: Moving Beyond Work Culture and Ethic. *Asian Social Science*, *9*(17), 111–19. doi.org/10.5539/ass.v9n17p111.

Universiti Kebangsaan Malaysia and UNESCO. (2016). *Harnessing Talent towards an Inclusive Malaysia. An Assessment of the National Policy on Science, Technology and Innovation (NPSTI) in Enhancing Social Inclusion in Research and Innovation*. Retrieved from www.unesco.or.id/download/Social_Inclusion_MalaysiaFinalReport.pdf.

Universiti Sains Malaysia. (2005). *Mahathir on Science and Technology: A Commemorative Volume in Conjunction with the Conferment of the Honorary Degree of Doctor of Science. The Launching of Academy of Sciences Malaysia*. (2nd ed.). Pulau Pinang: Penerbit Universiti Sains Malaysia.

World Bank. (2018). The World Bank in Malaysia. Retrieved from www.worldbank.org/en/country/malaysia/overview.

World Halal Forum. (2010, 22 June). Genetically Modified (GM) Crops and Halal Workshop Report. Kuala Lumpur Convention Centre, Kuala Lumpur. Retrieved from www.ask-force.org/web/Islam/GM-Crops-World-Halal-Forum-Kuala-Lumpur-2010.pdf.

Yusof, Z. A. and Bhattasali, D. (2008). Economic Growth and Development in Malaysia: Policy Making and Leadership. Working paper No. 27. The Commission on Growth and Development, World Bank.

Timeline

Event	Name	Date	Comments
First interactive science centre.	Pusat Sains Negara (PSN) [National Science Centre]	Nov 1996	
First national (or large regional) science festival.	Different organisations run events with different names	2010	NGOs and researchers run exhibitions, talks, workshops, fashion shows and roadshows
First university courses to train science communicators.	No formal courses in science communication	-	Master's and PhD candidates customise their research in bioethics, science communication, public perception
First master's students in science communication graduate.			There is a master's program in environmental journalism
First PhD students in science communication graduate.	The lead author, Mahaletchumy Arujanan	2013	Dr Arujanan is Malaysia's only PhD graduate in this area
National government program to support science communication established.	National Science, Technology and Innovation Policy	2011	Science communication is covered under this policy
National Science Week founded.	Theme was *Negaraku Berinovasi* [My Innovative Country]	2018	Launched by the Science, Technology and Innovation Ministry
First significant TV programs on science.			Used to have children's programs
First awards for scientists or journalists or others for science communication.	Special STEM Award presented by minister	2018	Awarded to Kuala Lumpur Engineering Science Fair and others
Other significant events.	*The Petri Dish*, first science newspaper	Feb 2011	

Contributors

Dr Mahaletchumy Arujanan is executive director of the Malaysian Biotechnology Information Centre (MABIC).

Noorshamira Shamsuddin is a project manager at the Malaysian Biotechnology Information Centre (MABIC) in Bandar Sunway, Selangor.

Farahana Nadzri is a project officer at the Malaysian Biotechnology Information Centre (MABIC).

24

MEXICO

From simple and centralised to expansion, diversity and complexity

Elaine Reynoso-Haynes, Susana Herrera-Lima, Ana Claudia Nepote and Lourdes Patiño-Barba

1. Introduction

The roots of public communication of science (PCS) in Mexico may be traced back to the early days of science in the 17th century. However, as in other parts of the world, the modern era of PCS in Mexico started in the 1960s, and was basically concentrated in the urban area of Mexico City. There were four main pillars for the development of science communication during those early years: UNAM (National Autonomous University of Mexico), CONACyT (National Council for Science and Technology), AMC (Mexican Academy for Science) and SOMEDICyT (Mexican Society for the Communication of Science and Technology).

Today, various PCS products and activities are dispersed throughout Mexico in different formats, media and spaces. Many institutions and persons with different profiles are involved: state and private universities; the various state councils for science and technology; and government and NGO agencies, private companies, science journalists as well as freelance science communicators. Other activities related to the field are research, evaluation, professional training, management, administration and commercialisation. This chapter tells the story of those early years of PCS and how such projects have evolved.

2. The early years

Although the modern era of PCS in Mexico started during the decade of the 1960s as a result of institutionalised efforts, the roots of this activity, as in most countries, can be traced back to the early days of science. Science in Mexico is deeply rooted in the knowledge of the native people whose lives were closely linked to the physical and natural world. The Mayans had a profound knowledge of natural phenomena, such as the cycles of water, natural harvest cycles and the capacity to predict astronomical events. Such knowledge was important to them not only for practical applications such as agriculture but also for religious and political reasons. In 1521, Tenochtitlan, the heart of the Aztec empire (the present site of Mexico City) fell to the Spanish conquerors. Spanish rule would last almost three centuries, producing a unique clash and fusion of European and local culture, knowledge, experience and beliefs. When the conquerors arrived in these remote lands they were very much impressed by the collections of natural specimens (flora and fauna) and botanical gardens owned by the Tlatoanis (the Aztec rulers) for study, preservation and reproduction. Soon after the conquest, the Spanish rulers organised expeditions throughout the territory they called New Spain with the purpose of collecting, registering and studying specimens of flora and fauna, as well as finding archaeological treasures. This practice continued into the early 19th century, with some items sent to Spain and the rest of the collection retained for research and display at the Museum of Natural History when it opened to the public in 1790 (Rico Mansard, 2007).

During this period, scientific knowledge was shared with the general public. Outstanding examples are two books about comets written by Carlos Sigüenza y Góngora, based on the ideas of Copernicus, Galileo, Descartes and Kepler. The purpose of these books was to reassure people about a comet that appeared in 1680. The popular belief was that comets were composed of exhalations from deceased bodies and human sweat and that they were precursors of calamitous events (Benítez, 1995).

Two other outstanding pioneers of popularisation of science in Mexico were José Antonio Alzate (1737–99) and Ignacio Bartolache (1738–90). Alzate was a priest who dedicated his life to research and writing articles for the general public on physics, chemistry, mathematics, astronomy, botany, archaeology, philosophy and literature. He was the founder of the first cultural newspaper in the New Spain (Enciclopedia de la literatura en México, 2019).

Bartolache wrote about medicine, astronomy and mathematics. One of his most important contributions was how to deal with smallpox, and in 1772 he started the first publication for the general public about medicine and physics called *Mercurio volante* [Flying Mercury]. In 1769, both Alzate and Bartolache studied the transit of Venus in front of the Sun (Moreno, 2013).

Institutionalisation of science began shortly after the independence of Mexico from Spain in 1821. The first scientific society in the American continent and fourth in the world, the Mexican Geographical and Statistical Society was founded in 1833 with the purpose of making a map of the young republic, collecting national statistics and publishing a journal for the general public (Azuela Bernal, 2012).

During the 19th century, several publications for the general public appeared, some published by the scientific societies and museums. One of the most popular was *Mosaico Mexicano* (1843–46), which included news about scientific discoveries and natural phenomena along with poetry, historical events, practical agricultural advice, biographies of celebrities, descriptions of exotic places, balloon trips and short stories (Cuevas, 2002).

The National Museum contained collections of natural history, documents, machines and objects of scientific, artistic and religious interest. In 1825, the director determined that the museum should have the double function of preservation and exhibition for the general public. Eventually, this collection would be split into several categories, including natural history. With this collection of natural specimens, the Museo de Tacubaya was opened in 1893 with the triple objectives of research, teaching and popularisation of science (Cuevas, 2002).

One of the most remarkable figures of the beginning of the 20th century was the astronomer Joaquín Gallo, head of the National Astronomical Observatory of Mexico for more than 30 years. He was very active in communicating science to the general public through talks and newspaper articles (Biro and Mateos, 2011).

The beginning of the 20th century was marked by the Mexican Revolution, a civil war from 1910 to 1920. The war began as an attempt to overthrow the 34-year regime of Porfirio Diaz. Although Diaz's regime was known for the development of science, industry, infrastructure and foreign investments in Mexico, it was also known for being a dictatorship that favoured a very small and outrageously wealthy group of individuals while the majority of the population was living in extreme poverty. The years after the revolution were devoted to a total and profound reconstruction of Mexico. In 1920, José

Vasconcelos, president of the National University of Mexico, proposed the creation of a federal ministry of education and launched an intensive literacy and educational campaign with several elements: an ambitious publishing effort, the creation of public libraries and the staging of cultural festivals for the population. His renovation project included what he called 'free discussions', seminars by the most important intellectuals and university professors, with the purpose of constructing a new identity for Mexico. One of the outcomes was the multifaceted movement known as 'Mexican Muralism'[1] with famous painters such as Diego Rivera, David Alfaro Siqueiros and José Clemente Orozco (Fierro, 2003).

Science and science communication also had a recovery process after the revolution. An example is the Mexican Society of Natural History, the second oldest society in Mexico, founded in 1868 and lasting until 1914. In 1936, Alfonso L. Herrera and Enrique Beltrán decided to revive this society with the purpose of promoting scientific and traditional knowledge of nature in Mexico. Its journal became the most important science communication product about natural sciences in the 20th century (Gío-Argáez et al., 2013).

The following decades are known as the era of modernisation and industrialisation of Mexico as well as the development of higher education (Aguilar and Serrano, 2012). The international and national contexts were appropriate for the process of the institutionalisation of science and technology. The National Council for Higher Education and Scientific Research was created in 1935 (Casas, 1985). Later, in 1952, UNESCO (United Nations Educational, Scientific and Cultural Organization) showed an interest in the development of science and technology in Latin America. In 1963, the 'United Nations Conference for the Application of Science and Technology for the Development of Less Developed Regions' launched an innovative educational movement that stressed the importance of improving science education as well as increasing the level of science education of the population (Massarani et al., 2015, p. 13). Mexico was an emblematic and inspiring country for the Latin American region due to its process of institutionalisation and consolidation of science, which triggered the process of institutionalisation of science communication (Sánchez-Mora et al., 2015).

1 Mexican Muralism was an artistic movement, starting in the 1920s, in which social, historical and political messages were portrayed on murals as part of the efforts to reunify the country and create a national identity under the post–Mexican Revolution government. See www.khanacademy.org/humanities/ap-art-history/later-europe-and-americas/modernity-ap/a/mexican-muralism-los-tres-grandes-david-alfaro-siqueiros-diego-rivera-and-jos-clemente-orozco.

3. The modern era of public communication of science in Mexico

The modern era of PCS started in the 1960s. The term used at the time was *divulgación de la ciencia*, which has a definite deficit model approach. Today the preferred term in Mexico is *comunicación pública de la ciencia*, which includes a wide range of approaches.

The pillars of the early stage of the modern era were UNAM, CONACyT, AMC and SOMEDICyT (Reynoso-Haynes, 2015a).

3.1. National Autonomous University of Mexico (UNAM)

Previous to this modern era of science communication, UNAM had two outstanding examples of science communication: the Geology Museum (1906) and the Botanical Gardens of the Institute of Biology (1959), which still exist today.

However, the first attempts at the institutionalisation of science communication also occurred in UNAM by the physicist Luis Estrada. In 1968, with the support of students and a small group of colleagues, mostly from a scientific background, he launched a journal called *Física* [Physics] for physics teachers and university students. Two years later, this journal was renamed *Naturaleza* [Nature] with the purpose of offering a wider scope of scientific topics. The elaboration of this journal became a practical school for science communicators, as well as an arena for theoretical discussions related to the field (Estrada et al., 1981). In 1970, Estrada and his colleagues founded a department within the UNAM with the purpose of communicating science to non-experts. In 1980 it became the Programa Experimental de Comunicación de la Ciencia (PECC) [Experimental Program for Science Communication]; and the following year became the Centro Universitario de Comunicación de la Ciencia (CUCC) [University Centre for Science Communication], the first university institution in Mexico devoted completely to science communication (Zamarrón, 1994). In 1997, CUCC became the present Dirección General de Divulgación de la Ciencia (DGDC) [General Direction for the Popularisation of Science] (Sánchez-Mora et al., 2015).

The DGDC has two science museums (Universum, opened in 1992, and the Museum of Light, opened in 1996) and communicates science using a full range of media such as the magazine *¿Cómo Ves?* It offers courses and workshops for teachers, children and the general public as well as postgraduate courses for training professional science communicators; organises and participates in a variety of PCS events and outreach programs; collaborates with research institutes and other museums within Mexico and abroad; and carries out studies and research in the field of PCS.[2]

Other PCS programs in UNAM include the publication of *Ciencias* [Sciences], a journal for university students published by the Facultad de Ciencias [School of Science] since 1980.

UNAM provides reliable information to news media with publications produced by different research institutes in different campuses in several states. It produces radio and television programs on scientific topics that are broadcast on commercial or cultural channels and stations as well as the UNAM's TV channel and radio station, and an increasing offering on the web (Sánchez-Mora et al., 2015).

Figure 24.1: The exhibit of the lunar rock in the science museum Universum. This rock was collected by astronauts from Apollo 17 on 19 December 1972.

Source: Universum.

2 See www.dgdc.unam.mx.

3.2. National Council for Science and Technology (CONACyT)

The second pillar of this modern era of PCS is CONACyT, founded in 1970. From the very start, this government agency communicated science to the general public as one of its duties. It produced two journals *Ciencia y Desarrollo* [Science and Development], aimed at the educated reader, and *Información Científica y Tecnológica* (ICyT) [Scientific and Technological Information] with a more popular science approach. CONACyT was a pioneer in offering a training course for science journalists (Sánchez-Mora et al., 2015).

CONACyT has always supported various PCS events, activities and products. In 2013 it launched a call for PCS projects with a multidisciplinary approach, to support the development of the field in different regions of the country.

3.3. Academia Mexicana de Ciencias (AMC) [Mexican Academy for Science]

The third pillar of the modern era is the AMC. In 1982, it started a form of science communication that became extremely popular: *Domingos en la ciencia* [Sundays in Science]. This series of informal talks by scientists and science communicators to the general public on a wide range of topics continues throughout the country to this day. Similar experiences are the Encuentros de divulgación científica [Encounters in Science Communication], an annual event since 1985 organised by the Sociedad Mexicana de Física [Mexican Society for Physics] in the city where the annual physics conference is held; and the 'Science and Technology Week' organised by the CONACyT every year since 1994 (Sánchez-Mora et al., 2015).

3.4. Mexican Society for the Communication of Science and Technology (SOMEDICyT)

The fourth pillar is SOMEDICyT, a network of professional science communicators founded in 1986. It started out with only 19 members located in the urban area of Mexico City, and today consists of full-time science communicators, scientists, teachers and journalists with 303 active members in 24 of the 32 states in the country.[3] Products and activities of SOMEDICyT include books on science topics for children, peer publications in the field, the development of exhibitions and science museums and products on the

3 See www.somedicyt.org.mx.

Internet. It has organised numerous national and international conferences and contributes to the professional development of the field by means of seminars, courses and workshops. SOMEDICyT offers an annual award to outstanding communicators and encourages young science communicators by organising contests for essays on different scientific topics (Reynoso-Haynes, 2015b).

Other publications appeared during this early stage such as *Chispa* [Spark] for children, *Avances y Perspectiva* [Advances and Perspectives] for the academic community, and *Cuadernos de Nutrición* [Notebooks on Nutrition] with an educational perspective. (Sánchez-Mora et al., 2015).

4. The expansion and diversification of PCS in Mexico

Most of the activities described above occurred in the urban area of Mexico City. Gradually, due to various factors, different projects appeared in other parts of the country.

The first factor is the creation of museums and science centres. Although traditional science museums have existed in Mexico since the 19th century, the first two hands-on science museums in Mexico were the Museo Tecnológico [Museum of Technology] in Mexico City founded in 1970 and the Centro Cultural Alfa [Alfa Cultural Centre] in Monterrey in 1978. In the 1980s, there was an international boom of interactive science museums and centres (Sánchez-Mora et al., 2015), and it reached Mexico between 1990 and 1996 with the opening of the first interactive museums in several cities as well as children's museums with an important component of science-related exhibits and activities. Most of these museums and science centres belong to the Asociación Mexicana de Museos y Centros de Ciencia y Tecnología (AMMCCyT) [Mexican Association of Science and Technology Museums and Centres] created in 1996. Through this network, its members collaborate, share exhibitions and stimulate professional growth (Sánchez-Mora et al., 2015; Padilla, 2000). AMMCCyT has 35 institutional members (Bonilla, 2016).

The second expansion factor is related to the incorporation of PCS activities into the agenda of REDNACECyT, the national network of state councils for science and technology, created in 1998 with the purpose of supporting the development of science and technology in Mexico's 32 states.[4] In 2012, a special fund was created within CONACyT to support and promote programs for the development of PCS in each of the 32 states. In some cases,

4 See www.rednacecyt.org.

these funds strengthened the development and infrastructure of existing efforts, for other states it was a starting point. Examples of projects that have been supported with these funds are science museums, travelling exhibitions, science fairs and training programs for science communicators. One of the main purposes is to reach marginal and rural communities in remote areas (Padilla and Patiño, 2012).

The third ingredient of the territorial expansion of PCS was the creation of the National Week of Science and Technology funded and organised by CONACyT since 1994. Although PCS is not the main purpose for many of the participating institutions (universities, research institutes, businesses and government organisations), this annual event for the general public and students has been an important learning experience in PCS. In some states, the demand for such events and activities is so great that one week is not enough. Similar events are offered throughout the year, promoting a strong relationship between schools and science institutions.

Today, many institutions are involved in PCS: CONACyT, state councils of science and technology; universities, research institutions, scientific societies, museums, zoos, NGOs, communication media and SOMEDICyT (Padilla and Patiño, 2010).

Figure 24.2: Outdoor science fair in the science centre Explora in León, Guanajuato.

Source: Lourdes Patiño-Barba.

In the last 10 years, several new institutions and groups dedicated exclusively to PCS activities and projects have sprouted—mostly civil associations, businesses and independent groups of professionals and students. State networks of science communicators have been formed with the intention of creating a community of full- or part-time professionals in PCS and promoting professional training in the field (Patiño, 2018).

5. Science journalism

Science journalists are becoming a strong and independent professional community. This movement began in 1979 when Mexico hosted the Third Iberoamerican Conference of Science Journalism in Mexico City. The conference was organised by the Asociación Mexicana de Periodistas Científicos (AMPECI) [Mexican Association of Science Journalists], CONACyT and UNAM. Science journalists from Spain, Latin America and Mexico participated. The topics were: a) media and science communication, b) assessment of science journalism, c) training science journalists, d) social projection of science journalism, e) environment and science journalism (Asociación Mexicana de Periodismo Científico, 1981).

Thirty-seven years later, in 2016 the Red Mexicana de Periodistas de Ciencia (RedMPC) [Mexican Network of Science Journalists] was formed for professional science journalists and students. This network has over 100 members. In 2017, it was incorporated as member 55 in the World Federation of Science Journalists, a non-profit international association. This achievement is important to the RedMPC because now it has a voice in the global panorama of science journalism with the opportunity of collaborating with other professionals on different international projects.[5]

6. Different profiles of science communicators in Mexico

PCS in Mexico started out as a volunteer 'missionary' free-time occupation, often at a personal cost and opposed by colleagues who considered communicating science to be a waste of time and a distraction from more important tasks of research and teaching. Today the PCS community is rich and diverse, with a wide range of professional profiles and experiences. For some, it is a full-time profession; for others, it is a secondary or complementary activity to their main

5 See redmpc.wordpress.com.

occupation. This second group includes scientists and persons who work in the media. Fortunately, the number of full-time specialised science communicators has increased significantly over the past two decades (Reynoso-Haynes, 2015a).

Due to its complexity as a multidisciplinary field, science communication professionals must specialise—for instance in a certain scientific discipline, or in audiences they address or the media they use. Usually this specialisation involves a combination of these elements. Therefore, within the group of full-time science communicators, we find people with many different backgrounds: writers, reporters, scientists, journalists, museographers, photographers, designers, computer scientists, engineers, artists and educators. There are others who support this activity through public relations, promotion, marketing, administration and management (Reynoso-Haynes, 2015a).

Studies performed by Padilla and Patiño (2010, 2013, 2016 and 2017) show that the different profiles are a consequence of the process of institutionalisation. Some research institutions, universities and science councils have departments devoted to PCS with professional science communicators, definite work plans and budgets. Unfortunately, this does not occur in several states in Mexico.

7. Official attitudes to PCS

Universities may appear to be the ideal places for the development PCS projects. The first obvious advantage is the proximity to a strong and diverse community with the latest scientific knowledge and a critical approach in practically any field: experts with various creative and technical abilities and artists who can participate in PCS as advisors, creators or evaluators. Other advantages are the infrastructure, laboratories and technical equipment as well as the financial benefits of being able to use all these facilities with much lower costs.

However, Patiño, Padilla and Massarani (2017) and Padilla and Patiño (2012) reveal that in many cases PCS is still not considered a priority in research and higher education institutions, and that it is often considered a minor activity by authorities and part of the academic community. The greatest disadvantage that science communicators working in universities face is that they are judged by 'academic standards and benchmarks' designed for other professions such as scientists. The situation becomes more critical when not only are the criteria used to evaluate science communicators 'imported' from other fields, but also the evaluators themselves, usually scientists who know nothing or very little about science communication and are not trained in the field. Sometimes these 'imported evaluators' look down on science communication as a minor activity compared to research and teaching. Therefore, the issue of how to evaluate science communicators is particularly important to those who work in universities and research institutions (Reynoso-Haynes and Tonda, 2013).

A group within the DGDC has been working on a proposal that includes peer evaluation, criteria and parameters for an adequate and just evaluation for a wide spectrum of profiles of science communication professionals. The first results of these discussions can be found in the proceedings of conferences and formal communications such as Delgado et al. (2003), Reynoso-Haynes (2008) and, more recently, in the XIII RedPOP (Latin American Network for the Popularisation of Science and Technology) and XIX SOMEDICyT conferences (Bravo, Reynoso-Haynes and Tonda, 2013). Such discussions conclude that the evaluation of products and their producers is inseparable. Any scheme proposed for this purpose is based on a specific conception of science communication, which includes the objectives that are pursued, the image of science portrayed and the relationship we seek with the recipient of our products.

Patiño and Padilla (2017) also found that financial support for PCS is still scarce. The study shows that science journalism is centralised in Mexico City, although there are several outstanding efforts in other parts of the country. Over 50 per cent of the institutions that perform PCS activities do not have a specialised department or a formal year plan. Budgets for science journalism tend to be quite low compared for those other functions of the institution. Most of the institutions studied do not have formal registers of the impact of their PCS activities. Another problem they found is the difficulty of keeping trained staff in PCS because a considerable number of the persons involved are students or professionals who have other activities. This is particularly true in science museums and centres where those involved in PCS activities (such as museum guides) usually remain only for a short period of time (six months or a year).

8. Professionalisation and training in PCS

In Mexico the number of programs for professional training in PCS is increasing. Courses, workshops, postgraduate courses, specialties, master's and PhD degrees in PCS are now available in different regions throughout the country. However, this is still insufficient due to the rate at which this activity is growing and the need to train professional science communicators.

Reynoso-Haynes (2009) proposed a classification of science communicators by generations depending on what kind of training they had received. The first generation of science communicators, the pioneers of the modern era who mostly started during the late 1960s and the 1970s, did not receive any kind of professional training because formal courses in science communication did not exist in Mexico or abroad at that time. The only school for this pioneer group was practical experience. Most of them had scientific backgrounds and the rest had training and experience in non-scientific fields such as communication, literature, humanities and journalism. Those whose initial background was science had to acquire experience and skills in some communication media, and those who started out with a non-scientific background had to learn the science they required in order to collaborate with scientists. The professional science communicator did not exist.

Then a second generation of science communicators emerged. Due to the lack of specific training programs in science communication, those who had an initial professional training in one of the fields mentioned above might enrol in formal postgraduate studies or specialised courses in a complementary field considered useful for their work. As the field of PCS started to grow stronger and more diverse, so too did discussions about what was required to be a professional science communicator. The need to plan and design specialised courses to train science communicators became evident. At UNAM, both generations combined their expertise to develop and teach these courses. The result was the first specialised course in PCS in Mexico, the Diplomado en Divulgación de la Ciencia [Science Communication Diploma], offered every year since 1995. This 240-hour course was designed with the purpose of providing the required theoretical and practical tools to enable graduates to communicate science to the general public (Reynoso-Haynes, 2009). The demand for this course increases constantly, as does the number of requests the DGDC receives to offer similar courses in other institutions within Mexico and abroad. Continuous evaluation has been a fundamental instrument for the planning and updating of the course.

Other options are graduate and postgraduate degrees as well as subjects in undergraduate programs. Not all of these programs are devoted exclusively to training professionals in the various areas of PCS, but this activity is a substantial part of the curricula. Such is the case at the Instituto Tecnológico y de Estudios Superiores de Occidente (ITESO) [Western Institute of Technology and Higher Education] in the city of Guadalajara, Jalisco, that has offered a master's degree in science communication and culture since 1998.[6]

Since 2003 the postgraduate program in Philosophy of Science offers a master's and a PhD with several terminal options, including one in science communication. Other examples of programs that include science

6 See posgrados.iteso.mx/maestria-filosofia-ciencias-sociales/.

communication are the course in science journalism at the School of Political Science of the UNAM and the postgraduate courses offered at the Universidad Autónoma Metropolitana (UAM) Metropolitan Autonomous University. Each of these programs has its entrance requirements, curricula, durations and graduate profiles depending on different needs and approaches. Both the ITESO and the UNAM programs have a strong emphasis in developing research skills in the field. Other programs, such as the *diplomados* (postgraduate courses) offered by the UNAM, SOMEDICyT and other universities are much shorter and have a more practical approach (Sánchez-Mora et al., 2015).

The new creative, intellectual and ethical challenges in the field of science communication increase day by day in complexity, richness and diversity. These include new theoretical and methodological contributions, a growing presence on the web, new proposals and approaches for communicating science, new media, as well as new professional options and needs. Based on the two UNAM experiences, the Diplomado en Divulgación de la Ciencia and the Science Communication branch of the Philosophy of Science Postgraduate Program (as well as present-day needs for professional development in the field in all its complexity), the Department of Training and Research within the DGDC is currently working on a project for a postgraduate Specialisation in Science Communication in collaboration with the School of Political and Social Sciences in UNAM. The purpose of this one-year, 640-hour program is to offer students a solid theoretical and methodological background as well as the opportunity for extensive practice in some area of personal interest (UNAM, 2018a).

Those who graduate from these specialised PCS programs constitute a third generation of science communicators.

9. Research and evaluation in PCS

Today PCS is considered an academic and professional field. According to Ana María Sánchez Mora (2010), the origins of institutionalised research can be traced back to 1988 when Luis Estrada coordinated a collective document with the title *Aspectos de investigación en comunicación de la ciencia* [Aspects of research in science communication]. It suggested that communication of science should be performed with a more professional and methodological approach and should have a multidisciplinary perspective. Certain research topics were proposed with the purpose of understanding and improving PCS products and activities by establishing a communicative bridge between the audience and scientific concepts through a process of re-creation of the initial scientific discourse. This includes a basic approach to the meaning of scientific culture, the problems related to language and the strategies used to communicate with non-specialists, and the need for evaluation of the effectiveness and originality of what is produced (Sánchez Mora, 2010, p. 119).

At the beginning of the 1990s a collection of books called *Divulgación para Divulgadores* [Science Communication for Science Communicators] emerged within the DGDC of UNAM with the purpose of sharing analysis, reflections and research related to PCS (Sánchez-Mora et al., 2015).

The proceedings of conferences of SOMEDICyT and RedPOP have published various studies in diverse fields of PSC, such as methodological proposals (Sánchez Mora, 1991) and the evaluation of the impact of journals (Tonda and Burgos, 2007). Research carried out in science museums and centres are the most numerous and have contributed considerably to our understanding of the role such environments and their activities play in science literacy as well as how and what people learn. A good part of these research projects has been carried out in UNAM's museums Universum and the Museum of Light (Sánchez-Mora et al., 2015).

This kind of research will significantly contribute to the development of improved frameworks for practice and evaluation in museums and will also provide a basis for future research. There are many examples of this kind of research in Reynoso-Haynes (2000, 2001, 2003), Lozano (2005), Rico Mansard (2009) and Sánchez-Mora (2002, 2006, 2009a, 2009b, 2012).

Initially research in PCS in Mexico arises from within the community of science communicators, but gradually professionals from other disciplines have become interested in the field. Today we can find research projects in PCS with different approaches such as communication, sociology, education, literature and social studies in science.

Another fundamental ingredient of the process of professionalisation and research in PCS is the emergence of international journals in science communication. The issues covered in these journals have had a definite impact on academic discussions in Mexico. For instance, the public communication of science approach, which leans towards the democratisation of knowledge and the empowering of citizens in scientific and technological matters has been significantly displacing that of the 'deficit model'. This can be seen in articles and theses written by professionals in Mexico and Latin America such as Merino and Roncoroni (2000), Caue (2002), Lozano (2005), Cevallos (2008) and Reynoso-Haynes (2012) (in Sánchez-Mora et al., 2015).

The two most important examples of postgraduate programs preparing future researchers in the field are the Master's in Communication of Science and Culture of ITESO in Guadalajara, Jalisco; and the science communication strand of the master's and PhD in Philosophy of Science at UNAM. The first one began in 1998 and approaches the study of science communication from a social and cultural perspective, considering the relationship between science, the media, the spaces, the institutions and the social interactions. The second one, the science communication strand of the postgraduate course in Philosophy of Science, began in 2003. The approaches in this program are philosophy and history of science; social studies in science; and science, technology and society.

Other examples of research with different disciplinary approaches are the projects developed by Ernesto Márquez, Jorge Padilla with Lourdes Patiño, and Elaine Reynoso-Haynes. Márquez and Tirado (2009) use a psychological approach to analyse the perceptions that Mexican teenagers have towards science and technology. Patiño and Padilla use different approaches to analyse the state of scientific culture in urban populations: first a psychological and sociological approach (Padilla and Patiño, 2011) and then one in which they consider the inclusion of science in the context of habits of cultural consumption (Patiño and Padilla, 2017). Reynoso-Haynes (2012) uses an educational approach to analyse the learning processes in museums and how these institutions can contribute to the construction of scientific culture, and proposes a working framework for the development and evaluation of science museums.

Another important line of research is that of diagnosis of PCS in different states in Mexico. These studies have been required and financed by CONACyT, the state councils for science and technology and SOMEDICyT. Using basically quantitative methodologies, these studies have provided useful information related to different aspects of PCS in Mexico (Padilla and Patiño, 2010, 2012, 2014, 2016). In 2017 in coordination with the RedPOP, Patiño and Padilla conducted a diagnosis of the popularisation of science in Latin America (Patiño, Padilla and Massarani, 2017).

The evaluation of products and research in PCS are closely related. The issue of evaluating science communication products and activities is an old, but at the same time contemporary, debate. A Latin American contribution to this discussion can be found in the proceedings of a workshop held in 2006 in Cartagena, Colombia, in which several experiences and proposals related to the evaluation of products and activities in this field were presented (Lozano and Sánchez-Mora, 2008).

10. PCS publications

There are several journals or popular science magazines published by universities, CONACyT and the state councils of science and technology. Most of these magazines are for readers with a high school or university educational level and a few are for children.[7]

As for books, we consider two categories of PCS publications: the PCS literature about different scientific issues for non-experts, and the literature about PCS issues for those interested in the field.

In 1986 the publishing house Fondo de Cultura Económica launched an ambitious editorial project: a collection of books on different scientific topics written by Mexican scientists and science communicators called *La ciencia desde México* [Science from Mexico]. In 1997, having reached 157 titles, the collection became international and its name was changed to *La ciencia para todos* [Science for Everyone], with the purpose of including authors from other Spanish-speaking countries (Farías, 2002). To date, the collection has more than 250 titles, and several have been re-edited (Torres, 2018).

The DGDC has several collections of books for the general reader that have served as a support to science students and professors of all educational levels due to the variety of scientific topics and issues.[8]

The SOMEDICyT also has collections for the general reader such as: *Colección Básica de Medio Ambiente* [Basic Collection of the Environment] and a collection of children's books about health (Reynoso-Haynes, 2015b)

As for peer publications for science communicators or those interested in the field, the two main contributors in Mexico are the DGDC of the UNAM and the Master of Science and Culture program of the ITESO. Both institutions publish collections of books that cover different professional aspects as well as trends in research issues in PCS.

7 Examples are: *Ciencia y desarrollo* (CONACyT), *Ciencias* (UNAM), *¿Cómo Ves?* (UNAM), *Elementos* (Autonomous University of Puebla), *Ciencia* (AMC), *Hypatia* (University of the State of Mexico) and *Conversus* (National Polytechnic Institute).

8 Examples are: the collection *Antologías ¿Cómo Ves,* which consist of articles originally published in the journal *¿Cómo Ves?* during its 20 years of existence with topics such as physics, mathematics, chemistry, astronomy and the environment. Other collections are *Divulgación para profesores* [Science Communication for Teachers], *Ojitos Pajaritos* for children, *Science and Art* , *History of Science*, *Agenda ciudadana de ciencia y tecnología* [Citizen Agenda for Science and Technology] a collection of 10 books with challenges in which science and technology has solutions that citizens should know and several ebooks and co-editions with other institutions (www.dgdc.unam.mx/libros).

Some of the topics addressed in the collections of the DGDC are research and evaluation in museums, scientific journalism, current debates in PCS and social, philosophical and historical studies related to science communication.[9]

The ITESO collections, created and coordinated by Susana Herrera and Carlos Enrique Orozco since 2000, are the result of research projects of their academic community. The titles of these collections are *De la Academia al Espacio Público* [From the Academia to the Public Space] and *Comunicar Ciencia en México* [Communicating Science in Mexico].[10]

There are also examples of international collaborations with other networks such as RedPOP (the Latin American Network for the Popularisation of Science and Technology) and books coordinated by Juan Nepote and colleagues from other countries.[11]

Elaine Reynoso-Haynes as president of the SOMEDICyT (2012 –14) coordinated a two-volume work with the title *Hacia dónde va la Comunicación Pública de la Ciencia* [Where is Public Communication of Science in Mexico Going?]. In these books the status of PCS in Mexico is presented as well as the main issues, discussions and proposals in the field. The first volume describes the origins and the institutions of PCS in Mexico and the second one presents the status of the professional field of PCS. These books, which were published in 2015, are the result of a collective effort of 27 authors covering a large range of experiences and the national challenges in PCS. The purpose of these books was to provide proposals for public policies (Reynoso-Haynes, 2015a, 2015c).[12]

9 See www.dgdc.unam.mx/libros/. Collections of books for science communicators are: *Divulgación para divulgadores* [Science Communication for Science Communicators], *Museos de la DGDC* [Musuems of the DGDC].

10 The titles of some of the chapters of these books represent a sample of the different research projects. For example, in 2012 professionalisation, public policies; magazines, films and museums; in 2015, communication and scientific culture; historical and social role of science; activism and science and social problems in marginal regions; in 2016, trends in PCS, audio-visual discourse and environment and health; and, in 2018, research in PCS and environmental communication and literature, science and theme parks.

11 In 2009 Juan Nepote and Paola Rodari wrote a book called *Más allá del océano. Ciencia y ciudadanos en Jalisco y Trieste* [Beyond the Ocean: Science and Citizens in Jalisco and Trieste], which discusses the similarities and differences in the construction of scientific culture in both countries. A second book is *Instrucciones para Contagiar la Ciencia* [Instructions to Infect you with Science] coordinated by Juan Nepote and Diego Golombek from Argentina, which contains articles where 29 Mexican and Argentinian authors share their stories about how they engage people in science in museums, schools, books, journals and several other projects.

12 These books were part of a large project launched in February 2012 by the AMC, CONACyT and the Advisory Council in Science for the President of Mexico with the purpose of analysing the state of science in Mexico in the international context with the intention of presenting specific proposals for the development of the country. More than 100 panels of experts of different branches of science took place in different cities. The results of all these discussion groups were published in a series of books with the title *Hacia dónde va la ciencia en México* [Where is Science in Mexico Going?].

11. The future and challenges of science communication in Mexico

In the article 'Public Communication of Science in Mexico: Past, Present and Future of a Profession' (Sánchez-Mora et al., 2015), several challenges and possible solutions were proposed. These were based on an analysis of how the field has evolved in Mexico and also internationally; the growing body of knowledge; the increasing need as well as the diversification of objectives for communicating science to different sectors of the population; the wide scope of objectives and the changing strategies and media; and the need to strengthen the professional field of PCS.

These proposed challenges were classified into three categories:

11.1. The challenges related to the national context

Although the need to incorporate science into the general culture of the population is recognised as urgent, the activities and programs to achieve this goal have not received an adequate level of acceptance and support from decision-makers and society as a whole. A greater presence in the media is necessary, coupled with more opportunities for encounters between experts and citizens so the latter can learn and participate in debates on matters related to scientific knowledge and how it impacts on their personal and collective lives. In order to reach this goal, collaboration between several sectors of society is required: researchers, teachers, industry, the media and decision-makers. The community of science communicators will act as intermediaries between all these sectors and the public.

The starting point for this collaboration is an analysis of the meaning of scientific culture and its link to society. Some of the issues to be tackled are the goals and objectives of a scientific culture for the population, the necessary basic knowledge and skills needed to fulfil these goals and objectives, and the attitudes and values that should be promoted when applying this knowledge (Reynoso-Haynes, 2007).

The balance between global and local issues is fundamental. The use of a 'glocal' model for science communication—based on combining global knowledge and its application to the local context, and considering local interests, problems, solutions and expertise—is highly advisable in cases in which citizen engagement is critical. This includes issues related to public health, environmental problems and climate change (Reynoso-Haynes, 2003, 2005).

11.2. The challenges related to the institutional context

Most science communicators in Mexico work in universities, higher education or state institutions. Unfortunately, due to the relative youth of the profession, programs, projects and groups in these institutions, they are vulnerable to political and institutional changes. A clear mission and strategies to guarantee their stability is required. Programs should not be attached to one specific administration or to political interests but to long-term plans framed within an institutional project. The project must be based on a collective analysis of the need to integrate science with the general culture of the population as well as the role of the community of science communicators in this task. Institutions need to have clear guidelines and plans for their development, including criteria for hiring, promotions, permanence and professional growth of its personnel. Different types of profiles for science communicators should be established with the purpose of facilitating these decisions. At the same time, the personnel hired must be evaluated using fair and relevant criteria based on the work they perform.

11.3. The challenges inherent to the activity

Issues such as the definition of the required scientific culture for the Mexican population and our social responsibility to and relationship with our audience must be dealt with. This area requires further contributions to the field of knowledge, with proposals of new theoretical and methodological foundations, studies, experimentation and evaluation.

Parameters must be established for evaluating products and activities for the purpose of learning and improving, not grading or legitimising. These tasks should not be viewed as 'intellectual luxuries' but as essential instruments for communicating science effectively, with quality and responsibility. Such activities should be seen as fundamental to the successful development of projects and be given full institutional support.

Last but not least, professionalisation requires the support of postgraduate studies, postgraduate courses, courses to learn new topics or skills, workshops, exchange programs and participation in forums and conferences.

12. Conclusions

Public communication of science in Mexico began in the 17th century as a consequence of the need to acculturate the 'New World' with the Western worldview introduced by the Spanish conquerors. Such efforts continued, with outstanding examples, well into the 20th century.

In the mid-20th century, an institutionalised government and academic effort took place with the purpose of strengthening science and technology throughout the country. As part of this movement it soon became evident that PCS was necessary with the objective of incorporating science into the general culture of the population. In the last couple of decades, PCS has flourished and expanded throughout Mexico with a wide and diverse range of programs, activities, science museums and centres, supported by professional networks and various programs for training science communicators. The growing PCS community is now composed of full-time science communicators, scientists, journalists and others from various backgrounds.

During the last decade, one of the main issues discussed by this community is how to better 'professionalise' the field. Diverse opportunities for specialised training are required as well as social and organisational structures that support and recognise PCS as a profession. PCS should be considered as a legitimate and therefore paid occupation, the same as other professional activities such as research and teaching. Although there are still a lot of people who participate in PCS activities as a part-time and voluntary job, there has been an increasing movement towards creating specialised PCS departments, mostly within universities and science and technology councils.

Terms such as 'science communication' or 'popularisation of science' are becoming quite common thanks to the participation of hundreds of science communicators in the mass media (TV, radio, internet, newspapers and journals) as well as massive events such as science fairs. The reasons for communicating science to the general public have also increased and become much more diverse. These commonly include an emphasis on the relationship between science and society, and also promote a culture that is compatible with science (as suggested by Cereijido (2016)).

Another reason for communicating science is to promote critical thinking in order to provide society with the criteria to identify fake news and to decide about pseudoscientific issues, such as whether or not to use 'miracle products' based on so-called scientific facts and popular practices that can be harmful to individuals or society as a whole.

The increasing interest for PCS, as well as the growing number of organisations, networks and people involved in this endeavour may appear to be a positive sign for the field; however, it is not devoid of challenges. Consensus must be achieved among all those involved based on shared knowledge, experience and resources with the purpose of creating alliances and collaborations that will result in greater impacts and better communication with society.

Evaluation of the impact of products, programs and activities is essential. Evaluation must be incorporated and extended throughout the country with the purpose of learning and improving our professional activity. Therefore, time and resources for evaluation must be considered as a necessary ingredient of every project. Last but not least, research and researchers in the field must be supported not only with the intention of contributing to growing field of knowledge but also with the purpose of providing new strategies and methodologies for science communication.

References

Aguilar, E. and Serrano, P. (2012). Posrevolución y estabilidad. Cronología 1917 –1967. México: Instituto Nacional de Estudios Históricos de la Revolución Mexicana (INHERM). Retrieved from inehrm.gob.mx/work/models/inehrm/Resource/437/1/images/posrevolucion_estabilidad.pdf.

Asociación Mexicana de Periodismo Científico. (1981). *Memorias del III Congreso Iberoamericano de Periodismo Científico*. México: AMPCI.

Azuela Bernal, L. F. (2012). La Sociedad Mexicana de Geografía y Estadística, la organisación de la ciencia, la institucionalisación de la Geografía y la construcción del país en el siglo XIX. *Investigaciones Geográficas, 52*. doi.org/10.14350/rig.30346.

Benítez, L. (1995). Carlos de Sigüenza y Góngora: Criollo, nacionalista y moderno hombre de ciencia. Instituto de Investigaciones Filosóficas, Universidad Nacional Autónoma de México. Retrieved from www.iifilologicas.unam.mx/pnovohispano/uploads/95sabernovo/art29_95.pdf.

Biro, S. and Mateos, G. (2011). Astronomía para todos: Joaquín Gallo en el Observatorio Astronómico Nacional (1915 – 1946). In J. Bartolucci (ed.), *La Saga de la Ciencia Mexicana* (pp. 189–212). Coordinación de Humanidades, Universidad Nacional Autónoma de México.

Bonilla, A. (2016). Asociación de museos y centros de ciencia renueva mesa directiva. CONACyT prensa. Retrieved from www.conacytprensa.mx/index.php/sociedad/museos/5039-ammccyt-renueva-mesa-directiva-apostara-por-divulgacion-de-la-ciencia.

Bravo, A., Reynoso-Haynes, E. and Tonda, J. (2013). La evaluación de los divulgadores de la ciencia: La experiencia de la Dirección General de Divulgación de la Ciencia (DGDC) de la Universidad Nacional Autónoma de México (UNAM). In *Proceedings of the XIII Reunión de la RED POP and XIX Congreso Nacional de Divulgación de la Ciencia de la SOMEDICyT* (pp. 79–88). Zacatecas, Mexico. Retrieved from www.redpop.org/redpopWeb/Contenido/Pagina/File/LIBRO%20 SOMEDICYT%20ok.pdf.

Calvo Hernando, M. (1999). *El Nuevo Periodismo de la Ciencia.* CIESPAL, Quito.

Calvo Hernando, M. (2005). *Ciencia y Periodismo Científico en Iberoamérica.* Conferencia de apertura en el II Congreso Iberoamericano de Comunicación Universitaria y I Reunión Iberoamericana de Radios Universitarias. España.

Casas, R. (1985). El estado y la política de la ciencia en México (1935 – 1970). *Cuadernos de Investigación Social, 11.* Instituto de Investigaciones Sociales, Universidad Nacional Autónoma de México.

Caue, M. (2002). Apresentação. In *Ciência e Inclusão Social, Estação da Ciência.* São Paulo, Brasil: Universidade de São Paulo, Fondo de Cultura e Extensão Universitária da Pro-Reitoria de Cultura e Extensão.

Cereijido, M. (2016). *Hacia una cultura compatible con la ciencia.* Retrieved from www.ccciencias.mx/es/ciencia-y-opinion/item/35-hacia-una-cultura-compatible-con-la-ciencia.html.

Cevallos, M. (2008). Monitoreo y evaluación de la divulgación científica. In L. Massarini and C. Polino (eds), *Los desafíos y la evaluación del periodismo científico en Iberoamérica: Jornadas iberoamericanas sobre la ciencia en los medios masivos.* Santa Cruz de la Sierra, Bolivia: AECI, RICYT, CYTED, SciDev.Net, OEA.

Cuevas, C. (2002). Historia y divulgación de la ciencia en México. In Dirección General de Divulgación de la Ciencia (ed.), *Antología de la divulgación de la ciencia en México* (pp. 121–29). México: Universidad Nacional Autónoma de México, México.

Delgado, S., Orta, A. and Heine, C. (2003). La evaluación de la divulgación científica: un reto por la calidad. In *XII Congreso Nacional de la SOMEDICyT y VIII Reunión de la Red POP.* México. Retrieved from www.somedicyt.org.mx/images/divulgadores/congresos/12/memorias/Memorias/resumenes/profesionalizacion/resumen_perez.htm.

Enciclopedia de la literatura en México. (2019). José Antonio Alzate. Retrieved from www.elem.mx/autor/datos/107001.

Estrada, L. (1988). Aspectos de investigación en comunicación de la ciencia (unpublished).

Estrada, L. et al. (1981). *La divulgación de la ciencia.* Cuadernos de Extensión Universitaria, México. Universidad Nacional Autónoma de México (UNAM).

Farías, M. (2002). *La ciencia para todos: 17 años de una aventura científica.* México: Fondo de Cultura Económica.

Fierro, R. (2003). *Templo Máximo de San Pedro y San Pablo. Museo de la Luz. 400 años de historia.* México: Dirección General de Divulgación de la Ciencia, UNAM.

Gío-Argáez, R. et al. (2013). *Revista de la Sociedad Mexicana de Historia Natural.* Sistema de Información Ciencias. Facultad de Ciencias, UNAM, México. Retrieved from repositorio.fciencias.unam.mx:8080/xmlui/handle/11154/139857.

Golombek, D. and Nepote, J. (2016). *Instrucciones para Contagiar la Ciencia.* México: Editorial Universitaria. Universidad de Guadalajara.

Herrera-Lima, S. and Orozco, C. (eds) (2012). *De la academia al espacio público. Comunicar ciencia en México.* México: ITESO. doi.org/10.2307/j.ctvjk2v1q.16.

Herrera-Lima, S. and Orozco, C. (2018). *Comunicar ciencia en México. Prácticas y escenarios.* México: ITESO. doi.org/10.2307/j.ctvjk2vtt.

Herrera-Lima, S., Orozco, C. and Quijano, E. (2015). *Comunicar ciencia en México. Discursos y espacios sociales.* México: ITESO. doi.org/10.2307/j.ctvjk2vsb.1.

Herrera-Lima, S., Orozco, C. and Quijano, E. (2016). *Comunicar ciencia en México. Tendencias y narrativas.* México: ITESO.

Ledesma, I. (2013). La Biología y los biólogos en México: ciencia, disciplina y profesión. In M. Kleiche-Dray et al. (eds), *La institucionalisación de las disciplinas científicas en México. Siglos XVIII, XIX y XX: estudios de caso y metodología* (pp. 99 –126). México: UNAM.

Lozano, M. (2005). Hacia un nuevo contrato social: la popularisación de la ciencia y la tecnología (Master's thesis). Universidad Nacional Autónoma de México UNAM, México.

Lozano, M. and Sánchez-Mora, C. (eds). 2008. Evaluando la comunicación de la ciencia. Una perspectiva latinoamericana, México D.F.: CYTED, AECI, DGDC-UNAM. Retrieved from www.redpop.org/wp-content/uploads/2015/06/Evaluando-la-comunicacin-de-la-ciencia.pdf.

Márquez, E. and Tirado, F. (2009). Percepción social de la ciencia y la tecnología de adolescentes mexicanos. *Revista Iberoamericana de ciencia, tecnología y sociedad, 2*, 1–19.

Massarani, L. (ed.) (2015). *Guía de centros y museos de ciencia de América Latina y el Caribe.* Brasil: RedPOP, Museu da vida, UNESCO, Biblioteca de Educacao e Divulgacao Científica Iloni Seibel.

Massarani, L. et al. (2015). RedPOP: 25 years of a Science Communication Network in Latin America. *Journal of Science Communication, 14*(03). doi.org/10.22323/2.14030406.

Merino, G. and Roncoroni, M. (2000). *La popularización de la ciencia y la tecnología, reflexiones básicas: un marco para la equidad* (Serie Pedagógica). La Plata, Argentina: FHCE.

Moreno, M. A. (2013). Ciencia y arte en dos publicaciones astronómicas novohispanas del siglo XVIII. *Anales del Instituto de Investigaciones Estéticas, 35*(102). 11–31.

Nepote, J. and Rodari, P. (eds) (2009). Más allá del océano: Ciencia y ciudadanos. México: Coecytjal, SISSA Medialab, CULagos Universidad de Guadalajara. Ediciones Arlequín.

Padilla, J. (2000). Desarrollo de los museos y centros de ciencia en México. In J. A. Chamizo (coord.), *Encuentros con la ciencia. El impacto de los museos y centros de ciencia*. México, D.F.: Ed. AMMCyT y Conacyt.

Padilla, J. and Patiño, M. D. L. (2010). Diagnóstico de la difusión y la divulgación de la ciencia en Michoacán. Consejo Estatal de Ciencia y Tecnología de Michoacán, Morelia, Michoacán, México.

Padilla, J. and Patiño, M. D. L. (2011). *Investigación sobre cultura científica, percepción y apropiación social de la ciencia en Michoacán*. Reporte técnico. Consejo Estatal de Ciencia y Tecnología de Michoacán. Morelia, Mich.

Padilla, J. and Patiño, M. D. L. (2012). *Fortalecimiento de la apropiación social de la ciencia y la tecnología en los estados – 2012*. Reporte de evaluación, Red Nacional de Consejos y Organismos Estatales de Ciencia y Tecnología (REDNACECYT) y CONACyT. México, D.F.

Padilla, J. and Patiño, M. D. L. (2013). *Diagnóstico de la divulgación de la ciencia en Guanajuato*. Consejo de Ciencia y Tecnología del Estado de Guanajuato, Guanajuato, Gto.

Padilla, J. and Patiño, M. D. L. (2014). *Diagnóstico de la divulgación de la ciencia en Querétaro*. Consejo de Ciencia y Tecnología del Estado de Querétaro, Querétaro, Qro.

Padilla, J. and Patiño, M. D. L. (2016). *Diagnóstico de la divulgación de la ciencia en Hidalgo*. Consejo de Ciencia, Tecnología e Innovación del Estado de Hidalgo, Pachuca, Hgo.

Patiño, M. D. L. (2015). *Motivos de afiliación y satisfacción de socios de Somedicyt*. Internal document of Somedicyt. México.

Patiño, M. D. L. (2018). Retos para la comunicación pública de la ciencia. Speech presented at Coloquio de Centros de Pensamiento, Ciudad de Panamá, Pan. 5 April.

Patiño, M. D. L. and Padilla, J. (2017). *Evaluación del impacto de las acciones de divulgación de la ciencia y la tecnología en el estado de Hidalgo*. Consejo de Ciencia, Tecnología e Innovación del Estado de Hidalgo, Pachuca, Hgo.

Patiño, M. D. L., Padilla, J. and Massarani, L. (2017). Datos de instituciones mexicanas. In *Diagnóstico de la Divulgación de la ciencia en América Latina: Una mirada a la práctica en el campo*. RedPOP-UNESCO y Fibonacci – Innovación y Cultura Científica, A.C. León, Gto – México.

Red Mexicana de periodistas de ciencia. (2018). *Red Mexicana de periodistas de ciencia*. Periodistas y comunicadores de ciencia. Retrieved from redmpc.wordpress.com.

Reynoso-Haynes, E. (2000). El Museo de las Ciencias: un apoyo a la enseñanza formal (Master's thesis). Facultad de Filosofía y Letras, Coyoacán, México, Universidad Nacional Autónoma de México UNAM.

Reynoso-Haynes, E. (2001). La evaluación de la divulgación de la ciencia. Speech presented at Rountable 'La evaluación de la divulgación de la ciencia', X Congreso Nacional de Divulgación de la Ciencia, Toluca, Estado de México, 7–9 March.

Reynoso-Haynes, E. (2003). La responsabilidad del divulgador en la formación de una cultura científica nacional. Speech presented at Rountable 'Cultura científica y cambio social', VIII Reunión de la Red Pop (Red de Popularisación de la Ciencia y la Tecnología de Latinoamérica y el Caribe), León, México, 26–29 May.

Reynoso-Haynes, E. (2005). Going glocal: UNAM's approach to global science. *Dimensions* (Bimonthly News Journal of the Association of Science and Technology Centres), *5* (September/October), 5–9.

Reynoso-Haynes, E. (2007). Entre lo global y lo local: hacia la construcción del contexto glocal para popu¬larisar la ciencia. In *Memorias de la X Reunión de la Red Pop*, Speech presented at X Reunión de la Red Pop (Red de Popularisación de la Ciencia de Latinoamérica y el Caribe). San José, Costa Rica, 9–11 May. Retrieved from www.cientec.or.cr/pop/2007/MX-ElaineReynoso.pdf.

Reynoso-Haynes, E. (2008). La evaluación de los comunicadores de la ciencia: una perspectiva mexicana. *Razón y Palabra: Primera Revista Digitalisada en Iberoamérica Especialisada en Comunicología. 65*. Retrieved from www.razonypalabra.org.mx/N/n65/actual/ereynoso.html.

Reynoso-Haynes, E. (2009). A graduate course for science communicators: A Mexican approach. *Journal of Science Communication* 8(1). doi.org/10.22323/2.08010304.

Reynoso-Haynes, E. (2012). La cultura científica en los museos en el marco de la educación informal (PhD thesis). Facultad de Filosofía y Letras, UNAM.

Reynoso-Haynes, E. (2015a). Introducción. Comunicación pública de la ciencia: Origen e instituciones. In *Hacia dónde va la Ciencia en México* (pp. 13–20). *17*. CONACYT, Academia Mexicana de Ciencias and Consejo Consultivo de Ciencias. México, D.F., México.

Reynoso-Haynes, E. (2015b). La Somedicyt y el desarrollo de la comunicación pública de la ciencia en México, in Comunicación pública de la ciencia: Origen e instituciones. *Hacia dónde va la Ciencia en México. 17,* p. 65. CONACYT, Academia Mexicana de Ciencias and Consejo Consultivo de Ciencias. Mexico, D.F., Mexico.

Reynoso-Haynes, E. (2015c). Comunicación pública de la ciencia: El oficio. *Hacia dónde va la Ciencia en México. 18.* México: CONACYT, Academia Mexicana de Ciencias and Consejo Consultivo de Ciencias.

Reynoso-Haynes, E., Monterrosa, C. and Macías, P. (2015). La formación de comunicadores de la ciencia en América Latina. In L. Massarani et al., *RedPOP: 25 años de popularisación de la ciencia en América Latina.* Museu da Vida, Casa Oswaldo Cruz, Fio Cruz-COC.

Reynoso-Haynes, E. and Tonda, J. (2013). Consideraciones generales sobre la evaluación de los divulgadores de la ciencia. Speech presented at Rountable La Evaluación de los Divulgadores de la ciencia. In *Proceedings of the XIII Reunión de la RED POP and XIX Congreso Nacional de Divulgación de la Ciencia de la SOMEDICyT* Zacatecas, Mexico. Retrieved from www.redpop.org/redpopWeb/Contenido/Pagina/File/LIBRO%20SOMEDICYT%20ok.pdf.

Rico Mansard, L. (2009). *Aportes a la museología mexicana.* Coyoacán, México: Dirección General de Divulgación de la Ciencia (DGDC), Universidad Nacional Autónoma de México (UNAM).

Rico Mansard, L. (2007). La historia natural tras las vitrinas. In *Museología de la ciencia: 15 años de experiencia* (pp. 37–66). Dirección General de Divulgación de la Ciencia, Universidad Nacional Autónoma de México, México, D.F.

Sánchez Mora, A. M. (1991). La mancuerna investigador-divulgador. In *Memorias del I Congreso Nacional de Divulgación de la Ciencia,* 18–20 April. SOMEDICyT, Morelia, México.

Sánchez Mora, A. M. (2010). *Introducción a la comunicación escrita de la ciencia.* México: Universidad Veracruzana. doi.org/10.25009/uv.2006.131.

Sánchez-Mora, C. (2002). El Museo como foro educativo. *Perspectivas Docentes. 27.* México: Universidad Juárez Autónoma de Tabasco.

Sánchez-Mora, C. (2006). Las visitas guiadas en los Museos de Ciencias. Museolúdica. *Revista del Museo de la Ciencia y el Juego, 8*(4–15).

Sánchez-Mora, C. (2009a). La evolución educativa de los museos de ciencia. *Ciencia y Desarrollo, 35*(236).

Sánchez-Mora, C. (2009b). La potencialidad educativa no formal de los museos universitarios. *Gaceta de Museos del INAH, 46,* 28–34.

Sánchez-Mora, C. (2012). Estudio de patrones de lectura de textos en museos: una contribución a la evaluación formativa de las cédulas. *Boletín Científico y Cultural del Museo Universitario de Medellín, Colombia, 25*, 81–90.

Sánchez-Mora, C., Reynoso-Haynes, E., Sánchez Mora, A. M. and Tagüeña, J. (2015). Public Communication of Science in Mexico: past, present and future of a profession. *Public Understanding of Science, 24*(1), 38–52. doi.org/10.1177/0963662514527204.

SOMEDICyT. (2018). Publicaciones dedicadas a la divulgación de la Ciencia y la Tecnología. Retrieved from www.somedicyt.org.mx/enlaces/publicaciones.html.

SOMEDICyT. (2020). Repositorio Divulga. Acervo de documentos digitales sobre la comunicación de la ciencia. Retrieved from www.somedicyt.org.mx/divulgadores/repositorio-divulga.

Tonda, J. and Burgos, E. (2007). Análisis y evaluación de las revistas de divulgación. In *Proceedings of the X Reunión de la Red de Popularización de la Ciencia y la Tecnología en América Latina y el Caribe (RED POP – UNESCO) y IV Taller 'Ciencia, Comunicación y Sociedad*, 9–11 May. San José, Costa Rica.

Tonda, J. et al. (2002). Antología de la divulgación de la ciencia en México. Universidad Nacional Autónoma de México, México, D.F.

Torres, I. (2018). La ciencia para todos del FCE supera los 250 títulos, *La crónica, 16*. October. Retrieved from www.cronica.com.mx/notas/2018/1097418.html.

UNAM. (2018a). Programa único de especialisaciones en ciencias políticas y sociales. Creación del plan de estudios de la especialización en comunicación pública de la ciencia. UNAM Posgrado, 25 May.

UNAM. (2018b). Posgrado en Filosofía de la Ciencia. Retrieved from www.posgrado.unam.mx/filosofiadelaciencia/programa/campos/comunicacion-de-la-ciencia.html.

Zamarrón, G. (1994). La divulgación de la ciencia en México: una aproximación. Serie Cuadernos de Divulgación 1. México: SOMEDICYT.

Timeline

Event	Name	Date	Comments
First interactive science centre established.	Museo Tecnológico de la CFE [Federal Commission of Electricity]	1970	Devoted to electricity, magnetism and technology
First national (or large regional) science festival.	National Week for the Popularisation of Physics	1985	Later the name changed to National Week for Science Communication

Event	Name	Date	Comments
An association of science writers or journalists or communicators established.	AMPECI (Association of Science Journalists)	1979	1986: Mexican Society for the Popularisation of Science and Technology (SOMEDICyt) for science communicators
First university courses to train science communicators.	Diplomado en Divulgación de la Ciencia, National Autonomous University of Mexico	1995	
First master's students in science communication graduate.	Maestría de Ciencia y Cultura offered by ITESO (Western Institute for Superior Studies)	1998	
First PhD students in science communication graduate.	Posgrado en Filosofía de la Ciencia [PhD in Philosophy of Science in Science Communication]	2003	
First national conference in science communication.	Organised by SOMEDICyT	1991	
National government program to support science communication established.	Government program devoted exclusively to support projects in science communication. Launched by CONACyT	2012	However, science communication had received support previously as part of scientific research projects
First significant initiative or report on science communication.	COECYT-Michoacán (J. Padilla and L. Patiño) diagnosis of science communication in the state of Michoacán	2010	Diagnosis of other states were performed in the following years
National Science Week founded.		1994	Organised by CONACyT
A journal completely or substantially devoted to science communication established.	*Ciencia y Desarrollo*	1975	Edited by CONACyT
First significant radio programs on science.	*Actividades científicas* [Scientific activities]	1972	Conducted by Juan José Morales
First significant TV programs on science.	*Problemas del mundo y del hombre* [Problems of the world and mankind]	1971	
First awards for scientists or journalists or others for science communication.	National Award for Science Communications in honor of Alejandra Jaidar	1991	Organised by SOMEDICyT

Event	Name	Date	Comments
Other significant events.	A book about comets for the general public	1681	Author Carlos Sigüenza y Góngora
	Centro Universitario de Comunicación de la Ciencia (UNAM) was created	1980	
	Launch of *Revista Chispa*	1981	Popular science magazine for children
	Sundays in Science progam begins	1983	Popular science talks
	La Ciencia desde México [Science from Mexico]	1984	Science communication editorial project is launched
	Foundation of SOMEDICyT	1986	
	Foundation of AMMCCyT	1996	
	First public policy for the support of science communication in the State of Michoacán. National Award for Science Communication and Journalism CONACyT	2010	

Contributors

Dr Elaine Reynoso-Haynes is director of education and research in science communication at the National Autonomous University of Mexico (UNAM).

Susana Herrera-Lima is professor in the Sociocultural Studies Department of ITESO University, Jalisco, México.

Ana Claudia Nepote is associate professor at the National School of Higher Studies at National Autonomous University of Mexico (UNAM) at Morelia, Michoacan, Mexico.

Lourdes Patiño-Barba is president of the Mexican Society of Public Communication of Science and Technology (SOMEDICyT) and senior consultant in Fibonacci Innovation and Scientific Culture.

25

THE NETHERLANDS

From the first science information officers to the Dutch Research Agenda

Anne M. Dijkstra, Frans van Dam
and Maarten van der Sanden

Science communication efforts in the Netherlands started with exhibitions in national history museums in the 19th century and popular articles about science and technology in the media in the 1930s. From the 1950s onwards, the Dutch government stimulated popularisation of science and technology as a way to foster the science–society relationship. Democratic, and later economic and cultural considerations were the reasons for setting up one-way and two-way science communication. This may explain why attitudes towards science and technology have largely been positive compared to most other European countries, but at the same time not all new technologies are accepted. Genetic modification is an example of a topic that raised a lot of debate in the 1990s; today, opposing views on vaccination show that acceptance of science and technology is not straightforward in the Netherlands. Science communication efforts are visible in many ways in Dutch society via organisations, events and activities. These are supported or organised by both private and public partners. Nowadays, Dutch researchers are increasingly stimulated to engage with society—for instance, via the Dutch Research Agenda. Science communication in the Netherlands can build on a rich variety of expertise and inputs.

1. Introduction: Dutch rationales for science communication efforts

In the Netherlands, science communication followed in the footsteps of Dutch public information campaigns in the agricultural sector. Immediately after World War II, the government focused on rebuilding Dutch society. In this process, science and technology played a role based on the economic principle that whatever is right for science and technology is also right for society. Although science communication did not yet exist as a field of study and practice in the Netherlands, attention to popularisation had been growing, with the aim of acquiring societal support for science and technology (Dalderup, 2000; Dijkstra, 2008).

From the 1950s onwards, various government programs supported science communication as a way to foster the science–society relationship. The first policy report that mentioned science communication was published by the government in 1957 (see Table 25.1 below for an overview of the most prominent government reports on science and society). The Bender Commission, established by an advisory committee of collaborating universities, argued that universities should systematically improve relationships with the groups in society they depend on, and try to gain public trust. This was basic public relations, but the commission preferred the label 'science information' (in Dutch: *voorlichting*). A democratic rationale emerged: everyone is entitled to have access to knowledge and information and should be able to use this to discuss matters of science and technology. Consequently, at the end of the 1950s, the first science information officials—as they were called—started working at the universities. Science communication in the Netherlands was still in its infancy (Dalderup, 2000; Dijkstra, Seydel and Gutteling, 2004; Wiedenhof, 1978).

Table 25.1: Overview of reports that discussed science and its relation to society and the role for science communication.

1957	*Commission Bender*
1974	*Nota Wetenschapsbeleid* [Report on Science Policy] by Boy Trip, the first minister of Research Policy
1982	*Wetenschap als gemeengoed* [Science as common good] by Professor Stappers
1984	*Integratie van wetenschap en techniek in de samenleving* [Integration of science and technology in society] by Wim Deetman, Minister of Education, Culture and Science
1990	*Wetenschap en techniek voor een breder publiek* [Science and technology for all] by PWT
1992	*Kabinetsstandpunt Publieksvoorlichting over wetenschap en techniek* [Cabinet's position on public information about science and technology]

2000	*Boeiend, betrouwbaar en belangrijk* [Fascinating, trustworthy and important] by the Ministries of Economic Affairs, of Education, and of Agriculture
2014	*Wetenschapsvisie 2025: keuzes voor de toekomst* [Science vision 2025: choices for the future] by the Minister of Education, Culture and Science
2019	*Nieuwsgierig en betrokken: de waarde van Wetenschap* [Curious and engaged: the value of science] by the Minister of Education, Culture and Science

1.1. Democratic rationales and the emergence of science shops

Science communication efforts received a boost when the first Minister of Research Policy, Boy Trip, took office in 1973. His report on research policy, *Nota Wetenschapsbeleid* (Trip, 1975), discussed extensively the background of both research policy and science communication. According to the minister, the pursuit of scholarly work should not take place (or be considered) separate from its societal context. Researchers should strive to come in close contact with the actors concerned. The minister believed that, in this way, the public would be able to develop their own opinions about scientific research, and public participation would be improved. In 1978, as a result of the report, the Office of Science Information was established. It championed the principle that citizens have the right to know and understand (Dijkstra, 2008; Stappers et al., 1983).

In the 1970s, when government influence on science and technology policy was strong, the first science shops were created at the universities. The concept originated in the Netherlands, and the first shops were run by students on a voluntary basis with support from employees. They were set up after debates on research policy concluded that universities should play a more prominent role in the solution of societal problems. They were based on democratic grounds, with a goal of supporting groups that could not afford to commission research (such as oppressed minorities and financially weak groups). From 1978 onwards, science shops received financial support from the universities (Lürsen, Mulder and Lieshout, 2000).

By about 2000, most universities hosted science shops, with the number peaking at 33. However, a few years later, several shops had to close as the universities stopped funding them. These funding cuts resulted from a combination of the economic downturn and a shift in policy as universities no longer considered societal support so important. Only a handful of science shops still exist in the Netherlands, but, interestingly, the concept has gained international support and can be found in universities all over the world (De Bok and Mulder, 2004; Lürsen et al., 2000; Mulder and Straver, 2015).

1.2. Economic rationales for science communication

At the beginning of the 1980s, several public debates emerged spontaneously in Dutch society, on topics such as nuclear energy and the environment. At the same time, in 1984, Minister of Education, Culture and Science Wim Deetman released a new policy report called *Integratie van wetenschap en techniek in de samenleving* [Integration of science and technology in society]. Key themes included the dissemination of information, the development of public opinion and social decision-making. The minister stated that it was necessary to intensify and diversify information dissemination efforts and that the public needed continuous science information to be able to follow developments. The economic rationale for science information started playing a more dominant role. Scientific as well as technological knowledge is considered indispensable for achieving economic progress.

Two new organisations were established in 1986 to enhance information dissemination efforts. The first, the Foundation for Public Information on Science and Technology (PWT, later the Dutch Science and Technology Association) replaced the Office of Science Information and dealt with informing the public about science and technology (Wiedenhof, 1995). The second organisation, the Netherlands Organisation for Technology Assessment (NOTA, renamed the Rathenau Institute in 1994) was commissioned to study societal and ethical aspects of science and technology, to inform policymakers about the outcomes, and to stimulate public debate about new developments. These tasks were partly inspired by the experiences from the US Office of Technology Assessment (Tuininga, 2000).

Five years later, in 1989, Minister Deetman again advocated the strengthening of public support for science and technology. He thought that fostering scientific literacy through increasing knowledge was important, since developments were moving so quickly that the gap between science and societal groups was widening. New initiatives such as the Science and Technology Week were organised and encouraged. From 1993, a series of six public debates on biotechnology topics were organised or commissioned by the government to raise public support for biotechnology. These culminated with a debate on genetically modified (GM) food in 2001 (Dijkstra, 2008). Activities were no longer organised solely on the basis of democratic or economic rationales—there is a growing awareness that science and technology are inherently connected to society. A cultural perspective entered Dutch thinking about the science–society relationship (Dalderup, 2000).

In his evaluation in 1995 of 10 years of science information campaigns, Wiedenhof concluded that the economic rationale had become more influential, but that democratic as well as cultural rationales also played a role. According to him, these rationales are one of the reasons why science information activities in the Netherlands were doing well compared to developments abroad (Wiedenhof, 1995, 2000). But in the following years, changes occurred as the government interfered in science communication more often—as science information had been renamed then to emphasise the change towards two-way transactions and dialogue—and demanded immediate and clear evidence of attitudinal effects. Science communication efforts were also aimed at science education, and the Dutch Science and Technology Association was dismantled in 2007. Dalderup (2000) considered that the economic rationale had become dominant by the beginning of the 21st century, as democratic or cultural motives for science communication were relegated to the background (cf. Dijkstra, 2007).

1.3. A more reluctant government

For many years a conservative government has been in office in the Netherlands, and it has been reluctant to stimulate science communication as it is not considered a core responsibility, particularly in light of the 2008 economic crisis. Economic profits have been the main drivers for science, technology and innovation, and, consequently, for science communication or public engagement. Despite this stance towards science communication, a few government initiatives are worth mentioning. In 2010, the government commissioned a societal debate about nanotechnology to tackle the public's experiences with biotechnology; in 2014, it started working on the so-called Dutch National Research Agenda, where priorities are driven by societal needs. In addition, universities started reconsidering their relationship with society.

A public debate about nanotechnology was organised in 2010 and 2011. Taking into account criticisms that an earlier debate on GM food in 2001 was biased in favour of GM products, this time a more bottom-up public engagement approach was used. The lesson learned from previous experience was that a societal dialogue should feed into decision-making. The responsible committee organised the societal dialogue in two phases, with the first aiming to provide essential information as not many people had heard of nanotechnology before, and the second phase aiming to establish a dialogue (Krabbenborg and Mulder, 2015). In practice, however, an evaluation of the activities arranged by the various organisations and selected and funded by the committee concluded that activities mainly focused on outreach and knowledge transfer, and that this was a missed opportunity for genuine dialogue and bottom-up approaches (Krabbenborg and Mulder, 2015).

Recently, the exercise on the Dutch National Research Agenda[1] has been relevant for science communication in the Netherlands. And, although the process has not yet been evaluated for its implications, it still can be considered an interesting example of a bottom-up approach aiming to include societal needs in research. In 2014, the government decided to ask Dutch citizens what research questions they considered important. The aim was to establish a Dutch National Research Agenda for the future, as outlined in a new policy report on science and its role in society (Ministerie Van OCW, 2014). The promise was that the responses would be taken seriously, and a budget would be allocated in a later phase to address these societal questions. The rationale was to connect science to society in a better way.

Everyone was quite surprised when about 12,000 questions were submitted, mainly by citizens—but representatives of various interest groups including researchers also handed in questions. Under the guidance of university professors, in the next step 25 so-called main 'routes' containing research areas were identified, which included 140 research questions to be addressed—for example, about climate change and sustainability. Meetings and deliberations with citizens were organised to discuss the questions and what researchers could or should do to answer them. In 2018, calls for large multidisciplinary research proposals were made available for researchers, with funding of €70 million. As a follow up in 2019, the Minister of Education, Culture and Science published a new policy report *Nieuwsgierig en betrokken: de waarde van wetenschap* [Curious and engaged: the value of science], which included a budget increase for the Dutch Research Agenda to €130 million from 2020 onwards (Ministerie Van OCW, 2019). In addition, the minister specifically allocated €1 million for science communication, for a pilot program to be organised by the Dutch Research Organisation to reward researchers who engage in dialogue with society. According to the policy report, the reasoning is that everyone should benefit from scientific findings, and these can be achieved by better connecting science to society—hence the emphasis on engaged researchers (Ministerie Van OCW, 2019).

At the same time (and independently from government efforts), Dutch universities have been reconsidering their relationship with society and are acknowledging that societal needs should be better incorporated in their research and policies. Following the example of the UK, some universities established offices of public engagement. This aligned with thinking about the science–society relationship in the wider world in which science communication is one aspect next to, for instance, science education, gender

1 See www.wetenschapsagenda.nl/?lang=en.

and ethics. These, including further engagement and open access, are also propagated at the EU level as guiding principles for responsible research and innovation.

To sum up, policies and their rationales have played a significant role in the Dutch science–society relationship and have strongly influenced efforts for science communication until the beginning of the 2000s. The next section discusses Dutch attitudes towards science and technology as well as towards specific technologies.

2. Attitudes towards science and technology in the Netherlands

In the Netherlands, public perceptions and attitudes towards science and technology have not been measured often. In 2000, the Social and Cultural Planning Office (SCP) and the Netherlands Organisation for Scientific Research (NWO) conducted a survey with 1,777 interviews (Becker and Van Rooijen, 2001). Respondents considered science trustworthy and prestigious and there was optimism about the ability of science to solve contemporary problems. Science and technology were both evaluated and the results were similar: both were considered good and beneficial for society (Becker and Van Rooijen, 2001). Five years later, in 2005, a Special Eurobarometer survey (Eurobarometer, 2005) showed that 97 per cent of the Dutch agreed that 'science and technology developments will help cure illnesses such as AIDs or cancer'. This was the highest rate in the EU. As well, 70 per cent agreed that 'science and technology make our lives healthier, easier and more comfortable'. Compared to other EU countries, the Dutch responses have been among the most positive in Europe, with approval levels similar to those of Sweden and Denmark.

This optimism had also been visible in attitudes towards biotechnology and related topics in the 1980s and 1990s (Gutteling et al., 2001). A more negative shift in media coverage occurring in countries such as Germany and the UK, in response to the birth of Dolly the cloned sheep and the marketing of GM soybeans by US-based agrochemical company Monsanto, did not occur in the Netherlands, as Einsiedel et al. (2002) pointed out. The Dutch government had invited the public to consider the risks of new technologies relatively early, when they commissioned six public debates on topics such as Herman the Bull (the world's first transgenic bovine born in 1990) and GM organisms for food applications. Despite these debates, however, in the 1990s and 2000s attitudes towards biotechnology gradually became more negative (but not as negative as in other countries) (Dijkstra, 2008).

The most recent Eurobarometer (2014) showed again that Dutch believe that science and technology innovation will have a positive impact. The Dutch perceptions (84 per cent) were the most positive, followed again by Sweden (83 per cent) and Denmark (82 per cent). Recent studies conducted when the public debate about nanotechnology was organised also showed positive attitudes towards this emerging technology, albeit for a specific audience that showed interest in science (cf. Dijkstra and Critchley, 2016). Finally, a recent study by Hanssen et al. (2018) found that Dutch attitudes towards genetic modification or genetically modified organisms are neutral, neither in favour of nor against the technology. Dutch attitudes are complex and related to general attitudes toward science and technology and to different aspects of trust. One conclusion from the studies was that the Dutch show a more active and engaged behaviour when their direct personal interests are involved.

3. How is science communication institutionalised in the Netherlands?

The previous sections considered the rationale for science communication efforts focusing on policies and government influence, and then described Dutch attitudes towards science and technology. Now we look at the way science communication has been institutionalised and is visible in organisations, events or education in the Netherlands. (See the timeline at the end of the chapter for an overview of significant science communication events and activities in the Netherlands.)

3.1. Museums and media

The first places in the Netherlands where science and technology were available for a broader audience were the natural science and history of science museums. The National Museum for Natural History opened in Leiden in 1820, long before any other similar institution. Several other museums opened their doors at the beginning of the 20th century. In 1904 the National Sciences Museum (renamed Museon) started its exhibitions in The Hague; in 1923, the Museum of Labour opened in Amsterdam. It subsequently went through several name changes before settling on NEMO Science Museum in 2016. In Leiden, the National Museum Boerhaave opened in 1928; it started with a series of interactive exhibits in the 1980s (Van Mensch, 2000).

Media reported on science early. The Dutch popular journal *Natuur & Techniek* [Science & Technology] has featured articles about science and technology since the 1930s. More science reporting emerged in newspapers, radio and television in the 1960s. A highlight is the reporting of the Apollo landing on the moon in 1969. Two reporters—Henk Terlingen, better known as mad-Henkie, and Chriet Titulaer—became well known for their enthusiastic broadcast of the landing (Dalderup, 2000). In 1966 the first television program showed The Young Researchers' Competition (Dalderup, 2000). However, for a long time, science communication on television was not considered important. Dalderup (2000) explains that in the year 1978/79 the total number of hours related to science communication was 10 hours out of a total of 3,000 hours, while in the neighbouring country Belgium this was 110 out of 2,800 hours.

In 1969, a separate Chapter of Science Journalism was created as part of the Netherlands Journalism Organisation. The conference 'Science in Journalism' in 1978 discussed the active role of journalism in science and technology, and the first science sections appeared in Dutch newspapers in 1981 (Volkskrant) and 1982 (NRC). In 1985 the Chapter on Science Journalism became the independent Association for Science Journalism in the Netherlands (VWN). However, as in other countries, times changed for science journalists, and in 2013 it was recognised that science communication had developed into a task that now also involved science journalists, and the name was changed to the Association for Science Journalism and Communication Netherlands. It had become almost impossible for science journalists to earn their income as independent journalists only.

The Dutch non-academic journal *Tijdschrift Wetenschap, Techniek en Samenleving* [Journal for Science, Technology and Society] provided a more reflective view on the science and society relationship from 1997 until 2005. It examined the relationship between science, technology and society, from the perspectives of both science communication and science and technology studies. Its readers were based in universities and government bodies. Until 2011 it continued as the Yearbook Knowledge Society, with its discussions focusing on a different societal theme each year, such as developments in surveillance and privacy issues.

3.2. Science centres and events

In the period 1960 to 1980, many other activities were initiated, including setting up the first science centre Evoluon by the company Philips in Eindhoven in 1966. The company wanted to show how science and technology, and mechanics and computerisation, lifted production levels with humanising technologies. The centre demonstrated how technology can solve societal problems. The first exhibition was a success, but the centre was not viable financially. It is noteworthy that Evoluon opened before the Exploratorium in the US, although it never served as an example for other science centres as the latter did (Van Mensch, 2000).

In 1986 the Dutch government commissioned advice about a National Centre for Science and Technology. Several proposals competed for the national centre, but the final report in 1987 was quite a surprise: the government decided that it would not fund a national centre. In 1989 there was another blow, when Philips decided not to proceed with the Cosmocentre Project as a replacement for the ageing Evoluon. Some of the original proposals for the national centre decided to further their plans anyway.

There were further discussions and a new plan was developed, also for a centre in Amsterdam. In 1992 the Ministers of Economic Affairs and Education and Science supported this plan, with promises of assistance from the city and industry. In 1997 NewMetropolis opened its doors in a building designed by the world-famous architect Renzo Piano. Unfortunately, the expenses were such that the museum suffered huge financial losses from the beginning. Therefore, in 1999, a restart was made as the science centre NEMO, later NEMO Science Museum, which is increasingly paying attention to science and hands-on activities (Van Mensch, 2000). Nowadays, NEMO is the best-known science museum in the Netherlands.

The creation of science centres and media coverage of science were complemented by setting up the first national science week. It was organised at the University of Utrecht in 1986 with funding from the national government. The next year, the Dutch Science Week Association, funded by the Ministry of Education (and later with additional funding from the Ministry of Economic Affairs) started organising an annual science week. The event had its ups and downs until it was discontinued in 2007. Currently, a smaller event takes place (without much government funding) to coincide with the European Science Night.

3.3. Science communication in other places

At the beginning of the 2000s, the liberal government gradually retracted most of its funding for science communication in the Netherlands, and activities and events started being organised by non-government actors. From 2010 onwards, for example, several music festivals in the Netherlands offer science lectures and demonstrations as well as music. The Zwarte Cross festival, a huge event, provides a whole tent to university researchers where they give lectures and demonstrations to festival participants who are willing to broaden their interest as well enjoy the music.

Another activity gaining ground in the Netherlands is the science café, with the first organised in Nijmegen in 2005. The organisers came across a *café scientifique* in France when their car broke down and they had to wait for repairs. The formula was an immediate success, and, by 2016, about 15 to 20 science cafés were successfully organised on a regular basis. Most of them are run by volunteers and free of charge and several are in cities where there is no university (Dijkstra, 2017).

3.4. From courses to master's programs and science communication research

Another outcome of science policy developments in the 1970s was the funding and development of courses in science communication, with the first ones starting in 1976. Jaap Willems, a biologist and former journalist, taught one in Nijmegen. He was the first PhD graduate in the field in the Netherlands and defended his thesis on science journalism and communication barriers in the same year (Willems and De Bekker, 1976). Only a handful of PhD students have graduated in science communication topics in the Netherlands. Niels Wiedenhof completed his thesis in 1978 on the development of science communication in the Netherlands, while more than 20 years later in 1999, Adriana Esmeijer analysed selection processes in science journalism (Wiedenhof, 1978; Esmeijer, 1999). The next batch of three PhD theses was defended in 2008 (Dijkstra, 2008; Van der Auweraert, 2008; Van der Sanden, 2008). Since then a handful of others have completed their theses.

Over this period the number of science communication educational programs at universities has risen slowly. At the end of the 1990s, the Minister of Education, Culture and Sciences announced that the general universities should offer science students the possibility of a track in science communication, science education or science management. This would fulfil societal needs, and the general universities in the Netherlands responded by developing science communication programs. In 2006, the technical universities successfully applied for science communication programs for their engineering students, and these started in 2007. At this moment, next to several single courses or modules, a science communication specialisation at the master's level is offered by about half the Dutch universities.

Research in science communication is conducted mainly at the universities that offer educational programs. Topics of research vary widely and are often connected to either the focus areas of these programs, the various backgrounds of the researchers involved, or externally funded projects in which the scholars participate. For instance, not only is the interaction between scientists and audiences studied, but research projects may also examine the role for scientists in the process of innovation (Van der Sanden and Flipse, 2016).

4. Organisational capacities

The Netherlands has always been a country where citizens organise themselves in associations. Most Dutch citizens are members of five or more associations either professionally, privately or on a voluntary or paid basis; and this holds for the Dutch science communication field as well (Riedlinger et al., 2018). The VWN exists as well as the association for public information officers (PWC). In 2013, SciComNL started, connecting practitioners in science communication with a 'community of practice' approach so that they could learn from each other.

National conferences for science communicators were organised soon after courses at universities were developed. Beginning at the end of the 1990s, the Dutch Science and Technology Association organised a few conferences to further knowledge about science communication, but never on a structural basis. Nowadays, with other partners, NEMO organises an annual Science Communication in Practice conference that attracts about 200 participants, mostly practitioners. Science communication students organise an annual student conference for all students in one of the Dutch programs. And last but not least, in 2022, Rotterdam will host the first PCST conference to be held in the Netherlands. This will be a collaborative effort of all universities and other organisations in the Netherlands involved in science communication.

4.1. Science communication in the Netherlands varies richly

The thinking on science communication as well as its practices are continuously changing in the Netherlands. Starting in the 19th century when the first natural history museum opened, the Dutch government was an important stimulator of science and society interactions. For a long time, science communication was considered a task of keeping the Dutch public informed. However, from the first policy report onwards, improving relationships between science and society by gaining public trust and making science available for everyone was an important driver for starting and stimulating efforts and providing funding. Many science communication efforts were established between the 1970s (when the first information officers started working at the universities) and the 1990s. Universities, as the case of the science shops has shown, played a prominent role in these activities. When governmental commitment became less prominent, other organisations slowly took over. Events such as promoting science and technology at music festivals have been funded by private parties, while science cafés offering more in-depth discussion of the newest scientific developments are mainly run by volunteers. In a more recent trend, the Dutch Research Agenda shows more government support for communication and demands that researchers play a more active role engaging in science communication.

5. Conclusion

When it comes to controversial topics, the Dutch start organising themselves sooner or later, as exemplified by the first spontaneous public debates on nuclear energy and the environment. Dutch citizens are becoming increasingly aware that science and technology are influencing our society, and the evidence lies in the many discussions in the public domain on topics such as privacy, climate change and robots. At the same time, citizens have to deal with new developments such as fake news or the trustworthiness of science. In recent years, Dutch citizens seem to engage more regularly in science-related activities, not only as receivers of information but also as active participants in scientific research. At home, in science centres and in informal education, they have engaged in and contributed to activities within the scientific process. In the Netherlands, these activities vary from regular participation by birdwatchers in bird population research to citizens who bring up questions about the living environment and team up with a university to become involved in research projects (see European Citizen Science Association, 2015). However, there is always tension, and despite memberships of associations and citizen science contributions, active engagement is often restricted to a few topics and to certain (often higher educated) publics.

Practices of science communication in the Netherlands are continuously changing. As this chapter shows, science communication practices started out mainly by informing people about science. The approach today is more inclusive and varied, as illustrated by the views of one prominent practitioner reported below.

Box 25.1: Changing practices from the perspective of a science communicator

Jac Niessen (1955) works for Wageningen University & Research (WUR) as the science information officer. His main task is to connect the media and the public at large with the experts at his university, as well as guiding scientific results towards the media. After his biology degree, he started as a science journalist for the popular magazines *Natuur en Techniek* and *Bionieuws*, and the agricultural magazine *Landbouwkundig Tijdschrift*. As science communicator for the science funding agency (NWO), he was an editorial board member for the annual national science quiz for Dutch TV.

He feels that over the years, the practice of science communication has changed substantially. Niessen has observed a shift from content to formats. Social media in particular has caused this shift: it is no longer about the story itself but more about the way it is told, who says what and to whom. The implication is that science communicators (like Niessen) do more with less content.

Before the age of social media, media coverage of announcements contained in press releases was the main medium for communicating about science. Communication was aimed at large audiences, such as those who read newspapers or watch TV. Nowadays, someone approaches the university with a question and that person receives an answer quickly, coming directly from an expert. On a daily basis, Niessen and his colleagues monitor what happens in society to find out what stories or developments dominate (social) media. They may then decide to publish a dossier on the topic, appoint an expert as spokesperson and use social media to attract attention. This approach has strongly accelerated communication, as well as the recycling of information.

Another major shift is the movement from transmission to dialogue. Debating issues as they are will mainly cause polarisation, according to Niessen. Instead, with the help of communicators, scientists engage in dialogue, and both parties may acquire better understanding of the various opinions.

In recent years, trust in public institutes such as universities has decreased. Now a story told by a person outside the university can often do a better job, and with more credibility, than a story told by a university researcher. For this reason, the university regularly asks prominent alumni to serve as ambassadors, who then may be asked if they are willing to use their networks on specific issues.

According to Niessen, the speed of communication will further increase in the future. Response times will come down and audiences will become smaller, perhaps in the end to a single individual.

References

Becker, J. W. and van Rooijen, P. M. (2001). *Het beeld van de wetenschap in Nederland. Opvattingen van de bevolking, wetenschappers, adviseurs en Kamerleden in 2000.* Den Haag: SCP/NWO.

Dalderup, L. (2000). Wetenschapsvoorlichting en wetenschapsbeleid in Nederland 1950–2000. Gewina: *Tijdschrift voor de Geschiedenis der Geneeskunde, Natuurwetenschappen, Wiskunde en Techniek, 23*, 165–92.

De Bok, C. and Mulder, H. (2004). Wetenschapswinkels in de kennissamenleving. *Tijdschrift voor Hoger Onderwijs, 22*, 123–39.

Dijkstra, A. M. (2007). Perspectief op het democratisch motief in de wetenschapscommunicatie. In J. Willems (ed.), *Basisboek wetenschapscommunicatie.* Amsterdam: Boom Onderwijs.

Dijkstra, A. M. (2008). Of publics and science. How publics engage with biotechnology and genomics (PhD thesis). University of Twente.

Dijkstra, A. M. (2017). Analysing Dutch Science Cafés to better understand the science–society relationship. *Journal of Science Communication, 16*(01), A03, 1–17. doi.org/10.22323/2.16010203.

Dijkstra, A. M. and Critchley, C. R. (2016). Nanotechnology in Dutch science cafés: Public risk perceptions contextualised. *Public Understanding of Science, 25*, 71–87. doi.org/10.1177/0963662514528080.

Dijkstra, A. M., Seydel, E.R. and Gutteling, J. M. (2004). Effectieve wetenschapscommunicatie: een communicatievraagstuk. Kennisdagen Communicatie 2003. Effectieve publiekscommunicatie: hints voor de Wetenschaps- en techniekcommunicatie. Papers, positiepapers en tips. Amsterdam: Stichting Weten.

Einsiedel, E., Allansdottir, A. et al. (2002). Brave new sheep – the clone named Dolly. In M. W. Bauer and G. Gaskell (eds), *Biotechnology. The Making of a Global Controversy* (pp. 313–47). London: Cambridge University Press.

Esmeijer, A. (1999). *Sluiswachters in de wetenschapscommunicatie. Een studie naar selectieprocessen bij het populariseren van (alfa)onderzoek*. Amsterdam, Vrije Universiteit.

Eurobarometer. (2005). Special Eurobarometer 224 'Europeans, Science & Technology'. European Commission.

Eurobarometer. (2014). Special Eurobarometer 419 'Public perceptions of science, research and innovation'. European Commission.

European Citizen Science Association. (2015). *Ten principles of citizen science*. London: European Citizen Science Association.

Gutteling, J., Midden, C., Smink, C. and Meijders, A. (2001). The Netherlands: controversy or consensus? In G. Gaskell and Martin W. Bauer (ed.), *Biotechnology 1996–2000. The years of controversy* (pp. 229–36). London: Science Museum.

Hanssen, L., Dijkstra, A. M., Sleenhoff, S., Frewer, L. J. and Gutteling, J. M. (2018). Revisiting public debate on Genetic Modification and Genetically Modified Organisms. Explanations for contemporary Dutch public attitudes. *JCOM, 17*(04), A01. doi.org/10.22323/2.17040201.

Koninklijke Nederlandse Academie van Wetenschappen. (1977). *Wetenschapsvoorlichting. Rapport over de instelling van een dienst wetenschapsvoorlichting*. Amsterdam: Koninklijke Nederlandse Academie van Wetenschappen.

Krabbenborg, L. and Mulder, H. A. J. (2015). Upstream public engagement in nanotechnology: Constraints and opportunities. *Science Communication, 37*, 452–84. doi.org/10.1177/1075547015588601.

Lürsen, M., Mulder, H. and Lieshout, M. (2000). Kronkelpaden en afslagen; ontwikkelingen in en om wetenschapswinkels. *Gewina: Tijdschrift voor Geschiedenis der Geneeskunde, Natuurwetenschappen, Wiskunde en Techniek, 23*, 207–13.

Ministerie Van OCW. (2014). *Wetenschapsvisies 2025: keuzes voor de toekomst.* Den Haag, Ministerie van Onderwijs, Cultuur en Wetenschap.

Ministerie Van OCW. (2019). *Nieuwsgierig en betrokken: de waarde van wetenschap.* Den Haag, Ministerie van Onderwijs, Cultuur en Wetenschap.

Mulder, H. and Straver, G. (2015). Strengthening community–university research partnerships: Science shops in the Netherlands. In B. Hall, R. Tandon and C. Tremblay (eds), *Strengthening community university research Partnerships: Global perspectives* (pp. 181–96). Victoria, British Columbia: University of Victoria and PRIA.

Riedlinger, M., Barata, G., Van der Sanden, M., Medvecky, F. and Joubert, M. (2018). PCST roundtable: The role of professional science communication associations in supporting science communicators and legitimising communication practices. Proceedings 15th International Public Communication of Science and Technology Conference, Dunedin, New Zealand, 2018.

Stappers, J. G., Reijnders, A. D., Möller, W. A. J. and Hesp, L. A. T. M. (1983). *Wetenschap als gemeengoed. Een studie van de wetenschapsvoorlichting in Nederland.* 's Gravenhage, Staatsuitgeverij.

Trip, F. H. P. (1975). *Nota wetenschapsbeleid.* 's-Gravenhage: Tweede Kamer, zitting.

Tuininga, E.-J. (2000). Sturing van wetenschap en technologie, 1965–2000. *Gewina: Tijdschrift voor de Geschiedenis der Geneeskunde, Natuurwetenschappen, Wiskunde en Techniek, 23*, 58–71.

Van der Auweraert, A. (2008). De onderzoeker als communicator. Een kwalitatief en verkennend onderzoek naar de determinanten van wetenschapscommunicatiegedrag (PhD thesis). Wageningen University and Research Centre.

Van der Sanden, M. (2008). Towards effective biomedical science communication (PhD thesis). Free University Amsterdam.

Van der Sanden, M. C. A. and Flipse, S. (2016). Science communication for uncertain science and innovation, *Journal of Science Communication, 15*(06), C05. doi.org/10.22323/2.15060305.

Van Mensch, P. (2000). Tussen object en samenleving: 25 jaar musea van natuurwetenschap en techniek in Nederland. *Gewina: Tijdschrift voor de Geschiedenis der Geneeskunde, Natuurwetenschappen, Wiskunde en Techniek, 23*, 193–206.

Wiedenhof, N. (1978). *Wetenschapsvoorlichting: een bijdrage tot beeld-, oordeels- en besluitvorming.* Eindhoven, Technische Hogeschool Eindhoven.

Wiedenhof, N. (1995). *Aarde, appel, ei -Wetenschaps- en techniekvoorlichting in Nederland en ook wat daarbuiten. 1985–1995.* Utrecht, Stichting PWT.

Wiedenhof, N. (2000). Wetenschaps- en techniekvoorlichting op zoek naar balans tussen Apollo en Dionysus. *Gewina: Tijdschrift voor de Geschiedenis der Geneeskunde, Natuurwetenschappen, Wiskunde en Techniek, 23*, 228–39.

Willems, J. and De Bekker, J. (1976). *Wetenschapsjournalistiek: klachten en communicatiebarrières*. Nijmegen: Universiteit Nijmegen.

Timeline

Event	Name	Date	Comment
First interactive science centre.	Evoluon, set up by the company Philips in Eindhoven	1966	Aimed to demonstrate how science and technology can solve societal problems
First national (or large regional) science festival.		Since 2000	Presence at science festivals, later also at music festivals
An association of science writers or journalists or communicators established.	Chapter of science journalists created in the Netherlands Journalism Organisation	1969	1985: The chapter becomes the independent Association for Science Journalism (VWN)
First university courses to train science communicators.	Science journalism courses	1976	Offered by Jaap Willems, for example
First master's students in science communication graduate.		Since 2000	Universities offer science communication programs after request by minister
First PhD students in science communication graduate.	Jaap Willems	1976	1978: Niels Wiedenhof 1999: Adriana Esmeijer
First national conference in science communication.	Organised by *Stichting WeTeN* [Association for Science and Technology Netherlands]	1990s	2013: Annual *Vakconferentie Wetenschapscommunicatie* [Science communication conference for professionals] organised by NEMO other partners
National government program to support science communication established.	Commission Bender	1957	Universities should strengthen their relationship with society
First significant initiative or report on science communication.	*Nota Wetenschapsbeleid* [Report on Science Policy]	1975	By Boy Trip, the first minister of research policy
National Science Week founded.		1987	1986: the first regional Science Week in Utrecht

Event	Name	Date	Comment
A journal completely or substantially devoted to science communication established.	*Tijdschrift Wetenschap, Techniek en Samenleving* [Journal for Science, Technology and Society]	1997–2005 (2011)	Followed up by the *Jaarboek Kennissamenleving* [Yearbook Knowledge society]. Both non-academic journals dedicated to science, technology and society and science communication
First significant radio programs on science.	Various programs	1960s	A highlight is the radio reporting on the moon landing in 1969
First significant TV programs on science.	The Young Researchers competition	1966	
First awards for scientists or journalists or others for science communication.	Various prizes and awards	1990s to 2000s	2009: *Boy Trip Fonds* [Boy Trip Fund/VWN Trip Fund] for science journalists
Date hosted a PCST conference.	PCST conference will be hosted in 2023	2023	
Other significant events.	The popular journal *Natuur & Techniek* [Science & Technology]	1930s	More science reporting emerges in the second half of 1960s; in newspapers, on radio and television
	First science shops emerge	1970s	Science shops originated in the Netherlands, with the first run on a voluntary base
	Platform *Wetenschap Communicatoren* [Association for Public Information Officers] formed	End of 1980s	Science communicators start organising themselves in various professional groups. 2013: ScicomNL network starts building a Community of Practice
	First science café Nijmegen	Since 2005	2016: 15–20 science cafés in various cities are organised on a regular basis

Contributors

Dr Anne M. Dijkstra is an assistant professor in science communication at the University of Twente.

Frans van Dam lectures on science communication and writing and presentation skills at Utrecht University, The Netherlands.

Dr Maarten van der Sanden is an associate professor of Communication Design for Innovation at TU Delft, Department of Science Education and Communication.

26

NIGERIA

Battling the odds: Science communication in an African state

Bankole Falade, Herbert Batta
and Diran Onifade

1. Introduction

Science and technology institutions, practices and policies abound in Nigeria but the phrase 'science communication' is not an entrenched one. At best, science communication practices can be described as fragmentary and incidental. A major issue is policy preferences, which do not elevate science communication to the same level as in Western societies. Closely related to this is education infrastructure and curriculum, which incorporates science communication as a subset of other disciplines.

Nigeria's plurality of religions in a strongly contested democratic setting between the old North and South also poses constraints to the spread of scientific ideas, and quite often this has to be factored into campaign programs. The polio vaccine controversy in the North shows the strength of religious beliefs as an obstacle to the spread of scientific ideas. The reaction of the religious leaders in the south to the Ebola virus epidemic also shows that religion, as an obstacle, is independent of affiliation. Religion can however be double-edged in having both adverse and complementary effects. Also, prevalence rates for the Human Immunodeficiency Virus (HIV) and alarmingly high annual death statistics from malaria tend to support the argument that more needs be done to encourage the citizenry to embrace modern scientific practices as premises for behaviour.

Notwithstanding, the country has made tremendous progress and the potential for science communication to accelerate the growth of science, technology and innovation (STI) indices, improve manufacturing, agriculture, health and wellbeing in Nigeria remains vast, awaiting more attention from both public and the private sector.

2. STI and Nigeria's growth and developmental challenges

Nigeria is Africa's most populated country with an estimated 186 million (2016) residents living on a land area of 923,768 km^2. It is the seventh most populous country in the world and although 14th in terms of land area in Africa; it accounts for 16 per cent of the continent's population. Nigeria is Africa's largest economy with gross domestic product (GDP) at $481 billion and per capita income around US$2,000 (current US$ for 2015). It is followed by South Africa at US$317 billion but with higher per capita income at about US$6,000 (National Bureau of Statsitics, 2020). According to the World Bank, between 2006 and 2016, Nigeria's GDP grew at an average rate of 5.7 per cent per year. Oil prices continue to dominate the country's growth pattern and their volatility imposes substantial constraints on planning and policy implementation. With a fertility rate of 5.5 births per 1,000 women, the population is expected to soar in the next few decades, raising developmental challenges (see World Bank, 2018). Science communication, embedded in STI performances, is critical to overcoming challenges in food production, health, manufacturing, the service sector and others.

3. Agricultural extension as precursor to science communication

Science communication in Nigeria has its origin in extension services, which date back to the early 20th century when agriculture accounted for over 90 per cent of the country's export earnings. Agriculture was soon overtaken by crude oil and although the petroleum sector is small, it is important for foreign exchange and fiscal revenues. Agriculture currently constitutes about 20–25 per cent of GDP; the oil industry accounts for 8–10 per cent; non-oil 13–14 per cent while the service sector is the highest at 52 per cent (Joseph-Raji and Timmis, 2018). Agricultural production thus remains very important to the economy.

Agricultural departments were established in Southern Nigeria in 1910 and Northern Nigeria in 1912, and a unified central body was formed in 1921 (Omotayo et al., 2001) after the amalgamation of both regions into present-day Nigeria in 1914. Constitutional reforms led to the creation of ministries of agriculture at federal and state levels (now 36 states) from where extension services were initiated for the commodity boards, the farm settlement schemes and the Agricultural Development Projects (ADPs), among others.

The farm settlement scheme was to train young school leavers in scientific innovations in farm practices with the expectation that these new technologies would diffuse into neighbouring farms and become linkages for the spread of future innovations. The ADPs, partly funded by the World Bank (World Bank, 2012) commenced in 1975, using the 'Training and Demonstration' method later replaced with the 'Training and Visit' method (Omotayo et al., 2001; Adebayo and Idowu, 2001). Subsequent programs such as the National Accelerated Food production program, the Green Revolution and Integrated Rural development programs were all very active examples of science communication, taking innovations to the farmers. The International Institute of Tropical Agriculture (IITA), established by the Ford and Rockefeller foundations, has also been active in research and innovation dissemination in Nigeria since it was founded in 1967. It is sited in Ibadan, the capital of Oyo State. With Africa producing more than half the world's cassava, IITA has been instrumental in the development and distribution of high-yielding disease-resistant cassava clones and continues to use genomic selection to improve cassava productivity.

Figure 26.1: ICT and agricultural extension practices.

Source: IITA.

Nigeria's 2016–20 agricultural policy framework called 'The Green Alternative' (Federal Ministry of Agriculture and Rural Development, n.d.) adopted a stakeholder approach, which seeks to build linkages across the value chain. Information and communication technology (ICT) platforms such as mobile phone and e-wallet schemes are now playing significant roles in the sector (Asogwa et al., 2015; Fadairo et al., 2015). The IITA also recently launched Nuru (IITA, 2018), an artificially intelligent assistant. Nuru (Swahili for light) uses machine learning to recognise leaves damaged by two important viral diseases of cassava (cassava mosaic disease and cassava brown streak disease), as well as damage by red and green mites. ICT platforms thus join traditional media forms such as radio, television and newspapers, which continue to play significant roles in the communication of improved farm practices.

The soap opera *Cock Crow at Dawn*, shown nationwide by the national broadcaster Nigerian Television Authority (NTA) in the 1980s, captivated the national audience for its portrayal of life in the rural areas. Co-sponsored by the United Bank for Africa and shot in a rural agrarian community, it was a mix of entertainment, agricultural extension and rural life. Several radio programs in English and the various local languages also aid the diffusion of innovations to farmers. They include *Aye Agbe* and *AgbeLere* in Yoruba; *Harama Manoma* and *Ina Manoma* in Hausa; *Onye Oruubi* and *Onye Oruugbo* in Igbo; and *Telefarmer*, *Country Farmer*, *Radio Farmer* and *Agribiotech* in English (see Oladele, 2006).

4. Science communication and health campaigns

Like agricultural extension services, science communication is also the basis of health campaigns in Nigeria. With a life expectancy at 55/56 years (WHO, 2016), Nigeria's health challenges are wide and varied. The country suffers the world's greatest malaria burden with approximately 51 million cases annually (Dawaki et al., 2016). In 2016, an estimated 3.1 million people were living with HIV/AIDS (WHO, 2016). Nigeria was one of three countries worldwide with active transmission of the wild poliovirus type 2. None has been detected since 2016, but there are still outbreaks of circulating vaccine-derived poliovirus type 2. Case studies below indicate the context and constraints to effective science communication in the health sector.

4.1. Pfizer drug trials

In 1996, Hopkins and Pfizer's medical team gave Trovan, an experimental meningitis drug, to about 200 children during a major outbreak of meningitis in Kano State, Northern Nigeria (Pfizer, 2009). While Pfizer claims the study plainly proved that Trovan helped save lives with a survival rate of 94.4 per cent, against less than 90 per cent for the current best treatment available in Nigeria at the time, the trial was seen as having fallen short of ethical guidelines and international best practices. The company made a US$75 million out of court settlement with the Kano state government (Okonta, 2014), but the event has continued to fuel conspiracy theories about Western health interventions.

4.2. Polio vaccine controversy

The international drive to eradicate poliomyelitis began in 1988. Nigeria's nationwide effort, led by the federal government, the World Health Organization, Rotary International, the United States Centers for Disease Control and Prevention, UNICEF and other international donor agencies provides a good case study of obstacles to science communication. It was a typical multimedia campaign but while it was very successful in the south, it met with resistance in the north, highlighting the negative effects of religious beliefs, the fragility of the relationship between the central government and the states and their combined effect on top-down science communication. The northern campaign was adversely affected by religious teachings against vaccination, rumours the vaccine had been contaminated by sterilising substances and reported deaths from taking the vaccine. It snowballed into a major vaccine revolt in July 2003 when two influential groups, the Supreme Council for Shari'ah in Nigeria (SCSN) and the Kaduna State Council of Imams and Ulamas, declared that the vaccine contained anti-fertility substances and was part of a Western conspiracy to sterilise Muslims (see Falade, 2015). Many northern states subsequently banned the distribution of the vaccine, fearing a backlash from their citizens.

The previous episode with Pfizer and ongoing wars in Afghanistan and Iraq added weight to the conspiracy theory of a plot to decimate Muslim populations worldwide. Opinion leaders in June 2004 eventually agreed to allow the campaign to resume: the death toll from polio had become unbearable and new findings led to conclusions that the polio vaccine-to-infertility rate may have been exaggerated. However, it was not until after a five-day immunisation tour of Egypt in 2007 that prominent Islamic scholars accepted the compatibility of vaccination with the Qur'an (Falade, 2015). Communicating the scientific approach to disease was not enough to change behaviour, religion had to accommodate science for lasting change.

4.3. Ebola in Nigeria

The campaign to prevent the transmission of Ebola to Nigeria from West Africa was led by the federal government through the Federal Ministry of Health. Posters (see Figure 26.2) were circulated in many local languages (Pidgin, Hausa, English, etc.) around the country, complementing radio and television announcements on symptoms and prevention. The virus was however transmitted to Nigeria by Patrick Sawyer, a Liberian-American who contracted the disease in Liberia before he flew into Nigeria (TheCable, 2015). Sawyer, who died of the disease, transmitted it to hospital staff; while some survived, others, including the consultant, Dr Ameyo Stella Adadevoh, died (more on Adadevoh below).

Like the polio vaccine controversy, the efforts to contain Ebola again raised the issue of a potential for conflict between science and religious beliefs in West Africa (Falade and Coultas, 2017). A Christian religious leader declined to postpone the church's annual convention, which attracted congregants from all over the world and told his pastors that those who had fasted for 100 days should have no fear of Ebola. A Muslim opinion said disease was with the 'permission of Allah'. When Nigeria was declared free of the virus by the World Health Organization, a pastor claimed that it was 'the hand of God on Nigeria' (Vanguard, 2014). However, some churches were active in the communication of prevention practices by suspending the practice of shaking hands during church services and serving Holy Communion in the mouth. Some churches also served as orphanages while others sourced medical equipment and consumables from abroad.

The epidemic also brought to public attention claims and counterclaims about local therapies. There were rumours the disease could be prevented by bathing in saltwater or eating bitter kola. Maurice Iwu, a professor of pharmacology, to whom the bitter kola claim was attributed, later clarified that he did not find a cure but findings from a 1999 research project showed that bitter kola can 'stop the replication of the virus' (Olawale, 2014). Another professor also claimed the virus could be cured using ewedu, a native vegetable (Vanguard, 2014).

Here again, as with polio, science communication was not sufficient. Religious beliefs and traditional medical practices about health also had to accommodate science.

4.4. HIV/AIDS

The first case of AIDS in Nigeria was reported in 1986 (FMH, 2014) and its prevalence increased exponentially until it peaked at 5.8 per cent in 2001 before progressively declining to 4.1 per cent in 2010. According to UNAIDS (2016), Nigeria has the second largest HIV epidemic in the world, with 3.4 million people living with HIV in 2014, including 380,000 children below the age of 14 (South Africa has the largest). A 2019 report titled 'Nigeria HIV/AIDS indicator and impact survey', however, shows that the prevalence of HIV among adults age 15–64 years is 1.5 per cent: 1.9 per cent among females and 1.1 per cent among males (NACA, 2019). Prevalence among children age 0–14 years is 0.2 per cent. The report, from a 2018 household survey, is regarded as the most comprehensive to date and involved 89,565 eligible households with 93.7 per cent response (NACA, 2019). HIV prevalence also varied by zone across Nigeria, with the highest prevalence in the South South Zone (3.1 per cent) and the lowest prevalence in the North West Zone (0.6 per cent). (Nigeria's 36 states and federal capital territory are divided into six zones: South South, South East, South West, North West, North Central and North East.)

As with Ebola, a Nigerian scientist also made claims for an effective treatment for HIV/AIDS. Dr Jeremiah Abalaka claimed to have developed a vaccine to prevent HIV infections. Before then, there were a plethora of claims in the media on instant cures for all manner of medical conditions such as AIDS, hypertension, diabetes, cancer, etc. paid for by herbalists and spiritualists. The difference with Abalaka's 1999/2000 claim was that he is a senior medical doctor and trained immunologist. The doctor became an instant celebrity with patronage from as high up as the military high command. One of the soldiers sent to him for treatment later said on national television that he was the only one of 30 referrals that survived (Ahmad, 2000). Abalaka's claims have not been independently verified using established testing methods and he has since been largely ignored by the media.

4.5. Malaria

Malaria is endemic in Nigeria and the country bears up to 25 per cent of the malaria disease burden in Africa (NMIS, 2015). It accounts for up to 11 per cent of maternal mortality, 25 per cent of infant mortality and 30 per cent of under-5 mortality. Added to this is person/hour losses for an average infection cycle to recovery, which may be up to a week.

Nigeria was one of the countries included in WHO's first large-scale multilateral initiative for malaria control between 1955 and 1969. The initiative, known as the Malaria Eradication Programme, relied on massive indoor spraying of dichlorodiphenyltrichloroethane (DDT). Although the goal was the complete eradication of malaria globally, it only succeeded in eliminating the disease from some regions, including southern Europe, the former USSR, and some countries of North Africa and the Middle East (Alilio et al., 2004). The use of DDT has been discontinued over environmental concerns. Also, the cheap antimalarial drug, Chloroquine, which in the past saved millions of lives, is no longer effective against new strains of malaria. The new drugs, combining artemisinin and its derivatives with other compounds, cost more than 10 times the price of Chloroquine, raising the cost beyond the reach of the poor (Gelband et al., 2004).

Science communication activities against malaria have thus focused on awareness campaigns on effective over-the-counter treatments and prevention strategies. Prevention strategies, which are often cheaper, target the mosquito that spread the malaria parasite by eliminating them using insecticides, or provide effective barriers with insecticide-treated nets, mosquito coils, etc. In recent years these public campaigns have seen the involvement of local and foreign NGOs, major international corporations and local artists and celebrities. The artists and celebrities feature in audiovisual messages.

5. Campaign for a clean environment

Nigeria's first significant environmental event was the Koko town toxic waste dump of June 1988. Prior to 1988, Nigeria responded to most environmental problems on an ad hoc basis (see Ogbodo, 2009). The attendant international outcry led to the creation of the Federal Environment Protection Agency in 1988, with responsibility for the administration and enforcement of environmental laws.

The second event was over crude oil extraction–related pollution. The Ken Saro Wiwa–led Movement for the Survival of the Ogoni People (MOSOP) had engaged in international awareness campaigns on the extent and dangers posed by oil exploration activities in the Niger Delta. MOSOP was brutally suppressed by the military, culminating in the killing of Ken Saro Wiwa and nine others in 1995 by the military government (see Boele et al., 2001). Former British Prime Minister John Major described the killing as 'judicial murder' (Rowell and Lubbers, 2010). Sadly, it is a reminder of the delicate balance between environmental concerns, public wealth in developing countries and politics. No major organised campaign against environmental issues has occurred since then in Nigeria—it appears that the effects of the silencing of MOSOP continue to affect the public.

Figure 26.2: Oil spill in the Niger Delta region.

Source: The Will Nigeria (thewillnigeria.com).

The federal government has, however, set up a committee to reconcile the Ogoni people. It commissioned the United Nations Environment Programme (UNEP) to do an environmental assessment study of the sites between 2009 and 2011. A stakeholder meeting was held in Geneva, Switzerland, in November 2014, and Nigeria's current president, Muhammadu Buhari, launched the clean-up of Ogoniland on 2 June 2016 in Bodo (Federal Ministry of Environment, n.d.).

6. The public and engagement with science

Examining public engagement with science in Nigeria's *Guardian* newspaper, Falade (2016) found that coverage increased in intensity over the period 2001–09 from 6.8 per cent to 8.3 per cent of total available space. More stories were written by science writers in 2009 (28 per cent) than in 2001 (18 per cent), and the paper devoted 7 per cent of science coverage to alternative/herbal medical practices. The broadcast media mirrored the general trend in terms of attention paid to the coverage of science. However, television stations tend to focus more on agriculture, environment, climate change and health. In this category are *Health Matters* on Channels Television, *The Environment* on NTA and *Our Environment* on Television Continental (TVC). *Agribiotech* is on TVC and focuses on Nigeria's emerging agricultural biotechnology sector. *The Voyage of Discovery,* which debuted on the NTA in 2006 as a 30-minute weekly show, specifically focuses on research. It later moved to Africa Independent Television. The program was conceived by the National Universities Commission to disseminate research findings using an easily digestible and entertaining media. Online, Nigerian issues feature frequently in the Africa section of SciDev.Net. SciDev.Net, the UK-based online portal (scidev.net), serves as an avenue for freelance journalists to publish science stories and get paid. Local websites devoted to science coverage such as AfricaSTI and EnvironNews are struggling from lack of funds.

A pilot public understanding of science survey (Falade and Bauer, 2018) using standard and modified questions from the Eurobarometer and National Science Foundation found that Nigerians have high levels of trust in scientists and religious leaders when compared with politicians, judiciary and the military. On the knowledge questions, issues where science and religion have similar interpretations produced interesting results. For example, 70 per cent of respondents agree 'father's gene decides sex of child' and 75 per cent also agree 'God decides sex of child'. Cross-tabulation shows that 82 per cent of respondents who agree it's 'Father's gene' are also happy it's 'God' (12 per cent disagree and 6 per cent don't know). When respondents were asked their first option for tackling health problems, 55 per cent selected Western medicine, 24 per cent prayers and 6 per cent traditional herbs. The second option for health was more revealing: 34 per cent selected prayers, 29 per cent Western medicine and 20 per cent traditional herbs. Only 12 per cent of the respondents chose Western medicine as first and second options. The results show that respondents consult science, religion and traditional medicine, albeit in different orderings. In addition, 84 per cent rated religion as playing a strong role in their lives, with 88 per cent believing in destiny.

7. Policy framework for science technology and innovation in Nigeria

7.1. Science education and communication

Nigeria's policy on science education, technology and innovation is predicated on its national aspirations. The report of the Vision 2010 Committee, announced in September 1997, sees science, engineering and technology as critical to the nation's development. The importance of science communication can be inferred from statements about the significance of health and environment, especially in the aftermath of the toxic waste dump in Nigeria from abroad. The National Policy on Education noted that radio and television educational broadcasting shall feature as part of the educational support service system (NERDC, 2004, 53). The section on science education stipulates that 'government shall popularise the study of the sciences and the production of adequate number of scientists to inspire and support national development' (p. 29). The National Mass Communication Policy (2004) seeks to promote Nigerian culture as the basis of creative expression and to facilitate the advancement of national unity, social co-existence, education, science and technology and the peaceful resolution of social problems and conflict.

The Federal Ministry of Science and Technology updated its National ST&I Policy in 2012.[1] Specifically, it seeks to popularise science through technology fairs, exhibitions, STI clubs and the mass media. Other strategies focused on human resources, biotechnology, research and innovation in natural (health) products, natural medicine and pharmaceutical research and energy, and ICT and emerging technologies such as nanotechnology. The country, it is argued, recognises the importance of STI to its development. However, the synergy among these policies and the extent to which science communication and popularisation components are implemented is not visible.

Perhaps it was in recognition of these shortcomings that the country came up with a new policy framework, the National Science, Technology and Innovation Roadmap (2017–30) (FMST, 2017). Nigeria's Minister of Science and Technology Ogbonnaya Onu said the new policy would coordinate and support the development of science and technology infrastructure for the country's socioeconomic development (FMST, 2018). Significantly, one of the new strategies is science literacy improvement and public/stakeholder engagement.

8. Science popularisation: Centres, museums and parks

8.1. Science centres and museums

Science centres, museums and parks lend themselves to the dissemination of scientific knowledge and the popularisation of science among lay publics. Cavalcanti and Persechini (2011) note the growing number and diversity of science museums in Brazil with scientists, educators, curators, journalists, etc. largely focusing on science popularisation. In comparison, Nigeria does not have a functioning science centre and a previous attempt by the Akwa Ibom state government, the Ibom Science Park, was abandoned in 2007. The state government, however, recently signed a Memorandum of Understanding to

1 The Federal Ministry of Science and Technology uses the term ST&I (science, technology and innovation) in policy statements. The term is sometimes abbreviated to 'STI' by other users. The terms are interchangeable.

complete the park (Dada, 2018). The first university-based science park was initiated at the University of Nigeria, Nsukka, in southeast Nigeria. The Lion Science Park project was inaugurated in July 2018 and is a partnership between the university, Ideon Science Park, scientists from Lund University in Sweden and LundavisionAB. Edward-Ekpu (2017) notes a recent effort to complete the natural history museum at the University of Ife, South West Nigeria.

8.2. Science in the public sphere

An initiative on Goal 17 of the United Nations Sustainable Development Goals for equitable access to science, technology and innovation took off in January 2018. The initiative involved a non-government organisation, Journalists for Social Development (Odogwu, 2018). The project seeks to provide a hangout for public dialogue on science and technology, a weekly radio and television talk show and a monthly science digest publication focusing on issues and developments in science and technology. The Science Communication Hub Nigeria also offers a platform where scientists, Nigerians and Africans engage with the public. The hub's website has articles, interviews, opinions and mentoring essays from scientists and provides a forum for meeting mentors, collaborators and other interested parties.

8.3. Technology Innovation Expo

Nigeria held its first Science, Technology and Innovation Expo in April 2017 in Abuja, Nigeria's capital territory. The Federal Ministry of Science and Technology (FMST, n.d.) said the purpose of the expo was to highlight the capacities of scientists, engineers and inventors in the country and to encourage researchers and investors to collaborate and market Nigeria's research results, inventions and innovations. The second edition was held in March 2018 and the government used the occasion to announce US$3 million support by the World Bank for emerging technological innovations and funding for setting up six technology and innovation hubs across the country (Soyombo, 2018).

9. Organisations engaged in communication activities

9.1. Federal Ministry of Science and Technology

The department charged with communication activities in the ministry oversees the formulation of policies and the promotion of science in rural areas and among women and children. It is also responsible for science fairs, exhibitions, workshops, conferences, mass media activities and the National

Science and Technology Museum (NSTM). The parastatals under the ministry, established for specific science and technology purposes, engage with the public directly. Such agencies include National Biotechnology Development Agency, National Centre for Technology Management, National Office for Technology Acquisition and Promotion and the Natural Medicine Development Agency. Nigeria's first science, technology and innovation policy was written in 2012.

9.2. Nigeria Academy of Science

The Academy is a leading self-accounting science body in Nigeria incorporated in 1986 but established earlier in 1977. Its mandate is to influence policies and strategies in science and encompasses the development and promotion of science, technology and innovation in Nigeria. The Academy's recent achievements include the development of a training manual for integrating research into policy and practice, accrediting agencies of the Ministry of Science and Technology, etc. The Academy publishes a journal, *The Proceedings of the Nigerian Academy of Science*, and hosts media roundtables and the NAS Science Media Awards. Instituted since 2010 for outstanding science reporting, the awards are intended to promote excellence in science and science-related journalism.

9.3. Science Association of Nigeria

The Association was formed in 1958 to provide a platform for Nigerian scientists to make their contributions to scientific and technological developments. Working in concert with UNESCO, the Association participated in the formulation of Nigeria's science policy. The *Nigeria Journal of Science* is published by the Association and reports original research outputs of scientists. The Association also organises annual science conferences.

9.4. NGOs in science communication

The Dr Ameyo Stella Adadevoh (DRASA) Health Trust[2] is a non-government organisation established in memory of the late Dr Adadevoh who treated Nigeria's first-ever Ebola patient, Patrick Sawyer, and contained the potential spread of infection, but lost her life in the process. DRASA's mission, according to Chief Executive Niniola Soleye[3] is to strengthen Nigeria's preparedness for future outbreaks. Health communication to drive behaviour-change is at the core of DRASA's work, and the approach is to get the public to understand 'why' and to dispel myths and rumours about infectious diseases. It uses several strategies

2 See drasatrust.org.

3 Personal communication, 19 August 2019.

to turn scientific evidence into digestible information tailored to the needs of different stakeholders. DRASA organises science communication activities for different groups: at-risk health workers; women as role models and influencers; secondary school students and local communities. DRASA trains health workers to understand why practising universal infection control precautions for all patients is key to protecting themselves and the wider society, and on the need to share data with government to inform future policies and guidelines. For these stakeholders, DRASA found that making infection control a personal issue is very effective: explaining how an infectious disease contracted from a patient while on the job can be carried into the home and potentially infect their family and community members. This strategy builds on the fact that self-preservation, a basic instinct, is as important as the natural impulse to protect those we love. They also run simulation programs to equip health workers for future outbreaks.

For communities, DRASA utilises a mix of strategies to ensure effective communication. One very important strategy uses community influencers, people who have been identified through the course of DRASA's work. For example, DRASA engages women as influencers because when their children, husbands or other relatives are ill, they tend to be decision-makers regarding whether to refer them to a health facility, or self-medicate, or practise traditional medicine. The NGO also has a Youth Ambassadors program where students are equipped and trained to drive positive health and hygiene-related behaviour change within their schools, families and wider communities. These individuals and groups are expected to propagate these messages to peers, family and other community members. DRASA is funded through donations and grants from the public. It receives no government funding for now.

10. Science communication education

10.1. Undergraduate education in science communication

Nigeria currently has about 158 universities, 120 polytechnics and other tertiary institutions recognised by the relevant government agencies offering numerous courses in science and technology. The University of Nigeria, Nsukka, and the University of Lagos commenced courses in communication studies in the 1960s under the title mass communication. Similar programs are now offered at universities and polytechnics nationwide with other titles such as communication arts, communication and language arts, media studies, communication studies and journalism studies. The establishment of the National Universities Commission, the regulator of tertiary education in Nigeria, led to the setting up of minimum benchmarks for the accreditation of courses.

There is no first-degree university program in science, health or environmental communication. However, an examination of the standards for undergraduate programs in Nigerian universities (NUC, 2011) shows a provision for a second-year, one-semester course in specialised reporting to cover these areas. It also provides for a third-year, one-semester course in the practice of writing popular science articles for magazines and newspapers. The emphasis is on the translation of scientific language, familiarisation with the literature of science and interviewing scientists. Some universities and colleges have increased the minimum standards by introducing additional credit units in health communication, development communication and environmental communication or journalism in the second, third or fourth year[4] to address the dynamics of science, technology and innovation in a digital age.

10.2. Postgraduate education in science communication

At the postgraduate level, the NUC (2011) guidelines provide for postgraduate diploma courses in specialised journalism, focusing on areas including environment, health and economics. However, the major focus is on reporting in areas such as sports, education, business and religion. For MSc and PhD programs, the NUC identifies areas of specialisation to include print journalism, broadcast journalism, advertising, public relations, behaviour change communication and health communication. In reality, students are awarded degrees in mass communication and they choose to develop their theses in any of these special areas. There is no postgraduate degree in science, health or environmental communication in Nigeria (Pate, 2018).

10.3. Review of communication programs

In an apparent recognition of the gaps in communication education in Nigeria, in September 2017, stakeholders (professional associations, regulatory bodies, tertiary institutions, civil society groups and United Nations agencies) convened at the Bayero University, Kano, to review the National Universities Commission draft Benchmark Minimum Academic Standards document (NUC, 2007). The conference recommended the study and award of bachelor's degrees in seven areas of communication studies: journalism and media studies, public relations, advertising, broadcasting, film and multimedia studies.

4 University of Uyo, Nigeria, Department of Communication Arts 2016 student handbook (in-house teaching manual).

10.4. Books, scholarly articles and conferences in Nigeria

Science communication programs in Nigeria have mostly relied on chapters in journalism textbooks that focus on science reporting or journalism. Curtis MacDougall's *Interpretative Reporting* is a prominent textbook in Nigeria (MacDougall and Reid, 1987). In the 1990s, the widely used reading material for science communication came through the effort of the African Council for Communication Education to develop books for African scholars and students. Kwame Boafo's edited module on specialised writing has three sections: business and finance; science, technology and health; and environment (Boafo, 1989). Nigerian authors have also developed science communication–related textbooks. Akinfeleye (1989) and Nwosu et al. (2008) focused on health communication and development, while Nwabueze (2007) and Soola et al. (2016) were on environmental communication. Wilson and Batta (2013) brought together the three main fields: science, health and environment. Notably the science communication section of Wilson and Batta included contributions from participants in Nigeria, India, China and Italy at the 12th Public Communication of Science and Technology (PCST) Conference in Florence, Italy, in 2012.

10.5. Scholarly articles on science communication by Nigerians

Articles on science communication in Nigeria are spread across high- and low-impact journals, open and pay-to-access journals, and others in non-indexed journals that are not online. Studies on the output of Nigerian scholars in science communication journals are yet to be conducted. A few examples here will show the spread of fields of research.

Soola (1988) examined agricultural communication and the African non-literate farmer; Olurundare (1988) assessed the role of science education in scientific literacy; Ehikhamenor (1990) examined informal scientific communication in Nigerian universities; Sanni et al. (2016) evaluated the quality of science, technology and innovation in Nigeria; Falade (2016) examined the role of religion in public understanding of and attitudes to science; Ekanem (2003) published communicating science information in a science-unfriendly environment.

In the field of health and environmental communication, Ajao and Ugwu (2011) studied the problems facing scientific medical information in Nigeria; Falade and Coultas (2017) studied the Ebola outbreak in west Africa; Nwabueze and Ekwughe (2014) analysed the coverage of Boko Haram's effects on the environment. Atinmo and Jimba (1998) conducted a longitudinal study of environmental reporting in Nigerian newspapers. Batta, Ashong and Bashir (2013) examined the coverage of climate change in Nigerian Newspapers; Batta (2012) examined newspaper information on traditional medicine in Nigeria; Falade (2016) studied the science content of a Nigerian newspaper; and Ashong and Batta (2013) evaluated the public communication of aesthetic genital surgery. The list is by no means exhaustive, and follows no particular order, but it does reflect the writings of Nigerian authors in science communication research.

11. Notable scientists and science journalists

11.1. Prominent Nigerian scientists

There are several notable Nigerian figures who have contributed in tremendous ways to the development of science communication, science education as well as science journalism. Peter Okebukola, professor of science education at the Lagos State University and former executive secretary of Nigeria National Universities Commission, the statutory body that oversees all the country's universities, specialises in higher education, science, computer and environmental education. He has published over 130 papers and books on science education and is on the editorial board of many national and international journals. He is also a member of several science popularisation committees and is the first African to win the prestigious UNESCO Kalinga Prize for the Communication of Science. He has also served on a number of international organisations including UNESCO, UNICEF, the World Bank and the UNDP.

A very recent face of science is Nigeria-born and educated Oluyinka Olutoye, MD, PhD, professor of pediatric surgery, Balfour College of Medicine and co-director of Texas Children Hospital, United States. Professor Olutoye specialises in foetal and neonatal surgery with specific interest in congenital diaphragmatic hernia and complex wounds. His pioneering surgery to remove a tumour from an unborn child at 23 weeks and restore the foetus to the mother's womb for delivery at 36 weeks drew worldwide attention and became a reference for scientists and health communication in Nigeria. Olutoye was born in Lagos and studied medicine at the Obafemi Awolowo University (OAU) Ile-Ife before moving to the United States.

Dr Elisabeth Rasekoala is a chemical engineer and founding president of the African Gong, the Pan African Network for the Popularization of Science & Technology and Science Communication. Rasekoala is a member of the African Union Commission Monitoring and Evaluation Committee on the

Science, Technology and Innovation Strategy for Africa (STISA – 2024). She is the founder and director of the UK-based African-Caribbean Network for Science & Technology, an NGO that focuses on race and gender equality science. She is a member of the UK Qualifications and Curriculum Authority, American Educational Research Association (AERA) and the European Educational Research Association (EERA). Rasekoala and the African Gong, along with other partners, have played prominent roles in the African Conference on Emerging Infectious Diseases and Biosecurity Series.

11.2. Science in the media and science journalists

The Nigerian media has many notable science journalists who have distinguished themselves in their chosen fields. Diran Onifade is one of Nigeria's most celebrated television journalists who worked for many years with the NTA. His most outstanding works were on unreported or under-reported diseases, and he was one of the first to write in-depth reports of the devastating impact of HIV/AIDS on Nigerian communities.

Toyosi Ogunleye, a former editor of *Sunday Punch* and now head of language services (West Africa) at BBC World Service, has won several awards for her reporting on health and environmental issues. One of her works focused on the impact of the gases released by a steel company on the health of the residents of a nearby estate in Lagos. It detailed the results of blood and urine tests and confirmed the presence of abnormally high doses of metals in their bodies.

Alex Abutu, founding president of the Nigeria Association of Science Journalists (NASJ) brought science writers from across Nigeria under one platform, before moving to become the secretary general of the African Federation of Science Journalists.

12. Science communication: Projecting to the future

Science communication (or mass communication of science) has an important role to play in overcoming the many developmental challenges facing the Nigerian state in agriculture, health, industry and environment. While progress has been made in many areas, there is room for improvement.

Improvements in STI indices is a worldwide marker of economic advancement, and there is no reason to expect Nigeria will be different. New policy implementation initiatives are important, going forward, to tackle the challenges of a developing economy faced with high levels of religious beliefs that may be antithetical to the spread of scientific ideas. We argue that if constraints to the diffusion of science are deliberately tackled by the relevant stakeholders—governments, religious leaders, science associations, science academics and institutes and civil society groups—modern science communication may yet emerge in Nigeria as a critical force in social and economic development.

However, science communication also needs to include debates about reducing the cost of treatment for malaria, HIV/AIDs and other diseases. This means facing up to established practices in the pharmaceutical industry and laws that protect them. We have to move from scaremongering statistics about how many have it, where they are, how they got it and when, to how to ensure access to cheap or free lifesaving medicine, where it is needed, when it is needed and ensuring availability for all those who need it. Currently, the cost of lifesaving medicine is beyond the reach of the world's poorest, the underinsured or those not insurable for 'pre-existing conditions', particularly in Africa. Modern drugs now treat HIV to the point where it is undetectable in the blood and the positive person can no longer transmit the virus (Rodger et al., 2016). The resurgence of measles in the United States, transmission of Ebola across Africa and the Zika virus disease to the Americas have shown that no country is free until all countries are free.

References

Adebayo, K. and Idowu, I. A. (2001). The aftermath of the withdrawal of the World Bank counterpart funding for the Ogun State Agricultural Development Programme (ADP) in Nigeria. *Journal of Sustainable Agriculture, 17*(2–3), 79–98. doi.org/10.1300/J064v17n02_09.

Ahmad, K. (2000). Public protests as Nigeria bans use of untested HIV vaccine. *The Lancet, 356*(9228), 493. doi.org/10.1016/S0140-6736(05)74166-0.

Ajao, O. G. and Ugwu, B. T. (2011). Problems facing scientific medical publications in Nigeria. *Journal of the West African College of Surgeons. January – March*, (1), 31–9.

Akinfeleye, A. (ed.) (1989). *Health communication and development.* Ibadan: Spectrum/Safari Books.

Alilio, M. S., Bygbjerg, I. C. and Breman, J. G. (2004). Are multilateral malaria research and control programs the most successful? Lessons from the past 100 years in Africa. *The American Journal of Tropical Medicine and Hygiene, 71*(2), 268–78. doi.org/10.4269/ajtmh.2004.71.268.

Ashong, C. and Batta, H. (2013). Sensationalising the female pudenda: An examination of public communication of aesthetic genital surgery. *Global Journal of Health Sciences*, *5*(2), 153–65. doi.org/10.5539/gjhs.v5n2p153.

Asogwa, S. C., Ugwoke, F. N. and Ozioko, F. E. (2015). ICT-based framework for improved food security in Nigeria. *West African Journal of Industrial and Academic Research*, *13*(1), 23–8.

Atinmo, M. I. and Jimba, S. W. (1998). Environmental reporting in Nigerian newspapers: A longitudinal study. *International Journal of Environmental Studies*, *55*(4), 333–43. doi.org/10.1080/00207239808711189.

Batta, H. E. (2012). Press coverage of traditional medical practice in Nigeria. *Journal of Communication*, *3*(2), 78–89.

Batta, H., Ashong, C. and Bashir, A. (2013). Press coverage of climate change issues in Nigeria and implications for public participation opportunities. *Journal of Sustainable Development*, *6*(2), 56–69. doi.org/10.5539/jsd.v6n2p56.

Boafo, S. K. (ed.). (1989). *Communication and culture: African perspectives*. Africa Church Information Service.

Boele, R., Fabig, H. and Wheeler, D. (2001). Shell, Nigeria and the Ogoni. A study in unsustainable development: I. The story of Shell, Nigeria and the Ogoni people – environment, economy, relationships: conflict and prospects for resolution. *Sustainable Development*, *9*(2), 74–86. doi.org/10.1002/sd.161.

Cavalcanti, C. and Persechini, P. (2011). Science museums and the popularisation of science in Brazil. *Field Actions Reports*. Special Issue 3.

Dada, B. T. (2018, 19 July). Gov. Emmanuel signs MOU with Australian firm on Ibom Science Park. *Leadership*. Retrieved from leadership.ng/2018/07/19/gov-emmanuel-signs-mou-with-australian-firm-on-ibom-science-park/.

Dawaki, S., Al-Mekhlafi, H. M., Ithoi, I., Ibrahim, J., Atroosh, W. M., Abdulsalam, A. M. and Ahmed, A. (2016). Is Nigeria winning the battle against malaria? Prevalence, risk factors and KAP assessment among Hausa communities in Kano State. *Malaria Journal*, *15*(1), 351. doi.org/10.1186/s12936-016-1394-3.

Edward-Ekpu, U. (2017). There are no science museums or centres in Nigeria and most of Africa. *SciTechAfrica*. Retrieved from scitechafrica.wordpress.com/2016/04/01/there-are-no-science-museums-or-centres-in-nigeria-and-most-of-africa/.

Ehikhamenor, F. (1990). Informal scientific communication in Nigerian universities. *Journal of the American Society for Information Science*, *41*(6), 419–26. doi.org/10.1002/(SICI)1097-4571(199009)41:6<419::AID-ASI4>3.0.CO;2-I.

Ekanem, I. A. (2003). Communicating science information in a science unfriendly environment: The experience of Nigeria. *Public Understanding of Science*, *12*(12), 203–9. doi.org/10.1177/09636625030122005.

Fadairo, O. S., Olutegbe, N. S. and Tijani, A. M. (2015). Attitude of crop farmers towards e-wallet platform of the Growth Enhancement Support Scheme for input delivery in Oke-Ogun area of Oyo state. *AgraInformatika: Journal of Agricultural Informatics, 6*(2), 62–71. doi.org/10.17700/jai.2015.6.2.184.

Falade, B. (2016). Genres of science news in the Nigerian press. *South African Journal of Science, 112*, 5–6. doi.org/10.17159/sajs.2016/a0157.

Falade, B. A. (2015). Familiarising science: A western conspiracy and the vaccination revolt in Northern Nigeria. *Papers in Social Representations, 24*, 3.1–3.24.

Falade, B. A. and Bauer, M. W. (2018). 'I have faith in science and in God': Common sense, cognitive *polyphasia* and attitudes to science in Nigeria. *Public Understanding of Science, 27*(1), 29–46. doi.org/10.1177/0963662517690293.

Falade, B. A. and Coultas C. J. (2017). Scientific and non-scientific information in the uptake of health information: The case of Ebola. *South Africa Journal of Science, 113*(7/8). doi.org/10.17159/sajs.2017/20160359.

Federal Ministry of Agriculture and Rural Development. (n.d.). *The Green Alternative*. Retrieved from fmard.gov.ng/green-alternative/.

Federal Ministry of Environment. (n.d.). The Ogoni Clean-up. Retrieved from www.environment.gov.ng/index.php/ogoni.

Federal Ministry of Health (FMH). (2014). *Integrated national guidelines for HIV prevention treatment and care*. National AIDS/STIs Control Programme. Retrieved from childrenandaids.org/sites/default/files/2017-05/Nigeria-Integrated-National-Guildlines-For-HIV-Prevention-treatment-and-care-2014.pdf.

Federal Ministry of Science and Technology (FMST). (2018). Science roadmap to facilitate rapid economic development – Onu. Retrieved from scienceandtech.gov.ng/2018/08/16/science-roadmap-to-facilitate-rapid-economic-development-dr-onu/.

Federal Ministry of Science and Technology (FMST). (2017). *National Science, Technology and Innovation Roadmap (NSTIR) 2030*. Retrieved from oer.covenantuniversity.edu.ng/sites/default/files/An%20Integrated%20Roadmap%202017-2030.pdf.

Federal Ministry of Science and Technology (FMST). (n.d.). Technology & Innovation Expo (5–9 March). Retrieved from www.scienceandtech.gov.ng/tech-expo.

Gelband, H., Panosian, C. B. and Arrow, K. J. (eds). (2004). *Saving lives, buying time: Economics of malaria drugs in an age of resistance*. National Academies Press. doi.org/10.17226/11017.

IITA. (2018, 24 June). African farmers get new help against cassava diseases: Nuru, their artificially intelligent assistant. Retrieved from www.iita.org/news-item/african-farmers-get-new-help-against-cassava-diseases-nuru-their-artificially-intelligent-assistant/.

Joseph-Raji, G. and Timmis, E. (2018). Nigeria biannual economic update: connecting to compete (English). Washington DC: World Bank Group. Retrieved from documents.worldbank.org/curated/en/769551524576691390/Nigeria-biannual-economic-update-connecting-to-compete.

MacDougall, C. D. and Reid, R. D. (1987). *Interpretative reporting.* Macmillan Publishing Company.

National Agency for the Control of Aids (NACA). (2019). Nigeria HIV/AIDS Indicator and Impact Survey, National Summary Sheet. Retrieved from www.naca.gov.ng/naiis-national-summary-sheet/.

National Bureau of Statistics (NBS). (2020). Nigeria population 2006 to 2016. Retrieved from www.nigerianstat.gov.ng/.

National Universities Commission (NUC). (2007). *Benchmark minimum academic standards for undergraduate programmes in Nigerian universities: Social Sciences.* Abuja: National Universities Commission.

National Universities Commission (NUC). (2011). *Benchmark minimum academic standards for undergraduate programmes in Nigerian universities: Social Sciences.* Abuja: National Universities Commission.

National Universities Commission (NUC). (2017). Draft of the Review of the National Universities Commission (NUC) Minimum Benchmark (BMAS) 2014 for Mass Communication Degree in Nigerian Universities. Retrieved from www.nuc.edu.ng/nuc-unbundles-mass-communication/.

NERDC. (2004). *Nigerian Education Research and Development Council, National Policy on Education* (4th edition). Lagos: NERDC Press.

NMIS. (2015). *Nigeria Malaria Indicator Survey.* Retrieved from dhsprogram.com/publications/publication-MIS20-MIS-Final-Reports.cfm/.

Nwabueze, C. (2007). *Environmental communication: Perspectives on green communication and information management.* Enugu: Daisy Press.

Nwabueze, C. and Ekwughe, V. (2014). Nigerian newspapers' coverage of the effect of Boko Haram activities on the environment. *Journal of African Media Studies, 6*(1), 71–89.

Nwosu, I. E, Soola, O. E. and Nwodu, L. C. (2008). *Communication for health and sustainable development in Nigeria.* Enugu: African Council for Communication Education, ACCE Nigeria.

Ogbodo, S. G. (2009). Environmental protection in Nigeria: Two decades after the Koko incident. *Annual Survey of International and Corporate Law, 15*(1), 2. Retrieved from digitalcommons.law.ggu.edu/annlsurvey/vol15/iss1/2/.

Ogundipe, S. (2014). Announce Ebola cure and go to jail, FDA Warns. *Vanguard.* Retrieved from www.vanguardngr.com/2014/09/announce-Ebola-cure-go-jail-fda-warns/.

Okonta, P. I. (2014). Ethics of clinical trials in Nigeria. *Nigerian Medical Journal: Journal of the Nigeria Medical Association, 55*(3), 188. doi.org/10.4103/0300-1652.132035.

Oladele, O. I. (2006). Multilinguality of farm broadcast and agricultural information access in Nigeria. *Nordic Journal of African Studies, 15*(2), 199–205.

Olawale, G. (2014, 22 August). Bitter kola, not a cure for Ebola virus – Iwu. *Vanguard.* Retrieved from www.vanguardngr.com/2014/08/bitter-kola-cure-ebola-virus-iwu/.

Olurundare, S. A. (1988). Scientific literacy in Nigeria: The role of science education. *International Journal of Science Education, 10*(2), 151–58. doi.org/10.1080/0950069880100203.

Omotayo, A., Chikwendu, O. D. and Adebayo, K. (2001). Two decades of World Bank assisted extension services in Nigeria: Lessons and challenges for the future. *The Journal of Agricultural Education and Extension, 7*(3), 143–52. doi.org/10.1080/13892240108438816.

Pate, U. (2018). Status of science and technology reporting in mass communication curriculum and training in Nigerian institutions. Presentation at the International Workshop on Science Reporting and Science Communication Curriculum Development at the National Press Centre, Abuja, 3–6 July.

Pfizer. (2009, 30 July). Pfizer, Kano State reach settlement of Trovan cases. Trovan fact sheet. Retrieved from www.pfizer.com/news/press-release/press-release-detail/pfizer_kano_state_reach_settlement_of_trovan_cases.

Rodger, A. J., Cambiano, V., Bruun, T., Vernazza, P., Collins, S., Van Lunzen, J. and Asboe, D. (2016). Sexual activity without condoms and risk of HIV transmission in serodifferent couples when the HIV-positive partner is using suppressive antiretroviral therapy. *JAMA, 316*(2), 171–81. doi.org/10.1001/jama.2016.5148.

Rowell, A. and Lubbers, E. (2010, 5 December). Ken Saro-Wiwa was framed, secret evidence shows. *The Independent.* Retrieved from www.independent.co.uk/news/world/africa/ken-saro-wiwa-was-framed-secret-evidence-shows-2151577.html.

Sanni, N., Oluwatope, O., Adeyeye, A. and Egbetokan, A. (2016). Evaluation of the quality of science, technology and innovation advice available to law makers in Nigeria. *Palgrave Communications 2,* Article number: 16095. doi.org/10.1057/palcomms.2016.95.

Sampathkumar, M. (2018, 16 January). Gates Foundation to pay $76 million of Nigeria's debt to help eradicate polio. The country already has one year without reported cases. *The Independent.* Retrieved from www.independent.co.uk/news/world/africa/bill-gates-nigeria-debt-pay-polio-eradication-foundation-money-pledge-a8163056.html.

Soola, E. O. (1988). Agricultural communication and the African non-literate farmer: The Nigeria experience. *African Media Review, 2*(2), 75–91.

Soola, E. O., Udoudo, A. and Ochonogor, C. I. (2016). *Issues and trends in environmental communication*. Oyo State, Nigeria: Kraft Books Limited.

Soyombo, D. (2018, 12 March). Nigerian government gets $3m from World Bank to create more innovation hubs. *TechpointAfrica*. Retrieved from techpoint.africa/2018/03/12/nigerian-government-innovation-hub/.

TheCable. (2015). Flashback: How Sawyer passed Ebola on to Dr. Ada Igonoh … and how she survived. Retrieved from www.thecable.ng/how-i-survived-Ebola-2.

UNAIDS. (2016). Investing in the AIDS response in Nigeria. Retrieved from www.unaids.org/en/resources/presscentre/featurestories/2016/february/20160210_Nigeria.

Vanguard. (2014, 23 October). Ebola – NMA calls for vigilance. Retrieved from www.vanguardngr.com/2014/10/Ebola-nma-calls-vigilance/.

WHO. (2016). Nigeria. Retrieved from www.who.int/countries/nga/en/.

Wilson, D. and Batta, H. (eds). (2013). *Science, Health, & Environmental Communication: Global Issues, & Local perspectives*. Ibadan: Ibadan University Press.

World Bank. (2012). Agricultural Development Projects in Nigeria. Retrieved from documents.worldbank.org/curated/en/658931468288625090/Agricultural-development-projects-in-Nigeria.

World Bank. (2018). World Bank Open Data. Retrieved from data.worldbank.org/country/nigeria?view=chart.

Timeline

Event	Name	Date	Comment
First interactive science centre established.	Ibom Science Park. Started 2007	2007	Ibom abandoned and yet to open. 2018: Lion Science Park
First national (or large regional) science festival.	STI Expo	3–7 April 2017	In Abuja, capital city of Nigeria
An association of science writers or journalists or communicators established.	Science Communicators Association of Nigeria (SCAN)	2005	In 2006, SCAN was succeeded by Nigerian Association of Science Journalists (NASJ)
First university courses to train science communicators.		1980	Science and technology reporting as electives in a university course
National government program to support science communication established.	The national ST&I policy	2012	

Event	Name	Date	Comment
First significant initiative or report on science communication.	The Basel Convention and the Barmako Convention	1988	Report into storage of 18,000 drums of hazardous waste in a small fishing village of Koko
National Science Week founded.	The National Festival	2001	Sponsored by Nigeria's oil industry
First significant radio programs on science.	*Agbe lere* (Yoruba)	1933	One of a number of programs
First significant TV programs on science.	A number of programs	1959	Programs on agriculture and health from 1959
First awards for scientists or journalists or others for science communication.	The Science Media Award	2010	Awarded by the Nigerian Academy of Science
Other significant events.	First science communication workshop involving journalists and teachers	3–6 July 2018	Nassarawa State University, SciDev.Net, Script and Robert Bosch Stiftung College workshop on science journalism and science communication curriculum
	Extension and communication in agriculture	1910	1910: Commenced Southern Nigeria 1912: Commenced Northern Nigeria
	International Institute of Tropical Agriculture (IITA) opened	1967	Located in Ibadan
	Global Polio Eradication Initiative	1988	2003–04: vaccine revolt in Nigeria
	Pfizer drug trials in Kano	1996	Experimental trial of trovafloxacin led to litigation and compensation
	Ebola virus disease	2014–16	A communication challenge
	HIV/AIDS Antiretroviral (ARV) guidelines released	2003	2014: HIV/AIDS: Consolidated guidelines
	Malaria control program with DDT	1955–69	2009–13: revised second strategic plan
	Koko toxic waste dump	1988	1990: MOSOP/Ogoni Bill of Rights

Contributors

Dr Bankole Falade is the South African Research Chair (SARChi) in Science Communication at the Centre for Research on Evaluation, Science and Technology (CREST), Stellenbosch University, South Africa.

Dr Herbert Batta is senior lecturer in the Department of Communication Arts, University of Uyo, Nigeria.

Diran Onifade is one of Nigeria's outstanding television journalists and now works as a consultant/trainer.

27

PAKISTAN

Changing landscape of science communication

Manzoor Hussain Soomro
and Khalil Raza

1. Background

Science and technology are linked with all aspects of society. Science has greatly advanced our understanding of the natural world and has brought countless creations of practical application and many meaningful advantages to the human species. The primary role of human society should be to promote and develop rational, logical ability and critical thinking amongst its citizens, so as to better understand the dynamics of nature, conserve it and benefit from technological developments on a sustainable basis. With regards to better understanding by society of scientific phenomena such as climate change, science communication plays an important role to raise awareness. Science journalism conveys scientific knowledge, concepts and processes to the public and policymakers, important for taking governing decisions. It bridges the gap between the scientific world and general public and exposes them to real world problems. However, the Islamic Republic of Pakistan lags far behind in science communication among the comity of forward-looking nations, even though the majority of its people are believers in Islam, a faith that encourages and strongly advocates education and learning 'from cradle to grave' for both men and women (Soomro and Tanveer, 2017).

Pakistan does not seem to have established the necessary framework to communicate science to its public effectively. The avenues for sharing of scientific thoughts and research outcomes among scientists, the general public and policymakers for decision-making are simply not enough. The fact is that interest in science is fizzling out among kids and the gap between science and society continues to widen in the country. Science communication and science journalism in Pakistan are still in their initial phases; Pakistanis are just getting familiar with the idea of 'science for all', but due to the interesting, engaging and magnetic nature of science, people are always keen to know more—and this has the potential to brighten the future of science and technology in Pakistan.

Pakistan is a developing country with a population of over 207 million (Ministry of Finance, 2018).

The country is blessed with enormous resources, human as well as natural, but its economic growth and development are not commensurate with its available natural endowment. Since its independence in 1947, Pakistan has gone through many serious challenges including geopolitical turbulence, internal instability, natural disasters and a changing political and democratic landscape. As a result, Pakistan today faces various development challenges ranging from a sluggish economy, illiteracy, water and food scarcity, an energy crisis and environmental issues.

The country's science and technology (S&T) is an emerging sector and has played an important role in the development of the country. Pakistan has been known for its exceptional and talented pool of scientists, engineers, doctors and technicians whenever they have been provided with a conducive environment in Pakistan or abroad. Becoming a nuclear power is an example. However, it is unfortunate that despite its demonstrated quality of human capital, Pakistan is underperforming in many sectors of its economy. The lack of consistent policies and weak governance have been the main challenges. As a result, the institutional framework to support science, technology and innovation (STI) has remained weak, and the talented people of the country have not been given the opportunity to shine.

There appears to be little recognition of the communication of science for the general public at the top, mainly because the basic understanding of the role of STI in economic growth has been somewhat lacking among the leadership. This situation can be attributed to the inability of the country's scientific community to educate its politicians and policymakers. We do not see concerted efforts on the part of government, aimed at changing the behaviour of the people at large. With the exception of the nuclear program, science institutions do not get adequate funding. (Krishna and Naim, 2005; Osama, Hassan and Chattha, 2015). Thus the country faces serious challenges in attracting the right talent and securing adequate funding, which has always been a challenge for researchers. Pakistan also lacks the conducive ecosystem and facilities needed for quality research in its universities and research institutes.

The Pakistani education system, like many developing countries, is influenced by sociocultural values, where questioning and inquiry are discouraged at schools as well as at home. There are arguments and counter-arguments on the relationship between science and religion: some people consider that science and religion are contradictory, while others believe in combining science with religion. At times, 'crackpots' denounce the globally accepted principles of biology, such as the theory of evolution. However, such perceptions exist in many developing and developed countries (Hoodhboy, 2016). In Pakistan, students are taught science by memorising some text rather than understanding the core scientific concepts, thus curtailing their cognitive abilities. Under such circumstances, it is difficult to develop and nurture scientific thinking and analytical skills (Soomro and Tanveer, 2017).

It is common for pseudoscientists and/or bureaucrats to become mainstream decision-makers or 'opinion builders' in Pakistani society; for example, premier institutions like Pakistan Science Foundation and Pakistan Council for Scientific and Industrial Research (PCSIR) have off and on been without a chief executive for long spells and looked after by bureaucrats with no science background. We also come across scams such as water-powered cars and doubtful treatment of human health ailments and dysfunctions. This is even though Pakistan is an Islamic state and functions in accordance with the Qur'an, a book of guidance with numerous indications and instructions to think critically and explore nature.

2. Historical perspectives

The 'Golden Age of Islam' in the 8th–13th centuries, emanating from the education and learning centres of the Middle East and well before the European renaissance, contributed significantly to modern science and technology and its communication (Faruqi, 2006). However, the need to communicate scientific knowledge to the public and to enlighten society about scientific discoveries in an organised way originated in the developed world. The developing world has the same communication needs, but they were only satisfied later, with the invasion of colonial powers in the developing regions. These include the Middle East, once cradle and bastion of education and knowledge. The debates around the potential benefit of science and technology in the public spheres emerged in post-colonial governments in the Indian subcontinent and Africa (Massimiano and Trench, 2008).

Science communication in present-day Pakistan originated well before Pakistan's creation and partition from India in 1947. Several scientific societies and publications began their work before independence. *Science* magazine was launched in the Urdu language in 1928 by Anjuman Tarraqi-e-Urdu in Delhi (Dawn News, 2011). It was the first popular science magazine in Urdu and, after independence, it continued publication from Karachi until 1956.

Pakistan recognised the importance of being able to communicate scientific concepts in local languages. In 1955 the Scientific Society of Pakistan was established to promote these discussions in the Urdu language. This society published Urdu journals *Science Bachoon Kay Liye* [Science for Children] and *Jadid Society* [Modern Society]; and their publications survived until 1972. The society used to organise annual science conferences where research papers were presented in Urdu and the entire proceedings were published in the Urdu language (Krishna and Naim, 2005).

The medium of instruction is of course a challenge when it comes to the teaching of science in Pakistan. It is commonly taught either in English or local languages, Urdu or Sindhi. Local languages are faced with the challenge of borrowing scientific terms from English. This can make it difficult to understand complex terms transcribed in Urdu and can lead young students to lose interest in science. One approach is to promote the national (Urdu) language in every walk of life including science. On the other hand, where science is taught in English, teachers at primary and secondary levels in public schools often do not have the capacity to employ English as a medium of instruction.

The Pakistan Association for the Advancement of Science (PAAS) was established in 1947 on the pattern of the British Association of Advancement of Science, for the promotion of science in Pakistan. Since its inception, PAAS has contributed towards communication and popularisation of science (International Science Council, 2007). PAAS provides a forum for scientific meetings, conferences and the publication of scientific papers for professionals, but not as much for public awareness. PAAS used to publish two journals, *Pakistan Journal of Science* and *Pakistan Journal of Scientific Research*, which were subsequently merged into the *Pakistan Journal of Science* (PJS) in 2008 (Pakistan Journal of Science, 2009). PAAS continues to organise an annual conference on science with a focus on key subjects ranging from health, agriculture, biological, physical to chemical sciences. It attracts over 300 participants each time and provides a space for interaction between the general public and scientists in the country. In 2018, PAAS held its 38th Annual Conference, sponsored by government agencies including the Higher Education Commission and Pakistan Science Foundation.

The PCSIR was established in 1953 to undertake and promote industrial and scientific research for the socioeconomic development of the country. Currently, it works under the Ministry of Science and Technology (MoST). PCSIR set up its publications branch in 1956 and launched a series of popular science publications, like *Science and Industry* and *Science Chronicles* in English, and in Urdu *Karwan-e–Science*, a popular science magazine (PCSIR, n.d.), all of which are not published anymore. PCSIR also launched a quarterly research journal, *Pakistan Journal of Scientific and Industrial Research*, in 1958, which continues to be published.

In 1974, a distinguished scientist of Pakistan, Professor Abdus Salam (who later won the Nobel Prize in Physics), proposed to establish an international forum so scientists from developing countries could interact with their peers from advanced countries. The Pakistan Atomic Energy Commission (PAEC) owned the idea and continues to hold the International Nathiagali Summer College (INSC), focusing on presentations of research by scientists (International Nathiagali Summer College, n.d.). In 2018, INSC held its 43rd event in the series. Since its inception, over 670 eminent scientists from abroad, including six Nobel Laureates, have participated in the summer college. The college has also facilitated the exchange of over 1,000 foreign scientists from 72 developing countries and benefited 7,000 Pakistani scientists.

The Sindh Science Society (SSS) was founded in 1971 to raise the public awareness of S&T through the local Sindhi language. SSS played a significant role in the promotion of science and provided numerous opportunities to the general public for interaction with the scientific community. SSS also published its monthly *Science Magazine* in Sindhi to disseminate the latest trends and developments in S&T. In the period 1976 to 1989, SSS faced some challenges and could not continue its monthly publication. It resumed publication in 1989 but ceased again in 1998.

Launched in 1981, the Urdu publication *Science Digest* continued until 2001 (Patairiya, 2006). Another popular science magazine, *Global Science*, was launched in 1998 and continued its paper-based edition until November 2016. The magazine could not sustain its print version but is now continuing with the web-based version. *Global Science* was Pakistan's most widely circulated Urdu-language magazine in S&T with an average monthly circulation of 4,500 copies.[1] Some of its special editions (including those that covered biotechnology reviews and the 2005 earthquake) peaked at 8,000 copies. It was a landmark achievement in terms of S&T journalism in Urdu.

1 See globalsciencemag.com.

During the course of history in Pakistan, a number of other popular science magazines and/or journals were launched for communication and dissemination of science to public. However, most of these magazines and journals have ceased to exist and only a couple of them have survived. This indicates that science journalism or communication has not taken off very well, despite the need, potential and available human capital in Pakistan.

3. The role of government and its policies

At the time of partition in 1947, Pakistan inherited a scanty infrastructure to begin its path towards socioeconomic development. Pakistan was an underdeveloped economy and many important sectors such as industry, transportation, trade and basic infrastructure were not sufficiently advanced, resulting in a very low standard of living. There were problems the new government had to address urgently, and the immediate challenge for Pakistan was the intense migration across the borders between India and Pakistan. The foremost priority for the newly established government was to provide food and shelter. Subsequently, the Government of Pakistan realised the significance of S&T for economic development and established a number of R&D institutions. Academic institutions and universities were also planned for provision of quality education in the country.

Liaquat Ali Khan, the first prime minister of Pakistan, undertook several initiatives to develop the S&T base of the country. He invited Professor Salimuzzaman Siddiqui, a renowned chemist of the Indian Subcontinent, to Pakistan by awarding him citizenship in 1951 (Dawn News, 2011). Professor Siddiqui was appointed as the first Science Advisor to the Government. He established PCSIR along with its 16 laboratories in different cities to support R&D and boost industrialisation. During the 1950s, Pakistan established numerous councils to carry out research and propose policy recommendations in emerging fields of S&T, such as medical, agriculture, nuclear, industry and forestry. This was the initial phase of institutionalisation of science and technology in the country, but formal development of policies for the S&T sector did not begin until the 1960s (Naim, 2001).

3.1. Science and technology policy development in Pakistan

In 1960, Professor Abdus Salam was entrusted by the then President Ayub Khan to formulate the National Science Commission (NSC) with a mandate to develop a plan of action for science in Pakistan. The NSC published its first report in 1960 with recommendations to establish R&D institutions and universities across the country (Naim, 2001). The NSC report served as a de facto first S&T policy in the early development phase of Pakistan. The key focus of the recommendations was to strengthen R&D organisations and universities with provision of adequate funding for research and human resource development. It emphasised developing a career service structure for science professionals and allocation of at least 2.5 per cent of the national budget for science (Osama, Hassan, & Chattha, 2015). These recommendations broadly covered promotion of scientific research but did not specifically address the public awareness of science or science communication.

During the 1960s, several important institutions were launched, and various councils were established to advise the government on policy matters related to defence, irrigation, flood control and housing research (Krishna and Naim, 2005). Pakistan saw phenomenal growth in agricultural productivity in the 1960s as the government embarked upon a plan to encourage farmers to use dwarf varieties of wheat. Transmission of improved technology to the farmers through an extension services program played an important role in the agricultural growth in Pakistan (Ahmed, Shah and Zahid, 2004). This had a profound impact on the national economy and is considered as the 'Green Revolution' for Pakistan, where agricultural production almost doubled during the 1960s and 1970s (Broughton, 2017). Over the years, the government deployed a number of extension services to encourage farmers to use new varieties of grains. A high-yielding cotton variety 'NIAB-78' was developed in 1983 by Pakistani scientists using nuclear technology, and an active extension program led to a revolution in cotton production and the agricultural economy of the country.

In 1972, Pakistan established MoST to plan, coordinate and direct efforts to ensure effective S&T governance and research programs. In 1973, the Pakistan Science Foundation (PSF) was established through an Act of Parliament as an autonomous organisation under MoST. PSF today serves as the premier funding agency for supporting scientific and technological research as well as for the promotion and popularisation of science in the country.

However, it was in 1984 that Pakistan adopted its first national S&T policy, after an extensive process of consultations. This policy was quite comprehensive and aimed to address challenges to upgrade the country's S&T landscape (Ministry of Science and Technology, 1984). For the first time in the history of Pakistan, S&T policy highlighted and provided guidelines to promote public awareness of science and technology. Chapter 8 of this policy provides clear guidelines for the promotion of public communication of S&T:

> Creation of widespread public awareness of the vital importance of science and technology is absolutely necessary if these are to be utilised as instruments for improving the quality of life of the people in Pakistan.

This policy placed special interest on inculcating widespread awareness of science in society, developing a scientific culture and emphasised increasing public awareness through radio, television, newspapers, popular science publications and the establishment of science museums and centres. However, not much has been done over the decades to implement the policy.

In the 1990s, the focus shifted towards developing and commercialising indigenous technologies. In 1994, the National Technology Policy and Technology Development Action Plan were adopted with a goal to commercialise research and boost industrial sectors.

In the first decade of the new century, the S&T sector underwent a phenomenal revival and made revolutionary progress during the military rule of General Pervez Musharraf. The budget for S&T was increased 60-fold and many programs and initiatives were launched to strengthen research and human resource capabilities, such as a free national digital library for universities, high-speed internet access and the provision of fully funded PhD scholarships to study abroad (Nature, 2009).

The University Grants Commission (UGC), set up in 1947 to provide grants to the universities, was entirely overhauled and transformed into the Higher Education Commission (HEC) in 2002. For the first time in Pakistan's history, universities began to receive substantial funding for research and infrastructure development. Funding for the universities increased more than 10-fold, empowering universities to undertake quality research comparable to international standards. At times this proceeded a little too fast and to some extent, beyond the absorption capacity of the universities, resulting in spillage of some resources (Osama, Hassan and Chattha, 2015).

Even though the S&T Policy of 1984 was never implemented as required, Pakistan adopted a new policy in 2012. It emphasised using STI as central pillars for socioeconomic progression of the country, but the S&T sector (other than defence research) remained low priority and under-funded. Subsequently, with the change of government, MoST launched a revised STI strategy 2014–18 for 'effective' implementation of the policy, but still the sector remained a low priority. Nevertheless, a new government in February 2019 has set up a task force on science and technology led by Professor Atta ur Rahman FRS (Fellow of the Royal Society), with the aim of revamping the S&T sector. Only time will tell what actions are planned and what outcomes can be expected.

It can be concluded, therefore, that historically successive governments undertook a number of initiatives to augment the S&T sector with the creation of new R&D organisations, universities and the formulation of STI policies. But all these measures have been unable to deliver socioeconomic benefits. Time and again, these policies have failed to create any significant impact because the framework for implementation of such policies lacks a conducive environment and requires consistency in the funding and implementation of such policies. Moreover, S&T has remained a rather low priority for successive governments over the last decade.

4. Institutional framework for science, technology and innovation

Pakistan has a three-tiered S&T ecosystem and overall the sector can be classified into three main divisions: 1) university-based S&T; 2) public sector; and 3) defence research. However, the success and progression of the S&T sector (except for the defence research) has been mainly linked to high-stature personalities rather than strong institutions.

The bulk of scientific research is conducted in public and some private universities as well as at R&D organisations, but none of them pursues any systematic science communication program. While universities in Pakistan do offer degrees in journalism, none offers any courses related to science journalism, although there is an elective course on environmental journalism (Higher Education Commission, 2013). The scope for promotion of public communication of science at universities has been very limited, though the government has recently allowed universities to launch public radio channels. Thus, several universities have launched FM radio stations, particularly for dissemination of agricultural and health services among adjacent communities (Express Tribune, 2015). Agricultural and animal husbandry universities also provide farmer advisory services.

MoST is the lead entity responsible for planning, coordinating and directing efforts to launch and carry out public sector S&T programs and projects aimed at Pakistan's economic development. One of MoST's principal aims is to build Pakistan's S&T capacity in the 21st century by ensuring effective S&T governance and enhancing the capacity of indigenous innovation systems.

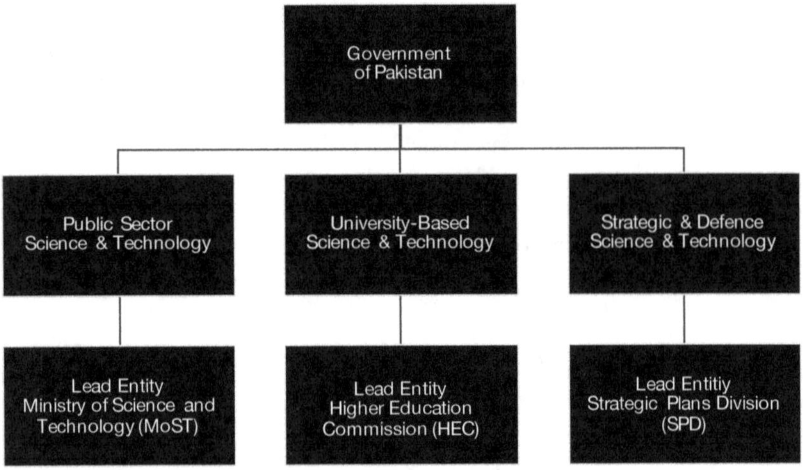

Figure 27.1: Pakistan's three-tiered S&T ecosystem

Source: After Osama, Hassan, and Chattha (2015).

4.1. Institutional support for popularisation of science in Pakistan

4.1.1. Pakistan Science Foundation

The Pakistan Science Foundation (PSF) was established in 1973 as an autonomous organisation under MoST. It is the apex body responsible for the funding and promotion of science education, research and communication in the country. The functions of PSF as laid down in its act through Article 4 (IV–VI) emphasise the promotion of public communication of science and technology, as follows:

i. The establishment of science centres, clubs, museums, herbaria and planetaria.
ii. The promotion of scientific societies, associations and academies engaged in spreading the cause of scientific knowledge in general or in the pursuit of a specific scientific discipline or technology in particular.
iii. The organisation of periodical science conferences, symposia and seminars.

PSF has established science societies, clubs, museums and S&T information centres for the communication of science to the scientific research community and the public. PSF organises and supports various conferences, science fairs and travelling expos, and training workshops and seminars throughout the country to engage the public and scientific community on emerging scientific and technological issues relevant to society.

In the wake of the 1984 S&T policy emphasising public awareness of science, PSF took several initiatives to popularise science and began its first science popularisation program in 1986 by hosting scientific film shows and a planetarium. Other PSF initiatives include the establishment of science caravans, science clubs in high schools, science film shows and popular science lectures.

4.1.2. Science caravans

To popularise science across the country especially in rural areas, PSF launched its Science Caravan program in 1988 (UNESCO, 2003). Up to 2019, nine science caravans have been established; they travel to far-flung areas of Pakistan and expose school children to some of the most fascinating scientific and technical concepts, processes and developments of modern science (Pakistan Science Foundation, 2017).

4.2. Financial support to scientific societies in Pakistan

Scientific societies work as non-government professional bodies for the promotion and development of scientific disciplines. Around 50 scientific societies are registered with PSF for financial support, but most of them provide forums for sharing scientific knowledge among their own communities through conferences, symposia, seminars and the publication of journals. Only occasionally do they hold open houses and science fairs or write popular articles for the general public. They do not normally interact with media but do send occasional press releases with research outcomes to raise awareness via wider dissemination to the public. As a consequence, Pakistanis are less aware of the potential of science and technology to address their local challenges.

4.2.1. National S&T fairs and travelling expos

In the 1990s, PSF organised a month-long mega National Science and Technology Fair every two years at the sports complex in Islamabad, where all R&D organisations, universities, defence R&D and production organisations and services as well as the manufacturing industry would join in. Around 1 million people, made up of the general public, students of all ages, families and the business community, visited those fairs each time. The last such fair was held in 1999; with the incident of 9/11 in 2001, the security situation changed quickly across the world and the mega fairs were stopped. However, in 2007, an indoor national S&T development fair was organised at the newly established Monument Museum Islamabad, which focused on S&T achievements of the past 60 years by Pakistan.[2]

2 Personal communication by senior author Soomro, who has been a member of the organising committee of science fairs in 1997, 1999 and 2007.

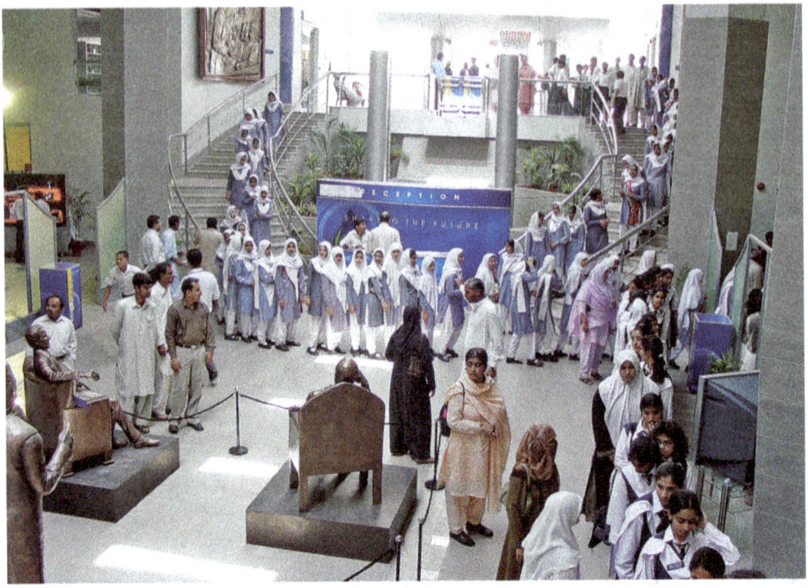

Figure 27.2: School kids during the Science & Technolgy Expo 2007: Shaping the Future.

Source: Manzoor Hussain Soomro, Pakistan Science Foundation (PSF).

Figure 27.3: Science and Technology Expo 2007: Shaping the Future.

Source: Manzoor Hussain Soomro, Pakistan Science Foundation (PSF).

Subsequently, PSF under the leadership of Professor Manzoor Hussain Soomro organised annual thematic travelling expositions on mathematics, water, environment, chemistry, biodiversity and energy, starting in 2008. They were held in 19 cities and towns across Pakistan and once in Afghanistan. These were visited by over 165,000 people (Pakistan Science Foundation, 2015). The expo series ended in 2013 as Soomro finished his services with PSF. The purpose of organising these travelling science expos was to stimulate public interest and illustrate the linkages between science, technology and economic/industrial development as well as the accomplishments of scientific discoveries.

4.2.2. Science centres

The PSF has a program for the establishment of science centres but its first and last science centre was set-up in 2000 through a public–private partnership at Faisalabad. This centre displays exhibits (only a few interactive) in the fields of natural history, biology, chemistry, physics and computer sciences, etc. with special emphasis on the application of science and technology in our daily lives. Subsequently, a proposal for setting up four more science centres was put to the government for funding but has not materialised yet.

4.2.3. Popular science lecture series

PSF as part of its science popularisation activities organises 'Popular Science Lectures'. These lectures address and highlight different scientific themes and issues of everyday life such as blood pressure and heart disease, floods, earthquakes, climate change, mosquito-borne diseases and viruses, etc. Eminent scholars, scientists, technologists and educationists are invited to lecture to diverse audiences.

4.2.4. Pakistan Museum of Natural History (PMNH)

Pakistan Museum of Natural History (PMNH) Islamabad, established in 1976 under the patronage of Pakistan Science Foundation, is mandated to provide current and historical specimens to researchers for their research and to raise awareness among students and the public, and to improve their understanding of the nation's natural history. PMNH plays an important role in public education to raise scientific awareness about fauna, flora and rocks and minerals, and shares the wonders of the natural treasures of Pakistan through dioramas, exhibits, lectures, workshops, seminars, posters and film shows (Rahim, 2016). PMNH currently holds nearly 1 million natural history specimens including a rare specimen of a whale shark, the skeleton of a blue whale and a real-size replica of one of the largest (5.5 metres high and 8 metres long) land mammals, the Baluchitherium.[3]

4.2.5. Pakistan Scientific and Technological Information Center (PASTIC)

PASTIC, like PMNH, is a subsidiary of PSF that mainly serves as an information repository of research by national R&D and S&T institutions and provides information services through modern IT tools and applications. PASTIC has undertaken initiatives to strengthen and promote the role of scientists and journalists, and to enhance interaction of the general public with science. PASTIC has organised training workshops and seminars for capacity building of journalists for effective communication of science to the public. PASTIC produced numerous science documentaries during the early 2000s, addressing local challenges and issues on environment, health, disaster risks and biotechnology (Pakistan Science Foundation, 2006). The documentaries were broadcast through local TV channels, but it all remained in 'project mode'.

4.2.6. National Museum of Science and Technology

The National Museum of Science and Technology (NMST) was established in Lahore, Pakistan, in 1965 by the Government of Pakistan in order to raise public awareness about science and technology and to inculcate interest among the masses. However, after the 18th constitutional amendment and devolution of numerous federal ministries in 2010, the NMST became a provincial institution with limited scope.

4.2.7. Urdu Science Board

Urdu Science Board, under the Federal Ministry of Information, Broadcasting and National Heritage, plays a vital role in promoting science and scientific culture in the country. It was established in 1962 as the Central Board for the Development of Urdu through a resolution to fulfil the constitutional requirement to develop Urdu as Pakistan's national language. It was renamed Urdu Science Board in 1984 with the objective of familiarising the masses with the new developments in the field of S&T across the globe. The board has so far published more than 800 books on different science subjects and a science encyclopedia in Urdu.

3 Baluchitherium (Paraceratherium) is an extinct rhinoceros that lived during the tertiary period about 20–30 million years ago. Its fossils were discovered in the Dera Bugti area of Balochistan province in the late 1990s by French and PMNH scientists.

4.2.8. Pakistan Academy of Sciences

Established in 1953, the Pakistan Academy of Sciences is a non-government scientific body of distinguished scientists in the country. The Government of Pakistan has given the consultative and advisory status to the academy on all national and international matters relating to the development of science and technology. It aims to promote science and technology in Pakistan, disseminate scientific knowledge and honour eminent scientists primarily through their election as fellows. It has played an important role in science popularisation and promotion of scientific research with its exchange programs with international scientific societies, academies and learned bodies. The academy also arranges popular science lectures, seminars, symposia, conferences and workshops at national and international levels.

5. Emerging trends in the promotion and popularisation of science in Pakistan

Recent trends are encouraging as there is an increasing level of realisation that advancement in STI is the way forward for Pakistan. In this context, the non-government entities and private sector are playing a lead role and taking a keen interest in promoting a culture of critical thinking and science in the country. These initiatives and private sector engagement are critical to raise the public awareness of science. With emergence of modern information tools and platforms, Pakistan has witnessed well-orchestrated and planned science popularisation activities throughout the country. Some of these encouraging trends and success stories on science communication are highlighted below.

Pakistan recently saw a very successful campaign for school education, called 'Alif Ailaan'. Launched in 2013 and finished in 2018, it was funded by the Department for International Development (DFID) of the UK Government (Alif Ailaan, 2017a). Its prime objective was to ensure that parents and communities demand good-quality education with a special focus on science subjects. They ran strong political and media advocacy activities for education at schools. The campaign published a three-volume document *Powering Pakistan for the 21st Century* (Alif Ailaan, 2017b), and recognised the significance of mathematics and science for the country's progress as well as for individual cognitive development.

After the Alif Ailaan finished, an education NGO called the Pakistan Alliance for Maths and Science (PAMS) was launched as its legacy. It is a collective effort of more than 70 civil society organisations, businesses and government bodies and individuals, aiming to provide Pakistani children with more and better opportunities to learn 21st century cognitive skills. PAMS and its member organisations during the past two years have had direct interactions with over 150,000 students across 30 districts of Pakistan (Pakistan Alliance for Maths and Science, 2018). These interactions have been through science fairs and festivals and other activities organised in government schools and Math-a-thon events organised in universities across Pakistan.

A number of scientific societies/associations (non-profit and for profit) have been established to further scientific culture in Pakistan's educational institutions and in the general public (Pakistan Alliance for Maths and Science, 2018).[4]

Khwarizmi Science Society (KSS) Lahore is one such example. It was established in 1997 to bring together scientists and scholars to work for the shared goal of making science accessible to the general public, especially children. Ever since, it has organised over 400 activities for the popularisation of science, including public lectures, symposiums, national workshops, the filming of scientific documentaries, science exhibitions and Melas (festivals). The Science Melas organised by the society are focused on simplifying rather complicated scientific phenomena to make them understandable to the general public. KSS inspired an enthusiastic young student to recreate Newton's telescope at the astronomical fair organised by the society (Farrell, 2011). The last Lahore Science Mela was organised at Lahore in October 2019.

Pakistan Science Club is also one of the emerging science clubs in Pakistan and is based in Karachi. It aims to promote a scientific research culture in society through interactive and innovative hands-on science activities, experiments and science fun events, as well as capacity-building workshops for children and their parents. In the wake of the Coronavirus pandemic (COVID-19), the club has begun 'Family Science Camps Online'.

Science Fuse Lahore is another interesting example, working as a social enterprise to promote science education among young learners. Science Fuse offers young students a unique opportunity to learn science by 'doing' at summer camps, extracurricular science workshops and after-school science clubs. Over the last few years, they have engaged almost 20,000 children through their informal science workshops and fun activities (Alveena, 2018).

4 Alif Ailaan, a nonprofit organisation working in the field of education in Pakistan since 2013, enlisted a number of emerging organisations dedicated to promoting scientific culture. Launched by a team of media and communications specialists, its program sought to highlight education on a priority basis in Pakistan and make the masses aware of the importance of education.

Similarly, a number of astronomical societies have been established to contribute towards the development of astronomy both as a science and as a fun activity for the public (Astronomy Without Borders, 2018).

With the emergence of the digital and online space, streaming services such as YouTube and online radios, are changing the course of broadcasting and telecasting. YouTube streaming has emerged as one of the critical disruptions for television. It is used as an effective tool for public lecturing and sharing popular science lectures. With the emergence of these technologies, Pakistanis are using digital applications to share news and popularise science.

The Eqbal Ahmad Centre for Public Education (EACPE) is an excellent online platform for the general public of Pakistan seeking to foster the use of science and reason to enable Pakistani citizens to participate fully in society as informed citizens (Razzaque, 2015). It runs an impressive series of online popular science articles and lectures by the prominent scientists and intellectuals of Pakistan covering various subjects of society, politics, culture, science and technology for the general public.[5]

Recently, many startups and initiatives have been spurred to seek out and leverage technology to promote scientific culture in Pakistan. One such example is an educational technology company called LearnOBots, launched in 2014 to promote science, technology, engineering and mathematics (STEM) education through robotics workshops and learning kits.

6. Science communication through media

Electronic, print and online media in Pakistan are vibrant sources of information. Television is considered to be the most popular medium for information and entertainment purposes, followed by radio and the print media. Almost all the media broadcasters have an online presence and web channels. Over the last two decades, the electronic media has made revolutionary progress with the emergence of cable and satellite TV networks (Baig and Cheema, 2015). Since then, the media industry has been considered one of the most influential forces in lawmaking and the political process in the country. However, science reporting or communication does not generally rank high among the priority areas for media outlets and newsroom hierarchies across the country. Pakistan has produced outstanding journalists in every field, including national politics, international affairs, entertainment and sports, but it would be very difficult to find a single 'true science journalist'

5 See eacpe.org.

in Pakistan. The obvious reason is that in the Pakistani media, whether print or electronic, journalists are usually obsessed with international and national politics and the demand for science journalism does not seem to be there. Those who are reporting science are not professional science journalists but rather general reporters.

Pakistan's media has a political bias where international and national political issues are given far more coverage than any other news. While countries of the developed world often have scientific issues like climate change high on the political agenda, it is often not part of political discourse in Pakistan and rarely discussed in the media. Pakistan is the seventh-most vulnerable country to climate change according to Global Climate Risk Index 2018 (Eckstein et al., 2020). Due to extreme weather events, thousands of people have been killed and millions have been displaced. Despite this grave situation, Pakistan's contribution to media coverage of climate change issues is rather negligible.

Sometimes English newspapers cover major science news or stories, such as 'Voyager 2 reaches interstellar space', but it is hard to discover such news in the Urdu media. There was hardly any coverage by Urdu channels on NASA's Voyager 2 when it entered interstellar space, despite this being one of the most remarkable events in the history of space travel.

There are over 83 TV channels in Pakistan, but none is dedicated to science and technology. Pakistan has produced only three popular science documentary series: *Ilm Kay Raastay* in 1980s, *Asrar-e-Jahan* in 1995 and *Bazm-e-Kainat* in 2003. The *Ilm Kay Raastay* program featured an interview of Professor Abdus Salam in late 1980s. All three documentary series were produced and developed by Professor Pervez Hoodbhoy, a Pakistani physicist who went on to receive the UNESCO Kalinga Prize for his efforts to increase public awareness of science and acquainting the public with the role of science in improving people's living conditions.

The lack of institutional support is considered to be a major contributing factor for underdeveloped science journalism (Ahmed, 2005). The business model of Pakistan's media outlets is such that they do not incentivise news reporting on science and technology. The universities in Pakistan do not offer any courses related to science journalism, which is one reason why science is never a topic of discussion among the public, or on electronic and social media. Scientists and researchers, both in public or private universities, are not given any incentives to share their scientific knowledge with the public for dialogue and policy making.

Radio plays a key role for public awareness of science in Pakistan, particularly in addressing agricultural and rural issues. Many rural farmers do not have an adequate access to communications technologies like TV or the web, but radio reaches a large rural population. Agricultural messages are broadcast at critical times of crop growth stages through radio channels year-round to educate the farming community about the latest and site-specific production technologies. Radio programs also include interviews with agricultural experts and progressive farmers.

The *Technology Times* is Pakistan's only newspaper devoted to the field of science and technology. On an average day, it receives about 80,000 hits. Their magazine is electronically distributed across Pakistan, particularly to universities and research institutes.

So far, only two Pakistanis have received the UNESCO Kalinga award for their exceptional work for increasing public awareness and popularisation of science: Misbah-Ud-Din Shami received his UNESCO Kalinga Prize in 1990 (UNESCO, 1990); and Pervez Amirali Hoodbhoy received his UNESCO Kalinga award in 2003 (UNESCO, 2003).

7. Conclusion

Science communication plays a significant role as it connects the general public with scientific arenas and helps them understand how science influences human lives. The country's science, technology and innovation fields are emerging sectors that have played an important role in the development of the country. However, the hard fact is that science and technology has never been a genuine priority for the Government of Pakistan, largely because of the many challenging issues the government has to deal with. Thus, science and technology have not really been successful in delivering social benefits to the country. At times, good policies for science and technology have been announced but the implementation of these policies has been weak and ineffectual.

Pakistan faces numerous development challenges ranging from a sluggish economy, poor literacy rate, water and food scarcity, to an energy crisis and environmental issues. The country must prioritise an adequate investment in STI to produce well-qualified and capable human capital to address these challenges. The country has to establish the necessary framework to communicate science to its public effectively to cultivate a science-literate society. With emerging digital technologies, the avenues for sharing scientific thoughts and research outcomes among scientists, the general public and policymakers will continue to grow. In this context, the institutional development of science communication would be critical to bridge the growing gap between science and society.

The fact is that science journalism or communication has not taken off, despite its potential, the existence of a valid demand and the availability of human capital in Pakistan. During the course of history in Pakistan, a number of popular science magazines and/or journals were launched but the majority of them have ceased to exist and only a couple of magazines have sustained publication. The national S&T policy adopted in 1984 highlighted and provided guidelines to promote public awareness of science and technology. It placed special emphasis on widespread awareness of science in society, developing scientific culture and emphasised increasing public awareness through radio, television, newspapers, popular science publications and the establishment of science museums and centres. The policy exists, but it needs to be implemented in its true spirit.

Overall, the Pakistani media including electronic and print are vibrant sources of information but science reporting does not rank high among the priority areas of media outlets and newsroom hierarchies across the country. However, with the emergence of online and digital platforms, many startups and non-government entities have spurred to leverage technology to promote scientific culture in Pakistan.

Over the decades, Pakistan has witnessed innovative and encouraging trends in the promotion of public awareness of science and technology. Momentous campaigns like Alif Ailaan, together with persistent efforts of various science societies and clubs, determined popular science publishers, and excellent and untiring efforts of some individuals, have created meaningful impact. However, in order to make a significant contribution to the public awareness of science and technology in Pakistan, the government needs to prioritise STI and focus on the promotion of public awareness of science and technology. There is a need for collective efforts and initiatives to bring positive change and help transform Pakistan (Soomro and Tanveer, 2017).

References

Ahmed, A. (2005, January 14). Barriers to science journalism in Pakistan. *SciDev.Net*. Retrieved from www.scidev.net/global/communication/opinion/barriers-to-science-journalism-in-pakistan.html.

Ahmed, I., Shah, A. S. and Zahid, M. (2004). Why the Green Revolution was short run phenomena in the development process of Pakistan: A lesson for future. *Journal of Rural Development & Administration*, *4*(35), 89–104. Retrieved from mpra.ub.uni-muenchen.de/9520/.

Alif Ailaan. (2017a). *Powering Pakistan for the 21st Century; Vol I of III: How Maths and Science Power Nations*. Pakistan Alliance for Maths and Science. Retrieved from www.mathsandscience.pk/wp-content/uploads/2017/01/How-Maths-Science-Power-Nations-1.pdf (site discontinued).

Alif Ailaan. (2017b). *Powering Pakistan for the 21st Century* (3 vols). Pakistan Alliance for Maths and Science. Retrieved from www.mathsandscience.pk/publications/ (site discontinued).

Alveena, J. (2018, 5 August). After more than 10 years abroad she moved back to Pakistan & is teaching Pakistani students all about science. *Mangobaaz*. Retrieved from www.mangobaaz.com/teaching-pakistani-students-lala-rukh-came-back.

Astronomy Without Borders. (2018). Lahore Astronomical Society. Retrieved from www.astronomerswithoutborders.org/my-awb/our-current-affiliates/network.html?id=523.

Baig, A. and Cheema, U. (2015). *Broadcast journalism in Pakistan: A hostage to media economics*. Islamabad: Centre for Peace and Development Studies.

Broughton, A. (2017, 8 September). The Green Revolution: Effects in Asia and implications for Africa. *Links International Journal of Socialist Renewal*. Retrieved from links.org.au/green-revolution-effects-asia-implications-africa.

Dawn News. (2011, 3 April). Anjuman Taraqqi-i-Urdu — the movement lives on. *Dawn News*. Retrieved from www.dawn.com/news/618112/anjuman-taraqqi-i-urdu-the-movement-lives-on.

Eckstein, D., Künzel, V. and Schäfer, L. (2020). *Global Climate Risk Index 2018*. Retreived from germanwatch.org/en/node/14987.

Express Tribune. (2015, 16 February). Lend your ear: New radio service takes healthcare to the people. *The Express Tribune*. Retrieved from tribune.com.pk/story/839329/lend-your-ear-new-radio-service-takes-healthcare-to-the-people/.

Farrell, J. (2011, 25 September). Pakistani recreates Newton's telescope. *Forbes*. Retrieved from www.forbes.com/sites/johnfarrell/2011/09/02/pakistani-recreates-newtons-telescope/#3e2c627b25c6.

Faruqi, Y. M. (2006). Contributions of Islamic scholars to the scientific enterprise. *International Education Journal, 7*(4), 391–99.

Higher Education Commission. (2013). *Curriculum of mass communication*. Islamabad: Higher Education Commission.

Hoodhboy, P. (2016). How Pakistan can develop a culture of science. Eqbal Ahmed Centre for Public Education. Retrieved from www.eacpe.org/content/uploads/2014/02/Developing-a-culture-of-science-in-Pakistan.pdf.

International Nathiagali Summer College. (n.d.). An Introduction. Retrieved from www.ncp.edu.pk/insc/index.php.

International Science Council. (2007). Pakistan Association for the Advancement of Science. Retrieved from council.science/member/pakistan-pakistan-association-for-the-advancement-of-science/.

Krishna, V. V. and Naim, S. T. K. (2005). *The Islamic Republic of Pakistan. A country report prepared for UNESCO*. Retrieved from citeseerx.ist.psu.edu/viewdoc/download?doi=10.1.1.494.2840&rep=rep1&type=pdf.

Massimiano, B. and Trench, B. (2008). *Handbook of Public Communication of Science and Technology*. New York: Routledge.

Ministry of Finance. (2018). *Economic Survey of Pakistan*. Islamabad: Government of Pakistan.

Ministry of Science and Technology. (1984). *National Science and Technology Policy*. Islamabad: Government of Pakistan.

Naim, S. (2001). Science and technology development in Pakistan. *Science, Technology & Society*, *6*(1) 97–132. doi.org/10.1177/097172180100600105.

Nature. (2009). Cash costs. *Nature*, 461, 11–12. doi.org/10.1038/461011b.

Osama, A., Hassan, S. and Chattha, K. (2015). *Pakistan science and innovation reviews*. Islamabad: Pakistan Innovation Foundation.

Pakistan Alliance for Maths and Science. (2018). *Capability Statement*. Islamabad: PAMS.

Pakistan Journal of Science. (n.d.). The Journal. Retrieved from www.paas-pk.com/SubPages/About/AboutUs.aspx?fcso=1&menuid=1001.

Pakistan Science Foundation. (2006). *PSF News*, *4*(3). Islamabad: Pakistan Science Foundation. Retrieved from www.psf.gov.pk/newsletters/july06.pdf.

Pakistan Science Foundation. (2015). International Traveling Expo on 'Water'. www.psf.gov.pk/expo_2013_2014.aspx.

Pakistan Science Foundation. (2017). Science caravans. www.psf.gov.pk/scienceCaravan.aspx.

Patairiya, M. (2006). *Science communication in South Asia: Challenges and prospects*. International Conference on Public Communication of Science and Technology, Seoul.

PCSIR. (n.d.). Scientific Information Centre, Karachi. Pakistan Council of Scientific and Industrial Research. Retrieved from www.pcsir.gov.pk/?pg=10208.

Rahim, H. (2016). *Role of Pakistan Museum of Natural History in imparting education in informal way*. International Conference on Earth Sciences Pakistan, Peshawar University, Peshawar.

Razzaque, A. (2015, 15 November). Science for the people. *The News*. Retrieved from www.thenews.com.pk/tns/detail/559791-science-for-the-people.

Soomro, M. H. and Tanveer, S. H. (2017). Ethics and education in Pakistan: Principles, policies and practice. In A. Battro, P. Lena, M. Sorondo, and J. V. Braun (eds), *Children and Sustainable Development- Ecological Education in a Globalized World* (pp. 385–96). Springer Nature. doi.org/10.1007/978-3-319-47130-3_29.

UNESCO. (1990). Kalinga Prize laureate. Retrieved from www.unesco.org/new/en/natural-sciences/science-technology/sti-policy/global-focus/science-popularization/prizes/kalinga-prize/kalinga-winners/kalinga-1990/.

UNESCO. (2003). Kalinga Prize laureate. Retrieved from www.unesco.org/new/en/natural-sciences/science-technology/sti-policy/global-focus/science-popularization/prizes/kalinga-prize/kalinga-winners/kalinga-2003/.

Timeline

Event	Name	Date	Comment
First interactive science centre established.	Faisalabad Science Centre	2000	The centre was established by Pakistan Science Foundation on a public–private partnership basis
First national (or large regional) science festival.	First was 1992 and the fourth was in 1999; all at Islamabad. Then the series ceased	17–30 Sept 1992	Organised by PSF and Ministry 'Science and Technology for Self-Reliance'. 2007: A smaller month-long National Science Fair, at the Monument Museum Islamabad
First National Conference in Science Communication.	'National Workshop on Science and Technology Reporting'	1991	Organised by the Ministry of Science and Technology, sponsored by Thomson Foundation and British Council
National government program to support science communication established.	Establishment of Science Media Cell in Pakistan Science Foundation (PSF)	2010	
First significant initiative or report on science communication.	Science Media Forum	2010	Organised by PSF

Event	Name	Date	Comment
National Science Week founded.	Proposal submitted to the government	None yet	10 November is World Science Day for Peace & Development, celebrated by students, teachers and scientists but not the general public
First significant radio programs on science.	*Tehqiq ki Duniya* [Research World]	Began late 1990s continued till 2004	2002–13: *Science ki Dunya* [Science World]
First significant TV programs on science.	*Bazm e Kainat* [Living in the Cosmos]	1994	Hosted by Pervez Amirali Hoodbhoy, of Quaid-e-Azam University, Islamabad 2000: *Asrar e Jahan* [Mysteries of the Universe]
First awards for scientists or journalists or others for science communication.	Dedicated awards for science journalists, by PSF	2017	Awards for scientists have been given ever since the establishment of Civil Awards in Pakistan
Other significant events.	A press briefing at the National Press Club in Islamabad	June 2017	Occasional press releases are issued by various organisations across Pakistan
	First museum established	1965	National Museum of Science and Technology, Lahore
	Kalinga Award to M. D. Shami	1990	2000: Kalinga Award to Pervez Hoodhboy

Contributors

Professor Dr Manzoor Hussain Soomro is the founder president of ECO Science Foundation (ECOSF), and was chair of the Pakistan Science Foundation (PSF).

Khalil Raza is the scientific officer at the ECO Science Foundation.

28

PHILIPPINES

From science then communication, to science communication

Garry Jay S. Montemayor, Mariechel J. Navarro and Kamila Isabelle A. Navarro

This chapter discusses the contexts surrounding the development of science communication (scicom) practices in the Philippines, tracing its history from the pre-Spanish colonial period onwards. Evidence shows that science was communicated in popular media during the Spanish colonisation period, which ended in 1898. Massive political and institutional changes happened during and after the American occupation, leading to the further development of science and scicom. As the Philippines is mostly an agricultural country, the practice of scicom was rooted in agricultural journalism, which might explain why initiatives until now focused more on popularising applied rather than theoretical science. As an academic track, scicom developed under the theory and practice of development communication, thus creating a distinct local style that is non-media centric and guided by participatory approaches. Case studies in scicom pertaining to agriculture, disaster management and health, as well as its challenges, are presented. The realisation of the role of scicom took a slow course, but innovative programs are now picking up after issues like Bt corn and eggplant, typhoon Haiyan and Dengvaxia triggered a greater appreciation of its value in everyday lives.

1. Introduction

The Philippines is a mere dot on the global map, a country of more than 7,600 islands in southeast Asia. Due to its abundant natural resources and biodiversity, it was a magnet for colonisation: by Spain, the United States and then by Japan. By the time the Philippines gained independence in 1946, colonialism had greatly influenced the institutional, cultural and general psyche of Filipinos.

It is in this environment that the development of science and technology (S&T) unfolded. Colonial policies from the 1700s to 1900s inhibited a relevant and nationalistic science community. Research and development (R&D) efforts focused on serving colonial objectives rather than the national interest. Instead of frontier research, investigations centred on taxonomy, discovering natural resources and identifying flora and fauna for by-products that would make a competitive mark in world trade. This perspective would influence the initial and even post-colonisation establishment of a national science agenda (Chamarik and Goonatilake, 1994).

Despite this initial dim scenario, the Philippine constitution in 1935 and the Science Act of 1958 articulated the role of S&T as a tool for national development. It took a series of reorganisations, downsizing of the bureaucracy and limited participation of the private sector to highlight government support to S&T. The role of scicom would thus take an even slower course, although various activities had already been started without being labelled as such. Once this field was given the attention it deserved, the appreciation for popularisation of S&T, knowledge-sharing initiatives, stakeholder engagement and the need to develop a core of science communicators through formal and innovative strategies were realised. These practical actions were accompanied by the emergence of an academic community from about 1985 that would offer courses in scicom. Its graduates would eventually influence minds not only in the Philippines but also in other countries in Africa and Asia and contribute to a global network of science communicators.

2. Brief pre-modern history of Philippine science communication

It is rather difficult to accurately recount scicom's history in the Philippines, as a unified, systematic study is yet to be done. There is no local academic journal title that focuses solely on scicom, although more recently articles about Philippine scicom have been published in various local and international communication journals.

It is clear that Filipinos during the pre-colonial times had deep knowledge of science and engineering, as evidenced by the richness of indigenous concepts and language in local astronomy (Ambrosio, 2010) and the design and creation of the Ifugao Rice Terraces (Conklin, 1980). Forms of scicom were present in pre-colonial times, with indigenous tribes passing down traditional knowledge of herbal remedies and agricultural practices. Some of this knowledge was written on perishable bamboo (Morrow, 2002), while other material was delivered through oral means, resulting in these traditions being overlooked and largely unrecorded until recently (Daoas, 1999). Some historical documents during the Spanish colonial period, however, indicate a more likely way science was communicated. Scientific advancements were led by priests of different religious orders (Velasco and Baens-Arcega, 1984; Anderson, 2007).

Later on, dissemination of agricultural information increased, especially during the time of Governor General Jose Basco y Vargas in the 1780s. During his term, the Sociedad Economica de los Amigos del Pais [Economic Societies of Friends of the Country] was formed as a private association of 'learned individuals' that aimed to improve agriculture in the country (Velasco and Baens-Arcega, 1984). This paved the way for communicating agricultural concepts and practices through technical and popular media. The Sociedad was also credited for establishing a comprehensive library and a museum of natural history (Anderson, 2007).

The final years of Spanish colonisation were marked by the establishment of science-related bureaus like the Manila Observatory in 1865, the founding of the University of Santo Tomas' Museum of Arts and Science in 1871 and advancements in medicine, such as anti-smallpox, cholera and leprosy campaigns (Velasco and Baens-Arcega, 1984; Anderson, 2007). At about this time, some vernacular newspapers like *El Ilocano* (founded in 1889) published science news articles. In its 28 June 1889 issue for example, an article featured topics including the origin of the earth, solar system, astronomy and geography. Some issues published articles about mathematics and proper counting (Montemayor, 2014).

By the American colonial period in the early 20th century, science had improved tremendously—many scholars were sent abroad for graduate studies through scholarships, and foreign grants and aid helped foster a stronger local science culture. In 1905, the Bureau of Science was established, evolving in 1987 into the current Department of Science and Technology (DOST) (Velasco and Baens-Arcega, 1984). Other bureaus, and eventually departments, were similarly set up for agriculture, health and natural resources. These departments would conduct their own R&D and have their own information and communication units.

3. Philippine science communication throughout the modern period

The advances made by the Philippines in science and its communication came tumbling down during the Japanese occupation in the 1940s. During this time, educational and scientific activities were practically halted. The capital Manila, the centre of such activities, was reduced to ruins during battles for liberation, virtually wiping out all previous research efforts and scientific collections (Caoili, 1987).

Come 1946, massive efforts were needed to reconstruct the newly independent Philippines, with the 1950s proving to be a tumultuous time for local scicom. Along with the massive efforts towards industrialisation mainly due to foreign aid, science and science education were seen as important drivers of socioeconomic progress. Thus, milestones occurred during the decade, such as the first Philippine National Science Week in 1951 (Official Gazette of the Philippines, 1951) and the creation of the now-defunct Science Foundation of the Philippines (SFP) in 1952,[1] a public corporation mandated to encourage the creation of school-based science clubs and societies. SFP was credited with starting the science club movement in 1956 (Antiola and Jose, 1982).

Scicom also existed in popular culture, like Teodorico Santos' alien invasion film *Exzur*[2] produced in 1956, which was one of the earliest, if not the first, sci-fi movies made in the Philippines (Santos, 2008a). It was around this period, however, that a report submitted by the Chairman of the Senate Committee on Scientific Advancement indicated the lack of science consciousness among Filipinos (Chamarik and Goonatilake, 1994). Since the nature of research then was basically a technology push rather than demand-driven, dissemination efforts were fragmented and few.

Science promotion and education thereafter became a priority area for S&T in the 1960s. The first science fair happened in 1960, and it became a national event in 1965 (Dagdayan, 1978). The first science quiz contest was spearheaded by SFP in 1961 and progressed into a national science quiz contest in 1969 (Reyes, 1978). To teach science using the vernacular, the Akademya ng Wikang Pilipino [Academy of the Filipino Language] started a

---

1 Republic Act No. 770, Congress of the Philippines, 20 June 1952.

2 Other notable classic local sci-fi movies include Richard Abelardo's *Zarex* (1958), about Filipino astronaut landing on the moon (Santos, 2008b); and also his *Tuko sa Madre Kakaw* (Gecko in Madre Kakaw, 1959), a story of a mad scientist who invented a serum turning animals into monsters (Santos, 2008c). Unfortunately, copies of these movies are now considered lost.

project to find words and/or translate science jargon into Filipino, the national language, for use in teaching science in schools. Their efforts resulted to the publication of an English–Filipino technical vocabulary dictionary in 1967.

It was also during the 1960s that experimental educational TV shows were commissioned through the Ateneo de Manila University's Educational Television Division (Rodrigo, 2006). Eventually, the drive to use mass media to educate youth through 'edutainment' initiatives paved the way for the classic 1980s children's program *Batibot* [Small, but Strong], which occasionally tackled science and maths subjects; and the mid-1990s hit *Sineskwela* [School on Air], a TV program intended to supplement the classroom-based elementary science education curriculum. The show reportedly reached about 14 million school children nationwide, with some research indicating that it resulted in improved science comprehension (Rodrigo, 2006). Many other child-oriented TV programs that tackled science topics followed later. Enhancing science education by training science teachers through TV programs was also initiated in the mid-1990s (Department of Education, 1997).

In terms of other popular media, local movies involving 'scientists' reached the height of their popularity during the 1970s and 1980s. Almost similar to the depictions of scientists in other countries, scientists in these films were usually portrayed as an evil expert, a mad intellectual, a helpless victim, a hermit prodigy, a foolish professor, a well-rounded genius or a heroic creator (Montemayor, 2013).[3]

The concept of development journalism, articulated in the late 1960s, was a response to Third World problems like poverty, low productivity and social inequality. This demanded writers go beyond mere reporting of facts to probing the 'depths of human drama'. The Philippine News Agency and the Philippine Press Institute had a team of writers concentrating on in-depth developmental news that covered, among others, population, agriculture, public health, environment and S&T (Jamias, 1987).

By the 1970s, science journalism had become a buzzword. Campus journalism became a regular component of SFP's annual Youth Science Camp Project since its first campus science journalism activity in 1971 (Ongoco, 1978). That same year the National Science Club of the Philippines, and the Science

3 Some examples of notable films that portray scientists as: (1) an evil expert: Ben Feleo's *Kalabog en Bosyo Strike Again* (1986); (2) a mad intellectual: Luciano Carlos' *Super Wan Tu Tri* (1985); (3) a helpless victim: Armando Garces' *Darna vs the Planet Women* (1975); (4) a hermit prodigy: Mike Relon Makiling's *Fly Me to the Moon* (1988); (5) a foolish professor: Tony Y. Reyes' *Fantastic Man* (2003); (6) a well-rounded genius: Ben Feleo's *The Crazy Professor* (1985); and (7) a heroic creator: *Bebong Osorio's Biokids* (1990).

Club Advisers Association of the Philippines, Inc. were formed (Dapul, 1978; Vergara, 1978). The first National Science Journalism Workshop for professional journalists, science club advisers and campus writers was implemented two years later, noting that science writing was a 'new Philippine frontier' (Bautista, n.d.). In 1976, the now-defunct Depthnews Science Service was launched in Manila. It provided weekly science features and radio scripts that would eventually reach about 250 newspapers and 300 radio stations in Asia and Pacific Islands (Amor et al., 1987).

Also worthy of mention is the role of UNESCO in accelerating science education in the country. For example, the SFP hosted the UNESCO-funded Asian Training Course for Leaders in the Promotion of Public Understanding of Science, Technology, and Environment (PUSTE) from 15 February to 14 March 1977. The training gathered 26 Asian leaders from 10 Asian countries to promote and 'institutionalise PUSTE through their out-of-school science, technology and environmental education (OSSTEE) program' (Science Foundation of the Philippines, 1978, Foreword). Two of the most important outputs of the training were: (1) the development of a framework for the promotion of PUSTE; and (2) the establishment of the Asian Coordinating Council on PUSTE.

The Annual DOST Media Awards commenced in 1988, giving recognition to practitioners from radio, television, print and, more recently, the cyber press. Other R&D institutions then started giving their own scicom awards. By mid-2001, the Philippine Science Journalists Association (PSciJourn) was formally organised as a non-profit association of media practitioners aiming to 'provide a dedicated network of people who realise the socio-economic transforming power of S&T'. It aimed to support the government's effort in developing an informed citizenry (Bautista, n.d.). Proclamation No. 437, s. 2003 (Official Gazette of the Philippines, 2003) declared every third week of July as Science Journalism Week. PSciJourn is mandated to conduct activities during this celebration.

Science centres were established by the government in Metro Manila. These include the National Planetarium in 1975, which also features astronomical myths and beliefs, and the Philippine Science Heritage Center in 1998, which details the history of local science. In response to the government's call for private sector support in science promotion, the Philippines' first interactive science centre—the Philippine Science Centrum—was established in 1984 in Marikina.

At this point, scicom was more unidirectional, reflecting the view that S&T information needed to be packaged in a form to encourage its appreciation and understanding. This would eventually progress into a more proactive process where communication had a significant role in creating an environment that enabled knowledge sharing and public engagement in an open and transparent dialogue.

4. The current state of science communication

Velasco (1998) reported that DOST stakeholders and communication practitioners felt that public appreciation of S&T was lacking, suggesting that science consciousness was low and bordering on apathy. However, the advent of biotechnology just before the start of the 2000s changed this scenario. Suddenly, a mere option in a scientist's toolbox generated diverse scientific, political, cultural and even religious viewpoints rendering it, as Liakopoulos (2002) noted, more of a social issue than a technological development. Projects were initiated and the hiring of science communicators commenced as funding support was made available by government, private sector and non-government organisations. A 2006 study indicated a shift in public awareness of the technology and increased use and access of information sources (Torres et al., 2006). During this period, an increase in Bt corn planting was documented.

Likewise, institutions started to adopt innovative methods in communicating science. Participative approaches like Cafe Scientifique were popularised in the Philippines by the Mind Museum in 2011. To further widen the scope of scicom, DOST initiatives like the Science Explorer Bus and NuLab (both are mobile science laboratories); and STARBOOKS (Science and Technology Academic and Research-Based Openly Operated Kiosks), a set of stand-alone digital S&T libraries, were deployed to geographically remote and economically disadvantaged areas with limited or no access to the internet. Indie-Siyensya, a science filmmaking contest, was initiated by DOST in 2016. Although all sub-agencies under DOST have their own scicom initiatives, the National Science and Technology Week (NSTW) celebration serves as a venue to showcase S&T through innovative scicom methods (Figure 28.1). The Department of Health's (DOH) various health campaigns, the Department of Agriculture's (DA) initiatives in communicating agriculture-related information to different stakeholders, and the Department of Environment and Natural Resources' (DENR) citizen science initiatives also regularly contribute to local scicom.

Given the country's susceptibility to natural disasters, weather and climate science topics have become increasingly prominent in various forms of the media, especially the internet. In 2009, the local broadcasting network GMA launched the nation's first dedicated science and technology online news section. Around this time, sub-agencies of the DOST, specifically those pertaining to disaster preparedness like PAGASA (Philippine Atmospheric, Geophysical and Astronomical Services Administration) and PHIVOLCS (Philippine Institute of Volcanology and Seismology), created popular pages on social media platforms for the quick and easy dissemination of important weather updates and safety information.

To combat the low profile of the DOST and science in the national consciousness, DOSTv, a weekday television program shown on the state-owned television channel, was launched in 2017. The program includes weather reports, interviews with scientists and officials, local and international science news and even S&T trivia (Burgos, 2017).[4] Homegrown resources like FlipScience, which prides itself as the first Filipino-made popular science website, have also materialised. Finally, individual science advocates, ranging from university-age science enthusiasts to established scientists and physicians, have built up a significant digital presence, enabling the unprecedented reach of S&T especially among younger Filipinos. A recent trend of fusing science and art also opens endless possibilities as a useful platform for public discourse.

Although the prominence of scicom efforts is at an all-time high, it is too early to tell if these have successfully embedded a culture of science in the national consciousness. However, as described later in the chapter, history has shown that initiating conversations on scientific topics can result in actionable change. Indeed, the ongoing story of scicom is an engrossing one, with a myriad of successes and even setbacks contributing to a robust scicom environment.

5. Public attitudes towards science and technology

Different nationwide surveys that aim to find the public's attitudes toward science and technology, such as that of the International Social Survey Programme (ISSP) and World Values Survey (WVS), have revealed a very

4 ABS-CBN's Knowledge Channel continues to air children-oriented science and mathematics programs, such as *Science Says* and *Sineskwela* re-runs. In July 2019, DOST launched *Siyensikat* over GMA News TV showcasing DOST-developed technologies.

interesting pattern: Filipinos' perception leans toward the negative side when asked about 'science' alone, but leans toward the positive one when asked about 'technology'.

For example, five ISSP surveys between 1993 and 2010 showed that there is an increasing trend in the number of Filipinos who agreed in the statements 'Overall, modern science does more harm than good', and 'We believe (trust) too often (too much) in science, and not enough in feelings and faith (religious faith)'. Respondents are becoming increasingly divided in their opinions in both statements.

The WVS survey in 2012 showed the same pattern. More Filipinos agreed than disagreed with the statements 'One of the bad effects of science is that it breaks down people's ideas of right and wrong', 'We depend too much on science and not enough on faith', 'Whenever science and religion conflict, religion is always right', and 'It is not important for me to know about science in my daily life'.

The general perception changes when Filipinos are asked about technology. Different WVS surveys revealed this finding. For example, those who believe that 'the scientific advancements we are making will help mankind' have increased by 50 per cent from 1996 to 2001. More Filipinos believe that it is good to 'give more emphasis on the development of technology' in the future, based on surveys done in 1996, 2001 and 2012. In 2012, the survey found that more Filipinos think that the world is better off because of technology than those who believe otherwise. More Filipinos agreed than disagreed in the statements 'Because of science and technology, there will be more opportunities in the next generation', and 'Science and technology are making our lives healthier, easier, and more comfortable'. Interestingly, two ISSP surveys found that more than 60 per cent of Filipinos feel proud about the country's scientific and technological achievements, and that the number has increased from 1995 to 2003.

Although these surveys seem to suggest that Filipinos are sceptical about science (e.g. Mangahas, 2018), the generally positive attitude towards technology shows their appreciation for more tangible outputs, or applications, of science.

6. Science communication as an academic discipline

Scicom as a local academic track is rooted in agricultural journalism, developed under the field of development communication (devcom). This enabled the flourishing of a distinct brand of scicom—one that is focused on applied science (particularly agriculture), which places a premium on the role of information science in scicom; is non-media centric; and is guided by participatory and public engagement approaches in communicating science to non-expert publics.

In 1954, the Office of Extension and Publication (OEP) was established in the then University of the Philippines College of Agriculture's (UPCA) Department of Agricultural Education in Los Baños, Laguna. OEP was mandated to help communicate and 'popularise' research results in agriculture to farmers. This office—where Nora Cruz-Quebral, the scholar who first defined devcom as a field, worked as a copyeditor upon earning her bachelor's degree—regularly published bulletins, pamphlets and brochures, and sent science news to major national broadsheets. Professor Juan F. Jamias, regarded as one of the earliest post-war agricultural journalists, worked here alongside Quebral. This experience of interfacing with non-expert publics, as well as the need to formalise communication training for agriculture students, put the OEP in the right position to offer a related academic course. The first course in agricultural communication was instituted in 1960, and a master's degree in agricultural communication—the first in the Philippines—in 1965. The OEP became a separate department, and after Quebral articulated 'development communication' as a field in 1971, it was renamed the Department of Development Communication in 1974 (Figure 28.2). In 1972, UPCA grew into an autonomous university, UP Los Baños (UPLB).

In the early 1970s, Professor Jamias articulated the idea of establishing scicom as a formal field of study within the context of devcom. A formal course on science reporting, essentially focusing on print, was instituted in 1974, although the course contents had already existed before under a different course title.

Although communicating research results is in itself a form of scicom, the scholarly roots of scicom 'UPLB-style' focused more on the implications of emerging information and communication technologies in information processing and access (Jamias, 1989). As early as 1983, the department formally discussed developing a scicom program (Librero, 2000; Jamias, 1984). Jamias initially conceptualised what he first called 'scientific communication' to refer to the communication of scientific and technical information, the idea of which was mainly based on his experiences in information technology as a visiting professor at the University of Sydney in 1984 (Jamias, 1984).

Impressed with what he saw, he went back to the Philippines in 1985, and proposed 'scientific communication' as an academic major in development communication, with a strong leaning toward information and library science.

Felix Librero, then department chair, strongly supported this idea. Together, they pushed for scicom to be included in the development communication curriculum. Although this idea was received favourably, the time was not yet ripe to proceed with the plan.

The first formal graduate course in scicom was instituted in 1985 (Jamias, 1997), and several students who enrolled in devcom eventually introduced scicom in their own university's communication courses. These universities include Xavier University in Northern Mindanao and Visayas State University in the Eastern Visayas, both of which had initiated fully fledged devcom undergraduate programs by 1986 (Visayas State University, 2016; Xavier University Development Communication Department, 2017). Other institutions have also instituted their own science journalism courses.[5]

Although the proposal for a scicom department was dropped in the mid-1980s, the department—now the Institute of Development Communication—continued the efforts to formally recognise scicom, beyond agricultural communication, as a distinct discipline. In 1988, Jamias proposed a scicom program with a threefold aim to: (1) promote science literacy; (2) promote science for human development; and (3) promote public understanding of science (Jamias, 1988). Although initial proposals gave emphasis to the agricultural sciences, it was clear in later proposals that scicom should also deal with other branches of science. The mid-1980s to mid-1990s proved to be a fruitful time to discuss scicom in devcom, as three scicom workshops happened in this decade. On 11–12 April 1985, a department-wide workshop on scientific communication was held. On 11–12 July 1989, a UNESCO-funded workshop explored the institution of a national academic program in science information. On 10–11 November 1993, a national scicom conference was held to 'flesh out the conceptual foundations for the science communication program' program' (Gomez, 1993, p. 3). This conference resulted in the establishment of the now-defunct Philippine Foundation of Science Communicators, Inc. (Montemayor, 2018).

During the early 1990s, several undergraduate courses in scicom were institutionalised, paving the way for the first undergraduate devcom batches to have a major degree in science communication. When the institute became an autonomous college in 1998, the Department of Science Communication was established, and the College of Development Communication (CDC)

5 For example, the then Asian Institute of Journalism (AIJ) created a course guide in science journalism in 1983. Velasco (1998) noted that, along with AIJ, many other universities—most of them located in Manila—already had scicom-related undergraduate courses during the time of her study, some of them have instituted scicom-related courses as early as the 1980s.

became the first to offer a clear scicom academic track among all universities in the country (Velasco, 1998). Since then, the CDC's Department of Science Communication has organised roundtable discussions on scicom, institutionalised additional scicom courses and collaborated with institutions in scicom activities. In a bid to further increase the capacity for S&T communication, DOST partnered with UPLB in 2016 to launch the first off-campus scholarship in Master of Science in Development Communication for its communication specialists.

CDC has recently opted to have a generalist orientation in its undergraduate curriculum, with all students taking scicom courses. Although other higher education institutions had incorporated scicom in some of their communication courses years ago, it was only in 2017 that the Commission on Higher Education (CHED) required all higher academic institutions offering devcom to include 'Risk, Disaster, and Humanitarian Communication' in their curriculum, and to add at least two scicom-related courses in their respective units, as mandated in CHED Memorandum Order No. 36, s. 2017 (Commission on Higher Education, 2017). The courses resemble the scicom courses in UPLB, where theory and skills on communicating science using participatory approaches and various media (including online platforms) to four publics—scientists, technicians, policymakers, and the general non-expert public—are emphasised.

Other universities are developing their own scicom-related courses beyond the scope of devcom, by following a transdisciplinary format. For example, Ateneo de Manila University's life sciences program has a new communications track that trains students to effectively engage the public in biological discussions across multimedia platforms. Its revamped communication program is offering 'Science and Risk Communication' as a core course, as well as an elective integrating philosophy and sociology to critically appraise the field of scicom. The De La Salle University is developing a minor in science communication for its organisational communication program, while the University of the Philippines Diliman campus is offering classes in science journalism and hopes to soon offer a corresponding undergraduate degree program (Dimacali, 2017).

While a majority of these newer initiatives are located within Manila, the strategic placement of devcom programs across the Philippines' three major islands assures the growth of scicom scholarship. They now contribute to a growing body of research on diverse issues relevant to Philippine society, like new strategies for the communication of health and environmental topics (including emerging tropical diseases, climate change and disasters) and evaluations of the effectiveness of these scicom initiatives.

Figure 28.1: Gabriela party representatives lead Dengvaxia-vaccinated children and their parents in a protest in front of the Department of Health's Manila office.

Source: Christian Yamzon, Defend Job Philippines (used with permission).

7. Case studies in Philippine science communication

7.1. Agriculture: Biotech crops

The introduction of the first biotech crop, Bt corn, in the Philippines in 2002 was a contentious process. Technological issues merged with religious, political, social and cultural issues, resulting in years of drama—like the uprooting of field trials, hunger strikes, boycotts and fearmongering in the media. These events were a baptism of fire for the science community, science communicators and government officials who had to deal with a diverse group of stakeholders from civil society groups, priests and nuns, and even politicians who all had something to say on the topic.

A multi-sectoral coalition of biotech advocates was set-up: scientists were suddenly asked to articulate the benefits to farmers in public fora; communications people were briefed on crop biology to respond to safety issues; and even bishops and priests were engaged in dialogue to understand their ethical concerns. This unfolding drama galvanised the combined efforts of the scientific community, science communication staff, media practitioners and other stakeholders to draw up a strategy to address a myriad of concerns and information requirements (Navarro et al., 2007; Panopio and Navarro, 2011).

Many initiatives were conducted by various government (DA, DOST, and UPLB) and non-government agencies (Biotechnology Information Center), farmers groups and private groups. These included dialogues with different stakeholders, creating field champions from the farming and local sectors and training the academic and scientific community to be actively involved in public briefings and engagement. Institutes inked a memorandum of understanding with PSciJourn to help establish a core of science writers from the different media platforms. Intensive media workshops, field visits and arranged interviews with scientists, farmers and local leaders were organised. Core experts who agreed to be key sources of information were trained on how to communicate for the layperson and engage in meaningful dialogue. Information briefs, story leads, event and institutional visits and identified experts were constantly given to the media practitioners in response to a perceived lack of 'interesting' pegs.

Farmers' acceptance and planting of Bt corn did not stop a similar process with the introduction of a second biotech product, Bt eggplant. Again, several challenges emerged, notably the premature termination of field trials because of the anti-biotech stance of local government and civil society groups. More damaging was the Supreme Court ruling in 2015 against further field trials based on a petition of civil society groups. Ironically, this event only heightened public awareness and interest in biotechnology and revitalised public dialogues and media articles. Eventually, the Supreme Court reversed its previous ruling, and science and industry have high hopes for its eventual commercial release (De Guzman, 2016). This time, the lessons learned from the first foray into biotech communication contributed to a more cohesive plan for public participation and engagement.

Interface with media and its outcome during this time may be validated by a 17-year media monitoring study by Navarro et al. (2011) and updated by Tome et al. (2017), which revealed a gradual progression of editorial perspective on Bt crops from a negative to neutral or positive tone over time. Uncertainty over the topic as well as lack of a tangible biotech product hampered the release of factual articles and opened the gates for their speculative and fear-based counterparts. In the succeeding years, science news using credible sources have become the norm. Efforts by the government and private sectors to provide media training proved to have significant impact as media practitioners now write about the topic on a regular basis, using less exaggerated metaphors and focusing on message frames relevant to consumers in general.

7.2. Disaster management: Typhoon Haiyan

Typhoon Haiyan (local name: Yolanda) was a landmark event of national significance that made the government and people realise the importance of science (and risk) communication. In November 2013, Haiyan made its first landfall in Guiuan, Northern Samar, with sustained winds of 315 kph (195 mph) and gusts up to 379 kph (235 mph), making it the strongest typhoon ever to make local landfall. Haiyan caused massive property damage, and significant human costs with about 6,300 dead and more than 28,000 injured (NDRRMC, 2014). Although some meteorologists claimed that these deaths were caused by the typhoon's unimaginable strength (Lagmay et al., 2015), several post-Haiyan analyses pointed out issues of interest to science communicators. Montemayor and Custodio (2014), in particular, described three problems that implicitly led to the widespread realisation of the role of scicom in everyday lives: (1) the public's (mis)understanding of the term 'storm surge';[6] (2) people's attitudes toward evacuation;[7] and (3) problems in institutional mechanisms in disaster mitigation.[8]

The Philippines has since seen notable improvements in scicom, especially in disaster reporting. For example, PAGASA finally formally translated 'storm surge' in Filipino as *daluyong bagyo*. Since then, storm surges have always been reported in media weather forecasts, and efforts to make other meteorological jargon more intelligible to the public have been consciously inserted in weather forecasts in subsequent typhoon events.[9]

6 Before Haiyan, the term was nearly non-existent in the consciousness of the ordinary, non-expert Filipino public. The argument was, had there been enough efforts to make the public understand the concept of a storm surge before the typhoon, casualties might have been reduced.

7 Many studies pointed out that people in the affected areas downplayed the early warnings and underestimated the strength of the typhoon. This still happened even though reports claimed that people in the affected areas had received evacuation training from their local governments, and that both the local and mainstream media had broadcast early warnings about the typhoon.

8 Studies had pointed out three problems that need to be addressed in disaster mitigation on the national scale: (1) poor disaster education given to local citizens; (2) lack of an efficient system in disseminating hazard information; and (3) lack of safe evacuation facilities.

9 Aside from Typhoon Haiyan, the experience brought about by Typhoon Ketsana (local name: Ondoy) in 2009, which flooded most of Metro Manila because of the unprecedented amount of rain it brought, and the release of the Metro Manila Earthquake Impact Reduction Study in 2010, which predicted the occurrence in the near future of a 7.2-magnitude earthquake in Manila and neighbouring areas, contributed to increasing scicom initiatives in the context of disasters.

7.3. Health: Yosi Kadiri and Dengvaxia

The DOH is mandated to guarantee equitable, sustainable and quality health for all Filipinos, especially the poor. Achieving this mandate is contingent on effective health promotion strategies. Currently, the DOH's campaigns typically involve door-to-door health promotion and dissemination of information, education and communication materials through traditional and digital media, coupled with catchy slogans.

One memorable effort is the anti-smoking campaign Yosi Kadiri [Smoke is Disgusting] in 1992. The campaign's centrepiece was the now-iconic mascot, a personification of the evils of smoking meant to counter the cool, masculine and handsome images of smokers cultivated by the tobacco industry, as exemplified by the Marlboro Man (Blanke, 2004). The mascot, along with anti-smoking messages and information, was plastered all over traditional media platforms and supplemented by billboards, stickers and comic books (Nieva, 2014). School visits and celebrity promotions further embedded the Yosi Kadiri's anti-smoking message in the national consciousness.

Yosi Kadiri had significant political impact, like the passage of the Tobacco Regulations Act in 2003. The law prohibited smoking in enclosed public places, banned the sales of cigarettes to minors and regulated cigarette advertisements. The Act paved the way for policies further restricting tobacco use, like the Sin Tax Reform Law that imposes higher taxes on tobacco products and the Graphical Health Warnings Law that stipulates that all tobacco products must carry graphic health warnings.

In contrast to Yosi Kadiri's success, the recent Dengvaxia immunisation program was a PR disaster, with stark consequences for the future of Philippine public health. Manufactured by Sanofi Pasteur, Dengvaxia is the world's first-ever dengue vaccine. Its sale was approved for the Philippines in December 2015. A mere four months later, a school-based immunisation program was launched by the DOH. The program would target over a million 9-year-old public school students in select regions of the Philippine island of Luzon.

In November 2017, Sanofi announced that Dengvaxia could cause 'severe dengue' in recipients who had never had the virus (termed 'seronegative patients'). At this point around 800,000 children had already been vaccinated, and 10 per cent of them were seronegative (CNN Philippines, 2018). The program was halted the following month, but sensationalised accounts of deaths started proliferating in national media, causing widespread panic and confusion (Figure 28.3). Local experts blamed the hysteria on Sanofi's lack of clarity on the term 'severe dengue'. Parents equated severe dengue with dengue shock syndrome, a far deadlier complication characterised by massive bleeding (Takumi, 2017). Moreover, parental consent forms were later found to lack information on vaccine risks and potential adverse effects (Torregoza, 2018).

Although no causal link between Dengvaxia and the children's deaths has been established so far, the damage was done. Hospitals subsequently documented significant drops in children's general vaccination rates and accounts of parents refusing even the DOH deworming program due to fears of another Dengvaxia-like scenario have been reported (Pazzibugan and Aurelio, 2018). Both the DOH and Sanofi are now challenged to restore public trust in their immunisation program and in their institutions as a whole.[10]

7.4. Institutional experiences: IRRI and ISAAA

Philippine-based international organisations like the International Rice Research Institute (IRRI) and the International Service for the Acquisition of Agri-biotech Applications (ISAAA) continue to be led by Filipino scicom experts. While rice research is the core business of IRRI, it plays an important role in developing and implementing strategic communication plans and activities, along with coordinating communications for the Global Rice Science Partnership. There is still much to be done to get people to understand the potential for its projects—like the Green Super Rice project that aims to develop stable high-yield cultivars that use less water, fertilisers and pesticides (IRRI, 2018). Though the Golden Rice project is still in the R&D stage, IRRI has had to respond to occasional attacks against this GM product, as people raise the same perceived fears as they did against Bt corn and Bt eggplant. An important process has been the conduct of public consultations as an integral component of the national biosafety regulatory process in the Philippines.

10 Other public health concerns such as influenza A(H1N1) virus, meningococcemia, Zika, avian flu and MERS-CoV in the past several years also contributed to the advancement of scicom initiatives in the health sector.

Figure 28.2: A group of science communicators from Africa and Asia assembled by ISAAA to help in knowledge-sharing initiatives in biotechnology.

Source: Eric John Azucena, ISAAA (used with permission).

ISAAA created a network of Biotechnology Information Centres in Asia, Africa and Latin America (Figure 28.4). A team of Filipinos transformed a proposal on knowledge sharing and science communication in 2000 into the Global Knowledge Center on Crop Biotechnology. These centres address communication challenges like bridging science and society (India), popularising genetic modification (China), understanding the saga of biotech papaya (Thailand), increasing biotech awareness for the masses (Malaysia) and strategising communication in biotech crop commercialisation (Bangladesh). Not only has scicom been institutionalised as a key component in the science arena but a cadre of scicom practitioners has also been developed (Navarro et al., 2013). A ripple effect for the appreciation and awareness of scicom in many countries began with home-grown Philippine initiatives.

8. Lessons learned, challenges in Philippine science communication and conclusion

Science communication had a slow start, but once it raised the interest of various stakeholders, it made a mark both locally and globally. There is now a core group of science communicators nurtured by an academic community and a thriving environment.

In a synthesis of different studies conducted by faculty members of UPLB's Department of Science Communication, Montemayor (2016) highlighted four good practices of different government and non-government agencies for scicom activities. Findings suggest that for a scicom initiative to be successful: (1) institutions should allot an adequate budget for their scicom activities to produce diverse outputs and achieve desired outcomes; (2) project teams should be composed of a mix of seasoned and young staff with smooth working relationships; (3) the scicom project should always have a unique selling point to establish its niche; and (4) the scicom project should receive strong support from top management.

Despite the idiosyncrasies brought on by the Philippine context, local scicom efforts continue to face similar challenges as their international counterparts (Navarro, 2018). While there are efforts to improve media coverage of science, particularly in the field of biotechnology, the research on both Filipino scientists (Ponce de Leon, 2011) and science journalists (Lacbayo, 2012) suggests the existence of a clash between the two groups, caused by a mutual lack of training and collaboration. The shared lack of formal scicom training was later attributed to the limited number of Philippine universities offering dedicated scicom classes and programs (Navarro, 2018). This has resulted in issues regarding the accuracy of science reporting in the media (Lacanilao, 2006). According to Ponce de Leon, Filipino scientists appear to subscribe to the outdated deficit model of scicom, affecting their communication approach to the public. The existence of a 'clash' between science and the media has long been debated, with surveys from the US (Hartz and Chappell, 1997) and Australia (Searle, 2013) suggesting the pervasiveness of this concept worldwide.

Challenges specific to Filipino science journalists include the absence of dedicated science reporters and experts in many newsrooms and low prominence of science stories in Philippine media (Congjuico, 2017). Due to rising costs in today's media landscape, even developed countries (Brumfiel, 2009; Ashwell, 2016) have had to contend with staff cuts and dwindling science news sections. The poor pay of Filipino journalists and limited access to funding hinders their search for science news leads and stories, further disincentivising the science beat and the science journalism career track.

Emerging scientific fields with their many issues have focused on developing a new form of public engagement that is more participative and dialogic. The number of actors involved in scicom is increasing, requiring new formats and modes of communication. Though a survey showed that while most Filipino scientists and academics strongly agree that they have a responsibility to communicate with the public, only 10 per cent of their time was devoted to this (Tome et al., 2014). Some 20 per cent have attended risk communication workshops. One major communication barrier is the difficulty of translating technical concepts into layperson's terms, and another is dealing with audiences with negative views about a scientific issue. Scientists felt that organisational support for scicom initiatives and establishing a community of practice would change mindsets.

To date, the reach of scicom efforts in the Philippines is viewed as insufficient by some practitioners (Navarro, 2018). This may be caused by barriers to access such as socioeconomic status, geography and language. One-fourth of the population live below the poverty line, potentially causing limited access to media-centric scicom efforts. Exacerbating this situation is the geographic spread of the population over the islands, making it a continuous challenge for science communicators to reach out to remote, poorly serviced and impoverished regions outside major metropolitan areas. The fragmented geography of the Philippines also introduces linguistic challenges, as various regions have distinct cultures and languages and there are approximately 180 different languages over the country (Simons and Fennig, 2018). For local scicom to be truly inclusive, future efforts must adapt to these factors or risk excluding a significant proportion of the population.

Although local scicom initiatives are still dominated by information delivery models (Brossard and Lewenstein, 2010), scicom initiatives guided by lay expertise and public engagement models are being carried out in the field, depending on the nature of the science topic to communicate, and the amount and source of project funding. It is hoped that more scicom initiatives in the future will focus on public engagement (Montemayor, 2016), and these challenges highlight the opportunities for growth in scicom. Navarro and Hautea (2011) listed the challenges to include capacity-building programs for different stakeholders; communication research to validate assumptions; identifying appropriate strategies to monitor and evaluate impact; developing a more responsive and relevant curriculum in secondary and tertiary education; and institutionalising the field in government and private sector initiatives.

Dr Gelia Castillo, a Philippine national scientist, has said that times have changed with different stakeholders now asserting their rights to know and participate in science-related decisions that affect their lives (Castillo, 2003). In the Philippines, it is no longer a case of science then communication but science communication.

References

Ambrosio, D. L. (2010). *Balatik: Etnoastronomiya, Kalangitan sa Kabihasnang Pilipino*. Quezon City, Philippines: University of the Philippines Press.

Amor, A., Icamina, P. and Laing, M. (1987). *Science Writing in Asia: The Craft and the Issues*. Manila, Philippines: Press Foundation of Asia.

Anderson, W. (2007). Science in the Philippines. *Philippine Studies*, *55*(3), 287–318.

Antiola, S. L. and Jose, L. E. (1982 December). Philippines. *UNESCO Bulletin Special Issue*. UNESCO Regional Office for Education in Asia and the Pacific. UNESCO Thailand, 158–76.

Ashwell, D. J. (2016). The challenges of science journalism: The perspectives of scientists, science communication advisors and journalists from New Zealand. *Public Understanding of Science*, *25*, 379–93. doi.org/10.1177/0963662514556144.

Bautista, V. M. (n.d.). History of science journalism in the Philippines. *PsciJourn: The official site of the Philippine Science Journalists Association Inc.* Retrieved from pscijourn.wordpress.com/about/.

Blanke, D. D. (2004). Lessons in legislation: Case studies from nine countries. In D. D. Blanke and V. D. C. Silva (eds), *Tobacco control legislation: an introductory guide*. Geneva, Switzerland: World Health Organization.

Brossard, D. and Lewenstein, B. V. (2010). A critical appraisal of models of public understanding of science: using practice to inform theory. In L. Kahlor and P. A. Stout (eds), *Communicating Science*. New York: Routledge.

Brumfiel, G. (2009). Science journalism: Supplanting the old media? *Nature Biotechnology*, *458*, 274–77. doi.org/10.1038/458274a.

Burgos, R. (2017). *Communicating science for the people: Considerations towards a strategic communication plan for DOST*. Philippines: Department of Science and Technology, Bicutan.

Caoili, O. (1987). *A history of science and technology in the Philippines*. Quezon City, Philippines: University of the Philippines Science Research Foundation.

Castillo, G. (2003). *Science communication whose time has come: An evaluation report on the Global Knowledge Center on Crop Biotechnology*. Los Baños, Laguna, Philippines: International Service for the Acquisition of Agri-biotech Applications (ISAAA).

Chamarik, S. and Goonatilake, S. (1994). *Technological independence – The Asian experience*. Tokyo, Japan: United Nations University Press.

CNN Philippines. (2018). Timeline: The Dengvaxia controversy. Retrieved from www.cnnphilippines.com/news/2017/12/09/The-Dengvaxia-controversy.html.

Commission on Higher Education. (2017). CHED Memorandum Order (CMO) No. 36, s. 2017: Policies, Standards, and Guidelines (PSGs) for Bachelor of Science in Development Communication (BSDevCom) Program. Retrieved from www.ched.gov.ph/wp-content/uploads/2017/10/CMO-36-s-2017.pdf.

Congjuico, T. S. (2017). Foundations for a responsive and relevant science journalism course. *Media Asia, 43*(3–4), 169–75. doi.org/10.1080/01296612.2017.1293319.

Conklin, H. C. (1980). *Ethnographic atlas of Ifugao: A study of environment, culture, and society in Northern Luzon.* New York: Yale University Press.

Dagdayan, C. B. (1978). PUSTE through science fairs. In *Science Foundation of the Philippines Asian training course for leaders in the promotion of public understanding of science, technology and environment* (pp. 150–51). Manila, Philippines: Science Foundation of the Philippines.

Daoas, D. (1999). *Efforts at protecting traditional knowledge: The experience of the Philippines.* Geneva, Switzerland: World Intellectual Property Organization.

Dapul, S. M. (1978). The Philippine Society of Youth Science Clubs: The moving factor. In *Science Foundation of the Philippines Asian training course for leaders in the promotion of public understanding of science, technology and environment,* (pp. 153–54). Manila, Philippines: Science Foundation of the Philippines.

De Guzman, P. (2016). The trial of Bt talong field trials. *BioLife Special Issue,* Quezon City, Philippines: J. Burgos Media Services.

Department of Education. (1997). DECS Order (DO) no. 53 s. 1997: Institutionalisation of the use of CONSTEL tapes for elementary and secondary school teacher. Retrieved from www.deped.gov.ph/1997/07/28/do-53-s-1997-institutionalization-of-the-use-of-constel-tapes-for-elementary-and-secondary-school-teacher/.

Dimacali, T. J. (2017, 19 January). UP to establish science journalism course. *GMA News Online.* Retrieved from www.gmanetwork.com/news/scitech/science/596297/up-to-establish-science-journalism-course/story/.

Gomez, E. D. (1993). Welcome remarks. Proceedings on the National Conference on Development Communication. Institute of Development Communication, University of the Philippines Los Baños, Los Baños, Philippines.

Hartz, J. and Chappell, R. (1997). *Worlds apart: How the distance between science and journalism threatens America's future.* Nashville, US: First Amendment Center.

International Rice Research Institute (IRRI). (2018). Institutional Brochure. Laguna, Philippines.

Jamias, J. (1984). Structuring communication education for the information society. Paper presented to the Department of Development Communication, University of the Philippines Los Baños, Los Baños, Philippines, 14 August.

Jamias, J. (1987). Humane goal for development journalism. *Philippine Communication Journal, 1*(2), 21–26.

Jamias, J. F. (1988). Towards a science communication program: A systems approach. Unpublished curricular program review report. Institute of Development Communication, University of the Philippines Los Baños, Los Baños, Philippines.

Jamias, J. (1989). Science communication and education in the development communication context. Diamond Jubilee Professorial Chair Lecture presented to the Institute of Development Communication, University of the Philippines Los Baños, Los Baños, Philippines, 27 September.

Jamias, J. F. (1997). The plain and simple tale of science communication as proposed in the Institute of Development Communication (IDC). In M. T. H. Velasco, A. G. Moran, J. C. Sison, and A. G. Flor (eds), *Communication of Scientific and Technical Information* (pp. 22–31). Diliman, QC: UP Open University.

Lacanilao, F. (2006). Problems with media and scientists. In G. P. Concepcion, E. A. Padlan and C. A. Saloma (eds), *Selected essays on science and technology for securing a better Philippines* (vol 1). Quezon City, Philippines: University of the Philippines Press.

Lacbayo, S. A. (2012). A comparative analysis of the attitudes and beliefs of scientists and science journalists toward science journalism in the Philippines (Bachelor's thesis). University of the Philippines Los Ba-os at Laguna, Philippines.

Lagmay, A. M. F., Agaton, R. P., Bahala, M. A. C., Briones, J. B. L. T., Cabacaba, K. M. C., Caro, C. V C. et al. (2015). Devastating storm surges of Typhoon Haiyan. *International Journal of Disaster Risk Reduction, 11*, 1–12. doi.org/10.1016/j.ijdrr.2014.10.006.

Liakopoulos, M. (2002). Pandora's Box or panacea? Using metaphors to create the public representations of biotechnology. *Public Understanding of Science, 11*(1), 5–32. doi.org/10.1088/0963-6625/11/1/301.

Librero, F. (2000). Science communication: A personal retrospective brief. *Proceedings on discoursing science communication: First roundtable discussion on science communication* (pp. 22–26). University of the Philippines Los Baños, Los Baños, Philippines.

Mangahas, M. (2018, 15 December). Skepticism about science. *Philippine Daily Inquirer*. Retrieved from opinion.inquirer.net/118181/skepticism-about-science.

Montemayor, G. J. S. (2013). Dissecting the genius behind the white lab coat: Constructing the images of Filipino scientists in films and biographies (Master's thesis). University of the Philippines Diliman, Quezon City, Philippines.

Montemayor, G. J. S. (2014). Communicating risks, risking (mis)communication: Mass media and the science of natural disasters. Public lecture delivered at University of the Philippines Diliman, Quezon City, Philippines, 3 September.

Montemayor, G. J. S. (2016). A study on the state of science communication practice in the Philippines in Scicom in the Philippines: Current practices and challenges. *Proceedings of the third roundtable discussion on science communication* (pp. 40–46). UP Los Baños, College, Laguna, Philippines.

Montemayor, G. J. S. (2018). Juan F. Jamias and the beginnings of science communication as an academic field in development communication. Paper presentation, 8th ADCEP National Conference, UP Los Baños, College, Laguna, Philippines.

Montemayor, G. J. S. and Custodio, P. A. (2014). Communicating risks, risking (mis)communication: Mass media and the science of disasters. *Kasarinlan*, *29*(2), 39–74.

Morrow, P. (2002). Baybayin – the ancient script of the Philippines. Retrieved from www.webcitation.org/5rpLR9aNS.

National Disaster Risk Reduction and Management Council (NDRRMC). (2014). Updates re: the effects of Typhoon 'Yolanda' or 'Haiyan'. Retrieved from www.ndrrmc.gov.ph/attachments/article/1329/Update_on_Effects_Typhoon_YOLANDA_(Haiyan)_17APR2014.pdf.

National Government Portal. (n.d.). About the Philippines. Retrieved from www.gov.ph/about-the-philippines.

Navarro, K. N. (2018). Challenges of communicating science: Perspectives from the Philippines (Master's thesis). The Australian National University, Canberra, Australia.

Navarro, M. and Hautea, R. (2011). Convergence in science communication: The tipping point. In M. Navarro and R. Hautea (eds), *Communication challenges and convergence in crop biotechnology*. Los Baños, Philippines: ISAAA and SEARCA.

Navarro, M., Escaler, M., Ponce de Leon, M. I. A. Z. and Tababa, S. (2007). The Bt Maize Experience in the Philippines: a Multi-stakeholder convergence. In D. Brossard, J. Shanahan, and T. C. Nesbitt (eds), *The Public, the Media, and Agricultural Biotechnology*. Wallingford, United Kingdom: CAB International. doi.org/10.1079/9781845932046.0327.

Navarro, M., Panopio, J., Malayang, D.B. and Amano, N. Jr. (2011). Print media reportage of agricultural biotechnology in the Philippines: a decade's (2000–2009) analysis of news coverage and framing. *Journal of Science Communication*, *10*(3), A01. doi.org/10.22323/2.10030201.

Navarro, M., Tome, K. and Gimutao, K. (2013). *From monologue to stakeholder engagement: The evolution of biotech communication*. Ithaca, New York: The International Service for the Acquisition of Agri-biotech Applications.

Nieva, R. (2014, 5 October). How the DOH used PR to create health and safety awareness. *Business Mirror*. Retreived from www.businessmirror.com.ph/how-the-doh-used-pr-to-create-health-and-safety-awareness/.

Official Gazette of the Philippines. (1951). Proclamation No. 279, s. 1951. Retrieved from www.officialgazette.gov.ph/1951/10/12/proclamation-no-279-s-1951/.

Official Gazette of the Philippines. (2003). Proclamation No. 437, s. 2003. Retrieved from www.officialgazette.gov.ph/2003/08/11/proclamation-no-437-s-2003/.

Ongoco, T. C. (1978). The Youth Science Camp: A Philippine innovation. *Science Foundation of the Philippines Asian training course for leaders in the promotion of public understanding of science, technology and environment* (pp. 156–57). Manila, Philippines: Science Foundation of the Philippines.

Panopio, J. and Navarro, M. (2011). Philippines: Drama and communication behind Asia's first commercialized Bt corn. In M. Navarro and R. Hautea (eds), *Communication challenges and convergence in crop biotechnology*. The International Service for the Acquisition of Agri-biotech Applications and SEAMEO Southeast Asian Regional Center for Graduate Study and Research in Agriculture.

Pazzibugan, D. Z. and Aurelio, J. M. (2018, 4 February). Panic over Dengvaxia hurting other immunisation, health programs. *Philippines Daily Inquirer*. Retrieved from newsinfo.inquirer.net/965935/panic-over-dengvaxia-hurting-other-immunisation-health-programs.

Ponce de Leon, M. I. A. Z. (2011). Science communication beliefs of researchers based in the Philippines and the United States: A qualitative analysis of research cultures and worldviews (PhD dissertation). Purdue University at West Lafayette, USA.

Reyes, V. F. (1978). PUSTE through the science quiz. *Science Foundation of the Philippines Asian training course for leaders in the promotion of public understanding of science, technology and environment* (pp. 152–53). Manila, Philippines: Science Foundation of the Philippines.

Rodrigo, R. (2006). *Kapitan: Geny Lopez and the making of ABS-CBN*. Quezon City: ABS-CBN Publishing, Inc.

Santos, S. (2008a). Pinoy Sci-Fi #1: *Exzur* (1956). Retrieved from www.video48.blogspot.com/2008/04/pinoy-sci-fi-1-exzur-1956.html.

Santos, S. (2008b). Pinoy Sci-Fi #2: *Zarex* (1958). Retrieved from www.video48.blogspot.com/2008/04/pinoy-sci-fi-2-zarex-1956.html.

Santos, S. (2008c). Pinoy Sci-Fi #3: *Tuko sa Madre Kakaw* (1959). Retrieved from www.video48.blogspot.com/2008/04/pinoy-sci-fi-3-tuko-sa-madre-kakaw-1959.html.

Science Foundation of the Philippines. (1978). *Asian training course for leaders in the promotion of public understanding of science, technology and environment*. Manila, Philippines: Science Foundation of the Philippines.

Searle, S. D. (2013). Scientists' engagement with the public. In J. K. Gilbert and S. M. Stocklmayer (eds), *Communication and engagement with science and technology, issues and dilemmas: A reader in science communication*. New York: Routledge. doi.org/10.4324/9780203807521.

Simons, G. F. and Fennig, C. D. (2018). *Ethnologue: Languages of the world.* Dallas: SiL International.

Takumi, R. (2017, 13 December). 'Miscommunication' over Sanofi definition of severe dengue blamed for panic over Dengvaxia – experts. *GMA News Online*. Retrieved from www.gmanetwork.com/news/news/nation/636456/miscommunication-over-sanofi-definition-of-severe-dengue-blamed-for-panic-over-dengvaxia-experts/story/.

Tome, K. C., Navarro, M. and Aldemita, R. (2014). Academics and scientists as biotech communicators: Perspectives, capabilities, and challenges. *Philippine Journal of Crop Science, 39*(2), 44–57.

Tome, K. G., Navarro, M., Mercado, S. and Villena, M. M. C. (2017). Seventeen years of media reportage of modern biotechnology in the Philippines. *Philippine Journal of Crop Science, 42*(1), 41–50.

Torregoza, H. L. (2018, 22 February). Parents were not informed of Dengvaxia's adverse effects – expert. *Manila Bulletin*. Retrieved from news.mb.com.ph/2018/02/22/parents-were-not-informed-of-dengvaxias-adverse-effects-expert/.

Torres, C., Suva, M., Carpio, L. and Dagli, W. (2006). *Public understanding and perception of and attitude towards agricultural biotechnology in the Philippines*. University of the Philippines Los Banos, International Service for the Acquisition of Agri-biotech Applications, and Southeast Asian Regional Center for Graduate Study and Research in Agriculture, Laguna, Philippines.

Velasco, J. R. and Baens-Arcega, L. (1984). *National Institute of Science and Technology 1901–1982: A facet of science development in the Philippines*. Geneva, Switzerland: National Science and Technology Authority.

Velasco, M. T. H. (1998). The role of science communication in the promotion of science and technology in the Philippine setting (PhD dissertation). University of the Philippines Los Ba-os at Laguna, Philippines.

Vergara, L. F. (1978). PUSTE through the Science Club Advisers Association of the Philippines. *Science Foundation of the Philippines Asian training course for leaders in the promotion of public understanding of science, technology and environment*. Manila, Philippines: Science Foundation of the Philippines.

Visayas State University. (2016). VSU DevCom celebrates 30th anniversary. Retrieved from www.vsu.edu.ph/articles/bulletin/events/459-vsu-devcom-celebrates-30th-anniversary.

Xavier University Development Communication Department. (2017). Who we are. Retrieved from www.xudevcomblog.wordpress.com/2017/10/10/xu-devcom/.

Timeline

Event	Name	Date	Comment
First interactive science centre established.	Philippine Science Centrum	1984	Private sector initiative with support from the Department of Science and Technology
First national (or large regional) science festival.		1951	Occurred during National Science Week
An association of science writers or journalists or communicators established.	Philippine Science Journalists Association (PsciJourn)	2001	
First university courses to train science communicators.		1960	Offered by the University of the Philippines College of Agriculture
First national conference in science communication.		1993	One of the first conferences resulted in the establishment of the Philippine Foundation of Science Communicators, Inc (PhilSciCom)
National government program to support science communication established.	Science and Technology Information Institute of the Department of Science and Technology	1987	
National Science Week founded.		1951	
First significant radio programs on science.	Science topics were discussed in *Mga Gintong Kaalaman* [Bits of Golden Knowledge]	1965	Hosted by Ernie Baron, a general radio program
First significant TV programs on science.	*Sineskwela* [School on Air]. May not be the earliest, but definitely the most popular science program	1994	Jointly produced by the ABS-CBN Foundation, the Department of Science and Technology, and Department of Education, Culture and Sports
First awards for scientists or journalists or others for science communication.	Annual Department of Science and Technology Media Awards	1988	
Other significant events.	First Youth Science Camp with Science Journalism workshop	1971	Through the Science Foundation of the Philippines

Contributors

Garry Jay S. Montemayor is the former chair of the Department of Science Communication, College of Development Communication, University of the Philippines Los Baños (UPLB), Philippines.

Dr Mariechel J. Navarro retired in 2015 as director of the Global Knowledge Center on Crop Biotechnology at the International Rice Research Institute (IRRI) in Los Baños, Laguna, Philippines.

Kamila Isabelle A. Navarro is an assistant professor lecturer at the Department of Communication at the De La Salle University and a lecturer at the University of the Philippines Diliman.

29

PORTUGAL

The late bloom of (modern) science communication

Marta Entradas, Luís Junqueira
and Bruno Pinto

1. Introduction

This chapter describes the emergence of modern science communication in Portugal. The chapter is organised in parts. Part 2 sets the context in which science communication activities emerged and flourished in the country during the mid-1990s, anchored by a top-down government policy. This story is an historical account of the social and political factors leading up to this important episode. Whenever possible, we situate national moves within academic and policy debates on the public understanding of science, which may have influenced them. Part 3 maps the main events, activities, group initiatives and moments in science communication since then and describes the emergence of a community of practitioners, and opportunities in the professionalisation of the field. Part 4, we consider the late blooming and rapid developments of today, and the overall impact of the top-down approach on the development of modern science communication.

2. The political context and the emergence of a government policy for 'scientific culture'

2.1. The pre-1990s

Modern science communication is relatively new in Portugal compared to its European neighbours, who have longer traditions of public understanding of science (PUS), or public participation in science policy. See, for example, the 'PUS movement' in the United Kingdom in the 1980s (Gregory and Miller, 1998) and publication of the internationally influential report *Public Understanding of Science* by the Royal Society in 1985, the Norwegian government policies in science communication since 1975 (Hetland, 2014), and the Danish consensus conferences organised since 1995 (Einsiedel et al., 2001). In the early 1990s, Portugal was a country with few modern scientific resources, public relations with science were weak, and the practice of science communication was scarce (Entradas, 2015). This was a consequence of a dictatorship and authoritarian state that ruled for more than 40 years[1] and kept scientific institutions and scientists away from society (Gonçalves and Castro, 2002). The second half of the 1990s saw, however, a turning point in science–society relationships, with 'scientific culture' and the 'promotion of science to the public'—as it was termed in our country—becoming an integral part of the science policy agenda. Since then, Portugal has quickly expanded its infrastructure for science communication, with political support, and continues to do so (Entradas, 2015).

During the 20th century, the university was an elitist space for the education of the few, based on the values of the New State (Rosas and Sisifredo, 2013). Research was confined mostly to the State Laboratories. Until the 1950s there were only four universities in the country—Coimbra, Lisbon, Porto, and Lisbon Technical University (Teixeira et al., 2007, p. 347)—and only 0.04 per cent of the Portuguese population completed a university degree. Today, there are 14 universities and 13 polytechnic institutes around Portugal, with 372,000 students enrolled (DGEEC, 2018), and 18 per cent of population has a degree (Instituto Nacional de Estatística, 2017).

With the fall of the authoritarian regime, overthrown by a military coup on 25 April 1974, the country focused on developing scientific infrastructure, expanding its scientific community and universities, and increasing the population's levels of education. Despite these developments, the science–society relationship was (still) distant from both the political agenda and university practices during the 1980s. The communication of science relied mostly on the activities of a few scientific authorities (Gago, 1990). There was no tradition of science journalism or science museums and exhibitions (Machado and Conde, 1988) and engaging with the public was not well regarded amongst the scientific community (Jesuíno and Diego, 2003). A study of the Portuguese scientific community in the early 1990s shows

1 The 'New State' was the far-right regime installed in Portugal from 1933 to 1974, created by Prime Minister Antonio de Oliveira Salazar, who ruled between 1932 and 1968, and continued under Marcelo Caetano, the last prime minister of the New State, ruling from 1968 until his overthrow in the Carnation Revolution of 1974.

that 'scientific dissemination' was regarded as an unimportant factor in the recognition of scientific authority and not a reputable activity for a scientist (Jesuíno et al., 1992; Machado and Conde, 1988). Still, it is during the 1980s that the first signs of a public dissemination culture in the country emerge; for instance, in 1982, the first national publisher Gradiva is created with an editorial profile oriented to science collections, probably a result of a growing public demand, and a community of science journalists begins to emerge. Central to these developments was the integration of Portugal in the European Union in 1986, which greatly advanced the economy in many sectors, including scientific and education infrastructures, and modernised the country more broadly, while also promoting stronger political and economic relationships with other member-states (e.g. Rodrigues, 2015; Soares, 2008).

2.2. The post-1990s

In the mid-1990s, we see a radical change in the science–society relationship in Portugal, which begins in the form of a top-down government policy (Entradas, 2015). In 1995, the Ministry of Science and Higher Education is created. José Mariano Gago, the first Minister for Science and Technology from 1995 to 2002, puts 'scientific culture' strongly on the political agenda, as part of a broader aim of building a scientifically literate society (Gago, 1990). Modern science communication thus emerges in a context of full political support, with the government becoming a major player in the promotion of initiatives to foster scientific culture in the country (Entradas, 2015). The 'policy for scientific culture' is perhaps the most significant event in the history of science communication in Portugal, having had positive impacts at many levels, and the turning point from which we can best trace the beginning of modern science communication in the country.

2.2.1. National 'policy for scientific culture'

The Portuguese national policy for scientific culture created in the 1990s was reflected in a series of actions by the government to encourage research institutions and scientists to increase their relations with society, and to widen public access to science. Two of the most pre-eminent actions were:

i. the formulation of legislation governing scientific research institutions, teaching and research staff, to expand and strengthen science communication. For example, the revised Legal Framework for Scientific Research institutions declared that all publicly funded research centres should communicate their scientific activity and allocate funding for this task (e.g. Legal Framework for Scientific Research Institutions, Article 13 of Decree Law No. 128/99, 17 April). Similarly, the higher education career statute establishes scientific dissemination as one of the duties of university teaching staff (Decree Law No. 205/2009, Article 4, revision of the Decree Law No. 448/79). The government's emphasis on science communication is seen in more recent examples, such as including researchers' communication activities in the assessment of their academic performance; requiring 'dissemination and public engagement' plans in project grants (e.g. Guidelines for FCT[2] Investigator 2016, Guidelines for Individual Stimulus 2017); and assessing science and society activities as part of the evaluation of the research and development (R&D) units for competitive funding.

ii. The second action was the creation in 1996 of Ciência Viva Agency (Science Alive! – National Agency for the Scientific and Technological Culture). This national non-profit public awareness association was funded by the government through the Ministry of Science and Technology to develop science communication infrastructure and activities in the country.

But there were other important government initiatives during these years, including the creation of fellowships (one to six years duration) in science and technology management (BGCT – Bolsa de Gestão de Ciência e Tecnologia) covering science communication. A second initiative was the addition of a new research area to the six areas already existing for individual fellowship applications at the postgraduate level. This was called PACT (Promotion and Administration of Science and Technology), and those intending to pursue science communication at the postgraduate level could apply for PhDs and postdoctoral fellowships. These fellowships were in place for almost a decade (2005–13). We do not have numbers for the ratio of management to science communication fellowships awarded during these years, but we believe it to have been split evenly. Importantly, this marks the early years in the emergence of a community of science communicators.

What we observed then is a growing panoply of opportunities to increase the presence of science in society. Science begins to be regularly presented in the media, the number of science museums and centres expands significantly, and scientific organisations create structures dedicated to outreach and training programs in science communication. These developments are the focus of Part 3 of this chapter. At the academic level, a body of social studies examining the science–society relationship, publics for science, and the scientific community emerges (e.g. Costa et al., 2002; Gonçalves, 1996).

2 FCT (Fundacao para a Cienciaea a Tecnologia [Foundation for Science and Technology]) is the Portugues natoinal funding agency for research.

2.2.2. Portuguese society and science

The effort by the Portuguese government to increase public scientific literacy is visible in the national surveys introduced in the 1990s (discontinued in the 2000s). The first survey was conducted in 1996/97 by the Science and Technology Observatory (OCT), part of the Ministry of Science and Higher Education, and the second in 2000 by the same institution.

It is perhaps not surprising that these studies portrayed a gap between the Portuguese population and science. Nevertheless, the 2000 survey saw an increase in public interest and positive attitudes to science and technology when compared to the 1996 survey. For example, 20 per cent of the respondents in 2000 versus 10 per cent in 1996 declared themselves very interested in scientific topics; there was a broader recognition that science could contribute to improving people's quality of life; and people had higher expectations about science and technology in general (OCT, 1996, 2000; Ávila and Castro, 2003). Yet the levels of 'scientific literacy' of the Portuguese have ranked low compared to European standards, as shown by the Eurobarometer surveys of knowledge conducted by the European Commission (1992, 2001, 2005). Portugal presents more similarities with the countries from the southern and eastern Europe than the northern European countries, which in general have stronger relations with science.

Despite the generally positive attitude towards science and an improvement in the science–society relationship during the 2000s, we also see signs of a decrease in trust in science among the Portuguese, indicated by a more negative view of the benefits that science brings to individual life and its role in solving societal issues. For example, in 2005, 77 per cent agreed that 'science and technology make our lives healthier, easier and more comfortable', compared to 62 per cent in 2010 (European Commission, 2005, 2010). This decrease in trust in science has been attributed to public controversies around scientific issues in the 1980s and 1990s, such as the bovine spongiform encephalopathy (BSE) or genetically modified foods (GMF), which were also felt by Portuguese society (Gonçalves, 1996). Perhaps most significant were local controversies around environmental impacts of incinerators (Lima, 1995; Gonçalves, 2003a) and the aborted construction of a hydro-electric dam in the Foz Côa Valley, interfering with one of the most important national Palaeolithic sites of rock art (Jesuíno, 2001; Gonçalves, 2001). The Côa Valley rock art site has been on UNESCO's world heritage list since 1998. Studies showed that the public remained a marginal actor in influencing policy and the scientific debates were highly politicised (Lima, 1995; Castro and Lima, 2003; Gonçalves, 2003a, 2003b).

This illustrates how the policy for 'scientific culture' was approached as a dissemination model rather than in dialogical contexts of public participation, which were at that time intensely under debate in Europe (House of Lords, 2000; European Commission, 2001; Wynne, 1996). The fact that modern science communication in Portugal was just in its beginnings may in part explain this. Although traditional deficit-style communication still predominates, public participation initiatives have emerged, such as citizen science initiatives and public participation labs (Laboratórios de Participação Pública) to engage local communities. These initiatives have often resulted from partnerships between municipalities and universities. An example is the Open Science Hub (2017), a partnership between Figueira de Castelo Rodrigo municipality and Leiden University, to engage local communities in the development of innovation products, through collaborations between schools, civil society, industry, universities and the broader community.[3] Another example is the initiative Participatory Budgeting for Science (2017) promoted by the Ciência Viva Agency and the Portuguese Foundation for Science and Technology (FCT), in which citizens get involved in decision-making on the Portuguese participatory budget for science through a voting process (Ciência Viva, 2017).

Mariano Gago (1948–2015) was the first Minister for Science and Technology in Portugal. During his mandate (1995–2002) in the XIII and XIV Constitutional Governments, he introduced science communication into the political agenda. He became an influential voice in the promotion of research and scientific culture through his tenure as president of the Junta Nacional de Investigação Científica e Tecnológica (JNICT), the precursor to the FCT, between 1986 and 1989, where he coordinated early efforts at modernising science policy. Not long after, he published his influential essay *Manifesto para a Ciência* (Gago, 1990), where he called for a change in academic institutions from their historic isolation to make Portuguese science more open to society. He asked for the renewal of scientific education and research, and the promotion of scientific culture in Portugal. Mariano Gago became Minister for Science, Technology and Higher Education of Portugal again in the XVII Constitutional Government, between 2005 and 2011.

Rómulo de Carvalho (1906–97) was an early promoter of scientific culture in Portugal. He was a physics and chemistry high school teacher and influential poet (under the name António Gedeão). He had an important role in promoting scientific culture in Portugal since the 1950s and is still a reference for science communication in Portugal—his birthday was officially named

3 See www.cm-fcr.pt/plataforma-ciencia-aberta/.

National Scientific Culture Day in 1996. He wrote several popular science book collections: *Science for young people* (10 volumes, 1952–62), *Physics for the people* (two volumes, 1968), *Notebooks of initiation to science* (18 volumes, 1979–1995), among other books and articles on science communication. He was founder and director of the first popular science periodical, *Gazeta da Física* [Physics Gazette], which was first published in 1946.

3. Science communication activities in modern Portugal

In what follows we offer a descriptive view of the evolution of science communication activities and emergence of a community of practitioners in the modern science communication period in our country.

3.1. When and what

3.1.1. Science museums and centres

The first signs of science being open to the public in Portugal can be traced back to the first museums and botanical gardens established at the end of the 18th century, associated with universities and based on private and royal collections (Fiolhais, 2011, 2014; Granado and Malheiros, 2015). The first were created in Lisbon—for example, the Royal Museum of Natural History (1768) and the Botanical Garden of Lisbon (1878) (today the National Museum of Natural History and Science)—and in Coimbra, the Cabinets of Natural History and Physics, and the Botanical Garden of Coimbra University (1772), currently the Science Museum and Botanical Garden of Coimbra University (Brigola 2003, 2010). In the second half of the 19th century, other institutions were established by professional groups such as geologists and naturalists, and scientific associations such as the Society of Geography of Lisbon (1875). Examples are the Geological Museum created in Lisbon in 1859 by pioneers in geology such as Carlos Ribeiro and Nery Delgado (LNEG, 2018), and the first zoological garden (the Lisbon Zoo) created in 1884 by three naturalists: Dr Pedro Van Der Laan, José Martins and the Baron of Kessler (Jardim Zoológico de Lisboa, 2018). Although these institutions were important spaces for people to access science, it is fair to say that their reach was limited, possibly as attractions for the educated few living in cosmopolitan areas. In the 1980s, the number of natural history museums, science museums, botanical gardens, zoos and aquariums was only 13 (Delicado et al., 2013).

This picture has changed with an increased number of science museums in the country. One of the most important is the Lisbon Oceanarium, built as one of the centrepieces of the 1998 Lisbon World Exposition, and the most visited cultural venue in the country with about 1 million visitors per year (Oceanário de Lisboa, 2015). Other science museums run by a diversity of actors (associations, companies, municipalities) are important spaces for public interaction with science. Examples include the Visionarium, an interactive science centre created in 1998; the Museum of Energy created by EDP (former state energy company) on the site of an old power plant in Lisbon in 1990; and the Museum of Pharmacy, maintained by the National Pharmacies Association since 1996. The number of science museums grew from 23 in 2000 to 40 in 2016, and of aquariums, zoos and botanical gardens from three in 2000 to 20 in 2011 (Instituto Nacional de Estatística, 2002, 2013, 2016).

3.1.2. Scientific associations

Scientific associations have been important players in science communication. The *Gazeta da Física*, one of the earliest science magazines for non-specialists, was founded in 1946 by Rómulo de Carvalho and a group of physicists, and in 1974 integrated into the Portuguese Physics Society as its official publication. Nowadays, in a context where national societies have lost some peer communication functions to their international counterparts, many scientific societies find outreach to be an important component of their activities: around 50 per cent say they regularly engage in public communication (Delicado et al., 2013). One of their best-known activities is the organisation of the national science Olympiads in mathematics (since 1983), physics (1985), chemistry (2000) and biology (2010) by their respective scientific societies. Besides the traditional scientific societies, there are associations created by researchers to promote citizen science. These include amateur astronomers' associations that organise skygazing events and associations for nature observation activities such as bird or butterfly watching. A survey by Delicado et al. (2013) found 62 of these associations in Portugal, 51 of which were created after 1990.

However, the most significant change within the realm of associations is seen with the creation of the Ciência Viva Agency, as described above. This has had a profound impact on the amount and diversity of science communication activities all over the country, allowing science to expand from the main cities to more peripherical areas. Ciência Viva rapidly became a nexus for science outreach (Costa et al., 2005) by promoting a national science communication program based on three main axes (Conceição, 2011). The first was to improve science teaching by funding experimental projects developed by schools. The second was the Ciência Viva no Verão [Science Alive in the Summer], a program of outdoor scientific activities directed at the general public. This had its first edition in 1997, Astronomy in the Summer, and expanded over the years to include other disciplines such as geology (1998), biology (2001) and engineering (2002). In these activities, citizens engage in astronomical observations, birdwatching, nature walks, spelunking and visits to technology sites (mines, factories, power plants, treatment plants), among others.

The third axis for Ciência Viva's activities was the creation of a national network of science centres and following a trending model of science exhibitions based on interactive modules and activities (Schiele, 2008), a novelty in science museology in Portugal at the time. This network of science centres has been built through partnerships between the agency and local actors including universities and municipalities, usually relying on a theme of local significance to organise the centre's activities. The first centre opened in Faro (Algarve) in 1997, a partnership between Ciência Viva, Albufeira and Faro municipalities, and the University of Algarve (Pinto and Amorim, 2018). The centre was installed on the site of a deactivated power plant with a focus on ocean sciences. The network also opened the Knowledge Pavilion in 1999 to serve as a flagship science centre under the agency's administration (Delicado, 2006). This is the largest and most visited science centre in the country and attracts about a third of the number of visitors for the whole network. Ciência Viva network centres received an average of 626,000 visitors per year between 2012 and 2015 (Garcia et al., 2016); for comparison, art museums had an average of 3 million visitors over the same period (Instituto Nacional de Estatística, 2013, 2014, 2015, 2016). Today, the Ciência Viva network has 20 science centres spread throughout the country (including the Azores Islands), with themes varying from astronomy and geology, to forestry, hydrology, biodiversity, energy, sustainability and navigation technology. Ciência Viva's initiatives have become very popular among universities and are among the main outreach activities in which universities participate (Entradas and Bauer, 2017). The program has been acknowledged as a successful model of science communication in Europe (Miller et al., 2002).

3.1.3. National events

National science events have also played an important role in public access to science, having grown in diversity and public reach over the last few decades. The earliest was the Science and Technology Week starting in 1998, promoted by the Ministry of Science (Conceição, 2011). Science Week activities are usually organised by universities and museums and include public lectures, exhibitions, visits to scientific institutions, open days and hands-on workshops. The European Researchers' Night and the FameLab promoted by the European Union (EU) in many European countries have become a staple of the universities' public outreach calendar. Similar events are organised by universities themselves, the most notable being Physics Week, started in 1996 by the Instituto Superior Técnico of Lisbon (IST-UL) and continued annually. During Physics Week, non-scientists participate in public lectures and an interactive exhibition of physics experiments called The Physics Circus is a core activity of the event. It is important to note the role of the Calouste Gulbenkian Foundation—a Portuguese institution established in 1956 to promote the arts, charity, science and education—in organising large science public exhibitions. Examples include the *At Einstein's Light* (2005) and the *Darwin's Evolution* (2009). The latter celebrated the 200th birthday of Charles Darwin and had 161,000 visitors in Lisbon (Delicado et al., 2010). The Institute Gulbenkian of Science (a research centre in the biomedical sciences, which is part of the Gulbenkian Foundation) has had a marked presence at one of the largest music festivals Nos Alive (Algés, Lisbon) since 2008 with a stand dedicated to public information about life sciences. For three days in July, about 600 participants per year (mostly young adults and teenagers) have engaged in outreach activities such as speed dating with scientists, experiments and science games in an informal environment (Leão and Castro, 2012).

3.1.4. Media science communication

In the national media, the 1980s are regarded as a landmark for an increase in science news in the most read national newspapers including the *Expresso*, the *Diário de Notícias* and *A Capital* (Fonseca, 2017; Machado and Conde, 1988; Mendes, 2003). These newspapers have published articles about science and technology since the establishment of democracy in the late 1970s, though irregularly. Dedicated sections to science in national newspapers came only later and not always as a regular feature in the papers. For example, the newspaper *Público* had a daily page on science news from the newspaper's creation in 1990 until 2007, when it was discontinued, returning in 2012 until the present day. *Diário de Notícias* had a daily science section between 1999 and 2003, and between 2007 and 2014, but today science news is published in the daily pages of this newspaper (Granado and Malheiros, 2015). Some of these newspapers had science supplements, which also often changed names and formats, sometimes being reduced and/or discontinued (Fonseca, 2017).

Despite what was on offer, a study in 2000 on the public consumption of newspaper articles and popular science magazines showed low readership rates of science news by the Portuguese public, below those of European counterparts (Freitas and Ávila, 2002). While this may be in part explained by the scant coverage of science in newspapers in the 1980s and 1990s, science was still fairly new for the Portuguese. As an attempt to increase science news in the media and public consumption, the Portuguese government under the rule of Minister Mariano Gago signed an agreement in 1998 with the national news agency Lusa to make science news, national and international, freely available to the national and regional press. This agreement ended in 2003, which might have contributed to the significant decrease in science news in Portugal in the last decade. Online newspapers have appeared in recent years: one example is the *Observador* (created in 2014), which often covers science and technology topics and policy. The decrease in traditional science media coverage, accompanied by the emergence of online newspapers, is a trend found in many countries and not just Portugal (Bauer et al., 2012).

Science media coverage can thus be characterised by a certain instability in the regularity of science news, sections and supplements in newspapers over the last two decades. Today, although most national newspapers including *Expresso*, *Correio da Manhã*, *Público* and *Diário de Notícias*, and cultural magazines such as *Visão* and *Sábado*, cover science topics regularly, *Público* is one of the few publications to include a science section.

The greatest change in publication of popular books on science happened with the establishment of the Portuguese science publishing company Gradiva in 1982, although some science collections from foreign authors had been translated into Portuguese much earlier. An example is the *Cosmos* collection of Portuguese titles, edited in the 1940s by the mathematician and science disseminator Bento de Jesus Caraça. Gradiva made a significant contribution by presenting new science authors to Portuguese audiences. More recently, other national editors such as Presença, Relógio D'Água and Europa-América have been publishing popular science books (Fiolhais, 2011; Granado and Malheiros, 2015).

Coverage of science on television and the radio has traditionally been low. A study on television newscasts in the four Portuguese public TV channels shows that in 2011 only 0.8 per cent of the news was about science and technology (ERC, 2012). There were only a few national TV productions such as the magazine *2001* (1996) or the program *MegaScience* (2004). *MegaScience* was broadcast on public TV with demonstrations of scientific experiments by presenters and guests. On radio the first long-term program was the *Antena 1 Science* (1996–2003), a forum where prominent scientists discussed scientific issues of public interest. Other examples of successful radio programs are *The days of the future* (2007) and *Antena 2 Science* (2009), which are still broadcast today on the public radio stations. The program *90 seconds of science* in *Antena 1*, produced by the New University of Lisbon since 2016 and featuring interviews with scientists about their research, has been very popular having reached around 600 episodes at the time of writing. Overall, despite the instabilities mentioned, science media coverage has grown significantly since the 1980s, due to an increasing availability of content in the editorial market, the press and a growth in public demand.

3.2. By whom? The community of practitioners

The community of science communication practitioners has traditionally been scarce and dispersed. Twenty years ago, it was mostly comprised of personnel working at science museums and a few popularising scientists and journalists, but the situation has changed quite considerably in recent years. This is visible in the increasing number of communication professionals and in the various attempts of professionalisation of science communication in the country, primarily in the shape of science journalism, and more recently PR staff at research institutions and universities (Entradas and Bauer, 2019).

The science journalist community has traditionally been small. The few journalists who reported on science in the 1980s considered themselves pioneers (Machado and Conde, 1988). It is likely that this community has decreased since the early 2000s. The number of journalists working regularly on science issues in the Portuguese media has been recently estimated as about 10 professionals (Granado and Malheiros, 2015), with one or two journalists working at one newspaper or magazine.

The increase in demand for these professionals in recent years is in great part driven by the establishment of the Ciência Viva Agency and its science centres, and the rise of PR offices/communication/marketing (under different names) at universities and research institutes (Entradas and Bauer, 2017). Although the number of science communicators in Portugal is unknown, we could expect a community of a few hundred, although the precise number might be difficult to predict without benchmarking the community. This number might, however, rise significantly if we consider within this spectrum professionals who, although they are not exclusively dedicated to science communication tasks, perform them as part of their jobs. We know from a nationwide study of the Portuguese research institutes conducted in 2015 that around 50 per cent of the research centres in Portugal employ personnel partly dedicated to science communication tasks who often combine their communication roles with administrative functions (Entradas, 2015).

Studies of members of this community in Portugal, although limited, point to an undefined professional identity of the community (Agostinho and Trindade, 2013) seen in the range of professional backgrounds, portfolios and skills, and temporary work contracts (Entradas and Bauer, 2017). This is not a singularity of our country, but rather a trend in many countries (Wellcome Trust, 2015; Buhler et al., 2007; Kohring et al., 2013).

Attempts at professionalisation (Evetts, 2003) are evident in the proliferation of platforms, networks and associations for public science communication, and in universities' efforts in offering training in science communication. A major step was the creation of the Portuguese Science Communication Association (the Rede SciComPT). This network has about 400 members, ranging from communications officers/PR and managers working in research centres and universities, science centres and museums, to science journalists, illustrators and scientists. This association was created in 2013 (and legally established in 2014) by a group of science communicators and science journalists, and aims to 'promote science communication in all its aspects, to enhance the collaboration between science communication professionals and to promote the participation of citizens in all matters involving science and technology' (adapted from Rede de Comunicação de Ciência e Tecnologia de Portugal, 2018). One of its main activities is the organisation of an annual conference normally attended by around 200 participants, though the first national congress of science communication took place in 2013, before the creation of this network (Granado and Malheiros, 2015). The SciComPT conferences have since been organised every year, taking place in science centres, museums and universities in different cities in Portugal, and serving as important meeting points for discussion among science communicators, practitioners and scholars (Rede de Comunicação de Ciência e Tecnologia de Portugal, 2018). Examples of other networks are the online social network group on Facebook SciCom Portugal (created in 2010), where more than 1,800 members interact on science communication topics; and the Finca-Pé discussion group, an informal forum where science managers and communication professionals meet six times per year in the greater region of Lisbon to discuss best practices and ongoing projects (Entradas and Bauer, 2019).

3.2.1. Initiatives for training and education in science communication

Formal training in science communication in Portugal was first directed at professional groups of journalists and scientists. For example, the Technical School for Journalists (Cenjor) developed and ran a three-month course on science journalism in 1999/2000 and in 2005/06 (Granado and Malheiros, 2015), and the Institute Gulbenkian of Science organised a series of science communication workshops for scientists in 2003, 2005, 2006, 2007 and 2010 (Lamas et al., 2007). At about the same time, the government funded the program Scientists in the Newsroom in collaboration with the daily newspaper *Público*, where scientists would spend three months writing news about science in the newspapers' rooms. These were important initiatives to strengthen the relationships between Portuguese journalists and scientists.

In terms of specialised education, the first master's courses in science communication were created in the early 2000s. In 2002/03, the University of Porto created a MSc degree in science communication, but this was discontinued in 2006/07, presumably due to lack of demand. In the following year, the University of Aveiro created a MSc degree in communication and education of science. That also ended in 2006/07. In 2011, the New University of Lisbon created a MSc in science communication, still running today due to the practical focus of the course. In 2017, 14 students completed a degree. The New University of Lisbon also promotes training modules and summer school courses in science communication (FCSH, 2018). In 2017/18, the University of Lisbon opened a MSc degree in scientific culture and dissemination of science; and in 2019/20, the University of Minho started offering a MSc in science communication and the University of Porto, a MSc in science education and dissemination. Workshops or short-term courses in science communication are offered by universities and larger research institutions (e.g. Iberian Nanotechnology Centre in Guarda). It is evident that the number of science communicators is increasing, that they perform a variety of jobs, and the field is beginning to take shape, in part catalysed by these important networks and training initiatives.

4. Final considerations: The impact of top-down initiatives

As we describe here, modern science communication emerged in Portugal over the last 25 years and can be attributed to top-down government initiatives, initiated in the mid-1990s under Mariano Gago's mandate. Despite its recent emergence, Portugal has quickly expanded its infrastructure for science communication and undergone remarkable changes. Some have followed models and trends of other European countries (e.g. measurements of scientific literacy, PUS models and interactive science centres, national initiatives such as the Science and Technology Week and the European Researchers' Night), but others are specific to the Portuguese political and social context, bringing singularities to modern science communication in Portugal in relation to other countries. The most significant is perhaps the national policy for scientific culture described above and the continuing role played by the government in supporting science outreach. This policy, which had initially been implemented through centralised initiatives such as the Ciência Viva Agency, has acquired more dispersed dimensions, with many actors assuming roles in the promotion of science in the country.

We can then ask what the impact has been of these government top-down initiatives on the development of science communication in our country. There are many indicators that point to a greater openness and accessibility of Portuguese science to society during recent years. The most prominent examples are the increasing number of initiatives for the public organised by universities and research centres (Entradas and Bauer, 2017), the network of Ciência Viva centres across the country, and the increased number of communication professionals. All these point to a national spread of science communication. However, we cannot attribute the national expansion entirely to national policies. Along the way, other factors have fostered the continued effort seen in Portugal in the field, particularly in more recent years. These include the resources allocated to public communication and professionalised staff in research institutions and universities; European demands and directives; and the overall international mobilisation for science communication. We can, nevertheless, say the national policy was the turning point and the motive for the beginning of a commitment to science communication in the country—the top-down efforts have certainly promoted scientific culture in Portugal.

This does not mean, however, that the field has become fully integrated in the scientific and societal spheres. It suffers from lack of resources and professionalisation, and public participation in research and policy is marginal (Entradas and Bauer, 2017). An explanation may lie in the national policies themselves, which foster a culture of increased scientific literacy, emphasising unidirectional 'deficit' approaches to communication—these may have inhibited a more intimate public involvement in science (Entradas, 2015). This raises questions such as to whether the dominant unidirectional practices are a response to the national policies, or a lack of understanding/interest in adopting mechanisms for public involvement, or national constraints such as lack of public interest or opportunities to participate and maintain a more decisive role in decision-making. Despite significant achievements over the years, much remains to be done to engage Portuguese society in science as required in modern societies. Science communication in Portugal could benefit from closer collaborations between the high diversity of professionals and stakeholders already involved in science communication and the broadening of bottom-up approaches to promote more dialogical communication—for example, setting up more structures to involve citizens in decision-making around science-related issues, and adopting successful models of public participation from neighbouring countries. To conclude, the initially adopted deficit model of communication has brought a certain amount of success, but it is now time to open modern science communication to other approaches such as dialogue and discussion in order to get a greater involvement and trust in science.

References

Agostinho, M. and Trindade, M. (2013). Research Management in Portugal: A Quest for Professional Identity. *Research Management Review, 20*(1), 1–8.

Ávila, P. and Castro, P. (2003). Compreender a Ciência: O Inquérito à Cultura Científica dos Portugueses. In M. E. Gonçalves (ed.), *Os Portugueses e a Ciência.* Lisboa: Publicações Dom Quixote.

Bauer, M. W., Howard, S., Romo, Y. J., Massarani, L. and Amorim, L. (2012). *Global science journalism report: Working conditions and practices, professional ethos and future expectations.* London: Science and Development Network.

Brigola, J. C. (2003). *Colecções, Gabinetes e Museus em Portugal no Século XVIII.* Lisboa: FCT Fundação Calouste Gulbenkian.

Brigola, J. C. (2010). *Os viajantes e o 'livro dos museus'.* Porto: Dafne Editora.

Buhler, H., Naderer, G., Koch, R. and Schuster, C. (2007). *Hochschul-PR in Deutschland. Ziele, Strategien und Perspektiven.* Wiesbaden: Deutcher Universitatserlag.

Castro, P. and Lima, L. (2003). Discursos sobre a Ciência num Debate Ambiental. In M. E. Gonçalves, (ed.), *Os Portugueses e a Ciência.* Lisboa: Publicações Dom Quixote.

Ciência Viva. (2017). Orçamento participativo Ciência Portugal. Retrieved from www.cienciaviva.pt/concurso/oppciencia/.

Conceição, C. (2011). Promoção de Cultura Científica. Análise Teórica e Estudo de Caso Do Programa Ciência Viva (PhD thesis). ISCTE – Instituto Universitário de Lisboa, Portugal.

Costa, A. F., Ávila, P. D. and Mateus, S. (2002). *Públicos da ciência em Portugal.* Lisboa: Gradiva

Costa, A. F., Conceição, C., Pereira, I., Abrantes, P. and Gomes, M. do C. (2005). *Cultura Científica e Movimento Social Contributos para a análise do programa Ciência Viva.* Oeiras: Celta Editora.

Delicado, A. (2006). Os Museus e a Promoção da Cultura Científica em Portugal. *Sociologia, Problemas e Práticas, 51*, 53–72.

Delicado, A., Cortez, A., Vala, F., Gago, M. do M. and Casaleiro, P. (2010). Comunicar ciência numa exposição: uma avaliação exploratória de A Evolução de Darwin através de PM. *Actas do I Seminário de Investigação em Museologia dos Países de Língua Portuguesa e Espanhola* (pp. 8–18). Porto.

Delicado, A., Rego, R., Conceição, C. P., Pereira, I. and Junqueira, L. (2013). *Ciência, Profissão e Sociedade – Associações Científicas em Portugal.* Lisboa: Imprensa de Ciências Sociais.

DGEEC. (2018). *Principais resultados do RAIDES 17 – Inscritos 2017/18.* Lisboa: DGEEC.

Einsiedel, E. F., Jelsøe, E. and Breck, T. (2001). Publics at the technology table: The consensus conference in Denmark, Canada, and Australia. *Public Understanding of Science 10*(1),83–98.

Entradas, M. (2015). Envolvimento societal pelos centros de I&D. In Maria de Lurdes Rodrigues e Manuel Heitor (org.) *40 Anos de Políticas de Ciência e de Ensino Superior*. Lisboa: Almedina.

Entradas, M. and Bauer, M. W. (2017). Mobilisation for public engagement: Benchmarking the practices of research institutes. *Public Understanding of Science, 26*(7), 771–88. doi.org/10.1177/0963662516633834.

Entradas, M. and Bauer, M. W. (2019). Kommunikationsfunktionen im Mehrebenensystem Hochschule. In *Forschungsfeld Hochschulkommunikation* (pp. 97–122). Wiesbaden: Springer VS. doi.org/10.1007/978-3-658-22409-7_5.

ERC. (2012). *Ciência no Ecrã – A divulgação televisiva da atividade científica.* Lisboa: ERC and IGC.

European Commission. (1992). *Special Eurobarometer 38.1. Europeans, Science and Technology*. European Commission.

European Commission. (2001). *Eurobarometer 55.2: Europeans, science and technology.* Retrieved from www.ec.europa.eu/public_opinion/archives/eb_special_en.htm.

European Commission. (2005). *Eurobarometer 63.1: Europeans, science and technology.* Retrieved from www.ec.europa.eu/public_opinion/archives/eb_special_en.htm.

European Commission. (2010). *Eurobarometer 340/Wave 73.1: Science and technology.* Retrieved from www.ec.europa.eu/public_opinion/archives/eb_special_en.htm.

Evetts, J. (2003). The sociological analysis of professionalism – Occupational change in the modern world. *International Sociology, 18*(2), 395–415. doi.org/10.1177/0268580903018002005.

Faculdade de Ciências Sociais e Humanas. (FCSH) (2018). *Comunicação de ciência.* Retrieved from fcsh.unl.pt/ensino/mestrados/comunicacao-de-ciencia.

Fiolhais, C. (2011). *A ciência em Portugal.* Fundação Francisco Manuel dos Santos.

Fiolhais, C. (2014). *História da ciência em Portugal.* Lisboa: Editora Arranha-Céus.

Fonseca, R. (2017). *A Ciência e a Tecnologia na Imprensa Portuguesa: 1976 – 2005: Entre jornais 'populares' e de 'qualidade'.* Novas Edições Acadêmicas.

Freitas, E. and Ávila, P. (2002). *Inquérito à Cultura Científica dos Portugueses (Final Report).* Lisboa: OCT.

Gago, M. (1990). *Manifesto para Ciência em Portugal: Ensaio.* Lisboa: Gradiva.

Garcia, J. L., Silva, P. A. and Ramalho, J. (2016). *O Público da Rede Nacional de Centros de Ciência Viva – Relatório Final.* Lisboa: Instituto de Ciências Sociais.

Gonçalves, M. E. (1996). Ciência e Politica em Portugal, o caso da doença das 'vacas loucas'. In M. E. Gonçalves (ed.), *Ciência e Democracia* (pp. 121–40). Lisboa: Bertrand.

Gonçalves, M. E. (ed.). (2001). *O Caso de Foz Coa, Um Laboratório de Analise Sociopolítica.* Lisboa: Edições 70.

Gonçalves, M. E. (2003a). Imagens Publicas da Ciência e Confiança nas Instituições: Os Casos de Foz Coa e da Coincineração. In M. E. Gonçalves (ed.), *Os Portugueses e a Ciência.* Lisboa: Publicações Dom Quixote.

Gonçalves, M. E. (ed.). (2003b). *Os Portugueses e a Ciência.* Lisboa: Publicações Dom Quixote.

Gonçalves, M. E. and Castro, P. (2002). Science, culture and policy in Portugal: A triangle of changing relationships? *Portuguese Journal of Social Sciences, 1*(3), 157–73. doi.org/10.1386/pjss.1.3.157.

Granado, A. and Malheiros, J. V. (2015). *Cultura Científica em Portugal.* FFMS, Lisbon. Retrieved from www.ffms.pt/FileDownload/54fca75d-9ddf-448c-b153-7c9c46753e58/cultura-cientifica-em-portugal.

Gregory, J. and Miller, S. (1998). *Science in public: Communication, culture and credibility.* London: Plenum Press.

Hetland, P. (2014). Models in science communication policy: Formatting public engagement and expertise. *Nordic Journal of Science and Technology Studies, 2*(2), 5–17. doi.org/10.5324/njsts.v2i2.2144.

House of Lords, Select Committee on Science and Technology. (2000). *Science and Society.* London: House of Lords, UK.

Instituto Nacional de Estatística. (2002). *Estatísticas da Cultura, Desporto e recreio 2000.* Lisboa: INE.

Instituto Nacional de Estatística. (2013). *Estatísticas da Cultura 2012.* Lisboa: INE.

Instituto Nacional de Estatística. (2014). *Estatísticas da Cultura 2013.* Lisboa: INE.

Instituto Nacional de Estatística. (2015). *Estatísticas da Cultura 2014.* Lisboa: INE.

Instituto Nacional de Estatística. (2016). *Estatísticas da Cultura 2015*. Lisboa: INE.

Instituto Nacional de Estatística. (2017). *Estimativas Anuais da População Residente*. Lisboa: INE.

Jardim Zoológico de Lisboa. (2018). *O Zoo: História*. Retrieved from www.zoo.pt/en/.

Jesuíno, J. (2001). O caso de Foz Côa: um fórum híbrido. In M. E. Goncalves (ed.), *O caso de Foz Côa: um laboratório de análise sociopolítica*. Lisboa: Edições 70.

Jesuíno, J. C., Amâncio, L., Avila, P. Carapinheiro, G, Costa, A. F., Machado, L., Patrício, T., Stoleroff, A. and Vala, J. (1992). *A Comunidade Cientifica Portuguesa nos Finais do Seculo XX*. Lisboa: Celta.

Jesuíno, J. and Diego, C. (2003). Estratégias de Comunicação dos Cientistas. In M. E. Goncalves (ed.), *O caso de Foz Côa: um laboratório de análise sociopolítica*. Lisboa: Edições 70.

Kohring, M., Marcinkowski, F., Lindner, C. and Karis, S. (2013). Media orientation of German university decision makers and the executive influence of public relations. *Public Relations Review*, *39*, 171–77. doi.org/10.1016/j.pubrev.2013.01.002.

Lamas, S., Araújo, S. J., Bettencourt-Dias, M. and Coutinho, A. G. (2007). Os Cientistas como Agentes na Comunicação de Ciência: Motivação, Formação e Iniciativas em Portugal. In L. Barbeiro (ed.), *Comunicação de Ciência*. Porto: Sete Pés.

Leão, M. J. and Castro, S. (2012). Science and Rock: How music festivals can boost the progress of science. *EMBO Reports*, *13*(11), 954–58. doi.org/10.1038/embor.2012.151.

Lima, L. (1995). *Estacão de tratamento de resíduos sólidos urbanos (Freguesia de S. João da Talha – Loures), Relatório de impacte social*, Lisboa: IPAMB.

LNEG (Laboratório Nacional de Energia e Geologia). (2018). Museu Geológico. Brief historical view. Retrieved from www.lneg.pt/museugeologico/.

Machado, F. L. and Conde, I. (1988). A divulgação científica em Portugal: do lado da produção. *Sociologia – Problemas e Práticas*, *5*, 11–38.

Mendes, H. (2003). Visibilidade da Ciência nos Mass Media: A tematisação da Ciência nos jornais Público, Correio da Manhã e Expresso (1990 e 1997). In M. E. Gonçalves (ed.), *Os Portugueses e a Ciência*, Lisboa: Publicacoes Dom Quixote.

Miller, S., Caro, P., Koulaidis, V., De Semir, V., Staveloz, W. and Vargas, R. (2002). *Report from the expert group-benchmarking the promotion of RTD culture and public understanding of science*. Brussels: European Commission.

Oceanário de Lisboa. (2015). EMAS 2015. IV Declaração Ambiental. Retrieved from www.oceanario.pt/content/files/emas_2015_oceanario_de_lisboa.pdf.

OCT (Observatório das Ciências e Tecnologias). (1996). Relatório do Inquérito a Cultura Cientifica dos Portugueses, Lisboa, OCT.

OCT (Observatório das Ciências e Tecnologias). (2000). Relatório do Inquérito a Cultura Cientifica dos Portugueses, Lisboa, OCT.

Pinto, B. and Amorim, I. (2018). 'A biodiversidade marinha nos museus de Portugal Continental: Uma introdução'. *Revista Museologia e Interdisciplinaridade* 7(14), 107–27.

Rede de Comunicação de Ciência e Tecnologia de Portugal. (2018). Rede SciComPT. Retrieved from www.scicom.pt/.

Rodrigues, C. (2015). Participation and the quality of democracy in Portugal. *Revista Crítica de Ciências Sociais, 108*, 75–94. doi.org/10.4000/rccs.6111.

Rosas, F. and Sisifredo, C. (2013). *Estado Novo e universidade: a perseguição aos professores*. Tinta da China.

Royal Society. (1985). *The Public Understanding of Science*. London: Royal Society.

Schiele, B. (2008). Science museums and science centres. In M. Bucchi and B. Trench (eds), *Handbook of public communication of science and technology* (pp. 2–40). New York: Routledge.

Soares, A. G. (2008). Portugal and the European Union: The ups and downs in 20 years of membership. *Perspectives on European Politics and Society*, *8*(4), 460–75. doi.org/10.1080/15705850701640835.

Teixeira, P., Cardoso, M. F., Sarrico, C. S. and Rosa, M. J. (2007). The Portuguese public university system: on the road to improvement? In A. Bonaccorsi and C. Daraio (eds), *Universities and strategic knowledge creation: Specialisation and performance in Europe*. Edward Elgar Publishing.

Wellcome Trust. (2015). *Factors affecting public engagement by researchers. A Study on behalf of a consortium of UK public research funders*. Wellcome Trust.

Wynne, B. (1996). Misunderstood misunderstandings: social identities and public uptake of science. In *Misunderstanding science: The public reconstruction of science and technology*, vol. 1 (pp. 19–46). United Kingdom: Cambridge University Press. doi.org/10.1017/CBO9780511563737.002.

Timeline

Event	Name	Date	Comment
First interactive science centre established.	Centro Ciência Viva do Algarve	1997	
An association of science writers or journalists or communicators established.	Rede de comunicação de ciência e tecnologia de Portugal [Science and Technology Communication Network of Portugal]	2014	Rede SciCom PT
First university courses to train science communicators.	Instituto Gulbenkian da Ciência (IGC)	2004–10	Workshops in science communication
First master's students in science communication graduate.	MSc in Science Communication, University of Porto	2002/03	
First PhD students in science communication graduate.	University of Coimbra and University of Minho	2015	Theses in science communication have been completed in PhD programs in sociology (ISCTE, ICS) and science education
First national conference in science communication.	At Pavilion of Knowledge in Lisbon	2013	
National government program to support science communication established.	Ciência Viva Program	1996	
First significant initiative or report on science communication.	*Públicos da Ciência em Portugal*	2002	
National Science Week founded.	Science and Technology Week	1998	
First significant radio programs on science.	*Antena 1 Science*	1996	One national radio channel produced by the Portuguese public broadcasting Rádio e Televisão de Portugal

Event	Name	Date	Comment
First significant TV programs on science.	*2001* (RTP2)	1996	A Portuguese 24-hour public service news channel owned by Rádio e Televisão de Portugal (RTP)
First awards for scientists or journalists or others for science communication.	Ciência Viva Montepio Prize	2012	For public communication work (big prize); educational projects (education prize); and science dissemination in the media (media prize)
Other significant events.	Ecsite Annual Conference	2017	Hosted by the Natural History and Science Museum, University of Porto

Contributors

Dr Marta Entradas is principal investigator at CIES, ISCTE-IUL, Lisbon University Institute, Portugal, and visiting fellow at the London School of Economics.

Dr Luís Junqueira is a research fellow at the Institute of Social Sciences, University of Lisbon, Portugal.

Dr Bruno Pinto is an assistant researcher in science communication at the University of Lisbon, Portugal.

30

RUSSIA

Russian pendulum: From glorious science propaganda to modest public engagement initiatives

Alexandra Borissova and Dmitry Malkov

1. Introduction

In 2017, Russia commemorated 100 years since the October revolution. As a matter of convenience, in this chapter we stick to the same time frame by summarising 100 years of evolution of modern science communication in Russia, from the space race–era science propaganda to the present-day Westernised approach. Not coincidentally, this evolution closely followed the train of political and ideological thought in the country, thus drawing a visible line between Russia and other world regions. To help our readers, we will highlight several historic periods unified by the same way of thinking about science communication. Throughout the chapter we will resort to different terminology to refer to science communication. Until quite recently, the prevailing term used to describe the field in Russia has been 'science popularisation', rather than communication. This heritage is still quite noticeable as the term enjoys widespread usage by the community and state officials. Even such terms as 'science propaganda' and 'enlightenment' have been somehow preserved and are used today. 'Science communication' is a relatively new term in Russia and was introduced to refer to public communication of science as a separate profession around 2010. This name is not as widely used by practitioners, but a growing number of people recognise it as a unifying term for both the professional field and the rising academic discipline of the same name. We do not plan to elaborate further on terminology but will attempt to trace the conceptual evolution of the field.

2. 'We will go another way' (1917–91)

The headline quote, attributed to Vladimir Lenin (in fact, it belongs to the poet Vladimir Mayakovsky), not only perfectly reflects the Soviet path in general, but also the development of science and science communication during the communist period. The objectives of science communication, or rather science propaganda as it was referred to at the time, were entirely politically driven. The young Soviet state proclaimed priority tasks of modernisation, industrialisation and urbanisation of the largely rural and illiterate country. Patriotic slogans declared the tsarist past to be scientifically backward, while the socialist future was seen as radiant and technologically advanced (Andrews, 2003). Therefore, alongside unprecedented strengthening of formal education (known as *likbez*, i.e. 'elimination of illiteracy'), massive efforts were made to introduce what is now known as informal science education or science popularisation. Implemented on the federal level, these programs enjoyed huge centralised support both politically and financially. This awareness was only reinforced by the Cold War as the Soviet Union strived to demonstrate its scientific and technological superiority over the United States and other countries.

Science propaganda via public lectures and discussions in the Soviet Union had an unimaginable reach by modern-day Russian standards, or by any standards for that matter. One of the flagships of this activity was the Znanie Society [Knowledge Society]. Signed into existence by Joseph Stalin in 1947 (Litvak, 2008), it was commissioned with the popularisation of scientific and political ideas—a combination that testifies to the predominantly political purpose of the Soviet-style science popularisation. By the 1970s, the society organised a variety of public lectures, themed nights, radio and television broadcasts and produced popular science films. In 1976 alone, the members of the society conducted more than 24 million public lectures (Nauka i Zhizn, 1977). The logic of the Soviet state assumed that every citizen had to undergo continuing education for the sake of the national economy and communism. Another example of the whopping outreach of the Znanie Society is the Soviet weekly newspaper *Argumenty i Fakty* [Arguments and Facts], which belonged to the society. In 1990, the newspaper entered the Guinness Book of Records for the largest weekly newspaper circulation in human history, namely 33.5 million copies (Fishman, 2005, p. 194).

Figure 30.1: One of many Soviet posters used to promote science. The messages on the poster testify to the political roots of popular science of that era: 'Science and communism are inseparable!' and 'October revolution opened pathway to space!'.

Source: Public domain (softsalo.com/sovet_45_poli/poli_53.html).

Science magazines also enjoyed extremely high circulations. The oldest and most prominent Russian science magazine is *Nauka i Zhizn* [Science and Life]. Founded in 1890, it promptly became one of the foremost popular science magazines in Soviet times and grew from a 16-page weekly digest to a monthly 64-page edition with a circulation of around 3.5 million copies in the 1960s–80s (Yakovenko, 2012). The magazine was, of course, politicised—its long-term vice editor-in-chief was Rada Adzhubey, daughter of Nikita Khrushchev, the First Secretary of the Communist Party, 1953–64. Apart from *Nauka i Zhizn*, this stage was also championed by *Znanie – Sila* [Knowledge is Power] with 700,000 print copies distributed monthly and *Khimiya i Zhizn* [Chemistry and Life] with 300,000 monthly copies. About a dozen other titles were widespread and could be found in any household. Conceived and perceived as a form of leisure for the entire family, these magazines often contained crosswords, food recipes and other information, which today we might deem irrelevant to science. Although they provided facts and figures, these periodicals suffered from the lack or even complete absence of any critical view or investigative journalism, as well as self-inflicted isolationism. Naturally, discussions about science policy were off the table entirely.

The immense circulations exercised by big popular science magazines required a large intellectual work force. Consistent science journalism education was not yet in place, but several schools did prepare science writers. Perhaps the most prominent and longest living science journalism studio was created early in the 1980s by the *Khimiya i Zhizn* magazine. Its target audience was mainly science and engineering students who gathered occasionally to test their science writing skills, discuss texts and sometimes just play the guitar. With time the science journalism studio became a unique source of writers for *Khimiya i Zhizn* and similar magazines.

Science was a subject of motion movies and TV programs. The flagship was *Ochevidnoe-neveroyatnoe* [Obvious-unbelievable], started in 1973 by a prominent Russian scientist and populariser Sergey Kapitsa. A son of the Nobel laureate physicist Pyotr Kapitsa and distinguished physicist himself, Sergey Kapitsa suffered from the Sagan effect[1] during his research career due to his public activity. Various sources report high officials from the USSR Academy openly telling his father that Sergey would not be elected to the Academy if he started a TV show. He was, indeed, never elected. Nevertheless, his TV program lasted for decades until his death in 2012, and Kapitsa Jr remains an icon of the modern-day science engagement in Russia (Borisova and Malkov, 2018).

Other forms of science popularisation kept pace. Science and technology museums were built as far as the eye could see to commemorate the glory of the Soviet technological progress. A considerable number of the 245 natural history museums and around 100 science museums we know today were built in the Soviet period (Russian Venture Company, 2016). Not surprisingly, more than half the science and technology museums were aerospace museums, a tribute to the country's historical developments in space exploration. Museums such as the Moscow Polytechnic Museum and the Memorial Museum of Cosmonautics were crown jewels of Soviet science propaganda, underpinned by the one-party system and fierce political support.

In parallel to these broad propaganda efforts, another way of thinking proliferated in Soviet society. Some scholars argue (Kukulin, 2017) that the seeds of the 'new age' thinking and various sorts of occultism were planted through popular science magazines of the Soviet era that freely published articles about paranormal phenomena, UFOs and other ideas of this kind alongside scientific pieces. Beliefs such as those of ancient astronauts started

1 The Sagan effect can be defined in various ways. One of these versions states that the scientific popularity of a scientist with the general public is considered to be inversely proportional to the quality and quantity of that scientist's scientific work (Baldscientist, 2013).

to spread from academic circles with both natural scientists and humanities scholars involved (Panchenko, 2018) and reached the audience of engineers and other parts of so-called Soviet intelligentsia.

3. Back to square one (1991–2001)

With the collapse of the Soviet Union at the beginning of the 1990s, the popularisation of science plummeted together with science itself. The grave economic situation (Milanovich, 1998) led to dwindling science funding and left virtually no room for any institutional support of science communication. In 1992, the total expenditures on science from all sources halved. By 1994 science funding in Russia as a share of GDP was about one-sixth of that of developed Western countries (Graham and Dezhina, 2008). Science and technology propaganda came to be perceived by the population as manipulative and patronising. With the influx of freedom in the 1990s, new age thinking spread from white-collar circles to a broader, less educated audience. Hence, the arrival of the 1990s meant a dramatic shift from the national dissemination of science towards other forms of leisure and interest areas. Foreign literature on culture, philosophy and spirituality cracked through the restrictions imposed by the Soviet government. Science was no longer appealing to the population and the country submerged in mysticism and obscurantism that still can be tracked in the modern Russia.

Under these circumstances, science communication naturally fell low in the list of priorities. While many Western countries were busy elaborating on novel approaches to science communication, including dialogue and participatory engagement models, there were no significant developments in this area in 1990s Russia whatsoever. Activities such as public lectures endured, of course, but their content switched to philosophy and humanities. The celebrated Auditorium of the Moscow Polytechnic Museum, which once hosted scientific luminaries such as Ilya Mechnikov or Niels Bohr, was now rented for public events on anything but science and technology. The Znanie Society crumbled and popular science lectures virtually disappeared. The society was partly brought back to life in 2015 under the Russian Ministry of Science and Higher Education, but it never came close to its past Soviet grandeur (State Legal Information System, 2015).

The fate of the once flourishing landscape of popular science magazines is documented in the history of the *Nauka i Zhizn* magazine. Once an icon of the Soviet Union with a circulation of millions of copies monthly, the magazine was reduced to a thick 144-page monthly edition and a circulation of 36,000 copies. *Nauka i Zhizn* managed to survive throughout the 1990s, but the print version never fully recovered. Later in the digital age, its online version managed to become one of the busiest in the market with a monthly audience of over 750,000 unique visitors (Rambler Top 100, n.d.). This case is quite unusual, however, as the vast majority of titles simply ceased to exist in the 1990s and never came back. What used to be a rich market in the USSR is barely a market at all in modern Russia (Yakovenko, 2012). Indeed, today there are print magazines with similar content, but there remain no strong players who are able to compete. Consequently, these print editions have been reduced to a niche product for a specific science-oriented audience and have lost their outstanding public role.

As the Russian economy started to recover after the default of 1998, some developments did occur in the late 1990s against the background of apathy towards science and science popularisation. For example, in 1998, *Khimiya i Zhizn* launched a news agency called InformNauka, which aimed at disseminating Russian science both within and beyond the country. To supply the agency with science writers, *Khimiya i Zhizn* relaunched its school of science journalism. The school was terminated in 2008, once management decided that they no longer needed to produce writers for the magazine and the agency, which itself was phased out a few years later. Several key radio programs also appeared at the end of this period. The oldest science program still active on the radio is *Granit nauki* [Granite of Science]. It started in 1997 as part of the then liberal radio station Echo of Moscow, which was owned by Gazprom-media. Radio Liberty in Russia also featured science in its programs since 1998, and later they strengthened this direction with their website broadcasts.

4. Privatisation of science communication (2001–11)

The situation slowly began to change around 2001. Irritated by the public contempt for science and the consequent rise in mysticism in Russia, several prominent figures stood up. Perhaps the most important of them was Dmitry Zimin, a Russian telecommunications tycoon with a profound respect for science, technology and education (Carnegie Medal for Philanthropy, 2013). Along with some other business owners, after the turbulent arrival of the market economy to the country, Dmitry Zimin adopted a more comprehensive model of corporate social responsibility. In 2001, he established the Dynasty Foundation, Russia's first private funder of science and science popularisation, which, among other things, began providing grant support to book publishing, science festivals, lectures and science museums. We will further focus on the activity of this foundation, as it was the first and for the larger part of its existence the only foundation of its kind in Russia and had an enormous impact both on science communication and philanthropy.

At first, the foundation focused on translating and publishing popular science books by foreign authors as Russian scientists at that time were not eager to engage in popularisation activities again. In fact, the academic community had come to despise this activity. The same was applicable to popular science lectures organised by the Dynasty Foundation in the early 2000s. Instead of inviting Russian scientists, the foundation mainly had to rely on foreign lecturers, frequently of significant calibre, such as various Nobel laureates (Dynasty Foundation, n.d.). Only by the mid-2000s did more Russian scientists begin to take up the role of public educators and lecturers. This allowed the foundation to organise entire science festivals (under the name Science Days) in all corners of the country. The Dynasty Foundation also introduced a variety of more relaxed and informal formats to Russia, i.e. science cafés. Most of the Dynasty-funded events were organised by enthusiasts and with little financial aid, but there is no doubt that the flow of private money throughout the 2000s helped popular science events and other activities regain some ground.

Figure 30.2: Dmitry Zimin, founder of the Dynasty Foundation.

Source: Maria Olendskaya (used with permission).

Independently from the Dynasty Foundation, groundwork was being laid for the largest present-day popular science event in Russia—the All-Russian Festival of Science NAUKA 0+. As the story goes, the festival owes its existence to European science journalists, who during a visit to the Lomonosov Moscow State University in 2005 asked the rector why there was no large-scale science festival in Russia (Ria Novosti, 2013). The university decided to take the risk and so the idea of a new national science festival was born. The first edition in 2006 was a local popular science event attended by 20,000 people and it quickly paved the way for a city-wide event the next year. But the true nationwide expansion occurred only later, as we discuss in the next section.

Another area of science popularisation transformed with the help of the Dynasty Foundation was science museums and centres. From 2006 to 2014, the Dynasty Foundation conducted eight grant calls for regional museums, which summed up to about 150 million rubles in funding to 90 museum projects across Russia. The foundation also provided grant holders with international internship opportunities at the Copernicus Science Center in Warsaw, eventually attended by almost 30 directors of regional Russian science museums.

This grant program was run by Irina Aktuganova, one of the pioneers of the modern revival of Russian science museums. According to her, when the program was launched in 2006, science museums in Russia were in a state of hopeless stagnation. Conventional science museums merely survived. At the same time, the idea of a science centre was represented by solitary bottom-up projects, such as the Experimentary in Irkutsk, which was created from scratch in 2005 by enthusiasts at the Siberian branch of the Russian Academy of Sciences. Curiously, it was inspired by the San Francisco Exploratorium. The idea of creating an interactive science centre occurred to the Russian biologist Konstantin Kravchenko when he visited the Exploratorium during his trip to the US in 1998 (Palshina, 2010). History repeated itself in 2012. After spending a year in San Francisco as a researcher, another young Russian scientist, Anton Sharypov, returned to his hometown of Krasnoyarsk (also in Siberia) to pursue the dream of creating his own science centre, Newtonpark (Burova, 2014). Given that back in the early 2000s, neither science museums nor science centres were of particular interest to anybody, the financial and expert support provided by the Dynasty Foundation made an invaluable contribution to the regrowth of the Russian science museum community.

Thanks to the Dynasty Foundation, the representation of researchers in the public domain—both politically and content-wise—expanded a great deal too. A popular science project called *Elementy* [Elements] got underway with foundation support in 2005. The website facilitated access to various resources that helped scientists get started with their non-academic writing. In 2008, another outlet called *Troitsky variant – Nauka* joined the Dynasty team as a political wing. Dubbed 'a newspaper of scientists and science journalists', it became (and in part remains) a place of vivid debate about science and education policy in Russia. Nearly everything became a subject of debate and criticism. As policymaking in the field of science suffered (or enjoyed, depending on the perspective) a great deal of change in the previous 25 years, *Troitsky variant – Nauka* provided mass media with a stable flux of stories (Graham and Dezhina, 2008).

While there were only a handful of dedicated science magazines, the 2000s was the era when science finally took off in general interest media outlets—independent, competitive and quickly growing in readership. Weekly magazines of political or business profiles were perhaps among the leaders in the popular science market during this period. Such periodicals as *SmartMoney*, *The New Times, Itogi* and *Russian Newsweek* featured elaborate science sections, produced by reporters with a predominantly scientific background. However, this did not last long as they fell victim to the paper press crisis in 2008–09 and, from 2012, they all were shut down, while the vast majority of authors left science journalism for good. Daily newspapers followed a similar trend. In 2000–06, Pleiades Publishing—a publisher of English translations of Russian scientific journals—ran a science communication project in the general interest media. According to its vice director-general, Nikolay Avanesov, the company financed the weekly supplements to *Literaturnaya gazeta* [Literary Newspaper], *Izvestiya* [News], *Parlamentskaya gazeta* [Parliamentary Newspaper] and *Rossiyskaya gazeta* [Russian Newspaper] that were dedicated to science, medicine and education.

When funding for this project came to an end, these outlets did not manifest independent interest in covering science. Science and technology sections were the first to be cut under economic pressures. At the same time, entire newspapers were being urged to go digital, but in online mode they had to compete with a new type of media—online newspapers. Three key players in this market were *Gazeta.ru*, *Lenta.ru* and *Polit.ru*, all of them founded from scratch as purely online media outlets in 1998–99 (Sapun, 2016). Their designated science sections appeared about five to six years after their creation, and survive to this day. The internet enjoyed the best of editorial freedom, and these outlets became a real school for a number of reporters who later created independent projects that shape contemporary Russian science journalism to a great extent.

Another important development during this period was the creation and growth of organised civil society groups related to science. Among the very first and most prominent is the Commission on Pseudoscience and Research Fraud, established in 1998 by the Russian physicist and Nobel laureate Vitaly

Ginzburg. The commission provided recommendations concerning arbitrary scientific issues and publicly criticised pseudoscience and paranormal beliefs. Another society called the Science Journalist's Club was founded in 2004 and brought together nearly all science popularisers in Russia at the time, including publicly active researchers. Note, however, that the plethora of science reporters of that period worked part-time in addition to their main, usually research-related occupation. Therefore, the line between science journalists and active researchers was quite vague. The club was in constant touch with other actors (e.g. club members initiated a series of translations of international popular science bestsellers that were then supported by the Dynasty Foundation as well as the *Elementy* website).

In 2009, in cooperation with the Commission on Pseudoscience and Research Fraud, the club initiated a broad public discussion that led to the withdrawal of public funding about to be allocated to the production of water filters promoted by the pseudoscientist Victor Petrik. The filters were advertised as a panacea from all the complications of city water supply and their inventor was publicly supported by top members of Russian Academy of Sciences and by Boris Gryzlov, back then the speaker of the lower chamber of the Russian parliament (Nazaretyan, 2010; White, 2010). However, an independent investigation launched by a member of the Commission and supported by many science journalists found that there was no evidence of the efficiency of these filters, as no proper tests had been performed. On the contrary, the filters were contaminating water with heavy metal ions. Civil society celebrated a big victory. Initially, the government planned to implement the filters country-wide, spending the equivalent of $500 billion by 2020, but after the debunking this amount went down to $5.5 million by 2017, and effectively even this money was only partially spent. However, the filters were installed in over 600 public institutions, including schools (Podorvanyk and Alexeeva, 2010).

Corporate science communication in universities and research institutes remained perhaps the largest lacuna in the market. In the Soviet Union, science and education were essentially separate entities. Most research was done in various institutes. Those of the Academy of Sciences were responsible for basic research, while others, supervised directly by the ministries, performed applied research and technology transfer. As for educational organisations, two types existed: *universitet* and *institute*. The former were fewer in number and gave basic education, preparing graduates for a research career. The second category provided purely applied training, thus preparing practitioners like teachers, engineers, doctors, etc. All *institutes* and nearly all *universitets* were teaching-only, and their academic staff were not involved in any research. For historical reasons this system is very similar to what we see in Germany with its universities and *hochschule*, as well as various societies of research institutes. Thus, Soviet universities were alien not only to their third, but also to their second mission (Scott, 2006). Recall that universities are not only responsible for developing human capital (education—the first mission) and producing new knowledge (research—the second mission), but they must also engage with societal needs and market demands by establishing links with the surrounding socioeconomic environment (the third mission). Today, most universities develop their strategies around these three missions. In the framework of Soviet society, centralised science popularisation functions along with the absence of competitiveness and civil society led to a model where communication offices just did not belong. Research organisations enjoyed funding settled by the government plan and did not need to perform any outreach activities, either for the sake of academic competitiveness or to engage with the public. This status quo remained and prevailed until the late 2000s with a couple of exceptions. Russian research organisations lagged behind in science communication and received a wakeup call much later.

The crisis in the mass media market and abrupt changes in the Russian political course hindered many of the projects discussed above, but their contribution did not slip through the cracks. On the contrary, this decade was a game changer for Russian science communication. The combination of private funding and extensive media coverage put science back on the public radar and brought science communication into the focus of the government.

5. The empire strikes back (2011–16)

The 2010s brought serious perturbations with respect to how science communication is sponsored and implemented. With the Russian budget enjoying higher oil and gas prices from the early 2000s (Sonin, 2004), the state finally became a bigger player in science and brought a greater asymmetry in the allocation of power and resources. This period is marked by a tectonic shift in the Russian academic system, which consequently affected science communication.

In 2013, the government initiated a long-expected reform of the Russian Academy of Sciences, accelerating the dismantling of the Soviet-style science management. The same year, a national academic excellence program called Project 5-100 was launched. The project was aimed at improving the position of leading Russian universities in international university rankings and, as a side effect, triggered a wave of important developments in corporate science communication. Bigger goals aided by additional funding allowed the universities to employ experienced and ambitious press officers and science communicators. This movement, however, was complicated by a lack of these specialists in the market. As a result, some of the newborn PIOs (public information officers) were former science journalists, others were researchers and lecturers, and yet others were public relations professionals in a broader sense. These people came out of their comfort zone to enter or even create an entirely new profession for the country. Project 5-100 thus became a locomotive of corporate science communication, ensuring that at least a dozen of the top Russian universities established and expanded their public information offices. Among the drivers of this process were the Moscow Institute of Physics and Technology (MIPT), Higher School of Economics (HSE), National University of Science and Technology MISiS, ITMO University in St Petersburg, Tomsk State University, Ural Federal University in Yekaterinburg, and several others.

In turn, the reform of the Russian Academy of Sciences (Clark, 2013) changed drastically an institution that was once famous worldwide as the flagship of Soviet science. Unlike European or American academies that are mostly expert bodies or honorary clubs, the Soviet (and later the Russian) Academy of Sciences was a powerful structure operating numerous research institutes across the country and managing a big chunk of state science funding. After the reform, the institutes and the funding fell under the control of a newly established Federal Agency for Scientific Organizations (FASO), and later under the direct control of the Ministry of Science, while the Academy started to function as an elite researchers' club with generous lifelong stipends. FASO stimulated competitiveness and demanded higher efficiency indicators from the institutes. Other factors, such as the establishment of the Russian Science Foundation (RSF) in 2013, fuelled competition for funding and other resources among Russian researchers. As a whole this setup gave institutes an impetus to start hiring public information officers, albeit with less vigour than universities. A recent assessment of the communication function in more than 1,500 institutes and 850 universities demonstrated that around 40 per cent had at least a communication representative (Russian Association for Science Communication, 2016). These figures are still low, but, importantly, in recent years there has appeared a group of organisations (mostly universities) that lead the pack. It should be mentioned, however, that for the institutes the distribution is quite even among fields of sciences and geography, while universities suffer strongly from the so-called Matthew effect[2]—a huge gap between the leaders from Project 5-100 and the weaker group of former *instituts* is observed.

2 The Matthew effect can be best summarised by the adage 'the rich get richer and the poor get poorer'.

This newly born community of science PIOs was given an additional stepping-stone to move forward. In 2014, the Russian Venture Company (RVC), a key government fund of funds,[3] launched a long-term project named Communications Laboratory intended to build bridges between researchers, journalists and PIOs. Heavily involved in the process of building the national science and innovation ecosystem, RVC was among the first government-backed structures to realise the existence of serious communication barriers in Russian science. The immediate goal of the Communications Laboratory was to organise educational and networking activities for researchers and communication professionals, as well as undertake some initial research on the state of science communication in Russia. The project stood out by virtue of its international outlook and deliberate reliance on the Western culture and terminology of science communication. As an example, inspired by Eurekalert and AlphaGalileo, Russia's first science news distribution website *Otrkytaya nauka* (Borissova and Koenig, 2017) was launched in the framework of the project in 2015.

The fact that some science journalists became eager to switch to a PIO job came from the changing situation in the mass media. In 2013–14 nearly all independent media outlets were put under some form of governmental control (Meduza, 2016), which delivered a blow to the already-dying freedom of speech (Kovalev, 2017). As a result, general interest mass media virtually lost their say in the field of science. As the public interest in science endured in Russian society (Vaganov, 2016), a number of reporters left their general interest magazines and started independent popular science websites that virtually took the place of the Soviet-era science magazines. Digital science magazines are, indeed, quite popular today. The leaders are *Popular Mechanics* (2 million unique visitors monthly, according to Liveinternet), *National Geographic Russia* (1.4 million unique visitors monthly, according to Rambler) and *N+1* (1.3 million unique visitors monthly, according to Rambler). The lack of competition from traditional print media, radio and TV where science is scarcely and poorly represented helps them attract relevant audiences and resources.

Modern Russian digital science magazines are technologically advanced and appealing to younger audiences—for example, through their use of handy mobile versions and mobile apps. A solid scientific background combined with digital acumen has inspired some of these magazines to explore international markets. For example, www.nmas1.org, a Spanish-speaking branch of the

3 A fund of funds (FOF) is an investment strategy of holding a portfolio of other investment funds, rather than investing directly in stocks, bonds or other securities.

Russian *N+1*, entered the top 5 most-read Spanish-speaking science media. Digital science magazines focus on research news rather than science policy and their reporters increasingly aspire to become part of the global science journalism community. The flipside here is that these outlets mostly rely on state and/or corporate funding. In other words, a good share of their financial support comes either from the Ministry of Science and Education or from leading Russian universities and research centres that influence the agenda.

The government control of science communication brought other dramatic changes. In May 2015, the entire research community in Russia was hit by a shockwave: all the grant programs of the Dynasty Foundation were discontinued along with the foundation itself as a consequence of it being labelled a foreign agent by the Russian government. This news caught Russian scientists flat-footed and rumbled over the world (Pokrovsky, 2015). The biggest private founder of science and science popularisation was kicked out of the game for reasons which people still speculate about. It is not far-fetched, however, to suggest that the Russian government saw political risks associated with the activity of the foundation. Some criticisms were heard earlier that the foundation politicised Russian science as it was providing financial support only to those researchers who voiced views against the government. After the Dynasty Foundation was forced to shut down, some of its projects were taken over by the much smaller independent Evolution foundation, which became the main body training semi-professional science lecturers (Rudneva, 2018). The ousting of independent funders meant that the government was preparing to enter the market itself with significant resources and its own view of how things should be done.

A good example of how the government re-entered the field of science communication is its sudden interest in science museums, which was, quite ironically, stirred by the Dynasty Foundation's museum program. Being the only grant program for science museums and centres, the closure of the Dynasty Foundation certainly hit hard. Nevertheless, it did set the wheels in motion. Despite persistent financial problems, the revitalisation of Russian science and natural history museums caught the attention of government officials even before the closure of the Dynasty Foundation. In April 2010, Russia's then president Dmitry Medvedev instructed the government to develop a new museum concept based on the Polytechnic Museum and its vast collections. This decision turned into an almost decade-long quest. Modernisation involved a massive reconstruction of the historic building by the Japanese architect Jun'ya Ishigami, a sweeping revision of all methodological and conceptual approaches, and major structural changes in how the museum was governed and operated. The very idea was largely lobbied for by Anatoly Chubais, an influential politician and former deputy prime minister in Boris Yeltsin's administration.

The complexity of the project was reinforced by a relatively modest overall professionalism in the field of science and technology museums in Russia at that time. Museum curators and specialists in Russia mostly came from arts backgrounds and had no scientific or technical background, let alone any understanding of how to create world-class science museums from the ruin. The widespread financial and political support that surrounded the modernisation process, however, assisted the development of the museum. By 2018 it was fair to say that the Polytechnic Museum had made enormous progress, both in terms of conceptual thinking and physical infrastructure.

According to the personal account of Natalia Sergievskaya, the deputy director of the Polytechnic Museum, the last 10 years have dramatically raised the bar on professionalism in the field of science museums and centres. International mobility and exchange made it possible for Russian museum experts to get access to best museum practices overseas. However, Sergievskaya acknowledges that the situation is far from balanced when it comes to Russia as a whole. Too much depends on political support and only a chosen few achieve it. This in part explains why the Moscow Polytechnic Museum, despite its apparent march towards a successful relaunch, has not excelled in community building. For example, the Russian Association of Science and Technology Museums (AMNIT), which is led by the Polytechnic Museum, currently remains in a state of dormancy. The sheer gap between the Polytechnic Museum and other regional science museums has made any well-meaning exchange of practices improbable. Nevertheless, through trial and error, the Polytechnic Museum is moving towards a scheduled opening in 2021 and hopes are high that after the launch it will provide more methodological support to regional counterparts.

Popular science events were also invaded by the government, although as we show later several private initiatives did make an enormous difference during this period. From the late 2000s and early 2010s, several state corporations, such as the State Atomic Energy Corporation (Rosatom) and the Russian Nanotechnology Corporation (Rusnano), fiercely began supporting popular science events. In 2008, Rosatom sponsored the creation of a network of Information Centers for Atomic Energy (ICAE) that were assigned the mission of science popularisation with a special emphasis on atomic energy. The centres currently exist in 17 Russian cities as well as in Minsk and Astana. Public events organised by ICAE are characterised by original and humorous formats, in part due to its key target audience—kids. For instance, the event called Trials Against Superheroes features a dramatised discussion where a prosecutor and lawyer try to prove whether a comic superhero violates the laws of physics. Rusnano, in turn, regularly supports its own series of popular science events and other activities, but with a focus on nanomaterials and technology.

In 2011, with increased support from the government, the science festival NAUKA 0+ attained the status of an all-Russian event with dozens of participating cities and hundreds of organisations involved. It had been organised since 2006 by the Moscow State University, and in 2017 the overall outreach of the festival was calculated to be 2.5 million people visiting around 6,000 events, organised throughout the year in 80 regions under the umbrella of the All-Russian Festival of Science. Despite this national conquest, the central stage for the event remains in Moscow, where each October the festival attracts the most visitors. For instance, during the 2018 edition, the Moscow event was attended by 870,000 people. A distinctive feature of the All-Russian Festival of Science is that each year the organising committee selects several central regional locations, where the event adopts a more centralised and ambitious character. Usually this role is taken by relatively big Russian cities, such as Krasnoyarsk or Murmansk.

The renewed role of the government in science popularisation was additionally accentuated by the establishment in 2014 of a prestigious award Za vernost nauke [For Commitment to Science], also sometimes referred to as 'True Science'. This symbolic initiative was launched by the Russian Ministry of Science and Education to honour popularisers in all fields, from science journalism to popular events.

Figure 30.3: Former Russian Minister of science and education Dmitry Livanov at the ceremony of the True Science award.

Source: Maria Olendskaya (used with permission).

Popular science events in Russia took a curious twist in the early 2010s when a German popular science project called Science Slam[4] took off in Russia. In retrospect, the success of Science Slam might be attributed to its informal and cheerful approach to science communication, which contrasted strongly with events organised by universities, museums and other learned societies in Russia. Public lectures were mostly given by academic scientists whose views on popularisation were influenced by the patronising image of Soviet science propaganda. In a striking contrast, Science Slam usually took place in bars and rock clubs (Zhitkova and Grishin, 2018). No surprise that they felt like a breath of fresh air, especially to younger audiences. Soon dozens of Russian cities embraced the idea, using local bars and clubs as a venue for this unorthodox format. In 2016, the scope of the movement became so staggering that a separate Science Slam Association was established. By 2018 the network included 38 cities, with more than 150 events and over 45,000 visitors.

The format is constantly evolving and is being adopted by companies, universities and even schools. Corporate Slams, in turn, are giving rise to thematic variations. Be it materials science, medicine or computer technologies, events for each of these subject fields have been organised at some point, all under the same brand and franchise. Notably, the Russian Nanotechnology Corporation (Rusnano) has been one of the biggest sponsors of this unfolding bottom-up movement. The corporatisation of Science Slam and several other similar projects, such as Smart Moscow, are a trend in Russia. Finally, a TV show inspired by Science Slam appeared on national television despite the otherwise weak presence of science on air.

Of all the activities described in this chapter, science communication training is perhaps one of the most recent phenomena. As discussed in the previous sections, in the 1990s Russian scientists were left out of the massive attempt of making every scientist a populariser of his or her own work, which did take place in the 1990s in the UK, for example, and presumably some other European countries (Bauer and Gregory, 2007). The idea that researchers should consider public communication as part of their professional duties was not around in Russia until recently and is far from being fully embraced even now. This realisation coincided with the advent of PR-oriented science communication in major Russian universities and other research organisations. Therefore, while in Europe the integration model of science communication (where public communication is an integral part of scientists' jobs) anticipated the specialisation model (where professional communicators are needed to

4 A Science Slam is a scientific talk where scientists present their own research work in a given time frame (usually 10 minutes) in front of a non-expert audience in an informal setting of a bar or a club.

serve as intermediaries between scientists and the audience) (Davies and Horst, 2016), in Russia both were rediscovered at roughly the same time. This produced a noticeable impact on the evolution of science communication training in Russia. While graduate and postgraduate communication training was considered a routine practice elsewhere, in Russia no such training was widespread until the mid-2010s, with several exceptions.

Russia's first science journalism master's degree was launched only in 2013 at the Moscow State Institute of International Relations (MGIMO) and was part of the government takeover of science communication. The initiative was backed by the Russian Academy of Science, the Moscow State University and the Ministry of Science and Higher Education. Essentially, it was the first step towards the formalisation of science journalism education, which at the time was only represented by various summer schools and short-term seminars. In 2014, the Moscow State University (MSU) decided to catch up and followed with another master's degree in science journalism and communication. Despite having a strong lobby, neither of the programs quite took off. The MGIMO program came to an end in 2015 and never made another admission, while the MSU program was put on hold later. In parallel to these developments, Saint-Petersburg State University, the second largest university in Russia, decided to launch another master's degree in popular science journalism in 2014. Based at the Faculty of the History of Journalism, the program had a more consolidated theoretical foundation. However, this orientation towards theoretical and historical aspects (rather than more practical aspects of modern journalism) was and is both the strength and weakness of the program.

6. Raising the bar on professionalism (2016–present)

The government attempt to take control over science communication training did not entirely pay off. What unified the higher education programs described above was their reliance on science journalism training as opposed to the broader domain of public communication of science. The very term 'science communication' was not widely used to refer to public communication of science and technology until the early 2010s, although it is hard to define any strict boundaries. Instead, the term was used perhaps more frequently in the context of scholarly communication. This balance changed dramatically with the unfolding of the RVC's Communications Laboratory project, which disseminated the public dimension of the term as well as a broader view of science communication as a professional field.

An attempt to launch a degree (BA this time) on this new premise was made in 2015 by the Moscow Polytechnic University, but it did not fly high either. Although the project was supported by RVC's Communications Laboratory and featured prominent Russian science journalists and communicators, the program was tormented by political winds at the university. Two years after the launch, the program management left the university, depriving the degree of its science and technology core. The students were transferred to a general PR and advertising track, thus rendering the project a failed experiment.

Since 2016, a new cascade of educational programs has been set in motion. Many of them were built on the basis of science communication as a separate and broad profession, encompassing all possible formats and variations of public engagement with science. These ideas were in part brought to Russia from overseas by internationally active members of the Communications Laboratory. In 2016, the race to build a full-blown science communication degree was joined by ITMO University, a top technical university located in St Petersburg and leading member of the abovementioned 5-100 Project. The new master's degree claimed an all-embracing view of science communication from science journalism and PR to science museums and visualisation. Most of the courses were designed specifically for the program and relied on job market leaders rather than academics. The program has already produced several waves of graduates. It is fair to say that among Russian higher education institutions, ITMO University has made the biggest bet on science communication research and education. On the sidelines of the master's degree, in 2016 ITMO University launched the first Russian MOOC on the same subject and continues to make periodic launches to the present day. Apart from the science communication degree, in 2018 the university launched a new English-speaking master's degree in art and science.

Numerous educational programs continue to appear at different levels, from summer schools to intensive seminars. Several part-time courses deserve to be mentioned (some appeared earlier than 2016), such as courses at the Moscow Higher School of Economics, Novosibirsk State University and the Moscow Institute of Physics and Technology.

An array of schools and short-term trainings also exists outside the universities, such as the Science and Journalism Summer School, which first appeared in 2011 and flourished in the years following. Traditionally organised as part of a large educational summer camp at the recreational compound of the Joint Institute for Nuclear Research, the school received much acclaim and support among leading scientists and journalists. Until now, it remains one of the most massive and oldest functioning science journalism schools in Russia. In higher education, the latest development occurred in 2018 with

Peter the Great St Petersburg Polytechnic University declaring the launch of another master's degree in public relations in the field of science and technology. It is too soon to make any judgements about the program itself, but its announcement marks another step towards the institutionalisation of science communication training in Russia.

The appearance of higher education programs was instrumental to a wider appreciation for science communication as a research field. Russian authors have been virtually absent from the international research literature on science communication, but the abovementioned educational programs have begun to build a core of communication practitioners who have made a switch towards research, or social scientists who chose science communication as their focus of interest. Still, research on science communication and engagement in Russia is a long way from becoming solid and internationally recognised. Most research projects were started recently and are still in progress or in the publishing phase. The Higher School of Economics (HSE) and the European University at St Petersburg are among the strongest research centres for social sciences in Russia and perform some studies of interest to science communicators.

HSE, along with Levada-centre, is a major provider of data on scientific literacy and public attitudes to science in Russia. HSE annually publishes the Science and Technology Indicators in the Russian Federation Higher School of Economics (HSE). This research repeatedly shows a certain degree of respect for the profession of scientist among the Russian population. Between 2003 and 2016, around one-third of Russians mentioned that they wanted their child to become a researcher, with a record high of 43 per cent in 2009 and a record low of 26 per cent in 2014. The number of people praising both the level of Russian science and the amount of its funding gradually grew from 1997 to 2016. In 1997 only 40 per cent of Russians said in their responses that research in Russia is stronger or at the same level as in developed countries, but by 2016 this figure grew to 58 per cent. In 1997 only 9 per cent of Russians thought that science funding is sufficient or surplus, but by 2016 this figure reached 44 per cent (Chernovich et al., 2016).

Little is known about public trust in science. The latest HSE data manifests strong support for both positive and negative notions about science and technology (Chernovich et al., 2016). Up to 75 per cent agree that they 'make our life healthier, easier and more comfortable' and 'future generations will have more opportunities thanks to science and technology'. But nearly the same percentage of the population thinks that 'science and technology change our life too fast' and 'advances in science and technology can lead to negative consequences for health and environment'. Another popular counter-scientific notion states that 'nowadays people draw too much attention to advances in science and technology and too little attention to the spiritual side of life'. Generally, there is no organised system of research in the field of public understanding of science, or any solid empirical work aimed at testing various approaches to science communication. This is, however, not a special case, but rather a manifestation of a generally poor institutionalisation of science and society studies in Russia.

The process of self-identification with the new profession and the international community was enhanced by the establishment of the Russian Association for Science Communication AKSON in 2016. It was established with no institutional support by five independent communication specialists: Alexandra Borissova (TASS Russian News Agency), Elena Brandt (Moscow Institute of Physics and Technology), Ekaterina Ivanova ('Teorii i Praktiki), Elena Zemtsova (Moscow Polytechnical University) and Dmitry Malkov (ITMO University).[5] Since then, the association has grown to more than 200 members from all over Russia and CIS countries,[6] bringing together both PIOs and science journalists. AKSON promotes and defends the interests of the professional community and engages in periodic networking and educational activities. Supported by RVC, AKSON gives out the annual Communication Lab award, which acknowledges prominent achievements in corporate science communication. Since 2018, AKSON operates the national competition for the European Science Writer of the Year award in cooperation with Rusnano. Both awards are presented during the annual Forum of Russian Science Communicators that has become a major networking site for Russian science PIOs and journalists.

Despite clear progress with educational programs and community building, it is premature to say that science communication is fully institutionalised in Russia. Another important feature is missing—namely, the inclusion of science communication in state policy. To this day, Russia has no separate state policy on science communication or engagement. However, several references are included in a key document that was officially published in December 2016: the Strategy for Development of Science and Technology (Russian Federation, 2016).

5 Affiliations are listed as of August 2016.

6 The Commonwealth of Independent States (CIS) was formed following the dissolution of the Soviet Union in 1991. It has nine members, plus two founding non-members.

Figure 30.4: Third Annual Conference of the Russian Association for Science Communication in 2018. Announcement of the inaugural winner of the Russian Sci&Tech Writer of the Year award: Shamil Troyanovsky.

Source: Shamil Troyanovsky (used with permission).

Sadly, the strategy barely addresses the lack of understanding of any science and society relationship in Russia. With respect to science communication, the document assumes the deficit model in urging the construction of an efficient system of communication in the field of science, technology and innovation to ensure the growth of economic and social susceptibility to innovation and high-tech entrepreneurship. The strategy additionally calls for adjusting the information policy so as to develop a technological culture and social susceptibility to the popularisation of important scientific and technological achievements, as well developing the role of outstanding researchers, engineers and entrepreneurs in ensuring the socioeconomic progress of the country. The conclusion is that the strategy must provide for the growth of influence of science on the Russian society.

Curiously enough, one sentence from the document hints at potential plans to step out from the deficit model and move towards enabling conditions for mutual influence of science and society and engaging the society by providing various feedback loops. It calls for 'conditions that contribute to reciprocal influencing between science and society by involving the society in shaping demand for the research outcomes' (Russian Federation, 2016). However, the gap between the deficit model understanding and these flirtations with participatory engagement is not bridged by any comprehensive recommendations or plan of action. Neither does the strategy call for the much-desired intensification of research in public communication of science. So far, these statements seem to be just buzzwords rather than real intentions. The general observation is that state priorities are mostly agenda-driven rather than evidence-based, and that the current strategy is poorly informed on available evidence and does not encourage any further research. Attempts are currently being made by the community to push for a full-blown science communication strategy, but they are nowhere near getting traction.

Acknowledgements

The authors extend their gratitude to Olga Orlova, Natalia Sergyievskaya, Lubov Strelnikova, Irina Aktuganova, Arina Pushkina, Larisa Bakulina, Andrey Kozhanov, Nikolay Avanesov and Alina Koroleva. Dr Alexandra Borissova expresses her special thanks to German Chancellor Fellowship (Alexander von Humboldt Foundation, Germany) and Professor Alexander Gerber (Rhine-Waal University, Germany) for their kind help and support during her research stay. Without it this work would not be possible.

References

Andrews, J. (2003). *Science for the masses: The Bolshevik state, public science, and the popular imagination in Soviet Russia, 1917–1934.* Texas A&M University Press.

Baldscientist. (2013, 10 February). The Sagan Effect. Retrieved from baldscientist.wordpress.com/2013/02/10/the-sagan-effect/.

Bauer, M. and Gregory, J. (2007). From journalism to corporate communication in post-war Britain. In M. Bauer and M. Bucchi (eds), *Journalism, science and society: science communication between news and public relations* (pp. 33–52). Routledge Studies in Science, Technology and Society. Abingdon, UK: Routledge. doi.org/10.4324/9780203942314.

Borisova, A. and Koenig, J. (2017). *Opit na nacyonalnom yazike* [Experience in the national language]. *Troitsky variant* (online). Retrieved from www.trv-science.ru/2017/04/11/opyt-na-nacionalnom-yazyke/.

Borisova, A. O. and Malkov, D. V. (2018). Five faces of Russian science communication. *Spokes Magazine.* Retrieved from www.ecsite.eu/activities-and-services/news-and-publications/digital-spokes/issue-39#section=section-indepth&href=/feature/depth/five-faces-russian-science-communication.

Burova, M. (2014). Interview with Anton Sharypov. Retrieved from www.sobaka.ru/krsk/city/city/30986.

Carnegie Medal for Philanthropy. (2013). Dmitry Zimin: Russian for philanthropy. Retrieved from www.medalofphilanthropy.org/dmitry-zimin-russian-philanthropy/.

Chernovich, E., Ditkovsky, K., Fridlyanova, S., Fursov, K., Gokhberg, L., Gorodnikova, N., Kotsemir, M., Kuznetsova, I., Lukinova, E., Martynova, S., Polyakova, V., Ratay, T., Rosovetskaya, L., Sagieva, G. and Sergeeva, V. (2016). *Science and Technology Indicators in the Russian Federation: Data Book.* Moscow: National Research University Higher School of Economics. . Retrieved from www.hse.ru/data/2018/02/08/1162213715/Science_and_Technology_Indicators_in_the_Russian_Federation_2016.pdf.

Clark, F. (2013). Reforming the Russian Academy of Sciences. *The Lancet, 382*, 1392–93. doi.org/10.1016/S0140-6736(13)62142-X.

Davies, S. R. and Horst, M. (2016). *Science communication: Culture, identity and citizenship*. UK: Palgrave Macmillan. doi.org/10.1057/978-1-137-50366-4.

Dynasty Foundation. (n.d.). Public lectures of the Dynasty Foundation. Retrieved from web.archive.org/web/20190214221207/http://www.dynastyfdn.com/programs/popular/lections.

Fishman A. M. (2016). Argumenty I Facty [Arguments and Facts]. *The Great Russian Encyclopedia*, vol. 2. Moscow.

Graham, L. and Dezhina, I. (2008). *Science in the New Russia: Crisis, aid, reform.* Indiana University Press.

Kovalev, A. (2017, 25 March). In Putin's Russia, the hollowed-out media mirrors the state. *The Guardian*. Retrieved from www.theguardian.com/commentisfree/2017/mar/24/putin-russia-media-state-government-control.

Kukulin, I. (2017). Periodicals for Engineers: Soviet Popular Science Journals and the Shaping of the Late-Soviet Scientific and Technical Intelligentsia's Interests. *Novoe Literaturnoe Obozrenie*, 145.

Litvak, N. (2008). *Information society: Permanent evolution*. Moscow: Kolos Press.

Meduza. (2016, 18 March). *12 redakzy za pyat let Razgony, blokirovki, uvolnenia glavredov, kak vlast reformirovala rynok SMI* [12 editorial offices in five years: Crackdowns, blockings, dismissals of editors-in-chief: how government reformed the mass media market]. Meduza.io. Retrieved from www.meduza.io/feature/2016/05/17/12-redaktsiy-za-pyat-let.

Milanovich, B. (1998). *Income, inequality, and poverty during the transition from planned to market economy* (pp. 196–202). Washington DC: The International Bank for Reconstruction and Development and The World Bank.

Nauka i Zhizn. (1977). Obshestvu Znanie 30 Let [The Znanie Society Turns 30]. *Nauka i Zhizn*, (5), 26. Izdatelstvo Pravda, Moscow.

Nazaretyan, K. (2010). Petrikgate. Saint Petersburg Branch of the Russian Humanist Society. Retrieved from www.humanism.su/en/articles.phtml?num=000084.

Palshina, I. (2010). Samiy krupniy v Rossii muzei zanimatelnoi nauki nahodtsiya v Irkutske [The Biggest Popular Science Museum in Russia is Located in Irkutsk]. *Argumenty I Facty, 43*. Retrieved from irk.aif.ru/science/science__details/78338.

Panchenko, A. (2018). The Age of Aquarius» for the Builders of Communism: New Age Culture in Late Soviet Society and the Problem of «Periods of Change. *Novoe Literaturnoe Obozrenie*, 149.

Podorvanyuk, N. and Alexeeva, O. (2010, 7 July). Voda iz filtra Petrika opasna dlya zdoroviya [Water from Petrik's filters is dangerous for heath]. Gazeta.ru. Retrieved from www.gazeta.ru/science/2010/07/07_a_3395369.shtml.

Pokrovsky, V. (2015, 9 July). Russia's only private science funder closes its doors. *Science magazine*. Retrieved from www.sciencemag.org/news/2015/07/russias-only-private-science-funder-closes-its-doors. doi.org/10.1126/science.aac8829.

Rambler Top 100. (n.d.). Retrieved from top100.rambler.ru/.

Ria Novosti. (2013). Interview with Viktor Sadovnichiy. Ria.ru. Retrieved from www.ria.ru/20131007/968271833.html.

Rudneva, V. (2018). Talking about the Big Bang: An exploratory study of how Russian science communicators use social media. Coursework material, Media and Communications Studies Programme, Department of Media Studies, Stockholms Universitet. Retrieved from www.diva-portal.org/smash/get/diva2:1212857/FULLTEXT01.pdf.

Russian Association for Science Communication. (2017). First report on the professionalisation of science communication in Russia. Russian Association for Science Communication (AKSON) (online). Retrieved from akson.science/news/2017/08/14/opublikovany-rezultaty-issledovanija-professionalizacija-nauchnoj-kommunikacii-v-rossii/.

Russian Federation. (2016). Scientific and Technological Development Strategy of the Russian Federation. Retrieved from online.mai.ru/StrategySTD%20RF.pdf.

Russian Venture Company. (2016). Analytical report on Russian science museums and centres. RVC official website. Retrieved from www.rvc.ru/upload/iblock/868/museums_study.pdf.

Sapun, A. (2016). Istoria razvitia internet-SMI v Rossiye [History of online mass media in Russia]. Dni.ru. Retrieved from www.dni.ru/academ/2016/8/1/347408.html.

Scott, J. (2006). The mission of the university: Medieval to postmodern transformations. *The Journal of Higher Education*, *77*(1) (online). doi.org/10.1353/jhe.2006.0007.

Sonin, K. (2004). The unbearable lightness of petrodollars. *Russia in Global Affairs*, *2*, 94–100.

State Legal Information System. (2015). *Ukaz Prezidenta Rossiskoy Federazii ot 11.12.2015 No. 617 "O sozdanii Obsherossiskoy obshesvenno-gosudarstvennoy organizatsii Rossiskoye obshestvo Znanie"* [Decree of the President of the Russian Federation of 11.12.2015 No. 617 'On the Creation of All-Russian Public Educational Organisation Russian Society Znanie']. Retrieved from publication.pravo.gov.ru/Document/View/0001201512110006.

Vaganov, A. (2016). Evolution of forms of science popularisation in Russia: 18th–21st centuries. *Science. Innovation. Education*, *3*(21).

White, G. (2010, 5 March). Russian inventor has friends in Kremlin, but skeptics outside it. *The Wall Street Journal.* Retrieved from www.wsj.com/articles/SB10001424052748704187204575101510173019130.

Yakovenko, I. (2012, 30 August). Rynok nauchno-populyarnich jurnalov. Analitichesky obzor [Market of popular science magaizes. An analytical review]. *Nazionalnaya tirazhnaya sluzhba.* Retrieved from www.pressaudit.ru/rynok-nauchno-populyarnyx-zhurnalov-analiticheskij-obzor/.

Zhitkova, V. and Grishin N. (2018). Bitva botanikov: skolko prinosyat komedyinie shou s uchenimi [Battle of geeks: How much money shows with scientists bring]. Rbc.ru. Retrieved from www.rbc.ru/own_business/26/04/2018/5ae052a19a7947223c1a8cbd.

Znanie Society. (n.d.). History of the Znanie Society. Retreived from www.znanie.org/OZR/history.htm.

Timeline

Event	Name	Date	Comment
First interactive science centre established.	Experimentary (Irkutsk)	2005	Experimentary was created by the Siberian branch of the Russian Academy of Sciences inspired by the San Francisco Exploratorium
First national (or large regional) science festival.	NAUKA 0+ established by Lomonosov Moscow State University	2005	Has grown from a local event attended by 20,000 people, to an all-Russian event with dozens of participating cities and 2.5 million participants at around 6,000 events
An association of science writers or journalists or communicators established.	Russian Association for Science Communication (AKSON)	2016	AKSON has more than 200 members from Russia and CIS countries. It brings together both PIOs and science journalists
First university courses to train science communicators.	Moscow State Institute of International Relations, MSc in science journalism	2014	This focuses on science journalism rather than science communication. It does not provide training for science communicators, but is the first master's degree to formally include science communication

Event	Name	Date	Comment
First master's students in science communication graduate.	ITMO University, MSc in Science Communication is considered to be Russia's first fully fledged degree on science communication	2018	The degree claims an overarching view of science communication from science journalism and PR to science museums and visualisation. Most of the courses were designed specifically for the program and relied on job market leaders, rather than academics
First national conference in science communication.	Russian Science Communicators Forum	2016	Organised by AKSON and since 2016 it has become a major networking site for Russian science PIOs and journalists
First significant initiative or report on science communication.	Communications Lab Project (Russian Venture Company)	2014	Launched to bridge the gap between scientists, media professionals and public information officers. A first national overview of the development of science communication in Russia
First significant radio programs on science.	*Granit nauki* [Granite of Science]	1997	Broadcast by radio station Echo of Moscow owned by Gazprom-media. Won a True Science award in 2016 from the Ministry of Science
First significant TV programs on science.	*Ochevidnoe-neveroyatnoe* [Obvious-unbelievable]	1973	Started by a prominent science populariser Sergey Kapitsa and lasted until his death in 2012. Since then there is no dedicated science TV program on any major TV channel
First awards for scientists or journalists or others for science communication.	True Science award was established by the Ministry of Science and Education	2014	True Science does not cover the work of PIOs or individual science journalists, but celebrates the best science media, radio and TV programs and science popularisers

Contributors

Dr Alexandra Borissova is the president of the Russian Association of Science Communicators AKSON and supervisor of Russia's first MSc in science communication at ITMO University in St Petersburg.

Dmitry Malkov is the founder of the Center for Science Communication at ITMO University in St Petersburg, and vice-president for international relations of the Russian Association of Science Communicators AKSON.

31

SINGAPORE

An evolving and increasingly complex relationship

Denise E. De Souza, Lieu Zi Zhao, Letchumi Mani, Glenn Toh and Benedict Lin

1. Introduction

The development of scientific communication in Singapore can be described in terms of an evolving and increasingly complex relationship between discourses *about* science and its role in society, and the communication *of* science and its ideas, through formal education and other forums. From early beginnings with top-down government-initiated promotion of science for pragmatic societal ends, discourses about science have broadened to now incorporate multiple voices, including more ground-up ones emerging from an established scientific community as well as members of the public. In tandem with this, the communication of science has also grown from early basic concerns with extrinsically motivated promotion of scientific literacy for economic needs to include more mature intrinsic concerns such as helping developing and established scientists to communicate their work more effectively. This chapter traces this evolution, highlighting the ways in which extrinsic and intrinsic motivations, as well as top-down and ground-up movements have shaped discourses about science, and the communication of science, in Singapore.

2. Historical background

Singapore's present commitment to science, technology and the achievement of excellence can be traced to overarching discourses built around the official histories of the founding of Singapore in 1819 by Sir Thomas Stamford Raffles and Singapore's former status as a British colony. Politically sanctioned discourses depict Singapore as a quiet fishing village that developed quickly into a thriving centre of *entrepot* trade,[1] made possible by forward-looking colonial policies that supported the growth of commerce. These early influences provide the backdrop of circumstances that helped to generate a rhetoric of change, vulnerability and uncertainty upon which the need to establish firm policies for Singapore's future growth, stability and economic prosperity were then based.

Singapore's early developmental trajectory, for over a century, drew on the benefits of its strategic geographical location and the trade and commercial activities this enabled, ostensibly with less emphasis placed on scientific and technological development and advancement (Tan, Wee and Subramaniam, 2017). However, the unfolding of political change—brought about by the traumas of the Cold War and decolonisation, leading to self-governance in 1959, the Malayan emergency and Singapore's separation from the Malaysian Federation in 1965—resulted in the need to address the urgent, pragmatic and complex constraints that followed. This was done under the leadership of the People's Action Party (PAP), which was voted into power in 1959. Constantly stressing the importance of political stability for 'existential matters of national survival and economic prosperity' (Tan, 2018, p. 1), the party has chosen to interpret this as an 'abhorrence of parliamentary Opposition' (Mutalib, 2004, p. 28). To date, while negotiations of political culture and ideology are ongoing, this narrative has enabled it to retain a strong grip on power. The PAP's nearly six decades of governance has enabled the fairly coherent and smooth implementation of policies and programs, with significant opposition stemming more from external rather than internal forces.

From the outset of independence, two ongoing constraints facing Singapore have been the heavy reliance on Malaysia for the provision of water (Tortajada, Joshi and Biswas, 2013) and the lack of access to a previously available supply of natural resources that resulted from the breakaway from Malaysia. At that time, this precipitated the need to look for alternative resources and discourses that could sustainably support an agenda of economic development, excellence and prosperity. Science and technology were identified as prospective areas to realise this quest for development (see Lau, 1998; B. T. G. Tan, 2017).

The importance of both science and technology was recognised in the early years of nationhood, as is evident from the formation of the Science Council of Singapore on 30 October 1967, just two years following Singapore's

1 *Entrepot* trade refers to a port, city or trading post where merchandise may be imported, stored or traded, usually to be exported again.

independence from Malaysia. The council was then assigned the significant role of overseeing research and development and promoting Singapore's indigenous capabilities in both science and technology (B. T. G. Tan, 2017).

3. Laying foundations (1965–80)

The term 'science communication', during the modern period, took on two meanings as it influenced two spheres of public life in Singapore: the communication of science within educational institutions as the formal learning of science; and the communication of science to the general public through less formal institutions and means (Tan, 2011). Prior to Singapore's independence, the establishment of the University of Singapore in 1905 and Nanyang University in 1956 focused on communicating science primarily through education and training, with less emphasis placed on research and development, or on communicating science to the general public (Tan et al., 2017). Tan (2011, p. 15), however, notes that 'This was recognised as a limitation since the majority of the population was not science literate when Singapore became a nation'. Measures were subsequently undertaken to remedy this situation.

The prevailing perception up to the 1980s was that Singapore, with its long history as a trading port, comprised a population accustomed to business and commercial activities more than science and technology. One expressed function of the Science Council of Singapore, therefore, was to make Singaporeans more comfortable with embracing the potential and imminent scientific and technological transformations (Glauberman, 1985), and to ensure that young people would, in the near future, be encouraged to take up careers in these fields.

In the early years of independence, this possibility was realised through establishing the Singapore National Academy of Science (SNAS), which promoted the advancement of science and technology. An informal science education centre was built to popularise science and generate awareness about advancements, and science-related educational entertainment in the mass media was introduced. These initiatives were influential in promoting science and fostering scientific literacy amongst the general public.

The SNAS was established by, and comprised, credentialed locally based scientists. Inaugurated at the end of July 1967, the SNAS organised its First Science Congress in August 1968. There, the important role of science and technology in Singapore's industrial, commercial and economic development was emphasised to the professional and academic community of scientists attending the congress. Dr Toh Chin Chye, then Minister for Science and

Technology, concluded his opening address by asking the delegates, 'How can science and technology be applied towards our economic development? That is the challenge we face today' (as cited in B. T. G. Tan, 2017, p. 14). The proceedings of this first congress, as well as the opening address, were subsequently communicated through the newly introduced, Singapore-run *Journal of the Singapore National Academy of Science*, which published its first issue in 1969 (B. T. G. Tan, 2017). In 1976, SNAS was reorganised as a broader umbrella organisation with the responsibility for overseeing all other scientific societies in Singapore. The task of promoting science and technology in Singapore was then transferred to the newly formed Singapore Association of the Advancement of Science (SAAS).

Apart from supporting the advancement of science and technology amongst the professional and academic community, the SAAS also made efforts to promote science to the general public. The proposal to establish an informal science education centre was sanctioned in the 1970s (Tan, 2011). From the literature available, the Science Centre was assigned two roles, both clearly motivated by concerns extrinsic to science itself. The first was to generate public interest that could support scientific and technological development for the sake of economic growth. In 1971, Dr Toh Chin Chye announced that Singapore's industries must be supported by a strong foundation and sustained interest in those two areas. An over-reliance on science and technological expertise imported from overseas was deemed undesirable and unhealthy, so the ministry and the Science Council of Singapore worked towards developing indigenous capabilities. Having seen similar developments taking place overseas, the Science Centre Board was established in November 1970 with the aim of promoting science and technology to the general public (Singapore Parliamentary Report, 1971).

The second role of the centre, targeted at school-going children and youth, was to highlight the relevance and applicability of science and technology to everyday life (Tan, 2011). The centre aimed to 'promote interest, learning and creativity in science and technology through an imaginative and enjoyable experience and contribute to the nation's development of its human resource' (Science Centre Singapore, 2018a). Although the idea of the Science Centre was mooted in 1967, it took 10 years of lobbying, fundraising and development before the centre opened its doors to the public officially on 10 December 1977. That same year, Singapore's first popular science magazine *Singapore Scientist* was launched with the goal of communicating interesting scientific issues, experiments and quizzes to school children (Dairianathan and Lim, 2014; Subramaniam, 2014). The writing in the magazine was pitched at the level of school students and was meant to communicate challenging topics in a way that they could comprehend.

Another key activity initiated by the Science Council of Singapore, with the help of the local media, was the Science and Industry Quiz. The quiz was introduced because of its potential effectiveness in educating and entertaining the public and thereby popularising science and technology (B. T. G. Tan, 2017). The program, publicly broadcast as a televised series by Radio Television Singapura, aired from 1972 to 1977. Its popularity reportedly 'grew to phenomenal proportions well beyond ... most optimistic projections' (p. 6). The annual quiz competition featured teams from Singapore's elite schools competing keenly 'in fierce battles' for the prestige of winning the top prizes and was viewed by large audiences (p. 7). Three other programs that followed after the series ended, *The Innovators* (1979), *Top of the Trade* (1977–79/1980, 1983) and *Science Challenge* (1980, 1989), similarly highlighted and legitimated the importance and prominence of science, technology and innovation to the public. This importance was clearly reflected in the involvement of both the national broadcasting authority and well-known academics from the then University of Singapore, who were the quizmasters and judges in these competitions (B. T. G. Tan, 2017).

The existence of these government-driven efforts at promoting science, however, does not mean that there was no pre-existing interest in communicating science to the general public. Pre-dating the Science Centre as an informal science education centre, though much less prominent, was the Raffles Museum, first conceived in 1823 as the library of the Singapore Institution (later the Raffles Institution) (Y.-L. K. Tan, 2017). The museum specialised in the study of fauna from the Southeast Asian region. Since Singapore's independence in 1965, the collection has moved from place to place before finding a permanent home in a corner of the campus of the National University of Singapore. In 2015, it appeared in its current reincarnation as the Lee Kong Chian Natural History Museum. Arguably, the museum can be said to have represented a less pragmatically motivated forum for the communication of science in the early period of Singapore's history.

In general, science communication in the modern period may be described as being shaped by the Singapore government's attitude towards, and interest in, science and technology in terms of the economic benefits that competencies in both could bring to Singapore in its early developmental phases. Science and technology were, and still are, accorded priority of place in Singapore, and this is reflected in the practices and discourses of the system implemented by the People's Action Party. Brown (2000) observed that in the early years of nationhood, an ideology of survivalism was set in motion and a siege mentality was inculcated through emphasising Singapore's economic vulnerability and susceptibility to threats due to its size and geographical location (Lee, 1996). As such, there has been an overwhelming tendency for the government to collocate science with technology. Official rhetoric tends to naturalise this collocation and associate the necessity of both for economic growth and for moving Singapore forward, thus serving also to advance the legitimacy of the ruling party.

4. Later developments (1981 onwards)

4.1. The 1980s: Addressing teething problems

In the early 1980s, it was observed that the effectiveness of Science Centre Singapore as an informal institution for communicating science to the general public was limited. The inaccessibility of the centre, resulting from the lack of proper roads to its then remote location, made visits difficult for local schools and the public. Professor Leo Tan Wee-Hin, then the CEO of the centre (1982–92) and a prominent and esteemed Singaporean scientist and educator, was tasked with remedying this situation. He contacted schoolteachers that he personally knew and offered to pay for chartered buses to ferry students to and from the centre and their schools. He subsequently initiated discussions with Singapore Bus Services to provide a bus service to improve access to the centre. However, a number of years passed before the area around the centre was developed and a public bus route was added, raising the centre's accessibility and profile. These and other efforts to address teething problems enabled and eventually led to more demotic, ground-up and less extrinsically driven attitudes towards science and the communication of science.

To attract more visitors, the first interactive science show in Singapore (performed by communicators from the Royal Institute, UK) was staged during Professor Tan's tenure. Demonstrating science through live entertainment was a novelty then, and these shows became extremely popular among students, popularising the use of similar demonstrations for both teaching science in the classroom and learning science in public spaces.

Like the Science Centre Singapore, the popular science magazine *Singapore Scientist* also suffered a variety of problems in its early years. These challenges included a shortage of funding, writers and contributors, the lack of a full-time editorial staff and the absence of keen interest from schools. To address this, Professor Tan persuaded and encouraged his colleagues at the Faculty of Science in the National University of Singapore to contribute to the magazine by writing about their research or current scientific issues of interest. He also sent the Science Centre staff to all schools as part of its outreach efforts, to promote and popularise the reading of this magazine. School libraries were encouraged to subscribe to the magazine to enable access to all students, regardless of socioeconomic status. The magazine was subsequently published by Science Centre Singapore for a total of 38 years. To refresh and re-align the content of the magazine with the current school curriculum, an overhauled and rebranded version of *Science Spy* was launched in 2015 in collaboration with a commercial publisher, Marshall Cavendish Education (Times Publishing Group, 2017), thus marking a step towards devolution of responsibility for science promotion and education away from the government.

Sustaining curiosity in science beyond the occasional Science Centre visit was also important, and the Young Scientist Badge Scheme was initiated to encourage young students to continue to engage in self-directed activities in an area of science that was of interest to them (Science Centre Singapore, 1983; Dairianathan and Lim, 2014). The badge scheme, launched in 1983, required students (with guidance from a teacher or a parent, if needed) to complete science activities listed in an activity card to earn a badge. To date, more than a million badges have been awarded since the launch of the scheme, a testament to the scheme's popularity and its success in nurturing interest in science. Some prominent scientists in Singapore have acknowledged that their interest in science began by participating in this scheme (Lim, 2017).

Overall, despite its early teething problems in the 1980s, under Professor Tan's leadership Science Centre Singapore established itself as a hub for the public to engage with science and technology and became a credible source for promoting the informal learning and communication of science both within and beyond the confines of the centre.

4.2. Mid-1990s and early 2000s: Mass media communicates science

In the first years following Singapore's independence, scientific communication centred on government-led efforts to promote science and technology and to advance scientific literacy among the general public as a key means of economic growth and survival. This led ultimately to more intrinsic interest in science, and arguably the beginnings of more ground-up participation in communicating science.

In the mid-1990s and early years of the 20th century, an interesting development was how scientific communication was used in the mass media to shape public opinion and attitudes towards events in which politics and advancements of science and technology were enmeshed. One event revolved around Singapore's ongoing problem of access to fresh water. To date, Singapore imports a large part (up to 60 per cent) of its fresh water from

Malaysia (Ghangaa, 2015; Law, 2003; Tortajada, Joshi and Biswas, 2013) and, based on four agreements signed in 1927, 1961, 1962 and 1990, can do so up to 2061 (Tortajada et al., 2013).

While the existence of the agreements highlights bilateral cooperation between Malaysia and Singapore, this relationship has not been free of disputes (see Tortajada et al., 2013, p. 152). Between 1997 and 2004 the Singapore–Malaysia water relationship was tested, in part, by Malaysia's desire to renew discussions about reviewing the price of water under the prevailing agreements and to increase it by at least 600 per cent. Singapore responded by announcing its intention to seek legal advice on whether Malaysia had a right to do this.

It was also during this time that Singapore publicly announced that alternative sources of water were being considered. While local research into water reclamation began in the 1970s, concerns about costs and technological reliability had hindered its development till the 1990s—when evidence that other countries were engaging successfully in water reclamation and treatment became available. Despite concerns about high costs (Srinivasan, 1997), during Singapore's water dispute with Malaysia in 1998, the NEWater Study on wastewater recycling was sanctioned and jointly undertaken by the Public Utilities Board of Singapore and the Ministry of Environment and Water Resources (Leong, 2010). Construction of the first desalination plant was also to begin in 1999 and was projected to be completed by 2003 (Lim, 1998a). It was eventually opened in 2005.

Apart from commissioning such projects to secure Singapore's long-term access to potable water (Lim, 1998b), a public communication plan was also developed. It strategised an approach to educate and convince the general public that it was safe to drink recycled water (Tortajada et al., 2013). One of the main challenges, when attempting to implement water-reuse policies has been helping the general public to overcome the 'yuck' factor associated with the notion of consuming 'sewage' or 'wastewater'. Leong (2010; also Lee and Tan, 2016; Tortajada et al., 2013) observes that, in Singapore's case, the media played an important role as a strategic partner in educating and garnering support and acceptance for NEWater by the time it was launched in 2003. This was done systematically by minimising the use of negative terminology associated with wastewater recycling during reporting. Instead, public attention was drawn to the treatment process rather than the water source, and the successful use and implementation of water recycling in other countries were also highlighted. The 'yuck' element was addressed by adopting 'a rational, scientific approach to the topic' (Leong, 2010, p. 124). In framing their science communication to the public, the media adopted the theme of 'Social Progress', foregrounding NEWater technology as a modern-day solution that would benefit Singaporean society and be a viable, self-sustaining alternative for replacing fresh water sources from Malaysia in the long term, thus diminishing Singapore's strong reliance on Malaysia's supply of water.

Through the media's communication of the science behind, and importance of, reused wastewater to water sustainability in Singapore, the process of getting Singaporeans to accept NEWater was made much smoother than experiences observed in similar projects by USA and Australia (Tortajada et al., 2013). This acceptance has not been an accidental phenomenon. It has been attributed, in part, to the very strong support the government has from the media in Singapore, efficient media management by the Public Utilities Board (Lee and Tan, 2016; Leong, 2010; Tortajada et al., 2013) and subsequent buy-in from Singaporeans who have embraced NEWater as one of four 'National Taps' that ensures the nation has a sustainable and diversified supply of water (Public Utilities Board, 2018). Science communication, in other words, was strategically used in the media to gain public support for addressing potential national difficulties.

5. From communicating science to nurturing science communicators (1990s to 2010s)

In comparison to how mass media used the theme of 'Social Progress' to frame science communication to the public, science communication in the field of education saw continuities of old discourses about science as well as newer developments. Broadly speaking, preoccupations over the teaching and learning of science, in this period, have been reflected in three dominant areas of discussion: (1) nurturing curiosity, nimble-mindedness and giftedness in students; (2) guiding students early on into 'the positive end of the attitude continuum' with regard to the importance of science; and (3) foregrounding the potential benefits of science to the future of Singaporean society and consequently, the need 'to be better "consumers" of science in the future' (Caleon and Subramaniam, 2008, p. 950). While the third continues a long-running narrative, the first two appear to have constituted emerging central discourses. To address these concerns educationally, what counts as science teaching and learning has progressively broadened to encompass more informal science teaching and learning events, incorporating applied activities, both within and outside of school and institutional contexts. Additionally, efforts have also begun to nurture effective science communicators, including not only teachers, but also students of science, scientists and professional science communicators.

5.1. Communicating science

The Science Centre's role as a hub for getting school children and the public to engage with science and technology informally was extended in the late 1990s. On the 20th anniversary of Science Centre Singapore, in 1997, the centre underwent a major revamping exercise. For this, it received S$34.7 million from the Ministry of Education to improve its facilities for school programs. This saw 95 per cent of Singapore schools signing up as institutional members to participate in the centre's science enrichment programs. The centre was consequently recognised as playing a key role in providing mass-based science education (Lim, 2017). During that time, the Science Centre Singapore also became a full member of the Asia Pacific Network of Science and Technology Centres (ASPAC), an organisation connecting science and technology centres in the region.

Apart from having a role in educating the public, the Science Centre has held regular workshops for educators to learn or upgrade their skills in science communication (Dairianathan and Lim, 2014). The workshops have been conducted to propagate the teaching of science using exhibition galleries and science show tricks. The teacher-participants learn and are encouraged to share creative strategies of communicating and teaching science, demystifying specific and challenging topics, and correcting common misconceptions.

In 2002, the Physics Demonstration Laboratory, located at the National University of Singapore's Faculty of Science (NUS FoS), was started by Professor Sow Chong Haur as another platform for science outreach. The laboratory, targeting students, aimed to demonstrate interesting phenomena through hands-on activities. In its initial stages, the Physics Demonstration Laboratory relied on using commercially available apparatus for demonstrations. It subsequently moved on to designing and building its own science demonstrations. Over the years, the laboratory presentations have evolved from standard 'show and tell' demonstrations to more 'engaging' participative demonstrations where audiences are asked to predict or explain the observed phenomena. This mode of science communication through participative demonstration has been adopted to increase student engagement, deepen learning and to allow common student misconceptions to surface and to be discussed. In 2008, the laboratory was renamed the NUS Science Demonstration Laboratory to include activities from the other science disciplines (NUS FoS, 2018a; Teng, 2016).

Communicating science in formal and informal educational contexts in Singapore during the 1990s up to 2010s, therefore, incorporated a variety of strategies that aimed to nurture curiosity, nimble-mindedness and a positive attitude towards science.

5.2. Nurturing science communicators

In comparison to the preoccupations surrounding the teaching and learning of science, Professor Tan (2011, p. 15), upon reflecting on the development of science communication in Singapore since 1965, noted that what was overlooked for some time was the nurturing of professional science communicators to work in the 'main-stream media, government agencies responsible for science-related matters, and the education service'. Despite the availability of overseas scholarships to pursue an MSc in science communication, offered in the 1980s by the National Science and Technology Board (now the Agency for Science, Technology and Research or A*STAR), Tan observed that 'both the media and educators did not think that was a priority' (p. 16).

This changed in January 2009 when the joint MSc degree program in science communication was introduced by the National University of Singapore (NUS) in collaboration with the Centre for the Public Awareness of Science (CPAS) at The Australian National University (ANU) (NUS FoS, 2018b). CPAS was responsible for developing one of the first specialised degree programs for science communication in 1987 (McKinnon and Bryant, 2017). Leveraging this expertise and combining the unique and complementary strengths of both ANU and NUS, the program at NUS targets both educators who wish to upgrade their skills in education as well as those working in areas of science policy, journalism and scientific writing. The program aims to achieve this through using a curriculum that focuses on communication skills and scientific content. Graduate students who enrol in the program are introduced to theories of communication, learn creative and innovative ways to display science and explore in-depth controversial and emerging science communication issues. They are also given opportunities to apply their learning to develop strategies to communicate science effectively to the public. Since its inception in 2009, the program has successfully trained 97 graduate students, with a yearly intake of 15–20 students.

While an MSc degree program in science communication is available, Singapore's tertiary institutions do not offer an undergraduate degree program. To promote public communication of science amongst undergraduates, Professor Sow, working in collaboration with the Ministry of Education, commenced the Young Educator in Science (YES) program in 2011. This program gives undergraduates opportunities to engage in activities to promote science to the general public and to younger students (NUS FoS, 2018c). YES members are often involved in conducting science demonstration workshops, holiday camps and public exhibitions to develop relevant skills in science communication. Since its inception in 2011, over 250 undergraduates have been trained under the YES program. The program has been very well received by the participants who get to develop and practice science communication skills while increasing their depth of knowledge about the subject matter at hand.

Also at the undergraduate level, since the mid-2000s, credit-bearing science communication courses have been embedded in the science degrees offered at both NUS and the Nanyang Technological University (NTU), two of six public universities in Singapore with the highest levels of undergraduate intake (Ministry of Education, 2017). For example, for all first-year undergraduates enrolled in the Faculty of Science in NUS, the SP1541 'Exploring Science Communication through Popular Science' is a compulsory module. The course aims to equip students with relevant knowledge and skills to communicate complex scientific content in ways that are comprehensible and accessible to non-experts, with a focus on audience-centric communication competencies (written and oral) (NUS FoS, 2018d). One unique feature of this course is the use of popular science texts as a teaching tool to highlight the techniques used to communicate effectively to a non-scientific audience.

Similarly, at NTU, undergraduates from different schools within the College of Science are required to take scientific communication courses (Bolton et al., 2018a, 2018b), which aim to develop students' academic and professional communication competencies. These courses emphasise the importance of designing written and spoken communicative events that are not only delivered appropriately to, but also received successfully by, the intended audience who may comprise students' scientific colleagues and members of the public. In the higher-level course, undergraduates learn about the structure, elements and language features of scientific texts and undertake a group research project, which takes them through the processes of proposing, designing and conducting research. They subsequently present their projects as a written scientific report and as an oral presentation targeted at a non-specialist academic audience.

In summary, the developmental efforts taking place within science communication in Singapore, up to the 2010s, include concerted efforts to develop a local pool of competent science communicators (professionally credentialed and non-credentialed in the subject) for both the private and public sectors.

6. Science communication branching out (2010–present)

Since 2010, science communication in Singapore has become more visible and has established a firmer footing, extending into previously unchartered spheres to meet the changing needs and demands of the times. The reach of science communication has now expanded so that it has become a viable career option to consider in Singapore and an area of growing importance in academic research. Efforts in communicating science to the public have brought science to different informal spaces and to a wider range of audiences. The growing influence and importance of science communication has been, perhaps, most evident from Singapore's role in hosting the inaugural Asia-Pacific Science Communication Conference 2018 and the recent establishment of a commercial entity to provide professional science communication services to private and public organisations.

6.1. Science communication as a career in Singapore

The increased interest in nurturing science communicators has opened up more avenues to work as a science communicator in Singapore. Graduates from the science communication program or science students with interests in communication can nowadays undertake specific roles in a variety of areas. These include, but are not limited to, roles in these four areas: (1) as corporate science communicators who work directly with researchers to translate the impact of an institute's scientific work to the public and the media, to train scientists to engage with the media and to manage media communications (e.g. responding to media inquiries, public relations, developing the corporate image); (2) as informal science educators who work in informal learning spaces such as nature parks, zoos, science centres and museums, and who develop and execute programs (e.g. walks, exhibition content, educational activities, the tour of exhibits etc.) to enhance interactions with the public; (3) as science and media communicators who create communication resources (e.g. publications in print, broadcast, electronic media) for science journalists, authors or documentary makers; and (4) as formal science educators such as teachers who communicate science content in a formal educational setting.

6.2. Science communication research in Singapore

Also gaining ground, though still in its infancy compared to other developed countries, has been interest in science communication research. Before 2000, studies on science educational pedagogy and informal science learning were loosely deemed as 'science communication' research. In Singapore, such investigations have been undertaken by practitioners involved in science literacy in education, or those involved in science educational and pedagogical research. Their findings have been published in peer-reviewed journals or archived in the digital repository of the National Institute of Education, Singapore (e.g. Lee, Hwang, Kim, and Wolff-Michael, 2009).

More recently, however, the number of academic research documents reporting on science communication in Singapore has been increasing. This may be attributed to the presence of the joint ANU–NUS master's of science in science communication program and the science communication research actively undertaken by academics such as Associate Professor Shirley Ho Soo Yee and her colleagues from NTU's Wee Kim Wee School of Communication and Information. To date, graduate students enrolled in the joint ANU–NUS master's program have been adding their research projects to a slowly expanding pool of locally based research into science communication. These graduate students must undertake a research project as part of their course requirements. The focus of research, in these projects, spans areas such as science pedagogy, public engagement and science literacy, science in the media and science on social media.

Research into the Singapore public's perceptions of science and technology, especially in controversial areas such as nanotechnology and nuclear power in Singapore have also been carried out by a range of Singapore-based researchers (Ho et al., 2018; Chuah et al., 2018; Lee et al., 2016; Ho et al., 2015b; Liang et al., 2015; George et al., 2014; Chua et al., 2008, Subrahmanyan and Cheng, 2000). A fairly recent and important contribution, by Associate Professor Shirley Ho and her colleagues, is a pathbreaking study on Singaporean attitudes towards science and technology and how these might shape the conversations of policymakers and communicators when discussing science and technology in the Singapore context (Ho et al., 2015a).

This survey of the general attitudes Singaporeans have towards science and technology finds that they believe that advancements in science and technology have improved their lives and have helped to sustain Singapore's economic competitiveness. The respondents also perceived that these benefits outweighed any harmful effects encountered. Their findings indicate that Singaporeans have a keen interest in scientific and technological knowledge and want to engage with policymakers about policies pertaining to these areas. The findings also highlight the fact that the Singaporean public does not have extreme attitudes (either positive or negative) towards science and technology issues. Other studies have noted that Singaporeans tend to adopt a 'perceived benefit and risk' viewpoint when assessing emerging issues (Chuah et al.,

2018; George et al., 2014) indicating that they are generally receptive to scientific and technological innovations so long as they are given sufficient information to make their own informed decisions about costs and benefits.

While the branching out of science communication within formal higher education contexts has seen important and promising developments in the training of professional science communicators and in science communication research, its extension within the public sphere has increasingly seen science being communicated to a broad spectrum of audiences through informal learning channels such as science shows, social media, public events and open days.

6.3. Science Centre Singapore

An important part of the Science Centre's mission under Professor Lim Tit Meng, the centre's CEO since 2010, has been to provide opportunities for all members of the public, regardless of age or educational background, to access science and to transition the centre's focus on science and technology to include engineering and mathematics as well.

Under Professor Lim's leadership, KidsSTOP—a specially dedicated space for pre-schoolers and early primary children under the age of eight to learn, explore and experience science—was opened in June 2014 (Science Centre Singapore, 2014a). KidSTOP comprises interactive physical exhibits that expose children to all-things-science in their formative years of development. The STEM Inc. program, an applied STEM learning program offered in collaboration with the Ministry of Education (Science Centre Singapore, 2014b), has also been introduced with the goal of working directly with schools to show students that STEM knowledge can be used to solve real-world challenges. Lessons and hands-on activities in the applied STEM learning program focus on bridging conceptual understanding in school curricula to applications in real-world scenarios, showing students the relevance of STEM in today's world and providing opportunities for them to apply their creativity and scientific knowledge to design and develop useful products.

Together with its partners, the Science Centre has also reached out to audiences who may not frequent its premises. This is done through the Singapore Science Festival, an annual festival that celebrates the best of science, technology and innovation in Singapore. Unlike other festivals typically set up in one central location, the Singapore Science Festival comprises multiple satellite events held over a period of three weeks in a variety of locations, such as shopping districts, heartland malls and even hawker centres. Such initiatives bring the Science Centre's activities to locals from all walks of life.

Beyond such show-and-tell initiatives involving science, another avenue for the general public to learn and keep up-to-date about innovations in science and technology is through experimenting for themselves. There are now co-creation spaces in the Science Centre in the form of a tinkering studio, the Einstein Room and the Eco-garden, which the public can access for a small fee to engage in hands-on activities, interact with objects and gain a deeper understanding of science. Tapping into the worldwide Makers Movement, which started in the USA in 2005, in 2012 the Science Centre hosted the first Mini Maker Faire in Singapore. In 2015, within a short span of three years, the Singapore Maker Faire was elevated to full Maker Faire status, giving agency to students and the local community to transform their science and technical knowledge into action.

6.4. Science communication in public spaces

Apart from the Maker Faire, science communication, over the years, has also taken place in public spaces such as the Housing Development Board (or public housing) heartland areas, community centres and shopping malls. These locations provide excellent spaces to communicate science to audiences who may not have had much exposure to, a strong interest in or inclination towards science. The events provide opportunities to those in this demographic to engage in activities and to interact with science exhibits. Examples of communication activities hosted in public spaces include The Pint of Science Festival, the Festival of Biodiversity and the Science Buskers Festival.

Other organisations have also participated in bringing the public to science. For a whole weekend in September, for example, as part of the One North Festival, the public is invited to experience research at major scientific hubs in Singapore such as the Biopolis and Fusionopolis. Organised by A*STAR and its partners, the festival is a celebration of research, innovation, creativity and enterprise. During this festival, members of the public are invited to attend talks by scientists, observe public science demonstrations and tour the research facilities. All the science communication activities at this event are helmed and hosted by A*STAR staff and their supporting partners.

Figure 31.1: A photograph at the local heartland showing Singapore science communicators demonstrating the power of lemon to produce electricity to interested members of the public. This was one of our recent science busking events organised by science communication students in the joint ANU–NUS master's degree program.

Source: NUS, Faculty of Science, Dr Lim Zhihan (used with permission).

A notable aspect of these events is that the impetus for the outreach is not always top-down. While a fair share of events such as Singapore Science Festival are organised by the country's leading institutions in research, ground-level initiatives such as Science Café SG exemplify events held for the common public by the people. Here, citizen groups arrange for scientists to present their work to interested members of the public. Table 31.1 highlights a sample of events where research carried out by scientists, scientific and technological advancements and educational campaigns are communicated in informal settings. The list is not exhaustive, but the events are representative of local, informal and public gatherings held to foster interest in science and technology, to enable scientists to interface with the general public to communicate their work and, perhaps most importantly, for members of the public to communicate their interest in science to others in interactive and entertaining ways.

Table 31.1: Science communication events in public spaces.

Public science communication events (year started)	Frequency	Description	Organised by
Omni Theatre (1985)	Daily	The general public can view movies and live shows related to science and technology.	Science Centre Singapore (SCS)
Science Centre Observatory (1989)	Weekly	The general public can participate in a variety of astronomy-themed events.	SCS
Buskers Festival (Part of Singapore Science Festival or SSF) (2008)	Annually	Participants use everyday tools and items to demonstrate scientific concepts in a show-and-tell competition.	SCS and A*STAR
Bucket Science Symposium (2009)	Annually	A thematic science show where performers use everyday tools and items to demonstrate scientific concepts.	NUS, Faculty of Science and SCS
NUS Science Demonstration Laboratory (2009)	Ad hoc	An experiential learning lab where students participate in guided demonstrations and hands-on experiments.	NUS, Faculty of Science
Science communication and STEM workshops / talks (2011)	Ad hoc	Informal/casual gatherings where scientists speak about their work to the general public.	*Asian Scientist* magazine
Festival of Biodiversity (2012)	Annually	This festival, organised by the National Parks Board in collaboration with the Biodiversity Roundtable, conducts workshops and exhibitions to promote conservation and create awareness about Singapore's biodiversity.	NParks with Biodiversity roundtable
Star Lecture (Part of SSF) (2012)	Annually	Informal/casual gatherings where scientists speak about their work to the general public. Conducted by Speakers from the Royal Institute of Science, London.	SCS and A*STAR
Science Café SG (2013)	Monthly	Informal/casual gatherings where scientists speak about their work to the general public.	Singapore Skeptics, Ground-Up Citizen group

Public science communication events (year started)	Frequency	Description	Organised by
Maker FAIRE SINGAPORE (Part of SSF) (2015)	Annually	Do-it-yourself technological event featuring hands-on activities, projects and tools that introduce the public to science, technology, engineering, art and maths.	SCS, A*STAR, and Defense Science Organisation
One-North Festival (Part of SSF) (2016)	Annually	A festival for the general public to engage in the latest science and technology–related activities in Singapore. The Xperiment science carnival, formerly a three-day event held in conjunction with the science festival in 2009, has now been incorporated into the One-North Festival.	SCS, A*STAR, and JTC Corporation
Tech Terik (2015)	Ad hoc	Casual gathering where the public can discuss up-and-coming science and technology innovations with a technology expert.	Ground Up Innovation Labs for Development
Pint of Science (Singapore edition) Festival (2018)	Annually	A festival where the general public can engage in the latest science and technology themed talks in Singapore.	A*STAR and SCS
Citizen Science Programs	Annually, biannually, or depending on group schedules	Public can participate in organised research efforts to collect data about particular research subjects.	NParks with various citizen groups
Science in the Café	Ad hoc	Informal/casual gathering where presenters and the public can discuss and share ideas on issues in STEM.	SCS
Show Us Your Science – Passion and Knowledge Exchange	Ad hoc	Opportunity to communicate with those passionate in science in SCS's exhibition gallery.	SCS

6.5. Building science communication networks in Asia Pacific

In 2018, NUS faculty, in conjunction with the 10th anniversary of the joint ANU–NUS master's of science in science communication, organised and hosted the inaugural Asia-Pacific Science Communication Conference. The conference aimed to establish connections with science communicators worldwide, especially those from the Asia-Pacific region, to share knowledge, experiences, techniques and innovations.

Figure 31.2: A group photograph of the speakers and participants of the inaugural Asia-Pacific Science Communication Conference 2018.

Source: NUS, Faculty of Science, Science Communication program committee (used with permission).

6.6. Science communication services by Wildtype Media Group

A milestone in the development of science communication in Singapore since the 2010s is the entry it has now made into the commercial sector. Within this sector, Wildtype Media Group Private Limited—headed by Adjunct Assistant Professor Juliana Chan, an award-winning, multi-talented biomedical scientist—has emerged as the first STEM-focused media company in Singapore that provides professional science communication services to government agencies, industry and academia in Singapore as well as the broader Asian region. Having a strong desire to communicate, and make more prominent, the valuable scientific research being conducted in Asia to the wider international lay and academic audience, Chan began the *Asian Scientist* magazine as a blog in 2011 (Chan, 2018). Two years later, Chan, as the founder and editor-in-chief of the magazine, launched and circulated a print version of the magazine. Its popularity has grown exponentially since then. The magazine continues to be very well received by scientists, students and the general public and, in May 2018, Asian Scientist Publishing expanded as Wildtype Media Group Private Limited, extending its marketing and communication services to the science, technological and medical sectors that want to make their work more accessible to the general public through the use of print, digital and social media campaigns. The media company has also contributed as a media partner in the One North Festival (Rohaidi, 2016) and has organised science public outreach programs such as a talk with Nobel laureate Sir Richard Roberts and a pre-med seminar (Chan, 2018).

7. Conclusion

The different developmental phases of science communication outlined in this chapter mark, very broadly, the various ways in which science and technology have been deployed and communicated, either intentionally or unintentionally, to support Singapore's national development. During Singapore's early years following independence, science communication comprised government-led discourses emphasising the importance of science and technology in securing Singapore's future as a nation. This communication was undertaken to garner public support and has ostensibly contributed to shaping the relatively high regard that Singaporean homes have for STEM (cf. Marginson et al., 2013, p. 56). Science communication from 1980 to 2000 broadened considerably and emerged from multiple sources—the government, formal and informal public institutions like schools and the Science Centre Singapore, the media and key individuals. The discourses during this period included examples of science being communicated in a manner that shaped and tapped public knowledge about science and its processes, as was seen in the case of NEWater. More recent developments have come to focus on developing a local pool of professional science communicators for different sectors in Singapore. Efforts have also been made to encourage the general public to participate in informal science communication sessions with professional scientists and with other members of the public.

These developments collectively highlight the firm foothold that science communication has in the political, educational, commercial and social spheres of life in Singapore. Singaporeans, thus far, have shown a keen interest in being informed about developments in science and technology. They have also expressed their desire for local policymakers to engage them when making decisions about policies pertaining to science and technology that will invariably affect their lives and the lives of future generations of Singaporeans (Ho et al., 2015a).

References

Australian Broadcasting Corporation. (2000). What's it all about? APEC Youth Science Festival. Retrieved from www.abc.net.au/science/singapore/what.htm.

Bolton, K., Kathpalia, S., See, E. K., De Souza, D. E., Lam, S. T.-E., Leong, A. and Ong, K. (2018a). *HW0128 Scientific Communication 1: Student's Course Guide* (second edition). The Language Communication Centre, Nanyang Technological University. Singapore: Routledge.

Bolton, K., Kathpalia, S., See, E. K., De Souza, D. E., Lam, S. T.-E., Leong, A. and Ong, K. (2018b). *HW0228 Scientific Communication 2: Student's Course Guide*. The Language Communication Centre, Nanyang Technological University. Singapore: Routledge.

Brown, D. (2000). *Contemporary nationalism: Civic, ethnocultural and multicultural politics*. London: Routledge.

Caleon, I. S. and Subramaniam, R. (2008). Attitudes towards science of intellectually gifted and mainstream upper primary students in Singapore. *Journal of Research in Science Teaching, 45*(8), 940–54. doi.org/10.1002/tea.20250.

Chan, B. (2018, 5 October). The science of effective communication. *The Straits Times*. Retrieved from www.straitstimes.com/singapore/the-science-of-effective-communication.

Chua, S., Giam, X., Yeo, C. and Tan, H. (2008). Attitudes towards biological conservation of one class of National University of Singapore undergraduate students. *Nature in Singapore, 1*, 9–20.

Chuah, A. S., Leong, A. D., Cummings, C. L. and Ho, S. S. (2018). Label it or ban it? Public perceptions of nano-food labels and propositions for banning nano-food applications. *Journal of Nanoparticle Research, 20*, 36. doi.org/10.1007/s11051-018-4126-5.

Dairianathan, A. and Lim, T. M. (2014). Science Centre Singapore as an alternate classroom. In A. L. Tan, Poon C-.L. and S. S. L. Lim (eds), *Inquiry into the Singapore Science Classroom*. Singapore: Springer. doi.org/10.1007/978-981-4585-78-1_14.

George, S., Kaptan, G., Lee, J. and Frewer, L. (2014). Awareness on adverse effects of nanotechnology increases negative perception among public: Survey study from Singapore. *Journal of Nanoparticle Research, 16*(12), 2751.

Ghangaa, M. (2015, 5 August). Singapore's water supply: Where does it come from? *The Straits Times*. Retrieved from www.straitstimes.com/singapore/environment/singapores-water-supply-where-does-it-come-from.

Glauberman, S. (1985, 1 March). Council out to change attitudes towards science. *The Straits Times*, p. 10.

Ho, S. S., Looi, J., Chuah, A. S., Leong, A. D. and Pang, N. (2018). 'I can live with nuclear energy if…': Exploring public perceptions of nuclear energy in Singapore. *Energy Policy*, *120*, 436–47. doi.org/10.1016/j.enpol.2018.05.060.

Ho, S. S., Yang, X., Liao, Y., Turner, D., Tan, R. and Chan, J. M. (2015a). A survey of public views and attitudes towards science and technology issues in Singapore. Singapore: Asian Scientist Publishing, Retrieved from www.asianscientist.com/wp-content/uploads/2015/11/A-survey-of-public-views-and-attitudes-towards-science-and-technology-issues-in-Singapore.pdf.

Ho, S. S., Yang, X., Thanwarani, A. and Chan, J. M. (2015b). Examining public acquisition of science knowledge from social media in Singapore: An extension of the cognitive mediation model. *Asian Journal of Communication*, *27*(2), 193–212. doi.org/10.1080/01292986.2016.1240819.

Lau, A. (1998). *Moment of anguish: Singapore in Malaysia and the politics of disengagement*. Singapore: Times Academic Press.

Law, I. B. (2003). Advanced reuse: From Windhoek to Singapore and beyond. *Water*, *30*(5), 31–6.

Lee, E. W., Shin, M., Kawaja, A. and Ho, S. S. (2016). The augmented cognitive mediation model: Examining antecedents of factual and structural breast cancer knowledge among Singaporean women. *Journal of Health Communication*, *21*(5), 583–92. doi.org/10.1080/10810730.2015.1114053.

Lee, H. and Tan, T. P. (2016). Singapore's experience with reclaimed water: NEWater. *International Journal of Water Resources Development*, *32*(4), 611–21. doi.org/10.1080/07900627.2015.1120188.

Lee, K. Y. (1996). Will Singapore be another slow-growing developed nation? Retrieved from www.nas.gov.sg/archivesonline/data/pdfdoc/lky19960314.pdf.

Lee Kong Chian Natural History Museum (LKCNHM). (n.d.) About the Museum. Retrieved from lkcnhm.nus.edu.sg/about/about-the-museum/.

Lee, Y. J., Hwang, S., Kim, M. and Wolff-Michael, R. (2009). Literacy in learning science: A Vygotskian approach. Retrieved from repository.nie.edu.sg/handle/10497/13584.

Leong, C. (2010). Eliminating 'Yuck': A simple exposition of media and social change in water reuse policies. *International Journal of Water Resources Development*, *26*(1), 113–26. doi.org/10.1080/07900620903392174.

Liang, X., Ho, S. S., Brossard, D., Xenos, M. A., Scheufele, D. A., Anderson, A. A., Hao, X. and He, X. (2015). Value predispositions as perceptual filters: Comparing of public attitudes toward nanotechnology in the United States and Singapore. *Public Understanding of Science, 24*, 582–600. doi.org/10.1177/0963662513510858.

Lim, R. (1998a, 2 March). Work on desalination plant likely to begin by mid-99. *The Business Times*, p. 2.

Lim, R. (1998b, 21 August). PUB embarking on various projects to ensure long-term water supply. *The Business Times*.

Lim, T. M. (2017). The formation and transformation of Science Centre Singapore. In B. T. G. Tan, H. Lim and K. K. Phua (eds), *Fifty Years of Science in Singapore* (pp. 471–88). Singapore, World Scientific Publishing.

Marginson, S., Tytler, R., Freeman, B. and Roberts, K. (2013). *STEM: country comparisons: International comparisons of science, technology, engineering and mathematics (STEM) education. Final report.* Melbourne: Australian Council of Learned Academies. Retrieved from dro.deakin.edu.au/view/DU:30059041.

McKinnon, M. and Bryant, C. (2017). Thirty years of a science communication course in Australia: Genesis and evolution of a degree. *Science Communication, 39*, 169–94. doi.org/10.1177/1075547017696166.

Ministry of Education. (2017). *Education statistics digest.* Retrieved from www.moe.gov.sg/docs/default-source/document/publications/education-statistics-digest/esd_2017.pdf.

Mutalib, H. (2004). *Parties and politics: A study of opposition parties and the PAP in Singapore* (second edition). Singapore: Marshall Cavendish Academic.

National Parks Board. (2002). *Annual Report 2002/03: Reaching out, reaching in.* Retrieved from www.nparks.gov.sg/-/media/nparks-real-content/about-us/annual-report/archive/2002-2003/ar03_3443_reaching-out-reaching-in.pdf.

National University of Singapore, Faculty of Science (NUS FoS). (2018a) Science Demonstration Laboratory. Retrieved from www.science.nus.edu.sg/outreach/sdl-main.

NUS FoS. (2018b). Joint Msc Science Communication Programme. Retrieved from www.science.nus.edu.sg/admissions/graduates/joint-msc-science-comm/.

NUS FoS. (2018c). Young Educators in Science Programme. www.science.nus.edu.sg/studentlife/student-development/yes.

NUS FoS. (2018d). SP1541 Exploring Science Communication through Popular Science. Retrieved from www.nus.edu.sg/nusbulletin/faculty-of-science/undergraduate-education/degree-requirements/curriculum-structure-and-graduation-requirements/sp1541-exploring-science-communication-through-popular-science/.

Public Utilities Board. (2018). NEWater. Retrieved from www.pub.gov.sg/watersupply/fournationaltaps/newater.

Rohaidi, N. (2016, 4 August). Science communication 101 with Asian Scientist's Editor-in-chief. *Asian Scientist*. Retrieved from www.asianscientist.com/2016/08/topnews/asian-scientist-juliana-chan-science-communication-masterclass-one-north-festival/.

Science Centre Singapore. (1983). Young Scientist Badge Scheme. Retrieved from www.science.edu.sg/for-schools/teaching-resources/young-scientist-badge-scheme.

Science Centre Singapore. (2014a). About KidsSTOP™. Retrieved from www.science.edu.sg/visit-us/kidsstop/about-kidsstop.

Science Centre Singapore. (2014b). About our applied learning programme. Retrieved from www.science.edu.sg/stem-inc/applied-learning-programme/about-our-applied-learning-programme.

Science Centre Singapore. (2018a). About Science Centre Singapore. Retrieved from www.science.edu.sg/about-us/about-science-centre-singapore.

Science Centre Singapore. (2018b). National Science Challenge. Retrieved from www.science.edu.sg/whats-on/singapore-science-festival/national-science-challenge-2018.

Singapore Parliamentary Report. (1971, 21 July). Addenda, Ministry of Science and Technology, Vol. 31, Column 15. Retrieved from sprs.parl.gov.sg/search/topic?reportid=006_19710721_S0003_T0012.

Srinivasan, V. (1997, 11 June). Self-sufficiency in water possible, but costly: BG Lee. *The Business Times*, p. 2.

Subrahmanyan, S. and Cheng, P. S. (2000). Perceptions and attitudes of Singaporeans towards genetically modified food. *Journal of Consumer Affairs, 34*, 269–90. doi.org/10.1111/j.1745-6606.2000.tb00094.x.

Subramaniam, R. (2014). Public education about science in Singapore: The role of science journalism via newspapers. In A. L. Tan, C. L. Poon and S. Lim (eds), *Inquiry into the Singapore Science Classroom* (pp. 273–96). Singapore: Springer. doi.org/10.1007/978-981-4585-78-1_15.

Tan, B. T. G. (2017). The Science Council and Singapore science in the 60s and 70s. In B. T. G. Tan, H. Lim and K. K. Phua (eds), *Fifty years of science in Singapore* (pp. 1–22). Singapore: World Scientific Publishing.

Tan, K. P. (2018). *Singapore: Identity, brand, power*. United Kingdom: Cambridge University Press. doi.org/10.1017/9781108561273.

Tan, W.-H. L. (2011). State of science communication and its development in Singapore. *International Journal of Science Education, Part B, 1*(1), 15–16. doi.org/10.1080/21548455.2011.552286.

Tan, W.-H. L. (2017). No conflict between faith and science for Professor Leo. *The Catholic News*, *67*(5), 5 March.

Tan, W.-H. L., Wee, A. T. S. and Subramaniam, R. (2017). The scientific society movement in Singapore. In B. T. G. Tan, H. Lim and K. K. Phua (eds), *Fifty years of science in Singapore* (pp. 531–49). Singapore: World Scientific Publishing. doi.org/10.1142/9789813140905_0029.

Tan, Y.-L. K. (2017). The Lee Kong Chian Natural History Museum. In B. T. G. Tan, H. Lim and K. K. Phua (eds), *Fifty years of science in Singapore* (pp. 489–529). Singapore: World Scientific Publishing. doi.org/10.1142/9789813140905_0028.

Teng, A. (2016, 6 June). Making science simpler and fun. *The Straits Times*. Retrieved from www.straitstimes.com/singapore/education/making-science-simpler-and-fun.

Times Publishing Group. (2017). Marshall Cavendish Education and Science Centre Singapore launch science magazine for children. Times Publishing Group, www.timespublishing.sg/join-us/career-opportunities/jobs/2015/07/24/marshall-cavendish-education-and-science-centre-singapore-launch-science-magazine-for-children.

Tortajada, C., Joshi, Y. K. and Biwas, A. K. (eds). (2013). *The Singapore water story: sustainable development in an urban city-state*. New York: Routledge.

Timeline

Event	Name	Date	Comment
First interactive science centre established.	Science Centre Singapore	December 1977	The idea for a science centre was first conceived in 1967 and it took 10 years to build
First national (or large regional) science festival.	National: Singapore Science Festival	2001	2000, Singapore hosted the Asia Pacific Economic Corporation (APEC) Youth Science Festival
First university courses to train science communicators.	Currently no formal undergraduate degree for science communication	–	Aspects of science communication may be covered in communication degrees. In addition, science faculties in Singapore universities mandate compulsory science communication modules for science undergraduates
First master's students in science communication graduate.	ANU–NUS joint master's degree of Science in Science Communication	July 2011	The first student was enrolled in January 2009
First PhD students in science communication graduate.	Currently no formal PhD program for science communication	–	Aspects of science communication may be covered in doctoral degrees in communication

Event	Name	Date	Comment
First national conference in science communication.	Inaugural Asia-Pacific Science Communication Conference 2018	November 2018	Aimed to allow experts and students to share knowledge, and held in conjunction with ANU–NUS master's of science communication
National Science Week founded.	No dedicated National Science Week	2001	But the Singapore Science Festival has a week of events before the satellite events happen
A journal completely or substantially devoted to science communication established.	Currently no academic journal has been established	–	The Science Centre has published a magazine targeted at schoolchildren known as *Singapore Scientist*
First significant TV programs on science.	The Science and Industry Quiz	1972	2004: The National Science Challenge is a televised competition for school students
Other significant events.	Lee Kong Chian Natural History Museum	April 2015	First natural history museum in Singapore. The collection has been inherited from predecessor museums that date back to 1878. 1976: Singapore Association for Advancement of Science was set up to promote public engagement of science and technology

Contributors

Dr Denise E. De Souza facilitates scientific communication courses at Nanyang Technological University, Singapore.

Dr Lieu Zi Zhao is a lecturer with the Special Programme in Science (SPS) at the National University of Singapore (NUS), Singapore.

Letchumi Mani heads the Outreach and Education Team at the Lee Kong Chian Natural History Museum.

Dr Glenn Toh is senior lecturer at the Language and Communication Centre, School of Humanities, Nanyang Technological University, Singapore.

Dr Benedict Lin is a lecturer at the Language and Communication Centre at the Nanyang Technological University, Singapore.

32

SOUTH AFRICA

Science communication throughout turbulent times

Marina Joubert and Shadrack Mkansi

1. Introduction

Science is practised within the contexts of societies. What's more, the public communication of science is sculpted by politics, culture and socioeconomic realities. As such, the relationship between science and society in South Africa is historically defined by our turbulent political past, and continues to be moulded by present-day challenges. Over the past 25 years, pivotal policy transformations have opened up new possibilities and inspired ambitions for developing a critically engaged knowledge society. However, some formidable obstacles remain in the pursuit of this goal.

2. The long shadows of colonialism and apartheid

In early African societies science and technology were integral elements of culture. This is demonstrated by archaeological evidence of remarkable ancestral abilities in prospecting and tool-making (Boshoff et al., 2000). Moreover, ancient fishing technologies, developed over a period of 10,000 years, are still in use in rural Africa (Bruton, 2016).

The colonial period brought social disruption, but also scientific exploration. During Dutch and British rule in South Africa (intermittently 1652–1910), the colonialists exploited local resources for the benefit of their imperial masters (Dubow, 2006; Huigen, 2009). Driven by intellectual curiosity, amateur scientists charted the interior of the country, while collecting mammals, birds, fishes, insects, fossils and plant specimens for collections housed in Europe. Colonial science served the interests of the local elite in navigation, astronomy and cartography. In later years, mining, plantations and medicinal plants also became important. Colonial intellectuals defined and described the indigenous people of Africa (Dubow, 2006) and many prominent naturalists visited the Cape. William Burchell, for instance, visited in 1822 and 1824 and popularised his explorations in two volumes, *Travels in the Interior of Southern Africa.* In 1836, the 27-year-old Charles Darwin spent almost three weeks in South Africa on the last leg of the five-year voyage of HMS *Beagle* (Thackeray, 2009). Between 1828 and 1833, Scottish surgeon and explorer Sir Andrew Smith led several expeditions into little-known parts of the country, resulting in a five-volume illustration of the zoology of South Africa. From the mid-1850s onwards, local scientists reported their discoveries to the science community in London via the *Cape Monthly Magazine*[1] (Dubow, 2006), but many field notes either remained unpublished or were destroyed when power changed hands at the Cape (Boshoff et al., 2000).

The notion of science as a tool to conquer and tame Africa and harness its resources lasted well into the 20th century (Tilley, 2011). Racial segregation and white privilege were well established and intellectual discussions focused mainly on matters of Western science and society, while legislation enforcing inferior education and work opportunities for black people suppressed the development of the majority of the population (Dubow, 2006; Du Plessis, 2015).

Political interference in science peaked during the Apartheid Era (1948–93), a period characterised by state control, censorship, strategic investments in science and secretive research (Dubow, 2006; Du Plessis, 2017). Science was viewed as a political tool, with scant regard for public accountability. The South African Broadcasting Corporation (SABC) supported apartheid (Tomaselli, 2002) and stringent laws clamped down on press freedom, with scientifically classified information tightly controlled by the state. For example, during the height of apartheid, official permission was needed to mention 'atomic energy' in the media and defiant journalists faced heavy fines and/or long-term imprisonments (Du Plessis, 2017). Consequently, the public, including most scientists, knew very little about the extent of South Africa's nuclear weapons capabilities.

1 One of the earliest examples of South African scientists being urged to disseminate their work more broadly dates back to an 1860 edition of the *Cape Monthly Magazine*. Editor Roderick Noble accused scientists of 'having a hardness about them' and presenting their work as a mere 'dry narration of facts' (Noble, 1860).

In 1953, the Bantu[2] Education Act[3] limited the education of black children to basic reading and writing. It was during this time that the apartheid architect Hendrik Verwoerd infamously said: 'There is no place for [the Bantu] in the European community above the level of certain forms of labour ... What is the use of teaching the Bantu child mathematics when it cannot use it in practice?'[4]

The close relationship between the National Party and the Dutch Reformed Church meant that religion shaped the values, norms and institutions of the ruling party (Ritner, 1967). With evolution banned from school curricula, curators at natural history museums were not allowed to refer to evolution in their displays, unless they indicated explicitly that evolution was presented as a theory rather than fact. In the 1970s, Transvaal Museum director Bob Brain bypassed this restriction by calling his two major exhibitions halls 'Genesis I' (for the evolution of life from microbes to mammal-like reptiles) and 'Genesis II' (for the exhibition hall dealing with mammals, including humans and their distant relatives). In both halls, the concept of evolution was reflected using evolutionary trees (phylogeny), without referring to evolution by name.[5]

Due to its apartheid laws, South Africa was politically isolated and faced stringent economic sanctions and cultural boycotts. Despite this, the South African science base strengthened and the country developed advanced facilities and expertise in fields such as geology, mining, energy, nuclear science, space science, medicine, agriculture, veterinary science and the natural environment (Sooryamoorthy, 2010).

Many academics defied apartheid and were actively involved in the struggle. They found a measure of protection from political meddling within the more liberal universities in the country, but even prominent South African scientists were occasionally 'tarred by an apartheid brush' and denied access to international societies and conferences (Wood, 2012, p. 40).

2 A collective term (incorrectly) used by the National Party to describe black people of African origin; the term later became a symbol of oppression. See South African History Online www.sahistory.org.za/article/defining-term-bantu.

3 Act No. 47 of 1953, Parliament of South Africa.

4 Dr Hendrik Verwoerd, South African Minister for Native Affairs (Prime Minister from 1958 to 1966), speaking about his government's education policies in the 1950s. See Apartheid Quotes About Bantu Education: www.thoughtco.com/apartheid-quotes-bantu-education-43436.

5 Personal communication with Francis Thackeray, October 2018.

In terms of science, the legacy of apartheid is ambivalent. Extensive investment in military and energy research resulted in cutting-edge innovations and sophisticated science-based industries, but it also created an inward-looking science with no regard for social justice and failed to provide basic services such as shelter and clean water to most South Africans (Mouton, 2006). Many (not all) scientific institutions in the country were an integral part of a system that promoted the (white) Afrikaner identity (Dubow, 2006). Public communication of science was mostly restricted to corporate communication efforts produced by scientific institutions (Du Plessis, 2017). Consequently, black South Africans remained entirely disconnected from science.

3. Pioneers of science communication in South Africa

Some of the earliest scientific institutions in South Africa were established by the British, including the South African Library (1822), the Royal Observatory at the Cape (1828) and the South African Museum[6] (1825). The Southern African Association for the Advancement of Science (also known as the S_2A_3) dates back to 1902 (Boshoff et al., 2000), while the Royal Society of South Africa, established in 1908, contributed significantly to intellectual vibrancy during the 20th century (Carruthers, 2008). Several scientific organisations currently leading science in South Africa were established shortly before or during the apartheid years, including the Council for Scientific and Industrial Research (established in 1945), as well as the Medical Research Council and the Human Sciences Research Council (both founded in 1969).

Despite the impediments presented by 20th-century politics, there were many gifted South African scientists[7] who were passionate about sharing their work with society. Some trailblazers and memorable milestones are listed below, with more details in footnotes.[8]

6 In its early years, the South African Museum was regarded as 'a private club for knowledgeable gentlemen' and it was not easy for the public to get access (Dubow, 2006, p. 37).

7 Early South African science popularisers include the writer and naturalist Eugene Marais (1871–1936); the first warden of the Kruger National Park James Stevenson-Hamilton (1867–1957); entomologist and author Sydney Harold Skaife (1889–1976) and fisheries scientist and oceanographer Cecil von Bonde (1895–1983), as well as the notable conservationists Thomas Chalmers Robertson (1907–89) and Ian Player (1927–2014). In the 1910s and 1920s, Sir George Cory became known for presenting spectacular chemistry shows to his students and the general public (Bruton, 2018).

8 This list is necessarily incomplete; if space allowed, many more names and events could be included.

Figure 32.1: The first human heart transplant on 3 December 1967 made a science celebrity of South African surgeon Christiaan Barnard and changed relationships between medicine, media and society globally.
Source: Heart of Cape Town Museum (used with permission).

The epic discovery of a living coelacanth off South Africa's coast, a fish thought to have been extinct for 65 million years, ignited global and local interest (Bruton, 2015, 2018). On 22 December 1938, museum official Marjorie Courtenay-Latimer found the unusual fish while inspecting a local catch. It was later identified by ichthyologist J. L. B. Smith. In December 1952, when Smith returned to South Africa with a (second) coelacanth caught in the Comoros, the SABC interrupted their regular programming to air a radio interview with him, recorded on the tarmac at Durban Airport. This interview was subsequently broadcast worldwide by the BBC and North American radio stations. Smith became a regular voice on radio and later television, sought after by journalists for his expertise on aquatic life and views on conservation. He wrote more than 800 popular science articles during 1940–60 period, and his scientific bestseller *Old Fourlegs: The story of the Coelacanth* sold over 800,000 copies and was translated into nine languages (Bruton, 2017).

In Cape Town on 3 December 1967, Christiaan Barnard (1922–2001) performed the first human heart transplant.

The dramatic nature of this medical milestone, along with the media flair of the charismatic surgeon himself, captured global attention and fundamentally changed the relationship between medicine, media and society around the world (Joubert, 2018a). At the time, the South African government saw an opportunity to improve South Africa's image and tried to co-opt Barnard as an ambassador for his country (Logan, 2003). When Barnard publicly spoke out against whites-only wards in hospitals, he was rebuked by then state president Nic Diederichs who warned that government would no longer protect him against his critics (Molloy, 1992).

The eminent paleoanthropologist Phillip Tobias (1925–2012) was another of South Africa's science icons. Based on his conviction that 'race is irrelevant in matters of the mind and spirit' (White, 2012, p. 423), Tobias fought against racism, specifically in universities and the scientific community. He became the first leading scientist in South Africa to present and narrate a television series. The six-part series *Tobias's Bodies* aired on SABC television in 2002 and took viewers on a fascinating journey of human evolution and its relevance to life in the 21st century. Notably, Tobias also used this science television series as a platform to combat racism.

4. Developments since democracy

Since 1994, democracy has fundamentally transformed South African society, including its science–society interface. The new government viewed public science engagement as a tool to help correct past imbalances and enhance socioeconomic growth. For the first time in the history of the country, the broad dissemination of science formed part of national science policy.

The White Paper on Science and Technology (DACST, 1996) underlined the transformational power of public engagement with science, and emphasised the importance of a society that valued science. It encouraged research organisations to participate in awareness-raising campaigns, and urged researchers to articulate clearly the benefits of their work to decision-makers and the public. Amidst this new enthusiasm for science communication, there were also significant challenges. Because of past political barriers most South Africans were disconnected from science; with few exceptions, the demand for increased public engagement was also new (and unwelcome) to many scientific institutions and scientists.

As one of its first major public engagement initiatives, the new government declared 1998 'The Year of Science and Technology' (YEAST'98). It was a signal that 'something has to be done to give South Africans a wake-up call', said then Minister of Arts, Culture, Science and Technology Lionel Mtshali (DACST, 1998). During this year-long nation-wide science communication campaign, exhibitions, science shows and public talks were organised in each of the nine provinces of South Africa. The aims were to demystify science as well as to create a special period during which much of the attention of the nation and the media would focus on science and technology. The campaign was also geared towards helping to realise Thabo Mbeki's (deputy president at the time) vision of an African Renaissance (DACST, 1998; Boshoff et al., 2000). Mbeki referred to YEAST'98 as 'a new movement which will ensure that our country is prepared for the challenges of the new millennium' (DACST, 1998). Based on the proclaimed success of this campaign, the government implemented an annual science week with the objective of 'taking science, engineering and technology to our people' (DACST, 1999, 2000). But the aims of YEAST'98 were not only overly ambitious, they were also too vague to be measured effectively. Therefore, a meaningful evaluation of YEAST'98 was not possible. The only indicator of success for these events was the number of people who participated, without regard for whether their participation made any difference to how they viewed or used science in their daily lives. Ironically, in many cases, this superficial approach to evaluating science engagement initiatives persists to this day (Weingart and Joubert, 2019).

In the early years of government-funded science engagement initiatives in South Africa, the objective of 'demystifying' science was a prominent goal, illustrating that science was perceived to be unknown, almost foreign, to most South Africans. Since then, terms such as 'awareness', 'outreach' and 'dissemination' have been used frequently. The prominence of 'engagement' in more recent policies and reports reflect the desire for more interaction and dialogue between science and society.

At grassroots level, those involved in the government-sponsored science awareness events in the early years of democratic South Africa experienced many challenges and obstacles. Despite efforts to spread activities around all nine of South Africa's provinces, it was impossible to reach rural areas where people were thinly distributed over vast distances. Additionally, in a mass media culture dominated by politics, crime and sport, it was not easy to secure media interest in science-related news and events. It also proved difficult to make science interesting and relevant to people who had to deal with unemployment and poverty in their daily lives. These challenges were compounded by the fact there were very few scientists who could speak to people in any of the country's indigenous languages. Collectively, the science communication community in the country had very little experience of how to communicate science effectively with diverse public audiences, and hardly any knowledge of catering for audiences who were mostly distant from science.

In many of the public events during YEAST'98, a disconnect between what was on offer (i.e. exhibitions and public talks) and the needs/interests of the audiences in attendance was painfully obvious.

5. Indigenous knowledge systems and pseudoscience—distinctions and overlaps

Since coming to power in 1994, the 'new' South African government has been committed to preserving, protecting and promoting indigenous knowledge systems (IKS). In 2004, the cabinet adopted an Indigenous Knowledge Systems Policy, and the 2016 Protection, Promotion, Development, and Management of Indigenous Knowledge Systems Bill was approved by parliament. These policies affirm African cultural values in the face of globalisation, support for services provided by traditional healers and recognition of their contribution to the economy, and the interface between indigenous knowledge systems and other forms of knowledge.

The National Research Foundation (NRF) provides funding for research into IKS-based innovations and the links between IKS and bioeconomy, astronomy, food security, farming, environment, storytelling and music, as well as for exploring women's roles in IKS and the IKS practices amongst specific communities such as the Khoi, Nama, Griqua and San.

A 2009 national survey demonstrated that the majority of South Africans favour indigenous knowledge over so-called 'Western science' (HSRC, 2010). Two-thirds (66 per cent) of respondents felt that IKS could offer lessons that could benefit everyone and 71 per cent felt that South Africans trusted too much in science and not enough in indigenous knowledge. However, claims that up to 80 per cent of South Africans regularly use traditional healers have been shown to be incorrect (Wilkinson, 2013). A 2008 survey shows that only 1.2 per cent of South African households consult traditional healers (Nxumalo et al., 2011), while a 2014 study reports that only 0.5 per cent of South Africans see traditional healers as a first point of call for health services (Statistics South Africa, 2015). Nonetheless, traditional healers are regarded as custodians of indigenous knowledge and form part of the local healthcare system. While some traditional healers collaborate with scientists to validate the healing properties of certain plants (Makunga et al., 2008; Ramchundar and Nlooto, 2017), others regard their declared healing powers as closely guarded secrets (Ndhlala et al., 2011). Some healers claim that their therapies cannot be scientifically tested, since their healing powers are at least partially of a spiritual nature. For these healers, subjecting these cures to scientific scrutiny could be conceived as being insensitive to their ancestors, who they consult in their healing practices.[9]

9 Presentation by Janice Limson at SciCOM100 Conference, Stellenbosch University, 5 November 2018.

A key concern around traditional healers relates to those who pose as herbalists but offer dubious remedies and propagate potentially harmful beliefs (Ndhlala et al., 2011; Nyundu and Naidoo, 2016). Examples of damaging myths and superstitions include trust placed in the magical powers of *muthi*[10] to protect during conflict (Nyundu and Naidoo, 2016); the belief that lightning is caused by witchcraft (Le Grange, 2007); the belief that sex with a virgin is a cure for HIV/AIDS (Pitcher and Bowley, 2002); and the stigma associated with people living with albinism (Baker et al., 2010). In extreme cases, human remains are used for *muthi* (Ndhlala et al., 2011). Traditional healers' use of animal parts, including lions, vultures and reptiles, are mostly for pseudotherapeutic effects such as making someone stronger, boosting luck or warding off bad spirits, but these practices raise conservation concerns (Ndhlala et al., 2011; Williams and Whiting, 2016). The problems that result from such myths and superstitions are widely recognised (Manzini, 2003), as is the need to eliminate outdated and potentially damaging practices (Du Plessis, 2017).

A notable illustration of the consequences of pseudoscientific beliefs can be found in an incident that occurred in South Africa around the year 2000 when Thabo Mbeki, South African president at the time, questioned the science of HIV/AIDS. Together with his health minister, Dr Mantombazana Tshabalala-Msimang, they promoted the use of the African potato and a concoction of garlic, beetroot and lemon as an HIV/AIDS treatment (Schneider and Fassin, 2002; Mbali, 2004; Nattrass, 2007). The subsequent delay in implementing an antiretroviral treatment regime in the public health system led to the deaths of more than 330,000 people, while about 35,000 HIV-positive babies were born during this period (Chigwedere et al., 2008).

Some scientists perceive a particular duty to speak out against these pseudoscientific beliefs (Joubert, 2018b), but it is a complex issue that requires sensitivity to local sociocultural dynamics. These beliefs are deeply connected to local cultures and their traditional way of life (Williams and Whiting, 2016) and may be linked to deep-rooted suspicions of so-called 'white' medicine (Batts, 2006).

Current debates on the 'decolonisation' of South African universities are closely related to an increasing awareness of the science information needs of local communities in the context of IKS. Scholars are currently exploring ways in which so-called 'Western' science and indigenous knowledge might be integrated. The implications and challenges of such an integration for public science engagement are obvious and numerous.

10 *Muthi* is a plant- or animal-based substance prepared by herbalists that is believed to heal, cleanse, strengthen or protect.

6. A new policy environment

Because of its defining role in policy and funding, government is a pivotal role player in public science engagement in South Africa. Current science policies, including new legislation, white papers and long-term strategic plans, demonstrate that policymakers see public engagement as a tool to democratise science, increase the societal impact of science and sustain public trust in science. In its 2019 White Paper on Science, Technology and Innovation (DST, 2019), the South African Department of Science and Technology (DST) reiterates the importance of a science-literate and science-aware society. Public engagement with science is regarded as a prerequisite for South Africa to become a knowledge-based society with a participatory mode of science governance. Citizens are no longer passive recipients of the products of science, but important contributors to the processes that shape science (DST, 2007, 2015). These policy intentions are consistent with global trends in science communication emphasising a transition from top-down, one-way communication towards a participatory mode of engagement. However, as elsewhere, the implementation of truly dialogic science engagement is complex and challenging. At present, most science engagement activities—with their emphasis on public talks, exhibitions and workshops—would fall into the category of providing information to a captive audience.

In 2015, the DST adopted a *Science Engagement Strategy* (DST, 2015), positioning science engagement as a way to enrich people's lives and empower them to reflect critically on issues rooted in science. This framework spells out wide-ranging plans to support, coordinate and evaluate science engagement at a national level, and proposes that public science engagement should become a mandatory activity for publicly funded researchers. These new policy intentions are confirmed in the 2019 White Paper (DST, 2019), including the intention of allocating a fixed percentage of public research funding to raising science awareness.

Three years since the announcement of this new engagement strategy, there is still a wide gap between these ambitious aims and the grassroots realities of public science engagement in South Africa. Key challenges relate to the multitude of target audiences, the dual nature of its objectives, the question of effective evaluation and the lack of tangible incentives for scientists' involvement in engagement activities.

It is evident that government paints with a very broad brush when identifying target audiences for public science engagement. The new strategy goes some way towards segmenting audiences. However, 11 target publics, namely learners, educators, industry, scientists and researchers, science interpreters, decision makers, journalists, students, tourists and indigenous knowledge holders, are still listed, with 'the general public' added as a catch-all category. Such broad and vague descriptions of audiences for science communication are problematic, since trying to engage everyone typically results in engagement programs that are superficial and impossible to evaluate, and may not be particularly effective for any specific audience (Borchelt, 2001). Scholars (e.g. Irwin and Horst, 2016) warn against the conflation of different publics into a generic whole. Furthermore, science communication activities aimed at 'the general public' are likely to connect with publics that are already interested, socially privileged and/or positively predisposed towards science at the expense of 'hard-to-reach' and marginalised groups in society (Kennedy et al., 2017; Dawson, 2018).

The current science engagement strategy of government proposes a blend of promotional, educational and engagement goals, using terms such as 'popularise', 'promote' and 'profile', which are reminiscent of a marketing/PR approach and deficit-style communication. In contrast with these aims, the strategy also specifies the development of a socially aware and critical society concerning matters of science and technology. The White Paper similarly frames the objectives of public science engagement activities as promoting science and 'enhancing its public standing' (DST, 2019, p. 33).

It is problematic to assume, as stated by government, that science will become popular amongst South Africans 'if all target publics participate in projects that make them aware and keep them abreast of key developments in science and technology' (DST, 2018, p. 10). In reality, more knowledge about science does not necessarily translate into more positive attitudes toward science (Cheng et al., 2008; Scheufele, 2014). In fact, scientific literacy and scientific ideology are negatively correlated in most countries, meaning that more knowledgeable citizens are more likely to reject new scientific ideas (Bauer, 2008).

Recognising the need for meaningful evaluation and impact measurement of its science engagement programs, the DST developed a comprehensive *Science Engagement Monitoring and Evaluation Framework* (DST, 2018). This framework proposes an extensive set of baseline measures and success indicators, along with the structures, processes and tools required for effective monitoring and evaluation. It will, no doubt, be challenging to implement these evaluation measures across a fragmented system where role players are not accustomed to being evaluated in this way. A key challenge is progressing from attractively simple measures such as counting numbers of visitors at public science events, towards measures capable of evaluating whether people have gained new understanding, insights or competencies through their engagement with science at these events.

In line with global trends, research funding instruments in South Africa are attributing increasing prominence to public science engagement. Local and international funders have introduced societal impacts as one of the criteria they consider when reviewing funding proposals, indicating their intention to position public engagement as an integral part of a research plan. However, it remains unclear whether these engagement plans have any effect on funding decisions and many scientists do not see any incentives for public science engagement from their employers. Consequently, calls for increased involvement in public engagement are often perceived as lip service rather than a genuine commitment on the part of funders and institutions (Joubert, 2018b). This apparent lack of rewards remains an obstacle when it comes to motivating scientists to become—and remain—actively involved in public science engagement activities.

7. The institutionalisation of science communication in present-day South Africa

The establishment of the South African Agency for Science and Technology Advancement (SAASTA) in 2002 is evidence of the democratic government's recognition of the need to coordinate public science engagement at a national level, and to invest in related activities. SAASTA is mandated to implement the science engagement initiatives of the DST.

The science engagement initiatives of government depend on a network of museums, science centres and festivals that were pioneered by dedicated individuals ardently committed to creating innovative spaces for informal science learning. The first science centre in South Africa was established as an 'exploratorium' at the University of Pretoria in 1977 by Lötz Strauss, while Brian Wilmot launched SciFest Africa in 1996. Other notable science centre pioneers include Shadrack Mahapa, Mike Bruton, Jan Smit and Derek Fish.

Figure 32.2: Held annually in Makhanda in the Eastern Cape Province of South Africa, SciFest Africa is the largest and oldest science festival in the country where young and old can enjoy and debate science.

Source: Water World – South African Institute for Aquatic Biodiversity (used with permission).

Figure 32.3: A network of science centres across South Africa forms an important part of South Africa's science communication ecosystem, particularly for engaging school learners and hosting public engagement events such as science weeks.

Source: Arcelor Mittal Science Centre (used with permission).

Science centres cater for diverse audiences and provide vital opportunities for informal science learning. They also play a key role in supporting the school science curriculum and providing spaces where learners from under-resourced schools can do practical science experiments. These centres[11] vary in size and sophistication, from small and rudimentary displays crammed into tiny spaces, to modern and well-equipped centres with interactive displays and custom-built auditoria, with some mobile outreach programs to peri-urban and rural communities. One challenge they have in common is that of financial sustainability in a country where the bulk of their visitors cannot afford to pay even a modest entry fee. In addition to funding from government, some science centres rely on industry support. A relatively common local model is that science centres are built on university campuses, with some financial support provided by the hosting university.

11 The SAASTEC website lists 37 science centres in South Africa in 2018, with 59 organisations listed as SAASTEC members. See www.saastec.co.za/membership/.

A few leading organisations on the local science landscape have adopted projects that focus on promoting public engagement with science. For example, the National Science and Technology Forum (NSTF) recognises top achievers in science at its annual 'NSTF-South32 Awards', with a special category for science communicators. The Academy of Science of South Africa (ASSAf) publishes a quarterly popular science magazine called *Quest*.

Several NGOs also make a significant contribution to engaging the South African public in science and conservation issues, including the Endangered Wildlife Trust (EWT), Birdlife Africa, the Wildlife and Environment Society of South Africa (WESSA) and the World Wide Fund for Nature South Africa (WWF SA). Citizen science projects scattered across the country allow thousands of volunteers to collect data on birds, butterflies, insects, spiders, frogs and mammals that feed into various science and conservation projects. For example, the Animal Demography Unit at the University of Cape Town uses citizen-generated data to create a series of atlases that feed into conservation policies, as well as a dynamic virtual (online) museum. Similarly, the National Biodiversity Institute and South African National Parks uses its network of botanical gardens and nature reserves as starting points for involving visitors in citizen science.

A notable development on the local science communication scene is the blending of science centres and tourism interests, with Maropeng and the Sterkfontein Caves, and the official visitor centres for the Cradle of Humankind World Heritage Site, as premier examples. Here, interactive displays chart the progress of humankind from our early beginnings in Africa, focusing on the science of palaeoanthropology and what it represents in terms of understanding our shared heritage and common humanity.

Today, South African scientists are increasingly visible on the local and international stage via their participation in science communication events and competitions such as FameLab, Three-Minute-Thesis and Pint of Science (Garrard, 2018), and at science cafés that regularly pop up on university campuses. Another creative science engagement platform that has spilled over to South Africa is Science & Cocktails organised at the Orbit Jazz club in Johannesburg since 2015. More examples of science-collaborations include Science Lens, a science photography competition organised by SAASTA since 2002, as well as the SKA (Square Kilometre Array) Shared Sky project that brought together South African and Australian artists in a collaborative exhibition on ancient cultural wisdom about the understanding of the night sky.

Figure 32.4: The potential of music and dance in engaging previously 'hard-to-reach audiences' with science is increasingly being realised. Hip hop in particular presents a unique opportunity to connect with underprivileged youth and make science relatable. This image showcases a Hip Hop U event facilitated by Jive Media Africa and the African Health Research Institute. Through such initiatives, science, hip hop and music are being combined in creative and entertaining ways.

Source: African Health Research Institute (used with permission).

8. Science journalism in South Africa

Nearly 70 per cent of South Africans rely on radio to find out about science; 65 per cent get science information from free-to-air television, while online sources are still relatively unimportant (Parker, 2017). In recognition of the importance of radio as a channel for public science engagement in the country, SAASTA has launched an initiative to place young, unemployed science and journalism graduates at community radio stations to increase good science content via community radio.

As in other countries, the number of specialist science reporters in South Africa has declined, raising concerns about the scope and quality of science journalism in the country. These include haphazard science reporting by untrained journalists and a blurring of science and pseudoscience in media reports (Claassen, 2011). Still, while science-focused pages in print media and programs on television and radio are rare, science does feature in mainstream print media as well as in actuality and investigative programs on radio and television. For example, *Carte Blanche*, a weekly, hour-long television program that celebrated its 30th anniversary in 2018, regularly features stories related to science, health and/or the environment. Local health journalism has been given a lifeline in the form of a donor-funded platform called Bhekisisa.

Furthermore, since its launch in 2015, the Africa edition of *The Conversation* offers a new intermediary platform for researchers to write for the mass media and general public.

Science media coverage in South Africa has some unique fingerprints that reflect issues of local importance and geographical relevance. For example, the media pays most attention to issues related to HIV/AIDS and education, topical environmental issues such as rhino poaching (Guenther, Weingart and Joubert, 2019) and large science infrastructure projects such as the SKA project (Gastrow, 2015).

9. Establishing science communication research and university training

Since the early 1990s, a handful of South African researchers started looking at public understanding of and attitudes to science—mostly attributing low levels of understanding and interest to the legacy of apartheid (e.g. Pouris, 1991, 1993, 2003; Blankley and Arnold, 2001). As Du Plessis and Masilela (2012) pointed out, these studies were primarily small and demarcated, highlighting the need for more nuanced and comprehensive assessments of the relationship between South African science and its publics. More recently, the Human Sciences Research Council (HSRC) has commissioned nationally representative surveys of the public's perceptions of science (Reddy et al., 2013), as well as public perceptions of astronomy and the SKA telescope (Roberts et al., 2014) and public perceptions of biotechnology (Gastrow et al., 2018). Investigations into representations of science in the media focused on biotechnology (Gastrow, 2010) and the SKA telescope (Gastrow, 2017). These studies reveal a complex mixture of perceived benefits and reservations in the way South Africans view science (Reddy et al., 2013; Guenther and Weingart, 2016) and highlight the influence of culture and cultural distance to science when interpreting public views of science, particularly for rural communities (Guenther and Weingart, 2018; Guenther et al., 2018). Research by Parker (2017) reveals a bleak overall picture of low interest in science and very few South Africans involved in activities where they could engage with science.

As local science communication policy changed in favour of public science engagement, a community of science communication practitioners emerged and expanded, but academic activity was still largely neglected (Du Plessis, 2017). In 2015, two research chairs in the field of science communication were established in South Africa, one each at Stellenbosch and Rhodes universities. Research at Stellenbosch focuses on the challenges of engaging a culturally diverse and educationally stratified society. The research agenda includes work on public perceptions and expectations of science, science as reflected in the mass media, the role of scientists and scientific organisations, and the influence of social media on science and science communication. The research team at Rhodes studies models of direct engagement between scientists and the public while exploring the benefits of science engagement to science students—and in particular their motivations for conducting research that has direct societal impact and co-creates knowledge with local communities. The work of these two research groups generates an evidence base for public science engagement within the local context, responding to the unique socioeconomic challenges and social stratifications that characterise South Africa.

South African science communication scholars and communicators participate in several international projects relevant to science communication. In 2011, the International Astronomy Union (IAU) decided to locate their international Office of Astronomy for Development (OAD) in Cape Town. South Africa joined the UK, Germany, China and India in the Mapping the Cultural Authority of Science (MACAS) project during 2012–15, with the aim of constructing a system of science culture indicators based on news analysis and public attitude data. The research outputs from this project are collected in a book, *The Cultural Authority of Science: Comparing across Europe, Africa, Asia and the Americas* (Bauer et al., 2019). South Africa is one of 14 countries participating in New Understanding of Communication, Learning and Engagement in Universities and Scientific Institutions (NUCLEUS), a Horizon 2020 project. Furthermore, the South African Institute for Aquatic Biodiversity (SAIAB), a business unit of the NRF, has been selected as an 'embedded nucleus' or test site for responsible research and innovation. The focus is on 'engaged' research—i.e. integrating meaningful and mutually beneficial dialogue with lay people into the research agenda.

10. Ongoing science communication challenges

South Africa is one of the most unequal societies in the world, with millions of people living in impoverished circumstances. For them, daily needs such as clean water and having enough food to eat are paramount, while debates about genetically modified organisms or nuclear energy seem superfluous (Lewenstein et al., 2002).

Figure 32.5: The construction of several components of the Square Kilometre Array in a central part of South Africa has provided a platform for engagement with local communities. Astrophysicist Dr Nadeem Oozeer explains how a radio telescope works to learners from Carnarvon Primary School.

Source: South African Radio Astronomy Observatory (used with permission).

Several of South Africa's 'big science' installations, such as the Southern African Large Telescope (SALT) and the SKA, are located in rural areas in close proximity to towns riddled with unemployment and other social problems. These super-sized telescopes are built to explore the evolution of the universe and the nature of black holes and dark matter. Understandably, people may question the need for these mega-investments from the public purse. Acknowledging that people in Africa have far more pressing challenges than exploring the universe, local scholars and communicators argue that astronomy is uniquely positioned to foster socioeconomic development and address developmental challenges in marginalised communities (McBride et al., 2018). Under these conditions, one can make a strong moral case for scientists to engage with disadvantaged communities (Manzini, 2003) and many local scientists perceive particularly strong duty to help improve people's lives (Joubert, 2018b).

In South Africa, school-level education[12] is plagued by problems such as underqualified and demotivated teachers, as well as poorly resourced or dysfunctional schools. In an effort to help address a huge science and mathematics education backlog, the bulk of publicly funded science communication initiatives target school-going youth and educators. For example, more than 70 per cent of participants in National Science Week and visitors to science centres are school learners (DST, 2015). This means that science engagement activities rarely cater for people beyond school-going age, limiting their ability to foster broad engagement with science. Language barriers cause a further disconnect between students and the science curricula taught at school. While the benefits of presenting science engagement activities in learners' home languages have been shown (Fish et al., 2017), South Africa lacks the capacity to deliver multilingual science engagement programs. Furthermore, the lack of a vocabulary that captures phenomena such as 'climate change' or 'genetically modified organisms' in the local vernacular is problematic.

11. Conclusion

The initial years of democracy were an optimistic time in South Africa, with high hopes for a unified 'rainbow nation' and prosperity that would spill over into a continent-wide African Renaissance. But, given the historical disconnects between science and the majority of South African citizens, along with huge socioeconomic disparities, the challenges of creating a scientifically literate society were vastly underestimated. Progress was derailed by inexperience with such challenges and misguided politics. Already monumental challenges were compounded by the intricacies of balancing the interests of indigenous knowledge systems (not including harmful superstitions) and modern science.

Despite these shortcomings, public communication of science in South Africa has made significant strides since the period when science was strategically isolated from the majority of its citizens. Currently, there is growing support for creating new connections between science and diverse audiences. Many role players in the local science ecosystem have taken up the challenge of making science publicly visible and accessible and engaging people in mutually beneficial dialogue. A solid start has been made with establishing science communication as a field of research and teaching in the country.

12 The dire state of mathematics education, for example, is revealed in the 2015 Trends in International Mathematics and Science Study (TIMSS) with Grade 4 learners in South Africa rated 49th out of 50 participating countries (Mullis et al., 2015).

Scientists are inspired to share their work by the country's unique fossil record, the ubiquitous access to the African night sky and rich biodiversity (Joubert, 2018b). It is therefore not surprising that palaeontologists,[13] astronomers[14] and health researchers[15] are amongst the most active science popularisers in the country.

It has to be acknowledged that science communication policies and strategies remain largely aspirational, with limited expertise and fragmented capacity in place to implement the government's ambitious plans. Even today, science engagement activities are mostly located in urban areas, usually limited to English and predominantly geared toward children of school-going age. They cannot hope to reach nearly 60 million citizens spread over vast rural areas, speaking 11 official languages and many more local dialects. Inadequate resources and the lack of appropriate metrics to evaluate communication programs remain key challenges, along with cultural and language barriers.

In present-day South Africa, there are specific factors that motivate scientists to engage with society. These include a desire to amend for past inequalities and improve the lives of people battling with poverty and disease.

Twenty-six years since the advent of democracy, it remains questionable what progress we have made in developing an appropriate science communication infrastructure that adequately responds to local needs and would be able to deliver a truly science-engaged knowledge society. We have made a solid start, but we have a very long way to go.

13 Palaeontologists who have made their mark in public communication include Robert Broom (1866–1951), Raymond Dart (1893–1988) and Bob Brain (1931–). More recently, Francis Thackeray has been instrumental in numerous exhibitions in South Africa and around the world, focused on South Africa's fossil heritage and human evolution. Another driving force in palaeontology outreach is Anusuya Chinsamy-Turan, who has written some of the first popular books on African fossils. Known for his innovative use of social media platforms and storytelling abilities, palaeontologist Lee Berger is currently one of the most visible scientists in South Africa (Joubert and Guenther, 2017).

14 Tony Fairall (1943–2008) and Mike Gaylard (1952–2014) stand out for their exceptional contributions to astronomy outreach in the past, while people like Claire Flanagan, Matie Hoffman and Kevin Govender continue the work of using astronomy to enlighten public audiences and promote social inclusion in science. Cosmologist Thebe Medupe played a key role in producing *Cosmic Africa*, a film that presents a panorama of the mythical and practical interaction of Africa's people and the African night sky.

15 Health research, and in particular HIV/AIDS and tuberculosis (TB), are hot topics in South Africa. Researchers known for their proactive media and public engagement, include Kelly Chibale, Tebello Nyokong, Bavesh Kana, Linda-Gail Bekker and Glenda Gray.

Figure 32.6: South Africans are keenly interested in the country's rich fossil heritage and palaeontologists play a leading role in engaging with society. Professor Francis Thackeray has a dream of putting a replica of 'Mrs Ples' in every classroom in the country. 'Mrs Ples', *Australopithecus africanus*, is a distant relative of all humankind, more than 2 million years old, from the Sterkfontein Caves in South Africa.

Source: Jose Braga.

Acknowledgements

The authors wish to thank the following people for their advice, comments and suggestions: Ina Roos, Francis Thackeray, Johann Mouton, Mike Bruton, Derek Fish, Beverley Damonse, Michael Gastrow, Nelius Boshoff, Lindsay Marshall and Karien Joubert.

References

Baker, C., Lund, P., Nyathi, R. and Taylor, J. (2010). The myths surrounding people with albinism in South Africa and Zimbabwe. *Journal of African Cultural Studies*, *22*(2), 169–81. doi.org/10.1080/13696815.2010.491412.

Batts, S. (2006). Traditional African potion as AIDS 'cure'? (No, just Woo). *Science Blogs*. Retrieved from scienceblogs.com/retrospectacle/2006/08/23/traditional-african-potion-as-1/.

Bauer, M. (2008). Paradigm change for science communication: Commercial science needs a critical public. In D. Cheng, M. Claessens, T. Gascoigne, J. Metcalfe, B. Schiele and S. Shi (eds), *Communicating science in social contexts: New models, new practices* (pp. 7–27). Berlin: Springer. doi.org/10.1007/978-1-4020-8598-7_1.

Bauer, M., Pansegrau, P. and Shukla, R. (2019). *The cultural authority of science: Comparing across Europe, Asia, Africa and the Americas*. London: Routledge. doi.org/10.4324/9781315163284.

Blankley, W. and Arnold, R. (2001). Public understanding of science in South Africa: Aiming for better intervention strategies. *South African Journal of Science, 97*(3/4), 65–69.

Borchelt, R.E. (2001). Communicating the future: Report of the Research Roadmap Panel for Public Communication of Science and Technology in the Twenty-first Century. *Science Communication, 23*(2), 194–211. doi.org/10.1177/107554700 1023002006.

Boshoff, S. C., Grebe, E., Mouton, J., Ravat, E., Ravjee, N. and Waast, R. (2000). *Science in South Africa: History, institutions and statistics*. Stellenbosch: Centre for Interdisciplinary Studies (CREST), Stellenbosch University, Stellenbosch.

Bruton, M. (2015). *When I was a fish: Tales of an Ichthyologist*. Johannesburg: Jacana Media.

Bruton, M. (2016). *Traditional fishing methods of Africa*. Cape Town: Cambridge University Press.

Bruton, M. (2017). *The annotated Old Fourlegs: The story of the coelacanth*. Cape Town: Penguin Random House.

Bruton, M. (2018). *The fishy Smiths: A biography of JLB & Margaret Smith*. Cape Town: Penguin Random House.

Carruthers, J. (2008). Scientists in society: The Royal Society of South Africa. *Transactions of the Royal Society of South Africa, 63*(1), 1–30. doi.org/10.1080/00359190809519207.

Cheng, D., Claessens, M., Gascoigne, T., Metcalfe, J., Schiele, B. and Shi, S. (eds) (2008). *Communicating science in social contexts: New models, new practices*. Berlin: Springer. doi.org/10.1007/978-1-4020-8598-7.

Chigwedere, P., Seage, G. R., Gruskin, S., Lee, T. H. and Essex, M. (2008). Estimating the lost benefits of antiretroviral drug use in South Africa. *Journal of Acquired Immune Deficiency Syndromes, 49*(4), 410–15. doi.org/10.1097/QAI.0b013e31818a6cd5.

Claassen, G. (2011). Science and the media in South Africa: Reflecting a 'dirty mirror'. *Communicatio, 37*(3), 351–66. doi.org/10.1080/02500167.2011.622288.

DACST. (1996). *White Paper on Science and Technology*. Pretoria: Department of Arts, Culture, Science and Technology. Government Printers.

DACST. (1998). *The South Africa Report, The Year of Science and Technology*. Pretoria: Department of Arts, Culture, Science and Technology. Government Printers.

DACST. (1999). *Annual Report*. Pretoria: Department of Arts, Culture, Science and Technology. Government Printers.

DACST. (2000). *Annual Report*. Pretoria: Department of Arts, Culture, Science and Technology. Government Printers.

Dawson, E. (2018). Reimagining publics and (non)participation: Exploring exclusion from science communication through the experiences of low-income, minority ethnic groups. *Public Understanding of Science, 27*(7), 772–86. doi.org/10.1177/0963662517750072.

DST. (2007). *Innovation towards a knowledge-based economy: Ten-year plan for South Africa (2008–2018)*. Pretoria: Department of Science and Technology. Government Printers.

DST. (2015). *Science Engagement Strategy*. Pretoria: Department of Science and Technology. Government Printers.

DST. (2018). *Science Engagement Monitoring and Evaluation Framework*. Pretoria: Department of Science and Technology. Government Printers.

DST. (2019). *White Paper on Science, Technology and Innovation*. Pretoria: Department of Science and Technology. Government Printers.

Du Plessis, H. (2015). The media and science communication: Exploring the complexity of communicating science in South Africa. *China Media Research, 11*(1), 46–54.

Du Plessis, H. (2017). Politics of science communication in South Africa. *Journal of Science Communication*, 16(3), A03. doi.org/10.22323/2.16030203.

Du Plessis, H. and Masilela, T. (2012). Report on the Human Sciences Research Council (HSRC) 2012 Conference – The Re-emergence of Astronomy in Africa – A Transdisciplinary Interface of Knowledge Systems. *Journal of Scientific Temper, 1*(3&4), 219–26.

Dubow, S. (2006). *A commonwealth of knowledge: Science, sensibility and white South Africa 1820–2000*. New York: Oxford University Press.

Fish, D., Allie, S., Pelaez, N. and Anderson, T. (2017). A cross-cultural comparison of high school students' responses to a science centre show on the physics of sound in South Africa. *Public Understanding of Science, 26*(7), 806–14. doi.org/10.1177/0963662516642725.

Garrard, C. (2018). Pint of Science: Bringing science to the public and highlighting African research. *South African Journal of Science, 114*(7/8), Art. #a0280. doi.org/10.17159/sajs.2018/a0280.

Gastrow, M. (2010). *The public understanding of biotechnology in the media: A report for the National Advisory Council on Innovation and the National Biotechnology Advisory Committee.* Pretoria: Human Sciences Research Council.

Gastrow, M. (2015). Science and the social media in an African context: The case of the Square Kilometre Array telescope. *Science Communication, 37*(6), 703–22. doi.org/10.1177/1075547015608637.

Gastrow, M. (2017). *The stars in our eyes: Representations of the Square Kilometre Array telescope in the South African media.* Pretoria: HSRC Press.

Gastrow, M., Roberts, B., Reddy, V. and Ismail, S. (2018). Public perceptions of biotechnology in South Africa. *South African Journal of Science, 114*(1/2), Art. #2017-0276. doi.org/10.17159/sajs.2018/20170276.

Guenther, L. and Weingart, P. (2016). A unique fingerprint? Factors influencing attitudes towards science and technology in South Africa. *South African Journal of Science, 112*(7), 8–11. doi.org/10.17159/sajs.2016/20160093.

Guenther, L. and Weingart, P. (2018). Promises and reservations towards science and technology among South African publics: A culture-sensitive approach. *Public Understanding of Science, 27*(1), 47–58. doi.org/10.1177/0963662517693453.

Guenther, L., Weingart, P. and Joubert, M. (2019). Bridging a gap? Mapping the science coverage of South African newspapers. In M. W. Bauer, P. Pansegrau and R. Shukla (eds), *The cultural authority of science: Comparing across Europe, Asia and the America* (pp. 122–36). New York: Routledge. doi.org/10.4324/9781315163284-8.

Guenther, L., Weingart, P. and Meyer, C. (2018). 'Science is everywhere, but no one knows it': Assessing the cultural distance to science of rural South African publics. *Environmental Communication, 12*(8), 1046–61. doi.org/10.1080/17524032.2018.1455724.

HSRC. (2010). *South African Social Attitudes Survey 2009.* Pretoria: Human Sciences Research Council.

Huigen, S. (2009). *Knowledge and colonialism: Eighteenth-century travellers in South Africa.* Leiden: Brill. doi.org/10.1163/ej.9789004177437.i-314.

Irwin, A. and Horst, M. (2016). Communicating trust and trusting science communication: Some critical remarks. *Journal of Science Communication, 15*(6), L01. doi.org/10.22323/2.15060101.

Joubert, M. (2018a). Reflections on the first human heart transplant and its impact on medicine, media and society. *Public Understanding of Science, 27*(1), 110–14. doi.org/10.1177/0963662517738619.

Joubert, M. (2018b). Factors influencing the public communication behaviour of publicly visible scientists in South Africa (PhD thesis). Stellenbosch University, Stellenbosch.

Joubert, M. and Guenther, L. (2017). In the footsteps of Einstein, Sagan and Barnard: Identifying South Africa's most visible scientists. *South African Journal of Science, 113*(11/12). doi.org/10.17159/sajs.2017/20170033.

Kennedy, E. B., Jensen, E. A. and Verbeke, M. (2017). Preaching to the scientifically converted: Evaluating inclusivity in science festival audiences. *International Journal of Science Education, Part B, 8*(1), 14–21. doi.org/10.1080/21548455.2017.1371356.

Le Grange, L. (2007). Integrating western and indigenous knowledge systems: The basis for effective science education in South Africa? *International Review of Education, 53*, 577–91. doi.org/10.1007/s11159-007-9056-x.

Lewenstein, B. V., Joubert, M. and Radin, J. (2002). Achieving public understanding of research in developing countries. PCST 2002 workshop report. Cape Town.

Logan, C. (2003). *Celebrity surgeon: Christiaan Barnard – A life*. Johannesburg: Jonathan Ball Publishers.

Makunga, N. P., Philander, L. E. and Smith, M. (2008). Current perspectives on an emerging formal natural products sector in South Africa. *Journal of Ethnopharmacology, 119*(3), 365–75. doi.org/10.1016/j.jep.2008.07.020.

Manzini, S. (2003). Effective communication of science in a culturally diverse society. *Science Communication, 25*(2), 191–97. doi.org/10.1177/1075547003259432.

Mbali, M. (2004). AIDS Discourses and the South African state: Government denialism and post-apartheid AIDS policy-making. *Transformation: Transformation Critical Perspectives on Southern Africa, 54*(1), 104–22. doi.org/10.1353/trn.2004.0023.

McBride, V., Venugopal, R., Hoosain, M., Chingozha, T. and Govender, K. (2018). The potential of astronomy for socioeconomic development in Africa. *Nature Astronomy, 2*, 511–14. doi.org/10.1038/s41550-018-0524-y.

Molloy, B. (1992). I went to the crucifixion. In D. Cooper (ed.), *Chris Barnard by those who know him* (pp. 240–53). Cape Town: Vlaeberg.

Mouton, J. (2006). Science for transformation: Research agendas and priorities in South Africa. In L. Box and R. Engelhard (eds), *Science and Technology policy for development, dialogues at the interface*. London: Anthem Press.

Mullis, I. V. S., Martin, M. O., Foy, P. and Hooper, M. (2015). *Trends in International Mathematics and Science Study (TIMSS)*. Boston: TIMSS & PIRLS International Study Center. doi.org/10.1007/978-94-007-2150-0_515.

Nattrass, N. (2007). *Mortal Combat: AIDS denialism and the struggle for antiretrovirals in South Africa*. Scottsville: University of KwaZulu-Natal Press.

Ndhlala, A. R., Stafford, G. I., Finnie, J. F. and van Staden, F. (2011). Commercial herbal preparations in KwaZulu-Natal, South Africa: The urban face of traditional medicine. *South African Journal of Botany, 77*, 830–43. doi.org/10.1016/j.sajb.2011.09.002.

Noble, R. (1860). Editorial. *Cape Monthly Magazine, 7*, 130.

Nxumalo, N., Alaba, O., Harris, B., Chersich, M. and Goudge, J. (2011). Utilization of traditional healers in South Africa and costs to patients: Findings from a national household survey. *Journal of Public Health Policy, 32*(Suppl. 1) S124–S136. doi.org/10.1057/jphp.2011.26.

Nyundu, T. and Naidoo, K. (2016). Traditional healers, their services and the ambivalence of South African youth. *Commonwealth Youth and Development, 14*(1), 144–55. doi.org/10.25159/1727-7140/1799.

Parker, S. (2017). Development of Indicators for the Measurement of the South African Publics' Relationship with Science (PhD thesis). Stellenbosch University, Stellenbosch.

Pitcher, G. and Bowley, D. (2002). Infant Rape in South Africa. *The Lancet, 359*(9303), 274–75. doi.org/10.1016/S0140-6736(02)07532-3.

Pouris, A. (1991). Understanding and appreciation of science by the public in South Africa. *South African Journal of Science, 87*(7), 358–59.

Pouris, A. (1993). Understanding and appreciation of science among South African teenagers. *South African Journal of Science, 89*(2), 68–69.

Pouris, A. (2003). Assessing public support for biotechnology in South Africa. *South African Journal of Science, 99*(11–12), 513–16.

Ramchundar, N. and Nlooto, M. (2017). Willingness of African traditional healers to collaborate with researchers to further develop their traditional medicines in Kwazulu-Natal Province of South Africa. PULA: *Botswana Journal of African Studies, 31*(1), 163–79.

Reddy, V., Gastrow, M., Juan, A. and Roberts, B. (2013). Public attitudes to science in South Africa. *South African Journal of Science, 109*(1), 1–8. doi.org/10.1590/sajs.2013/1200.

Ritner, S. R. (1967). The Dutch Reformed Church and Apartheid. *Journal of Contemporary History, 2*(4),17–37. doi.org/10.1177/002200946700200404.

Roberts, B., Struwig, J., Ngungu, M. and Gordon, S. (2014). Attitudes towards astronomy and the Square Kilometre Array (SKA) in South Africa. Tabulation report based on the 2013 round of the South African Social Attitudes Survey (SASAS) prepared for Education and Skills Development (ESD) Research Programme, Human Sciences Research Council, Pretoria.

Scheufele, D. A. (2014). Science communication as political communication. *Proceedings of the National Academy of Sciences*, *111*(Suppl. 4), 13585–92. doi.org/10.1073/pnas.1317516111.

Schneider, H. and Fassin, D. (2002). Denial and defiance: a socio-political analysis of AIDS in South Africa. *AIDS*, *16*(Suppl. 4), S45–S51. doi.org/10.1097/00002030-200216004-00007.

Sooryamoorthy, R. (2010). Science and scientific collaboration in South Africa: Apartheid and after. *Scientometrics*, *84*(2), 373–90. doi.org/10.1007/s11192-009-0106-y.

Statistics South Africa. (2015). *General household survey 2014*. Pretoria.

Thackeray, J. F. (2009). Darwin's interest in the natural history of the Cape: from beetles to antelope, plants and granite. *Transactions of the Royal Society of South Africa*, *64*(1), 79–81. doi.org/10.1080/00359190909519241.

Tilley, H. (2011). *Empire, development, and the problem of scientific knowledge, 1870 – 1950: Africa as a Living Laboratory*. Chicago: University of Chicago Press. doi.org/10.7208/chicago/9780226803487.001.0001.

Tomaselli, K. G. (2002). A brief history of South African journalism, mass communication and media education. *African Communication Scholars Association*, *4*(1), 111–35.

Weingart, P. and Joubert, M. (2019). The conflation of motives of science communication – causes, consequences, remedies. *Journal of Science Communication*, *18*(03), Y01. doi.org/10.22323/2.18030401.

White, T. D. (2012). Phillip V. Tobias (1925–2012). *Science*, *337*(6093), 423. doi.org/10.1126/science.1225988.

Williams, V. L. and Whiting, M. J. (2016). A picture of health? Animal use and the Faraday traditional medicine market, *Journal of Ethnopharmacology*, *179*, 265–273. doi.org/10.1016/j.jep.2015.12.024.

Wilkinson, K. (2013). Do 80% of S. Africans regularly consult traditional healers? The claim is unproven. *Africa Check*. Retrieved from https://africacheck.org/reports/do-80-of-south-africans-regularly-consult-traditional-healers-the-claim-is-false/.

Wood, B. (2012). Phillip Valentine Tobias (1925–2012). *Nature*, *487*(7405), 40. doi.org/10.1038/487040a.

Timeline

Event	Name	Date	Comment
First interactive science centre established.	The Exploratorium, University of Pretoria	1977	Later renamed Sci-Enza
First national (or large regional) science festival.	SASOL Scifest	1996	Now SciFest Africa
Association of science writers or journalists or communicators established.	South African Science Communicators' Network (SASCON)	1998	No longer active
First university courses to train science communicators established.	MPhil (Science and Technology Studies); specialisation in public science engagement	2015	Stellenbosch University (the university also has a journalism program going back further)
First master's students in science communication graduation.	MPhil (Science and Technology Studies); specialisation in public science engagement, Stellenbosch University	2017	
First PhD students in science communication graduation.	Science and Technology Studies, Stellenbosch University	2018	
First national conference in science communication.	Promoting Public Understanding of Science and Technology in Southern Africa	1996	At the University of the Western Cape
A national government program to support science communication established.	Year of Science and Technology	1998	
First significant initiative or report on science communication.	Report on 'Year of Science and Technology 1998'	1999	
National Science Week founded.		2000	A government-initiated flagship science promotion project
A journal completely or substantially devoted to science communication established.	*Archimedes*, a science magazine distributed mainly to white schools, published by the (former) Foundation for Education, Science and Technology (FEST)	1959–2003	2004: *Quest*, a quarterly popular science magazine published by the Academy of Science of South Africa
First significant radio programs on science made.	*New Science* (later *Science Matters*)	1994	Presented by Christina Scott (1961–2011) on the public radio station SAFM

Event	Name	Date	Comment
First significant TV programs on science made.	*Die Brandkluis* [The Safe]	1970	Afrikaans-language science program hosted by Marinus Wijnbeek
First awards for scientists, journalists or others awarded for science communication.	Science Communication Award	2006	Awarded annually by the National Science and Technology Forum (NSTF)
PCST conference hosted in the country.	PCST-7, Cape Town	2002	
Other significant events	First Science Centre Network Conference	1995	Hosted by Unizulu Science Centre at Mtunzini, KwaZulu-Natal
	Science Centre Network formed	1996	SAASTEC (Southern African Association of Science and Technology Centres)
	First Science Centre World Congress hosted	2011	Hosted by Cape Town Science Centre in Cape Town

Contributors

Dr Marina Joubert is a senior researcher at the Centre for Research on Evaluation, Science and Technology (CREST) at Stellenbosch University.

Shadrack Mkansi is a science awareness platforms manager at the South African Agency for Science and Technology Advancement.

33

SOUTH KOREA

A different exemplar

Hak-Soo Kim

1. Introduction

South Korea is no longer a developing country. Its gross domestic product (GDP) ranks just outside the top 10 countries in the world. Its technological advancement is well demonstrated in Korean industries and products such as electronics, ICT, automotive, oil refining, steelmaking and shipbuilding. As of 2016, Korea's gross R&D investment occupied 4.24 per cent of the nation's total GDP, the top rank in the world (joint with Israel). In that year, Korea's private companies made 77.7 per cent of the nation's R&D investment and employed 69.7 per cent of the nation's R&D personnel (Ministry of Science and ICT, 2018a). This is quite different from other countries, with the private sector, rather than the public one, being the major player in R&D in Korea.

Compared with Japan and China, Korea was very late in accessing Western science and technology. Scholars of the Realist School of Confucianism called Silhak brought back from Beijing Chinese-translated science books from the mid-18th century, and re-translated some into Korean (Park, 1986). Japan, on the other hand, had, from the mid-16th century, imported Western science and technology through trade with Portugal and later through systematic encounters with the Netherlands. China was exposed to Western science by European Catholic priests from the early 17th century. Therefore it is natural that many basic science-related terms being used in those three countries originated from Japanese translations with Chinese characters, including 'science' per se as 科學 (literally meaning 'study of classification').

Today, South Koreans take pride that Korea is the only country in the world that has fully accomplished both political democratisation and advanced industrialisation from a zero base after World War II. Germany and Japan were already highly advanced countries before WWII, but Korea was not. During the Japanese occupation (1910–45), Korea was absolutely plundered for Japanese interests and benefit. This situation worsened in June 1950, when North Korea invaded South Korea with the support of China and Soviet Russia. By the time the armistice was signed in 1953, the Korean peninsula was completely ruined. South Korea's annual per capita income for the early 1960s was less than US$100. South Korea was one of the most impoverished countries in the world, having a large population and no natural resources except coal. It was even poorer than North Korea, which had abundant resources of coal and other minerals but only half the population of South Korea.

When Korea gained its independence after Japan's defeat in 1945, approximately 80 per cent of the Korean public were illiterate and by 1961, 70 per cent remained so. The general public had little exposure to, interest in or knowledge of science. Major General Park Chung-hee had seized power in a military coup in 1961, and he recognised the importance of science in national development because he had observed the advancement of Japan and America during his military training in both countries. His political leadership focused on making Korea an economically well-to-do country through developing and mobilising science and technology above all (Hyun, 2005).

President Park began a political campaign for making the public aware of the importance of science. He implemented four conditions needed to advance science and technology that Dedijer (1963) suggested:

- science policy should be included and emphasised as a key in national development policy
- the political elite should recognise that science and its successful implantation are necessary for national progress
- a central research organisation should be established to foster research, to demonstrate science's role in society's major decision-making, and to support growth of science in universities
- a scientific community of researchers and teachers should be organised to cultivate science in society.

These developments are familiar to the author of this chapter, and parallel his involvement in developing Korea's national policies on science communication. When the author went to the USA for graduate study in 1978, Korea was still poor enough to demand a national examination of qualification for overseas study due to severe lack of foreign currency. In 1977, Korea's gross national income (GNI) per capita was less than US$1,000 (US$960 in 1977, US$27,600 in 2016).[1] But the Korean people were at last liberated from hunger and the basic need for household electricity was satisfied. Professor Kim's initial exposure to the philosophy of science was in a graduate communication class taught by Professor Richard F. Carter at the University of Washington (Seattle) in 1978–82. This exposure was a shock, a chance to contemplate science and technology and the policies for developing them. In 1982, Korea was politically still under the highly oppressive regime of another general: President Chun Doo-hwan. Economically, however, the country was booming, exporting labour-intensive products such as textiles and shoes and some moderate tech products such as electronic goods, steel and ships. The nation was desperate for political democratisation and scientific and technological advancement.

The author describes how he became involved in science communication:

> I can't forget one day in early winter in 1986 … in which I began to serve as a professor of communication in Sogang University after having moved from Hanyang University in Seoul. I saw a public notice in a daily newspaper that a symposium on science popularisation would be held in a conference room of the Press Center in downtown Seoul. I decided to attend there just to look at what would be talked about. A few natural scientists were found to deliver normative arguments for popularising science toward the general public. They stressed that science popularisation was urgent for national progress into further high-technology industry. But they made little mention of specific communication or media strategies for it. At the close, I commented on some potential strategies. Although I had to leave immediately to go home, several persons followed me to an elevator and wanted to meet me at my office the following day. They turned out to be public officials of the Ministry of Science and Technology who were anxious to make national policies for science popularisation. This is how I became involved in Korea's national policymaking for, first, science popularisation, and later, science communication from more of the public's perspective.

In principle, functional needs precede structure, although function may follow structure later (e.g. 'role'). This principle is more constructive (Carter, 2020; Kim et al., 2014) than the structural-functionalism often cited (e.g. Parsons, 1977). In Korea, the policy on science communication came from the functional need for national development. Thus, many of its derivatives, whether they were institutions, activities or studies, were the *first* such ventures in Korea, and the

1 IndexMundi (2019) using the World Bank Atlas method.

author became closely engaged in starting (or re-starting) them, as the first US-trained PhD in communication who showed serious interest and expertise in science communication in Korea. He became the first social scientist to be awarded one of the Orders of Science and Technology Merit (Woong-bi-jang) (Korea's equivalent to a US National Medal of Science) bestowed by the President of the Republic of Korea in April 2001, for his contributions to the development of science communication in Korea.

The story that follows segments Korea's history of science communication into three themes: 1) public familiarisation with science; 2) professionalisation of science popularisation; 3) specialisation of science communication. This division reflects Korea's unique history of science communication development. The conclusion suggests a new direction for effective science communication in a rapidly changing era.

2. Public familiarisation with science

On 15 August 2018, President Moon Jae-in delivered the 72nd anniversary address for Korea's Independence Day. In the address, he mentioned five deceased patriots who had not been well remembered for their great contributions to Korea's independence from Japanese rule. One was Yong-Gwan Kim who in 1934 created the first Science Day in Korea. The date he and 31 colleagues chose was 19 April, the anniversary of Charles Darwin's death (Park, 2017).

Yong-Gwan Kim was the pioneer in exposing the Korean general public to science and technology during Japanese colonial rule. At college in Seoul he majored in ceramic engineering before moving to Tokyo in 1918 for another year's study. This overseas study completely changed his career. He believed Japan's remarkable modernisation came from infusing people's ordinary way of life with science and technology, and he returned to Korea to establish the Invention Society in 1924 and publish the first science magazine *Science Joseon* ('Joseon' means Korea) in 1933.

The first Science Day in Seoul included a car parade, lectures, a radio talk, experiments, science movies and visits to science-related institutes. It was so successful that it continued for a week. It was also held in Pyongyang, then the biggest city in the north of the Korean peninsula. Contemporary intellectuals supported Yong-Gwan Kim and his activities, and this momentum led to the founding on 5 July 1934 of the Society of Disseminating Scientific Knowledge, for which he served as the executive director.

However, the Japanese colonial government in Korea came to regard the Science Day festival as being like a Korean independence movement and moved to restrain it after 1937. Yong-Gwan Kim was arrested and imprisoned in 1938. All movements to increase public familiarisation with science and technology were completely abandoned. He was released from prison in 1942, three years before Korea's liberation from Japan, but nothing remained and most of his former fellows had changed into pro-Japanese collaborators. A disappointed man, he moved to Manchuria and vanished from sight.

The first movement to familiarise the Korean public with science and technology had taken place in 1883, about 50 years before the first Science Day. Korea had signed a treaty of commerce and friendship with Japan, USA, Britain, Germany and Russia between 1876 and 1884, and now confronted a new era of civilisation and enlightenment. During the late 'Joseon Kingdom', public emissaries were sent to Beijing and Tokyo and they saw new advanced systems that might be adopted in Korea. One of their conclusions was that newspapers seemed to be a key to enlightening the general public in those developed countries.

Yeong-Hyo Park, an official emissary of King Gojong to the Japanese government, returned to Korea in January 1883 after having spent five months in Japan. Advised by the leader and symbol of Japan's modernisation, Fukuzawa Yukichi, Park brought three Japanese journalists and a printing machine to Korea to establish the first modern newspaper. He persuaded the government to publish the *Hanseong Sunbo*, a newspaper issued every 10 days. The first issue came out on 31 October 1883, announcing that it aimed to enlarge public knowledge of foreign as well as domestic news. It was intended to overcome underdevelopment by introducing developed countries' civilisations and systems (Cha et al., 2001).

From the first issue, it reported on electricity, trains, steamships, the telegraph and even astronomy. The newspaper was printed in Chinese instead of Korean, because the former was then the government's official language. Its circulation was estimated to be about 3,000. This means the newspaper was accessible to a very limited number of educated intellectuals, not to the majority of citizens. The newspaper stopped publication a year later when Park and his followers attempted the Reformist Revolution on 4 December 1884. The revolution failed and he fled to Japan.

The Korean peninsula became independent of Japan in 1945, but soon after was again engulfed in severe ideological and political conflict. Following the Korean War in 1950, it was divided into South Korea and North Korea. South Korea, a free and democratic country, had many newspapers that reported widely on science and technology, especially in 1957 when the Soviet Union launched Sputnik. The *Hankook Ilbo,* a daily, established the first science desk in 1958, an independent Department of Science and Technology News (Korea Science Journalists Association, 2014). Citizen exposure to science and technology news, however, was still very limited, given national literacy and the limited number of subscriptions.

How, then, did Koreans become fully familiar with science and technology? When Korea was liberated from Japanese rule in 1945, only about 130 scientists were available (Hyun, 2005). By 2016, Korea's research workforce has increased to 361,292 (full-time equivalents), ranked just behind China, USA, Japan, Russia and Germany (Korea Institute of S&T Evaluation and Planning, 2017). This leap started with the regime of President Park Chung-hee who seized power in a military coup in 1961.

In his first Five-Year Economic Development Plan (1962–66), President Park set six national directions for realising the vision of modernisation: industrialisation oriented to exporting; consecutive five-year planning for economic growth; projects to increase the income of people in farming and fishing; inducements to foreign capital and technology; development of infrastructure facilities; and the government's guarantee for supporting corporate development projects. He was determined to change Korea's main business from exporting light industrial goods to exporting heavy industry and chemical products by advancing the level of science and technology (Hyun, 2005).

Science and technology were now considered the backbone of economic growth. However, few R&D infrastructure facilities were available outside a handful of universities. The government thus established public institutions accountable for science and technology development: the Korea Institute of Science and Technology (KIST) in 1966 as the nation's major R&D institute, modelled on the US Battelle Memorial Institute; the Ministry of Science and Technology (MOST) in 1967 as the government's top decision-making agency; the Korea Academic Institute of Science and Technology (KAIST) in 1971 as the nation's major producer of advanced degrees (MS, PhD) in science and engineering; the Seoul National Science Museum in 1972, the second grand opening following a major renovation; and the Korea Science and Engineering Foundation (KOSEF) in 1977 as the nation's research funding agency.

These newly established institutes and agencies demanded many engineers and scientists, but Korea had produced only 196 domestic PhDs in science and engineering between 1945 and 1970; and in October 1970, only 309 graduate students were enrolled in the nation's graduate programs of science and engineering. There were, however, 1,220 Korean graduate students in similar programs in the US, and approximately 1,400 Korean natural and engineering scientists with advanced degrees in science and engineering were working in America. So President Park's government invited these high-quality Korean scientists to return to Korea. They were offered very special treatment: for instance, KIST provided them with modern housing and a salary three times that of a major national university professor, in addition to relocation expenses from overseas.

Figure 33.1: President Park's stone monument (12 February 1973).

Source: KOFAC 50-year history.

President Park's impact and that of these new institutions extended the government's political motto 'Scientification of the Whole Nation', which is inscribed on his 1973 stone monument (see Figure 33.1) in the Seoul National Science Museum, currently National Children's Science Center. He was assassinated in 1979 but his legacy survives.

2.1. Professionalisation of science popularisation

In the 1980s, South Korean citizens began to enjoy a moderate level of economic prosperity, but they still suffered political oppression under the military regime of General Chun Doo-hwan. Protests against his dictatorship were so strong and persistent that the government was forced to take a big step toward political democratisation and constitutional reform, and full democracy came in 1987.

The new economic prosperity demonstrates the success of President Park's five-year economic plans. The government had bred and promoted Korea's Chaebol (unique private conglomerates) so that Korean industries could compete internationally. These conglomerates took advantage of the so-called catching-up strategy to learn and imitate the science and technology used in advanced countries' products (Kim, 1998; Lee and Lim, 2001). Scientists and engineers used their graduate training in advanced countries to make effective use of the catching-up strategy to improve the quality of Korean industrial products.

However, Korean industry needed innovations to survive in the world's competitive markets. This demanded more capable college graduates in science and engineering. The government encouraged students to undertake majors in science and engineering instead of law and medicine, and industry supported students with scholarships and jobs. The government introduced national policies to bridge science and society so that the public could better understand science and technology, support R&D, and appreciate scientists and engineers. This shift was a final blow to Korea's traditional Confucian view that civil servants, literary scholars and farmers are socially preferable to artisans, craftsmen and tradesmen or merchants.

The term 'science popularisation' started to be commonly mentioned among scientists and was regarded as their public mission. This demanded mobilisation of scientists not only for R&D but also for science popularisation. Some professors became more interested in writing popular science than in conducting research.

The Korea Science Writers Association (KSWA) was founded in 1977 by 10 prominent scientists and science writers, including Professor Moon-Hwa Hong (pharmacologist) and Professor Jung-Hum Kim (physicist) (Korea Science Writers Association, 2008). One year later, the Korean Science & Technology Publication Association (KSTPA) was established by publishers of science and technology books. The two associations aimed to promote science writing and the publishing of science books, hoping to make science more popular with the general public. They instituted awards and, in 1984, KSWA awarded the first Science Writer Prize to Professor Myung-Ja Kim, an active science writer and broadcaster. KSTPA awarded its first Science Book Prize in 1983 (Korean Science & Technology Publication Association, 2018).

As Korean industries became more competitive, the government pushed public R&D institutes to make their research results more available to industry. Those R&D outcomes produced a plethora of scientific and technological information. Major newspaper and broadcasting media came to employ science-specialised journalists to satisfy public curiosity and were ready to organise the Korea Science Journalists Association (KSJA). On 15 December 1984, 51 journalists working at 13 media outlets gathered at KIST, Korea's biggest national R&D centre and established KSJA.

The inaugural declaration of KSJA ended with a commitment to play a part in developing the nation's science and technology through expansion and improvement of science journalism (Korea Science Journalists Association, 2014). The government welcomed KSJA's interest in promoting national policies on the development of science and technology and President Roh Tae-woo (1988–93) delivered the keynote speech to the 1991 KSJA meeting. In Korea today, the journalist's role of promoting science and technology seems stronger than reporting on science with a critical eye.

The government was interested in the efforts that other advanced countries were making to expand and improve science popularisation. Officials from the Ministry of Science and Technology visited the author after the 1986 symposium and invited him to use his international experience to advise them on a national policy for science popularisation. Consequently, he conducted policy research on utilising mass media for science popularisation, examining the state-of-the-art techniques of other advanced countries. The research project commissioned by the government was the first full-scale undertaking dealing with national science communication policy in Korea.

The research, *Considerations and Policies for Science Communication Media*, was completed in November 1987 (Kim, 1987). It introduced established arguments that science popularisation and science communication were needed not only for national modernisation but also for further development of democracy (e.g. by using scientific information to improve rational decision-making). It also analysed the state of the art in organisations (e.g. the American Association for the Advancement of Science (AAAS), the British Association for the Advancement of Science (BAAS) and the National Association of Social Workers (NASW)), systems (e.g. awards, fellowships), media (e.g. *Science*, *Nature*) and science news in US, UK, Japan and Korea. It presented a comprehensive picture of potential policies to advance science popularisation and science media in Korea.

The ministry then funded the author for a series of research projects through the Korea Science and Engineering Foundation (KOSEF): in 1988, for policy development on educating and training social leaders to enhance their understanding for the importance of science, science policy and science popularisation; in 1990, for building an effective system of publicising science-related information to facilitate science journalism; in 1991, on activating 'Science Month of April' for science popularisation; in 1992, on long-term plans for science popularisation with an analysis of US examples for the science popularisation movement. A collection of these produced the first book on science popularisation policies in Korea: *Studies on Policies for Science Popularization in Korea* (Kim, 1993). It introduced to government officials and the scientific community new strategies for extending science to the general public.

In 1996 the Korea Research Foundation appointed the author as principal investigator of a three-year research project to diagnose the state of scientific culture in Korea. Six researchers analysed the Korean public's understanding of science and technology, the contents of science-related news reports in major dailies, the contents of science-inclusive advertisements in mass media, the reflections on science in plays and movies, the implications of science in recreation/leisure activities and facilities, and the potential of mutual development between science and society. This resulted in the first comprehensive book about scientific culture in Korea: *Understanding of Scientific Culture – Communication & Comparative Analysis* (Kim et al., 2000).

The 1991 research on 'Science Month of April' introduced to Korea the world's major science festivals: UK's BAAS summer festival, Edinburgh International Science Festival, USA's National Science and Technology Week and Japan's Science and Technology Week. Korea had briefly celebrated 'Science Day' under Japanese rule, but after independence, it regularly commemorated the establishment of the Ministry of Science and Technology in 1967. However, people did not see Science Day as a 'festival'. They could not imagine that science and technology, seemingly rigid and formal, might be entertaining until the Korea Science Foundation (KSF), a scientific-culture promotion agency, held the first week-long Science Festival in April 1997.

The Science Festival was extended into the 'First APEC Science Festival' held in Seoul in August 1998. Korea was scheduled to host the Asia-Pacific Economic Cooperation (APEC) conference of Science Ministers in 1998. The author was asked to suggest a science-related project on which APEC countries could cooperate beyond R&D collaboration by the science officer of then President Kim Young-sam. He came up with the idea of an 'APEC Science Festival' for youths, with the host rotating among member countries. Because APEC countries had few commonalities regarding culture, history, nature, race, science and youth were a suitable common ground for mutual development in APEC countries.

This produced in Korea a new vocation of science-specialised public promotions through exhibitions, events and entertainment.

Science popularisation tries to improve science and technology literacy. It is based on the learning-theory model: more interest leads to more knowledge, which results in attitude change, hopefully more positive. Communication is supposed to serve two functions: information transmission and persuasion. The former is assumed to increase interest and knowledge; the latter to change attitude. Thus, popularisation effects have been measured as to how much interest the public has in science, how much knowledge they have and how positive their attitude is toward science (Miller, 1983; Durant et al., 1989). Further variables are often also measured, such as political knowledge (e.g. Bauer et al., 2000).

In 1991, the KSF surveyed the public understanding of science for the first time, and then again in 1995. Korea Gallup designed and executed both as face-to-face interviews. Survey results were reported in international forums (Kim and Yoon, 1993, 1995). Many questionnaire items were modelled, for comparison and in light of competition, on the US and the EU surveys of public understanding, literacy and perceptions of science. (These surveys would continue into the 21st century.) They assessed interest in science, knowledge of scientific facts, perceptions of science's effects on living conditions, health, economic development and environment, as well as media-related sources of scientific information (see Appendix II, Kim et al., 1996).

These surveys, based on the learning-theory model, take the information provider's point of view and perspective. But there was a growing view that Korea needed a new, innovative model that could reflect the public's point of view and perspective about science and communication. Eventually, the ministry's subsidiary organisation, the Science and Technology Policy Institute (STEPI), decided to fund the author in a joint research proposal with a US team that included two University of Washington faculty members, Professors Richard F. Carter and Keith R. Stamm. The full report of this research, which focuses on public engagement with science, was published in a book (Kim et al., 1996).

Public engagement starts from relevance brought forth by vexing problems and/or by issues in which available, often conflicting, solutions compete. The theory is that problems and issues therefore govern involvement with science and technology, insofar as science and technology are conceived to contribute to solving those problems and issues. Communication is considered effective if it achieves public exposure to and focused attention on problems and/or issues, and subsequently relating them to science and technology, recognising that this might *not* accomplish knowledge gain and attitude change about science. This process of engagement was suggested as a new model for measuring public understanding of science (PUS) after two pilot tests in Korea and US.

This new model (see Kim, 2007b) was anticipated to mark a turning point for the establishment of science communication and for traditional PUS studies. Science communication practitioners could improve their effectiveness by starting with contemporary problems and/or issues. This might enable the public to construct meaningful impressions of science and technology, irrespective of knowledge of or attitude toward them. Korea was set to start a new era of science communication, not depending upon the limited learning-theory model.

3. Specialisation of science communication

Although South Korea's industrialisation and economy seemed to have become much stronger in the 1990s, its international competitiveness was falling. Korean companies had long been indifferent to R&D investment and the export of their products was declining. In late 1997, Korea encountered a financial crisis due to lack of foreign currency and had to ask the International Monetary Fund (IMF) for a bailout. In return, the government had to restructure its national economic system, which gave rise to massive company closures and sweeping employee dismissals in 1998–99. The so-called IMF crisis, one of the biggest economic disasters in Korea's modern history, did give momentum to R&D investment and scientific and technological advancement.

The Korean government grew much more concerned with advancement of science and technology, now even more firmly believed to be the engine of economic growth and prosperity. The government became more committed to involving the public and to encouraging talented youths to major in science and engineering, greatly increased the national R&D budget, and induced the general public to recognise the importance of science and technology. The Ministry of Science and Technology and its subsidiary organisation, the KSF, established the Academy for Scientific Culture, an educational institution for science communication at Sogang University, promising their full funding for the academy's operation. The academy was headed by the author who, with Professor Deok-hwan Lee of the Department of Chemistry at Sogang University, constituted an operating committee. Lee had earned a high reputation as a traditional science populariser and science communicator, writing numerous columns in newspapers and often appearing on TV as a science commentator. In 2003, the Academy for Scientific Culture was formally founded as an affiliated institution of the Sogang University Graduate School of Mass Communication.

The academy focused on science communication training primarily for scientists and employees of science-related institutions. Its program consisted of an eight-week training course (four courses per year; one three-hour night class per week). Those eight classes covered the following themes: importance of scientific culture, principles of science communication, science magazines, science speech, science journalism, science online media, science broadcasting and science policy. Program graduates received a Certificate in Science Communication Leadership. The Academy for Scientific Culture trained about 1,400 science-related personnel before it closed in March 2012.

KSF urged Sogang University to establish an independent master's degree program focused on producing specialists in science communication. This first graduate program specialising in science communication in Korea was to be fully funded and, in 2004, a new MA program of science communication was founded within the Sogang University Graduate School and headed by Professor Lee. This second academy program was a two-year MA degree program requiring a master's thesis. It emphasised theories and research methodology of science communication (Sogang University and Korea Federation for the Advancement of Science and Creativity, 2011). The MA program has produced 33 graduates as of 2017 and continues to exist, although it is no longer funded by the government. KSF also used to fund KAIST master's program of science journalism that admitted science-related media professionals as students from 2010. Its main focus was on helping them to keep up with advances in science and technology.

There is no historical record of undergraduate classes in science communication in Korea. However, the contribution of Won-Bok Hyun, a former science journalist, is noteworthy. A pioneer of science journalism in Korea, he studied advanced science reporting as a one-year fellow of the Columbia University School of Journalism in New York in 1967–68. Working as a science journalist and even after retirement in 1975, Hyun taught an undergraduate class of science journalism at the Department of Journalism, Hanyang University, in 1969–85 and at Sungkyunkwan University in 1976–84 (Song, 2011).

The first PhD in science communication from Korean Schools was produced in 2004 in the Sogang University Graduate School. In 2004 Dr Seong-Cheol Park completed a doctoral dissertation entitled 'Cognition of Scientific and Technological Topics on the Media'. He discovered people were more engaged (i.e. interest and cognition) with science and technology in respect to relevant problems. His work did not use the traditional research model of a relationship between scientific messages and knowledge gain or attitude change.

Even as Korea began its economic renewal, overcoming the so-called IMF crisis, President Roh Moo-hyun, who took power in 2003, remained committed to advancing the level of science and technology. His government upgraded the position of the Minister of Science and Technology to that of Deputy Prime Minister. It authorised the National Science and Technology Council to control and coordinate the government's total R&D budget across all the ministries. In addition, it adopted 'ScienceKorea' as a key political slogan for improving public understanding of science and its contribution to economic growth. KSF was authorised to implement diverse activities for the ScienceKorea movement (see Cho, 2012; Cho and Kim, 2012, for KSF's/KOFAC's (since 2008) activities).

To start a social movement for ScienceKorea, the government promoted Professor Woo-Suk Hwang, a stem-cell researcher, as a national hero for pioneering breakthroughs in human stem-cell cloning research. However, his research team's publications in *Science* (2004–05) were found to have fabricated evidence, resulting in a world-famous fraud scandal (Kim, 2007a). This might have damaged the cultural authority of science in Korea (see Gauchat, 2012).

Still obsessed with the potential contribution of science to economic growth, the Korean government was interested in how much the general public supports and understands science and technology. KSF began to conduct a national survey every two years starting in 2002. Like the earlier surveys of 1991 and 1995, it was designed and executed by Korea Gallup or other private poll companies. The questionnaire was modelled after the traditional PUS measures such as interest in and understanding of diverse scientific concerns (e.g. scientific and medical discoveries, new invention of technology, environmental pollution, economy, education and agricultural problems, military and international policies) and attitudes toward the scientist's attributes (e.g. endeavour to solve future problems, work for the benefit of mankind, effort to contribute to society, working alone, being unable to get enjoyment of life).

These longitudinal data (2002–16) were recently analysed by two master's students of Professor Martin Bauer at the London School of Economics (Lee, 2017; Chae, 2017). They confirmed that the Confucian instrumental attitude toward science, though fading away in the younger generation, still influenced the Korean public's utilitarian sense of science, and a media event such as the match between AlphaGo (a computer 'Go' player powered by artificial intelligence) and a Korean 'Go' player promoted public knowledge of artificial intelligence.[2]

The PCST Network, an international network for people active in studying and practising the public communication of science and technology, was launched in 1989. During the seventh conference held in Cape Town, South Africa, in 2002, Korea submitted a proposal to host the ninth conference in Seoul in 2006 with the theme 'Scientific Culture for Global Citizenship' and the promise that the KSF would support the event. It would be an historical occasion for science communication in Asia: the first time the PCST conference was hosted by a non-Western country. Competing with China's proposal, Korea's was accepted in a vote of the Scientific Committee of PCST Network.

2 'Go' is an abstract strategy board game for two players, in which the aim is to surround more territory than one's opponent.

Figure 33.2: PCST-9 notice.

PCST Network allowed a 'host' country to have two members on the Scientific Committee and the author was joined on the committee by Dr Sook-Kyoung Cho, a historian of science (especially science museums) and a senior staff member of KSF. She dedicated herself to getting the full support of not only KSF but also other related organisations, including the Ministry of Science and Technology. The Local Organizing Committee based in Korea composed the program in close coordination with the international Scientific Committee. The conference (see Figure 33.2) attracted 463 participants from 31 countries (PCST Network, 2018). Through this international conference, the Korean government and public got to recognise the significance and potential contribution of science communication to the public understanding of science and, further, to economic growth and national prosperity.

The field of science communication had long been unfamiliar even among communication scholars and the associations that represented them. But there was growing interest in the field, and in 2007–08 the Korean Society for

Journalism and Communication Studies (KSJCS) moved to establish a new sub-group, the Division of SHER (Science, Health, Environment and Risk) Communication. The author served as chair of the preparatory committee, which led to KSJCS approving the SHER Communication Division in 2008. Professor Sung-Kyum Cho of the Chungnam National University was elected the first chair of that division. Since then, the division has been very active, providing a platform for presentation of academic research papers in the spring and autumn annual conventions of KSJCS.

Science communication in practice began to branch out. The Ministry of Science and Technology was eager to have a science-specialised public television channel. Although a few public television channels such as KBS and EBS sometimes produced excellent science programs, the government thought the general public should have access to more science programs to enhance scientific literacy. KSF organised the preparatory committee to begin a science TV channel and YTN, a 24-hour public cable television news channel, was chosen to accommodate another science-specialised channel with KSF's financial support. *YTN Science* was launched in 2007. In Korea, if a cable television channel is selected as being in the public interest by the Korea Communications Commission, every cable television firm is required to carry that public-interest channel. *YTN Science* is now running as such a public channel. Although it is questionable how much the science channel has widened public exposure to science and technology in this internet-rampant era, the channel has created a new cohort of professional program producers.

The National Science Museum had existed in Seoul since 1926. In 1990, it moved to Daejeon, located in the middle of South Korea and embracing several nearby science research institutes. Daejeon would host the 1993 Daejeon World Exposition. According to a 2008 National Science Museum analysis, its approximately 3,000 exhibits consisted of mostly eyes-on materials, artefacts, visual images, dioramas and hands-on displays whose main focus was on transmitting scientific knowledge, accompanied in most cases by printed labels and history. Only 26 of the exhibits were related to or connected with Korea's contemporary problems and issues (Lee and Kim, 2008).

As South Korea's industrialisation and economic levels grew, the government decided to build an advanced national science museum in Gwacheon near Seoul, and so organised a special bureau within the Ministry of Science and Technology, with a preparatory committee in 2001. The special bureau was to deal with practical business affairs for constructing the new museum and the preparatory committee was its advisory body. The committee was composed of diverse experts including an architect, a science historian, an exhibition centre builder, a curator, a science educator and a science communication specialist. However, it did not take long before the different intentions and goals of the bureau and the preparatory committee came to the surface. The former wanted to emulate or copy a science museum or centre from advanced countries, while the latter wanted to create a world-unique, Korean-styled science museum. However, at that time, the Korean government's 'catching-up' stance and strategy prevailed everywhere and so the special bureau was able to thwart the preparatory committee's hopes.

A member of the preparatory committee, the author proposed a basic principle that a science display in the new museum needed to enhance public engagement by being related to relevant problems that were presently or potentially threatening our lives and society. The committee's architect strongly suggested a creative design of the museum building. But those points did not fit major concerns of public officials that were, rather, to expedite the process of decision-making and to seek a world-level resemblance to advanced countries' museums. The Gwacheon National Science Museum opened in 2008 and it could in effect be said to be the first modern, considerably interactive science museum in Korea. It made full use of hands-on, audio-visual and simulation techniques in style, while concentrating on transmission of scientific knowledge.

Following the opening of the Gwacheon National Science Museum, the government decided to build national science museums in three other big cities: Busan, Daegu and Gwangju. The Gwangju National Science Museum opened in October 2013, the Daegu National Science Museum in December 2013 and the Busan National Science Museum in December 2015. Elsewhere in Korea, the number of public and private science museums or centres increased greatly, from 60 in 2008 to 130 in 2017.

Commemorating its 10th anniversary in 2018, the Gwacheon National Science Museum is renovating major halls such as the Traditional Science Hall and Basic Science Hall. Their exhibition is going to use more storytelling techniques so that the public might hopefully feel more connection (Gwacheon National Science Museum, 2018), but it is unfortunate that the exhibits still tend more towards the informative than the engaging.

4. Conclusion and discussion

Private industry and commerce have become very powerful in Korea. The government is no longer the leader of the nation but just one leading actor, especially in the area of science and technology. Private corporations such as Samsung, LG, Hyundai and SK, in order to survive in the global market, are charged with advancing the cutting-edge levels of science and technology via their huge R&D centres.

Korea has changed into a highly S&T-friendly country in a relatively short period of time. Science and technology are strongly believed to be the main engine for economic growth and social change by the public. But what about democratic change? What about public engagement? What about the problem-solving capability of science and technology that might constitute the cultural authority of science (see Kim, 2019)?

As Korea has experienced rapid modernisation, the most important thing seems to be the commitment of both the political elite and the public to solving national problems through science and technology. As long as they are fully engaged with solving problems, they cannot avoid orienting to science and technology in clarifying problems and constructing their solutions. President Moon Jae-in's current government has put emphasis on public interests and social responsibility of science and technology, and the Ministry of Science and ICT pays new attention to solving problems closely related to the public's daily life, such as unhealthy foods, chemical hazards, cyber misconduct, transportation-related problems, epidemic diseases, environmental pollution and natural disasters (Ministry of Science and ICT, 2018b). To clarify these problems, the government pursues active communication with the public and seeks to fund R&D to develop technologies to solve them.

Science communication needs to enable people to be more engaged with the multifaceted interdependent nature of problems. A solution (e.g. cars) for the transportation problem brought about a more complex problem of climate change (Kim, 2012c), which demands interdisciplinary teamwork (Kim et al., 2016). Science and technology, scientists and technologists too, require effective interdependence. Thus, for interdisciplinary problem-solving, science communication might need to contribute, first and foremost, to constructing teamwork prior to and along with using science and technology to contribute to producing an innovative solution (Kim, 2020). With effective interdependence we could get super-charged engagement.

Today, we can access scientific knowledge almost without limits of space and time through internet communication technology. Why do we still need public science campaigns, science festivals and exhibitions as well as science museums and centres (and so many educational institutions) that mostly aim to pour and push scientific knowledge into the public's memory? Many might be a waste of public money.

There is an argument for a new approach, a new paradigmatic model of science communication for effective problem solving, individual or team, disciplinary or interdisciplinary. The traditional learning-theory model for knowledge gain and attitude change presumes just a persuasion function for communication—a presumption that has all too often been the information provider's wishful thinking. An alternative (Kim, 2007b, 2012a, 2012b) relates science to a problem-solving situation that is the key to bringing forth public engagement. And then it expects people to construct some impression about science, whatever the changes in scientific knowledge or attitude change: an impression relevant to where they are coming from. But the government and science institutions are resistant to this suggested new approach, as are students of science communication, who continue to assume that if scientists transmit scientific knowledge, the public will learn it and cultivate a positive attitude toward science. They still hold the typical notion that the next best (competitively) is to try harder rather than to try better.

Acknowledgements

The author thanks Professor Emeritus Richard F. Carter of the University of Washington and two anonymous reviewers for their helpful comments. One reviewer in particular provided not only many detailed, earnest comments for revision but also a nice mention: 'It is a great story and we do not want to lose all the excitement in the way the chapter unfolds.'

References

Bauer, M., Petkova, K. and Boyadjieva, P. (2000). Public knowledge of and attitudes to science: Alternative measures that may end the 'science war'. *Science, Technology, and Human Values, 25*(1), 30–51. doi.org/10.1177/016224390002500102.

Carter, R. F. (2020). Behavioral foundations of effective problem solving. Retrieved from bfeps.org.

Cha, B.-K., Oh, J.-H., Jung, J.-S., Lee, K.-J., Lim, J.-S. and Shin, I.-S. (2001). *Our Newspaper 100 Years*. Seoul: Hyeonamsa.

Chae, J. (2017). A study of techno-movement and hype cycle through the AI 'Go-match' in Korea (MS thesis). London School of Economics and Political Science, UK.

Cho, S.-K. (2012). An experience of science communication in Korea: The space-sharing project with mass media. In B. Schiele, M. Claessens and S. Shi (eds), *Science Communication in the World: Practices, Theories and Trends*. New York: Springer. doi.org/10.1007/978-94-007-4279-6_11.

Cho, S.-K. and Kim, O.-T. (2012). From science popularization to public engagement: The history of science communication in Korea. In B. Schiele, M. Claessens and S. Shi (eds), *Science Communication in the World: Practices, Theories and Trends*. New York: Springer. doi.org/10.1007/978-94-007-4279-6_12.

Dedijer, S. (1963). Underdeveloped science in underdeveloped countries. *Minerva, 2*(1), 61–81. doi.org/10.1007/BF01100781.

Durant, J. R., Evans, G. A. and Thomas, G. P. (1989). The public understanding of science. *Nature, 340*(6 July), 11–14. doi.org/10.1038/340011a0.

Gauchat, G. (2012). Politicization of science in the public sphere: A study of public trust in the United States, 1974 to 2010. *American Sociological Review, 77*(2), 167–87. doi.org/10.1177/0003122412438225.

Gwacheon National Science Museum. (2018). The 2018 business plan of the Gwacheon National Science Museum (internal document issued on 25 May 2018).

Hyun, W.-B. (2005). *President leadership of science and technology*. Seoul: Gwahaksarang.

IndexMundi (2019). Korea – GNI per capita – IndexMundi – Country Facts. Retrieved from www.indexmundi.com/facts/korea/gni-per-capita.

Kim, H.-S. (1987). *Considerations and policies for science communication media*. Final report submitted to the Korea Science and Research Foundation.

Kim, H.-S. (1993). *Studies on policies for science popularization in Korea*. Seoul: Iljinsa.

Kim, H.-S. (2007a). The scandal of Hwang Woo-Sok. In M. W. Bauer and M. Bucchi (eds), *Journalism, Science and Society*. New York: Routledge.

Kim, H.-S. (2007b). PEP/IS: A new model for communicative effectiveness of science. *Science Communication, 28*(3), 287–313. doi.org/10.1177/1075547006298645.

Kim, H.-S. (2012a). Measuring PEP/IS, a new model for communicative effectiveness of science. In M. Bauer, R. Shukla and N. Allum (eds), *The culture of science: How the public relates to science across the globe*. New York: Routledge.

Kim, H.-S. (2012b). Engagement: The key to the communicative effectiveness of science and ideas. In B. Schiele, M. Claessens, and S. Shi (eds), *Science communication in the world: Practices, theories and trends*. New York: Springer. doi.org/10.1007/978-94-007-4279-6_18.

Kim, H.-S. (2012c). Climate change, science and community. *Public Understanding of Science, 21*(3), 268–85. doi.org/10.1177/0963662511421711.

Kim, H.-S. (2019). The authority of science in Korea: A new explication and measurement. In M. Bauer, P. Pansegrau and R. Shukla (eds), *The cultural authority of science: Comparing across Europe, Asia, Africa and the Americas*. London: Routledge.

Kim, H.-S. (2020). Realizing interdisciplinarity among science, humanism, and art: A new paradigmatic explication on community problem solving (manuscript under review).

Kim, H.-S., Carter, R. F. and Stamm, K. R. (1996). *Developing a standard model of measuring the public understanding of science and technology*. Seoul: Science and Technology Policy Institute (STEPI).

Kim, H.-S., Kim, D.-K., Kim, C.-H., Kim, Y.-S., Roh, B.-S. and Kim, K.-M. (2000). *Understanding of scientific culture – communication & comparative analysis*. Seoul: Iljinsa.

Kim, H.-S., Kim, S. and Kim, C. (2016). Realizing unifying/unified technology: A new social science paradigm. *Hyonsang-gwa-Insik* (the journal of Korean Association of Humanities and Social Sciences), *40*(3), 135–60.

Kim, H.-S., Shin, D.-C. and Rhu, J.-H. (2014). *Development of models for interdisciplinary research among science, humanism, and art*. Seoul: The Korea Academy of Science and Technology.

Kim, H.-S. and Yoon, Y.-H. (1993). The Korean public understanding of science and technology: A 1991 survey report. Paper presented to the 1993 International Conference on the Public Understanding of Science and Technology, the Chicago Academy of Sciences, USA.

Kim, H.-S. and Yoon, Y.-H. (1995). Does the Korean public understand science and technology? Paper presented to the 1995 International Conference on the Public Understanding of Science and Technology, Beijing, China.

Kim, L. (1998). Crisis construction and organizational learning: capability building in catching-up at Hyundai Motor. *Organization Science, 9*(4), 506–21. doi.org/10.1287/orsc.9.4.506.

Korea Institute of S&T Evaluation and Planning. (2017). Comparison of R&D manpower between Korea and other major countries. *KISTEP Statistics Brief*, no. 24.

Korea Science Journalists Association. (2014). *The 30 Years' History of the Korea Science Journalists Association 1984–2014*. Seoul.

Korea Science Writers Association. (2008). *The 30 Years' History of the Korea Science Writers Association 1977–2007*. Seoul.

Korean Science & Technology Publication Association. (2018). History. Retrieved from www.kstpa.or.kr/main.php?mm=s13.

Lee, E.-A. and Kim, H.-S. (2008). Effective exhibition communication of science: An analysis of the South Korea's National Science Museum. *Journal of Technology Innovation, 16*(2), 95–123.

Lee, H. (2017). Is there support developing for basic science in Korean between 2002–2016? Instrumentalism & Confucianism (MS thesis). London School of Economics and Political Science, UK.

Lee, K. and Lim, C. (2001). Technological regimes, catching-up and leapfrogging findings from the Korean industries. *Research Policy, 30*(3), 459–483. doi.org/10.1016/S0048-7333(00)00088-3.

Miller, J. D. (1983). Scientific literacy: A conceptual and empirical review. *Daedalus, 112*(2), 29–48.

Ministry of Science and ICT (2018a). News release. Brief report issued on 18 January 2018 for Minister Mr. Yoo's visit to the Korea Industrial Technology Association.

Ministry of Science and ICT (2018b). Workshop for preliminary exploration by the Science Advisory Group for the Public's Daily Life. Proceedings prepared for the workshop on 27 August 2018, Seoul.

Park, S.-J. (2017,17 April). Do you know the original 'Science Day' held in Korea? *The Hangyoreh Shinmun*. Seoul.

Park, S.-R. (1986). Scientific thought of the 'Practical Thought' school. In M.-K. Cho et al. (eds), *Depth research of Korean thoughts*. Seoul: Woosuk.

Parsons, T. (1977). The present status of 'structural-functional' theory in sociology. In T. Parsons (ed.), *Social systems and the evolution of action theory*. New York: Free Press.

PCST Network (2018). Scientific culture for global citizenship: A final report. Retrieved from www.upf.edu/pcstacademy/_docs/final_report_pcst9.pdf.

Sogang University and Korea Foundation for the Advancement of Science and Creativity. (2011). *The Academy for Scientific Culture* (printed brochure for introducing programs).

Song, S.-Y. (2011). A brief memoir about the deceased Mr. Won-Bok Hyun. Retrieved from kast.or.kr/kr/member/memoir.php?bbs_data=aWR4PTkzNyZzdGFydFBhZ2U9MjAmbGlzdE5vPTkmdGFibGU9Y3NfYmJzX2RhdGEmY29kZT1odWkmc2VhcmNoX2l0ZW09JnNlYXJjaF9vcmRlcj0=||&bgu=view&idx=937.

Timeline

Event	Name	Date	Comment
First interactive science centre established.	Daejeon National Science Museum	1990	2008: Gwacheon National Science Museum
First national (or large regional) science festival.	Korea Science Foundation (currently KOFAC)	1997	
An association of science writers or journalists or communicators established.	Science writers	1977	1984: Science journalists 2008: Science communication scholars
First university courses to train science communicators.	Sogang U Academy for Scientific Culture	2003 (full-scale)	
First master's students in science communication graduate.	Sogang University MA Program of Science Communication	2006	An independent program
First PhD students in science communication graduate.	Sogang University Graduate School	2004	
First national conference in science communication.	Organised by a division of SHER (Science, Health, Environment, Risk) Communication	2008 (academic)	
National government program to support science communication established.	Academy for Scientific Culture	2003 (full-scale)	
First significant initiative or report on science communication.	First comprehensive research report	1987	By author of this chapter, Hak-Soo Kim
National Science Week founded.	Science Day (old)	1934	1967: Science Day (new) 1997: Science Festival
First significant TV programs on science.	*YTN Science* TV	2007 (full-scale)	
First awards for scientists or journalists or others for science communication.	Science Book Prize	1983	1984: Science Writer Prize
Date hosted a PCST conference.	PCST-9, Seoul, South Korea	17–19 May	

Event	Name	Date	Comment
Other significant events.	First PUS survey	1991	
	New PUS Measurement Model	1996	
	New Science Communication Model PEP/IS	2007	

Contributor

Professor Hak-Soo Kim is distinguished professor, College of Transdisciplinary Studies, Daegu-Gyeongbuk Institute of Science and Technology (DGIST), in Daegu, and Professor Emeritus of Communication, Sogang University, Seoul.

34

SPAIN

Evolution and professionalisation of science communication

Gema Revuelta, Vladimir de Semir and Carolina Llorente

1. Introduction

The development of modern science communication (SC) in Spain is the result of numerous individual events, both great and small, participated in by many people, some famous and others completely unknown, and countless activities and actions, some meticulously planned and others completely random. This chapter will not be exhaustive. Our aim is to improve the overall understanding of the evolution of SC over recent decades (modern SC or public communication of science and technology), and how the narrative of Spain complements our understanding of SC in other countries in the world. The evolution of scientific communication in Spain cannot be understood without paying attention to its relationship with the rest of European Union countries, especially its neighbours France and Portugal, as well as its close connections with Latin American countries, with whom Spain shares a language and a huge number of cultural elements. After a brief historical introduction, this chapter reviews the sociopolitical context in Spain during the 20th century and how changes in that context influenced the emergence and evolution of 'modern science communication'. It is difficult to set a precise starting date for this, although there was more attention to scientific issues in the media in the 1970s and 1980s. This coincides—and not by chance—with an intensification and an internationalisation of science itself, as well as with an opening of the country to the world.

We review the respective roles played by mass media, museums, universities, professional associations and administrations (national, autonomous and local). Although Spanish scientific communication has experienced growth and recession phases since the 1970s, it has never stopped professionalising. Good examples are the offer of specific university studies on science communication (mainly master's and postgraduates courses); the growth of professional associations in number of partners and in their activities; the creation of collaborative networks (such as units of scientific culture, museums and science centres, etc.); the emergence of research groups dedicated to the analysis of this field; and the consolidation of major professional events.

2. Science and SC in Spain before the 1970s

The need to recount great happenings is something inherent in the human condition and even before printing had been invented, the public communication of news or events was already taking place. Epidemics, plagues, weather forecasts, environmental disasters, wars and their technological implementations, and quarrels between wise men (or among wise men, mages and monks) are examples of issues that have been present in the 'public arena' throughout history. Nevertheless, SC was transformed in Spain with the printing press and technological advances, as well as scientific revolutions and the institutionalisation of science. Scientists themselves (engineers, doctors, astronomers, naturalists etc.) for centuries acted as disseminators in Spain (López-Ocón Cabrera, 2000).

The amassing of artefacts and the 'culture of the curio' (Bolaños, 2008, p. 44), common to all countries with a colonial past or that commissioned great scientific expeditions, formed the basis for some of the first science museums in Spain. Examples include the country's National Museum of Natural Sciences, founded in 1771 on the basis of the collections of Pedro Franco Dávila or, towards the turn of the century, today's Museum of Natural Sciences in Barcelona, based on the collections of Francesc Martorell.

Science news in the press is as old as the press itself. SC historians have found records of science news dating back to the 17th century in France and England. Unfortunately, research into the history of SC in Spain is so scant that we have nothing on record regarding these periods, but there is every reason to think that Spain also published this kind of article in the same period. This is particularly so given the fact that the country was at its cultural zenith (the so-called 'Golden Age' of Spanish hegemony in Europe, spanning the 16th and 17th centuries), with writers such as Cervantes and painters like Velázquez. Spain has some of the oldest universities in the world and in this period Spain enjoyed significant development in science and technology in areas associated with naval activity, geographical expansion and medicine, partly as a result of the legacy of Arab culture.

Studies of the 18th century give some insight of the coverage of science issues in newspapers (Martín Melero, 2008). Guillamet i Lloveras has mentioned the efforts of the editor of the newspaper *Diario de Barcelona* to introduce science subjects on a regular basis during 1792–1810; from the very first day, these subjects were announced as a means of contributing to 'general instruction and common utility' (Guillamet i Lloveras, 1998, p. 119). Other examples over the course of the 19th century include *La Crónica Científica y Literaria*, edited by José Joaquín Mora and later renamed *El Constitucional* (Cruz Seoane and Saíz, 1983). Mora, like other Spanish liberals of the 19th and 20th centuries, ended up in exile in Latin America. In the second half of the 19th century, the reciprocal influence between Spain and Latin America, particularly between intellectuals and artists, was quite strong. Publications such as *La América, Crónica Hispano-Americana* and *El Museo Universal* mirrored this desire for communion between Spain and the countries of Latin America (López-Ocón Cabrera, 1990). These journals reflected Latin American cultural and political life, and also featured scientific content (Martín Melero, 2008; Graiño Knobel, 2014).

The common language made it easy for Spain and Latin American countries to share stories across media outlets. This has increased in the present era, since digital media in Spanish has a great diffusion across the Spanish-speaking world, regardless of the country where it originated.

We have an even greater understanding of the individuals and the media that ensured science coverage in the press of the first half of the 20th century. Worthy of note is the work of Odón de Buen who edited the weekly journal *El Mundo Científico* from 1903 (Calvo Roy, 2013); and the regular contributions of Josep Comas I Solà (Roca Rosell, 2004) and Miguel Masriera to the newspaper *La Vanguardia* (de Semir, 2014).

Spanish politics took a dramatic turn in the first half of the 20th century, and this impacted every sphere including SC. Although the country did not take part in either of the two world wars, events in Spain impoverished and isolated the country for decades. The century began with a recently restored constitutional monarchy, but 1923 saw a coup d'état and the monarchy joined the dictatorship of Primo de Ribera until 1930. After a year of provisional government, the Second Republic was installed in 1931, only to be followed by a bloody civil war from 1936 until 1939. In 1939 the dictator Francisco Franco took control and dominated the country's history for nearly

40 years. The countries that participated in World War II had difficulties in recovering, but for Spain things were even worse. The country was on its knees and its people divided by the Spanish Civil War. The political regime was based on 'autarchy' (self-sufficiency) but implemented no serious plans for the country's recovery, at least in its early decades, and the international community ceased providing support. The country was ravaged by poverty and famine, and most of its people were illiterate. Repression (both moral and political), curtailments of rights (both social and labour) and international isolation marked this period, especially during the dictatorship's early years. Many scientists emigrated, mainly to France and Latin America.

During the autarchy, scientific institutions founded before the Spanish Civil War were reopened, but research barely opened up to the new knowledge and disciplines that were emerging on the international scene. The same was the case with the country's universities. Management of scientific and academic institutions, as with all other strategic fields, was dominated by the military, certain families (those on the winning side of the Civil War) and the religious movement known as Opus Dei, of which Franco was a supporter (Sanz Menéndez and López García, 1997).

Towards the end of the 1950s, the regime began opening up to the outside world. Its last three Development Plans (between 1964 and 1975) were the key to jump-starting Spain's economic recovery. Amongst other strategies, they included clear investment in science, technology and industry, albeit only in some disciplines, industrial sectors and cities. In the 1970s, the country began to reap the economic benefits of tourism. It experienced some economic growth (in a period known as *desarrollismo* or 'developism'), even if it was at a rate far below that of countries such as the United Kingdom, Germany or France.

Nevertheless, Spain's science system had to wait for five closely interrelated events in order for great change to occur:

1. the definitive end to the dictatorship in 1975 following Franco's death, and the start of democracy
2. the transfer of some powers from the Spanish State to the country's autonomous communities, which had begun to be constituted towards the end of the 1970s
3. the country's entry into the European Union in 1986
4. the first Law on Science that for the first time regulated science activities and researchers' careers
5. Spain's first National Research and Innovation Plan, in 1988.

The effect was significant. Democracy facilitated entrance to the European Union, and consequently Spanish science became internationalised and able to access European funding. The Spanish autonomous communities implemented their own research, development and innovation (RD&I) plans. The Ley de la Ciencia [Spanish Law of Science] contributed to the professionalisation of research and access to regular funding. Spanish science entered the international circuit, becoming increasingly competitive. Between 1981 and 2003 the number of scientific publications indexed in ISI with Spanish authors increased from 3,382 to 24,737; and the investment in RD&I rose from 0.43 per cent to 1.1 per cent of GDP (Gómez et al., 2006). Research centres and universities, more productive in research but also competing more strongly among themselves, increased their communication to the media, through press releases and interviews, and mass media disseminated the achievements of Spanish scientists.

The 1990s and the first five years of the 21st century represented a period of great expectations for Spanish science: large science infrastructure was built, more resources were invested in research, and good working conditions attracted high-level scientists back to the country after they had left in search of greener pastures abroad, as part of the country's own 'brain drain'. This budgetary growth was halted by the economic crisis that began in 2007 and that has persisted in Spain, and its impact on the system was soon felt (Regalado, 2010). The new Law on Science of 2011, proposed a raft of actions to ensure that research is given enough resources and autonomy, but many of the key points have not been complied with and the Spanish science system remains economically weakened (Pain, 2013).

The history of SC in Spain closely mirrors that of science itself. Although the 1960s and 1970s had some journalists and communicators dedicated to the subject and the (sole) television station broadcast some science and nature programs, terms such as 'science communication' and 'the public communication of science' were practically unknown in the country. Expressions such as 'science journalism' and 'the museology of science' were on few lips prior to the last decade.

3. The role of mass media in the origin of modern science communication

Well before the communication of research was an everyday activity of Spain's universities and research centres, the mass media were covering science news and the persona of the science journalist began to appear in the press and on television.

3.1. Newspapers and magazines

Ever since the press has existed it has featured news on science. For many years, scientists themselves commented upon scientific discoveries, and discussed conferences and visits by leading personalities. Occasions such as Einstein's visit of 1923, the 1929 Barcelona Universal Exposition and the arrival of humans on the moon were of particular importance in the dissemination of science news. In 1955, Manuel Calvo Hernando (1923–2012), later the undisputed pioneer of scientific journalism in Spain and Latin America, discovered his vocation for science by covering the world conference Atoms for Peace in Geneva. Calvo Hernando was known for his prolific journalistic work, but also for being one of the first who, with his books and lectures on scientific journalism, contributed to identifying it as a profession and an academic discipline. Later he played a role in professional associations in Spain and Latin America. Manuel Toharia, another pioneer of Spanish scientific communication, has said of Calvo that he was the 'inventor of scientific journalism' (at least in Latin America) (Toharia, 1999, p. 197).

The pace of developments picked up when newspapers created their first regular (usually weekly) science sections and supplements. In the 1970s, for example, a well-known television personality Manuel Toharia edited the science section of the newspaper *Informaciones*. In 1982, *La Vanguardia* published its first supplement dedicated to science, making it the first of the many Spanish dailies to do so in the 1980s and 1990s (Lopez and Olvera-Lobo, 2017). The supplements clearly showed how SC is the sum of events both great and small, of grand plans and coincidences. More specifically, they are closely intertwined with the history of *The New York Times*, the computerisation of the press in the 1980s and the personal trajectories of two childhood friends: de Semir and Wagensberg.

Between 1970 and 1975, *The New York Times* saw a severe decline in its readership and a significant fall in its advertising revenue due, amongst other reasons, to competition from television. The business initiatives taken to halt this decline included the creation of themed supplements to increase the news draw for potential readers, build loyalty bridges and open up new advertising markets. This included the creation in 1978 of *Science Times,* a weekly science section published every Tuesday. The supplement not only helped the paper overcome its crisis but also became a model to be followed by many other dailies around the world, making it one of the cornerstones of the consolidation of science journalism. In Spain, the first newspaper to follow the model of *The New York Times* was *La Vanguardia*.

La Vanguardia's vocation for scientific dissemination could be seen almost from the newspaper's beginnings. Its first issue was published in 1881 and Camille Flammarion made his first contributions in 1892 (Voltes Bou, 1988). Traditionally this daily, like others, commissioned famous doctors and scientists to comment on science (Roca Rosell, 2004). From 1962, it featured a weekly section on medicine, also edited by doctors. Such a section came as no surprise in a city boasting a long tradition in the medical and pharmaceutical sectors (Duran and Piqueras, 2006). At the beginning of the 1980s, *La Vanguardia* embarked upon the process of modernising its production system. Among those responsible for the renovation of the newspaper was the journalist Vladimir de Semir (one of the authors of this book chapter). In a providential conversation with his childhood friend Jorge Wagensberg, who directed the Science Museum of La Caixa Foundation (created in 1981), he understood that science could be a topic of great interest to the public. Wagensberg encouraged him to think about the idea of a specific section or supplement on the subject. De Semir convinced the newspaper's directors and 1982 saw the birth of *La Vanguardia* science supplement (de Semir, 2014). For the very first time, the person responsible for coordinating the paper's science news was not a scientist, but a journalist (Morales, 2007). The supplement was published until 1997 and its work in consolidating the position of science journalism in Spain has been widely acknowledged. In 1986, the Spanish National Research Council (CSIC) created the first Science Journalism Prize in Spain; *La Vanguardia* won the inaugral award.

Other leading newspapers had their own science sections or supplements in the 1980s. The *Futuro* [Future] section of *El País* was one of the most widely read, and at one point had a large group of staff journalists and outside contributors. It enjoyed widespread recognition (being awarded the CSIC Science Journalism Prize in 1987) and a long life. *ABC*, which, alongside *La Vanguardia*, is one of Spain's oldest dailies, had a science supplement and boasts a long history in covering science news (Lopez and Olvera-Lobo, 2017). Others, like *El Mundo*, also covered science and, particularly, health news. The boom in science supplements took place between the 1990s and the first five years of the 21st century. However, the financial crisis, exacerbated by the press's poor response to the growth of the internet, put an end to most science supplements in Spain as in other countries. With the closing of these sections, journalists covering science were moved to other sections or made redundant (some moving to institutional communication jobs). However, science sections and supplements had already had a decisive effect on the recruitment and consolidation of readers. Some were attracted by science and others became used to reading the information offered in these supplements weekly: it was well worked, had good sources and were often accompanied by infographics with very good graphic support. This interest made the science news cease to be something strange and allowed them to progressively get into daily affairs, in the regular sections of the newspaper (de Semir and Revuelta, 2017).

Today, even though science (and especially medical and technology) news receives a lot of coverage in the press, neither native online newspapers nor those that began life offline boast amongst their workforces the number of specialist journalists they had at the end of the last century and the beginning of this one. Therefore, it is worth highlighting the case of the supplement *Tercer Milenio* [Third Millennium] of *El Heraldo*, which has survived for more than 25 years, first with weekly updates—on Tuesday, like *The New York Times*—and now with daily ones and a presence on social media networks (Perla Mateo, 2018). Their editor, Pilar Perla, was one of the first women journalists in Spain to head science sections in the press (previously, Malen Ruiz de Elvira had been responsible for *El País*'s *Futuro*).

Science magazines also fostered the dissemination of science in Spain. The situation in the country is paradoxical. On the one hand, the number of science magazines and their readerships have always been smaller than countries with a greater tradition in the field, such as France and the United Kingdom, something that has led to the failure of a number of magazines that enjoyed a fleeting existence for a few years. On the other hand, Spain boasts a magazine with a presence comparable to that of the most popular magazines (such as *Hola*), something truly exceptional not only in this country but, indeed, anywhere in the world. We are talking of *Muy Interesante* [Very Interesting], which has not only survived for well over 30 years, being first published in 1981, but is also a leading voice on Twitter with more than 8 million followers. In a 2004 interview, its editor explained that the basis of the magazine's success was its use of suggestiveness, humour, inquisitiveness and even 'naughty' takes on science news (Islas, 2004).

Some of Spain's science magazines enjoying a longer life and now benchmarks in modern SC are *Mundo Científico* [Science World] (the country's version of France's *La Recherche*), published from 1981 to 2003, and *Investigación y Ciencia* [Research and Science], the country's oldest (first published in 1976). Some emblematic magazines have not only been sold in Spain, but also in Latin America. This reflects their authors' origins, often from both sides of the Atlantic. For example, the journal *Investigación y Ciencia*, in addition to content from *Scientific American* and its international editions, includes articles and original sections written by researchers from Spain and Latin America.

3.2. Television

Science has had little presence on television either during the dictatorship or the Spanish democratic era (Toharia, 1990). The first 'face of science' for many Spaniards was Luís Miravitlles who, between 1959 and the end of the 1970s, directed and presented science programs on TVE (including *Visado para el futuro* [Visa for the future], from 1963 to 1965). Radio experienced a similar trend: there were very few examples of radio programs specialising in science. It is difficult to find documentation about the pioneer science radio programs, but one of them was *Los progresos científicos* [Scientific Advances], conducted by Manuel Vidal, and broadcast for first time in 1941 (Guions de Ràdio Barcelona, n.d.).

The most successful television program was *El Hombre y la Tierra* [Man and the Earth], a wildlife and nature program broadcast weekly from 1974 to 1980 in a good timeslot. Directed by Félix Rodríguez de la Fuente, the program became so successful that many schools used it as teaching material for their natural science classes. It was also sold abroad, something unheard of for a program produced in a Spain that had only recently emerged from its international insolation (Salcedo, 2011; Alberich-Pascual and Aguirre Salmerón, 2015). But the director's accidental death put an abrupt end to the series. The program's success and its director's popularity are truly exceptional, and all 181 episodes are still viewed today on the TVE website (RTVE, 1975).

The early 1980s saw the start of programming by the 'autonomous broadcasters' (i.e. the public television undertakings of the autonomous communities into which Spain is divided), and in 1988 the country permitted private television channels (from the end of the 1990s the number of channels, including pay television ones, would explode). TVE no longer enjoyed a monopoly. It has been said that 'the appearance of competition in the Spanish television market encouraged the neglect and marginalisation of science programming aimed at the general public' (Gutiérrez Lozano, 2002). However, some autonomous broadcasters, such as those of Catalonia (TV3) and Andalusia (Canal Sur), featured science programming from the very start (Toharia, 1990). Canal Sur's environmental information program *Espacio Protegido* [Protected Area] is one of the best-quality and longest-lived programs, on air for more than 20 years. One of the best-known television science programs was *Redes* [Networks], broadcast by TVE between 1996 and 2014 in a low-audience timeslot. Although the program dealt with important issues and enjoyed the input of top-class researchers, it also often featured pseudoscientific content, making it the butt of wide-ranging criticism amongst the scientific community and science communicators (Carmena, 2002). In all these years, science has had very little presence in television news (León, 2008; Francescutti, 2010).

The arrival of the internet in the mid-1990s and the advances in information and communication technologies have radically changed the ecosystem of scientific communication in Spain. Blogs, video platforms and other initiatives in social networks have as many followers as 'traditional' media. But the main change has to do not only with the competition from these initiatives, but also with the change that social networks have brought about in the distribution of information. The commercial algorithms on which this new distribution is based have affected the visibility of articles and journalistic pieces produced by the media (native or traditional), so these are in a continuous adaptation to the new rules of the game.

Here again is the connection between Spain and Latin America in the distribution flows that mark social networks and large digital platforms (Google, YouTube, etc.). Many of the major Spanish media have greatly increased the number of readers resident in Latin American countries, which further narrows the cultural divisions between these two worlds.

4. Science centres and science museums

Aside from the mass media and popular science books, the 1970s also saw SC finding a place in museums. Generally these were associated with former natural science collections (dead or inert) or with scientific/technical devices. The field was rounded off by the zoos and botanical gardens. Some had begun as a private initiative based on collections and curios, but had at some stage become public property and the responsibility of municipal governments or science societies and institutions, such as the CSIC. These museums have often been dubbed 'showcases', since their function was more to conserve and preserve heritage, and particularly to display it, rather than share knowledge or interact with visitors.

The first 'interactive' science museum (or 'science centre') in Spain was the Barcelona Science Museum, founded in 1981. It would later be rechristened CosmoCaixa to highlight the profile of the La Caixa Foundation, the bank foundation that created it and has ensured its position for almost 40 years as one of the city's leading cultural centres and one of the world's top science museums. Barcelona's was similar to other 'science centres': opening at the end of the 1970s (and over the 1980s and 1990s in other parts of the world) with the emphasis on interactivity and a 'please *do* touch' ethos. The science centres, with the San Francisco Exploratorium as a precursor (inaugurated in 1969), also had in common their focus on the explanation of scientific concepts, so they did not need to contain collections or specimens

(Páramo Sureda, 2009). Instead, they contained exhibition modules created specifically with didactic and playful functionality, often based on the incipient computer and audiovisual technologies then proliferating.

However, Barcelona is characterised by its minimal reliance on information technology and its use of real objects, some highly valuable, in addition to those museum resources typical of a science centre (Wagensberg, 1997). Jorge Wagensberg (1948–2018), a physicist, directed the museum for decades and was also responsible for its renovation and expansion in 2004. He had a large influence on the design and planning of science centres that sprang up in the 1990s in Spain, Latin America, Europe and other parts of the world. His ideas on the role of museums in society led him to affirm that they should be the cathedrals of the 21st century (Wagensberg, 2001). Some idea of his international influence can be gained from the fact that for the first issue of the journal *Public Understanding of Science*, he was invited to be a guest author on science museums (Wagensberg, 1992). On his death in March 2018, the Association of Science and Technology Centers stated that 'Jorge Wagensberg was not only a remarkable physicist, but a visionary in science communication who was able to see the importance and beauty of real objects and how to make them part of an interactive museum' (Staveloz, 2018).

The year 1983 saw the creation of the Casa de las Ciencias [House of the Sciences] in A Coruña, an initiative of the city council, but under the clear leadership of Ramón Núñez. He was director for many years and was also behind the science centres created there in subsequent years (La Domus, the Acuarium Finisterrae and the MUNCYT Coruña), making this small Galician city a focal point of scientific communication.

Wagensberg and Núñez had differing visions of what science centres should be, but both were a great influence on the development and expansion of science museology from the 1980s. Just 15 years after the first interactive museum, those of Alcobendas, Tenerife, Murcia, Malaga, Las Palmas, Cuenca, Extremadura and Valladolid had opened or were about to open, along with Granada's Science Park and the planetariums of Pamplona, Castellón and Madrid (Núñez, 1997). In 2009, the expansion of science centres had become a 'gray tide', where (as the Science Park of Granada Director explained) the term 'tide' refers to a trawling movement and 'gray' (instead of black in the popular expression 'black tide') indicates that what moves in this process is gray material—that is, intelligence, knowledge (Páramo Sureda, 2009, p. 250). This boom was not limited to Spain: other countries saw a similar tale, showing the cross-fertilisation typical of the modern globalised world. In the same period, traditional museums reacted, modernising themselves and introducing interactivity and more participative activities in their programming. Today, the boundaries between the 'old-school' traditional museums and the 'new' interactive ones have become blurred.

The year that saw the start of the economic crisis also, paradoxically, marked the greatest celebration of science that Spain has ever seen: '2007, the Year of Science', an initiative that witnessed the start of the country's Science and Technology Museums Network. By 2016, this network was made up of 24 entities of varying types (López García-Gallo, 2016), a figure that shows the scale of the activities in the sector. Since the start of the crisis, growth has slowed and investment fallen off, despite the fact that many centres have become a core asset in the teaching of science and become some cities' key cultural and even tourist hotspots (Páramo Sureda, 2003; Pérez and Gómez, 2011; Revuelta, 2014).

5. Key science communication events and associations

The first SC meeting in Spain was held in Madrid in 1958, as part of the XXIV Luso-Spanish Congress for the Progress of Sciences (Avogadro, 2005). The first big SC encounters to take place in Spain were closely associated with Manuel Calvo's trips to Latin America and his relationships with science journalists. By the 1960s there had already been debates on science journalism in countries like Chile and Ecuador. Manuel Calvo and Aristides Bastida founded the Asociación Iberoamericana de Periodismo Científico [Iberoamerican Association of Scientific Journalism]. After the first conference of this association in Caracas in 1974, Madrid hosted the second conference in 1977. In 1973 the Asociacion Española de Periodismo Científico was created (also presided over by Manuel Calvo). The organisation, now named the Asociación Española de Comunicación Científica (AECC) [Spanish Science Communication Association], has seen impressive growth over the last 10 years, and now (as of 2018) boasts more than 400 members. Its president is currently Antonio Calvo Roy and it is a member of the European Union of Science Journalists Association (EUSJA) and the World Federation of Science Journalists (WFSJ). The year 2018 saw the association returning to the Ibero-American spirit that characterised its beginning.

Despite these pioneering encounters of the 1970s and 1980s, real growth in SC in Spain did not take off until the 1990s and, above all, in the 21st century. Barcelona, La Coruña and Granada, apart from Madrid, were particularly active in this field. The year 1990 saw the creation of the

Associació Catalana de Comunicació Científica (ACCC) [Catalan Science Communication Association], founded by a large group already very active in the field in Catalonia (media journalists, science popularisers and institutional communicators, in the main). ACCC has had six presidents since then, two of them women (Mercè Piqueras and Cristina Ribas), something always worth highlighting in a world where much work is still to be done in terms of gender equality. Again in 1990, Barcelona hosted the International Symposium on Science Journalism, organised by the Dr Antonio Esteve Foundation, a non-profit organisation that has always played a very active role in promoting science communication.

In 1991, the CSIC headquarters in Madrid hosted the second conference of what would formally become known as the Network for the Public Communication of Science and Technology, or PCST Network. At the time, the group did not have the statutes or the organised structure it has today but was rather an informal network of professionals and academics interested in sharing experiences, highlighting the value of SC and, above all, looking to the future. The first PCST Network event took place in Poitiers in 1989 (Fayard, Catapano and Lewenstein, 2005), and the idea behind the initiative came from Pierre Fayard, then a professor at the city's university. This made Fayard, whose doctoral thesis was entitled 'L'émergence médiatique et la professionnalisation de la communication scientifique à destination des non-spécialistes' [The emergence of the media and the porfessionalisation of scientific communication aimed at non-specialists], one of the network's founding fathers. Vladimir de Semir would, after the Madrid conference and, above all, that in Montreal in 1994, play an essential role in the network's consolidation and in its promotion in Spain and Latin America (the presence of Spanish speakers at all its conferences is notable). Spain is the only country to have hosted PCST conferences twice: Madrid in 1991; and, in 2004, Barcelona welcomed more than 700 participants to the eighth conference, hosted by the Barcelona City Council and Pompeu Fabra University (de Semir and Revuelta, 2004).

The last international meeting of the 1990s was the first conference on the Social Communication of Science, held in Granada in 1999 and organised by the Parque de las Ciencias, UNESCO, the University of Granada, the Government of Andalusia and CSIC. In this case, participants hailed from Spain and many other Latin American countries. The conference was a great success, leading to its being held again in different cities in subsequent years. The AECC has, since the 10th conference (in 2017 at the University of Cordoba), been the entity responsible for organising this international event. Other SC conferences held regularly in Spain and dealing with the profession are Campus Gutenberg, organised by the Pompeu Fabra University –

BSM (since 2011); Ciencia en Redes [Science in Social Media] (since 2012), currently organised by the AECC; and the Jornadas de Divulgación Innovadora [Innovative Outreach Meetings], organised by Zaragoza Foundation of Knowledge (since 2013).

One cannot close this section without mentioning the events and celebrations of SC directed to the public. One of the oldest and longest-lasting celebrations is the Biology Week of the University of Murcia, launched for the first time in 1987 and celebrating 31 years in 2018. As in other European countries, in the 1990s some autonomous communities (in Catalonia since 1995) started celebrating Science Week. In 2001, the Spanish government extended the celebration throughout the country, coinciding with the creation of the Spanish Foundation for Science and Technology (FECYT). FECYT is a public agency dependent on the Spanish government with two missions: to encourage Spanish research and its communication. Both the FECYT creation and the nationalisation of the coordination of science week occur in a political context in which dissemination and communication were being promoted by the Spanish government itself, an administration that until then had shown little interest in science communication promotion. The ministry on which science depends (sometimes the Ministry of Science, at other times the Ministry of Education) had not clearly engaged with scientific communication until then. In 2004, the Spanish government issued a public call for scientific communication and dissemination actions for first time, in the form of a national program like the rest of the competitive calls for research grants (Comisión Interministerial de Ciencia y Tecnología, 2003). This new funding source stimulated the proliferation of communication and outreach activities in the country, especially from universities and research centres. In 2007, the impact of such funding was decisive, as we will see below. The Spanish government, therefore, joined SC late, but when it did, its contribution was decisive.

The biggest celebration on science that has taken place in Spain was named '2007, El Año de la Ciencia' [2007, the Year of Science]. Here again, coincidences and personal stories would have a lot to do with the development of events. During his time as councillor of the City of Knowledge, in the City Council of Barcelona, Vladimir de Semir with Gema Revuelta (also one of the authors of this chapter, and at that time Director of Scientific Culture at the Barcelona City Hall) devised the initiative '2007, Barcelona Year of Science'. De Semir proposed to the Secretary of State for Science of the Spanish government that the initiative be extended throughout Spain. Both had known each other for years, since they directed the two pioneering scientific communication master's degrees (Revuelta, 2007). The celebration was endowed by the Spanish government with an unprecedented budget for universities and research centres to carry out scientific dissemination activities throughout the year. In addition, for the first time a call was opened for these entities to hire personnel for communication and science promotion in what were called Units of Scientific Culture and Innovation, or UCC + I (units that, since then, have been fundamental in the dissemination of Spanish science). The Spanish Network of Science Museums was created during the Year of Science, as well as a network of local administrations active in SC. The Year of Science is considered a moment of inflection in the country's scientific communication (Lopez and Olvera-Lobo, 2017). This celebration, together with the initiatives led by the FECYT and the national funding programs for the promotion of scientific culture, added to the activities that had been developing since the previous decades, acting as a trigger for those who had not yet started and as a multiplier for those who had long careers.

From 2007 on there are numerous initiatives in scientific dissemination that focus on a mass-event format. The Naukas science festival in Bilbao, related to the blog with the same name organised from the project of Miguel Artime, Antonio Martínez and Javier Peláez (since 2011), is one of the most popular events. The Spanish version of the Famelab contest, organised by the FECYT (since 2012) and the performance of a group of comedian scientists called Big Van Ciencia (since 2013), are also widely publicised.

6. The role of universities and research centres

Universities have played an essential role in the expansion and consolidation of scientific communication. Their influence has been exercised at three levels: research, training and communication.

Scientific communication as a field of research does not appear until the 1990s. However, since the 1970s we can find books and research papers on scientific journalism. Manuel Calvo Hernando is the forerunner in the analysis of this topic (Moreno Castro, 2002) and probably the most cited author in Spain and Latin America in the early decades. Despite his influence on academic research, Calvo Hernando published most of his work outside the university (Calvo Hernando, 1977, 1982, 1999), since it was not until the mid-1980s that he began teaching at San Pablo University-CEU.

In the bachelor's and postgraduate degrees in communication and journalism, it was rare to include subjects of scientific journalism. Research groups focused on these topics have not abounded either, and until the 1990s there is very little literature on SC. The first doctoral thesis in scientific journalism is defended by María Alcalá Santaella Oria de Rueda at the Complutense

University of Madrid in 1992. Two other PhD theses followed at the same university, in 1992 and 1994 (Moreno Castro, 2002). During the second half of the 1990s several theses were presented in different universities. Manuel Calvo himself defended his in 1999, at the age of 75 years.

The first university centre in Spain dedicated specifically to scientific communication was the Observatory of Scientific Communication of the Pompeu Fabra University (UPF), founded in 1994. The centre soon became a reference for SC in Spain, publishing the first academic journal in Spanish specialised in SC: *Quark, Science, Medicine, Communication and Culture* (1995–2007). The UPF also launched the first SC master's degree. It played a fundamental role in the international projection of Spanish SC, being the first university group in the country to participate in European SC projects (The ENSCOT Team, 2003) and to publish in the first international journals of the field (Ribas and Cáceres, 1997). The Observatory of Scientific Communication inspired the creation of other similar centres in Europe, and was an initiative of Vladimir de Semir, professor of scientific journalism at the UPF since 1992. De Semir directed the centre for almost two decades with Gema Revuelta, the sub-director. In 2014 both left the observatory and created the Science, Communication and Society Studies Centre, also at the UPF (CCS-UPF). This time Gema Revuelta directs the centre, while de Semir is the chair of both the Social Council and Scientific Council of the Center of Studies on Science Communication and Society. The CCS-UPF continues and extends the research and training projects that both have led for two decades.

In the University of Valencia, a specific course in science and technology for students of journalism was launched at the beginning of the century (Moreno Castro and Gómez Mompart, 2002). Carolina Moreno is responsible for the course and is one of the first professors specialised in scientific journalism in Spain (along with Carlos Elías, from the Carlos III University of Madrid). The University of Valencia has also been a reference in the SC for other reasons. Thus, if *Quark* had been the first journal on SC in Spanish, the magazine *Mètode* from this university was the first one of its type in Spain (although originally it was only published in Catalan). Directed by the biologist and writer Martí Domínguez, it celebrated 25 years in 2017 and is currently published in Catalan, Spanish and English. At the beginning of the 21st century, the first professorships dedicated to the SC emerged. The first ones started at the University of Valencia (2002), followed by the University of Valladolid (2005), the University of Girona (2008), the University of Zaragoza (2009) and the University of Basque Country (2010) (Lopez and Olvera-Lobo, 2017).

Universities have also played a fundamental role in training in scientific communication for new generations of professionals. UPF launched the first master's degree in scientific communication in Spain in 1995, a program that has been offered continuously since then and that now has two simultaneous versions (on-site and one online) (de Semir, 2009). Another of the pioneering universities is the University of Salamanca, whose master's degree, directed by Miguel Ángel Quintanilla, was offered for the first time in 1998. During the first decade of the 21st century, other universities offered similar training (master's or postgraduate courses), such as the University of Valencia, the Carlos III University of Madrid, the National University of Distance Education, the University of Oviedo etc. (Lopez and Olvera-Lobo, 2017). However, after a period of boom, excessive competition and the economic crisis caused a number of these training programs to close. At the moment, only six universities offer SC training in the form of master's degrees.

During PhDs or master's degrees on science topics, it is not unusual to include workshops or courses on communication skills, relations with the media or other issues related to SC. However, it is unusual in undergraduate studies. In this sense, UPF was also a pioneer for including a subject of scientific communication in the curriculum of a bachelor's degree in science. Since 1998 the bachelor's degree in human biology includes the subject 'scientific communication', in which students not only learn how to improve their communication skills but also undertake public engagement activities and learn to design, organise and evaluate them. Some other bachelor's courses offer communication courses, but not many.

Finally, it is necessary to explain the role that universities have played in communicating their own research. This experienced a first impulse from the internationalisation of Spanish science at the end of the 1980s, and also later when universities were competing to attract the best students. The universalisation of schooling in Spain, coinciding with the generation of the baby boomers, produced a massive influx of students to the university classrooms in the 1970s and 1980s. One of the first bachelor's degrees to impose an access limit was medicine (since 1979 only students with very good grades can register). During the 1980s this practice was extended to all areas of knowledge, so that at the beginning of the 1990s all universities competed to be chosen by the best students (the system is mainly public). Faced with this new situation, universities were forced to improve their communication strategies. They soon discovered that research carried out by the professors is one of their main assets to 'sell' their institution to the general public. The large universities expanded their institutional communication teams to include people specialised in research dissemination, and graduates trained in scientific communication with master's degrees had an excellent reception. The scientific journalists affected by the staff reductions of mass media also found new jobs in institutional communication. Small universities were encouraged to join this trend when the Spanish government created a fund for the establishment of Units of Scientific Culture and Innovation in 2007. This transformed the landscape of science communication in Spain.

At present, practically all Spanish universities and research centres carry out scientific communication in one way or another. Some are limited to sending press releases and publishing their research news through their different channels (websites, social networks, university digital publications or even radio stations and TV channels). Others, in addition, organise outreach, public engagement or authentic participatory activities. Many research groups also carry out their own public engagement actions (especially those that participate in European and international projects).

The growth of communication that emerges in universities and research centres, together with the experimentation around new more participative formats (such as citizen science initiatives, hackathons, social labs, etc.) have been one of the main transformations of communication of science in Spain in the last decade.

It is difficult to anticipate what will happen in the future, but if the trends that we observe today continue, we can expect that universities and research centres will intensify and professionalise their science communication activities, gaining more presence in the public sphere both by their online communication and by the organisation of face-to-face events. In parallel, some other sources of information (pseudotherapy and fake-news promoters, for instance) could also expand their visibility because algorithms that regulate social networks and big online platforms (i.e. Google, YouTube) tend to favour them.

Paper-only newspapers and magazines will tend to disappear, while it is not possible to predict what will happen with online newspapers and television channels. They need to solve their business models without damaging their editorial content quality, but it is a hard task with a distribution highly dependent on commercial algorithms.

Finally, we can expect that science communication research will continue growing and that Spanish researchers in this field will be more present on the international sphere.

References

Alberich-Pascual, J. and Aguirre Salmerón, M. (2015). Changes in television nature documentaries in Spain. From Félix Rodríguez de la Fuente to Frank Cuesta (1974–2011). *Revista de Comunicación Vivat Academia. Universidad Complutense de Madrid, 130*, 1–34. doi.org/10.15178/va.2015.130.1-34.

Avogadro, M. (2005). Periodismo de la Ciencia: Aproximaciones y Cronología. *Razón y Palabra, 43*. Retrieved from www.razonypalabra.org.mx/anteriores/n43/mavogadro.html.

Bolaños, M. (2008). *Historia de los museos en España: memoria, cultura, sociedad.* Gijón: Trea.

Calvo Hernando, M. (1977). *Periodismo científico*. Madrid: Paraninfo.

Calvo Hernando, M. (1982). *Civilizacion tecnologica e informacion*. Barcelona: Editorial Mitre.

Calvo Hernando, M. (1999). *El Nuevo Periodismo de la Ciencia*. Quito: Editorial 'Quipus', CIESPAL.

Calvo Roy, A. (2013). *Odón de Buen: Toda una Vida*. Zaragoza: Ediciones 94.

Carmena, E. (2002). Premio a un dudoso divulgador científico. *El Escéptico* (verano), 6–7. Retrieved from www.rtve.es/tve/b/redes.

Comisión Interministerial de Ciencia y Tecnología. (2003). *Plan Nacional de Investigación Científica, Desarrollo e Innovación Tecnológica 2004–2007*. Madrid. Retrieved from www.ciencia.gob.es/stfls/MICINN/Investigacion/FICHEROS/ResumenPlan_Espanol.pdf.

Cruz Seoane, M. and Saíz, M. D. (1983). *Historia del periodismo en España*. Madrid: Alianza.

de Semir, V. (2009). Master in Scientific, Medical and Environmental Communication. *JCOM: Journal of Science Communication, 8*(1), 1–5. doi.org/10.22323/2.08010302.

de Semir, V. (2014). *Decir la ciencia: divulgación y periodismo científico de Galileo a Twitter*. Barcelona: Universitat de Barcelona, Publicacions i Edicions.

de Semir, V. and Revuelta, G. (2004). Scientific knowledge from, for and through cultural diversity. *Science Communication, 26*(2), 211–18. doi.org/10.1177/1075547004270784.

de Semir, V. and Revuelta, G. (2017). *Periodistas científicos : corresponsales en el mundo de la investigación y el conocimiento*. Barcelona: Editorial UOC.

Duran, X. and Piqueras, M. (2006). *Walks around the scientific world of Barcelona*. Barcelona: Ajuntament de Barcelona. Retrieved from ajuntament.barcelona.cat/barcelonallibres/es/publicacions/walks-around-scientific-world-barcelona.

Fayard, P., Catapano, P. and Lewenstein, B. (2005). The international public communication of science and technology network. A brief historical overview. In G. Revuelta and B. Bonmati (eds), *PCST International Conference (Barcelona 2004)* (pp. 27–35). Barcelona, Spain: Pompeu Fabra University. Retrieved from www.upf.edu/pcstacademy/_docs/fayard.pdf.

Francescutti, L. P. (2010). La información científica en los telediarios españoles. *Cuadernos de la Fundación Dr Antonio Esteve, 21*.

Gómez, I., Sancho, R., Bordons, M., and Fernández, M. T. (2006). La I + D en España a través de publicaciones y patentes. In J. Sebastián and E. Muñoz (eds), *Radiografía de la investigación pública en España* (pp. 275–302). Madrid: Biblioteca Nueva.

Graiño Knobel, S. (2014). La evolución de los argumentos justificadores de la divulgación y el periodismo científico. *Prisma Social, 12*, 232–97.

Guillamet i Lloveras, J. (1998). Pedro Pablo Husón de Lapazarán i els Inicis del Periodisme Cultural i Científic. *Treballs de Comunicació, 10*, 119–29. Retrieved from www.raco.cat/index.php/TreballsComunicacio/article/view/242790/325459.

Guions de Ràdio Barcelona [Radio Barcelona Scripts] (n.d.). Dipòsit Digital de Documents de la UAB. Retrieved from ddd.uab.cat/record/33214?ln=ca.

Gutiérrez Lozano, J. F. (2002). La divulgación científica en la programación de las televisiones generalistas. *Comunicar, 10*(19), 43–48. Grupo Comunicar. doi.org/10.3916/C19-2002-08.

Islas, K. (2004). Se puede informar sobre ciencia entreteniendo y divirtiendo al lector (entrevista a José Pardina). *Quark, 34*, 95–97. Retrieved from www.raco.cat/index.php/quark/article/viewFile/55074/63216.

León, B. (2008). Science related information in European television: A study of prime-time news. *Public Understanding of Science, 17*(4), 443–60. doi.org/10.1177/0963662505607308 9.

López-Ocón Cabrera, L. (1990). Las relaciones científicas entre España y la América Latina en la segunda mitad del siglo XIX. *Revista de Indias, 50*(188), 305–34.

López-Ocón Cabrera, L. (2000). Momentos y lugares de la ciencia Española siglos XVI-XX La contribución de ingenieros, médicos y naturalistas a la divulgación científica: Ciencia Burguesa. *Historia, 16*(291), 33–53.

López García-Gallo, P. (2016). Los museos y centros de ciencia en España. *ICOM España, 13*. Retrieved from issuu.com/icom-ce_librovirtual/docs/icom-ce_digital_13.

Lopez, L. and Olvera-Lobo, M. D. (2017). Public communication of science in Spain: A history yet to be written. *Journal of Science Communication, 16*(03), Y02. doi.org/10.22323/2.16030402.

Martín Melero, C. (2008). *Periodismo científico en España: una aproximación histórica a la divulgación de la ciencia en las publicaciones periódicas de los siglos XVIII y XIX*. Universidad Complutense de Madrid.

Morales, P. (2007). *La ciencia en los medios de comunicación. 25 años de contribuciones de Vladimir de Semir*. Barcelona: Fundación Dr Antoni Esteve.

Moreno Castro, C. (2002). La investigación universitaria en periodismo científico. *Ámbitos, 9–10*, 121–41.

Moreno Castro, C. and Gomez Mompart, J. L. (2002). Ciencia y Technologia en la formacion de los futuros comunicadores. *Comunicar Journal 19: Comunicación y divulgación de la ciencia, 9*(12), 19–24. doi.org/10.3916/C19-2002-04.

Núñez, R. (1997). Los centros de divulgación científica en España: Panorama a finales del siglo XX. *Quark, 9*(8), 29–38.

Pain, E. (2013). Spain's research council approaches bankruptcy. Science. *341*(6143), 225. doi.org/10.1126/science.341.6143.225.

Páramo Sureda, E. (2003). El conocimiento puede ser contagioso. El papel de los museos. *Quark, 28–29*, 118–23. Retrieved from quark.prbb.org/28-29/028118.htm.

Páramo Sureda, E. (2009). Comunicar la ciencia desde los museos. In A. González Valverde and A. Pérez Manzano (eds), *Contar la ciencia. Comunicar la ciencia desde los museos*. Fundación Séneca. Retrieved from fseneca.es/web/contar-la-ciencia.

Pérez, L. and Gómez, A. (2011). *Efectos económicos y sobre el empleo del parque Territorio Dinópolis de Teruel*. Editado por Fundación Economía Aragonesa.

Perla Mateo, M. P. (2018, 16 October). Especial 25 aniversario de Tercer Milenio: 25 años de Tercer Milenio (Divulgar en 3D). *Heraldo*. Retrieved from www.heraldo.es/noticias/suplementos/tercer-milenio/divulgacion/2018/10/16/25-anos-tercer-milenio-divulgar-3d-1271972-2121028.html.

Regalado, A. (2010, 12 March). Spain turns to science for stimulus. *Science*. Retrieved from www.sciencemag.org/news/2010/03/spain-turns-science-stimulus.

Revuelta, G. (2007). Barcelona Ciencia 2007 La crónica. *Quark, 39–40*, 21–33. Retrieved from quark.prbb.org/39-40/039021.pdf.

Revuelta, G. (2014). Impacts of science communication on publics, cities and actors. *Journal of Science Communication, 13*(1). doi.org/10.22323/2.13010301.

Ribas, C. and Cáceres, J. (1997). Perceptions of science in Catalan society. *Public Understanding of Science, 6*(2), 143–66. doi.org/10.1088/0963-6625/6/2/003.

Roca Rosell, A. (2004). *Josep Comas i Solà, astrònom i divulgador*. Barcelona: Ajuntament de Barcelona. Retrieved from ajuntament.barcelona.cat/barcelonallibres/ca/publicacions/josep-comas-i-solà-astrònom-i-divulgador.

RTVE (1975). Presentación de la serie ibérica de 'El hombre y la Tierra' por Félix Rodríguez de la Fuente. Retrieved from www.rtve.es/alacarta/videos/el-hombre-y-la-tierra/presentacion-serie-iberica-hombre-tierra-felix-rodriguez-fuente/4507801/.

Salcedo, M. (2011). El antropomorfismo como herramienta de divulgación científica por televisión: estudio de El Hombre y la Tierra. *Communication and Society, 24*(1), 217–46. Retrieved from www.unav.es/fcom/communication-society/en/resumen.php?art_id=386.

Sanz Menéndez, L. and López García, S. (1997). Continuidad y Cambio en las Políticas de Ciencia y Tecnología durante la Autarquía y los Inicios del Desarrollismo. *Quaderns d'Història de l'Enginyeria, II* (pp. 70–98). Retrieved from upcommons.upc.edu/bitstream/handle/2099/673/politica_tecnologica.pdf?sequence=1.

Staveloz, W. (2018). In memoriam: Jorge Wagensberg. Association of Science and Technology Centers. Retrieved from www.astc.org/member-news/memoriam-jorge-wagensberg/ (site discontinued).

The ENSCOT Team. (2003). ENSCOT: The European Network of Science Communication Teachers. *Public Understanding of Science, 12,* 167–81. doi.org/10.1177/09636625030122003.

Toharia, M. (1990). La ciencia en televisión. *Arbor, 136*(534), 123–36.

Toharia, M. (1999). Manuel Calvo H.: Inventar el periodismo científico. *Chasqui. Revista Latinoamericana de Comunicación, 9*(66), 7–11.

Voltes Bou, P. (1988). La Col·laboració de professors universitaris a l'antiga Vanguardia. In *I Simposium Història de la Universitat de Barcelona* (pp. 741–43). Barcelona: Universitat de Barcelona.

Wagensberg, J. (1992). Public understanding in a science centre. *Public Understanding of Science, 1*(1), 31–35. doi.org/10.1088/0963-6625/1/1/007.

Wagensberg, J. (1997). La vitrina: Reflexión en siete horas. *Quark, 9*(8), 20–28. Retrieved from www.raco.cat/index.php/Quark/article/view/54654.

Wagensberg, J. (2001). A favor del conocimiento científico (los nuevos museos. *ÉNDOXA: Series Filosóficas, 14,* 341–56. UNED. Madrid.

Timeline

Event	Name	Date	Comment
First interactive science centre established.	Barcelona Science Museum (now CosmoCaixa)	1981	Later rechristened CosmoCaixa in 2004
First national (or large regional) science festival.	Feria de la Ciencia is the largest and longest science festival in Spain (apart from Science Week)	2003	Organised by Sociedad Andaluza para la Divulgación de la Ciencia and Fundación Descubre

Event	Name	Date	Comment
An association of science writers or journalists or communicators established.	Asociacion Española de Periodismo Científico [Spanish Association of Scientific Journalism]	1974	Now named the Asociación Española de Comunicación Científica (AECC) [Spanish Science Communication Association]
First university courses to train science communicators.	Master's degree in scientific, medical and environmental communication, at Pompeu Fabra University – BSM (UPF BSM)	1995	A professional master's degree for graduates with a multidisciplinary background (communication, science, etc.) and a one-year degree (400 teaching hours)
First master's students in science communication graduate.	Master's degree in scientific, medical and environmental communication of Pompeu Fabra University – BSM (UPF BSM)	1995	A professional master's degree for graduates with a multidisciplinary background (communication, science, etc.) and a one-year degree (400 teaching hours)
First PhD students in science communication graduate.	María Alcalá Santaella Oria de Rueda	1992	In scientific journalism, at the Complutense University of Madrid
First national conference in science communication.	The first documented SC meeting in Spain was held in Madrid	1958	Part of the XXIV Luso-Spanish Congress for the Progress of Sciences
National government program to support science communication established.	Spanish Foundation for Science and Technology (FECYT). Public agency dependent on the Spanish government with two missions: to encourage Spanish research and its communication	2001	2004: For the first time the Spanish government issued a public call for proposals for scientific communication and dissemination, and in the same manner as other disciplines
First significant initiative or report on science communication.	A series of books and papers on scientific journalism, primarily by Manuel Calvo Hernando	1970s	Manuel Calvo Hernando is probably the most cited author in Spain and Latin America in the early decades
National Science Week founded.	Science Week in Catalonia	1995, 2001	Late 1990s: The week was celebrated by other autonomous communinites 2001: Spanish government made it a national event
A journal completely or substantially devoted to science communication established.	*Quark*, *Science*, *Medicine*, *Communication* and *Culture* published by the Pompeu Fabra University (1995)	1995	

Event	Name	Date	Comment
First significant radio programs on science.	Los progresos científicos [Scientific Advances] from Radio Barcelona	1941	Conducted by Manuel Vidal
First significant TV programs on science.	Visado para el futuro [Visa for the future]	1963	From 1963 to 1965
First awards for scientists or journalists or others for science communication.	Science journalism prize awarded by the Spanish National Research Council (CSIC)	1986	*La Vanguardia* newspaper received the first award because of its supplement dedicated to science
Date hosted a PCST conference.	Spain is the only country to have hosted PCST conferences twice: 1991 Madrid, 2004 Barcelona	1991, 2004	Madrid hosted the second conference (organiser CSIC) 2004 Barcelona hosted the eighth conference (organiser Pompeu Fabra University)
Other significant events.	First subject of scientific communication in the official curriculum of a scientific degree	1998	Human Biology, Universitat Pompeu Fabra
	The biggest natoinal celebration on science was 2007, El Año de la Ciencia [2007, the Year of Science]	2007	This began at the City Council of Barcelona and was later supported by the Spanish government and then extended throughout Spain

Contributors

Dr Gema Revuelta is director of the Studies Center on Science, Communication and Society at Pompeu Fabra University and director of the master's degree in scientific, medical and environmental communication at UPF – Barcelona School of Management.

Vladimir de Semir is a science journalist, and president of the social and advisory councils of the Studies Center on Science, Communication and Society at the Pompeu Fabra University.

Carolina Llorente is associate professor of Science Communication and Responsible Research and Innovation at the Department of Experimental and Health Sciences at Pompeu Fabra University.

35

TAIWAN

From nationalising science to democratising science

Chun-Ju Huang, Yuh-Yuh Li
and Yin-Yueh Lo

1. Background

Taiwan, officially the Republic of China (ROC), is a country located in East Asia, with neighbours that include China to the west, Japan to the northeast and the Philippines to the south. With one main island and smaller ones around its periphery, Taiwan has a population of about 23.6 million people in an area of about 36,000 km^2, which is close to the size of the Netherlands.

Previously known as Illa Formosa ('Beautiful Island' in Portuguese), the island was mostly inhabited by aborigines before the 17th century, when the Netherlands and Spain set up colonies, paving the way for mass immigration later by the Han people of China. The island was initially annexed in 1683 by the Qing Dynasty, the last imperial house of China, and was ceded to Japan in 1895 after the Qing Dynasty lost the Sino-Japanese War. While Taiwan was under Japanese rule, the Republic of China (ROC) was established on the mainland of China in 1912 after the fall of the Qing Dynasty. World War II ended in 1945 and, following Japan's surrender to the Allies, the ROC regime took control of Taiwan. However, the resumption of the Chinese Civil War led to the ROC's loss of the mainland to the Chinese Communists, with the ROC government and its remnant army fleeing to Taiwan in 1949. Such a complicated colonial history has nurtured Taiwan's characteristics as a small but tough country that has sought constant survival under various challenging international circumstances and balances of power among global leaders.

After the new regime from mainland China forcefully put down various protests in Taiwan, the domestic construction sector was let loose. In the early 1960s, Taiwan began a prolonged period of rapid economic growth and industrialisation. The tag 'Made in Taiwan' has been successfully applied to products such as textiles, plastic toys and bikes in the 1980s, and personal computers and computer chips in the 1990s. At one time, four out of every five notebook computers produced globally were designed in Taiwan. Since the 2000s, value-added innovation has been a highlight of its economic progress. Taiwan is now the 23rd largest economy in the world, and its high-tech industry plays a key role in the global economic supply chain. It also ranks quite high in terms of freedom of the press, health care, public education, economic freedom and human development. With education as one example, Taiwan was 4th in 2006, 12th in 2009, 13th in 2012 and 4th in 2015 in the PISA science assessment. The country benefits immensely from a highly skilled workforce, is among the most educated countries in the world, and has one of the greatest percentages of citizens holding an undergraduate education degree or higher.

Taiwan has worked hard to develop high-quality technology and education, striving to maintain its position amidst global competition. This chapter offers a brief overview of science communication in Taiwan through three different periods of evolution: 1945–80, 1981–2000 and 2001 to the present. The first period 1945–80 concentrated on 'useful science'. The second period highlighted the importance of science education, science popularisation and the emergence of civil awareness of one's environment, and sheds light on the emergence of science communication. Taiwan (under the Kuomintang 'KMT' ruling party, which led the occupation of Taiwan in 1949) formally ended almost 40 years of martial law in 1987, opening the way for its citizens to participate in free public debates, engagement and discussions about civil society. This newfound freedom spurred public debates on science and technology. The last period deals with developments from 2001, including systematic efforts to improve science communication. This outline gives us a chance to look both backward and forward and offers an example for other countries.

2. Period of national reconstruction (1945–80): Import of Western sciences

2.1. Science as a means of modernisation

The introduction of modern science in Taiwan can be attributed to Japan's colonisation until 1945 along with the government of the KMT that followed. Modern science served as the key to industrialisation and modernisation.

Under the nearly 60 years of colonisation, the Japanese government launched many infrastructure projects for education and research, such as the first university in Taiwan. The colonial government introduced modern agricultural science, medicine and engineering science into higher-level education. During this period, Japan was the first choice when Taiwanese sought higher education opportunities overseas. Western modern science was thus transferred into Taiwanese elites via Japan. The Japanese provided the first contact with modern science and, despite 'useful science' being at the forefront, a strong cornerstone for the development of scientific research was laid down. Japan was an important agent of knowledge transfer for Taiwan, and it was not until the 1960s that the number of students studying overseas began to focus on other countries such as the United States and those in Europe.

The call for modern science in mainland China emerged after the rejection of traditional values during the May Fourth Movement in 1919, or seven years after the Republic of China was founded. During the Movement, participants wanted to adopt the Western ideals of 'Mr Science' and 'Mr Democrat' to construct a modern nation (Jiang, 2016, pp. 68–70). However, after the KMT fled from the civil war in mainland China to Taiwan in 1949, its ruling government set up the island as a military base. 'Mr Science' served to form this new country through scientific endeavours such as geographical surveys, infection control, reducing infant mortality and improving agriculture productivity; while the ideals of 'Mr Democrat' were suppressed in Taiwan, with no public debates allowed until 1987 when martial law was lifted.

To demonstrate the government's commitment to science, the National Taiwan Science Education Center was established in 1956 with the target of complementing school education. The target groups were pupils below ninth grade. It provided equipment for scientific experiments at schools and, from 1960, organised yearly national science and technology exhibitions for schools. These exhibitions cultivated many science talents.

Under the conflicting relationship across the Taiwan Strait and the international context of the Cold War, the United States provided a broad array of resources to equip Taiwan as fortification against the Soviet Union. With billions of US dollars in financial aid and soft credit provided by the US over 20 years, Taiwan had the capital for reconstruction. The focus of early scientific research was on military technology, and any discoveries of this research were kept highly secret (Lu, 2018, p. 15). As the KMT government's hopes of returning to the mainland faded, attention turned to the domestic side. The Ten Major Construction Projects[1] (1974–79) provided Taiwan with the transportation facilities and factories for the build-up of its heavy industry.

With the US National Science Foundation as a reference, the Taiwanese National Science Council (NSC)—the first governmental organisation for the promotion and funding of science and technology research—was established in 1959, eventually transforming into the Ministry of Science and Technology (MOST) in 2014. The NSC drafted an agenda for the development of science and technology, and played a crucial role in setting up the Industrial Technology Research Institute (ITRI) in 1973 and Hsinchu Science Park in 1979 (Taiwan's version of Silicon Valley). ITRI and Hsinchu Science Park formed the cornerstones in transforming Taiwan into a high-tech island that produced almost 80 per cent of the world's notebook computers in the late 1990s and early 2000s under the moniker 'Made in Taiwan'.

2.2. Press coverage of science

Newspapers made up the most common form of media during this first period. Media coverage of science rarely reflected on knowledge values and, if it did, the tone was typically a state press release instead of a news report. Journalists focused on the political contexts of science rather than the scientific findings. From one science journalist's observation, the media coverage of science was either politically orientated or person-orientated (Jiang, 1985). Jiang's observation is that scientific knowledge was rarely the focus of the public media at that time.

Compared to high press circulation, only a few magazines were targeted at scientific topics for the public. The first magazine for the popularisation of science was *Public Science* (1951–74), which aimed at 'sparking public interest in scientific knowledge and popularising science education' (cited from Lu, 2018, p. 128). One of the most influential Taiwanese popular science journals is *Science Monthly* (1970–present), which was initiated by a group of Taiwanese students and scholars who had studied abroad (mostly in the United States) and established the journal. *Science Monthly* was expected to:

> serve as a good source of extracurricular reading and an effective public platform for information exchange. As a means of introducing and popularising new scientific knowledge, this platform should enlighten the public, cultivate scientific approaches, and establish a sound foundation for society (Lin, 2010, p. 90).

1 The Ten Major Construction Projects were major projects completed in Taiwan during 1974–79, including the Taiwan Taoyaun International Airport (formerly known as Chiang Kai-shek International Airport), the nuclear power plant and the first national highway.

These two magazines strongly reflect a belief in a knowledge-deficit between science and society. From the 1970s to the 1990s, *Science Monthly* played a crucial role in the popularisation of science in Taiwan, and not just through its printed issues: Y. C. Hsieh, one of the editors serving *Science Monthly* from 1976 to 1978, was later the first-known Taiwanese PhD graduate in science communication (1984, University of Illinois at Urbana-Champaign, the United States). Furthermore, the publishing company initiated and organised several science and technology conferences for the public in the 1980s.

During the Cold War era, Taiwan's society called for economic progress, leaving no place for ecology and the environment. Equipped with science and technology, Taiwan successfully transformed itself into a modern society. Science was generally viewed as hardware to advance everyday life and modern infrastructure, but rarely as software that could shape culture and the worldview. Furthermore, government financial aid and many returning students who had studied abroad in the United States created a strong public impact so that science and technology soon became entrenched in Taiwanese culture. Scientific development during this period barely responded to the environmental issues facing Taiwan. It was not until the rise of the environmental movement in the 1980s that the domestic scientific community finally turned its attention inward (Lin, 2010).

3. The Taiwan Miracle (1981–2000): Rooting science through education and communication

3.1. Science education

The rapid industrialisation and economic growth of Taiwan is widely known as the Taiwan Miracle. With an urgent need for fostering future scientists and technicians, the Department of Science Education (DSE) was established in 1982 under NSC. DSE played a crucial role in laying the cornerstone for science communication and aimed to strengthen, support and popularise leadership in science education. Since 1982, DSE funded research projects in science education and cooperated with the Ministry of Education to improve it in many ways—for example, science curriculum in higher education, teacher cultivation for high school science, special education programs for scientifically gifted students and the design of a national science exam. Starting in 1989, it began to fund science outreach programs such as a book series introducing key technologies, films on space science and weekly technology news. For the purpose of promoting people's understanding of science and technology and enhancing basic scientific literacy, in 2000 DSE began to raise funding for its 'popular science education project' to support activities such as national or regional science outreach, exposition programs, hands-on science and creative science teaching plans. For example, aiming to promote science in local schools, National Science Week provides opportunities for pupils to have personal contact with scientists. Started in 2006, it combined with Taiwan's railway system and became the Taiwan Railways of Popular Science. During National Science Week, the 'popular science' train travels around the island, introducing scientific knowledge and promoting the importance of science. The project continues to this day and is regarded as the birthplace of Taiwanese's systematic science communication, though the target audience are mostly students. The spirit of science education nowadays shapes people's understanding of public science communication, which was primarily viewed as only informal science education at the beginning.

The first doctoral program at the Graduate Institute of Science Education was established at National Taiwan Normal University in 1986 to cultivate professionals in science education research. In 1994, the first master's program was established at the same institute. Through the 2010s, these programs have gradually expanded research topics into informal science education, science popularisation, science activities and science communication. The National Museum of Natural Science in Taichung City, the country's first science museum, opened in 1986 as part of a grand 12 cultural construction projects scheme in the 1980s. This museum's exhibition area is composed of a science centre, a space theatre, a life science department, a human culture department, an earth environment department and botanical garden. The aim is to inspire the public's interest in science, to help all levels of schools achieve their educational goals and to lay the groundwork for the long-term development of the natural sciences. In 1988, the Association of Science Education in Taiwan was formed by a group of scholars to promote academic research in science education and to enhance communication between academic institutions and among researchers. The first professional journal of science education, the *Chinese Journal of Science Education*, was published by this association in 1993. Many of its members later became the initiators for science communication in Taiwan.

Due to the demands of national construction and social development, the progression of science communication during this period was largely associated with science education, with no clear boundary between science communication and science education. Society's imagination of science communication strategies reflected the assumption of the deficit model, where a knowledge gap was assumed to exist between scientists and the general public. Be it an educational unit or a science museum, their main purpose was to fill this knowledge gap. Such a top-down movement of science education turned out well. Scientific knowledge transmission methods were indeed very rapid and efficient, successfully fostering skilled workers and technicians in a short period of time and creating the economic miracle in Taiwan that so impressed Asian countries over the 1980s.

3.2. Popular science

On television, *Mr Ko and Ms Chi* became an important popular science program in 1983–84 (*Ko* is the phonetic sound for science in Mandarin Chinese and *Chi* represents technology). The program introduced useful science and technology to its audience. For example, the first episode was about home computers and the second about agricultural technology in Taiwan. Mr Ko was a scientist who explained various scientific knowledge, while the other moderator, Ms Chi (a retired stewardess), helped contextualise the knowledge and to make it relevant to daily life. The first governmental investment on a popular science film was *Space Exploration* in 1989. The video was produced by NASA in the United States. After its copyright was sold to Taiwan, the script was re-written and the video edited and dubbed.

Science soon became very popular. A well-known Japanese science magazine, *Newton,* was translated in 1983, changing the concept of introducing science to the masses. The Chinese version of *Scientific American* had been published since 2002, and the understanding of popular science broadened from explaining science knowledge in an understandable way, into viewing science as a distinct culture. In 1991, a series of books about integrating knowledge ranging from science to social science and the humanities was translated and published with the hope of redefining science culture in Taiwan. A member of the publisher's commission behind the series recalled that at that time, 'the popular science books we had in mind not only introduced scientific concepts with lay language, but also promoted popular science as a culture' (Lin, 2010, p. 230). The series consisted of six books, beginning with *Chaos: Making a new science*, by James Gleick. This series encouraged Taiwanese scientists to contribute efforts to popular science. Written by journalist Yu-Ling Yang and biologist Sze-Cheng Lo and published in 1996, *The story of snake venom research in Taiwan* illustrates the development of life science in the framework of snake venom research.

Translation has played an important role in popularising science to date. Aside from science magazines and books, comics from Japan and South Korea relating to science and technology are important for public science education. Translation in Taiwan has two meanings: the first is the *language* translation from other languages into Chinese, and the second is changing *scientific* language into public language. Translation became an effective way for introducing science into Taiwan, but the strong dependency on foreign material might have hindered society and the scientific community in Taiwan from reflecting upon science in their own cultural context.

While popular science takes its root in Taiwan, science and technology journalism failed to seize any opportunities to construct their identity as professionals. Taiwanese journalists for science and technology formed an informal association in 1983, but the association did not transform into a professional community and remained as an annual get-together.

3.3. Criticism of science

People in Taiwan enjoyed the convenience of new technology without questioning the incompatibility between science and Taiwan's traditional beliefs/culture. It was not until the rise of environmental movements that a critical voice on science burst forth in the late 1980s. Taiwan's national policy targeting heavy industry and striving for strong economic growth came at the expense of its environment. It was common to see factory wastewater pollute farmlands and poison crops. The unfortunate result was doubt placed upon the correct application of science and technology, and even ambivalent public attitudes toward science and technology. Environmental initiatives developed from about 1985. Demonstrations erupted against controversial technological applications, such as those against nuclear power and against the building of a chemical factory by DuPont, revealing public concerns about threats from science and technology. Eventually, DuPont withdrew its investment plan in 1986, making it the first time in Taiwan that a foreign company had cancelled its investment due to environmental protests. In response to public pressure, the central government established the Environmental Protection Administration in 1987, followed by similar agencies that were set up at the local level.

These environmental movements and demonstrations contributed to public debates on the benefits and costs of science and technology. Taiwanese society no longer took science and technology for granted.

4. Sustainable development (2001–present): Increasing public engagement with science

4.1. Public participation in science and civic scientific literacy

From the 2000s to the present, the range of civil movements in Taiwan has broadened and diversified. In 2000, Taiwan experienced its first peaceful political transition of presidential leadership when the Democratic Progressive Party (DPP) replaced the KMT as government. Taiwanese people are now much more confident in the democratic system under which public engagement could result in positive social changes. With this shift in the political climate, public participation in social issues is now common, and social scientific issues are no exception.

The 'Survey of Civic Scientific Literacy' project funded by MOST collects regular nationwide surveys on the public's understanding of science and technology, starting from 2008. The findings have helped bring Taiwan into the realm of other countries like the US, those in the EU, China and elsewhere that possess empirical data about their citizens' attitudes toward science and technology. So far, this project has collected four representative samples: in 2008, 2012, 2015 and 2018.

Despite generally positive opinions toward science and technology, the public's *utility* attitudes (in answer to questions such as 'will science and technology make our future lives better') have been negatively impacted by social events such as illegal food additives and environmental pollution, while *worrying* attitudes (such as 'technological development creates an artificial inhumane way of living') remain constant over time. Attitudes differ across generations—the cohort analysis of Li (2019) reveals that the generation born after 1980 has a more negative view of science and technology compared to previous generations. The different attitudes between generations born before and after 1980 is a reflection of their views on the environment.

Despite its outstanding performance scores in PISA science assessment, in countries like Taiwan where scientific concepts are 'imported', it is common that an embedded cultural belief in the paranormal can co-exist with a culture of science (Bauer and Durant, 1997; Needham, 1956). Fortune-telling and astrological television programs are popular (Chiu, 2006; Tsai et al., 2012; Tseng et al., 2014). At first glance, it seems incompatible that Taiwanese on the one hand can pursue the newest technology, while on the other hand place their belief in fortune-telling. These survey results indicate the unique scientific and cultural phenomena that have taken root in Taiwan. While Western science and technology are the newcomers, the obvious instrumentalisation of science and technology leads to a general view in Taiwan that science is understood as a means of spurring economics and connecting with the global world. But traditional Chinese science (like Chinese medicine, lunar calendar, Yijing, etc.) is still alive in people's daily life culture. Although Western science and traditional Chinese science seem to have few interactions, the two systems have a phenomenal co-existence, and each serves its own purposes in everyday life of Taiwanese people.

4.2. From the budding to blooming of science communication academia and practice

The atmosphere of public engagement has brought many changes in academia, especially in the new field of interdisciplinary research. The Taiwan Science, Technology & Society (STS) Association was established in 2008 and aims to promote research related to science and technology and seek creative solutions to problems. Members of Taiwan STS Association cover professions including historians, sociologists, philosophers, educators, scientists and engineers. While this association shares something in common with the Association of Science Education in Taiwan, due to the similar disciplinary make-up of their members, it has a different emphasis: more on the interdisciplinary perspectives of the overall social sciences and humanities, and more on research and promotion of policymaking and practical strategies for public participation in science (Huang, 2016).

Universities have seized opportunities to offer STS-related programs and courses, and new institutions have been established. The first Institute of Science, Technology and Society was formed at National Yang-Ming University in 2008. The year 2014 was a landmark for science communication. Feng Chia University announced the first bachelor program of science communication, while National Chung Cheng University offered the first master's program of science, technology communication and society. National Pingtung University then established the first department of science communication based on its graduate institute of Mathematics and Science Education Department. Today, about five to six related programs exist at different universities in Taiwan.

In 2007, MOST launched 'Development of Taiwan Science Communication Industry', which promotes science communication by integrating scholars, experts and media companies through a funding budget. Under the assistance and guidance of the project, subsidised units produce science news programs, science TV programs and science education films. With the government heavily promoting this project, the Golden Bell Awards—the most important awards for Taiwanese television production—officially established the Science Program Award in 2012. The goal is to encourage more excellent science TV stories and to cultivate science into people's daily lives.

Acknowledging that expanding scientific and technological development has generated amazing results as well as many problems, MOST established a new discipline to deal with science communication in 2010: 'Education of Science Popularisation and Communication' (renamed in 2016 as 'Science, Technology, Society and Communication'). Six research domains are addressed: 1) science and technology governance; 2) technological research and development and society; 3) science and technology risk; 4) scientific literacy and education; 5) science and technology communication; and 6) culture, history, science, and technology. Since then, government funds have been steadily invested in research on science communication, as well as related subjects such as science education, history and philosophy of science, public health, health education, law, etc. This initiative can be regarded as an important milestone in deepening science communication theory and practice, and has also led to the burgeoning development of science communication.

With the government increasing the importance of communicating science to the public under its agenda, an article of the Fundamental Science and Technology Act was adopted in 2016, which ensures certain proportions of research budgets go to popularising science. Furthermore, using the UK Science Media Centre as a role model, MOST funded the first science media centre of Taiwan in 2017.

These implementations symbolically present the government's promise of advancing science communication, complementing the launch of new communication channels for the public by publicly funded scientific research organisations. The most significant one is by Academia Sinica, which in 2017 launched a website with a whole new perspective to engage the public with science. So far it has received a very good response. In the private sector, a science blog project launched in 2011 has since turned into an important website (PanSci) for popular science in Taiwan. PanSci is now the biggest and most significant knowledge community of science communicators, and a for-profit company PanMedia has created a whirlwind that has driven other scientific-related websites. Aiming to establish a knowledge ecosystem, the company promotes both online and offline courses and meetings related to science.

5. Challenges and opportunities

The field of modern science communication has rapidly developed in the last few decades in diversity and academic institutions. The idea of science communication in Taiwan has gradually transformed from 'science education' to 'public science education', and now to 'public communication of science'. The traditional top-down model and the belief of knowledge deficits remain predominant, yet the call for public engagement with science is becoming more intensive under the extension of democracy. Still, several obstacles need to be overcome should we look to further develop this issue.

In response to the wave of 'public engagement with science and technology' that arrived in Taiwan around the 2010s, public surveys show that public engagement remains much more talking than doing. The great majority of Taiwanese adults believe that people should participate in the decision-making process of any science and technology policy, but active participants in public activities—such as attending public meetings or hearings, donating money, or assisting with fundraising campaigns for scientific researchers—are only minor (Tsai, 2015; Li, 2019). Equipped by online technology, scholars and a few public initiatives (such as 'vTaiwan' and 'g0v') aim at pushing information transparency and civic participation. Technology has revolutionised public demonstrations, like the Sunflower Movement—a movement driven by a coalition of students and civic groups in 2014 to protest against the economic integration policies of the former KMT government. Civil initiatives now address science and technology policy. They believe that applying digital technology and social media can help encourage better communication, thus making public governance possible.

Viewed under a microscope, the poor-quality coverage of science by mainstream media is another big challenge. Issues include a lack of professional science journalists and coverage of science having to depend on translations of tabloid newspapers from Western countries. The dearth of professional science journalists makes it difficult to frame scientific knowledge properly and makes it hard to cultivate any form of scientific culture in society. Depending on news from scientific-competent Western countries, inadequate translations of foreign science news often lead to a 'double-distortion' effect (Huang, 2014) as well as detachment from scientific knowledge within society. Furthermore, the mainstream media tends to politicise science, even as science-related issues that attract public attention are highly politically ambivalent. It is thus a challenge to maintain public discussions of science in an open and reasonable tone without being distracted.

The rise of the internet has had an enormous impact on journalism's traditional ecosystem and has pushed mainstream media into an even deeper corner. Journalists are no longer the monopoly source for scientific information. Traditional media organisations can no longer afford to offer positions for science journalists. Thus, when seeking information about science and technology, the general public now typically first goes to individual blogsites or scientific-related websites. Issues such as fake news and misinformation in this online era make it even more crucial to have access to credible sources for information and opinion. The current general media environment leaves little room for anyone to be positive.

Science and technology together are regarded as an 'imported culture' for any in-depth image of the Taiwanese, and many factors, such as national dignity and self-esteem, are involved in disseminating scientific knowledge (Huang, 2016). As a country, Taiwan is proud that in such a short period it has caught up with the developmental pace in the world with regards to infrastructure and institutionalisation of science communication. New ideas, such as democratisation of science, open science or citizen science, are coming forth and making Taiwan rethink its understanding of science communication and the boundary between the autonomies of science and civil society. It is a never-ending struggle between scientists and science communicators, between reporting science accurately and reporting science attractively, as well as arguing about what should be compromised. We look forward to a day when science can be communicated in a reflective way, while at the same time different parties are shown their proper respect. Taiwan's own history has already aptly presented the interdependency between democracy and the modern development of science communication.

Acknowledgements

We would like to thank the officials of the Ministry of Science and Technology of Taiwan, Kuo Yun-Wen and Mei Chia-yu for their key information, which helped greatly in completing this article.

References

Bauer, M. and Durant, J. (1997). Belief in astrology: A social-psychological analysis. *Culture and Cosmos: A Journal of the History of Astrology and Cultural Astronomy, 1*, 55–72.

Chiu, H.-Y. (2006). *Religion, occultism, and social change in Taiwan.* Taipei: Laureate Book.

Huang, C.-J. (2014). Double media distortions for science communication – an analysis of 'compiled science news' transforming in Taiwan. *Asian Journal of Communication, 24*(2), 128–41. doi.org/10.1080/01292986.2013.835328.

Huang, C.-J. (2016). Public communication of science and technology in Taiwan. In M.-H. Chiu (ed.), *Science Education Research and Practices in Taiwan: Challenges and Opportunities* (pp. 279–97). New York: Springer.

Jiang, C.-J. (1985). What kind of science news should be covered? Retrieved from resource.blsh.tp.edu.tw/science-i/content/1985/00110191/0001.htm.

Jiang, C.-J. (2016). Science popularisation in a casual view. *Scientific American* (Taiwanese Edition Magazine), *178*(12), 68–70.

Li, Y.-Y. (2019). Status quo, change and social differences of public attitude of science and technology in Taiwan. In H.-S. Lin (ed.), *Report of 2018 Taiwan civic science literacy survey.* Kaohsiung: Research Center for Promoting Civic Literacy.

Lin, C.-C. (2010). *Forty years legend of Taiwanese science communities – The natural and engineering scientists and Science Monthly in 1970s and 1980s.* Hsinchu: National Chiao Tung University Press.

Lu, I.-M. (2018). *Taiwan's years toward science and technology: The keypersons and events.* Taipei: Chu Liu Book Company.

Needham, J. (1956). *Science and civilisation in China: Volume 2, History of scientific thought.* Cambridge: Cambridge University Press.

Tsai, C.-Y. (ed.). (2015). *Overview of 2015 Taiwan public science literacy.* Kaohsiung: Center for Promoting Civic Scientific Literacy.

Tsai, C.-Y. et al. (2012). Effects of exposure to pseudoscientific television programs upon Taiwanese citizens' pseudoscientific beliefs. *International Journal of Science Education, Part B, 2*(2), 175–94. doi.org/10.1080/21548455.2011.610132.

Tseng, Y.-C., Tsai, C.-Y., Hsieh, P.-Y., Hung, J.-F. and Huang, T.-C. (2014). The relationship between exposure to pseudoscientific television programs and pseudoscientific beliefs among Taiwanese university students. *International Journal of Science Education, Part B, 4*(2), 107–22. doi.org/10.1080/21548455.2012.761366.

Timeline

Event	Name	Date	Comment
First interactive science centre established.	National Science Education Center	1956	
First national (or large regional) science festival.	Science and Technology Week	2000	
An association of science writers or journalists or communicators established.	Science and Technology Journalists Association in Taiwan	1983	It is not officially registered, rather an informal association
First university courses to train science communicators.	Science Communication in National Chengchi University (NCCU) and opened by Dr Y.-C. Hsieh	1984	2014: Feng Chia University announced the first bachelor program of science communication
First master's students in science communication graduate.	National Chung Cheng University offered a master's in science, technology communication and society	2014	First graduate 2018
First PhD students in science communication graduate.	Journalism PhD program at National Chengchi University	1983 (journalism)	1984: First graduate Y.-C. Hsieh 'Science and the press in Taiwan', University of Illinois. 1986: Science education program at National Taiwan Normal University
First national conference in science communication.	Annual Conference for Science Communication	2007	Organised by the Development of Taiwan Science Communication Industry project
National government program to support science communication established.	DSE's Popular Science Education Project	2000	
First significant initiative or report on science communication.	Advisory book for science communication policy	2010	Kwan, S. J. (2010). *Taipei: Development of Taiwan Science Communication Industry*
National Science Week founded.	Science and Technology Week	2000	The theme of the first week was earthquakes, to commemorate the 921 earthquake in 1999

Event	Name	Date	Comment
First significant radio programs on science.	*The Secrets of Nature*	1983	Via the Police Broadcasting Service, hosted by Chang, Jhih-Jie
First significant TV programs on science.	*Mr Ko and Ms Chi*	1983	In Chinese pronunciation, Mr Ko refers to Mr Democracy, Ms Chi refers to Ms Technology
First awards for scientists or journalists or others for science communication.	Wu Ta-You Award for Science Writing	2002	
Other significant events.	Establishment of the Science Media Center Taiwan	2017	

Contributors

Chun-Ju (Jerome) Huang is a full professor at the General Education Center of National Chung Cheng University, Taiwan.

Yuh-Yuh Li is a research fellow and an adjunct assistant professor at the Research Center for Promoting Civic Literacy at National Sun Yat-sen University in Kaohsiung, Taiwan.

Dr Yin-Yueh Lo is an assistant professor in the Department of Communications Management of Shih Hsin University, Taiwan.

36

THAILAND

From temples and palaces to modern science communication

Ganigar Chen, Wijitra Suriyakul Na Ayudhya and Chanin Suriyakul Na Ayudhya

1. Introduction

In Thailand, the history of science communication has not been well documented. As in many other countries, science education has been given a high priority by the Thai government with tremendous investment in the national education system. Science is considered to be an important subject at school and for placement exams rather than something that is used in daily life. The notion of communicating science to the general public is not very widespread or something there is public demand for, although there have been efforts by some academics and groups who see its value to society. In recent years, with exchanges and learning with global communities, science communication has become a topic of interest, and more universities and research agencies recognise its importance. This chapter will present the history of science communication in Thailand during the past 150 years in three different sections: from the first record of science communication in Thailand when the Thai monarchy became involved and exerted a strong influence on science; the modern science communication era with the development of science museums and science festivals; and possible trends and developments into the future based on current policy and the national development agenda.

The education system in Thailand began in temples and the royal palace. The temple was a place where all activities take place for Thais from birth till death; it could be considered as the first school in Thailand. The word for science in Thai, *vidyasart*, was derived from the Sanskrit word meaning 'knowledge'. The word was used in Thailand to represent modern scientific knowledge stretching back not more than 200 years. Although the idea of science is quite new, Thais have invented and used simple technologies for more than 700 years—technologies such as irrigation systems and rice machines since Sukhothai was the capital city from the year 1430 (although the word 'technology' was first included in the Thai dictionary only in 1982). The first two scientific instruments that came from the Western world were the telescope and the sextant. King Narai the Great had received them as presents from Louis XIV of France in 1685 during the age of the Ancient City, Ayudhaya (1350–1767).

In the 18th century, during the Rattanakosin era, modern science was first introduced to Thailand. The first group of Thai students went to study science in European countries during the reign of King Rama III (1724–1850). However, the first science communication activity recorded in Thailand dates back 150 years to King Mongkut (also known as Rama IV of Thailand, and the fourth king of the Chakri Dynasty from 1782 to the present). The King was a passionate scientist and, in 1866, from his studies in Western and Indian manuscripts in science and astronomy, he calculated when and where a total solar eclipse would occur. He announced and explained the eclipse to the public two years before the event occurred on 18 August 1868. He encouraged all diplomats and government officers to observe the eclipse at Warkor in the southern part of Thailand. Politically, this strengthened the international reputation of the Siamese King (Science Society of Thailand, 2018b). Because of this, 100 years later, he was honoured as the Father of Science in Thailand.

2. First record of science communication

Stories of science and science communication in Thailand started during the reign of King Mongkut, at a time when Thailand was called 'Siam'. Many foreigners may not recognise King Mongkut (Figure 36.1), but they will know the musical play *The King and I*—the King in this play was based on him. He was one of the great Thai leaders and played an important role in bringing the country out of the political turmoil of colonisation during his rule from 1851 to 1868 through his wisdom and knowledge.

Figure 36.1: King Rama IV (King Mongkut) on his throne, the Father of Science in Thailand.

Source: Science Society of Thailand under the Royal Patronage of His Majesty the King. © National Science Museum, Thailand (used with permission).

This was marked as the most important event in Thai history in terms of science, science communication and politics. King Rama IV made other announcements, using the royal bulletin to the public (similar to government newspapers today) to describe astronomical phenomena such as asteroids and comets. People in the old days were afraid of these phenomena and considered them bad luck. The King reassured the people that they were natural events and of no harm, contradicting most myths. These announcements could be considered as first attempts at science communication in Thailand.

King Rama IV died from malaria not long after the visit to Warkor, but his story and influence did not end then. One hundred years later, on the commemoration of 200 years of Bangkok as the capital city of Thailand, the Science Society of Thailand proposed to the Ministry of Science that the country honour King Rama IV as the Father of Science in Thailand for his knowledge of astronomy, and in leading the nation to be free from colonisation through his outstanding vision and talents in science and politics. Therefore, 18 August, the date that diplomats and royal officers were called to Warkor to observe the full solar eclipse (Figure 36.2), was approved by cabinet as the National Science Day for the modern science community. Warkor has now been developed into a historic science park with science centres and aquariums offered as lifelong learning science centres for schools and the public. Every year in August, National Science Week is celebrated throughout the country in schools, universities and science centres.

King Rama IV brought in and promoted several new technologies during his reign: the printing machine, photography, the automobile and astronomical equipment. Thais became more acquainted to the possibilities these technologies offered, but only small groups of people in the city had the opportunity to use them. Modernisation, in terms of technology, continued during the reign of King Rama V (1868–1910) when royal family members educated in Western countries brought back technologies for national development: the rail system, ships, medicine and architecture. Despite this, very little evidence related to public science communication can be found. After Thailand became democratised in 1932 and during the reign of King Rama IX (King Bhumibol, 1946–2016), science and science communication were again introduced, not for the sake of science but for improving people's lives and well-being in terms of health awareness, agriculture and land use, and environmental improvements.

3. Modern science communication era in Thailand

Efforts to communicate science by the scientific community and government bodies began around 1934/35 when a group of science scholars from the Faculty of Arts and Science, Chulalongkorn University (the first university in Thailand), started the first science club or science association. The first chairman of the club was Dr Charoen Thampanich, but it was when Dr Klum Watcharobol, a well-known young scientist who in 1935 had just returned from the United Kingdom, joined Thongsuk Pongsathat (the second chairman) that the science club became more active in science communication activities. The aim of the two co-chairs was to ensure scientific activities were made available to the public. The major activity of the club in 1936 was to organise scientific talks by both visiting professors and young scientists, and to promote and sell science books as part of fundraising efforts for the club. Later, the club was handed over from the Faculty of Arts and Science to the Chulalongkorn University Alumni Association. The club continued to organise meetings for members who were mostly scientists and scholars in the university, aiming to exchange ideas, communicate scientific knowledge and organise networking among Thai scientists. A few years later, a formal committee was formed and the club reached an important milestone: the launch the first science magazine in Thailand, the bi-monthly *Thailand Science Club Magazine*. The magazine covered stories of science both for academics and the public, as well as stories about the university's alumni, faculty and students. The first issue in 1947 is considered the first publication of modern science communication in Thailand. Amazingly, the magazine has continued to be published to this day (Figure 36.3) (Science Society of Thailand, 1998).

Until 27 January 1948, the science club was officially registered as a legal entity under the name Science Society of Siam (SSS), and later became the Science Society of Thailand with open membership. Members included people in science and science educators from universities all over the country, mostly scholars and academics. Membership to the society was open to associate members, science teachers and university students. In 1951, the royal patronage of King Rama IX was granted to the society. Currently there are about 3,000 members nationwide. The society has operated as the backbone of science activities in Thailand, and this year it celebrated its 70th anniversary. The activities of the Science Society of Thailand cover a wide range of areas to promote science to the nation's youth and the public, to strengthen the recognition of scientists and to create a scientific network. Science popularisation activities include issuing monthly science magazines, holding science lecture tours in schools and universities, promoting youth science competitions, sponsoring science radio programs, promoting scientific talent, supporting academic research exchange, encouraging the emergence of new communities of scientific disciplines, empowering science teachers and organising national and international science symposiums and exhibitions.

The Science Society of Thailand is involved in communicating with policymakers to drive science policies important to Thailand. It drives many important science promotion events, including the development of a nation-wide youth science project competition with the support of the Ministry of Education and the Ministry of Science and Technology; the initiation of the Science Olympiad of Thailand; and the creation and administration of science recognition awards such as the Annual Thailand Outstanding

Scientist Award, the Annual Science Teacher Award, and the Outstanding Senior Scientist Award. The keys to the success of the society are its genuine interest in science popularisation and its openness in involving stakeholders. It has been effective in uniting various partners, and this has contributed to a significant and long-term positive impact of science to Thai society. Although the Thai government has now established new agencies to be directly in charge of several science promotion activities, the legacy and the role of the society as a community of people with a common interest in promoting science are still important for the country's science activities.

Around the year 2000, with a national policy that encouraged science and technology, the concept of modern science communication was gradually introduced to Thailand by Thai academics and policy agencies. International experts in science communication were invited to give talks and professional training through university collaborations. Professor Susan Stocklmayer from the Centre for the Public Awareness of Science of The Australian National University and Professor Mike Gore from the Questacon Science and Technology Centre were among the first academics to introduce science communication theory and practice to Thailand through two workshops held at Chulalongkorn University and Khonkaen University. At that time, the training was offered to university academics, teachers and science museum professionals.

In 2005, at a seminar of the 31st Congress on Science and Technology of Thailand organised by the Science Society, the National Science and Technology Development Agency (NSTDA) raised the importance of the role of the media and the engagement of scientists in science communication. NSTDA proposed that in addition to public relations, science communication should be promoted to connect scientists and the public, disseminate scientific and technological knowledge to the people (especially those who lack access to information) and strengthen public understanding of science. NSTDA also proposed drafting a science communication curriculum for science students at university as a long-term partnership project between the Ministry of Education, the Ministry of Science and Technology and the Ministry of Information and Communication Technology, saying it would take 10 years before the curriculum could be implemented. In the meantime, short courses on science communication were planned for journalists and related professions. In 2007, NSTDA opened the Science Media Center as the active body in promoting science communication and bridging scientists, especially in NSTDA, to the public through science talks, media communication and other activities. This enthusiasm for these proposals has encouraged a lot of activities in science communication and inspired a number of people interested in science journalism and informal science education to work in different ways to promote science communication. This includes people working for museums and science centres, who later took very active roles in promoting science communication activities and developing science communicators.

Even with the tremendous optimism and enthusiasm shown by various organisations, after almost 20 years a curriculum on science communication is not yet available for a full degree at most universities. In 2014, Srinakharinvirot University announced a master's degree of arts in science and health communication and began the course in 2016. Other universities have started to offer science communication as a three- to six-credit elective in their faculty of science or education or other faculties. Chulalongkorn University, for example, offers science communication in the Faculty of Science, Department of Environment Science; whereas Rajamongkol University offers science communication as an elective for science and education students at its Thanyaburi Campus. Attempts have been made by various universities to open full programs on science communication but there remain two main barriers. Although many universities are interested in offering a science communication degree, the first barriers are regulations set by the Office of Higher Education Commission that require highly qualified professors for a new curriculum or degree. This has delayed attempts to offer this degree and, as of 2020, although many universities have become more interested in science communication, it is still only offered as a three-credit course in either the faculty of education or faculty of science. Other programs more or less cover science communication, including the "Media and Communication Innovation Program" in the Faculty of Education, Kasetsart University, or the "Bachelor of Technology Program in Medical and Science Media" in the Faculty of Medicine, Mahidol University. The second barrier is uncertainty about the work and career path of a science communicator. Where could graduates find employment, and what work would they be doing? At the moment the number of students enrolled even in elective courses is still low. However, most students pursuing a teaching career will be more interested in the subject.

Short courses in science communication seem to be more successful. They are usually organised by institutions rather than universities and run by experts in practice rather than academic institutions due to the lack of professionals in the area. Organisations that are regularly involved in offering such training include NSTDA, the National Science Museum (NSM) and the Science Society of Thailand. Trainers often come from other countries such as the United Kingdom (British Council), Germany (Goethe Institute), Australia (Australian Embassy) or the European Union project on science awareness program. At this moment, we estimate that less than 100 Thai scholars in science communication, with degrees from the United Kingdom, France, Australia, New Zealand and the United States, currently work in Thai universities and the Ministry of Science and Technology.

4. Development of science museums and science centres in Thailand

The first science museum in Thailand was initiated with the involvement of the Science Society of Thailand. Field Marshal Plaek Phibunsongkhram, then prime minister of Thailand, presided at its opening on 25 June 1954. The museum had a mission to promote science and education for Thai society, and began as a natural history museum where specimens and collections of plants and animals and archives of science books were displayed and stored. In an effort to make it sustainable, the Science Society transferred the operation of the museum to the Faculty of Science, Chulalongkorn University, in 1956. Unfortunately, resourcing continues to be an issue, and although the museum contains a valuable collection of natural science exhibits, the faculty does not have the resources to promote the museum to a wider audience but instead caters more to small groups of visitors (Science Society of Thailand, 1998).

In 1962, the Ministry of Education established the Bangkok Planetarium, the first planetarium in the middle of Bangkok; and in 1975, the Center for Education Museum was established next door. The aim of the Education Museum is to promote lifelong learning through informal experiences, and the exhibition gallery displays a collection of scientific equipment, basic science exhibits and astronomy equipment. In 1992, the Planetarium and the Center for Education Museum were combined and collectively named as the Science Center for Education, operating under the Department of Non-Formal Education in the Ministry of Education. The combined centre focuses on lifelong learning, both non-formal and informal, to serve the people of Thailand. Taking a holistic approach to learning, the centre focuses on science, technology, natural science, environmental science and astronomy. Many scientists have shared their experiences of visiting the planetarium and the centre when they were young, and how this inspired them to pursue their careers in science. Until now, more than 15 small science centres for education had been developed in Thai provinces. These provincial science centres play important supportive roles to local schools, providing extra-curricular experiences in science and extra classrooms for students in the non-formal education system.

In 1995, the NSM was established by the Ministry of Science and Technology. The organisation is a state enterprise with the mission to promote public awareness of science through exhibitions, research and collections. NSM's mission focuses more on science for the public and the relevance of science to daily life for the development of the nation. NSM operates a series of science and science-related museums, including the National Science Museum (shown in Figure 36.4), the Natural History Museum, the Information Technology Museum and the Rama IX Ecology and Environment Museum. The National Science Museum has become a key player to connect various research and science agencies under the Ministry of Science and Technology as well as other ministries, universities and societies to drive science popularisation and science communication in Thailand.

Figure 36.2: Night view of the iconic cube building of the National Science Museum, located in Pathum Thani province, 40 km north of Bangkok, which was opened to the public in 2000. The building was especially designed to present a new image of a modern science museum.

Source: Science Society of Thailand under the Royal Patronage of His Majesty the King. © National Science Museum, Thailand (used with permission).

This became the turning point, allowing science communication in Thailand to leap ahead. NSM provides many attractive channels of science communication to communicate science to the public, such as interactive exhibitions, science activities, shows and dramas, laboratory programs and walk rallies (see Figures 36.5 and 36.6).

Even though science communication centres in Thailand have been well spread geographically, many people in the countryside cannot access these essential facilities and tools due to travel limitations. It is difficult for people from remote areas to get to the museums. Therefore, in 2002, the National Science Museum decided to launch the Science Caravan Project. It includes interactive exhibitions, activities, games, shows, laboratories and fun activities for children. Moreover, in some places, the caravan offers a science camp, science training workshops, and teacher-training programs. By using various types and styles of science communication at the same time and place at the caravan site, information can be relayed to a wider target audience. This project is one of the most effective ways to communicate science to people in rural areas and around the country.

NSM promotes science communication and trains science communicators in Thailand by providing a number of workshops to researchers, educators, museum staff and students. It also organises contests and competitions on science communication, such as drama contests, film contests, writing contests and university student programs in science communication. The Young Thai Science Ambassadors program has been continuously carried out since 2004 and this is considered a prestigious program that has inspired and created a network of young science communicators in Thailand.

5. Science festivals in Thailand

The first science festival in Thailand was initiated with the aim of promoting science popularisation and scientific culture. In 1968, on the 100th anniversary of King Rama IV's Solar Eclipse Event, the Science Society of Thailand started a large-scale public science event with new technology and modern displays. This was organised at the Science Center for Education and presided over by His Majesty King Bhumibol (King Rama IX). This festival also provided communication between Warkor and Bangkok to emphasise the connection of this special event with the important historical astronomical phenomenon.

In 1982, the Thai government accepted a recommendation from the Science Society to set 18 August—the day of the eclipse predicted by King Rama IV—as National Science Day. The first science festival in 1982 had all sorts of activities to commemorate the 'Father of Science': the opening ceremony, the presentation of an award for outstanding scientists and science teachers, the Talents in Science program, a science exhibition by various science agencies, science project competitions and science lectures. The event was well-covered by the media in both Thai and English. The success of the first festival has led to sponsorship and support from the private sector to fund awards for youth science programs. In 1983, the Minister of Science, Technology and Energy joined the National Science Day event and saw it as a good opportunity to promote science to the public. He appointed the Ministry of Science to extend the event from one day to a week (National Science Week, 18–24 August), made it an annual event and allocated a budget so that the activity could be more widely implemented (Chen, 2014).

In 1997, the event moved to the Sirikit Convention Center, where it could accommodate more people and, for the first time, activities were organised in 23 provinces nationwide. This has significantly changed the atmosphere of science popularisation in Thailand and increased our ability to reach a wider audience. Since 1982, August has become the month of science both for academics and the public. The Ministry of Education, through the Science Society of Thailand, supports universities to organise youth science activities during Science Week, while the Ministry of Science and Technology provides additional funding to approximately 18 major universities to assist in the organisation of an open house for local science events under the national theme. All science education centres, science research agencies and faculties of science in universities have become the central points in organising science events during the week. Science Week activities are also implemented in schools throughout the country.

Figure 36.3: The National Science and Technology Fair, the largest public science event organised annually every August.

Source: Science Society of Thailand under the Royal Patronage of His Majesty the King. © National Science Museum, Thailand (used with permission).

In 2008, the Ministry of Science and Technology appointed the National Science Museum to organise the National Science and Technology Fair, a mega-size science festival (40,000 square metres) in the Central region. It had cutting-edge technology and thematic exhibitions relevant to global issues such as climate change, energy, environment and health (Figure 36.8). The festival included international participants organised through embassies and has been recognised as a flagship project in science popularisation. In 2018, the Ministry of Science and Technology decided to promote science to support its initiatives in creating innovation. This was known as the Thailand 4.0 policy in creating innovation, and it was given a special budget to organise further mega-scale science festivals in four other locations outside Bangkok, at Chiangmai, Khon Kaen, Songkhla and especially Warkor, where the big celebration returned Thai science graciously to its origin. The science festival has reached more than 1.5 million people nationwide and is organised by the Ministry of Science and Technology and the Ministry of Education (National Science Museum, 2018).

6. Policy and future of public science communication in Thailand

In early 2000, the NSTDA with the National Statistical Office funded a study on public attitudes towards science and technology. The results were published in 2006 and, to no one's surprise, television was the most influential and effective mode in communicating science to the public. According to the report, only 20 per cent of survey participants could name a Thai scientist (National Statistical Office, 2006). Another study, by Hathayatham (2005), reported that most Thais could only name the past two kings of Thailand (King Bhumibol and King Mongkut) as scientists. This reflected the limited interest of the public in scientific research and the lack public information about Thai scientists. However, while most Thais still consider science and technology to be important to themselves and the country, pursuing a science career is not promoted or encouraged by parents for their children.

The NSTDA also funded a study conducted by a team from the Science Society of Thailand to evaluate science promotion activities and develop a national policy proposal on science awareness. The study concluded that in the past 20 years, most science activities were intended to promote public understanding and interest in science, with less effort placed on long-term human resource development in science and technology (Mongkolkul, 2006). The report proposed guidelines for enhancing public awareness of science as follows:

1. Target: more scientific activities should be planned to engage youths 18 years and older, people from medium-income families, and policymakers.
2. Tools: various tools in science popularisation should be employed such as print media and mass media, science museums and science festivals during the National Science Week, national science awards, national youth science awards, public science in science curricula, and 'scientists meet policymakers' sessions.
3. Infrastructure: basic infrastructure to help with the popularisation of science needs to be established. This includes the development of science communicators, creation of science media centres, engagement of science associations, and funding for science awareness on research and development.
4. Policy: to ensure the successful implementation of these guidelines, the government should support a system to include science awareness policies, the restructuring of management or responsible agencies, and budget allocation. In addition, science literacy surveys should be conducted on a regular basis to monitor change and set future directions.

At the same time as the study (and for the first time) the Ministry of Science and Technology included a strategy on 'promoting public awareness of science' as one of the key five strategies for the ministry's action plan for 2004–13 (National Science and Technology Development Agency, 2004). The strategy set goals to enhance public awareness and understanding about science and technology, ensure public use of science in daily life, and provide access to science learning resources in local communities. There were four measures under this strategy: to enhance learning and creativity among youths and the public; to promote the engagement of scientists and policymakers in communicating science; to develop more learning centres and resources in science; and to increase public access to science news and information.

After several governmental restructures, the National Science Technology and Innovation Policy Office (STI) became the key player in developing the national strategy. Science awareness and science communication were considered to be of less priority and mentioned only as a part of the action plan. The National Science Technology and Innovation Policy and Plan for 2012–21 places more emphasis on increasing competitiveness through research and development, preparing for change due to globalisation, supporting social equality and security through investment in science infrastructure and research funding, and developing human resources.

Science awareness and communication fall under human resource development but were given a lower priority than formal education or workforce capacity building (National Science Technology and Innovation Policy Office, 2012). Regardless of all the established policies, a recent report by the Organisation for Economic Co-operation and Development (OECD) still ranked science literacy in Thailand to be below the OECD average. The report blamed this on the quality of science education offered in schools and the limited availability of qualified science teachers in rural or difficult areas. With regards to government policy in promoting science, since 2015 the government has pushed forward the policy of Thailand 4.0 to advance the nation through science, technology and innovation. This has driven a demand for educational reform and the active participation of science agencies, as well as increased involvement by the private sector to promote science and science education. To date, there are science schools, special science classrooms and scholarships available. Informal science education and science communication, which at one point were considered trivial, are now more in demand, and agencies from both private and public sectors are moving in this direction and developing more programs to accommodate the demand.

An example is the International Science Film Festival launched in 2004 as a cooperative effort by the Goethe Institute with the Institute for the Promotion of Teaching Science and Technology, the NSM, and several universities and science centres. The operation extended to provincial schools and universities and had reached out to 600,000 students in 2017.

During the past decade, it has been encouraging to see many research agencies becoming more interested and involved in science awareness and science communication. There are fewer than 15 agencies under the Ministry of Science and Technology, more than half of which have recently established their own awareness program units. These include agencies such as the Geo-Informatics and Space Technology Development Agency, the National Science and Technology Development Agency, the National Astronomical Research Institute of Thailand and the National Science and Technology Policy Office. The National Science Museum has played a major role on behalf of the Ministry of Science and Technology as the national platform in coordinating all agencies within the Ministry of Science and Technology, as well as cross-ministries to organise national science communication and science popularisation programs all year round. These programs include the National Science and Technology Fair that reaches out to approximately 1 million people annually, science caravans and science and technology competitions.

A recent policy of the Ministry of Science and Technology encourages the engagement of young scientists in public science communication, especially those who had received national scholarships and are working in rural areas.

The National Science Technology and Innovation Policy Office provides funding to awareness programs from various agencies, to prototype science communication models adopted from the United Kingdom, the United States and Australia based on their international experience, models such as FameLab (Figure 36.9), SchoolLab, Maker Space, Science Idol, Science Ambassadors and others. These programs promote scientists in science communication and empower future scientists with communication skills.

Getting scientists involved in these programs is difficult. A report by Chen in 2017 revealed barriers to the engagement of scientists in science communication, including policy, time, confidence and skill. The government has to make tremendous efforts to encourage young scientists to contribute to science communication, but these efforts are unlikely to be successful unless the barriers are unlocked systematically.

In addition to agencies related to science and technology, communication of science and environment issues has been carried out extensively by other ministries: the Ministry of Education works to promote science, technology, engineering and mathematics (STEM) learning and STEM careers; the Ministry of Natural Resources and Environment publicises issues of biodiversity and environmental protection; the Ministry of Public Health and the Ministry of Energy work in the areas of their responsibility. Numerous universities have started to encourage their researchers to be more engaged in science, and medical universities, such as Mahidol University, have launched online and science channels on cable television for the public.

The employment of professionals or trained science communicators is still in its early stage in Thailand. The role and importance of being a science communicator is unclear due to the lack of understanding of the profession. Most people working in the field are involved in public relations, as scientists or educators, as shown in Figure 36.10. Very few agencies are aware of this profession and the differences between being science communicators and science educators.

7. Science communication in the future

Due to the government's strong will to promote science, technological research and industry, on 2 May 2019, the Department of Higher Education merged with the Ministry of Science and Technology, giving birth to a new ministry called the Ministry of Higher Education, Science, Research and Innovation. This signalled the national need to strengthen and integrate research units both in academic and other public sectors. The growing number of science and technology–related agencies provides a signal that science communicators will be in demand. While the infrastructure to produce more science communicators domestically is still questionable, there are scholarships for Thai students returning from abroad (for example, from the United Kingdom and Australia), who are trained and ready to kick off the professional development of science communicators. The group of students who returned between 2001 and 2018 could be the hope for the future of science communication in Thailand.

Figure 36.4: Research show by a naturalist, one of the most popular programs of the National Science Museum where scientists of the museum communicate science to the public.

Source: Science Society of Thailand under the Royal Patronage of His Majesty the King. © National Science Museum, Thailand (used with permission).

Since the government's policy in 2018 was aimed at promoting science and innovation to raise the competitiveness of the nation, high levels of investment have been made into building new science research facilities and industrial incubators. The government has invested more on learning resources such as museums, learning centres and online media. It seems inevitable that science communication will be in greater demand in the future.

Although science activities have been promoted in the past 10 years, it was obvious that the biggest effort was still focused on student engagement. To really reach out to the general public, increased effort, better techniques and more resources will be required in order to be successful. Five barriers have been identified:

1. The public perception of science is that science is not for adults. Some people find science difficult but others recognise the importance of science, especially for students to study in schools and get high grades in examinations so they can have good future careers.
2. In order to communicate science to adults in a stimulating and interesting manner, it is important to gain knowledge from scientists themselves as they are aware of interesting scientific facts and evidence. Scientists from universities or research institutes can share their knowledge by storytelling to science journalists, the media and museums, which makes it critical to involve scientists in science communication.
3. Thailand needs more science communicators and better-trained science communicators if it is to make science interesting to the public. This can be a weakness and an opportunity at the same time. Currently, it is obvious that there are not enough well-trained science communicators due to the non-availability of study programs at both undergraduate and graduate levels in Thai universities. Therefore, the attempts made by research agencies to communicate science have been ineffective. As the need of science communicators increases, this will provide capacity building and career opportunities in this area.
4. Science communication at the national and organisational levels is still not strongly encouraged. In Thailand, even though the government is generous in terms of providing adequate budgets and investing in infrastructure, various research organisations should allow their scientists to be more freely engaged and involved in public science communication. In countries such as China and Korea, the policy on science communication is strongly supported as part of a national strategy of promoting scientific innovation in a technology-based society. In China, science popularisation is identified to have the same priority as research support.
5. There is limited involvement of journalists and public media in science communication. Journalists do not feel that there is enough support from the scientific community. The lack of cooperation or communication with scientists and the limited scientific information that is generally available make it difficult for journalists and the media to present scientific information of enough interest to attract public attention. Moreover, the high cost and complicated production of scientific content has discouraged the media from getting involved in science communication.

Despite all the barriers, the future of science communication in Thailand looks promising. Due to future changes in technology, society and environment, it is likely that science communication will become an emerging topic of interest. Possibilities include having online platforms to communicate science as well as having decentralised or localised science communication modes. This will ensure that the majority of Thais would have access to science on demand, especially scientific information pertinent to their everyday lives. In order to make science more relevant and valuable, and to move the nation forward through science, technology and innovation, Thailand needs to create an ecosystem with the involvement of all stakeholders, so that science is integrated in people's lives.

References

Chen, G. (2014). National Science Festival of Thailand: Historical roots, current activities and future plans of the National Science Fair. *Journal of Science Communication*, *13*(4). doi.org/10.22323/2.13040304.

Chen, G. (2017). Approach in Engaging Scientist in Science Communication. Paper Presentation on Academic Forum 'Dual Upgrading of Theory and Practice in Science Communication', Beijing, China.

Chen, G. (2018). Engaging scientists in science communication (poster). Public Communication of Science and Technology Conference. Dunedin, New Zealand.

Hathayatham, A. (2005). The Public Awareness of Science in Thailand: A Case Study in Biotechnology (PhD thesis). The Australian National University.

Mongkolkul, K. (2006). *Promoting Public Awareness of Science and Technology for the Thai Society.* Bangkok: National Strategy, National Science and Technology Development Agency.

National Science Museum (2018). *National Science Museum Annual Report 2018, Thailand.* Retrieved from www.nsm.or.th/images/Annual_Report_2018.pdf.

National Science and Technology Development Agency (2004). *National Strategy in Science and Technology 2004–2013.* Bangkok.

National Science Technology and Innovation Policy Office (2012). *Science, Technology, and Innovation Plan 2012–2021.* Bangkok.

National Statistical Office, National Science and Technology Development Agency & Science and Technology Awareness Subcommittee. (2006). *Youth Interest in Science and Technology Report Year 2006.*

Office of Science Technology and Innovation Policy. (2012). *Science Technology Innovation Policy and Plan I:2012–2021.* 1st edition. Bangkok, Thailand: Office of Science Technology and Innovation Policy.

Science Society of Thailand. (1998). *50 Years Science Society of Thailand under Royal Patronage of His Majesty The King.* Bangkok, Thailand.

Science Society of Thailand. (2018a). *70 Years Science Society of Thailand under Royal Patronage of His Majesty The King*. Bangkok, Thailand.

Science Society of Thailand. (2018b). *Warkor: Past-present-future*. Bangkok, Thailand.

Timeline

Event	Name	Date	Comment
First interactive science centre established.	The Center for Education Museum	1975	1992: name was changed to Science Center for Education
First national (or large regional) science festival.	Science festival, organised at the Science Centre for Education and presided over by His Majesty King Bhumibol	1968	National Science Day established 18 August 1982 and celebrated with festival
An association of science writers or journalists or communicators established.		1947	
First university courses to train science communicators.	Master's degree of arts in science and health communication at Srinakharinvirot University	2016	Short courses offered by institutions rather than universities are more successful
First national conference in science communication.	Science communication workshops	2000	2008: Hhosted the ASPAC (Asia Pacific Network of Science and Technology Centres) conference
National government program to support science communication established.	The National Science Museum Project	2000	
First significant initiative or report on science communication.	National policy on the development of public awareness of science and lifelong learning.	2006	While there has not been a report specifically on science communication, the 2006 report did refer to these matters
National Science Week founded.	National Science Week celebrated with festival, lectures, exhibitions	18 August 1982	National Science Day was established in 1982, and was extended to a National Science Week in 1983
A journal completely or substantially devoted to science communication established.	*Science*, a monthly journal established by the Science Society of Thailand	1948	*Science* communicated news, information and events to the public and members of the Science Society

Event	Name	Date	Comment
First significant radio programs on science.	Science Society of Siam (later Science Society of Thailand) established a science radio program	1950	This program was produced as a talk with questions. The first few stories related to oil, scientific disciplines, seasons in Thailand, etc.
First significant TV programs on science.	*Asset of the Land* was the first science film produced by the Science Society of Thailand	1960	It was screened in a movie theatre
First awards for scientists or journalists or others for science communication.	The first award for scientists was announced on the first national science day	18 August 1982	
Other significant events.	The first Science Lecture Tour in 1965	1965	

Contributors

Ganigar Chen is vice president of the National Science Museum (NSM), Thailand.

Dr Wijitra Suriyakul Na Ayudhya is the director of Research Division, Information Technology Museum (ITM) at the National Science Museum (NSM), Thailand.

Chanin Suriyakul Na Ayudhya is the head of the Science Communication Division at the National Science Museum (NSM), Thailand.

37

TURKEY

From the Ottoman Empire to the Republic

Gultekin Cakmakci and
Sevinc Gelmez-Burakgazi

1. Historical development of science communication in Turkey: Summarising crucial milestones

The scientific and industrial revolutions that emerged from the mid-18th century onwards divided countries into two streams: the developed and the less developed. Science communication became a key feature of developed countries. Partly due to other priorities, Turkey joined this trend rather late. This section presents the history of science communication in Turkey by dealing separately with the Ottoman Empire and the entity that succeeded it, the Republic of Turkey. The main focus will be on the Republic, which began in 1923.

In the Ottoman Empire's policies, applied sciences (engineering, medicine) were given priority in order to meet military, medical and agricultural needs, followed by physical sciences (chemistry, physics, astronomy, etc.) and mathematics to support the applied sciences (Dursun, 2009). The social sciences had a later stage of development than the applied and physical sciences (Dursun, 2009).

The Empire period spanned more than 600 years starting in 1299 as a multinational reign including *millets*,[1] like Muslim, Armenian, Syriac and Roman. During the reign, especially in the early establishment years, studies

1 In the Ottoman Empire, a *millet* was an independent court of law pertaining to 'personal law' under which a confessional community (a group abiding by the laws of Muslim Sharia, Christian Cannon law or Jewish Halakha) was allowed to rule itself under its own laws.

of science were encouraged. Istanbul Observatory (1557–60) and nearly 300 *medreses* (learning centres) were established and welcomed students from all over the world to study and conduct research in mathematics, astronomy, philosophy and religious sciences (Ihsanoglu, 2007). The first higher education institute, Mühendishane-i Bahri Hümayun, was established in 1773 to train engineers and soldiers.

From the 17th century onward, the Ottoman Empire began to lose its power. In an attempt to regain this power, a modernisation project with a European influence was put into practice: Tanzimat (1839–76). In the Tanzimat period (meaning reorganisation) many social, military, political, economic and educational reforms similar to those in the West were promulgated. Examples of social and educational ones are as follows: the first public education system, the first modern university (Darülfünun), establishment of a private press sector, the first telegraph. From the beginning of 1800s to the 1900s there was a remarkable investment in scholarships program that sent students to Europe for education (Gencoglu, 2008; Sisman, 2004). Most of these students, especially those sent in Tanzimat period, were studying basic science.

Despite all these efforts at reform and modernisation, the Ottoman Empire came to an end after World War I, when it was replaced by the Republic of Turkey in 1923. The new state, formed according to the nation-state model, would build new institutions and renovate old ones. The new government, under the presidency of Mustafa Kemal Atatürk—a towering figure of the 20th century—carried out crucial reforms in different spheres, including science. In the field of education, many reforms were introduced to inform the public in various fields and to teach them to read and write. These include alphabet reform, secular education, women's rights, village institutes and *millet mektepleri* (society schools). The new era with Atatürk began with high hopes for modernisation, a society with a strong science-based foundation, and a science-literate community. He fought to build a democratic, strong and modern state based on science: 'Science is the most reliable guide for civilisation, for life, for success in the world. Searching a guide other than science means carelessness, ignorance and heresy' (from Atatürk's speech to teachers in Samsun in 1924) (Atatürk Research Center, 2019).

In these early years of the Republic, educational reforms were made so that science and technology could develop. Reforms in industry and the economy followed education.

The first systematic movement in education was the Maarif Kongresi (Congress of Education) in 1921. The aim of this congress was giving 'a national direction to education' (Akyuz, 1983). The second step was Tevhid-i Tedrisat Kanunu (Law on Unification of Education). By means of the law, in 1924, all institutions (e.g. religious, secular, foreign) were gathered under the umbrella of the Ministry of National Education. In these times, the economy was mostly based on agriculture. The steps to be taken for the purpose of development were determined in 1923 at the first economic congress in Izmir. In this congress, Atatürk and deputy minister of economy Esat Bozkurt underlined the necessity of economic development and growing the national economy. In a public speech in Alaşehir in 1923, soon after the Izmir Economy Congress, Atatürk said:

> Following the military triumph we accomplished by bayonets, weapons and blood, we shall strive to win victories in such fields as culture, scholarship, science, and economics … the enduring benefits of victories depend only on the existence of an army of education. (Atatürk Research Center, 2019)

In parallel with these thoughts, in the early years of the Republic, there was an emphasis on the principles of science and technological transformation as never before seen in the history of Turkey (Bahadir, 2001; Inonu, 1999). By means of technology transfer, sugar and cement factories were built. In 1924 the High School of Mining Engineering in Zonguldak was established. Due to the worldwide 1929 economic crisis, the Second Economy Congress convened in 1930 adopted a statist model (where the state has substantial centralised control over social and economic affairs). Despite the worsening economy in the crisis, in the 1930s Turkey sent students (mainly engineering students and technical personnel) to Europe for education, to train them to work in industry.

One can say that few nations in the world have ever experienced the massive changes that Atatürk triggered in Turkey in such a short period of time. The Atatürk era began with high hopes for science and a science-literate community. However, these attempts were not enough to match modern Western science and other issues and turbulent times pushed science off the national agenda. The young state of Turkey had to focus on the need to build itself quickly. During the Republic's early years, science and technology transfer was seen as an easy and practical way to help achieve this aim. Yet, unfortunately, the importance and meaning of technological production based on science was not sufficiently comprehended even in the 21st century. Science was generally a low priority and despite the good intentions contained in many five-year development plans, national investment in science has continued to be low. It was not until 1963 that a national agency TUBITAK (the Scientific and Technological Research Council of Turkey) with a responsibility for science and science communication was created. Science coverage in the media has been weak and there are only few programs to train science journalists and science communicators. Nonetheless, there are promising developments: the number of science centres has increased rapidly over the years, and research in science communication is a new but active field. In the following sections, these issues will be discussed in detail.

1.1. Science policies and funding bodies in science communication

In order to talk about the emergence of science communication in the Republic of Turkey (after 1923), one might begin a chain with the establishment of the first university. The University Reform Act in 1933 can be considered the first link. After this Act, science in Turkey flourished (Inonu, 1999). In the same year, Istanbul University (formerly Darülfünun) was established. These were the first steps in the institutionalisation of science. Following Istanbul University, Gazi University (formerly Gazi Teacher Training Institute) (1926) and Ankara University (1946) were founded as initiatives of Atatürk. Education was considered important as a transformation agent to develop Turkey into a 'modern' society.

With the 1946 law on universities, administrative autonomy and legal status were given to universities and their duties were documented. This documentation gave a framework to the universities in their responsibilities for communicating science. At this time, Turkey was lagging behind many other Western countries in its institutionalisation of science. The institutionalisation of science in Turkey was clearly established with the formation of TUBITAK in 1963. This was the second link. In this manner, science was coordinated by a structure separate from universities. For Türkcan (2001), the establishment of TUBITAK is the start of science policy in Turkey. The purpose of TUBITAK was described as follows:

> In Turkey, to develop science-based research and development activities according to the priorities of the country's development, to encourage, organise and coordinate the activities and to access the available scientific and technical information and ensure availability (Official Gazette, 1963, p. 7).

Another important link in the chain of science policy in Turkey was the OECD Science and Development: Pilot Teams Project. The aim of this project was to highlight scientific activities as an important factor in economic growth and promote the idea that these activities should be the subject of a planned policy at the national level (OECD, 1967). Historical and economic analysis indicated that Turkey needed to formulate research facilities to manage the dynamics between science, technology, manufacture and development, with important implications for communication.

1.2. Media and science journalism

The first newspapers of the republican era, *Emel* (published in Amasya in 1920), *İntibah* (published in Bursa in 1921), and *Küçük Mecmua* (published in Diyarbakir in 1922), mainly covered political news rather than science. When science journalism of the Republic of Turkey is considered, given the decrease that occurred in the number of pages in newspapers in the 1930s, it appears that science news was mostly affected negatively (Kologlu, 1997). In 1945, with the transition to multi-party life on the political scene, science and technology news was neglected (Kologlu, 1997). In any case, the media is not designed to build public engagement in science.

Ten months after the announcement of the Republic in 1923, *Muallimler Mecmuasi* was the only scientific journal (Bahadir, 2001). The contents included subjects such as atoms, heat and mechanics. Over the next few years of the Republic, between 1923 and 1928, six science journals were published: *Muallimler Mecmuasi*, *Darülfünun Fen Fakültesi Mecmuasi*, *Mühendis Mektebi Mecmuasi*, *Kimya ve Sanayi Mecmuasi*, *Fen Alemi Dergisi* and *Tabiat Alemi Dergisi*. *Darülfünun Fen Fakültesi Mecmuasi* was the first new scientific journal of the Republican era. Among these journals, *Fen Alemi Dergisi* (January 1925 – December 1926) was the first popular science journal in Turkey. Subjects covered in the first issue of this journal included airways, electricity in the houses, white coal and new style ships. In the same year, 1925, a second popular science journal was published in Turkey: *Tabiat Alemi Dergisi*. In the following decades there were no other popular science journals besides these two until the TUBITAK publication *Bilim Teknik* [Science Technology] started in 1967. Today, TUBITAK, with four high-circulation popular science journals, continues its successful dissemination activities.

Media tools are important in communicating science. Radio broadcasting began in 1921 and TV was introduced to Turkey in 1952. The first broadcasts were basically culture and art programs. In the following years, with an increasing number of channels, science news—mostly translated from international news agencies—appeared on TV. When it comes to science communication in the media, there are now many popular science journals (for example, *Herkese Bilim ve Teknoloji*, *Bilim ve Teknik*, *Bilim Çocuk*, *Merakli Minik*, *Bilim ve Ütopya*, *Bilim ve Gelecek*, *Eğlenceli Bilim Dergisi*) and also science, technology and innovation news in newspapers, but most newspapers do not have separate section for science. Dursun, Becerikli and Dursun (2010) investigated the visibility and representation of science news in printed media in Turkey between 1993 and 2008. They analysed a total of 4,568 science news items in three high-circulation Turkish newspapers. According to the results, science news in the newspapers mainly covered medicine, biotechnology and health issues followed by astronomy, technology and nutrition news. Besides traditional media, another important initiative in social media channels is the 'Science Communication Platform of Turkey' and on the internet, bilimiletisimi.com.

1.3. Universities and public talks

When it comes to academic studies, there are not any specific departments to train professional science journalists (Becerikli, 2013), nor a national association for science journalists, nor a science news agency. There is not a specific department called 'science communication' in universities. However, there are some master's and PhD programs on issues related to science communication such as Science and Technology Policy Studies in the Middle East Technical University (METU), and a non-thesis program at Ankara University, called Science and Society Studies. The mission of the Science and Technology Policy Studies PhD program is explained in the following way on the department webpage (METU, 2018):

> PhD Program in Science and Technology Policy Studies is supported by various disciplines such as economics, administrative sciences, engineering, sociology, history, philosophy, communication and cultural studies … Recent developments in the knowledge-intensity of economic activity and rapid technological advancement have significant socio-economic repercussions at the level of nation states, regions, industries, markets, and firms. In this context the program aims to confront the challenges by providing several concentration areas for policy making.

The first international science communication conference in Turkey, the 14th Conference of the Network for the Public Communication of Science and Technology (PCST), co-chaired by Prof. Dr Gultekin Cakmakci and Brian Trench, was hosted in 2016 in Istanbul, Turkey. More than 400 international science communication scholars, researchers and practitioners participated in the conference. This was a unique experience for the host country and an important initiative to increase the awareness of science communication in Turkey. Prof. Dr Erkan Yuksel, Anadolu University, with his research team have been running an annual Health Communication Symposium[2] since 2015; however, there is not yet a national conference on science communication.

2 saglik-iletisimi.org.

Lastly, there are the Cafe Scientifique talks, first conducted by TUBITAK in 2014, which focus on young people at school-level. These activities—where expert speakers give information and are then asked questions by the participants—include elements of both the 'dialogue' and the 'deficit' models (Trench, 2008); nonetheless, elements of the 'participation' model also need to be considered in public engagement initiatives in Turkey.

2. Science communication policies and funding bodies in science communication

Turkey's longstanding aim, as articulated by Atatürk, has been 'to reach the level of contemporary civilisation'. In order to accomplish this aim, science and technology is important, as well as a science-aware community. The importance of the development in science and technology is appreciated and valued in Turkey; nonetheless, the importance of science communication in this process needs greater recognition (Bursali, 2000). Turkey has made plans for science (and, by implication, for science communication) but it has lacked the political will and the resources to carry out these plans effectively.

In the 1960s, science and technology policies were added to the agenda for economic and social development during the second Five-Year Development Plan period (1968–72) (State Personnel Presidency, 1969). The establishment of TUBITAK in 1963 was an important step toward the institutionalisation of science and technology. TUBITAK is an advisory agency to the government and responsible for science and technology policymaking together with the Supreme Council for Science and Technology. Its missions and responsibilities are:

1. to advance science and technology
2. conduct research and support Turkish researchers
3. promote, develop, organise, conduct and coordinate research and development in line with national targets and priorities (TUBITAK, 2018).

With the Division of Science and Society in TUBITAK, the following activities are coordinated:

1. to promote scientific literacy among the public
2. to raise awareness of science
3. to instil a culture of science, technology, and innovation (TUBITAK, 2018).

To this end, TUBITAK organises and supports activities by governmental institutions and universities that will engage the public with science. The other activity field of the Science and Society Division is popular science publications, as stated earlier.

Development plans including national and international goals and objectives are important in the history of Turkey. From the third Five-Year Development Plan (State Personnel Presidency, 1972) onward, the emphasis on technology is of note. In the fifth Five-Year Development Plan (State Personnel Presidency, 1984), 'keeping up with the developments in science and technology to mirror the rapid changes occurring in the world' (p. 15) is included. TUBITAK and universities were made responsible for achieving this aim.

In 1983, the Supreme Council for Science and Technology was established. The role of the council has an implied responsibility for science communication, which was explained in the following way:

> [the council] is the highest-ranking Science and Technology and Innovation policy-making body in Turkey with decision-making power and the role of identifying, monitoring and coordinating policies in Science and Technology areas in accordance with national goals for economic and social development and national security (TUBITAK 2010, p. 6).

The Supreme Council did not achieve its expected outcome. Minister of state Dr Nimet Ozdas (2000, p. 40) explained its lack of effectiveness:

> It was expected that this system would gain momentum as Science and Technology would enter the political agenda of the country with the effective operation of such a board. Unfortunately, this board held its first meeting in 1989, six years after its establishment. Thus, in Turkey is considered spent in vain one of the world's most precious resources in terms of both science and technology: time.

The first official policy document on science, the Turkish Science Policy 1983–2003 (TUBITAK, 1983), is a well-prepared and important document with the participation of various stakeholders. But the committee organised by the Turkish Supreme Council for Science and Technology to implement the policy efficiently failed to meet for many years and the report has never, in fact, been applied thoroughly. This negative situation regarding implementation can be blamed on reasons such as the failure of stakeholders to carry out their roles properly, different government priorities and problems in political stability, media indifference (as the main topic of those years was elections) and limited dissemination. Over the years, the various plans proposed a combination of deficit and dialogue model approaches.

During the 30-year period from 1963 to 1993, Turkey lacked an active, efficient and systematic science and technology policy (Yalcin and Yalova, 2005). In 1960s and 1970s, the science and technology policy in Turkey was mainly based on the promotion of research in the natural sciences (Saritas, Taymaz and Tumer, 2006).

To overcome these problems regarding the implementation of the plan, the Supreme Council for Science and Technology decided to prepare a project called Vision 2023 Technology Foresight. The project began by examining education, and the education system in Turkey was put under the spotlight in terms of quality and quantity (TUBITAK, 2005b). The report highlighted educational issues and also considered extending the involvement of the wider society and sectors in the economy, the level of political support, the integration of science and technology policies and other sectoral policies (Saritas, Taymaz and Tumer, 2006).

Like the *Turkish Science and Technology Policy 1993–2003* report (TUBITAK, 1993), the 1983 plans functioned to a limited extent. The stated aim in the 1993–2003 report is to catch up with world technology. In order to achieve this aim:

> the full and special production factors in the country should be made available to the educational system, and the scientific and technological research-development system should be improved with the most efficient use of country resources (TUBITAK, 1983, p. 8).

In terms of the emphasis on education, it is possible to trace the 'deficit model' effect and education orientation in the report. In the 'deficit model', education is considered the most fundamental way to eliminate the lack of knowledge. Besides education, the report also emphasises the importance of research and development facilities.

In 2004, TUBITAK published another document, *Vision 2023: Science and Technology Policy*. In this report, public awareness and public participation were given official emphasis: 'In every segment of society, awareness-raising ... should be coordinated, and systems should be established to ensure wide participation in such activities' (TUBITAK, 2004, p. 32). Within this aim, attention is drawn to the mission of education. Therefore, the report has signs of the 'deficit model' and 'dialogue model' as discussed by Bucchi (2009) and Trench (2008).

In 2005, TUBITAK published its *Science and Technology Policies Implementation Plan*. In this plan, some actions are identified as being of considerable importance in developing an awareness of science and technology and 'encouraging the active participation of social actors in decision-making processes' (p. 5). This kind of active participation might be evidence of the 'dialogue model'.

One of the most recent plans, TUBITAK's *2018–2022 Strategic Plan* (2018), highlights the importance of expanding the culture of science, technology and innovation in society and increasing awareness. Not surprisingly, it is possible to see traces of common elements in science/technology policy and education policy. For instance, it is noteworthy that in the Ministry of National Education *2015–2019 Strategic Plan* (2015), the idea of forming science classes and organising science fairs so that students will be able to evaluate events and facts from a scientific point of view shares common ground with the public focus in recent national science and technology plans and reports.

In today's Turkey, there is no report that specifically uses the terminology of 'science communication'. Instead, we can trace science communication as 'science and society' in the policy documents. Although there are some awards for science writing (for example, the Sedat Simavi Award) and funding bodies like foundations and associations, government and public support for science communication is limited. Nevertheless, TUBITAK encourages scientific publications and research with a program initiated in 1993 (Arioglu and Girgin, 2003), and Bilim Akademisi (Academy of Science) and TUBA (the Turkish Academy of Sciences) have awards, grants and projects. TUBA defines part of its mission as:

> to give direction to the science policies of our country, to give all stakeholders science-based consultancy service, to encourage science and scientists, to make people adopt scientific thinking, to work for making 'Turkish' a science language, and to fortify international scientific collaboration representing our country internationally.[3]

3. Science communication in informal environments

3.1. Science centres

Science and technology centres were included in development plans in 2001, with the eighth Five-Year Development Plan (State Personnel Presidency, 2001). The following statement from the plan was important in explaining public engagement in science: 'Interactive Science and Technology Centers will be established and developed in such a way as to support formal education, in order to make science, technology, and society come closer together.'

Looking at the nature of museums in Turkey, there has been a wide variety of museums in different fields, but the first science centre, Feza Gürsey Science Center, was established in 1993 in Ankara. This science centre was established and run by

3 See www.tuba.gov.tr/tr.

Ankara Municipality. Partly based on this successful initiative, other municipalities in different cities and towns started to do public engagement activities in science and technology. Table 37.1 shows the science museums, science centres, observatories and planetariums established thereafter in different cities.

Table 37.1. Main science centres in Turkey.

Name	Established	City	Mainly funded/run by
Feza Gürsey Science Center	1993	Ankara	Ankara Municipality
Deneme Science Center	1998	Istanbul	Science Center Foundation / Istanbul Technical University
Bekirpaşa Municipality Science Center	2008	Kocaeli	Bekirpaşa Municipality
Istanbul Museum of the History of Science and Technology in Islam	2008	Istanbul	Ministry of Culture and Tourism, Turkish Academy of Science (TUBA), TUBITAK
Karşiyaka Municipality Science Museum	2009	Izmir	Karşiyaka Municipality
Gaziantep Planetarium and Science Center	2010	Gaziantep	Gaziantep Municipality
Ödemiş Municipality Science Center	2011	Izmir	Ödemiş Municipality
Eskişehir Science Experiment Center	2012	Eskişehir	Eskişehir Municipality
Bursa Science and Technology Center	2012	Bursa	TUBITAK and Bursa Municipality
Karaman Municipality Science Center	2012	Karaman	Karaman Municipality
Avcilar Science Center	2013	Istanbul	Avcilar Municipality
Sancaktepe Science Center, Observatory and Planetarium	2014	Istanbul	Sancaktepe District Governorate, Sancaktepe Municipality and Istanbul Development Agency
Konya Science Center	2014	Konya	TUBITAK and Konya Municipality
Kocaeli Science Center	2014	Kocaeli	TUBITAK and Kocaeli Municipality
Elaziğ Science Center	2015	Elaziğ	TUBITAK and Elaziğ Municipality
Kayseri Science Center	2017	Kayseri	TUBITAK and Kayseri Municipality
Üsküdar Science Center	2018	Istanbul	TUBITAK, Üsküdar Municipality, Turkish Technology Team Foundation

The Turkish Supreme Council for Science and Technology (SCST) plays a critical role in setting the agenda and policies in science and technology. At its 23rd meeting on 27 December 2011, the SCST set a roadmap to promote science and technology among the public (SCST, 2011). In this meeting, TUBITAK, in cooperation with local authorities, was given the main role to establish science centres around the country to enhance children's interest and curiosity towards science and technology. TUBITAK aimed to complete a science centre in all 16 metropolitan cities by 2016, and in all 81 cities by 2023. This target was subsequently modified and rescheduled. Bursa Science and Technology Center, Konya Science Center, Kocaeli Science Center, Elaziğ Science Center, Kayseri Science Center and Üsküdar Science Center were funded by TUBITAK and their respective local municipalities. The local municipality is mainly responsible for the establishment of the science centre and TUBITAK is mainly responsible for the development of exhibitions, training of explainers and providing academic consultancy. Afterwards, the local municipality runs the centre. Kalkan and Turk (2017) argue that this model of partnership has some problems—for example, the municipality sees the science centre as a division of their municipality and political arena. Thus, the quality and quantity of explainers and administrators could be an issue. Rather than hiring staff and experts in different areas, the municipality may prefer to select staff locally, leading to a possible lack of quality and diversity.

3.2. Science and art centres

The first science and art centre in Turkey, Yasemin Karakaya, was opened in Ankara in 1995. Partly due to high demand from the public and national priority policies of the SCST, currently there are 124 science and art centres in 81 cities with a population above 25,000 students (SCST, 2011). These centres are designated for gifted and talented students. There is a huge demand for these centres; therefore, through an examination, primary school students (up to Grade 4) are placed in three fields (music; a 'general talent' field including science history, geography, etc.; and visual art) according to their talent. The nature of diagnostic tests for the selection of students has been criticised, as has the fact that these centres are only for gifted and talented students. There have been public demands for science and art centres for all children no matter what interests and abilities they have.

These centres are run by the Ministry of Education and are free of charge for students who pass the entrance exam. These students can attend their centres until they graduate from high school. They take courses and extra-curricular activities and projects at the centre for around eight hours per week during school terms. They attend the centre at weekends or weekdays based on their school timetable.

3.3. Science festivals and STEM enrichment programs

The 24th meeting of the SCST on 7 August 2012 focused on the dissemination of a science culture, spirit of research and research skills among students through science fairs. TUBITAK Science and Society Division is responsible for coordinating the public engagement initiatives.[4] TUBITAK gives funds to primary and secondary schools to run science fairs. Funding for each school is around US$1,000 and almost all applicant schools are successful. Although the funding is quite modest, the impact can be immense (Sontay et al., 2019). In particular, it has enhanced collaboration among schools, universities and industry, and it has also created social inclusion in science education. TUBITAK Science and Society Division also has other funding (around US$20,000 per event) for organising large-scale science festivals. These science festivals have attracted children to participate in many hands-on science activities.

There are non-profit initiatives in this field. The Turkish Technology Team Foundation,[5] Turkish STEM Alliance,[6] STEM & Makers Fest/Expo[7] and Maker Faire[8] are among them. The Turkish Technology Team (T3) Foundation, founded by several entrepreneurs, supports educational projects and technology start-ups. They organise science engagement activities for primary and high school students and provide several support programs for university students and grants for young tech start-ups. The T3 Foundation aims to support 1001 Technology Teams and 1001 Technology Ventures by 2023. On 20–23 September 2018, the T3 Foundation organised the biggest tech festival at new Istanbul Airport.[9] The Teknofest Istanbul Aerospace and Technology Festival aimed to showcase Turkey's technological advancement and promote cutting-edge technological products. The festival received good publicity in the media and had a great impact on the public's awareness about technological innovations in Turkey.[10]

The Turkish STEM Alliance (founded in 2015), a member of the EU STEM Coalition, is an independent body of networks for promoting public engagement with STEM. It unites STEM practitioners, researchers, policymakers and the public to enhance the quality of STEM education and broaden participation in

4 See www.tubitak.gov.tr/tr/destekler/bilim-ve-toplum/ulusal-destek-programlari.
5 See turkiyeteknolojitakimi.org/en.
6 See www.stemalliance.center.
7 See www.stemandmakers.org.
8 See turkiye.makerfaire.com.
9 See www.teknofestistanbul.org.
10 See www.bilimgenc.tubitak.gov.tr/makale/teknofest-istanbul-havacilik-uzay-ve-teknoloji-festivali-icin-geri-sayim-basladi.

STEM. The Turkish STEM Alliance consists of members from science centres, science museums, professional development centres, NGOs, STEM centres, companies, research centres and public organisations. It has organised STEM & Makers Fest/Expos[11] in different cities, such as Ankara, Kocaeli, Konya, Antalya, Gaziantep, Diyarbakir, Malatya, Adiyaman and Mersin since 2015. In 2020, Van, Kastamonu and Bolu were added to these cities for STEM & Makers Fest/Expos. With these cities, more than 200,000 participants have engaged with STEM & Makers activities. STEM & Makers Fest/Expo, a member of the European Science Engagement Association, has been organised in collaboration with universities, schools, local authorities and industry. Makers Faire is also quite popular in Turkey and attracts many young people.

4. Science communication in non-formal environments

Science communication in non-formal environments such as the media need significant improvements, (Becerikli, 2013; Cakmakci and Yalaki, 2012, 2018; Trench et al., 2014). Although there are several newspapers in Turkey, most of them do not have a science section, and news related to science is not reported by science reporters but rather by generalist journalists. Nonetheless, there are a few science reporters, such as Esra Oz. She covers news and articles on health-related issues on CNN Turk.[12] She also has books on informed decision-making about health issues (Oz, 2018).

Becerikli's 2013 study with 73 journalists revealed that most have limited knowledge and expertise in stories on science and technology (Becerikli, 2013; Erdogan, 2007). Environmental issues, ecology, historical texture and heritage, archaeology and evolution theory were among the areas they found difficult (Becerikli, 2013). Many public and private universities have faculties of communication but none has a science communication division. There are pioneering scholars who offer courses on science communication for students, scientists and journalists: Irfan Erdogan (Erdogan, 2007), Erkan Yuksel (Yuksel et al., 2014), Ciler Dursun (Dursun, 2010) and Ahmet K. Suerdem (Veltri and Suerdem, 2013).

The SCST plays an important role in science communication in non-formal environments, with TUBITAK responsible for promoting, funding and carrying out cutting-edge scientific research and making the findings available.

11 See www.stemandmakers.org.

12 See www.cnnturk.com/yazarlar/guncel/esra-oz.

It publishes popular science books as well as science magazines for children and the public. TUBITAK has three popular science magazines: *Curious Puppy* targets pre-school children, *Science and Children* targets primary school students (ages 7–14) and *Science and Technology Magazine* targets high school students. TUBITAK has an online science news platform, *Bilim Genç*,[13] for youth and the public. Besides these there is an independent news aggregator, bilimiletisimi.com (in Turkish, 'science communication'), which aggregates science, technology, health, education and business news content from a variety of sources (e.g. online newspapers and popular magazines).

5. Research in science communication

Although science communication activities are carried out in Turkey, studies on science communication as a discipline are limited (Gelmez-Burakgazi, 2017; Veltri and Suerdem, 2013). Research on science communication in the national context could be categorised under three main headings: descriptive studies, science journalism and media, and interdisciplinary studies (science education, public relations). There are some dissertations and theses (Arca 2004; Arslanoglu, 2014; Erdem, 2011; Gelmez-Burakgazi, 2012; Guzeloglu, 2012; Utma, 2015); journal articles (Becerikli, 2013; Dursun, 2010; Gelmez-Burakgazi, 2017; Gelmez-Burakgazi and Yildirim, 2014); and books/book chapters (Erdoğan 2007; Utma, 2017) in the field.

Most studies conducted on science communication in Turkey concern science journalism and the media. In her thesis, Arca (2004) examines the contribution of popular science magazines and scientific journalism in Turkey. Similarly, Erdogan (2007) explores the organisational and contextual structure of scientific journalism in Turkey through science communication. This study is pioneering as it is the first science communication study supported by TUBITAK. Other studies are concerned with nationalist discourse in science news (Erdem, 2011); scientific journalism and the profile of science journalists (Becerikli, 2013); science journalists' views on science news (Arslanoglu, 2014); and the process of scientific journalism and analysis of science news in a university newspaper (Utma, 2015). The common result of these studies is to say that the importance of science, technology and innovation news and science journalism is ignored, and that most science news is not produced by professional science journalists.

13 See www.bilimgenc.tubitak.gov.tr.

Elsewhere, Dursun (2010) portrays the development of science communication and different approaches, and comparing these with other developed countries. Guzeloglu (2012) conducts an interdisciplinary study investigating the role of public relations in creating awareness of the consumption of the products of nanotechnology. Another interdisciplinary study was conducted by Gelmez-Burakgazi (2012, 2013). She examines how fourth- and fifth-grade students use science information and the effective uses of sources and processes in communicating science to students, with a major focus on bridging science education and science communication.

Science communication is a fertile field of study in Turkey. Compared to other countries, science communication research and practices are conducted in a limited range of places (science centres, universities, etc.) by a limited number of science communicators and researchers. Nevertheless, we believe that, in time, with the initiatives of these communicators and researchers the visibility and importance of the field will develop. We need to emphasise that well-structured policies and investment on science and technology and on science communication will catalyse development of the field.

References

Akyuz, Y. (1983). *Ataturk and 1921 Education Congress, Education in Republic Time* [in Turkish] (pp. 89–104). Istanbul: Ministry of National Education.

Arca, E. (2004). Science communication model of Turkey: The importance of the role of scientific journalism (Master's thesis). Istanbul Technical University, Unıversite Louis Pasteur.

Arioglu E. and Girgin Z.C. (2003). Critical overview of the publication ranking of our country [in Turkish]. *Journal of Science and Utopia, 105*, 38–41.

Arslanoglu, O. B. (2014). The gatekeepers of science: Science journalism from the view of journalists in Turkey [in Turkish] (Master's thesis). Istanbul Bilgi University, Istanbul.

Atatürk Research Center. (2019). *Talking in Alasehir* [in Turkish]. Retrieved from www.atam.gov.tr/ataturkun-soylev-ve-demecleri/alasehirde-halkla-konusma.

Bahadir, O. (2001). *The first popular science magazines of the republic and modernisation* [in Turkish]. Istanbul: İzdüşüm Publications.

Becerikli, S. (2013). Thinking on the profiles of Turkish science, technology and innovation reporters and their news making practices [in Turkish]. *İletişim Fakültesi Dergisi, 45*, 1–18.

Bucchi, M. (2009). *Beyond technocracy: Science, politics and citizens.* Dordrecht: Springer. doi.org/10.1007/978-0-387-89522-2.

Bursali, O. (2000). *Science in the 75th year of the Republic, Results 1923–1998*, national meeting [in Turkish]. Istanbul: TUBA.

Cakmakci, G. and Yalaki, Y. (2012). *Promoting student teachers' ideas about nature of science through popular media.* Trondheim, Norway: S-TEAM/NTNU.

Cakmakci, G. and Yalaki, Y. (2018) Promoting pre-service teachers' ideas about nature of science through science-related media reports. In O. Tsivitanidou, P. Gray, E. Rybska, L. Louca and C. Constantinou (eds), *Professional Development for Inquiry-Based Science Teaching and Learning* (pp. 137–61). Dordrecht: Springer.

Dursun, Y. (2009). From past to present science and technology in Turkey [in Turkish]. Ankara University, *Journal of Social Sciences, 1*(1).

Dursun, Ç. (2010). Development of science communication in the world and different approaches: From science to society, to society in society [in Turkish]. *Kurgu Online International Journal of Communication Studies, 2*, 1–31.

Dursun, Ç., Becerikli, S. and Dursun, Y. (2010). *The reception of the science, technology and innovation news and to develop science journalism* [in Turkish]. TUBITAK Project.

Erdem, P. (2011). Nationalist discourse in science news in the written press in Turkey [in Turkish] (Master's thesis). Ankara University, Ankara.

Erdogan, I. (2007). *Journalism and science communication in Turkey: Structural features, problems and suggested solutions* [in Turkish]. Ankara: Pozitif Publications.

Gelmez-Burakgazi, S. (2012). Connecting science communication to science education: A phenomenological inquiry into multimodal science information sources among 4th and 5th graders (unpublished PhD dissertation). Middle East Technical University, Ankara.

Gelmez-Burakgazi, S. (2017). Science communication in Turkey in the light of critical events, political documents, reports and studies [in Turkish]. Selçuk University, *İletişim Dergisi, 10*(1), 232–61. doi.org/10.18094/josc.303022.

Gelmez-Burakgazi, S. and Yildirim, A. (2014). Accessing science through media: Uses and gratifications among fourth and fifth graders for science learning. *Science Communication, 36*(2), 168–93.

Gencoglu, M. (2008). Sent from the Ottoman Empire to the West (1830–1908) – A group biography research [in Turkish] (PhD dissertation). Hacettepe University, Ankara.

Guzeloglu, E. B. (2012). The role of public relations in the process of creating social awareness for the consumption of nanotechnological products [in Turkish] (PhD dissertation). Ege University, İzmir.

Ihsanoglu, E. (2007). Lessons from the Ottoman Empire. *SciDev.Net.* Retrieved from www.scidev.net/global/indigenous/opinion/lessons-from-the-ottoman-empire.html.

Inonu, E. (1999). *An overview of science in Turkey in the republican period and expectations about the Future*. TUBA Conference on the 75th Anniversary of the Republic of Turkey I. Ankara: TUBA Publications.

Kologlu, O. (1997). *Science to public: Science journalism in Turkey*. İstanbul: Türk Bilim Tarihi Kurumu Publications.

METU. (2018). Science and Technology Policy Studies (PhD Program). Retrieved from stps.metu.edu.tr/phd-program.

Ministry of National Education. (2015). *2015–2019 Strategic Plan*. Retrieved from sgb.meb.gov.tr/meb_iys_dosyalar/2015_09/10052958_10.09.2015sp17.15imzasz.pdf.

OECD. (1967). *Pilot Teams' Project on Science and Economic Development (Turkey)*, DAS/SPR/67.8.

Official Gazette. (1963). Retrieved from www.resmigazete.gov.tr/arsiv/11462.pdf.

Oz, E. (2018). *Mysterious tips on healthy eating – Detective Duru in pursuit of truth-1* [in Turkish]. Istanbul: Pupa Yayinlari.

Ozdas, M. N. (2000). *Science and technology policy and Turkey* [in Turkish]. Ankara: Bilimsel ve Teknik Araştirma Kurumu.

Saritas, O., Taymaz, E. and Tumer, T. (2006). *Turkey's National Technology Foresight Program – A contextualist description and analysis*. Retrieved from ideas.repec.org/p/met/wpaper/0601.html.

Sisman, A. (2004). *Ottoman students sent to France during the Tanzimat period, 1839–1876* [in Turkish]. Ankara: Türk Tarih Kurumu.

Sontay, G., Anar, F. and Karamustafaoglu, O. (2019). Opinions of secondary school students attending 4006-TUBITAK Science Fair about science fair [in Turkish]. *International e-Journal of Educational Studies, 3*(5), 16–28. doi.org/10.31458/iejes.423600.

Sontay, G., Yetim, H., Karamustafaoglu, S. and Karamustafaoglu, O. (2019). Developing an Entrepreneurship Scale for 5th Grade Students. *International Journal of Educational Methodology, 5*(2), 203–20.

State Personnel Presidency. (1969). *Second 5-Year Development Plan*. Retrieved from www.sbb.gov.tr/wp-content/uploads/2018/11/İkinci-Beş-Yıllık-Kalkınma-Planı-1968-1972%E2%80%8B.pdf.

State Personnel Presidency. (1972). *Third 5-Year Development Plan*. Retrieved from www.sbb.gov.tr/wp-content/uploads/2018/11/%C3%9C%C3%A7%C3%BCnc%C3%BC-Be%C5%9F-Y%C4%B1ll%C4%B1k-Kalk%C4%B1nma-Plan%C4%B1-1973-1977%E2%80%8B.pdf.

State Personnel Presidency. (1984). *Fifth 5-Year Development Plan*. Retrieved from www.sbb.gov.tr/wp-content/uploads/2018/11/Be%C5%9Finci-Be%C5%9F-Y%C4%B1ll%C4%B1k-Kalk%C4%B1nma-Plan%C4%B1-1985-1989.pdf.

State Personnel Presidency. (2001). *Eighth 5-Year Development Plan*. Retrieved from www.sbb.gov.tr/wp-content/uploads/2018/11/Eight-Five-Year-Development-Plan-2001-2005.pdf.

Supreme Council for Science and Technology (SCST). (2011). 23rd meeting of the SCST [in Turkish]. Retrieved from goo.gl/Vz8s18.

Trench, B. (2008). Towards an analytical framework of science communication models. In D. Cheng, M. Claessens, T. Gascoigne, J. Metcalfe, B. Schiele and S. Shi (eds), *Communicating science in social contexts: New models, new practices* (pp. 19–35). Netherlands: Springer. doi.org/10.1007/978-1-4020-8598-7_7.

Trench, B., Bucchi, M., Amin, L., Cakmakci, G., Falade, B., Olesk, A. and Polino, C. (2014). Global spread of science communication: Institutions and practices across continents. In M. Bucchi and B. Trench (eds), *The Routledge handbook of public communication of science and technology* (2nd edition)(pp. 214–30). New York: Routledge.

TUBITAK. (1983). *Turkish Science Policy: 1983–2003* [in Turkish]. Ankara.

TUBITAK. (1993). *Turkish Science and Technology Policy: 1993–2003* [in Turkish]. Ankara.

TUBITAK. (2004). *Vision 2023: Science and Technology Policy* [in Turkish]. Retrieved from www.tubitak.gov.tr/tubitak_content_files/vizyon2023/Vizyon2023_Strateji_Belgesi.pdf.

TUBITAK. (2005a). *Science and Technology Policies Implementation Plan* [in Turkish]. Retrieved from www.tubitak.gov.tr/tubitak_content_files/BTYPD/strateji_belgeleri/BTP_UP_2005_2010.pdf.

TUBITAK. (2005b). *Vision 2023 Technology Foresight Project*. Retrieved from bit.do/eV4eE.

TUBITAK. (2010). *Science, Technology and Innovation*. Retrieved from www.tubitak.gov.tr/tubitak_content_files/BTYPD/arsiv/STI_in_Turkey_2010.pdf.

TUBITAK. (2018). *2018–2022 Strategic Plan*. Retrieved from www.tubitak.gov.tr/sites/default/files/2204/tubitak_2018-2022_stratejik_plani.pdf.

Türkcan, E. (2001). Turkey's place in the age of technology: Science and education policies that will move Turkey into the modern age [in Turkish]. *Yeni Ekonomi Sempozyumu*, 173–80.

Utma, S. (2015). Bilim İletişimi ve Bilim Gazeteciliği: Ege Üniversitesi Haber Ajansi Örneğinde Üniversitelerde Bilim Haberlerinin Üretilmesine Yönelik Bir İnceleme (PhD dissertation). Ege University, İzmir.

Utma, S. (2017). Concept of science communication and relationship of science-communication [in Turkish]. In A. Guney (ed.), *Science center with all angles: A conceptual reading on science centers* (pp. 103–21). Istanbul: Çizgi Publications.

Veltri G.A. and Suerdem A. K. (2013). Worldviews and discursive construction of GMO-related risk perceptions in Turkey. *Public Understanding of Science*, *22*(2), 137–54. doi.org/10.1177/0963662511423334.

Yalcin, C. and Yalova, Y. (2005). *Turkey in the light of science and technology policies* [in Turkish]. Ankara: Nobel Publications.

Yuksel, E., Kocak, A., Aydin, S. and Yalcin, K. (2014). *Check Up: Health Communication*. Konya: Literatürk Academia.

Timeline

Event	Name	Date	Comment
First interactive science centre established.	Feza Gürsey Science Center	1993–95	See www.fezagurseybilimmerkezi.com
First national (or large regional) science festival.	Science festivals promoted by TUBITAK	1990s	2018: The Teknofest Istanbul Aerospace and Technology Festival www.teknofest.org
An association of science writers or journalists or communicators established.	Türkiye Gazeteciler Cemiyeti [Turkish Journalists' Society]	1946	See www.tgc.org.tr
First university courses to train science communicators.	First field study on science communication and science journalism (Prof. Dr Ciler Dursun, Ankara University; funded by TUBITAK)	2008–10	2011–13: Second field study was in science journalism in Turkey, Prof. Dr Ciler Dursun, Ankara University, funded by TUBITAK) 2010: Science, Technology and Health Communication, Prof. Dr Erkan Yüksel, Anadolu University 2010: Science Communication, Dr Hacer Erar, Atilim University
First master's students in science communication graduate.	Science Journalism, Ankara University	2004	Supervised by Prof. Dr Ciler Dursun
First national conference in science communication.	Health Communication Symposium	2015	Organised by Prof. Dr Erkan Yuksel, Anadolu University, www.saglik-iletisimi.org
National government program to support science communication established.	The Scientific and Technological Research Council of Turkey (TUBITAK)	1963	See www.tubitak.gov.tr/en

Event	Name	Date	Comment
First significant initiative or report on science communication.	First field study	2010	Prof. Dr Ciler Dursun, Ankara University
National Science Week founded.		1970s	See www.bilimgenc.tubitak.gov.tr/makale/bilim-ve-teknoloji-haftasi-etkinlikleri-turkiyenin-dort-bir-yaninda
A journal completely or substantially devoted to science communication established.	*Bilim ve Teknik*, TÜBİTAK science and technology magazine	1967 2007	See www.bilimteknik.tubitak.gov.tr/dergimiz/hakkimizda
First significant radio programs on science.	Programs featuring Prof. Dr Celal Sengor	2010s	Professor Sengor of the Istanbul Technical University was a guest speaker for different radio channels
First significant TV programs on science.	Programs featuring Prof. Dr Celal Sengor	2010s	Professor Sengor of the Istanbul Technical University was a guest speaker for different TV channels
First awards for scientists or journalists or others for science communication.	Awards administered by the Turkish Academy of Science (TUBA)	1993	See www.tuba.gov.tr/en
Date hosted a PCST conference.	PCST-14 in Istanbul	2016	
Other significant events.	*Fen Alemi Dergisi*, first popular science magazine	1925	2010: A second magazine *Eğlenceli Bilim Dergisi*, editor Dr Hacer Erar www.atilim.edu.tr/tr/eglencelibilim/page/2942/eglenceli-bilim-dergisi
	Most influential books on science communication	2007	*Journalism in Turkey and science communication*, by Prof. Irfan Erdoğan
		2011	*Understanding communication*, by Prof Irfan Erdoğan
		2009	Senturk, E. (2009). The effect of science centres on students' attitudes towards science. Middle East Technical University, Ankara
		2012	Gelmez-Burakgazi, S. (2012) 'Connecting science communication to science education: A phenomenological inquiry'. Middle East Technical University, Ankara
		2012	Cakmakci, G. and Yalaki, Y. (2012). *Promoting student teachers' ideas about nature of science through popular media*. Trondheim, Norway: S-TEAM/NTNU

Contributors

Gultekin Cakmakci is a professor of science education at Hacettepe University and teaches courses on STEM education and public engagement with STEM.

Dr Sevinc Gelmez-Burakgazi is a faculty member on the Curriculum and Instruction Program at Hacettepe University, Turkey.

38

UGANDA

Cultural values and modern media as drivers of science communication

Ivan Nathanael Lukanda

The chapter focuses on the evolution of science communication in Uganda. The chapter starts with the influence of cultural values on the science communication field but ends by emphasising the impact of modern media on the field. It states the role of the central government and (international) non-government organisations (NGOs) in sharing scientific information. It describes the genesis of science communication training and research at universities, and how associations have formed over the years. Some of the key science communicators are identified, and opportunities for employment of science communicators highlighted. The chapter ends with a discussion of some of the challenges in communicating science, the controversies generated by the various channels, and the manoeuvres to counter the barriers to sharing scientific knowledge. The terms adult literacy, sensitisation and engagement tend to be associated with this field in Uganda.

1. Introduction

Science communication is vital in modern societies as economies try to find solutions to the challenges of energy, food, water, climate change and treatment of diseases such as HIV/AIDS, cancer and malaria. Scientific knowledge is meant to help societies develop efficient technologies that can change society for the better: it is important in decision-making if individuals and policymakers are to make informed choices in democracies; and societies depend on science in health care, food production, communication technologies, transportation, and preservation (Davies and Horst, 2016). In the case of Uganda, science communication shapes what people know about issues such as climate change, food security, cancer and childcare. Science communication combines with aspects of cultural values and helps us understand the contemporary society we live in. Therefore, the development of the field marks an advancement in science–society relations and reflects a collaboration 'between institutions and between the cultures of science and of institutions and the culture of the wider society' (Bucchi and Trench, 2016, p. 151).

Since the 1970s, governments around the world have recognised the importance of the field and formed ministries of science, technology and/or research in 'anticipation of positive spin-offs' to national development (Schiele, 2018, p. 17). Politicians, then, either hired scientists as advisors or appointed them as ministers to mainstream science into the political systems. Uganda has had several physicians as ministers of health, the same way it has had lawyers as ministers of justice and constitutional affairs. The 'visibility' and influence of such scientists in public affairs, including the media, could be a 'relevant dimension to analyse a country's scientific culture' in terms of the public understanding of science and the respect accorded to such knowledge and its agents (Bucchi and Trench, 2016, p. 161). Thus, scientific culture must find a place in the general culture of a society if science is to be relevant to people who can put it to good use and help society manage its lived challenges. The interaction between the cultures and institutions and the general society is usually reflected or framed through the media—both traditional as well as online/ social media.

2. Culture and science communication

The development of science communication cannot be divorced from the diverse cultural values and symbiotic meanings communities attribute to nature since time immemorial. Plants, animals, rocks, mountains and objects in the universe feature prominently in the anthropology, ethnobotany, ethnomedicine, geography, history and language of many societies in Uganda. For instance, 89 species of plants that could be used in '26 cultural and social purposes' have been identified (Kakudidi, 2004). These purposes include wedding and religious ceremonies, prevention and treatment of illnesses, and sources of food. Many Ugandan tribes have plants, animals and insects as their totems. A totem signifies a mythical bond between individuals and communities. Totems symbolise religion, social cohesion and environmental protection. Totems could have formed the basis for laws and regulations as hunting, killing and exploiting of totem animals, insects and plants was/is forbidden because such actions border on hurting the ancestral spirits. Thus, cultural values create an ethical (spiritual) link between people's culture and the environment. In many Ugandan communities a clan is the genealogy of biological ancestry as traced through paternity lineage and the totem is the emblem of that socialisation. In some communities, names are given according to clan and it is a taboo to eat your totem. It is also a taboo for individuals from the same clan to marry or have any form of sexual relationship as a way of preventing incest and inbreeding. As a result, 'totemism can lead to environmental protection due to the fact that many tribes have multiple totems' with thousands of animals, plants and insects considered as totems (*emisiro*) among the Basoga, Baganda, Banyoro and other tribes in Uganda (Bamuturaki, 2011). There is a belief that totems and humans are 'related' and this biocultural heritage is communicated from generation to generation. Communities that have functional clan and customary land systems could be in a better position to protect the environment as exploitation has to conform with ecology (Epilla, 2014).

Indeed, kingdoms have been central to environmental protection. For close to a decade, the Uganda Wildlife Authority (UWA) has been using the Buganda clan system to raise awareness about environmental protection in the central region. For instance, the Queen Mother of the Buganda kingdom (Nabagereka) visited UWA in 2013 and laid the foundation stone for the construction of a monument for her totem, the cane rat (*omusu*) (UWA, 2013). This action made her a UWA ambassador on environmental protection. During the more than two-decade armed conflict from 1987 in northern Uganda, the indigenous woody plants conservation strategy involved using 'cultural law and local bye-law' (Epilla, 2014, p. 1). Rim-Rukah et al.'s study (2013) underscores the importance of beliefs in environmental protection. The study shows that 'particular ecosystems or habitats (such as sacred groves and sacred rivers/pond); and … animals or plant species (such as totem and tabooed species) were not as endangered as other species' (Rim-Rukeh, Irerhievwie and Agbozu, 2013, p. 426). A more recent study in Zambia reveals that totemism, taboos and 'traditional authority' forbid the use of fruit trees for firewood among the Tonga (Kanene, 2016, p. 3). Thus, the clan is not only an effective component in social, political and economic organisation, but also in communicating conservation directly or indirectly. For instance, a village in Burkina Faso is reportedly living with crocodiles because the majority of the people in the locality consider the animal their totem and believe the animals and humans cannot hurt one another (Adams, 2008). These traditional beliefs are hardly respected by the scientific establishment in Uganda as many scientists are trained in and follow Western cultural norms. However, it is common for scientists to use local proverbs to explain conservation/scientific issues to ordinary people, an indicator that they believe in some of the traditions.

Several kings, including the Kabaka of Buganda, the Omukama of Bunyoro, the Omukama of Tooro, the Kyabazinga of Busoga, the Emorimori of Teso, and other cultural leaders have been key agents, especially in mobilising people for immunising against and fighting HIV. Their efforts are complemented by the religious leaders who often use their pulpits to mobilise the public for immunisation and to sensitise communities about HIV prevention, treatment and management.

2.1. Use of storytelling, drama and music for enforcing belief systems

Science communication is dependent on common understanding within the audience and its success depends on understanding the cultures, beliefs, perceptions, attitudes, fears and promises of the target population. This makes the 'arts an excellent tool for health communication' as they encourage community engagement and behavioural change through social learning (Sonke et al., 2018, p. 401). The power of the arts derives from the fact that it is a homegrown communication tool ingrained in the local cultures. Such efficiency can be achieved through music, dance, drama, recitals, rituals, storytelling, orators and village criers in enforcing the belief system since culture determines how a message is decoded. These forms of communication are embedded in many Ugandan societies, as tools of social cohesion. They are recognised, accepted, valued and trusted. In Ankole, 'rap music' locally known as *okwivuga* continues to be used to pass on messages (Mushengyezi, 2003, p. 116). In Busoga and Buganda, drums were and are used to mobilise people, and the sound of the drum could indicate the purpose for mobilisation—danger, joy, community work or just gathering to receive important information.

The mobilisation involved an official going around the village drumming at every residence and announcing whatever message the local council had for the residents—usually to do with council meetings or else with cholera awareness, immunisation campaigns, road repair or constructing a water spring. This placed responsibility for action and response onto each individual resident. It was effective in its outcome because this 'interpersonal' nature of indigenous forms not only allowed for easy encoding, decoding and feedback, but also made it incumbent on the 'message-receiver' to respond to the message because of the shared relationship with the 'senders' (Mushengyezi, 2003, p. 112). Drumming as a mode of mobilisation worked when societies were still homogeneous, speaking a common language and with less urbanisation. The increase in urbanisation and its related challenges, and the emergence of new forms of media, means that drumming is no longer an ideal form of mobilisation. New forms of mass mobilisation are expensive and require government support if science messages are to reach the intended audience.

Figure 38.1: Black-and-white colobus monkey, the totem of the Ngeye clan in Buganda.

Source: Phionah Katushabe (used with permission).

Figure 38.2: A hippopotamus, the totem of the Envubu clan in central Uganda.

Source: Phionah Katushabe (used with permission).

3. Government's support toward science communication

Deliberate science communication started in 1948 when the government launched adult literacy using music, dance and drama (Kiwanuka-Tondo, 1990). The history of deliberate science communication in Uganda is synonymous with the history of broadcasting in the country. Broadcasting started in 1954 following a recommendation by a committee chaired by the Earl of Plymouth to the British Colonial Office to explore the option of setting up radio stations in the colonies, including Uganda. Among other reasons, broadcasting was necessary to communicate issues related to education, health and agriculture (Great Britain, 1937). Following the 1945 nascent nationalist struggles in which the governed demanded fair working conditions from their employers and the Buganda kingdom, disbandment of price controls on the export of cotton, abolition of the Asian monopoly over cotton ginning and participation in electing representatives to the local government, the colonial government realised the need for a broadcast infrastructure to explain its programs and actions to the local people. Based on the Plymouth report, the Uganda Broadcasting Services was set up to promote colonial interests and suppress the rising pro-independence wave. A later committee chaired by Gervais Huxley established the feasibility of broadcasting in Uganda (Huxley, 1958). Thus, 'the colonial government's tacit recognition of the role of both the broadcast media and the indigenous languages in influencing public opinion and political consciousness' (Chibita, 2006, p. 113) can be extended to science communication as agriculture, health and other scientific issues were later to be relayed in local languages as access increased. Indeed, Kiwanuka-Tondo (1990, p. 50) argues that in societies where education facilities are insufficient, the media, especially through instructional programming, have been seen as a 'substitute for this formal education'. He argues that by the 1980s, drama had become an important component of radio content, and was used in the field of reproductive health as a driver of science communication.

4. Education system and science communication (1960–70)

The establishment of television (Uganda Television) in 1963 was a forerunner to the launch of education broadcasting in 1964 (Sekeba, 2016). The integration of broadcasting into education was preceded by the construction of the national theatre. The Government of Uganda opened the national theatre in 1959. Apart from the premises being a venue where amateur groups entertained city revellers, the national theatre has remained a venue for national inter-school competitions. The competitions are modelled along the Makerere College Inter-hall competitions, which started in 1947 (Twesigye, 2001), and often follow an annual theme selected by the Ministry of Education and Sports.

The Ministry of Education and Sports, in conjunction with the Ministry of Information, initiated the intensive model. The model was meant to integrate instructional radio and television broadcasting into school and college curricula. Broadcasting acted as 'audio-visual aids' that could even 'substitute for teachers' as the number of teachers was insufficient to match the growing demand for education at a time when the country needed patriotic citizens to understand the 'challenges and aspirations' of the 'young' country (Kiwanuka-Tondo, 1990, pp. 55–6). Weekly programming prioritised science, mathematics, history, geography and literature, as these were considered vital for a sovereign Uganda. At the same time, Makerere University through its college of education and external studies (formerly the extra-mural department) started sporadic adult and continuing education programs.

Further, the author avers that in the 1960s, the Ministry of Information helped the Ministry of Agriculture to promote the coffee and cotton industries, then the predominant cash crops of Uganda. This period marked the birth of farm radio and television. The content included audio and film recordings of how coffee is planted, pruned, picked, dried and packed in sacks; and how cotton is planted, pruned, harvested and ginned. The documentaries did not include processing because ordinary Ugandans were concentrating on producing the crops, and not marketing or processing the products. Marketing was the work of the Coffee Marketing Board and the Lint Marketing Board for cotton. Audios were particularly important because most Ugandans could only afford radio, and very few could afford the available black-and-white television sets (colour television was only introduced on the eve of the 1975 summit of the Organisation of African Unity (now African Union).

5. Science communication under Idi Amin (1971–79)

When Idi Amin overthrew the Milton Obote government in the 1971 coup, broadcasting to the rural folk in selected local languages continued with minimum interruption for his first two years (Kakooza, 2012). Farmers remained a priority in Radio Uganda's programs, and tuned into the *Calling Farmers* program, along with the *Coffee Club*, a program running in conjunction with the Uganda Coffee Marketing Board to promote coffee growing in the country. These programs attracted extension workers, co-operative officers and parastatal staff as guests. However, instructional broadcasting for school and programs on agriculture started declining with the economy following the expulsion of Asians in 1972. Broadcasting succumbed to the brain drain at this time as professionals were wary of the security situation in the country, thereby virtually ending relevant science communication.

It is worth noting that during the liberation war that overthrew Idi Amin in 1979, several booster stations were destroyed. These included the Masaka television booster in Central Uganda and the cross-border service booster station of Radio Uganda at Bobi in Northern Uganda. Other stations were neglected. Such destruction interrupted not only science communication but broadcasting in general.

6. The post-Amin era (1980–85)

This period was marked by dramatic changes at the top of government. Uganda had three presidents in a space of roughly two years. It was only in 1981 that the new government of Milton Obote (Obote II) started rehabilitating the state broadcaster. In 1982, the government established the directorate of education broadcasting (Kiwanuka-Tondo, 1990); and at this time programming for children was introduced on both radio and television as a way of sensitising youngsters to the importance of academics in general and health in particular.

Broadcasting played a crucial role in health information campaigns during this period. Campaigns focused on preventing waterborne diseases, especially typhoid and cholera. The campaigns emphasised improving sanitation, drinking boiled water, eating cooked food and covering food to prevent contact with flies, which are agents of germs causing such diseases. Dramatisation of information messages was the key model.

A childcare campaign was started by government agencies such as the Uganda National Expanded Programme for Immunisation (UNEPI) and UNICEF in partnership with the directorate of educational broadcasting. It was important to inform, educate and mobilise the public, especially parents, to take children for immunisation against what were then the six child-killer disease—measles, whooping cough, polio, tuberculosis, tetanus and diarrhoea—and to encourage parents to utilise vaccines to immunise their children.

These collaborations were usually preceded by workshops and conferences to which print and broadcast journalists were invited—not only to be oriented about the campaign, but also to ensure coverage of the campaigns. The media messages were supplemented by posters at dispensaries, health centres, health departments and immunisation centres, schools and trading centres.

In 1987, UNICEF supported the Ministry of Education to revise the primary school curriculum to include immunisation, especially against the six killer diseases. This health education succeeded largely due to a combination of factors: the Resistance Councils (village chiefs), religious leaders, and the scouts and girl guides who supplemented health officials in disseminating information about HIV. The multiple-information-platform approach was a precursor to the fight against HIV/AIDS, which started around the same time. Prevention and basic management of HIV/AIDS skills are taught in primary and secondary schools. HIV prevention and management programs have been incorporated into the activities of local councils, religious organisations, women, youth, schools, health departments and virtually all government departments.

7. Science communication under liberalised media system

Broadcasting has been a key component of science communication from colonial times to date. Following the liberalisation of the media landscape in Uganda in 1992, several radio and television stations were opened. Currently, Uganda has about 300 radio stations and more than 30 television stations (Uganda Communications Commission, 2015). Although science is not a major component of their content, they often cover breaking stories on disease outbreaks, natural disasters and climate change. The national environmental management authority (NEMA) is working with NTV Uganda to run articles about protecting the environment in Uganda. The program, *NTV Green,* an intrinsically public relations forum for NEMA, involves the sponsor taking journalists to well-protected sites or places where NEMA is doing a good job and showing them to the public. Several broadcasting stations have programs on agriculture, which implicitly include a scientific approach to both the environment and agriculture. It is important to note that all the radio and television stations carry advertisements on malaria prevention and management. They also carry advertisements on immunisation, especially of children. It should be remembered that most of the radio stations have programs on health, where a medical practitioner is often hosted, say once a week, to talk about particular health issues. Just like the broadcast outlets, the newspapers often publish stories on health and the environment.

Online versions of newspapers also run science stories. On special days to mark international efforts in fighting diseases such as malaria, HIV/AIDS, sickle cell and cancer, the newspapers often publish pull-outs about the respective diseases. Key in all these science communication endeavours is that science must make news for it to be covered; or it must have a sponsor for the programs to run. The sponsors are usually government departments and local and international NGOs. In some cases, science is covered as business and economics. Notable among the media houses is local tabloid television *Bukedde*, which runs a science and technology segment in its popular *Agataliko Nfuufu* [Dustless News] at 10 pm, usually focusing on both local and international inventions.

Several radio stations run programs on astronomy. These include Radio Simba, Prime Radio and Central Broadcasting Services (CBS). A key science communicator of the mid-1990s and early 2000s was CBS's (Kalabalaba) Ernest Lule Basajjakabwe, whose 'strange but true stories' about American guns that fired bullets to target the enemy from crowds of people and a description of the rarely used drones to trail enemies before shooting at them are still memorable long after his death. Although he often used a sensational approach to trigger the minds of the audience to imagine the positioning and movement of the bullets and the stars, his radio programs were an attempt toward science communication.

8. Formation of the Ministry of Science and Technology

In 2016, the Government of Uganda established the Ministry of Science, Technology and Innovation (MOSTI). Headed by a cabinet minister, the ministry is mandated to plan and coordinate science and technology innovations in academic institutions, industries, agriculture, commerce and the informal sector (MOSTI, 2018). Although the ministry runs on a paltry budget compared to defence and works, its existence is a sign that Uganda considers science vital in its development, at least on the website.

9. Formation of the Uganda Media Centre

In 2005, the government created a propaganda hub known as the Uganda Media Centre to 'effectively facilitate communications of government policies, programs, and projects to the public through the media' (Uganda Media Centre, 2018). In addition to being home to the government 'wordsmen', the media centre allows cabinet ministers, ruling party members of parliament, and heads of government departments to address the media on key issues in the country. These issues include those of a scientific nature. For instance, health ministers have used the media centre to address issues related to outbreaks of diseases and to explain why the radiotherapy machine at the National Referral Hospital was not functioning. Officers from the Ministry of Agriculture and Uganda National Meteorological Authority (UNMA) have addressed the media on issues related to climate change, especially the element of when and where the drought and floods or landslides are expected.

10. Government intervention in spraying tsetse flies

Through health education in the 1980s, communities in Busoga in southeastern Uganda were mobilised in a campaign against tsetse flies. They were taught and involved in making mono-screen traps using old car tyres and plant materials, identifying sites where the flies are common and placing traps to prevent outbreaks that have ravaged the area for more than a century (Lancien et al., 1990; Okoth et al., 1991). The mono-traps were occasionally supplemented by aerial spraying, and radio was used to warn people against eating vegetables during the days when helicopters were operating. Although initially a top-down approach, the participation of the people made the use of mono-traps a sustainable technology to minimise the effects of the deadly tropical flies that cause *nagana*[1] among animals and sleeping sickness among humans. Unfortunately, this disease, first detected in 1900 in the Buganda region, has spread through the country, largely as a result of animal movement from one region to another. The failure by some farmers and herdsmen to spray their animals, and poor sensitisation, means the disease keeps recurring (Yolisigira, 2015).

The government has attempted to put in place mechanisms for communicating key messages to individuals who may not have access to the mainstream media. According to Article 50 of the *Local Governments Act 1997* (amended in 2015), the village committee, headed by the village (Local Council I) chairperson, is expected to serve as the communication channel between the government, the district or higher local council and the people in the area. It is supposed to help the government to implement its programs. In practice, it is largely a top-down government conduit, although occasionally the reverse happens.

1 A disease of cattle, antelope and other livestock in southern Africa, characterised by fever, lethargy and oedema, and caused by trypanosome parasites transmitted by the tsetse fly.

11. NGOs as key players in science communication

Non-government organisations (NGOs) and inter-governmental organisations (IGOs) are key players in communicating science in Uganda. For instance, Advocates Coalition for Development and Environment (ACODE), a pro-poor public policy think tank, often conducts research and engages several stakeholders before providing alternative policies to government (ACODE, 2018). The Centre for Research in Energy and Energy Conservation (CREEC) trains communities in off-grid areas to generate renewable energy and provide appropriate sources of energy for cooking and lighting (CREEC, 2018).

The Program for Accessible Health Communication and Education (PACE) Uganda and the NGO FHI360 partner with the Ministry of Health and local governments to develop and implement behavioural change programs, including access to contraceptives, family planning services, child health, reproductive health, sex education, prevention of HIV, malaria, tuberculosis and malnutrition, among others (FHI360, 2018; PACE, 2018). Reach Out Mbuya and Reach A Hand Uganda mostly educate communities about HIV (Reach A Hand Uganda, 2018; Reach Out Mbuya, 2018). The Uganda Water, Sanitation and Hygiene NGO network collaborates with its partners through sharing information (UWASNET, 2018). The Volunteer Efforts for Development Concerns (VEDCO) supports the Uganda government's food and nutrition security through knowledge exchange with its partners (VEDCO, 2018). The involvement of all these organisations signifies the importance of health as a driver of science communication in Uganda. The United Nations Development Programme (UNDP), World Health Organization (WHO), United States Agency for International Development (USAID) and ChildFund are some of the prominent organisations sponsoring such communication.

12. Schools as platforms of science communication

Schools and universities organise annual exhibitions, where students showcase what they do in and outside class. It is common at such fairs for students to show 'tourists' what happens when two chemicals are mixed. In some cases, students showcase digital applications, locally made robots, value-added agricultural products and rudimentary cars.

Several institutions, especially those under the National Agricultural Research Organisation, organise workshops for scientists. These workshops are often instructed by communication consultants and senior journalists, with the intention of training scientists in how to disseminate their research to the public through the media. The Uganda Biotechnology Information Centre (UBIC) has organised annual essay competitions on a specific topic for secondary schools since 2016, with the aim of consolidating bioscience in the formal education system. In 2016, students wrote essays to demonstrate their knowledge about biosciences. Selected schools participate in the finals, where prizes are awarded for the winning essay and top schools (UBIC, 2016).

13. Agricultural shows, science festivals and communication

These are common in Uganda. The most popular is the annual Source of the Nile trade and agricultural show in Jinja, which has been organised by the Uganda National Farmers' Federation (UNFFE) since 1993. The show is a platform for demonstrating and identifying innovations made locally and by foreign companies to improve production and marketing (UNFFE, 2018). It is a stage for learning modern technologies in the agricultural value chain and a potential market for products. In recent times, similar shows have been organised by the Buganda kingdom in its Mengo Palace Farm Clinics by the *Daily Monitor* newspaper in conjunction with NARO at regional agricultural research institutes, and the Harvest Money Expos by *New Vision* newspaper in conjunction with the Dutch embassy in Uganda at Namboole Stadium. These shows have a component of entertainment and are accompanied by circuses and musicians to entertain revellers, considering that music is an instrument of social interaction and knowledge dissemination.

At the annual Taxpayers Appreciation Week, organised by the Uganda Revenue Authority, different government agencies offer information and provide services at subsidised prices or for free. Many government agencies share leaflets, brochures and pamphlets, and their personnel explain to the public what they do. The agencies explain to the public issues related to their health (HIV, cancer, diabetes), the environment (planting trees and managing plastics), new seed varieties, preventing accidents, and government policy on education, among other issues. The government agencies also take complaints from the public during this event.

There is also real-time moon viewing at Makerere University: Benon Fred Twinamatsiko, a lecturer at the Department of Physics at Makerere University has, since January 2014, been using a telescope to show the public how the moon moves, and how its movement affects the weather since January 2014 (Alina, 2014). The monthly event is open to members of staff at the university, children and the general public.

14. Training and research beginning at universities

Formal training in science communication started in 2002 following Makerere University's proposal to the Swedish International Development Aid (SIDA) to sponsor a regional training program—the Eastern Africa postgraduate diploma in environmental journalism and communication (Böklin, 2004). The diploma attracted students from Burundi, Kenya, Rwanda, Tanzania and Uganda.

The International Health Sciences University (now Clark University) has been involved in organising dialogues with communities living in slum areas south of the capital Kampala. Based on such dialogues, usually coordinated by graduate students, the university organises medical camps as interventions to better the lives of disadvantaged people.

In 2014, the theme of science communication was incorporated into the specialised journalism course unit taught to fourth-year bachelor of journalism and communication students at Makerere University. This theme raised science communication to the level of other themes covered in that course unit, such as politics, parliament, business, gender, climate change and conflict. Makerere's Master of Arts in Communication curriculum includes science communication not as a formal component, but through the dynamism of the lecturers/facilitators. Communication (general principles of communication) is a cross-cutting course at the university.

Attempts have been made to integrate science communication into the secondary school curriculum. In September 2018, UBIC organised a nationwide workshop for teachers to acquaint themselves with modern agricultural biotechnology research at NARO (ISAAA, 2018). Teachers were expected to pass on information to their students and communities, as a way of increasing awareness and enhancing appreciation of modern biotechnology in the country. The workshop was attended by NARO scientists, the Uganda National Examination Board and the National Curriculum Development Centre. Such moves are aimed at not only increasing knowledge, but also improving the teaching of science in the country.

In October 2018, the Department of Journalism and Communication at Makerere University in collaboration with Science and Development Network (SciDev.Net) organised a course for the training of trainers in a science communication workshop for journalism lecturers under SciDev.Net's Script program. The Script program had been launched on the sidelines of the Next Einstein Conference in Kigali, Rwanda, in March 2018 (Deighton, 2018). The aim was to integrate science journalism into the department's curricula. Through its Nairobi office, SciDev.Net has been operating in Uganda since 2002. SciDev.Net worked with the Department of Journalism and Communication at Makerere University to organise the first science communication conference in April 2019. Prior to that conference, students from humanities, social sciences and natural sciences participated in an online science communication course under the supervision of journalism lecturers.

In November 2018, the Africa Union Commission through the Platform for Aflatoxin Control in Africa concluded a two-day training course for about 30 journalists on aflatoxin communication and reporting (aflatoxin is a plant fungal disease). The training preceded the launch of a National Aflatoxin Journalist Network for public awareness about aflatoxin contamination of food and feeds (Anyango, 2018).

15. The birth of associations, conferences and meetings on science

The Uganda Science Journalists' Association (USJA) was established in 2005 as a non-profit organisation for journalists, communicators and researchers interested in advancing science communication in Uganda. Like many NGOs in Uganda, the membership numbers are unknown as mobilisation is usually activity-focused. However, there is photographic evidence that USJA has been involved in organising upcountry trainings in reporting climate change as part of mentorship in science journalism (USJA, 2018). A Facebook chat with its founder Odinga Balikuddembe revealed that the association has a membership of about 80 journalists. Although its funding is not clear, the association is a member of the World Federation of Science Journalists (WFSJ).

Health Journalists Network in Uganda (HEJNU) is another non-profit organisation committed to enabling public understanding of health care in Uganda. Its membership is drawn from journalists working in print, radio, television and online publications. With a declared membership of about 70, the association often partners with government, especially the Ministry of Health, and civil societies to conduct training (HEJNU, 2018). Most members of the association also report about other subjects, including politics and agriculture, and could be members of associations in these areas. HEJNU uses its network to help members find sources for their stories. The association partners with local and international NGOs, such as AVAC (Global Advocacy for HIV Prevention), to organise science cafés, where health writers meet a guest scientist(s) in informal settings (AVAC, 2018). The partners usually provide the funding for its activities.

In a nutshell, the associations are still weak with fluid membership. Moreover, it is common for an association to declare as a member any participant who turns up for its event, even without subscribing. Thus, the impact of associations is limited to science streams where the project sponsors have budgets to facilitate journalists' activities. The agenda of associations is largely driven by their funders.

16. Who are the science communicators?

There are jobs in science communication in several areas, including journalism, public relations and communication officers in hospitals, environmental protection agencies, and research institutes and science communication trainers. Communicators fall into a number of different categories, as listed below.

Researchers: In partnership with their universities or funders, researchers often address the media about their findings. Scientists working at the meteorological centre may talk to the media when they anticipate severe weather conditions, such as drought or floods. Scientists at NARO often address the media about new seed varieties and biotechnology related issues.

Policymakers: Sometimes ministers and government departmental directors share research findings with the public. The politicians sometimes simply refer in passing to scientific results outside their line of duty during public gatherings, yet their prominence makes journalists quote them on science issues.

Journalists: These are major sources of information on science. They often meet scientists by appointment or by invitation to discover the latest results. Journalists are targeted when major science journals publish impactful results likely to make global news. They can be invited to attend the launch of scientific reports, receive emails or search for this information to fill gaps in their stories.

Development partners (USAID, John Hopkins School of Public Health, PACE, Uganda Health Communication Alliance): These institutions often sponsor scientists to do research and own the findings. They regularly publish findings on their websites as accountability to their funders, but also for the public to access the information.

NGOs: Just like IGOs, NGOs will pay researchers to do work, but the NGOs publish it as their work. These findings give credibility to such institutions.

Communication officers: Part of their job is to attend exhibitions and spend long hours explaining what the scientists did. They can act as gatekeepers to the scientists as they direct journalists to the appropriate scientist(s). Communication officers may appear on talk shows and write opinions on behalf of the science organisations.

17. The terminology of 'science communication'

Science communication in Uganda goes by several terminologies: adult literacy, training, sensitisation, education, engagement, participation, information sharing and knowledge exchange. Although there are no clear demarcations between the terminologies, the use of the term 'adult literacy' seems to be phasing out. The involvement of stakeholders is seeing a surge in the use of other terms to symbolise science communication.

'Adult literacy' is used in reference to people who missed school or for people who want to acquire new knowledge and skills outside the formal education system. In the context of science communication, this issue is addressed by workshops where adults learn about new agricultural technologies and prevention of hygiene-related diseases.

The concept of 'sensitisation' is commonly used in relation to teaching communities about immunisation, condom use, HIV and avoidance of unnecessary use of antibiotics.

'Training' is a term used by many NGOs in their outreach. For instance, CREEC uses the term to refer to the diffusion of information in making renewable energy accessible. PACE and FHI360 use the term 'support' to refer to their outreach activities. Reach Out Mbuya and Reach A Hand Uganda use the term 'educate' to denote contact with some of the beneficiaries. Other terms used by the NGOs include 'engagement', 'participation', 'information sharing' and 'knowledge exchange'.

18. Challenges, controversies and manoeuvres

The biggest challenge is that the country has few scientists. Coupled with this challenge is the fact that very few are willing to talk to the public and the media. This means that most of the scientific findings remain in the laboratories and academic journals to the disadvantage of the taxpayers who are supposed to benefit from such research. Many journalists have established working relationships with scientists for purposes of easing the tension between the two groups, and for knowledge sharing to the benefit of the public.

The political instability of the 1970s and 1980s affected the dissemination of messages. A change in government had meant a change in programming, and often health (science) programs were affected. In some cases, the government was unable to operate the national broadcaster according to the planned schedule. Although there is a fear that such a scenario may occur again (Uganda has never changed power peacefully), the availability of multiple FM radio stations, private television channels and online publications suggests that the effect can be minimised in a situation of a similar political disturbance.

Radio in the period 1970s and 1980s faced a problem of credibility as propaganda was the key component of program content when the country had a single broadcaster. This problem has metamorphosised as the new sources of information, especially social media, are associated with fake news. Many science journalists try to use scientific sources of information, but the general scribes are not under any obligation to stick to laboratory findings and they often use political sources in their stories, even those on science.

Superstition persists among many communities. There are still individuals opposed to immunisation and women who insist on delivering babies at home and not in hospitals. Some people associate HIV and mental illnesses with witchcraft. Such beliefs make people reluctant to adopt scientifically proven ideas. There is irrational use of communication channels to promote quackery and unscientific options in the place of science. For instance, Ugandan media platforms carry a lot of messages from herbalists trying to explain scientific phenomena. This compounds the superstition around scientific explanations to issues. There are also religious controversies. Uganda is a religious country and every weekend the faithful voluntarily congregate. Unfortunately, some religious cults discourage their members from seeking medical services, taking children out of immunisation programs and sending them to school.

A further challenge is the lack of funding. Most projects do not budget for dissemination, although some science institutions have started factoring in the aspect of dissemination in their project proposals. This can be used to fund media coverage. The costs of disseminating information in a commercial environment are high, and it is expensive to design high-quality messages that meet the demands of the scientists. This issue is exacerbated by the proliferation of newspapers, radio and television stations and online publishers, which makes it hard to justify the selection of a few in the face a stringent budget. Uganda has many news outlets. While multiple platforms are a blessing, they provide a challenge when it comes to choosing the channel to use for a specific message. While channels tend to be associated with certain audience profiles, editorial manoeuvres demand that news outlets capture the biggest possible size of the inelastic audience. Often, media houses adjust to eat into the audience of both related and unrelated channels. The challenge for science communication is that the difficulty in choosing an appropriate channel combines with other social, economic, political and cultural factors to make the message ineffective.

Jargon and cultural attitudes pose a further challenge. Because most scientists are not trained in communication, they tend to use technical terms that are hard for journalists to translate to laypeople. Many journalists are taking it upon themselves to learn how scientists work and are then translating their terminology to ordinary people in their stories, but this learning must match the journalists' passion for the subject and scientists' passion for public media engagement. Scientists can be reluctant to participate in the work of dissemination. Many consider that they are hired to do laboratory work, thereby making science outreach a charity exercise.

Training would help overcome part of the difficulties, encouraging scientists to use simpler terminology and helping bridge the gap between scientists and journalists, but there is a general inadequacy of trainers as science communication has not been mainstreamed in the curricula of universities. There are attempts, however, to incorporate modules in journalism and communication at Makerere University. Some agencies such as SciDev.Net often organise workshops for journalists to improve the way they report, but such training is still rare.

Finally, there are cases of politicians hijacking technical subjects such as environmental protection and genetically modified organisms (GMOs). Politicians determine whether a forest should be degazetted or not. Recently they have hijacked the subject of GMOs, politicised it and pushed the scientists into silence. As a result, the law to legalise or ban GMOs has been shelved because members of parliament and some NGOs have succeeded in making the issue of GMOs emotive.

19. Conclusion

Science communication in Uganda has evolved alongside cultural values attached to health and the environment. The participation of cultural and religious leaders in backing key science messages, and the use of music, dance, drama and comedy, has enhanced science messages. The government broadcaster, now UBC, and later other forms of media, has significantly supported the sharing of scientific findings and learning about such issues and events. Initially, the deficit approach was common, particularly as the technology did not allow serious participation before mobile phones became numerous. Now new forms of communication have democratised science communication to the level of engagement. NGOs have largely influenced the move toward multiple-way communications. Increasingly, the field is providing jobs to researchers, journalists and communication officers, although the possibilities for employment are still limited. Science communicators are researchers, policymakers, journalists and communication officers. However, with public relations entering science institutions and training of trainers intensifying the future of the field is promising.

References

Adams, S. (2008, 5 November). Crocodiles and humans live side-by-side in African village. *The Telegraph*. Retrieved from www.telegraph.co.uk/news/earth/earthnews/3394964/Crocodiles-and-humans-live-side-by-side-in-African-village.html.

Advocates Coalition for Development and Environment (ACODE). (2018). Who we are. Retrieved from www.acode-u.org/about.

Alina, M. (2014, 10 September). The woman in the moon: Full moon viewing. *Makerere University News*. Retrieved from news.mak.ac.ug/2014/09/woman-moon-full-moon-viewing.

Anyango, J. (2018, 12 November). Ugandan journalists trained on Aflatoxins communication and reporting. *Makerere University News*. Retrieved from news.mak.ac.ug/2018/11/ugandan-journalists-trained-aflatoxins-communication-and-reporting.

AVAC. (2018, 29 September). Health journalists link up with scientists at science cafés. Retrieved from www.avac.org/blog/health-journalists-link-scientists-science-cafés.

Bamuturaki, M. (2011). Totems immortalized for conservation and education. Retrieved from bamuturakimusinguzi.blogspot.com/2011/03/totems-immortalized-for-conservation.html.

Böklin, G. (2004). *Regional training programme in environment journalism and communication in the eastern African region*. Stockholm: Swedish International Development Co-operation Agency.

Bucchi, M. and Trench, B. (2016). Science communication and science in society: A Conceptual Review in Ten Keywords. *Technoscienza. Italian Journal for Science and Technology Studies*, *7*(2), 151–68.

Centre for Research in Energy and Energy Conservation (CREEC). (2018). CREEC to start solar pumping training course. Retrieved from creec.or.ug/.

Chibita, M. B. (2006). Indigenous language programming and citizen participation in Ugandan: An exploratory study broadcasting (unpublished PhD thesis). University of South Africa.

Davies, S. R. and Horst, M. (2016). *Science communication: Culture, identity and citizenship*. London: Palgrave Macmillan. doi.org/10.1057/978-1-137-50366-4.

Deighton, B. (2018, 29 March). SciDev.Net launches Script training course. *SciDev.Net*. Retrieved from www.scidev.net/sub-saharan-africa/capacity-building/news/scidev-net-launches-script-training-course-1.html.

Epilla, R. (2014). Institutions, utilisation and conservation of indigenous woody plants in Pader internally displaced persons' (IDP) camps, Northern Uganda (PhD dissertation). Makerere University.

FHI360. (2018). Communication for healthy communities (CHC). Retrieved from www.fhi360.org/projects/communication-healthy-communities-chc.

Great Britain. (1937). *Colonial Office: Interim Report of Broadcasting Committee ('Plymouth Report')*. Colonial no. 139. London: HMSO.

Health Journalists Network in Uganda (HEJNU). (2018). About us. Retrieved from hejnu.ug/about-us/.

Huxley, G. (1958). *Report of the committee of inquiry into the organisation, policy and operation of the government's information services*. Uganda Committee of Inquiry (Chairman: G Huxley). Kampala: Government Printer.

International Service for the Acquisition of Agri-biotech Applications (ISAAA). (2018). Uganda biotechnology information centre initiates efforts to integrate biosciences into Uganda's education system. Retrieved from www.isaaa.org/kc/cropbiotechupdate/article/default.asp?ID=16869.

Kakooza, M. (2012). *Reality check: Revisiting the media freedom debate at Uganda's independence*. Kampala: Uganda Media Development Foundation.

Kakudidi, E. K. (2004). Cultural and social uses of plants from and around Kibale National Park, Western Uganda. *African Journal of Ecology*, *42*(s1), 114–18. doi.org/10.1111/j.1365-2028.2004.00472.x.

Kanene, K. M. (2016). Indigenous practices of environmental sustainability in the Tonga community of Southern Zambia. *Jàmbá -Journal of Disaster Risk Studies, 8*(1), 2–7. doi.org/10.4102/jamba.v8i1.331.

Kiwanuka-Tondo, J. (1990). Educational broadcasting in Africa: The case of Uganda. *The African e-Journals Project, 4*(2), 48–63.

Lancien, J., Muguwa, C., Lannes, C. and Bouvier, B. (1990). Tsetse and human trypanosomiasis challenge in South Eastern Uganda. *International Journal of Tropical Insect Science, 11*(3), 411–16. doi.org/10.1017/S1742758400012832.

Ministry of Science, Technology and Innovation (MOSTI). (2018). Mandate. Retrieved from mosti.go.ug/.

Mushengyezi, A. (2003). Rethinking indigenous media: rituals, 'talking' drums and orality as forms of public communication in Uganda. *Journal of African Cultural Studies, 16*(1), 107–17. doi.org/10.1080/1369681032000169302.

Okoth, J., Kirumira, E. and Kapaata, R. (1991). A new approach to community participation in tsetse control in the Busoga sleeping sickness focus, Uganda. A preliminary report. *Annals of Tropical Medicine & Parasitology, 85*(3), 315–22. doi.org/10.1080/00034983.1991.11812567.

Program for Accessible Health Communication and Education (PACE). (2018). PACE – Interventions for health impact. Retrieved from pace.org.ug/.

Reach A Hand Uganda. (2018). Engaging youth through sexual health education and fun. Retrieved from reachahand.org/programs/.

Reach Out Mbuya. (2018). The vision of Reach Out Mbuya. Retrieved from www.reachoutmbuya.org/vision-mission/.

Rim-Rukeh, A., Irerhievwie, G. and Agbozu, I. E. (2013). Traditional beliefs and conservation of natural resources: Evidences from selected communities in Delta State, Nigeria. *International Journal of Biodiversity and Conservation, 5*(July), 426–32.

Schiele, B. (2018). Cultures of scientific culture. *Culture of Science, 1*(1),15–23. doi.org/10.1177/209660831800100103.

Sekeba, D. S. (2016). *The media bullets in Uganda: A reference guide to history of newspapers, their role in church and politics of Uganda.* Kampala: Angel Agencies Ltd.

Sonke, J., Pesata, V., Nakazibwe, V., Ssenyonjo, J., Lloyd, R., Espino, D., Nieves, M., Khandakji, S., Hahn, P. and Kerrigan, M. (2018). The arts and health communication in Uganda: A light under the table. *Health Communication, 33*(4), 401–08. doi.org/10.1080/10410236.2016.1266743.

Twesigye, C. (2001). Uganda. In *The world encyclopaedia of theatre: Africa.* 3rd edn. (pp. 323–26). London: Routledge.

Uganda Biosciences Information Centre (UBIC). (2016). *Ten facts about biotech and biosafety by 2015*. Kampala, Uganda: Uganda Biosciences Information.

Uganda Communications Commission. (2015). Access and usage of communication services across Uganda. Kampala, Uganda. Retrieved from www.ucc.co.ug/files/downloads/20150130-UCC-access-usage-dissemination.pdf.

Uganda Media Centre (UMC). (2018). Our Mandate. Retrieved from www.mediacentre.go.ug/about/about-umc.

Uganda National Farmers Federation (UNFFE). (2020). Uganda agricultural Show 2020. Retrieved from www.facebook.com/pg/UNFFE/posts/.

Uganda Science Journalists Association (USJA). (2018). Membership. Retrieved from www.facebook.com/pg/usjaug/posts/.

Uganda Water and Sanitation NGO Network (UWASNET). (2018). Objectives and mandate. Retrieved from uwasnet.org/about/.

Uganda Wildlife Authority (UWA). (2013). Queen routes for increased investment in tourism. Retrieved from ftp.ugandawildlife.org/component/k2/queen-routes-for-increased-investment-in-tourism.

Volunteers Efforts for Development Concerns (VEDCO). (2018). Food and nutrition. Retrieved from www.vedcouganda.org/.

Yolisigira, Y. (2015, 21 October). Government rolls out drive against tsetse flies. *Daily Monitor*. Retrieved from www.monitor.co.ug/News/National/Government-rolls-out-drive-against-tsetse-flies/688334-2922348-7wgbr3z/index.html.

Timeline

Event	Name	Date	Comment
First interactive science centre established.	Open Forum on Agricultural Biotechnology (OFAB)	14 December 2007	Coordinated by Uganda National Council of Science and Technology
An association of science writers or journalists or communicators established.	Uganda Science Journalists Association	2005	2014: Health Journalists' Network in Uganda
First university courses to train science communicators.	No direct course on science communication	None	2014: Makerere University introduced this topic in its specialised journalism workshop

Event	Name	Date	Comment
First national conference in science communication.	Makerere University 19th Annual Media Convention	April 2019	Organised by the Department of Journalism and Communication, with funding from SciDev.Net's Script Project
National Science Week founded.	No national event but individual universities/ research institutes organise exhibitions		
First significant radio programs on science.	Radio Uganda program focused on coffee and cotton.	1960	This was the birth of farm radio in Uganda
First significant TV programs on science.	Programs focused on agriculture	1963	Uganda Television, now Uganda Broadcasting Corporation
First awards for scientists or journalists or others for science communication.	National awards in categories of agriculture, environment, and health organised by the African Centre for Media Excellence (ACME)	2013	2017, 2018: The Open Forum on Agricultural Biotechnology organised competitions for journalists with the best winning a trip
Other significant events.	First Science Communication Conference hosted at Makerere University	25 April 2019	

Contributor

Dr Ivan Nathanael Lukanda is a research fellow at Stellenbosch University in South Africa.

39

UNITED KINGDOM

The developing relationship between science and society

Melanie Smallman, Simon J. Lock
and Steve Miller

The 1950s British Prime Minister Harold Macmillan is reported to have replied to a journalist's question as to what was likely to blow his government off course with the words 'events, dear boy, events'.

The development of science communication in the UK from the mid-1980s onwards is one of the best-documented stories in this field, punctuated by a series of reports from both the scientific community and the government itself—and by a number of 'Macmillian events' that blew science's relationship with the wider world hither and thither. Here we offer a series of episodes that changed how we think about the relationship between science and society. We describe these largely chronologically and imply that they heralded new eras of science communication. This does not mean, however, that previous approaches simply disappeared: many old ideas were buried momentarily or continued as an undercurrent, less visible but ready to resurface as and when conditions allowed and required them to do so.

1. The advent of the 'modern' era: From 'mad cows' to the 'crisis' in science communication (1985–97)

During the 1980s, the Conservative government of Margaret Thatcher adopted a policy for scientific research that prioritised near-market, applied projects over those of a more fundamental, 'pure science' nature, as part of its efforts to harness science as a driver of economic growth (Agar, 2011; Guise,

2014). The scientific community was concerned and a feeling grew that this (as they saw it) misguided and short-termist approach was due to a failure of politicians and the wider public to understand science (understood widely to include science, technology, engineering and medicine (STEM)). More 'public understanding of science' (PUS) was needed.

In 1985, the Royal Society (RS, founded in 1660), the UK's premier scientific society, published a report on 'The Public Understanding of Science' (Royal Society, 1985), referred to as the Bodmer Report[1] after Sir Walter Bodmer, the chair of the committee that drew it up. This was pivotal: amongst other things, it stated that scientists should consider it their duty to communicate to their fellow citizens about their work and its importance, galvanising and authorising researchers to communicate with the 'general public'.

Bodmer resulted in the establishment of the Committee on Public Understanding of Science (COPUS), with representatives from the RS, the British Association for the Advancement of Science (founded in 1831, and now the British Science Association (BSA)), and the Royal Institution (RI). COPUS organised funding schemes for PUS activities, and handed out prizes for initiatives, individuals and science popularisation books—of which there followed quite a boom, led by Stephen Hawking's 1988 *Brief History of Time*. A national science week was established in 1994; typical of the UK it managed *not* to coincide with its continental European counterpart.

In 1989, John Durant (the first UK Professor of Public Understanding of Science at Imperial College), Geoffrey Evans and Geoffrey Thomas published the first major survey of public understanding of science in the UK. They found high levels of interest, but that only some 14 per cent of British citizens could be called 'scientifically literate' (a term imported from the United States) according to tests of knowledge of scientific terms and processes. They concluded: 'If modern science is our greatest cultural achievement, then it is one of which most members of our culture are largely ignorant' (Durant, Evans and Thomas, 1989, p. 13). To the scientific institutions involved in PUS activities this confirmed a deficit in the public understanding of science, and the battle was on to fix that deficit before it was filled by charlatans and soothsayers (see Richard Dawkins' 1996 *Richard Dimbleby Lecture* on BBC TV, for example). This approach, which assumed that more knowledge would build a public more supportive of science, became known as the 'Deficit Model' (e.g. Gross, 1994).

1 In what follows, we use 'Bodmer' to refer to the report and the PUS movement it engendered.

Bodmer had also called for more science to be carried in the media, a call taken up enthusiastically by the British Broadcasting Corporation (BBC), which already had several TV and radio programs. Independent broadcasters also increased their science offerings. Newspapers responded by appointing science journalists, correspondents and editors, and some introduced special science sections into their regular pages. Particularly influential was (now Lord) Melvin Bragg, who had made his name in arts programming: his BBC Radio 4 program *Start the Week* regularly featured science themes and leading scientists. Debates around the notion of 'The Two Cultures', first identified as such by C. P. Snow (1959, 2012), flared into life again: was science overweening and over powerful or was it looked down upon as being less worthy by the media and political elite who were mainly arts and literary educated? (Weldon, 1991; Wolpert, 1991; Gregory and Miller, 1998; Lock, 2016).

Political legislators and representatives were also Bodmer targets. It had been suggested that MPs and Members of the House of Lords would benefit from independent scientific advice, and in 1989 the Parliamentary Office of Science and Technology (POST) was set up with charitable funding. Parliament adopted POST three years later (and made it a permanent institution in 2001). The government set up its own Cabinet Office of Science and Technology (OST) to handle policy on science, and its report of 1993—*Realising our Potential*—led to UK research councils being charged with ensuring that the work they supported was communicated to taxpayers, and with setting up advisory committees to help them (UK Government, 1993).

In the higher education sector, courses in science communication were started at master's level (Imperial College, London, 1992) and undergraduate level (University College London, 1990), and graduate training was started at several universities and colleges, some paid for by the Wellcome Trust, the UK's largest funder of biomedical research. Research into PUS also gained momentum via a dedicated program by the Economic and Social Research Council. In 1992, Durant and his Science Museum colleague Jane Gregory set up the peer-reviewed journal *Public Understanding of Science*, which joined the renamed *Science Communication* to give the PUS community two outlets for scholarly research. PUS, with its twin aims of improving scientific literacy and public attitudes towards science was well on its way: the outlook was bright with just a few clouds on the horizon.

One of those clouds, however, was bovine spongiform encephalopathy (BSE) or 'mad cow disease'. At the same time that Bodmer's committee was writing its report, farmers in the UK were reporting cattle collapsing from a disease that left them staggering and slobbering across the farmyard and dying in considerable discomfort (see Jasanoff, 1997; Millstone and van Zwanenberg, 2001; Reeves, 2002; Frewer et al., 2002, for examples of scholarly approaches to BSE and the crisis it caused in the UK). The Central Veterinary Laboratory first recognised BSE as a novel cattle disease in 1986. At that point, the medical/scientific jury was out on just what was happening. BSE looked like scrapie, well known for affecting sheep, and officials at the Ministry of Agriculture, Food and Fisheries (MAFF) speculated that cattle being fed with protein-supplement feed containing the remains of scrapie-infected sheep were somehow catching a similar disease.

The problem was that no bacterium or virus could be found to be carrying the infection and the idea of 'prions' (Prusiner, 1982) had yet to be accepted by mainstream vets and doctors (Reeves, 2002). As concerns for human health from eating BSE-infected beef started to grow, along with media interest in what was happening down on the farm, scientists were able only to tell politicians that the chance that BSE could infect humans was 'most unlikely' (MAFF/DoH, 1989, advice cited in Beck, Asinova and Dickson, 2005), although they did advise measures to destroy infected cattle and ensure they did not enter the food chain.

In May 1990, however, the (now-defunct) middle-brow newspaper *Today* published two articles, one of which linked the death of a cat to BSE and one of which proclaimed: 'Scientific proof: mad cow link to humans' (Wilenius, 1990; Brough, 1990). A more serious blow was struck by the heavyweight *Independent on Sunday* (Nicholson-Lord, 1990), which reported that based on medical advice from its own Fellows, beef was no longer being served at Magdelene College Cambridge. The market for British beef was already in decline, and its banning from High Table was clearly a major blow.

In response, Agriculture Minister John Gummer told *The Independent on Sunday* that his whole family ate beef (Cannon et al., 1990), and posed with his young daughter taking bites out of beef burgers (although it is clear from the photos that it was Gummer who had bitten into both burgers). Gummer's second-in-command David McLean poured scorn on the non-beef-eaters: 'If there are some people who do not want to believe that it is safe, God help them. But let them … not scare the vast majority of us who have common sense' (Craig and Francis, 1990; War, 1990).

While Prusiner's prion-infection theory of diseases like kuru, scrapie and BSE gradually gained traction amongst the medical community, opinions on the safety of British beef amongst scientists, politicians and the media remained divided. The 'settlement' came in 1996: on 20 March, the left-wing tabloid *Daily Mirror* published a scoop under the full-page headline 'MAD COW CAN KILL YOU', with an official-looking stamp on it. Later that day, Health Minister Stephen Dorrell announced to Parliament that a cluster of cases of a disease known as variant Creuzfeld-Jacob Disease in young people was probably linked to eating BSE-infected beef. Government advice that British beef was 'safe' had been turned on its head. Confidence in the government—seen as having misled the public for a decade over the safety of British beef—collapsed. According to the official Phillips report into the whole affair: 'When on March 20, 1996 the Government announced that BSE had probably been transmitted to humans, the public felt they had been betrayed' (Phillips, Bridgeman and Ferguson-Smith, 1997).

Just before Dorrell's announcement to parliament, Astronomer Royal Sir Arnold Wolfendale (1996) had carried out a review of PUS/science communication activities for the government. It was largely positive with few recommendations for major changes. 'Steady as you go', then. Viewing the debacle from across the Atlantic Ocean immediately after the government's *volte face* on BSE, Sheila Jasanoff came to a rather different conclusion: she saw a state of 'civic dislocation' in which there had been an 'unprecedented breakdown of communication' between British citizens and their government. In the dislocated state 'trust in government vanished and people looked to other institutions—the high street butcher, the restaurant, the media, the supermarket—for information and advice'. For her, British society had changed 'in profound ways that call for new forms of engagement between citizens and their government' (Jasanoff, 1997).

So how did BSE play into the era of the Deficit Model? Parallel to (but separate from) 'official' efforts to increase science communication, and hence public literacy in matters scientific, historians and sociologists of science in the UK had been pointing out that relations between citizens and research communities were much more complex than the idea that 'the more you tell people about science, the more they will know and the better they will like it'. The Bodmer Report had suggested that research into the efficacy of science communication activities should be funded by the Economic and Social Sciences Research Council, although their actual program owed more to pre-existing for science and technology studies (STS) than Bodmer itself. Many researchers had looked at how science was appreciated and understood by people in their everyday lives (e.g. Wynne, 1991) and how they could cope with and implement scientific advice in real-world situations (e.g. Irwin, 1995). These researchers stressed the importance of the context in which information was imparted and by whom (Layton et al., 1993). As part of this parallel discussion, Alan Gross (1994) proposed a 'contextual approach' to PUS that stressed the science we need to live plus appreciation of local knowledge and culture. 'The contextual model explores the ramifications of the interaction between science and its publics … it depicts communication as a two-way flow,' he explained (p. 6), unlike the deficit approach that stressed scientific *suf*ficiency as against public *de*ficiency.[2]

2 See also Layton et al. (1993) for further sources of these ideas.

Looking to put some of this research into practice, in 1994 the Biotechnology and Biological Sciences Research Council (BBSRC) sponsored the Science Museum in London to run a 'consensus conference' on plant biotechnology that attempted to get more citizen involvement in science policy issues (UK Consensus Conference on Plant Biotechnology, 1994; Joss and Durant, 1995). The panel of 25 lay citizens reached a consensus that they were not against technologies to improve crop qualities and yields so long as they were well regulated and labelled. When the 1996 UK social attitudes survey found that, after a decade of Bodmerism and all the efforts it engendered, there had been little change in British scientific literacy, the deficit approach seemed clearly to be failing (see Miller, 2001, for a discussion of this). But just what was PUS/scientific literacy anyway and how might it work in the BSE situation?

In 1993, Durant had published a paper asking just that question. He looked at three possible definitions of scientific literacy: knowing a lot of science, knowing how science worked, and knowing how science *really* worked. Questions probing the first two definitions had made their way into the surveys used in 1989 and 1996 to see how scientific literacy was progressing. For the first definition—knowing a lot of science—the problem in the BSE case was that the 'science' in question was far from the tried-and-tested, simple textbook science that we are supposed to learn at school. Right from the first identification of BSE as a new cattle disease, there was controversy as to just what the science was—about the source of BSE, the course of BSE, and its infectiousness to humans. Knowing a lot of science might be nice, but not too helpful in this case.

The second definition—knowing how science works—might have been more helpful. But the understanding of prion diseases and their infectious agents was at the stage of 'science in the making' and government officials were faced with immediate decisions to be made, without the benefit of long periods whilst the scientific/medical community tested one hypothesis against another. In any case, many researchers (e.g. Collins and Pinch, 1993) had shown that the 'hypothesis testing' account of the way science works often departed considerably from how it *really* worked, it tended to leave out much of the social context of the scientific community and how it came to judge what was and what was not good and relevant science. Understanding how science *really* worked was clearly not an unproblematic task; scientific literacy *per se* simply was not the 'answer'. Although the BSE scandal brought the (pre-)dominance of PUS/scientific literacy to a close in the UK (see Bauer et al., 2007), it did not do away with the need for good science and clear communication (Miller, 2001): it simply showed that this alone was not enough.

2. GM foods, the New Labour Government and the participatory turn (mid-1990s – mid-2000s)

In the wake of the BSE scandal, 'New' Labour under Tony Blair won the May 1997 election with a huge majority (418 MPs out of a House of Commons of 659); more than half of all sitting Conservative MPs, including several ministers, lost their seats. This political sea-change marked a watershed in UK science–society relations and approaches to science communication. Drawing strongly on the ideas of Anthony Giddens and the think tank Demos,[3] New Labour saw public participation as an important way to address the perceived democratic deficit, which was seen to be causing citizens to feel increasingly distant from and disillusioned with traditional decision-making structures, leading to a drop in voter turnout in elections and increasing cynicism in politics and government (Barnes et al., 2007).

Citizen participation was also seen as an important way to raise standards in public services and to find the best possible fit between user needs and service capacity. As a consequence numerous public and citizen participation opportunities and exercises arose. These ranged from the best value tendering regime, which placed a duty on local authorities to involve local citizens in reviewing services (Barnes et al., 2007), to the 'New Deal for Communities' program that involved local people in decisions about community regeneration (Smallman, 2016b). Various participatory methods such as citizens juries, deliberative polling and citizens panels were also introduced to inform local health care priority-setting decisions (Abelson et al., 2013). Giddens' influential book *The Third Way* described the need for more public participation in science, arguing that given the increasingly complex relationship we have with science and technology '[d]ecision making in these contexts cannot be left to the "experts", but has to involve politicians and citizens' (Giddens, 1998, p. 59). The democratisation of science and technology in this context was very much part of the process of the modernisation of government in the UK (Smallman, 2016b).

While the spectre of BSE cast a 'long shadow' (Stilgoe, 2007), the Labour government was keen to signal it had learnt lessons and moved on in its approach to dealing with the communication of public science. As Sir Robert

3 Sociologist Anthony Giddens was a pioneer of the 'Third Way' view of welfare and participation that regarded the job of government as being to give citizens a 'hand up' rather than a 'hand out', making them active citizens rather than passive recipients of welfare. His views were very influential in the Labour government under Prime Minister Tony Blair. The Demos think tank was a persuasive advocate of public participation in (scientific) decision-making, rather than leaving everything to scientific experts.

May, the Chief Scientific Adviser at the time of the Phillips inquiry, described in his evidence, the prevailing instinct in these situations up until this episode had been 'to hold the facts close' so that a 'simple message can be taken out into the market place'. BSE, however, suggested to him that the 'full messy process whereby scientific understanding is arrived at, with all its problems, has to be spilled out into the open' (Phillips, Bridgeman and Ferguson-Smith, 2000).

Many of the themes emerging from BSE, in particular the debate about scientific uncertainty, risks and how best to communicate science to the public, were also put under the spotlight at the beginning of 1999 when genetically modified (GM) crops became a controversial issue, drawing government, scientific institutions, the public, the media and industry back into public debate.

GM crops had been rapidly introduced in the UK market from the mid-1990s, prompting unease amongst green activist groups. Media coverage grew, and GM became a political issue in the House of Commons. On 12 February 1999, *The Guardian* published a letter from 12 scientists supporting the unpublished research of Dr Arpad Pusztai on the harmful effects of GM potatoes fed to rats (Rhodes et al., 1999). The matter became front-page news for almost two weeks. Many of the media outlets initiated high-profile campaigns against GM crops. The episode echoed the BSE controversy in several ways, not least when Prime Minister Tony Blair was quoted in a national newspaper as being happy to eat 'Frankenstein Food' and to feed it to his children; he was frustrated that the potential benefits of GM food were being ignored in the escalating row (Daily Mail, 1999).

It was in this questioning climate that the Science and Technology Select Committee of the House of Lords appointed Lord Jenkin of Roding to conduct a study into society's relationship with science (House of Lords Select Committee on Science and Technology, 2000). An earlier Lords report into the management of nuclear waste had devoted a chapter to considering the public acceptability of the issue. That report had acknowledged the complexity of public attitudes and values with respect to science and technology, the influence of the media and the importance of public trust in institutions (House of Lords Select Committee on Science and Technology, 1998). With the ongoing controversy over GM foods, and the changing nature of the scientific advisory processes within government in response to the BSE affair, Jenkin's team examined both the sources of information that shaped public attitudes to science and the mechanisms for facilitating dialogue between scientists and the rest of society. Notable in the production of this report was the appointment of Durant and STS academic Brian Wynne, who had been openly critical of the prevailing approaches to science communication for many years, as special advisors.

The final report, *Science and Society*, was published in February 2000, and concluded that society's relationship with science was in a critical phase. A 'crisis in confidence' on the part of the public, was how the committee characterised the relationship between science and society; and it put forward many different recommendations as to how this relationship could be improved, not least by advocating a shift away from 'simply giving information' to 'engaging the wider public in dialogue about what science could and should be doing' (p. 13). There was, however, a tension within the report: dialogue and engagement meant assigning a level of legitimacy to the public, allowing public values to be considered, and opening up the science policymaking process; in other senses, it appeared more to do with listening to the public's attitudes, and then, having done so, science could be trusted to get on with scientific research. As the report emphasised, dialogue was intended to secure science's licence to practice, not to restrict it—thus the scientific experts maintained authority and responsibility over science.

The effects of these recommendations were both quick and widespread, with science communication and scientific institutions in the UK (for example, the RS and the Research Councils) shifting the focus of existing programs in PUS or science communication to 'science in/and society' programs, prioritising dialogue exercises over more traditional one-way transmission activities. Perhaps the most obvious of all institutional changes, and a sign of how much the House of Lords report had changed both the landscape and language used in a short space of time, was the reformulation of COPUS. Following extensive reviews of the committee in the previous two years, it was agreed by all the partners that COPUS should be remodelled in this new climate as 'an inclusive partnership between the many sectors now involved in communicating science' (Lock, 2011). COPUS was now no longer to be used as an acronym, but as a brand 'Copus', with a new expanded council reflecting a broader range of stakeholders in science communication than the original three founding bodies. Chaired by Bridget Ogilvie, previously chief executive of the Wellcome Trust, the new body was intended to oversee science communication at a national level. But it was made clear that this was not to be one-way science communication; instead, it was to be focused on 'supporting ways of increasing public engagement with the issues and processes of science' (p. 25). The COPUS grant schemes were also reformulated under this new agenda to fund efforts at dialogue with the public.

Commitment to dialogue and debate was soon put to the test: in 2002, the UK government set up the first nationwide public dialogue on the heated issue of GM foods (Rowe et al., 2005). The Agriculture and Environment Biotechnology Commission had previously concluded that the development of GM crops had 'suffered as a result of the lack of opportunity for serious debate about the full range of potential implications of GM agriculture, on the basis of clear understandings of what is involved, away from concern that had been created by campaigning elements of the media'. Consequently, they argued that the government needed to 'encourage comprehensive public discussion of the ecological and ethica—including socioeconomic—issues which now have arisen' (AEBC, 2001).

Comprising a series of open public meetings, a dedicated debate website and a series of closed discussions (which acted as control groups), the GM Nation debate ran over the summer of 2003 and is believed to have reached more than 20,000 people (Rowe et al., 2005). Its stated aims were twofold:

> To promote an innovative, effective and deliberative programme of debate on GM issues, framed by the public, against the background of the possible commercial production of GM crops in the UK and the options for possibly proceeding with this; and through the debate provide meaningful information to Government about the nature and spectrum of the public views, particularly at grass roots level, on the issue to inform decision-making (PDSB, 2003).

'GM Nation?' was an 'unprecedented', and therefore experimental, national event in public dialogue. From a number of perspectives, however, it 'failed'. For the government and some scientists, it was seen to do little to take the heat out of the GM debate (Gaskell et al., 2004; Horlick-Jones et al., 2006). Others argued that it was impeded by lack of time and money (House of Commons, Environment, Food and Rural Affairs Committee, 2003), that it failed to engage with a sufficiently wide array of people, that it was primarily a legitimatory exercise and that it lacked focus (Irwin et al., 2012; Council for Science and Technology, 2005; House of Commons, Environment, Food and Rural Affairs Committee, 2003).

The GM debate was also seen to be foreshadowing a controversial emerging science—nanoscience. Although the UK was leading the way in this promising new field, would public opposition curtail it? In *See through Science* (Wilsdon and Willis, 2004), Demos used the launch of a report on nanoscience from the RS and Royal Academy of Engineering (Royal Academy of Engineering and the Royal Society, 2004) as the backdrop against which to make the case for 'upstream engagement', using this as an opportunity to correct the mistakes that were made with GM. In part to forestall any potential public opposition to this new area of science, the UK government launched the ScienceWise program in 2004, to encourage, fund and support policymakers to involve the public in decisions around science and technology. At the same time COPUS and its grant schemes was wound down and retired.

The ScienceWise scheme was launched to 'build the capacity of citizens, the science community and policymakers to engage in the dialogue necessary to establish and maintain public confidence in making better choices about critical areas in science and technology' (HM Treasury 2004, p. 108). The focus of PUS and related science communication activities had been firmly on increasing the amount of scientific knowledge and understanding that the public held, yet the effect of the BSE and GM crops controversies had shifted the focus onto issues of scientific advice and public trust in science and scientists; and dialogue and consultation were seen as the new orthodoxy in science policy and communication circles.

The switch of funding from 'traditional' science communication to processes that involved the public over policy-relevant issues had had its effect on the community. For example, the BSA was struggling to change itself to fit with the new dialogue and policy-oriented approach to science–society relationships. A review of strategy had developed a new purpose for the association—'to create a positive social climate in which science, and organisations dependent on it, advances with public consent, involvement and active support'—moving it away from its previous focus on science communication and public understanding (British Association for the Advancement of Science, 2005). Their 2005 annual Science Communication Conference was a fractious affair with a clear division between those individuals and organisations there to discuss practical efforts to improve science communication and those interested in discussing dialogue and influencing government policy.

As the drift from deficit to dialogue became a surging tide, criticism began to emerge. There was much discussion about the representativeness of the participants and whether dialogue was meant to represent existing views or to help to form new ones (Lezaun and Soneryd, 2007). Evaluations of dialogue activities highlighted how little impact they had on policy and that the objectives of science often went unchallenged at these events. Critics argued that the events themselves worked on the assumption that science is an inherent good, and limited public participation to voicing aspirations and concerns, rather than discussing the type of world that the particular science or technology was building (Macnaghten, Kearnes and Wynne, 2005; Wynne, 2006; Smallman, 2017).

3. MMR to post truth

While the period from the late 1990s to the mid-2000s has been characterised here and elsewhere (for instance, Stilgoe et al., 2014; Irwin, 2014; Smallman, 2014) as a move from deficit to dialogue, it is important to recognise that other, more 'traditional' science communication activities, aiming to explain science to the general public, have nonetheless continued to take place—and to flourish. The British Association for the Advancement of Science (BA) *has* continued to run its annual meeting in September of each year and its National Science Week (renamed British Science Week in 2014) in March; the Edinburgh Science Festival *has* taken place each spring. In 2002, Frank Burnet and Kathy Sykes, science communication academics and practitioners based in Bristol, launched the Cheltenham Science Festival. This became well known for originating the global FameLab contest, which invites scientists to compete in the style of TV's *X Factor* to give the best public talk. Science festivals have since mushroomed across the UK—from Brighton to Dundee—such that there were more than 30 science festivals involved in the UK Science Festivals Network in 2018.

Activities to encourage school children to take up careers in science and technology have also remained an important feature of the UK's science communication landscape. The RS continues to open its doors to the public every July for its annual Summer Exhibition, attracting several thousand school visitors in 2017. In 2006 Engineering UK organised the first Big Bang Fair, which aimed to show young people 'the exciting and rewarding opportunities out there [in science] for them'. Claiming to be the 'largest celebration of STEM for young people in the UK', the fair attracted 80,000 visitors in 2018 (Big Bang: UK Young Scientists and Engineers Fair, 2018). Since 2013, however, the event has been subject of criticism for the heavy involvement of arms manufacturers and fossil fuel companies (e.g. Bell, 2013).

In this spirit of explaining science to the public and sharing scientists' enthusiasm for their subjects, the citizen science movement also grew significantly in the UK during the 1990s and early 2000s. While Alan Irwin's original conception of citizen science as a way to create more active 'scientific citizenship' by bring the public and science closer together through dialogue and decision-making around environmental risks (Irwin, 1995), it has become a term to refer to involving citizens in the 'doing' of science—at least in part to help share scientists' enthusiasm for their subjects. Typically, citizens are involved in gathering or analysing large amounts of data. For example, the BBC and *The Daily Telegraph* set up the MegaLab in 1995, which, led by Prof. Richard Wiseman of the University of Hertfordshire, used national television, radio and press (BBC1's *Tomorrow's World*, BBC Radio One and *The Daily Telegraph*) to test whether it is easier to detect lies in print, radio or TV (Wiseman, 1995). In 2014, 'citizen science' was added to the Oxford English Dictionary (Bonney et al., 2016) and today is the subject of activities from scientific establishment organisations ranging from the BSA to the National Endowment for Science, Technology and the Arts (NESTA).

Reflecting the dual purpose—democratising science and promoting science—that has arisen within science communication in the UK in the early 21st century, the term 'public engagement' came into popular use amongst the community in the early 2000s (Suerdem et al., 2013). This term—which implies both capturing the public's attention and involving them in decisions about science without actually specifying either—was a compromise that dissipated tensions between these two different viewpoints within the community. A debate that took place on the psci-comm mailing list, titled 'the Importance of Public Understanding of Science' in May 2001—around the time of the 'move' from deficit to dialogue—illustrates this well.[4] Key figures from science and society discussed how on one hand the term 'engagement' showed the new bottom-up approach to 'science and society' relations, while on the other hand it disguised the PR nature of many science communication activities. However, no one disagreed that engagement was a better term than PUS.

Cementing engagement as the agreed compromise—and burying any possible tensions between the two approaches to science and society even deeper—in 2008, the Higher Education Funding Councils, Research Councils UK and the Wellcome Trust funded a series of six 'Beacons of Public Engagement' and a National Coordinating Centre for Public Engagement (NCCPE). The six beacons—based in Wales, Edinburgh, East Anglia, Manchester, North East England and UCL—were 'university-based collaborative centres that were set up to support, recognise, reward and build capacity for public engagement' (NCCPE, 2018). Funded for four years, the beacons aimed to encourage a culture change in UK universities, to open them up to engagement with the wider public.

The beacons adopted a broad definition of public engagement, to encompass 'a whole family' of types of engagement. Reflecting this, the beacons—and the subsequent 'Catalysts for Public Engagement'—have driven forward a wide variety of science communication activities ranging from Steve Cross's UCL-based 'Bright Club',[5] where UCL academics were trained to perform stand-up comedy about their work, to Swansea University's 'Little Voices Shouting Out / Lleisiau Bach yn Galw Allan' project that supports children to carry out research on issues that matter to them and to present their findings to policymakers (Little Voices, 2018). The NCCPE itself has also developed best practice guidance for universities, the annual 'Engage' conference and a journal.

4 See www.jiscmail.ac.uk/cgi-bin/wa-jisc.exe?A1=ind01&L=PSCI-COM.

5 See www.ucl.ac.uk/culture/projects/bright-club.

Over time, a number of these traditional science communication events have changed shape and focus, often reflecting debates taking place in the more dialogic arm of science in public. In 2018, for example, under the leadership of Chief Executive Katherine Mathieson, the BSA launched a new mission 'to transform the diversity and inclusivity of science; to reach under-served audiences, and increase the number of people who are actively engaged and involved in science' (British Science Association, 2018). Similarly, Muki Haklay at UCL developed a new format of citizen science that moves closer to Irwin's original democratising idea. In Haklay's 'extreme citizen science', scientists and non-scientists (or professional and non-professional scientists, as he terms these two groups) work together to decide the scientific problems to work on and how to collect and validate data (Haklay, 2013). Nevertheless, the purpose of sharing scientists' enthusiasm for science has remained.

4. The pushback against dialogue

Despite these moves to bridge the divide between science promotion and democratisation, many scientists continued to hold the view that the media was misrepresenting science and influencing public opinion inappropriately. These tensions came to the surface around the time of public debates about the safety of combined MMR (measles, mumps and rubella) vaccines, GM foods and mobile phones in the early 2000s. In particular, media-driven concerns about the safety of the combined MMR vaccine—concerns unsupported by the majority of the biomedical community—led to large numbers of (particularly middle class) parents refusing to have their child vaccinated. As a result, there were a series of local epidemics of measles, in which a number of young children died.

In January 2002, Liberal Democrat Peer Lord Dick Taverne and former Wellcome Trust Director Brigit Oglivy had launched the organisation 'Sense about Science', which aimed to 'put evidence at the heart of public discussion'. On the launch of Demos' 2005 *See Through Science* report, Taverne made clear his opposition to moves to democratise science, writing a letter to the journal *Nature* subtitled 'We must face the fact that science—like art—is not a democratic activity'. His letter asserted: 'You do not decide by referendum whether the Earth goes round the Sun' (Taverne, 2004). 'Sense about Science' has since developed an 'Evidence Base': a database of experts who want to put good science at the heart of public discussion; and a series of 'making sense of …' briefings that guide the public on what questions to ask about scientific issues. It has intervened in a number of public controversies including MMR and GM.

Around the same time (2002), the Science Media Centre (SMC) was launched. Housed at the Royal Institution, yet independent from both it and the government and led by Director Fiona Fox, the SMC proposed to take a proactive stance by providing science stories, and scientists, to the media where they felt there was 'a public interest or a developing controversy'. Its mission statement claimed to 'provide a focal point for scientists to explain the nature of their work, discuss its consequences, and engage in public discussion over the benefits and risks'. The website of the SMC claimed to trace its roots back to the desire of the House of Lords' *Science and Society* report (2000), 'to renew public trust in science'. (The Lords' report had largely concluded that the scientific community should deal with the media as it was, rather than seek to change how it operates.)

The establishment of the SMC was, on the one hand, a sign that members of the scientific community were following this and trying to deal with the media on its own terms, through effective communication. On the other hand, the move could equally be interpreted as an indication that some within the scientific community were still angling for more control over media messages about science—for example, by being able to field the 'right' scientists to communicate particular messages. Communication of accurate science was seen as the means of improving the relationship between the two, as the SMC's current aim proposes 'supporting them [scientists] to engage with the media; creating more opportunities for them to get their voices heard on the big science, health and environment stories of the day' (SMC, n.d.). For all the 'Sense about Science' and SMC efforts, paradoxically it was to be a journalist, Brian Deer, who did most to debunk the MMR scandal in the public mind, exposing Andrew Wakefield, the researcher who first suggested there was evidence to question the safety of the vaccine, as being in the pay of lawyers acting for families suing the National Health Service (see Deer, 2009, for example).

Moves to take hold of the agenda did build some momentum in the mid-2000s: Ben Goldacre wrote a popular 'Bad Science' column in the *Guardian* from 2003 to 2011, highlighting instances of pseudoscience and the misuse of science. In 2012, biology student Elise Andrews set up the Facebook Group 'I f***ing Love Science', attracting more than a million 'likes' in its first year—ostensibly from its posts presenting interesting and amusing facts about science, but also because sharing posts from the site allowed people to demonstrate their scientific and rational identities and their belonging to this 'tribe' (Marsh, 2018). The former *Times* science writer Mark Henderson authored the popular book *The Geek Manifesto – Why science matters* in 2013.

It drew again on media scares around GM crops, vaccines and nuclear power. *The Geek Manifesto* advocated for more scientific thinking in public life, this time giving it a more activist edge, claiming:

> Something is stirring among those curious kids who always preferred sci-fi to celebrity magazines. As the success of Brian Cox and Ben Goldacre shows, geeks have stopped apologising for an obsession with asking how and why, and are starting to stand up for it instead. The Geek Manifesto shows how people with a love of science can get political, to create a force our leaders can no longer afford to ignore (p. 12).

5. Responsible research and innovation

In parallel to the 'fightback' initiatives, however, there have been efforts to make science and scientific research agendas even more, not less, public property, areas of legitimate interest and concern for 'ordinary' citizens. The concept of responsible research and innovation (RRI) developed in the late 2000s under the auspices of the European Commission (see Smallman, 2018, for a discussion on the relationship between citizen science and RRI). In 2011, the Commission developed and adopted a concept of RRI that built upon the earlier ideas around public participation and dialogue, but with the aim of involving all actors (not just citizens or experts) throughout the process of innovation such that science could be more firmly rooted in society and society's needs and ambitions (Owen et al., 2012). This heralded a move from 'science in society' to 'science with and for society' (Laroche, 2011) and RRI was introduced as a cross-cutting theme of the Horizon2020 program. In RRI, science communication was seen as one of six key strands rather than a standalone activity.

Several UK institutions took up the challenge of RRI, most notably, the Societal Issues Panel of the Engineering and Physical Sciences Research Council (EPSRC). This panel had been set up in the wake of the GM debate, with the aim of identifying future issues of concern. The panel invited Richard Owen, then Professor of Responsible Innovation at the University of Exeter, to help them develop a more general framework for researchers. In 2013 their framework for Responsible Innovation, based on the principles of Anticipation, Reflection, Engage and Act (AREA) was published (Miller, 2016). Shortly after publishing this framework, the EPSRC (with the BBSRC) put out a call to fund a series of synthetic biology research centres that would include specific work on RRI. Those on the assessment panel, which included Owen, reported that the proposals had taken up the

AREA approach and embedded RRI (and public dialogue) within their sense of 'excellent research', producing more imaginative and interesting research proposals (Miller, 2016).

While it is very much an approach to governing emerging technologies, RRI has had significant implications for science communicators involved in public participation. The concept has arguably shifted the role of the science communicator from one who explains science to the public, to one who helps scientists and technology developers understand society. Arguably the objective of helping science to succeed remains, but it is achieved by helping science do more socially acceptable research.

6. Brexit and the post-truth era

On 2 June 2016, in the run up to the UK European Referendum, former education minister Michael Gove was challenged on television to name any economic experts who supported his position to leave the European Union. Refusing to answer the question, he replied: 'People in this country have had enough of experts (Gove, 2016).' This comment sent a chill down the spine of many scientists, leading many once again to seek better science communication, and the reassertion of expertise and reason, as the answer to many of the world's problems. *New Scientist* ran an editorial in June 2016, summing up this sentiment with the headline 'Post Brexit, experts need to reassert their value to society'. In it they called for scientists to improve their communication skills so that they can 'speak the emotional language of the victors' (New Scientist, 2016).

Others offered more nuanced positions—for instance, Tracy Brown from 'Sense about Science' argued that there was no evidence to support claims of falling support for experts (Brown, 2016). Others still suggested that Michael Gove's comment resonated with the public only because such a gulf had arisen between expert accounts of the world—including the impacts of science and technology—and the day-to-day experiences of ordinary people. Listening to the public more and involving more diverse perspectives in scientific and technological developments would be the answer to avoiding future troubles (Smallman, 2016a).

7. Conclusions

The timeline for science communication and science and society relations in the UK shows that this is far from a linear story. There are repeating cycles of activity as one or other of the various strands we have identified come to the fore or retreat into the background: at no time does any theme we have identified come to a complete end; at no time does the prevalent 'model' completely supersede what has gone before—there is no 'death of the dinosaurs' event that consigns a whole ecosystem of science communication to the fossil record. Instead, ecosystems evolve both independently and interactively at different times, at different rates of 'progress'.

In many ways, the UK is an exemplar of—sometimes even a driver of—changing science communication landscapes further afield than its own national borders. This is particularly the case across the European Union, where programs on Raising Public Awareness of Science and Technology (1999–2004, essentially PUS-plus) were replaced by science-and-society and then science-with-and-for-society as dialogue and debate and other engagement-oriented approaches came forward in the UK.

Although many of the main strands exist, develop and co-exist with one another, there are changes nonetheless. In the UK science communication training for researchers—workshops, courses, programs—has developed alongside demands in government for science to have demonstrable—and measurable—impact. Research council (and other) funding has helped to provide incentives for this. To an extent these efforts have been institutionalised in, for example, the National Coordinating Centre for Public Engagement, which allowed the beacons projects to develop and the Research Excellence Framework (REF) that gives additional funding to institutions that can demonstrate 'impact'. But it is also—typical of the UK—a story of many activities by individuals and individual groups and institutions.

If we are to draw just one lesson from our study of science communication in the UK it is that there is no one 'right way'. Having a genuine landscape inhabited by a multi-faceted ecosystem means that models, approaches and activities can come to the fore and fade into the background as the circumstances demand.

References

Abelson, J., Blacksher, E., Li, K., Boesveld, S. and Goold, S. (2013). Public deliberation in health policy and bioethics: Mapping an emerging, interdisciplinary field. *Journal of Public Deliberation, 9*(1), Article 5.

Agar, J. (2011). Thatcher: Scientist. *Notes Rec. 65*, 215–32. doi.org/10.1098/rsnr.2010.0096.

Agriculture and Environment Biotechnology Advisory Board. (AEBC). (2001). *Crops on Trial: A Report by the AEBC.* London: AEBC.

Barnes, M., Newman, J. and Sullivan, H. C. (2007). *Power, participation and political renewal: Case studies in public participation.* Bristol: Policy Press. doi.org/10.2307/j.ctt9qgrqs.

Bauer, M., Allum, N. and Miller, S. (2007). What can we learn from 25 years of PUS survey research? Liberating and expanding the agenda. *Public Understanding of Science, 16*, 79–95. doi.org/10.1177/0963662506071287.

Beck, M., Asinova, D. and Dickson, G. (2005). Science and Risk Assessment: A case study of the UK Bovine Spongiform Encephalopathy Crisis. *Public Administration Review, 65*(4), 396–408.

Bell, A. (2013, 18 March). The Big Bang Fair: A depressing vision of science and engineering. *The Guardian.*

The Big Bang: Young Scientists and Engineers Fair. (2018). 80,000 visitors attend The Big Bang Fair 2018. Retrieved from www.thebigbangfair.co.uk/news/media-and-press/press-releases/80-000-visitors-attend-the-big-bang-fair-2018/.

Bonney, R., Phillips, T. B., Ballard, H. L. and Enck, J. W. (2016). Can citizen science enhance public understanding of science? *Public Understanding of Science, 25*(1), 2–16. doi.org/10.1177/0963662515607406.

British Association for the Advancement of Science. (2005). *Connecting Science: What we know and what we don't know about science in society.* London: The British Association for the Advancement of Science.

British Science Association. (2018). *Revolutionising people's relationship with science.* Retrieved from www.britishscienceassociation.org/news/introducing-our-new-strategy.

Brough, G. (1990, 16 May). Scientific proof: Mad cow link to humans. *Today,* p. 1.

Brown, T. (2016, 19 September). The idea of a post-truth society is elitist and obnoxious. *The Guardian.* Retrieved from www.theguardian.com/science/blog/2016/sep/19/the-idea-post-truth-society-elitist-obnoxious.

Cannon, G., Connor, S., Hague, H. and Nicholson-Lord, D. (1990). Mad cows and Englishmen. *Independent on Sunday,* May 20, p. 19.

Collins, H. and Pinch, T. (1993). *The Golem: What you should know about science.* Cambridge: Cambridge University Press.

COPUS. (2001). A new direction for COPUS. Retrieved from www.copus.org.uk/copus_councilpaper.htm.

Council for Science and Technology. (2005). Policy through dialogue. Retrieved from webarchive.nationalarchives.gov.uk/20130705054945/http://www.bis.gov.uk/assets/cst/docs/files/whats-new/05-2180-policy-through-dialogue-report.pdf.

Craig, O. and Francis, W. (1990). We were frightened then Gummer rang. *Today*, May 16, p. 5.

Daily Mail. (1999). The Prime Monster. Fury as Blair says: I eat Frankenstein food and it's safe. February 16, pp. 1–2.

Daily Mirror. (1996). Mad Cow can kill you. March 20, pp. 1–2.

Dawkins, R. (1996). Richard Dimbleby Lecture, BBC TV. Retrieved from www.youtube.com/watch?v=7hEwYhMhdO8.

Deer, B. (2009, 8 February). MMR doctor Andrew Wakefield fixed data on autism. *The Times*. Retrieved from www.thetimes.co.uk/article/mmr-doctor-andrew-wakefield-fixed-data-on-autism-mgj82qsk50g.

Department of Health, and Ministry of Agriculture, Fisheries and Food (DOH/MAFF). (1989). *Report of the Working Party on Bovine Spongiform Encephalopathy ('Southwood Report')*. London: UK Department of Health.

Durant, J. R. (1993). What is scientific literacy? In J. R. Durant and J. Gregory (eds), *Science and culture in Europe*. London: Science Museum.

Durant, J. R., Evans, G. A. and Thomas, G. P. (1989). The public understanding of science. *Nature, 340*, 11–14. doi.org/10.1038/340011a0.

Frewer, L., Miles, S., Brennan, M., Kuznesof, S., Ness, M. and Ritson, C. (2002). Public preferences for informed choice under conditions of risk uncertainty. *Public Understanding of Science, 11*, 363–72.

Gaskell, G., Allum, N., Wagner, W., Kronberger, N., Torgersen, H., Hampel, J. and Bardes, J. (2004). GM foods and the misperception of risk perception. *Risk Analysis: An International Journal, 24*, 185–94. doi.org/10.1111/j.0272-4332.2004.00421.x.

Giddens, A. (1998). *The Third Way: The renewal of social democracy*. Cambridge: Polity Press.

Gove, M. (2016, 3 June). Interview with Faisal Islam on Sky News. Retrieved from en.wikiquote.org/wiki/Michael_Gove.

Gregory, J. and Miller, S. (1998). *Science in public: Communication, culture and credibility* (pp. 45–51). New York: Perseus/Basic Books.

Gross, A. G. (1994). The roles of rhetoric in the public understanding of science. *Public Understanding of Science, 3*, 3–23. doi.org/10.1088/0963-6625/3/1/001.

Guise, G. (2014). Margaret Thatcher's influence on British science. *Notes Rec. 68*, 301–09. doi.org/10.1098/rsnr.2013.0067.

Haklay, M. (2013). Citizen science and volunteered geographic information: Overview and typology of participation. In D. Sui, S. Elwood and M. Goodchild (eds), *Crowdsourcing Geographic Knowledge* (pp. 105–22). Dordrecht: Springer. doi.org/10.1007/978-94-007-4587-2_7.

Henderson, M. (2013). *The Geek Manifesto – Why science matters.* London: Corgi Books.

HM Treasury. (2004). Science & innovation investment framework 2004 – 2014. London: HMSO.

Horlick-Jones, T., Walls, J., Rowe, G., Pidgeon, N., Poortinga, W. and O'riordan, T. (2006). On evaluating the *GM Nation*? Public debate about the commercialisation of transgenic crops in Britain. *New Genetics and Society, 25*, 265–88. doi.org/10.1080/14636770601032858.

House of Commons, Environment, Food and Rural Affairs Committee. (2003). *Conduct of the GM Public Debate, Eighteenth Report of Session 2002–3.* London: HMSO.

House of Lords Select Committee on Science and Technology. (1998). *Management of Nuclear Waste, 3rd Report.* Session 1998–1999, HL Paper 41. London: HMSO.

House of Lords Select Committee on Science and Technology. (2000). *Science and Society.* London: HMSO.

Irwin, A. (1995). *Citizen Science: A study of people, expertise and sustainable development.* London, New York: Routledge.

Irwin, A. (2014). From deficit to democracy (revisited). *Public Understanding of Science, 23*(1), 71–6.

Irwin, A., Jensen, T. E. and Jones, K. E. (2012). The good, the bad and the perfect: Criticizing engagement practice. *Social Studies of Science, 43*(1), 118–35. doi.org/10.1177/0306312712462461.

Jasanoff, S. (1997). Civilization and madness: The great BSE scare of 1996. *Public Understanding of Science, 6*, 221–32. doi.org/10.1088/0963-6625/6/3/002.

Joss, S. and Durant, J. (1995). The UK National Consensus Conference on Plant Biotechnology. *Public Understanding of Science, 4*, 195–204. doi.org/10.1088/0963-6625/4/2/006.

Laroche, G. (2011). Responsible Research and Innovation from an EU Commission's perspective. Retrieved from www.ambafrance-uk.org/IMG/pdf/Gilles_LAROCHE.pdf.

Layton, D., Jenkins, E., McGill, S. and Davey, A. (1993). *Inarticulate science: Perspectives on the public understanding of science and some implications for science education*. Leeds: Leeds Media Services.

Lezaun, J. and Soneryd, L. (2007). Consulting citizens: Technologies of elicitation and the mobility of publics. *Public Understanding of Science, 16*, 279–97. doi.org/10.1177/0963662507079371.

Little Voices. (2018). Little Voices Shouting Out / Lleisiau Bach yn Galw Allan. Retrieved from www.publicengagement.ac.uk/do-engagement/inspire-me/case-studies/little-voices.

Lock, S. J. (2011). Deficits and dialogues: science communication and the public understanding of science in the UK. In D. J. Bennett and R. C. Jennings (eds), *Successful science communication: Telling it like it is* (pp. 17–30). Cambridge: Cambridge University Press.

Lock, S. J. (2016). Cultures of incomprehension? The legacy of the Two Cultures debate at the end of the twentieth century. *Interdisciplinary Science Reviews, 41*, 148–66.

Macnaghten, P., Kearnes, M. and Wynne, B. (2005). Nanotechnology, governance, and public deliberation: What role for the Social Sciences? *Science Communication, 27*(2), 268–91. doi.org/10.1177/1075547005281531.

Marsh, O. M. (2018). Nah, musing is fine. You don't have to be 'doing science': Emotional and descriptive meaning-making in online non-professional discussions about science (PhD thesis). University College London.

Miller, S. (2001). Public understanding of science at the crosssroads. *Public Understanding of Science, 10*, 115–20.

Miller, S. (2016). Training Showcase: The UK's Engineering and Physical Sciences Research Council's Framework for Responsible Innovation. *RRI Tools*. Retrieved from www.rri-tools.eu/documents/10184/193151/2_RRITOOLS-EPSRC_Showcase_final.pdf/2189f09b-4d83-4425-92c6-3ff8e8129639.

Millstone, E. and van Zwanenberg, P. (2001). Politics of expert advice: Lessons from the early history of the BSE saga. *Science and Public Policy, 28*, 99–112. doi.org/10.3152/147154301781781543.

National Co-ordinating Centre for Public Engagement (NCCPE). (2018). *Beacons for Public Engagement*. Retrieved from www.publicengagement.ac.uk/nccpe-projects-and-services/completed-projects/beacons-public-engagement.

New Scientist. (2016, 29 June). Post Brexit, experts need to reassert their value to society. Retrieved from www.newscientist.com/article/mg23130803-500-post-brexit-experts-need-to-reassert-their-value-to-society/.

Nicholson-Lord, D. (1990, 30 September). It's a mad, mad, diet of MAFF-speak. *Independent on Sunday*, p. 4.

Owen, R., Macnaghten, P. and Stilgoe, J. (2012). Responsible research and innovation: From science in society to science for society, with society. *Science and Public Policy, 39*(6), 751–60. Oxford University Press. doi.org/10.1093/scipol/scs093.

Phillips N., Bridgeman, J. and Ferguson-Smith, M. (1997). The BSE Inquiry Report: Volume 1. London: HMSO.

Phillips, N., Bridgeman, J. and Ferguson-Smith, M. (2000). The BSE Inquiry Report: Evidence and Supporting Papers of the Inquiry into the Emergence and Identification of Bovine Spongiform Encephalopathy. London: HMSO.

Prusiner, S. B. (1982). Novel proteinaceous infectious particles cause scrapie. *Science, 216*, 136–44. doi.org/10.1126/science.6801762.

Public Debate Steering Board (PDSB). (2003). *GM Nation? The findings of the Public Debate*. Report of the Public Debate Steering Board. London: Department of Trade and Industry.

Reeves, C. (2002). An orthodox heresy: Scientific rhetoric and the science of prions. *Science Communication, 24*, 98–122. doi.org/10.1177/107554700202400105.

Rhodes J. et al. (1999, 12 February). Letter. *The Guardian*, p. 7.

Rowe, G., Horlick-Jones, T., Walls, J. and Pidgoen, N. (2005). Difficulties in evaluating public engagement initiatives: Reflections on an evaluation of the UK GM nation? Public debate about transgenic crops. *Public Understanding of Science, 14*, 331–52. doi.org/10.1177/0963662505056611.

Royal Academy of Engineering and the Royal Society. (2004). *Nanoscience and nanotechnologies: Opportunities and uncertainties*. London: Royal Society.

Royal Society. (1985). The Public Understanding of Science (Bodmer Report). London: Royal Society.

Science Media Centre (SMC). (n.d.). About us. Retrieved from www.sciencemediacentre.org/about-us/.

Smallman, M. (2014). Public understanding of science in turbulent times III: Deficit to dialogue, champions to critics. *Public Understanding of Science, 25*(2), 186–97. doi.org/10.1177/0963662514549141.

Smallman, M. (2016a). Make innovation more equitable to restore trust in experts. Research Fortnight. London.

Smallman, M. (2016b). What has been the impact of public dialogue in science and technology on UK policymaking? (PhD thesis). University College London. Retrieved from discovery.ucl.ac.uk/1473234/.

Smallman, M. (2017). Science to the rescue or contingent progress? Comparing ten years of public, expert and policy discourses on new and emerging science and technology in the UK. *Public Understanding of Science, 27*(6), 655–73. doi.org/10.1177/0963662517706452.

Smallman, M. (2018). Citizen science and responsible research and innovation. In S. Hecker, M. Haklay, A. Bowser, Z. Makuch, J. Vogel and A. Bonn (eds), *Citizen science: Innovation in open science, society and policy*. UCL Press: London. doi.org/10.2307/j.ctv550cf2.24.

Snow, C.P. (1959/2012). *The two cultures*. Cambridge: Canto Classics.

Stilgoe, J. (2007). The (co-)production of public uncertainty: UK scientific advice on mobile phone health risks. *Public Understanding of Science*, *16*, 45–61. doi.org/10.1177/0963662506059262.

Stilgoe, J., Lock, S. J. and Wilsdon, J. (2014). Why should we promote public engagement with science? *Public Understanding of Science, 23*(1), 4–15.

Suerdem, A., Bauer, M. W., Howard, S. and Ruby, L. (2013). PUS in turbulent times II – A shifting vocabulary that brokers inter-disciplinary knowledge. *Public Understanding of Science*, *22*(1), 2–15. doi.org/10.1177/0963662512471911.

Taverne, D. (2004). Let's be sensible about public participation. *Nature*, *432*, 271. doi.org/10.1038/432271a.

UK Consensus Conference on Plant Biotechnology. (1994). Final Report. London: Science Museum.

UK Government. (1993). *Realising our potential: A strategy for science, engineering and technology*. London: HMSO.

War, S. (1990, 9 June). Loophole allows export of 'Mad Cow' beef. *The Independent*, p. 3.

Weldon, F. (1991, 2 December). Thoughts we dare not speak aloud. *The Daily Telegraph*, p. 14.

Wilenius, P. (1990, 11 May). Mad Cows' Disease kills cat. *Today*, p. 1.

Wilsdon, J. and Willis, R. (2004). *See-through science: Why public engagement needs to move upstream*. London: Demos.

Wiseman, R. (1995). The megalab truth test. *Nature*, *373*(6513), 391. doi.org/10.1038/373391a0.

Wolfendale, A. (1996). Report of the Committee to review the contribution of scientists and engineers to the public understanding of science. London: HMSO.

Wolpert, L. (1991, 9 December). So much for artistic license. *The Daily Telegraph*, p. 12.

Wynne, B. (1991). Knowledges in context. *Science, Technology and Human Values*, *16*, 111–121. doi.org/10.1177/016224399101600108.

Wynne, B. (2006). Public engagement as a means of restoring public trust in science-hitting the notes, but missing the music? *Community Genetics*, *9*(3), 211–20. doi.org/10.1159/000092659.

Timeline

Event	Name	Date	Comment
First interactive science centre established.	The first purpose-built independent science centre in the UK was Techniquest in Cardiff	13 November 1986	1931: Interactive exhibits in the Children's Gallery at Science Museum in London from 1931. 2000: New science centres—including the National Space Centre in Leicester, and Dynamic Eart in Edinburgh—were opened with funds from the Millennium Foundation
First national (or large regional) science festival.	This event in York, turned out to be the first annual meeting of the British Association for the Advancement of Science (BAAS) (now the British Science Association, BSA)	26 September 1831	1943: BAAS ran its annual meeting specially themed on 'Science for the Citizen', later the title of a bestselling book by Lancelot Hogben. 1989: Edinburgh Science Festival, first regional science festival
An association of science writers or journalists or communicators established.	The Association of British Science Writers was founded by J. G. Crowther and Maurice Goldsmith	3 March 1947	Crowther was a science journalist on the *Manchester Guardian* and Goldsmith of *Reynolds News* (last published 18 June 1967)
First university courses to train science communicators.	MSc in science communication at Imperial College under John Durant	1990/91 academic year	1989–90: Undergraduate BSc module in science communication at University College London by Steve Miller
First master's students in science communication graduate.	Graduates from the Imperial College MSc in Science Communication	1993	
First PhD students in science communication graduate.	Jane Gregory, University of London	1998	Thesis title 'Fred Hoyle and the popularisation of cosmology'
First national conference in science communication.	BAAS 'Science and the Citizen' conference (mainly for science enthusiasts)	1943	1990: 'Politics and Publics for Science and Technology' conference organised by the Science Museum and EASST. Mainly for academics. 1993: 'Building Bridges' conference organised by BAAS. 2002: First national 'Science Communication' conference organised by BAAS/BSA. 2013: First national 'Engage' conference organised by the NCCPE

Event	Name	Date	Comment
National government program to support science communication established.	The Committee on the Public Understanding of Science (COPUS), instituted its first funding scheme	1986	COPUS had representatives from the Royal Society, the BSA and the Royal Institution. Funded by UK Government's Office of Science and Technology
First significant initiative or report on science communication.	Report of the Royal Society on 'The Public Understanding of Science' led to the setting up of COPUS and the foundation of university SciComm courses	1985	1993: The first government report in this area was 'Realising our potential', which put a duty on government-funded research councils to communicate science
National Science Week founded.	The UK 'Week of Science' launched	1994	Parallel to the European Week of Science and Technology but held at different time of year
A journal completely or substantially devoted to science communication established.	*Public Understanding of Science* founded as a peer-reviewed SciComm research journal	1992	Founded by the Science Museum (published by the Institute of Physics)
First significant radio programs on science.	*The Stream of Life* on the BBC by Julian Huxley	1925	1942: BBC Home Service *Man's Place in Nature*
First significant TV programs on science.	BBC TV *Science Review*	1952	1954: Zoo Quest 1957: The Sky at Night 1964: First edition of BBC's flagship science TV program *Horizon*
First awards for scientists or journalists or others for science communication.	Royal Society Faraday Award to Charles Taylor	1986	1998: Susan Greenfield first woman to win Faraday Award. 2005: Royal Society Kohn Award for early-career scientists for public engagement given to Colin Pulham
Date hosted a PCST conference.	PCST-16, Aberdeen	2020	Postponed to 2021 because of COVID-19
Other significant events.		1066	Bayeux Tapestry includes the 1066 passage of Halley's comet in its woven narrative
		1802	Evening public lectures established at The Royal Institution (founded in 1799)
		1993	London 'Public Understanding of Science' seminar series established

Event	Name	Date	Comment
		1994	First UK consensus conference on 'Plant Biotechnology' held at the Science Museum
		2008	Six regional 'Beacons of Public Engagement' plus a National Coordination Centre set up
		2014	UK Research Excellence Framework includes 'Impact Statements' so universities explicitly receive funding for science communication efforts

Contributors

Dr Melanie Smallman is lecturer in science and technology studies and co-director of the Responsible Research and Innovation Hub at University College London's Department of Science and Technology Studies.

Dr Simon J. Lock is associate professor in science communication and governance in the Department of Science and Technology Studies, University College London.

Steve Miller is Emeritus Professor of Science Communication and Planetary Science at University College London.

40

UNITED STATES OF AMERICA

Science communication in the USA: It's complicated

Bronwyn Bevan and Brooke Smith

1. Introduction

Science communication in the United States, as a field, has a complicated history and contested definitions. In part this is due to the decentralised nature of the country's scientific research agencies and professional societies. Unlike many countries, there is no single governmental scientific research agency or policy group that oversees the endeavour of science communication. But this is not the only reason for the field's lack of clear definitions and parameters. A look into the past reveals a broad array of historical forces and stakeholders that have intervened, over the years, to produce divergent purposes, strategies and outcomes in science communication here in the United States. In this chapter we apply a critical lens to understanding how these forces have forestalled the development of a coherent field of practice, while at the same time producing a vibrant (if jostling) set of opportunities to engage the public with science.

In the US, science communication can be conceptualised as consisting of multiple constituent parts or dimensions. It involves communication skills, the ability to connect with people or to tailor messages that resonate with various non-scientific audiences for specific objectives. It involves communication tools, such as news articles, podcasts, films and live productions (e.g. science cafés). It involves a variety of science communication actors, including scientists, science outreach communicators, journalists, informal science educators and others. Science communication is provided by a range of institutions such as universities, science museums, professional societies, media outlets and small non-profits as well as the public relations and communications arms of industry and academia. It takes place in many settings, from café to auditorium, classroom to smart phone screen, congressional office to town hall. As a field of inquiry it has generated several subfields, such as the science of science communication (with its attendant research journals, conferences and collateral), science communication training programs (including institutes, courses and even graduate degree programs) and science communication professional organisations and networks (Gascoigne et al., 2010). Researchers in science communication span a range of disciplines, including sociology, psychology, neuroscience, learning sciences, political science, science and technology studies (STS), communication and more. There are a range of funding organisations and scientific societies that have a stake in advancing the field. These disparate efforts are largely uncoordinated, even within given regional communities, leading to what some have described as a largely fragmented science engagement ecosystem across the US (Lewenstein, 2001).

But the multiple dimensions of science communication are only part of the complexity. Its goals and objectives also vastly differ (Sugimoto and Weingart, 2015). A recent study by the National Academies of Sciences, Engineering, and Medicine (2016) synthesised the literature to name five distinct goals for communicators of science: to make the public aware of science, to build their appreciation of it, to develop their understanding of it, to inform their actions with it and, in turn, to have science itself become more informed by the public. This report did not directly call out other purposes we can see in the field, such as increasing the diversity of the STEM workforce, securing public support and funding for science as an enterprise, informing the production of science itself or developing scientific reasoning for citizenship in a democracy. One study revealed that scientists' motivations for online engagement include (in ranked order) to defend science, inform the public, excite the public, build trust between the public and science, and tailor messages for particular publics (Dudo and Besley, 2016).

Compounding this diversity of motivations and purposes, science itself can be conceptualised in multiple ways: as a body of knowledge, a process of inquiry, a professional practice and an avocation. Audiences could be differentiated by age (e.g. kids, adults, a mixture), degree of specialisation or purpose (e.g. total novices, policymakers, experts from other fields), type of interest (e.g. uninterested, community welfare, personal hobby, professional) and an infinite variety of other categorisations. They attend science communication events with a range of preparation and prior knowledge. Sometimes they may be motivated by a specific need-to-know, other times by more general curiosity. Some may identify closely with science and others may not. Some may be active participants who seek out science, some may stumble upon science presentations in public settings such as music festivals, and others (such as students on field trips) may experience engaging with science/ scientists as a form of coercion.

The decentralised nature of science communication in the US means that different organisations and actors define, pursue and describe science communication along some but not all of the above purposes and dimensions. In this somewhat cacophonous context, the field of practice is challenged to align its purposes, actors, skills, tools and settings with the goals and motivations of its audiences. Moreover, this range of overlapping approaches and foci leads to many instances of professionals in the field using the same words to describe their practices of science communication (e.g. communication, public engagement, science outreach) but sometimes meaning different things. Often, one-directional communication (e.g. science briefs shared with policymakers or science journalism) leads to engagement (e.g. scientists and policymakers collaborating in meetings or scientists engaging in the public discourse), further blurring the lines of what is communication or engagement (Smith et al., 2013). Consequently, people talk past or misunderstand one another, which limits knowledge-building in the field, whether research- or practice-based (Lavery, 2018).

In short, science communication in the United States (ironically) suffers from a communication—and coordination—problem. This lack of coordination and centralisation means that telling its story is not easy. In this chapter, we zoom out to examine the history of science communication in the context of the history of the US writ large. In so doing, we posit that different institutional stakeholders in the US—e.g. scientists, industrialists, policymakers, educators—have historically sought to use science as a means of exerting social control at times of social disruption and change (e.g. urbanisation, immigration, militarisation or social atomisation in the age of the internet). These different historical moments, in turn, have led to the emergence of disparate, sometimes misaligned, purposes and approaches to science communication, which today we experience as a cacophony. This cacophony can be positive, in that it promotes new thinking, innovation and creativity in the field and practice. But it can also lead to inefficiencies, misalignment and conflicting messages that can alienate or anger some audiences. In this chapter, we provide an overview of these historical forces before discussing both tensions and possibilities for the work moving forward.

2. The history of science communication in US history

We use a critical lens to describe how science communication as a field has developed in the US. Specifically, we examine the ways that science communication has been used to exert and to challenge existing power structures and to advance particular agendas.

Historians and sociologists of science have long applied a critical lens to science as an enterprise (see, for example, Haraway, 1988; Harding, 1998; Latour and Woolgar, 1986). Some studies in science communication, particularly on the subject of climate change, have also used this lens (e.g. Russill, 2018); however, to date, critical perspectives have not permeated the mainstream practices and journals of science communication in the US (Rauchfleisch and Schäfer, 2018).

There are ways to tell the history of science communication in the US without respect to larger historical currents or contested power relations. However, we believe that looking at the intersections of social change and science communication affords particular insights about the production, experience and evolution of science communication that can help the field today as it seeks to reconcile competing visions of what science communication is and can be. We recount this history via perspectives from science communication and policymaking, science journalism, the learning sciences and informal science education. We aim to be representative, not exhaustive, in our historical account.

For the purposes of this chapter, we define science communication broadly as a diversity of activities, with a variety of purposes, that strengthen the connections between the scientific research enterprise (its history, processes, people and products) and public audiences. We understand moments of science communication to be moments of cultural production, where different social and personal histories, interests and motivations converge to shape experience and meaning-making (Nasir et al., 2014). In these moments the discourse adopted by science communicators—the language, images, and representations they use—is always ideologically saturated (Bakhtin, 1981), as is the meaning made by their audiences. By ideologically saturated we mean that words, images, stories and representations used in science communication do not exist independent of history, culture, disciplinary epistemologies and social stances; as such they constitute particular Discourses of science (Gee,

2004).[1] That is, they are imbued with and invoke, explicitly or implicitly, specific positionalities and ideological commitments as to what science is and why we the public should care about it. Surfacing the ways in which science communication advances powerful interests is critical to any examination of its history. However, it's also important to note that despite its historical alliances with economic and social power structures, science communication as a practice is never monolithic or ultimately hegemonic but instead creates sites of exchange where power, and powerful ideologies, can be both exerted and challenged (see Foucault, 1990).

2.1. Acculturation and citizenship in the context of rapidly changing demographics

As in many nations, in the US science communication as a field began in the 19th century with the second industrial revolution (Lewenstein, 2016). Large-scale industrialisation, urbanisation and immigration dramatically transformed social organisation in communities across the country. A primary goal among the educated and upper middle classes was to acculturate new urban populations to 19th-century progressive American ideologies related to the role of science and economic prosperity. At the heart of this ideology were Enlightenment views on science as a source for rationality and democratic citizenship (Dewey, 1916). In this context, science communication involved creating new public spaces—such as the American Museum of Natural History in New York, the Franklin Museum in Philadelphia and the Museum of Science and Industry in Chicago—that opened up private collections to the public. Public lectures proliferated, as did guided tours of exhibitions, and short classes.

The creation of scientific societies—targeting science professionals and dedicated amateurs, modelled on the British Royal Society—extended this ideology of science as a tool of Enlightenment rationality and social progress to more privileged social sectors. Societies included the American Association for the Advancement of Science (founded in 1848), the National Geographic Society (1888) and the American Chemical Society (1876). As in Britain, these societies were private, and intended for both practising professionals and dedicated amateurs of particular social classes (Kohlstedt, Sokal and Lewenstein, 1999; Rhees, 1987).

1 Gee contrasts 'little d' discourse and 'big D' Discourse. Whereas discourse with a lower case d refers to the ways in which individuals discuss or represent ideas, Discourse with a capital D refers to the much broader social, cultural and historical conceptualisation, significance and representation (e.g. science as culturally neutral, science as a means for global prosperity) that may shape how little d discourse develops.

Figure 40.1: Science demonstrations have been a mainstay of science communication. Here a wary audience closely observes a cow's eye dissection.

Source: © Exploratorium (used with permission).

As such, the goal of science communication at this time of social change was education and *enlightened citizenship*, towards building a democratic society (Terzian, 2013). Science was presented as a mode of inquiry, a collection of artefacts, and a tool for—and promise of—rational, democratic progress. This Discourse links to a strand of science communication in the US today commonly associated with informal science education. This sector involves many large and established institutions—primarily museums, but also media outlets such as public television and radio—and receives significant funding from entities concerned with science education. In turn, this strand of science communication conducts research and assessment focused on learning outcomes, where learning is conceptualised broadly to include the development of audiences' science interests, identities, skills, concepts, epistemologies and practices (National Research Council, 2009).

2.2. Instrumentalism in the context of rapid economic growth

In the first half of the 20th century, science communication efforts broadened (i.e. were not replaced but expanded) to include more instrumental views of science. Specifically, the rapid growth of manufacturing and agricultural industries led to a perceived need to ensure that new scientific knowledge and technologies were put into practice. In 1914, the Smith Lever Act established a partnership between the US Department of Agriculture and a number of land-grant universities charged with applying research to agricultural practices. Extension agents were arguably some of the earliest science communicators in the US, connecting farmers and researchers to allow them to benefit from each other's knowledge and craft. In this context, science communication was motivated by a desire to achieve efficiencies (of all types) to advance economic prosperity.

This Discourse of science was accompanied by efforts to win the hearts and minds of the public, to encourage them to support investment in the scientific enterprise. Along these lines, the birth of the field of science journalism occurred. Science journalists sought to persuade the public to accept science as a force for ongoing social improvement (Lewenstein, 1992). From Pluto to penicillin, scientific discoveries covered in the news elicited awe and promise during this time. By 1934 there were enough science journalists to launch the National Association of Science Writers (NASW), dedicated to improving their craft.

Aided by the proliferation of radio and later television, public science evangelists began to appear. These figures mediated between professional scientists (employed in the burgeoning scientific industry) and the public. Figures such as Mr Wisard, who had a television show that ran from the early 1950s into the mid-1960s, engaged children and families in the science of everyday objects and phenomena (LaFollette, 2012). These shows both advocated for science as an important enterprise and emphasised the applicability and usability of science. Such views were taken up by Warren Weaver, an influential figure who served on the board of the American Association for the Advancement of Science (AAAS) in the 1950s and took a strong interest in the public understanding of science. The launch of Sputnik, and the doubling down of federal investments in science education (Rudolph, 2002) propelled science communication efforts aligned with a Discourse of science focused on the practical, utilitarian and instrumental uses of science in an increasingly technological society.

Such views were not unconnected to broader social and historical forces, particularly the role of science in prosecuting World War II and later the Cold War arms and space races between the US and the Soviet Union. In this instrumentalist Discourse, uses of science were entwined with rhetorical positioning of the US as a bastion of political liberty, which was closely connected with the expansion of the military industrial complex, putatively to defend those liberties. As such, science communication was focused on advocating for further state investment in science and scientists in the interest of nationalism.

Today, such Discourses exist in efforts to have science communication excite people about the potential future and current utility of science, partly because State funding depends on that support. From 1969 photos of Neil Armstrong walking on the moon to the Twitter account of today's Mars Rover, NASA's science communication exemplifies science communication as public relations. DuPont's 'Better Living Through Chemistry' campaign is an industry-driven example.

It is notable that instrumental approaches—because they tout science as authoritative and essential to economic prosperity—lend themselves to deficit framing. There is a literature spanning decades that demonstrates the weaknesses of deficit-based models relative to asset-based approaches in which the interests, experiences and cultural patterns of interaction are positioned as the means for productive engagement with, for instance, science; this is especially true for efforts to engage individuals from racially or socioeconomically marginalised communities (Bartolomé, 1994; Gonzalez, Moll and Amanti, 2005; Simis et al., 2016). Yet, a recent consensus study found that in the US the use of deficit models is most common in efforts to engage the public with what are considered to be controversial social issues—those same issues that may be most urgent in terms of decisions, direction and behaviour changes, such as vaccination, climate change and food safety (National Academies of Science, Engineering and Medicine, 2017).

2.3. Empowerment and actualisation in an increasingly mechanising society

As the Cold War continued, and wars in Korea and Vietnam were waged, the dominant Discourse of science in the US became more divergent, with the space race providing a sort of fulcrum for competing views. On the one hand, with NASA's massive investment in science communication and public relations and efforts by a range of other industrial players (from pharmaceuticals and General Electric to the military), there was an ongoing advocacy for the instrumental role of science in advancing social, technological and economic prosperity. After all, this was the era that led to the polio vaccine and a man on the moon. At the same time, the publication of Rachel Carson's *Silent Spring* in 1968, soon followed by Apollo 17's first images of Earth taken from space (the 'blue marble'), sparked environmental movements that joined with other liberation movements of the time (civil rights, feminism, etc.) to challenge existing power structures, including that of science. Some public science evangelists began to shift their lens from the instrumental dimensions of science for advancing economic prosperity to the relationship of science to topics of social concern. In 1969, at the height of the Vietnam War, the Union of Concerned Scientists was founded in response to the scientific community's concerns about the ways in which science was being used to inform military, environmental and social policies.

These trends towards challenging existing power structures triggered reflection in the science journalism community about their role. Was it the responsibility of science journalists to cover the remarkable science that led to pesticides, or was it their responsibility to cover the health concerns over the use of pesticides? Was the goal of these communication efforts to excite people about the potential of science, or was it to connect content to social issues and decisions (Nelkin, 1987)? Science journalists increasingly tried to address adverse results of science, as well as demonstrate how and where science was being used to address adverse social and environmental conditions. In 1974, the second episode of the renowned American television show NOVA focused on the devastating effects of land and water mismanagement on the health of the Colorado River. Five of the first 11 NOVA episodes adopted critical perspectives on how human progress and tools of science had put the planet and its inhabitants in various types of danger.

Figure 40.2: The Exploratorium, an early interactive science museum founded in 1969, greatly expanded the field by making its exhibit designs and teaching materials widely available and free of cost.

Source: © Exploratorium (used with permission).

Along with calls for 'power to the people' emerged a novel form of public science museum that operated with a motto akin to 'science for the people': the interactive science centre. In the US, this movement was launched in the late 1960s with the founding of the Exploratorium and the Lawrence Hall of Science, both in the San Francisco Bay Area (which had just concluded the 'Summer of Love'). The goal of these new institutions was to empower rather than strictly enlighten (as in the 19th century) the public to observe, question and engage with the natural world (Cole, 2009). Visitors were invited to participate in designed experiences (mostly interactive exhibits) where they could observe and develop their own questions about scientific phenomena—e.g. resonance, shadows, waves, linguistics. In this sense, a major goal of these early science centres was not to propagate knowledge but to stimulate questions about the natural and social world. Exploratorium founder, Frank Oppenheimer—who often touted a perhaps apocryphal story of a visitor to the museum writing him a note to tell him of her feelings of immense empowerment when after her visit to the museum she had gone home and re-wired a lamp on her own—said of that institution 'the whole point … is for people to feel they have the capacity to understand things' (Cole, 2009).

This emerging Discourse of science as empowerment—adding to continuing propagation of the Discourses of science as enlightened citizenship and as instrumental to a nationalist and economic agenda—aimed to support the populace to pay attention, to ask questions, to see science as something that we all can do, and to use science to challenge authority. The interactive science centre movement has not always or solely advanced this Discourse, oftentimes resorting to more instrumentalist or enlightenment models. But this empowerment Discourse has continued in pockets—including new community-driven models of science centres such as the Science Gallery in Dublin—as well as in the development of community-based science programs such as the COASST (Coastal Observation and Seabird Survey Team) project or the Learning in Places project in Washington State.

COASST works collaboratively with Aleutian fishing communities to collect data on fish, mammal and seabird populations as a means of tracking the effects of sea temperature rise as a tool for political action. Science communication activities include workshops about climate change and trainings for data collection, in order to empower local individuals and communities to advocate for mitigation efforts to preserve their indigenous practices and villages. Learning in Places is a collaboration among local schools, a community-based garden program and a university to engage students with socioecological justice issues (e.g. food sustainability and water usage) through investigations involving ecological reasoning and decision-making. Many programs that provide informal science education programs in prisons, such as Nalini Nadkarni's Oregon program, position science engagement as a tool for personal empowerment, where prisoners develop skills and understanding, but moreover personal meaning, through conducting pond studies or cultivating mosses for local industries. Community-based programs such as Guerrilla Science design interactions between the public and scientists so that their audiences can engage with science in ways that empower them to take action, whether to apply dream science to their dreams or neurosciences to their love lives. Science as empowerment is commonly measured through constructs such as activist uses of science, and pursuit of science in everyday and informal contexts indicating a personal commitment and sense of self-efficacy in science.

3. Staying relevant in an age of social atomisation and science communication solidification as a field

At the turn of the 21st century, in the context of the ubiquitous smartphone and an increasingly on-demand society, science communication in the US has grown more atomised. On the one hand, empowerment approaches have been embraced by advocates for topics ranging from HIV/AIDS research and popular epidemiology (Brown, 1992) to GET City in Lansing Michigan, where teenagers use the tools of science to investigate issues of community importance, such as the effects of urban heat islands on community health (Calabrese and Tan, 2010). Science as enlightenment continues in a broad array of informal science efforts, and science as instrumental is also advanced in many public engagement events, such as TED talks, public lectures and films.

On the other hand, science communication is being forced to play out against a backdrop of misinformation, a competitive media environment, identity politics and modern-day political communication (which some might call post-truth). In an age when science communicators are facing waning attention and growing scepticism the field continues to explore the dialogic, cultural and value-laden nature of the exchange and meaning-making of ideas (Kahan et al., 2013; National Academies of Science, Engineering and Medicine, 2017). This builds on the public understanding of science field's reflections of its evolution from science literacy from public understanding to an era more focused on the relationship between science and society (Bauer et al., 2007). The subfield of informal science education first began to engage with such issues in the wake of the formal education system's attention to multiculturalism in the 1990s and the development of a large body of work focused on the cultural (and therefore value-laden) dimensions of how people engage with science and mathematics in everyday settings, such as playing dominoes, scoring basketball or shopping in supermarkets (National Research Council, 2009). In science communication, this 'cultural turn' has led to more targeted and niche approaches to communicating science, including conceptualising the two-way nature and mutual learning and benefit that can come through science engagement. Though even this new term has multiple meanings (Lewenstein, 2016), and we might increasingly question the role of science communicators in social contexts where values outweigh facts and data (Kahan et al., 2013).

Figure 40.3: Guerrilla Science stages live interactions with scientists, mathematicians and artists. Here science communicators discuss probabilities at the National Mathematics Festival.

Source: © Guerrilla Science International (used with permission).

More science communication efforts today adopt two-way and asset-based approaches to engaging public audiences with science. For example, more of this work takes place beyond university lecture halls and museum floors, where cultural scripts that imbue knowledge with the scientist predominate. From Nerd Nites at nightclubs to Guerrilla Science's work in music festivals, community events and county fairs, most of these events seek to connect cutting-edge science to specific interests of the target audiences (e.g. what neurosciences can tell us about sexual attraction at speed-dating events). Science Discourse in these contexts is related to social transformation relevant to the questions and needs of the public audiences as individuals and as subcommunities. Where is science already a part of your life, your values, your priorities? How do you or could you use it to advance your goals? Much of this Discourse places the public audience member—not science itself—at the centre of the interaction.

Social transformation is also present in how the role of scientists is conceived today. As science communication and journalism have increasingly moved to more niche and participatory outlets—such as blogs and social media—who is doing the science journalism has changed too. Scientists have found themselves in new positions as the number of paid science journalists has declined. In this context, the international media outlet *The Conversation* has emerged as a leading place for scientists (and other academics and experts) to share their work beyond traditional journal articles, reports and books. This shift in role and responsibility has occurred just as awareness of shared communal issues, such as climate change—along with concomitant political resistance in the US to taking action on it—have raised a sense of urgency of the need to enlist the public as allies to advocate for political change.

In her AAAS Presidential Address, Dr Jane Lubchenco challenged the scientific community to enter into a 'social contract with science' (Lubchenco, 1998). Lubchenco called on scientists to not only prioritise research questions that inform the most pressing questions of our time, but to communicate their knowledge, insights and findings with the public. Indeed, studies have found that scientists are highly interested and active in communicating directly with the public (Besley, Dudo and Yuan, 2018). Scientists describe a variety of objectives for meeting their science communication goal: most scientists cite their top objectives as defending science from misinformation and informing the public about science, but others communicate and engage in order to strengthen the public's trust in science and shape policymaking (Dudo and Besley, 2016).

As more scientists are communicating and engaging directly, there is increasing focus to ensure that their communication is based on research about how best to communicate science. To this end, in 2012 the US National Academy of Sciences launched the Science of Science Communication Colloquium, which ultimately led to the development of a synthesis of the research called *Communicating science effectively: A research agenda*. As the study's committee co-chairs state: 'Fortunately, a growing body of scientific evidence can help inform the most effective ways of communicating with the public under different circumstances' (National Academies of Sciences, Engineering and Medicine, 2017). However, many practitioners, like science communication trainers, do not access or use research to inform their practice (Besley and Dudo, 2017). With this, we might infer that many science communication practitioners did not access or influence this report developed by researchers in the field. Indeed, a study by Bevan et al. (2019) found a disparity between researchers and practitioners in science communication—for example, researchers expressed far less engagement with issues of equity and inclusion than did practitioners, and researchers were less likely to indicate that they had strategies for staying up with new practices than practitioners reported for staying up with research. In contrast, there was much greater overlap among researchers and practitioners who identified with the subfield of informal science education.

Parallel to the rise of scientists engaging directly with the public has been an increase in the subfield of science communication trainers (Besley and Dudo, 2017), professionals who train and support scientists to be effective in their communication skills and messaging. Groups like the Alan Alda Center for Communicating Science, Seattle-based COMPASS and AAAS have pioneered approaches to supporting scientists' communication skill development, while hundreds of other courses and sessions are being developed (often by scientists themselves). The objectives and motivations of science communication trainers similarly run the gamut, ranging from wanting to ensure scientists inform policies to helping scientists engage in their local education systems. Frequently, trainers are unable to articulate specific goals for their programs (Besley and Dudo, 2017).

There can be a strong deficit orientation to some of this work. That is, scientists who see issues such as the role of vaccines and genetically modified organisms in meeting health and nutrition needs for an exploding world population are intent on public audiences embracing their analysis of the challenge and the solution. In this context, scientists are enlisted 'to set the record straight', particularly in the context of the internet-led spread of alternative, non-scientific views about (for example) the safety or efficacy of vaccines; and to advocate for particular, science-informed policies in a transforming world. This maps to Dudo and Besley's 2016 finding that 'defending science from misinformation' is a primary motivator for many scientists to engage the public.

To help scientists avoid falling back on deficit orientations, and in so doing vitiating the impact of their work, many new programs have been developed (led by scientists and science communicators) to re-imagine science communication as a means of promoting social justice and broadening participation in science. This extends the Discourse of science for social transformation to take on issues of social justice. For example, recent economic and natural disasters in Puerto Rico have resulted in Ciencia Puerto Rico's work to position science as a tool for economic development and social justice, including diversifying the scientific workforce. We Act, a group located in New York, collaborates with low-income people of colour to build healthy communities. These approaches seek to measure change by understanding who is participating in, who is contributing to, and who is benefiting from science as an endeavour.

4. Discussion: Tensions and possibilities in science communication

This brief historical overview is necessarily superficial and does not provide space for other approaches (e.g. a rhetorical analysis) or to the many counterexamples that we might have discussed. For example, a 1950s kind of instrumentalism—preparing for the new economy—drives much of the current focus of 2019 informal science education related to coding, though much is also underpinned by a discourse of equity and inclusion for young people from communities historically excluded from science. TED Talks, referenced earlier, are contemporary means of communication that rely on traditional lecture-based, one-way transmission models of engagement delivered by individuals who, notoriously, are primarily white, male and highly educated. At the same time, TED Talks are also heralded as breaking down walls of the academy to democratise current science (and other issues).

Figure 40.4: Sensory speed dating. Neuroscientists are teamed with stand-up comedians as they guide public audiences at music festivals through inquiries into the science of attraction.

Source: © Guerrilla Science International (used with permission).

Despite the existence of these and plenty of other counterexamples, we hope that examining how larger social and historical forces may have shaped the development of different science communication goals and their attendant strategies can help science communicators adopt a more critical, and therefore inclusive, perspective on what science communication seeks to accomplish and how it goes about its work.

We can see how different Discourses of science—enlightened citizenship, instrumentalism, tool for personal and community empowerment, and for social transformation—have emerged at particular times and also persisted over time. As new Discourses of science have emerged in response to their historical moment, old ones have continued. Moving into the future, as the public is increasingly bombarded with information—ranging from 24/7 news cycles, 23andMe[2] and similar DYI genetics testing, and internet misinformation trolling—new Discourses will undoubtedly emerge that may for example position science as a resource in a teeming sea of information.

Harkening back to our introduction about the different dimensions of science communication, it is clear that in the US science communication field different goals are being pursued by different institutional actors adopting different engagement strategies—citizen science, podcasts, science cafés, etc.—towards different ends. The resulting cacophony is one of the drivers for the current push for the development of a science of science communication and for an infrastructure to support science communication.

Developing 'a science' is a move made by many different fields seeking to establish political, academic and social legitimacy (e.g. learning sciences, network science and organisation science). Fundamentally, it refers to the development of a systematic body of knowledge and theory that can be used to guide and analyse practice. The fact that this field of study is beginning to form reflects both the growth and the disorder of the field: a 'science' suggests that order will be instilled. Some leading science communication researchers come from the larger communication field, which can include strategic communication, journalism, public relations persuasion and advocacy, perhaps leading with (but not exclusively so) an instrumentalist Discourse of science. Researchers in the informal science education field tend to lead with both enlightened citizenship views as well as individual empowerment, with some drawing from the broader education research, especially related to learning. Of course, there are many examples of researchers pursuing other Discourses of science. We do not mean to essentialise but to make sense of

2 23andMe is a privately held personal genomics and biotechnology company based in Sunnyvale, California.

the emphases in the literature that show that science communication research often stresses how science communication practices lead to changes in public beliefs, decision-making, and behaviour. Whereas informal science education research often stresses how the practices contribute to dimensions of learning, writ large (see Bevan et al., 2019).

Similarly, as the field—driven by funding agencies that invest in scientific research such as the Kavli Foundation, the Gordon and Betty Moore Foundation, the Rita Allen Foundation, the Simons Foundation and others—seek to develop a national infrastructure for science communication, competing Discourses of science may suggest different directions. For example, training programs are underpinned by ideologies of science that dictate how science communicators are prepared. To date, many have adopted instrumentalist Discourses—focused on helping scientists excite and engage audiences about the work that they do in order to build support for science, or to 'defend science' (Dudo and Besley, 2016), rather than to position science as a tool for social transformation (e.g. as efforts such as Ciencia Puerto Rico or GET City pursue). If preparation programs seek to adopt Discourses of science as empowerment, they would focus their trainings primarily on helping scientists to develop relationships with and understanding of the communities and individuals they seek to empower. Less attention would be focused on messaging and storytelling, and more on listening and collaborating. Again, we do not suggest that this is an either/or, but we do posit that there are unacknowledged emphases and ideologies at work in the field in the US.

Organisational theory suggests that new fields rapidly tend towards isomorphism: institutions and institutional actors and routines become like one another over time (DiMaggio and Powell, 1983). A case in point is the way in which the field of interactive science centres has developed. The number of such fully interactive museums has mushroomed around the globe since they were first founded 50 years ago. Although the first of these organisations tended to focus on science as empowerment, the radical (playful) way in which they engaged people with science—building on a laboratory for inquiry model familiar to scientists—was so novel, in a society that understood science learning to be about books and demonstrations, that it led to the creation of a new Discourse of the science centre as a playground, where science is 'fun'. This is the Discourse and model that the science centre field largely consolidated around in the 1990s and 2000s, with notable outliers, many of which have older histories as early science and industry museums (e.g. Boston's Museum of Science, Philadelphia's Franklin Institute, Chicago's Museum of Science and Industry), which tend to keep a close connection with scientists and the scientific community. These outliers may more closely associate themselves with science education and science communication than entertainment.

We raise this issue because as the science communication field consolidates it is essential that it is careful about what is in and what is out. Early interactive science centres, while arguing for an approach echoing 'science for the people' did not adequately articulate which people. Efforts to engage publics from communities historically marginalised from science—by social forces such as racism, sexism, hetero-normativism, etc.—were not explicit in their histories, their mission statements or even their design choices. As a result, the field has drifted towards serving mostly young people, on school trips or with their parents, and mostly young people from educated middle-class sectors. Again, there are exceptions to this rule. The Science Gallery International model seeks to engage adults with contemporary science and society issues. Some of the earliest maker spaces were developed in community settings, such as the San Francisco Mission neighbourhood, or Watsonville California or Detroit Michigan, to support and engage local audiences with science in the context of addressing immediate interests or needs (e.g. fixing a bike, building a bird house, fabricating a tortilla maker).

Is cultural relevance and responsiveness at the centre of the discussion in science communication as it begins to develop and solidify as a field? What does this mean in an age of too much information with an undercurrent of misinformation? Until we see cultural relevance and inclusiveness addressed in a comprehensive or widespread way there is a risk that the field will be developed by and for existing science stakeholders (with histories of deficit model thinking). Organisational theory suggests that later efforts to map social inclusion and cultural relevance back into the structures, the research and the work will be an uphill climb.

5. Conclusion

Rather than eschewing the multiple and competing goals of science communication, we hope that this cursory historical view of how such goals emerged reflect how values underpin work in the field. In that light, rather than pushing scientists and science communicators towards one set of common goals, we suggest that a careful articulation of goals and the subsequent alignment of audience, strategy and evaluation techniques are essential for both strengthening practice and working towards coherence across the STEM engagement ecosystem.

We know diversity is a strength, from natural ecosystems to our social structures: the more diverse things are, the more effective and resilient they can be. A science of science communication can create the structures, the theory and the connectivity to allow for diverse approaches to thrive (see Trench and Bucchi, 2010). But science communication practitioners must recognise themselves, their questions and their challenges in the research that is developed. The communities we seek to engage must also see their concerns and questions addressed in both research and practice. The field will best serve those who construct it. As it begins to solidify, we need more members of the public, and especially those publics long excluded from science, to inform its formation. For some of us this may mean that we need to 'lean out' so that others—those representing and working with communities who have not been an active or visible part of the history of science communication—can 'lean in' and create a more inclusive, and therefore more impactful, field of practice. We hope that this history, as retold a decade from now, will be able to recount such a new chapter for the field.

Acknowledgements

This chapter was supported in part by a grant from the US National Science Foundation (no. 1612739). The opinions expressed here are those of the author and not of the foundation. The authors thank Bruce Lewenstein and John Besley for their comments and feedback on the development of the chapter.

References

Bakhtin, M. (1981). Discourse in the novel (C. Emerson and M. Holquist, trans.). In M. Holquist (ed.), *The dialogic imagination* (pp. 259–422). Austin: University of Texas Press.

Bartolomé, L. I. (1994). Beyond the methods fetish: Toward a humanizing pedagogy. *Harvard Educational Review, 64*(2), 173–94. doi.org/10.17763/haer.64.2.58q5m5744t325730.

Bauer, M. W., Allum, N. and Miller, S. (2007). What can we learn from 25 years of PUS survey research? Liberating and expanding the agenda. *Public Understanding of Science, 16*(1), 79–95. doi.org/10.1177/0963662506071287.

Besley, J. and Dudo, A. (2017). Landscaping overview of the North American science communication training community. Retrieved from informalscience.org/sites/default/files/Communication%20Training%20Landscape%20Overview%20Final.pdf.

Besley, J., Dudo, A. and Yuan, S. (2018). Scientists' views about communication objectives. *Public Understanding of Science, 27*(6), 708–30. doi.org/10.1177/0963662517728478.

Bevan, B., Crowley, C., Risien, J. and Storksdieck, M. (2019). *Where informal STEM education and science communication meet: Two studies chart the intersection of ISE and SciComm*. Washington DC: Center for the Advancement of Science Education.

Brown, P. (1992). Popular epidemiology and toxic waste contamination: Lay and professional ways of knowing. *Journal of Health and Social Behavior, 33*(3), 267–81. doi.org/10.2307/2137356.

Calabrese Barton, A. and Tan, E. (2010). 'We be burnin!': Agency, identity, and science learning. *Journal of the Learning Sciences, 19*(2), 187–229. doi.org/10.1080/10508400903530044.

Cole, K. C. (2009). *Something incredibly wonderful happens: Frank Oppenheimer and his astonishing Exploratorium*. New York: Houghton Mifflin Harcourt.

Dewey, J. (1916). *Democracy and education: An introduction to the philosophy of education*. New York: Macmillan.

DiMaggio, P. J. and Powell, W. W. (1983). The iron cage revisited: Institutional isomorphism and collective rationality in organizational fields. *American Sociological Review, 48*(2). doi.org/10.2307/2095101.

Dudo, A. and Besley, J. (2016). Scientists' prioritization of communication objectives for public engagement. *PLoS One, 11*(2). doi.org/10.1371/journal.pone.0148867.

Foucault, M. (1990). *The history of sexuality* (vol. 1). New York: Vintage Books.

Gascoigne, T., Cheng, D., Claessens, M., Metcalfe, J., Schiele, B. and Shi, S. (2010). Is science communication its own field? *Journal of Science Communication, 9*. doi.org/10.22323/2.09030304.

Gee, J. (2004). *Situated language and learning: A critique of traditional schooling*. London: Routledge.

Gonzalez, N., Moll, L. C. and Amanti, C. (2005). *Funds of knowledge: Theorizing practices in households, communities and classrooms*. New York: Lawrence Erlbaum Associates. doi.org/10.4324/9781410613462.

Haraway, D. (1988). Situated knowledges: The science question in feminism and the privilege of partial perspectives. *Feminist Studies, 14*(3), 575–99. doi.org/10.2307/3178066.

Harding, S. (1998). *Is science multi-cultural?: Postcolonialisms, feminisms, and epistemologies*. Bloomington, Indiana: Indiana University Press.

Kahan, D. M., Peters, E., Dawson, E. and Slovic, P. (2013). Motivated numeracy and enlightened self-government. *Behavioural Public Policy, 1*, 54–86; Yale Law School, Public Law Working Paper No. 307. doi.org/10.2139/ssrn.2319992.

Kohlstedt, S. G., Sokal, M. and Lewenstein, B. V. (1999). *The establishment of science in America: 150 years of the American Association for the Advancement of Science.* New Brunswick, NJ: Rutgers University Press.

LaFollette, M. C. (2012). *Science on American television: A history.* Chicago: University of Chicago Press. doi.org/10.7208/chicago/9780226922010.001.0001.

Latour, B. and Woolgar, S. (1986). *Laboratory life: The construction of scientific facts.* Princeton, NJ: Princeton University Press.

Lavery, J. V. (2018). Building an evidence base for stakeholder engagement. *Science, 361*(6402), 554–56. doi.org/10.1126/science.aat8429.

Lewenstein, B. V. (ed.). (1992). *When science meets the public.* Washington DC: American Association for the Advancement of Science.

Lewenstein, B. V. (2001). Who produces science information for the public? In J. Falk, E. Donovan, and R. Woods (eds), *Free-choice science education: How we learn science outside of schools* (pp. 21–43). New York: Teachers College Press.

Lewenstein, B. V. (2016). Views: Public engagement. *Informal Science.* Retrieved from informalscience.org/news-views/public-engagement.

Lubchenco, J. (1998). Entering the century of the environment: A new social contract for science. *Science, 279*(5350), 491–97. doi.org/10.1126/science.279.5350.491.

Nasir, N. S., Rosebery, A. S., Warren, B. and Lee, C. D. (2014). Learning as a cultural process: Achieving equity through diversity. In R. K. Sawyer (ed.), *The Cambridge handbook of the learning sciences* (second ed.) (pp. 686–706). New York: Cambridge University Press. doi.org/10.1017/CBO9781139519526.041.

National Academies of Sciences, Engineering and Medicine. (2016). *Science literacy: Concepts, contexts, and consequences.* Washington DC: The National Academies Press.

National Academies of Sciences, Engineering and Medicine. (2017). *Communicating science effectively: A research agenda.* Washington DC: The National Academies Press. doi.org/10.17226/23674.

National Research Council. (2009). *Learning science in informal environments: People, places, and pursuits.* Washington DC: The National Academies Press.

Nelkin, D. (1987). *Selling science: How the press covers science and technology.* New York: W. H. Freeman.

Rauchfleisch, A. and Schäfer, M. (2018). Structure and development of science communication research: Co-citation analysis of a developing field. *Journal of Science Communication, 17*(3). doi.org/10.22323/2.17030207.

Rhees, D. J. (1987). *The chemists' crusade: The rise of an industrial science in modern America, 1907–1922.* Philadelphia, PA: University of Pennsylvania.

Rudolph, J. L. (2002). *Scientists in the classroom: The cold war reconstruction of American science education.* New York: Palgrave. doi.org/10.1057/9780230107366.

Russill, C. (2018). The 'danger' of consensus messaging: Or, why to shift from skeptic-first to migration-first approaches. *Frontiers in Communication, 3.* doi.org/10.3389/fcomm.2018.00037.

Simis, M. J., Madden, H., Cacciatore, M. A. and Yeo, S. K. (2016). The lure of rationality: Why does the deficit model persist in science communication? *Public Understanding of Science, 25*(4), 400–14. doi.org/10.1177/0963662516629749.

Smith B., Baron N., English C., Galindo H., Goldman E., McLeod K. et al. (2013). COMPASS: Navigating the rules of scientific engagement. *PLoS Biology, 11*(4).

Sugimoto, C. and Weingart, S. (2015). The kaleidoscope of disciplinarity. *Journal of Documentation, 71*, 775–94. doi.org/10.1108/JD-06-2014-0082.

Terzian, S. (2013). *Science education and citizenship: Fairs, clubs and talent searches for American youth, 1918–1958.* New York: Palgrave Macmillan. doi.org/10.1057/9781137031877.

Trench, B. and Bucchi, M. (2010). Science communication: An emerging discipline. *Journal of Science Communication, 9*(3). doi.org/10.22323/2.09030303.

Timeline

Event	Name	Date	Comment
First interactive science centre established.	New York Museum of Science and Industry	February 1936	Many of the early museums had a blend of interactives and objects, and the earliest museums that we now think of as interactive (e.g. Exploratorium, Pacific Science Center) had more objects than interactives when they opened
First national (or large regional) science festival.	Possibly Flagstaff Festival of Science	1990	Suggested as the first one by Ben Wiehe, manager of the national Science Festival Alliances
An association of science writers or journalists or communicators established.	National Association of Science Writers	1934	
First university courses to train science communicators.	Possibly New York University	1960s	

Event	Name	Date	Comment
First master's students in science communication graduate.	Several master's degree programs started in the mid-1960s, although some students graduated with science communication specialties in the 1940s and 1950s		1946: MS in Journalism and Mass Communication, with Science & Technical Writing specialty, Iowa State University 1950: MS in Agricultural Journalism, University of Wisconsin-Madison 1966: MA in Journalism, with Science Writing specialty, University of Missouri 1966: MS in Science Communication, Boston University 1967: MA in Mass Communication, with specialty in Science Journalism, University of Minnesota
First PhD students in science communication graduate.	A cluster of PhD students graduated from Indiana University after studying science communication topics in the mid-1970s	1970s	The cluser included Sharon Dunwoody, Edna Einsiedel, Marcel LaFollette and Holly Stocking. Also Rae Goodell at Stanford
First national conference in science communication.	Possibly the conferences of the Association of Science Technology Centers (ASTC)	1970s	AAAS in the US held sessions around the early 1970s. ASTC was founded in 1973
National government program to support science communication established.	The National Science Foundation created a program on public understanding of science in 1958	1958	The National Aeronautics and Space Administration and the Atomic Energy Commission supported science communication activities before that
First significant initiative or report on science communication?	*When Doctors Meet Reporters*. Kreighbaum, Hillier. New York: New York University Press	1957	
National Science Week founded.		1985–99	A National Science Foundation initiative
A journal completely or substantially devoted to science communication established.	*Science Communication*	1979	The original title until 1994 was *Knowledge: Creation, Diffusion, Utilisation*

Event	Name	Date	Comment
First significant radio programs on science.	Science Service's *Science News of the Week*	1920s	
First significant TV programs on science.	*Serving through Science*	1946	
First awards for scientists or journalists or others for science communication.	Westinghouse-AAAS Science Journalism Award	1946	
Other significant events.	Establishment of Science Service	1920s	Publisher of *Science News Letter* (now *Science News*), and ran the 'Things of Science' mail order system
	National survey	1957	First full survey about public knowledge of and attitudes toward science
	Incorporation of Council for the Advancement of Science Writing	1960	The first NGO committed to science communication
	Founding of the Center for the Advancement of Informal STEM Education	2009	

Contributors

Dr Bronwyn Bevan is director of research at the Wallace Foundation.

Brooke Smith is the director of public engagement at the Kavli Foundation.

www.ingramcontent.com/pod-product-compliance
Lightning Source LLC
LaVergne TN
LVHW052011200926
848481LV00005B/1462